Reciprocating Machinery Dynamics

Reciprocating Machinery Dynamics

A.S. Rangwala

PUBLISHING FOR ONE WORLD

NEW AGE INTERNATIONAL (P) LIMITED, PUBLISHERS

New Delhi • Bangalore • Chennai • Cochin • Guwahati • Hyderabad
Jalandhar • Kolkata • Lucknow • Mumbai • Ranchi

Visit us at **www.newagepublishers.com**

Published by New Age International (P) Ltd., Publishers
First Edition: 2006
Reprint: 2010

Branches:

- No. 37/10, 8th Cross (Near Hanuman Temple), Azad Nagar, Chamrajpet, **Bangalore**- 560 018. Tel.: (080) 26756823, Telefax: 26756820, E-mail: bangalore@newagepublishers.com
- 26, Damodaran Street, T. Nagar, **Chennai**- 600 017. Tel.: (044) 24353401, Telefax: 24351463 E-mail: chennai@newagepublishers.com
- CC-39/1016, Carrier Station Road, Ernakulam South, **Cochin**- 682 016. Tel.: (0484) 2377004, Telefax: 4051303 E-mail: cochin@newagepublishers.com
- Hemsen Complex, Mohd. Shah Road, Paltan Bazar, Near Starline Hotel, **Guwahati**- 781 008.Tel.: (0361) 2513881 Telefax: 2543669, E-mail: guwahati@newagepublishers.com
- No. 105, 1st Floor, Madhiray Kaveri Tower, 3-2-19, Azam Jahi Road, Nimboliadda, **Hyderabad**- 500 027. Tel.: (040) 24652456, Telefax: 24652457, E-mail:hyderabad@newagepublishers.com
- RDB Chambers (Formerly Lotus Cinema) 106A, 1st Floor, S.N. Banerjee Road, **Kolkata**- 700 014. Tel.: (033) 22273773, Telefax: 22275247, E-mail:kolkata@newagepublishers.com
- 16-A, Jopling Road, **Lucknow**- 226 001. Tel.: (0522) 2209578, 4045297, Telefax: 2204098 E-mail: lucknow@newagepublishers.com
- 142C, Victor House, Ground Floor, N.M. Joshi Marg, Lower Parel, **Mumbai**- 400 013. Tel.: (022) 24927869 Telefax: 24915415, E-mail: mumbai@newagepublishers.com
- 22, Golden House, Daryaganj, **New Delhi**- 110 002. Tel.: (011) 23262370, 23262368, Telefax: 43551305 E-mail: sales@newagepublishers.com

ISBN (10) : 81-224-1813-9
ISBN (13) : 978-81-224-1813-2

Rs. 525.00

C-10-04-4533

Printed in India at Glorious Printers, Delhi.
Typeset at Goswami Associates, Delhi.

PUBLISHING FOR ONE WORLD
NEW AGE INTERNATIONAL (P) LIMITED, PUBLISHERS
4835/24, Ansari Road, Daryaganj, New Delhi-110002
Visit us at **www.newagepublishers.com**

PREFACE

Considerable interest in the application of the theory of structural dynamics to the design of crankshaft, connecting rod and ram associated with reciprocating machines has existed for over 150 years. Since the advent of the earliest water-driven power saw mills and textile-making looms to steam engines for railroad locomotives, problems of deformed and broken crankshafts have plagued the operators and manufacturers of the machines. Problems associated with the kinematics of relative motion between parts and lubrication were so extensive that little effort was expended in understanding the impact of pistons and connecting rods on the dynamic characteristics of the crankshaft. In the 1920's, with the introduction of multi-cylinder internal combustion engines in mass produced automobiles, greater emphasis was placed on the design of balance masses to counter the dynamic forces created by the reciprocating masses. And with the availability of digital computers and structural dynamics software, the degree of reliability expected and achieved from crankshafts is attributable in no small measure to a thorough understanding of the torsional and flexural vibration characteristics of the crankshaft and its bearing system.

The need for a comprehensive textbook on the dynamics of the crankshaft, connecting rod and piston, incorporating the most recent developments of the subject, has been strongly felt for some years. Although a very extensive amount of research has been done, and an even greater number of publications exist on the subject, little effort has been spent in putting together in one volume topics including conceptual design, dynamic analysis, fluid flow in manifolds, design optimization, vibration measurement and dynamic balance. The book is primarily written to meet the needs of practicing engineers in a large variety of industries where reciprocating machines are used, although all of the material is suitable for college undergraduate level design engineering courses. It is expected that the reader is familiar with basic to medium level calculus offered at the college undergraduate level.

The first chapter of the book deals with classical vibration theory, starting with a single degree of freedom system, to develop concepts of damping, response and unbalance. The second chapter deals with types and classification of reciprocating machines, while the third chapter discusses detail design aspects of machine components. The fourth chapter introduces the dynamics of slider and crank mechanism, and provides explanation of the purpose and motion of various components. The fifth chapter looks into dynamic forces created in the system, and methods to balance gas pressure and inertia loads. The sixth chapter explains the torsional vibration theory, and looks at the different variables associated with it. Chapter seven analyzes flexural vibrations and lateral critical speed concepts, together with journal bearings and their impact on a rotating system.

Advanced analytical techniques to determine dynamic characteristics of all major components of reciprocating machinery are presented in chapter eight. Methods to mitigate torsional vibrations in a crankshaft using absorbers are analyzed in close detail. Various mechanisms of flexural excitation sources and their response on a rotor-bearing system are explored. Stability of a rotor and different destabilizing mechanisms are also included in this chapter.

Techniques in vibration measurement and balancing of reciprocating and rotating systems are presented in chapter nine. Chapter ten looks at computational fluid dynamics aspects of flow through intake and exhaust manifolds, as well as fluid flow induced component vibrations. Chapter eleven extends this discussion to pressure pulsations in piping attached to reciprocating pumps and compressors. Chapter twelve considers the interaction between the structural dynamics of components and noise, together with methods to improve sound quality. Optimized design of components of reciprocating machinery for specified parameters and set target values is investigated at length in chapter thirteen.

Practicing engineers interested in applying the theoretical model to their own operating system will find case histories shown in chapter fourteen useful. Computer programs provide complete methods in evaluating the performance of components and system. Electronic spreadsheet type of calculations are shown here to consider several aspects of the overall problem. Program modules are arranged to calculate the full array of design parameters, dimensions and dynamic characteristics. As a further guide, actual example problems using the program modules are provided within the programs as well as in the text. Using a step-by-step procedure a complete dynamic behavior and response of a system can be obtained through these program modules, thus providing the reader with a full design of the system to satisfy a given set of requirements. A suite of program sub-routines for dynamic balancing of rotating systems in the shop or in the field is also included.

I gratefully remember and appreciate past students of the course on this topic who have sent in comments and reported errors, and express my hope that those who work with this treatise will do likewise. I am indebted to Mr. Mark Belloni of General Electric Company for performing a vast amount of computational work in finite element analysis, and to Ann Pulido for valuable advise on the text and layout of the book. I greatly appreciate comments provided by Dr. Weize Kang of TK Engineering Associates, Dr William Marscher of Mechanical Solutions, Inc., Dr. Raj Subbiah of Siemens-Westinghouse, Mr. Dean Nelson of Kohler Company and Mr. John Cagney of Navistar International for checking the problems and reading the proof.

Orlando, FL **A. S. Rangwala**
U.S.A.

LIST OF SYMBOLS

The symbols provided here are mostly to facilitate identification with commonly used parameters in the equations and the associated text. However, because of the considerable number of topics discussed in the text the relevant variables are defined within each section, and may even be redefined for convenience and better understanding of the subject matter.

a, A = cross sectional area, Mach number

ao = amplitude of support

an = Fourier coefficient of $Sin(nwt)$

bn = Fourier coefficient of $Cos(nwt)$

c = damping constant, clearance

C = condenser capacity

cc = critical damping constant

$C1, C2$ = constants

d, D = diameters

e = eccentricity

e = amplitude of pendulum support

E = modulus of elasticity

Eo = maximum voltage, $Eo\text{Sin}(wt)$

f = frequency = $\omega/2\pi$

fn = natural frequency

f and g = numerical factors

F = force in general or dry friction force in particular

F = frequency function

g = acceleration of gravity

G = shear modulus

h = height

i = station number, $\sqrt{(-1)}$, imaginary unit of complex number

I = moment of inertia

J = polar mass moment of inertia

j = $\sqrt{(-1)}$, imaginary unit of complex number

k, K = spring constants, kinetic energy

Δk = variation in spring constant

l, L = length

ln = distance from nth station

L = inductance

m, M = *mass*

M = moment or torque

= angular momentum vector

= magnitude of angular momentum vector

n = station number, number in general, gear ratio
p = real part of complex frequency s, pressure
$p1$, $p2$ = parameters
Po = maximum force, Po Sin(wt)
P = force, potential energy
q = natural frequency of damped vibration
R = electrical resistance
s = complex frequency = $p \pm jq$
r, R = radius of circle
q = load per unit length on beam
Q = condenser charge
t = time
T = period of vibration = $1/f$
To = maximum torque ToSin(wt)
T = Torque, tension
V, V = velocity
v, V = volume
W = weight, work or work per cycle
X = displacement
Xo = maximum amplitude
xst = static deflection, usually, Po/k
$y = yo\text{Sin}(wt)$ = amplitude of relative motion
y = lateral deflection of string or bar
Y = response
Z = impedance

α = angle, bypass factor
α_n = nth crank angle in reciprocating engine
α_{mn} = influence number, deflection at m caused by unit force at n
β_n = angular amplitude of vibration of nth crank
δ = small length or other parameter in general
δ_{st} = static deflection
ε = eccentricity, parameter
λ = length, multiplier
μ = mass ratio m/M
μ_1 = mass per unit length of strings, bars
ξ = damping coefficient
ρ = radius of gyration, density
θ = angle
ϕ, φ = phase angle or some other angle
φ_n = phase angle between vibration of nth crank and first crank
ψ = an angle
ω = circular frequency = 2pf
ω = angular velocity
Ω = large angular velocity
ω_n, Ω_n = natural circular frequencies

CONTENTS

1 Vibration Theory Fundamentals

1.1 INTRODUCTION

Mechanical vibrating systems consist of elements such as a spring for storing potential energy, mass and inertia for kinetic energy, and damper for dissipating mechanical energy. The vibration process alternatively converts energy between its potential and kinetic forms. In its general sense the vibration is a periodic motion that repeats itself in all its details after a certain interval of time, called the period of vibration. Some energy must be replaced in each cycle of vibration from an external source to maintain the vibration.

In a linear spring the change in length is proportional to the force acting along its length, the constant of proportionality referred to as the spring's flexibility, which is the reciprocal of its stiffness. The ideal spring has no mass, and so the forces acting at its opposite ends are equal and opposite. A mass is a rigid body; so its acceleration, according to Newton's second law of motion, is proportional to the force acting on the mass. In a viscous damper the applied force is proportional to the relative velocity of its connecting points. The damping coefficient provides the proportionality constant, and is the characteristic parameter of the damper. The ideal damper also is massless, and so the force at one end is equal and opposite to the force at the other end.

Besides translational motion, a vibrating system may execute rotational motion. The elements of a mechanical system executing pure rotation of the parts are analogous to the elements executing pure translation. In a rotational system inertia stores kinetic energy, while stiffness and damping parameters are defined with reference to angular rotation and velocity. The analogy between translational and rotational motion of a vibrating body carries on to the mathematical equations describing the motion of the system; force used in linear motion is replaced by torque for rotational motion.

A plot of the linear displacement, or angular rotation, against time may be a complicated curve. The simplest kind of periodic motion is a harmonic motion, with the displacement expressed as a harmonic function of time t and angular velocity ω, also referred to as the circular frequency, and is measured in radians per second. The maximum value of the displacement, x_0, is called the amplitude of vibration. The period of vibration, T, measured in seconds, is the reciprocal of the frequency of vibration, f, measured in cycles per second. The relation between ω, f and T are as follows. A full cycle of vibration takes place when ωt has passed through 360°, or 2π radians. Then the sine function resumes its previous path. Thus, when $\omega t = 2\pi$, the time interval t is equal to the period T, or $T = 2\pi/\omega$ seconds. Since f is the reciprocal of T, $f = \omega/2\pi$ cycles per second. For reciprocating machinery the frequency is often expressed as cycles per minute, so $f = 30\omega/\pi$. In a harmonic motion displacement is given by

$x = x_0 \sin(\omega t)$. Velocity is determined by differentiation with respect to time, so $dx/dt = x_0\omega \cos(\omega t)$. Thus, velocity is also harmonic, with a maximum value of $x_0\omega$. Acceleration is given by the second derivative of x with respect to time t:

$$d^2x/dt^2 = -x_0\omega^2 \sin(\omega t) \tag{1.1}$$

also harmonic, with a maximum value of $x_0\omega^2$.

A vibrating system is said to have one degree of freedom if its geometrical position can be expressed at any instant in time by one number only. For example, a piston moving in a cylinder can be specified by giving the distance from the cylinder end, thus classifying it as a single degree of freedom system. Another example is a crankshaft in rigid bearings; the system is fully described by the angle between any one crank and the vertical plane. In general, if it takes n numbers to specify the position of a mechanical system, the system has n degrees of freedom. A disk moving in its plane without restraint has three degrees of freedom: the x and y displacements of its center of gravity and the angle of rotation about the center of its gravity. A circular disk rolling down an inclined plane has only one degree of freedom; if it partly slides and rolls, it has two degrees of freedom, one of translation and one of rolling. A rigid body moving freely in space has six degrees of freedom, three of translations and three of rotation. Thus, it takes six coordinates to express its position, usually denoted as x, y, z, θ, φ and ψ.

A single degree of freedom system is not necessarily a simple one. For example, a 12-cylinder gas engine with a rigid crankshaft and a rigidly mounted cylinder block has only one degree of freedom with all its components such as pistons, connecting rods and camshaft. This is because a single parameter, such as the crank angle through which the shaft has turned, determines completely the position of all its moving parts. But if the cylinder block is mounted on flexible springs so that it can freely move in all directions as in an automobile engine, it has seven degrees of freedom, six in reference to the motion of the block as a rigid body in free space and crank angle as the seventh coordinate.

1.2 VECTOR METHOD OF REPRESENTING RECIPROCATING MOTION

The motion of a vibrating particle can be conveniently represented by means of a rotating vector, x_0 (Fig. 1.1). When time is reckoned from the horizontal position of a vector rotating at a uniform angular velocity, the horizontal projection of the vector is $x_0 \cos(\omega t)$, and the vertical projection will be $x_0 \sin(\omega t)$. Either projection can be taken to represent harmonic motion of a reciprocating component. This representation gives rise to the name circular frequency for ω.

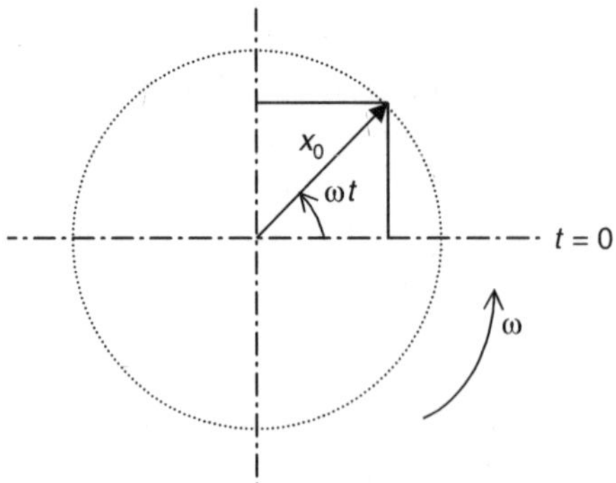

Fig. 1.1. Horizontal Projection of a Rotating Vector

This method of representing reciprocal motion is convenient. If a point is simultaneously subjected to two motions of the same frequency which differ by the phase angle ϕ, namely a cos (ωt) and b cos $(\omega t - \phi)$, the geometric sum of the two vectors gives the total motion (Fig. 1.2). Parallelogram a, b rotates in a counter-clockwise direction with uniform angular velocity ω, and the horizontal projections of all the vectors represent displacement as a function of time. Line $x - x$ represents a specific instant in time for which the vector diagram is drawn. Displacement of the sum is the sum of the ordinates for a and b.

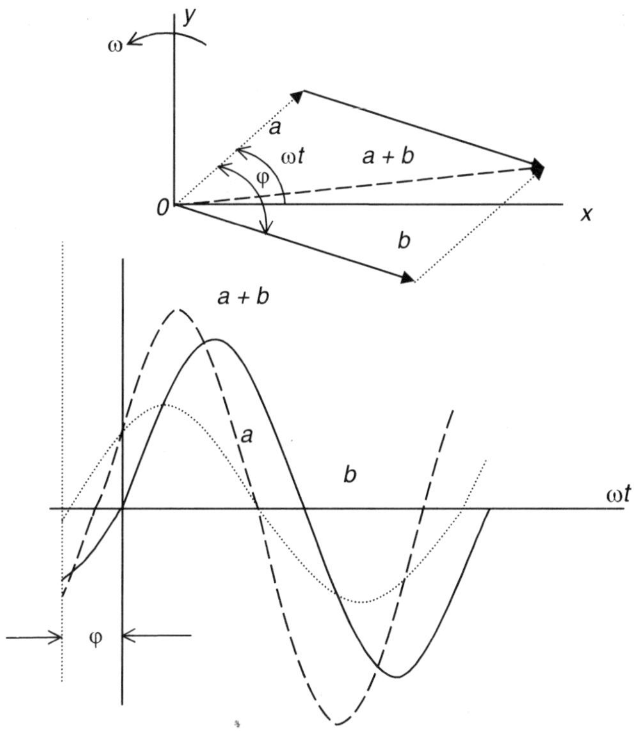

Fig. 1.2. Two Vibrations Added by their Vectors

Addition of the two vectors is permissible only if vibrations are of the same frequency. Motions a sin (ωt) and a sin $(2\omega t)$ can be represented by two vectors, the first rotating at an angular speed ω, the second at twice that speed. The relative position of these two vectors is changing continuously, thus a geometric addition of the vectors has no physical implication.

A special case of vector addition often encountered is the addition of sine and cosine waves of different amplitudes, a sin (ωt) and b cos (ωt). The two vectors are perpendicular (Fig. 1.3), so $a \sin(\omega t) + b \cos(\omega t) = [(a^2 + b^2)^{1/2}] . \sin(\omega t + \varphi)$, where $\tan \varphi = b/a$. For example, consider the sum of two motions 10 sin $(25t)$ and 20 sin $(25t + 1)$. At time $t = 0$, the first vector points vertically up, with no horizontal projection. The second vector is turned 1 radian in a counterclockwise direction from the first vector. The graphical vector addition shows the sum vector to be 26.76 units long.

1.3 COMPLEX NUMBERS METHOD

The vector method offers visualization of harmonic motions that is simpler than the consideration of the sine wave by itself. For numerical calculations, however, it is not well adapted. Complex numbers employing imaginary numbers offers a simpler method. A complex number can be represented graphically by a point in plane where real numbers are plotted horizontally and imaginary numbers are plotted along the vertical axis. Using $j = (-1)^{1/2}$, imaginary numbers are j, $2j$, $3j$, etc. By joining that point with the origin, the complex number can be made to represent a vector (Fig. 1.4). If the angle of the vector is α with the horizontal axis, and of length a, then the vector is $a(\cos \alpha + j \sin \alpha)$. Since harmonic motions are represented by rotating vectors, the variable angle ωt may be substituted for fixed angle α to obtain a rotating vector. Its horizontal projection will then be a reciprocating motion. The sum of the two motions considered earlier by the vector method can be solved using complex numbers. The first vector is $-10j$ and the second vector is $-20j \cos(1) + 20 \sin(1) = -10.8j + 16.8$. The sum of the two vectors is $16.8 - 20.8j$, representing a vector of length $(16.8^2 + 20.8^2)^{1/2} = 26.7$ units.

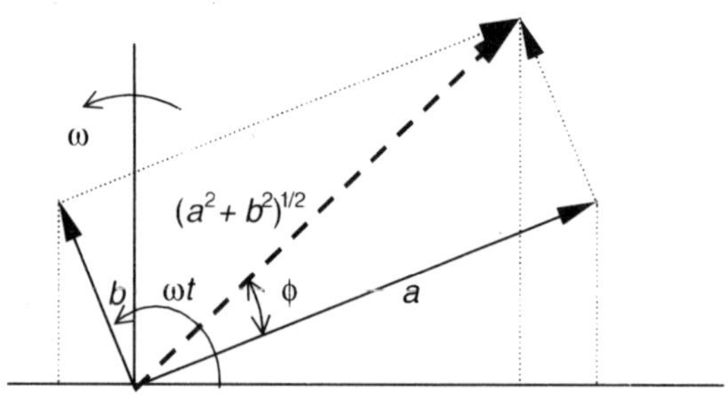

Fig. 1.3. Addition of Sine and Cosine Waves

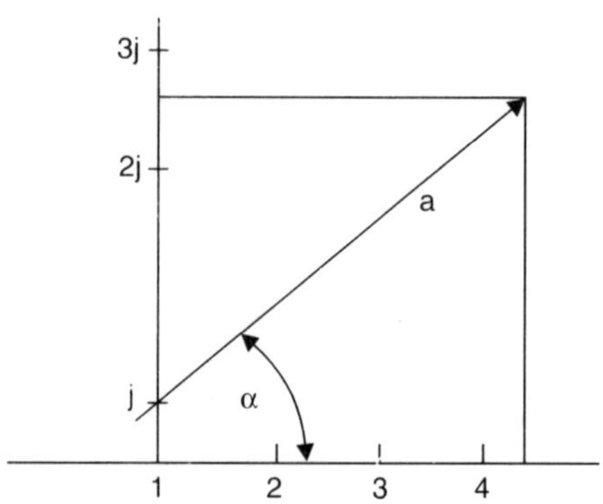

Fig. 1.4. Vector Representation in a Complex Plane

Differentiating $a\{\cos(\omega t) + j \sin(\omega t)\}$, gives $j\omega a\{\cos(\omega t) + j \sin(\omega t)\}$. Note that by definition, $j.j = -1$. Thus, differentiation of a complex number is equivalent to multiplication by $j\omega$.

In vector representation differentiation multiplies the length of the vector by ω and turns it ahead by 90°. Thus, multiplying a complex number by j is equivalent to moving the vector by 90° without changing its magnitude. In making extended calculations with complex numbers the ordinary rules of algebra are followed. Algebraic calculations are performed without much recourse to their physical meaning, only the final answer is interpreted using the real part for the vector magnitude and phase angle from both the real and imaginary parts.

A concept of importance in many applications is that of work done by a harmonically varying force on a harmonically varying motion of the same frequency. Consider a force $P = P_o \sin(\omega t + \varphi)$ acting upon a body for which the motion is given by $x = x_o \sin(\omega t)$. The work done by the force during a small displacement dx is Pdx, which may be written as $Pdt(dx/dt)$. During one cycle of vibration ωt varies from 0 to $2\pi/\omega$. After integration, the work per cycle is $W = \pi P_o x_o \sin(\varphi)$.

In order to understand this expression for work a graphical approach is useful. In Fig. 1.5*a* the ordinates are the displacement x and the in-phase component of the force. Between A and B the force is positive, say upward, and the body is moving in the same direction, so the work done is positive. Between B and C, on the other hand, the body moves downward toward the equilibrium point while the force is still positive, so that negative work is done. The work between A and B cancels that between B and C, and over a whole cycle the work done is zero. Thus, if a harmonic force acts on a body subjected to a harmonic motion of the same frequency, the component of the force in phase with the displacement does no work. Since the velocity vector is 90° ahead of the displacement, it follows that force does work only with that component which is in phase with the velocity.

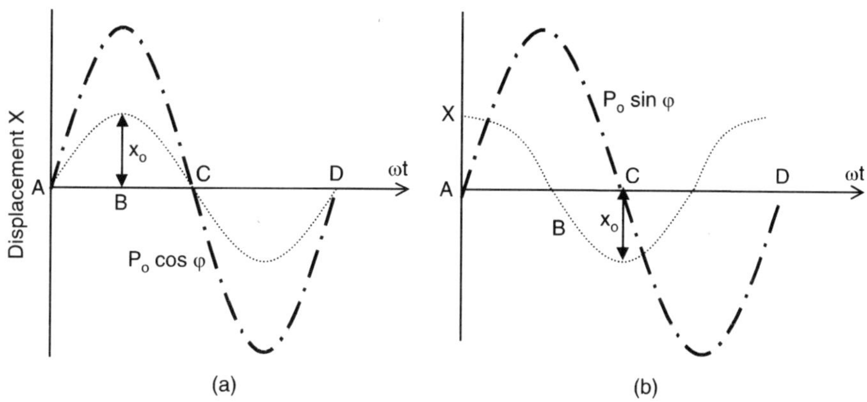

Fig. 1.5. Work Done on Harmonic Motion, (a) In Phase and (b) 90° Out of Phase

1.4 FREE VIBRATIONS WITHOUT DAMPING

Consider a mass m suspended by a spring from a fixed location, as shown in Fig. 1.6. Stiffness of the spring is defined by k, the force required to extend it by 1 unit. A damper is also placed between the mass and the fixed wall. The damper does not transmit force to the mass when at rest, but as soon as it moves it exerts a damping force of cdx/dt.

It is proportional to the velocity and directed opposite to the mass. c is the coefficient of viscous damping. An external alternating force $P_o \sin(\omega t)$ is also acting on the mass. The problem consists in determining the motion of mass m due to this external force. The equation of motion can be obtained by applying Newton's second law of motion, which states that force is equal to mass multiplied by acceleration. All forces acting upward will be considered positive when acting downward. Spring force has the magnitude $-kx$, and is negative since it pulls upward. Damping force acting on the mass is also negative, $-cdx/dt$. The downward forces acting on the mass are:

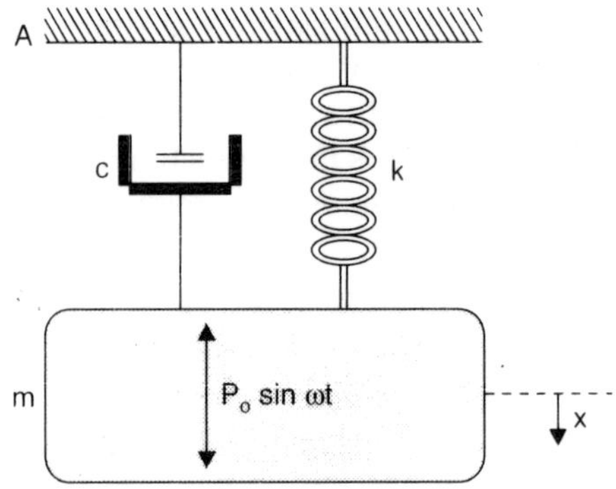

Fig. 1.6. Single Degree of Freedom Vibrating System

$$-kx - c\frac{dx}{dt} + p_o \sin(\omega t) \tag{1.2}$$

Newton's law gives:

$$md^2x/dt^2 + cdx/dt + kx = P_o \sin(\omega t) \tag{1.3}$$

This equation is known as the differential equation of motion of a single degree of freedom system. Before developing a solution of the general equation, simplified cases will be considered first. If there is no external applied force and no damping, the equation reduces to:

$$md^2x/dt^2 + kx = 0 \tag{1.4}$$

The most general solution of this equation is:

$$x = C_1 \sin\sqrt{\frac{k}{m}}\,t + C_2 \cos\sqrt{\frac{k}{m}}\,t \tag{1.5}$$

where C_1 and C_2 are arbitrary constants. Physically, the result as it stands is indefinite, since the constants can take any value. This is because the problem itself was not fully defined. If it is specified that the mass is pulled out of its equilibrium position to $x = x_o$, then released without initial velocity, two conditions are specified:

At $t = 0$: $x = x_o$ and $dx/dt = 0$.

The first condition gives $C_2 = x_o$. For the second condition, (1.5) must be differentiated, then get $C_1 = 0$. Substitution into (1.5) leads to the specific solution:

$$x = x_o \cos\sqrt{\frac{k}{m}}\,t \tag{1.6}$$

This represents an undamped vibration. Denoting the period of vibration by T, then:

$$T = 2\pi\sqrt{\frac{m}{k}} \tag{1.7}$$

Customarily, $(k/m)^{1/2}$ is denoted by ω_n, called the natural circular frequency. The natural frequency f_n is given by:

$$f_n = \frac{1}{T} = \frac{1}{2\pi}\sqrt{\frac{k}{m}} = \frac{\omega_n}{2\pi} \tag{1.8}$$

measured in cycles per second.

In the derivation of the differential equation of motion the effect of gravity has been neglected. The amplitude of vibration is measured from the position where the downward force mg is held in equilibrium by an upward force $k\delta$, δ being the extension of the spring due to gravity. To measure x_1 from the unstressed position of the spring, then $x_1 = x + \delta$. In the differential equation of motion, then, x must be replaced by x_1, and on the right hand side of the equation a force mg must be added.

If it is assumed that the motion is harmonic, the frequency can be calculated from an energy consideration. In the center of the swing the mass has maximum kinetic energy, while at either extreme position it has no kinetic energy. At the same time the spring is stretched (or compressed) to the fullest at the extreme positions, and has elastic, or potential, energy stored in it. Between the middle and extreme positions the spring-mass system has both kinetic and potential energy, the sum of which has to be constant since there is no external work done on the system. Thus, the kinetic energy at the center position must equal elastic energy at the extreme position. For a linear spring the potential energy when stretched over a distance x is $1/2kx^2$. The kinetic energy at any instant is $1/2mv^2$. If the motion is assumed to be $x = x_o \sin(\omega t)$, then $v = x_o\omega \cos(\omega t)$. At the extreme position the potential energy is $1/2kx_o^2$, and the kinetic energy at the center position where the velocity is maximum is $1/2m\omega^2x_o^2$. Equating the two energies, $1/2kx_o^2 = 1/2m\omega^2x_o^2$, from which $\omega^2 = k/m$, independent of the amplitude x_o. This method is of significance when dealing with problems of greater complexity, where the generalized energy method, known as the method of Rayleigh, leads to a result.

1.5 TORSIONAL PENDULUM

Fig. 1.7 shows a disk of moment of inertia I attached to a shaft of torsional stiffness k, defined as the torque necessary to produce one radian twist at the disk. The twisting motion of the disk is under the influence of an externally applied torque $T_o \sin(\omega t)$. Since only the torsional displacement φ is required to describe its position from the equilibrium position, this is a single degree of freedom problem. For a rotating body, Newton's law states that torque is equal to moment of inertia times angular acceleration, or:

$$T = Id^2\varphi/dt^2 \tag{1.9}$$

The three torques acting on the disk are the spring torque, damping torque and external torque. The spring torque $-k\varphi$, where φ is in radians, is negative, just as the spring force in the previous case was $-kx$. The stiffness k will depend on the shaft cross-section geometry and its material shear modulus. The damping torque is $-cd\varphi/dt$, caused by the shaft material's internal damping. The damping constant c is the torque on the disk caused by an angular speed of rotation of 1 radian per second.

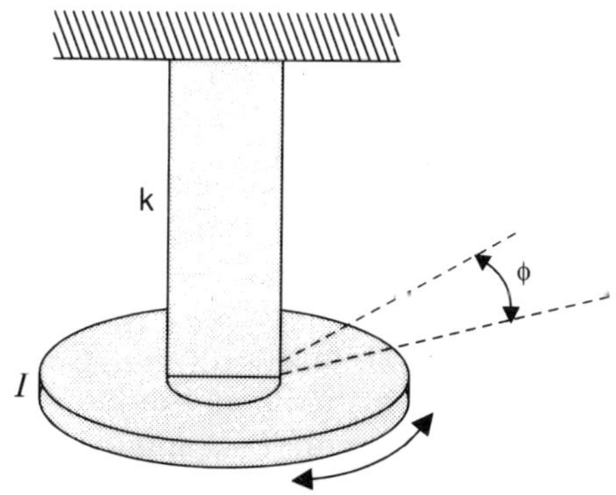

Fig. 1.7. Torsional Pendulum

Application of Newton's law yields the differential equation of motion:

$$Id^2\varphi/dt^2 + cd\varphi/dt + k\varphi = T_o \sin(\omega t) \quad (1.10)$$

This equation is similar to the translational case, and the solution for the undamped free condition can be readily obtained by substituting I for m, T_o for P and φ for x in the linear case.

Any system with inertia, elasticity and damping proportional to the velocity, where the displacement can be described by a single quantity, can be defined in this manner. Consider two disks of moment of inertia I_1 and I_2 joined by a shaft of torsional stiffness k, (Fig. 1.8). On the first disk torque $T_o \sin(\omega t)$ is acting. Both disks can assume an angular position independent of the other by twisting the shaft. However, if the quantity of interest is angle of twist in the shaft, it is possible to express the motion in terms of this quantity alone. Let φ_1 and φ_2 be the angular displacements of the disks, then $\varphi_1 - \varphi_2$ is the shaft twist, $k(\varphi_1 - \varphi_2)$ is the spring torque, $cd(\varphi_1 - \varphi_2)/dt$ is the damping torque. Applying Newton's law to the first disk, then the equation of motion is:

$$T_o \sin(\omega t) = I_1 d^2\varphi_1/dt^2 + cd(\varphi_1 - \varphi_2)/dt + k(\varphi_1 - \varphi_2) \quad (1.11)$$

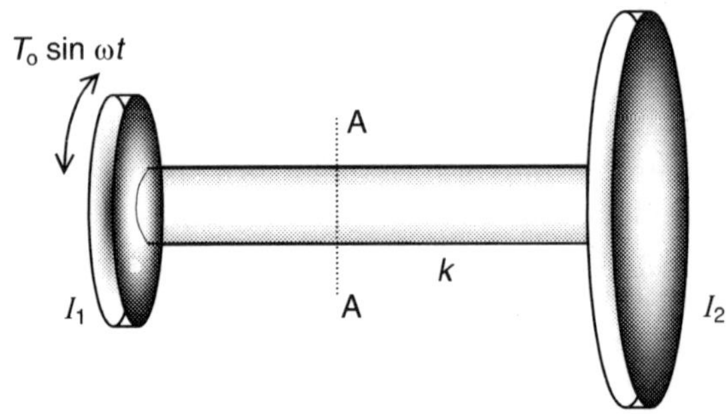

Fig. 1.8. Two Disks Mounted on an Elastic Shaft

and to the second disk,

$$0 = I_2 d^2\varphi^2/dt^2 + cd(\varphi_2 - \varphi_1)/dt + k(\varphi_2 - \varphi_1) \quad (1.12)$$

Rearranging the variables and subtracting the results from the two equations yields:

$$(T_o/I_1)\sin(\omega t) = (d^2\varphi_1/dt^2 - d^2\varphi_2/dt^2) + (c/I_1 + c/I_2)$$
$$d(\varphi_1 - \varphi_2)/dt + (k/I_1 + k/I_2)(\varphi_1 - \varphi_2) \quad (1.13)$$

Defining the twist angle $(\varphi_1 - \varphi_2) = \psi$, then multiplying the equation by the expression $I_1I_2/(I_1 + I_2)$ gives the equation of motion with a single variable ψ,

$$I_1I_2/(I_1 + I_2)d^2\psi/d^2t + cd\psi/dt + k\psi = T_0 I_2/(I_1 + I_2)\sin(\omega t) \quad (1.14)$$

The solution of this equation gives information on the angle of twist, but angular rotation of individual disks cannot be obtained. A variation is shown in Fig. 1.9 where a gear and pinion system is attached to the shaft. Assume gears G and P to be without any inertia. Also assuming the gear teeth to be rigid, torsional flexibility is limited to the two shafts. Gear ratio is n. The differential equations of motion for this system could be derived from Newton's law, but the system could be reduced to that of Fig. 1.8 by omitting the gears and replacing I_2, k_2 and ψ by equivalent parameters. If I_2 is clamped and torque T_o is applied to I_1, then I_1 will rotate through an angle φ_o, so $k = T_o/\varphi_o$. Due to the gears the torque in the shaft k_2 is T_o/n, and the angle of twist of k_2 is T_o/nk_2. Since I_2 is fixed, this is the angle of rotation of the pinion P. Angle of the gear G is n times smaller, or T_o/n^2k^2. Adding angle T_o/k_1 for shaft k_1 gives angular rotation of I_1. Thus, the equivalent k is obtained from the expression:

$$\frac{1}{k} = \frac{\varphi_o}{T_o} = \frac{1}{k_1} + \frac{1}{n^2 k_2} \tag{1.5}$$

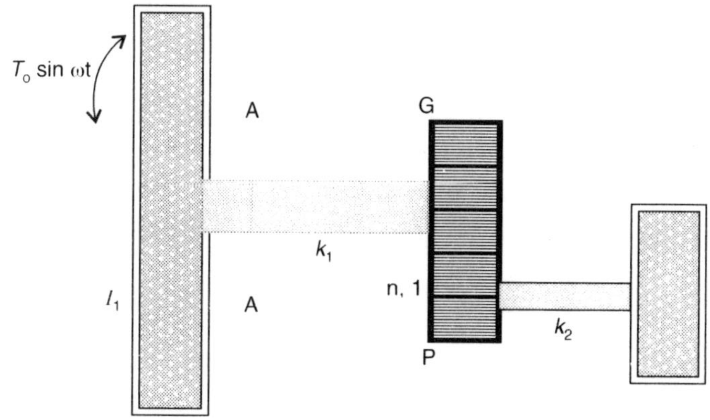

Fig. 1.9. Geared System

Now consider the inertia. Angular acceleration a in k_1 and G becomes $n\alpha$ in k_2 and P. Hence the torque in k_2 is $n\alpha I_2$. This is also the torque at the pinion P. Gear G sees it n times larger, so the torque at A is $n^2\alpha I_2$ and the equivalent of I_2 in the gearless system is $n^2 I_2$. In general, then, a geared system such as that of Fig. 1.9 can be reduced to an equivalent non-geared system (Fig. 1.8) in the following manner.

Divide the system into separate parts each of which has the same speed within itself. Figure 1.9 has only two parts, but there may be several. Choose one of these parts as the primary and assign numbers n to each of the other parts so that n is the speed ratio with respect to the primary. $n > 1$ for speeds higher than the primary part's speed, while the speed ratio of the primary part is unity. Next, remove all gears and multiply all spring constants k and inertias I by the factors n^2. The differential equation of the reduced gearless system is then the same as that of the original geared design.

In still another variation of Fig. 1.6, instead of the sinusoidal force acting on the mass, the upper end of the spring is made to move up and down with an amplitude a_o, the motion of A being $a_o \sin(\omega t)$. It can be shown that this motion of the top of the spring is equivalent to a force on the suspended mass. If x is the downward displacement of the mass, spring extension will be $x - a_o \sin(\omega t)$, spring force is $-k\{x - a_o \sin(\omega t)\}$ and damper force is $-c\{dx/dt - a_o\omega \cos(\omega t)\}$. Newton's law then gives:

$$md^2x/dt^2 + cdx/dt + kx = ka_o \sin(\omega t) + ca_o\omega \cos(\omega t) \tag{1.16}$$

As mentioned earlier in section 1.2, the sum of a sine and cosine waves of the same frequency is also harmonic, so:

$$md^2x/dt^2 + cdx/dt + kx = [(ka_o)^2 + (ca_o\omega)^2]^{1/2} \sin(\omega t + \varphi_o) \tag{1.17}$$

Thus, the motion of the top of the spring with amplitude a_o is equivalent to a force of amplitude $\{(ka_o)^2 + (ca_o\omega)^2\}^{1/2}$. Terms ka_o and $ca_o\omega$ in this expression are the maxima of the spring and damper forces, while the expression inside the parenthesis provides the peak magnitude of the force when the mass is clamped.

1.6 FREE VIBRATIONS WITH VISCOUS DAMPING

Undamped free vibrations persist indefinitely, which cannot happen; all free vibrations eliminate with time. The term viscous damping is associated with the expression cdx/dt since it represents the conditions of damping due to the viscosity of oil in a dashpot. Other types of damping exist, but their solution cannot be determined with as much ease. Consider the function $x = e^{st}$, where t is the time and s an unknown constant. Upon differentiation the same function multiplied by a constant results. This function when substituted in the equation:

$$md^2x/dt^2 + cdx/dt + kx = 0 \tag{1.18}$$

permits division by e^{st} and leads to an algebraic equation instead of a differential equation. Thus, assuming e^{st} as the solution, then the above equation becomes:

$$(ms^2 + cs + k) \, . \, e^{st} = 0 \tag{1.19}$$

This is a quadratic in s, with two values that will satisfy the equation:

$$s_{1,2} = -\frac{c}{2m} \pm \sqrt{\left(\frac{c}{2m}\right)^2 - \frac{k}{m}} \tag{1.20}$$

The most general solution of the problem is:

$$x = C_1 e^{s_1 t} + C_2 e^{s_2 t} \tag{1.21}$$

where C_1 and C_2 are arbitrary constants. The physical significance of this equation will depend on whether the expressions for s are real or complex. When $(c/2m)^2 > k/m$, both values of s will be real. Also, they are both negative since the square root is smaller than the first term $c/2m$. Thus (1.21) describes a solution consisting of the sum of two decreasing exponentials, as shown in Fig. 1.10. The case of $C_1 = 1$ and $C_2 = -2$ is shown as a dashed line.

This figure shows that the motion is not oscillatory but rather a creeping back to the original position. This is due to the fact that when $(c/2m)^2 > k/m$ the damping c is large. For smaller values of c pertaining to more practical cases, (1.20) gives complex values for s. The damping c at which this transition occurs is called the critical damping c_c:

$$c_c = 2m\sqrt{\frac{k}{m}} = 2\sqrt{mk} = 2\,m\omega_n \tag{1.22}$$

If the damping is less than critical damping, the solution consists of two factors, the first a decreasing exponential and the second a sine wave. The combined result is an exponentially decreasing sine wave, lying in the space between the exponential curve on both sides of the phase angle axis (Fig. 1.11). The smaller the damping constant c, the flatter will be the exponential curve and the more cycles it will take for the vibrations to be eliminated.

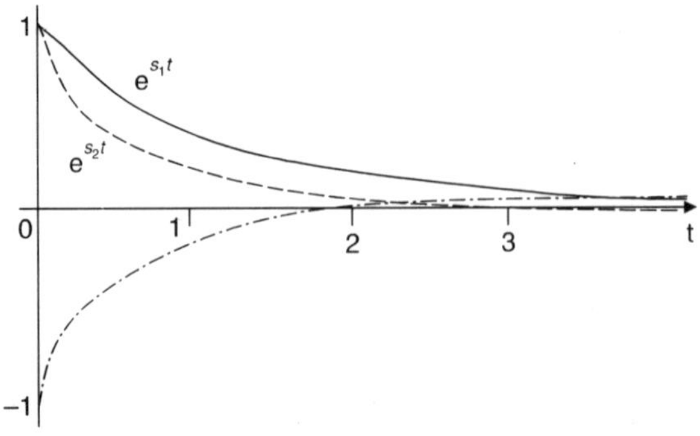

Fig. 1.10. Motion of Single Degree of Freedom System with $c/c_c > 1$

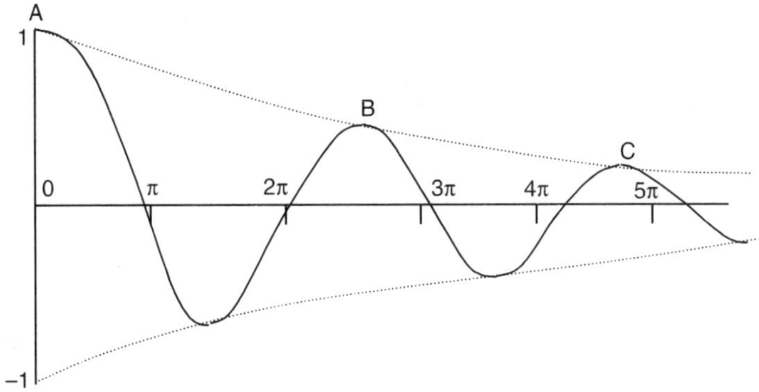

Fig. 1.11. Free Vibration of a System with $c/c_c < 1$

The rate of this dying down of the vibrations is of interest, and can be calculated by considering two consecutive maxima of the curve, $A - B$, $B - C$, and so on. During the time interval between any two such maxima, the amplitude of vibration diminishes from:

$$e^{-\frac{c}{2m}t} \text{ to } e^{-\frac{c}{2m}\left\{t+\frac{2\pi}{q}\right\}} \text{ where } q = \sqrt{\frac{k}{m} - \frac{c^2}{4m^2}} \tag{1.23}$$

The latter of these two expressions is equal to the first one multiplied by the constant factor:

$$e^{-\frac{\pi . c}{mq}} \tag{1.24}$$

which factor is smaller than unity. This factor is the same for any two consecutive maxima, and is independent of the amplitude or of time. Thus, the amplitudes decrease in a geometric series.

The frequency of vibration diminishes with increasing damping. If written in a dimensionless form with the help of (1.23), it becomes:

$$\frac{q}{\omega_n} = \sqrt{1 - \left\{\frac{c}{c_c}\right\}^2} \tag{1.25}$$

This equation is plotted in Fig. 1.12, where the ordinate q/ω_n is the ratio of the damped to the undamped natural frequency, and the abscissa is the ratio of the actual to the critical damping. The figure is a circle; for critical damping the natural frequency is zero. Due to the horizontal tangent of the circle at $c = 0$, the natural frequency is nearly constant and equal to $\sqrt{(k/m)}$ for all typically encountered values of damping ($c/c_c < 0.2$).

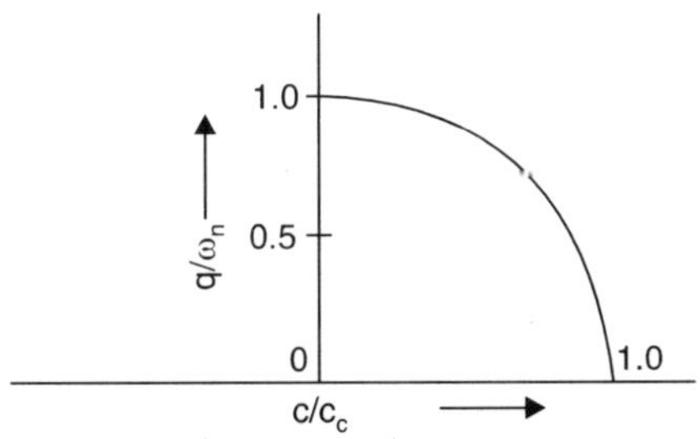

Fig. 1.12. Natural Frequency as a Function of Damping

1.7 FORCED VIBRATIONS WITHOUT DAMPING

Another particular case of (1.3) is the one where the damping term is made zero:

$$md^2x/dt^2 + kx = P_o \sin(\omega t) \tag{1.26}$$

Substituting a harmonic solution of the type $x = x_0 \sin(\omega t)$ into this equation yields:

$$-m\omega^2 x_o \sin(\omega t) + kx_o \sin(\omega t) = P_o \sin(\omega t) \tag{1.27}$$

producing the result:

$$x = \frac{P_o/k}{1-(\omega/\omega_n)^2} \sin(\omega t) \tag{1.28}$$

The expression P_o/k has the same physical significance of static deflection of the spring under a constant load P_o. Thus, if $P_o/k = x_{st}$, the solution becomes:

$$\frac{x}{x_{st}} = \frac{1}{1-(\omega/\omega_n)^2} \sin(\omega t) \tag{1.29}$$

Including the homogeneous solution containing the two integration constants, the total solution of the problem becomes:

$$x = C_1 \sin(\omega_n t) + C_2 \cos(\omega_n t) + \frac{x_{st}}{1-(\omega/\omega_n)^2} \sin(\omega t) \tag{1.30}$$

The first two terms represent the undamped free vibrations, the third term is the undamped forced vibration. Examining the implications of this result, x/x_{st} is a sine wave with an amplitude of $1/\{1-(\omega/\omega_n)^2\}$. For values of $\omega/\omega_n > 1$ the amplitudes are negative, and of $\omega/\omega_n < 1$ the amplitudes are positive. When the amplitudes are negative, it only means that the amplitudes are positive but with a phase angle lag of 180°. It also means that for $\omega/\omega_n < 1$ force and motion are in phase, and they are out of phase for $\omega/\omega_n > 1$. For $\omega/\omega_n < 1$ the mass is below the equilibrium position when the force pushes downward; the mass is above the equilibrium position when the force pushes downward for $\omega/\omega_n > 1$. At $\omega/\omega_n = 1$ the forced frequency coincides with the natural frequency, so the force is pushing the mass at the right time in the right direction. The amplitude then increases indefinitely. In the case of a pendulum, it is pushed slightly in the direction of motion every time it reaches the end of a swing; a small force is required to produce a large amplitude. This phenomenon is known as 'resonance', and the natural frequency is called the 'resonant frequency'.

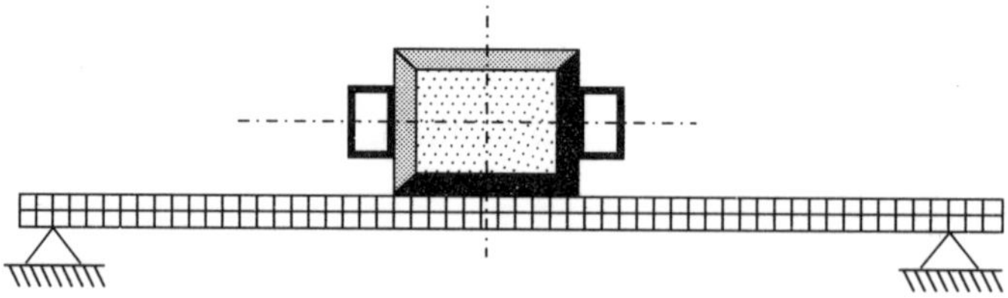

Fig. 1.13. Unbalanced Rotating Machine on a Beam

An important application of this theory is when the amplitude of the impressed force varies with ω^2. Fig. 1.13 shows a beam on two supports carrying a rotating machine with an unbalanced centrifugal force of $m_1\omega^2 r$, where m_1 is the unbalance mass and r is its distance from the center line of the shaft. This force can be resolved into a vertical component $m_1\omega^2 r \sin(\omega t)$ and a horizontal component $m_1\omega^2 r \cos(\omega t)$. If it is assumed that the beam is rigid for displacements in the horizontal direction, and flexible for displacements in the vertical direction, then it becomes a single degree of freedom system. Here the motor mass m and beam stiffness $k = 48EI/l^3$ are acted upon by the frequency dependent vertical disturbing force of amplitude $m_1\omega^2 r$. Considering the motion of the motor and beam system, it is equivalent to motor mass inertia force of $m\omega^2 a_o$ acting on the beam. The solution for this problem can be found directly from (1.28) by substituting $m\omega^2 a_o$ for P_o. Then:

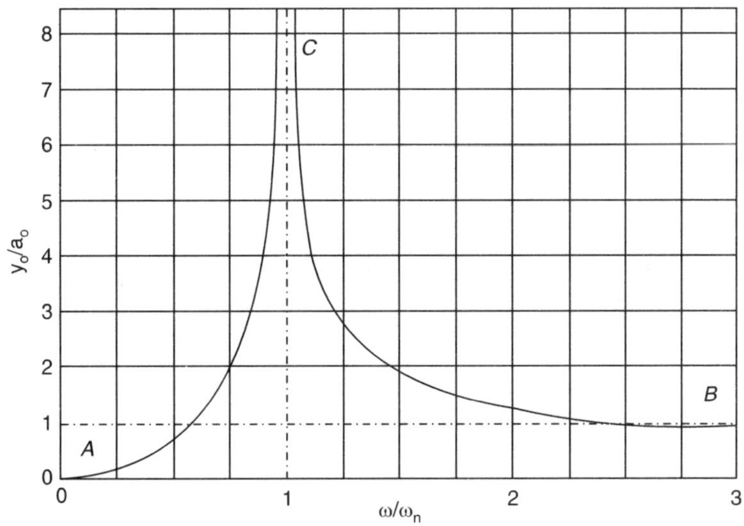

Fig. 1.14. Resonance Diagram for Rotating Machine on a Beam Example

$$y_o = \frac{m\omega^2 a_o/k}{1-(\omega/\omega_n)^2} = a_o \frac{(\omega/\omega_n)^2}{1-(\omega/\omega_n)^2}$$

or

$$\frac{y_o}{a_o} = \frac{(\omega/\omega_n)^2}{1-(\omega/\omega_n)^2} \qquad (1.31)$$

If y_o/a_o is plotted as a function of ω/ω_n, the result is shown in Fig. 1.14. At C there is resonance, and so spring extension is infinitely large in the absence of damping. Since this result is not in agreement with observations, it is necessary to consider damping. It shows the relative motion of a system in which the end of the spring is subjected to an alternating motion of constant amplitude a_o. The plot also shows the absolute motion of a system in which the mass experiences a force of variable amplitude $m\omega^2 a_o$. The ordinates of the three points A, B and C representing (1.31) can be physically explained as follows. At A the frequency ω is nearly zero, so the top of the spring moves at a slow rate, the mass follows this motion and the spring does not stretch, and so $y_o = 0$. At B the motion of the top of the spring is very rapid, so the mass cannot follow and stands still in space.

1.8 FORCED VIBRATIONS WITH VISCOUS DAMPING

The complete equation of motion is:

$$md^2x/dt^2 + cdx/dt + kx = P_o \sin(\omega t) \qquad (1.32)$$

The full solution consists of the complete solution of the equation with the right hand side set to zero and a particular solution of the whole equation (1.32), or:

$$x = e^{-\frac{c}{2m}t}\{C_1 \sin(qt) + C_2 \cos(qt)\} \text{ particular solution} \tag{1.33}$$

Thus, only the particular solution needs to be determined. Assuming a solution of the type $x = x_o \sin(\omega t - \varphi)$, the particular solution can be determined by finding x_o and φ. Newton's law requires that the sum of all the forces be zero at all times, implying that the sum of the vertical and horizontal components must also be zero. From these conditions the unknowns x_o and φ are solved:

$$x_o = \frac{P_o/k}{\sqrt{\left\{1 - \frac{\omega^2}{\omega_n^2}\right\}^2 + \left\{2\frac{c}{c_c}\frac{\omega}{\omega_n}\right\}^2}} \tag{1.34}$$

$$\tan(\varphi) = \frac{2\frac{c}{c_c}\frac{\omega}{\omega_n}}{1 - (\omega^2/\omega_n^2)} \tag{1.35}$$

The expressions for the amplitude and the phase angles are in terms of ratios of frequency and damping, where *cc* is the critical damping. P_0/k may be interpreted as the static deflection of the spring due to P_0. These relations are plotted in Fig. 1.15. The amplitude diagram contains a family of curves, one for each value of damping. All curves lie below the one for zero damping, since the amplitude of forced vibration is reduced by damping. The maxima of the different curves do not occur any more at $\omega/\omega_n = 1$ but at a smaller frequency. In fact, three different frequencies need to be distinguished, all of which coincide for $c = 0$:

- $\omega_n = \sqrt{(k/m)}$ = undamped natural frequency
- $q = \sqrt{\frac{k}{m} - \left\{\frac{c}{2m}\right\}^2}$ = damped natural frequency
- The frequency of maximum forced amplitude, or the resonant frequency.

The phase angle diagram is also of interest. For no damping it was seen that the force and displacement are in phase below resonance and 180° out of phase above resonance. So the phase angle shows a discontinuous jump at the resonance point. For damping values different from zero the other curves in the phase angle diagram are plotted. In general, damping tends to smooth out the sharpness of the undamped diagrams for both amplitude and phase.

Energy relations involved in this process also give a deeper understanding. For very slow motions $\varphi = 0$, and little work is done over a whole cycle, or no mechanical energy is converted into heat. Starting from the equilibrium position, the external force moves through a certain distance before reaching the extreme position, when it does work. But that work is used to store potential or elastic energy into the spring. During the next quarter cycle the motion goes against the external force and the spring gives up its stored energy. At slow speeds, then, the work of the external force is transferred into elastic energy and nothing is converted into heat. At resonant frequency $\varphi = 90°$, and work dissipated per cycle is $\pi P_o x_o$. The external force is equal and opposite to the damping force in that case, so the work is dissipated by the damper. The spring force and the inertia force are equal and opposite, and also in phase with displacement. Each of these forces individually perform work during a quarter cycle, but store the energy, returning it during the next quarter cycle. The work is stored periodically as elastic energy in the spring and as kinetic energy in the mass.

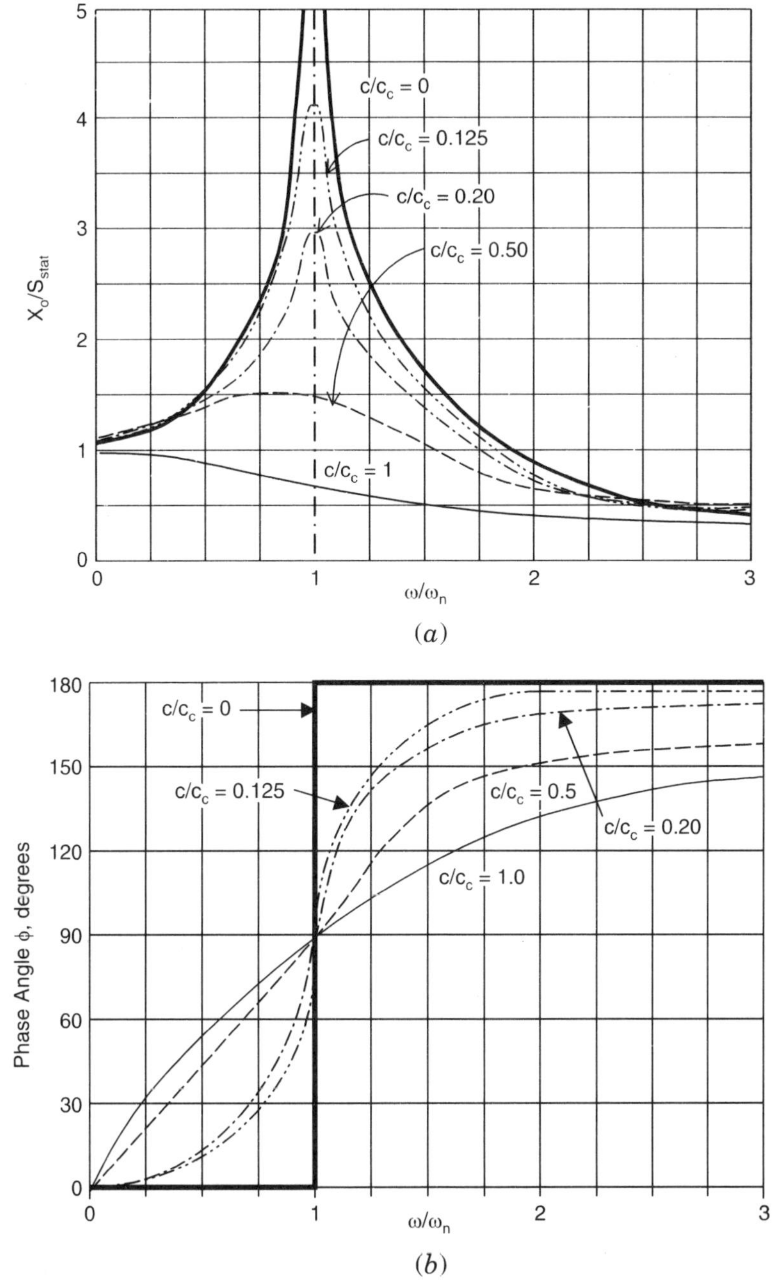

Fig. 1.15. (*a*) Amplitudes of Forced Vibrations, (*b*) Phase Angle Between Force and Displacement as a Function of Frequency for Various Damping Values

Returning to the assertion pertaining to the particular solution, the general solution consists of the damped free vibration superposed on the forced vibration. After a short time the damped free vibration disappears and the forced vibration alone persists. For this reason the forced vibration may also be referred to as sustained vibration and the free vibration becomes transient vibration. The values of the constants C_1 and C_2 depend on the conditions at the start and can be calculated from these conditions using the analytical process described earlier.

It is possible to describe the whole process by physical reasoning. For example, consider a mass suspended by a spring and acted upon by an external harmonic force of frequency eight times as slow as the natural frequency of the system. The mass is clamped while the external force is acting. Suddenly the clamp is withdrawn. Assume the damping in the system is such that the free vibration decreases by 10 percent for each cycle. Assume also that the clamp is released when the forced vibration has its maximum amplitude. From the initial conditions it follows that at the instant of the release the mass has no deflection and no velocity. The prescribed forced vibration starts with $x = x_o$ and $dx/dt = 0$. These two conditions can only be satisfied by starting a free vibration with $x = -x_o$ and $dx/dt = 0$. Then the combined motion will start at zero with zero velocity. Fig. 1.16 (*a*) shows the free vibration, 1.16 (*b*) shows the forced vibration and 1.16 (*c*) the combined motion.

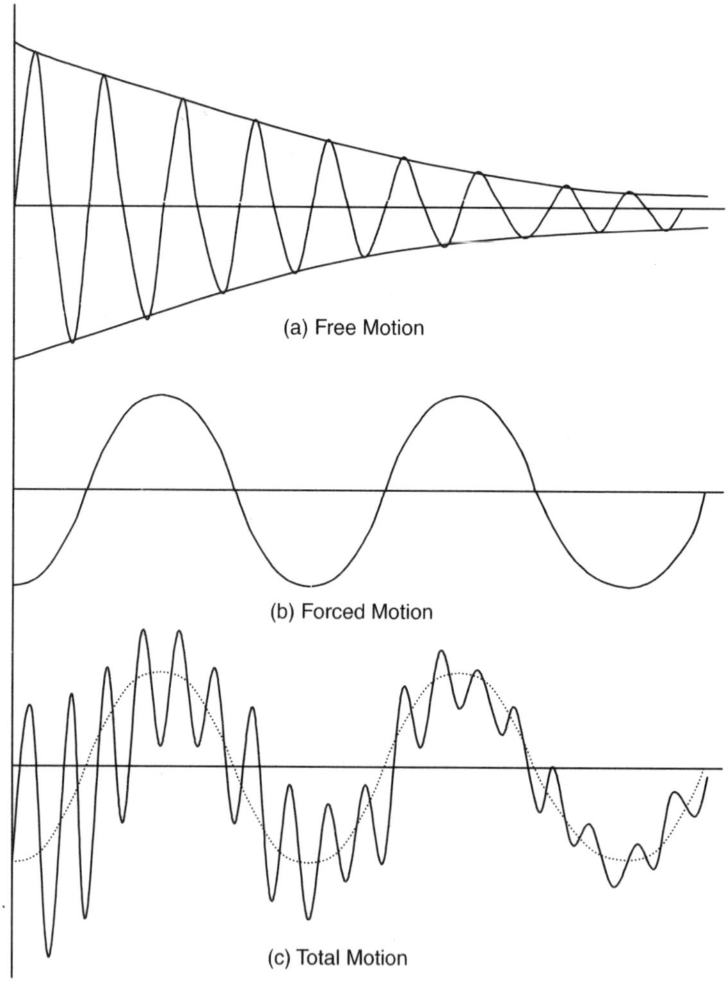

Fig. 1.16. Starting Transient Motion of a Forced Vibration

Note that the transient vibration disappears quickly and that the maximum amplitude at the start is nearly twice as great as the sustained final amplitude. If the difference between free and forced frequencies is small and damping is also small, the free and the forced vibration vectors would retain nearly the same relative motion during one revolution, the included angle between them changing only slightly. The vectors can then be added geometrically, and during one revolution of the two vectors the motion will practically be a sine wave of frequency $\omega_1 \approx \omega_2$ and amplitude the arithmetic sum of the two amplitudes. After a large number of cycles, the relative position of the vectors varies, and the magnitude of the sum vectors also changes. The resulting motion can then be described as an approximate sine wave with frequency ω_1 and an amplitude varying slowly between the sum and the difference of the amplitudes of the free and forced vibrations. This phenomenon is known as beats. Note that the free vibrations will be eliminated after a while due to energy dissipation from the damping. In order to have sustained beats it is necessary to have two sustained, or forced, vibrations.

1.9 VIBRATION DUE TO ROTATING ECCENTRIC WEIGHT

A classic example of forced vibrations with viscous damping is that of a rotating machine of mass m with a rotating eccentric mass m_u mounted in a shaft and bearings, Fig. 1.17. The rotating eccentric mass m_u follows a circular path of radius e with respect to the bearings. Considering only the motion in the vertical direction, the displacement of mass m_u relative to m is $x_2 - x_1 = e \sin(\omega t)$, where x_2 and x_1 are the absolute displacements of m_u and m and ω the angular speed of the machine rotor.

The differential equation of motion of the system is:

$$m_u d^2x/dt^2 + m_1 d^2x/dt^2 + cdx_1/dt + kx_1 = 0 \tag{1.36}$$

Differentiating $(x_2 - x_1)$ with respect to time and substituting in (1.36),

$$(m_u + m_1)d^2x/dt^2 + cdx_1/dt + kx_1 = m_u e\omega^2 \text{ sint } (\omega t) \tag{1.37}$$

This equation is of the same form as (1.32): by substituting $(m + m_u)$ for m and $m_u e\omega^2$ for P_o the resulting displacement, velocity and acceleration of the rotating machine mass m can be readily obtained. Note that the magnitude of eccentric mass m_u will be insignificant compared to the machine mass m.

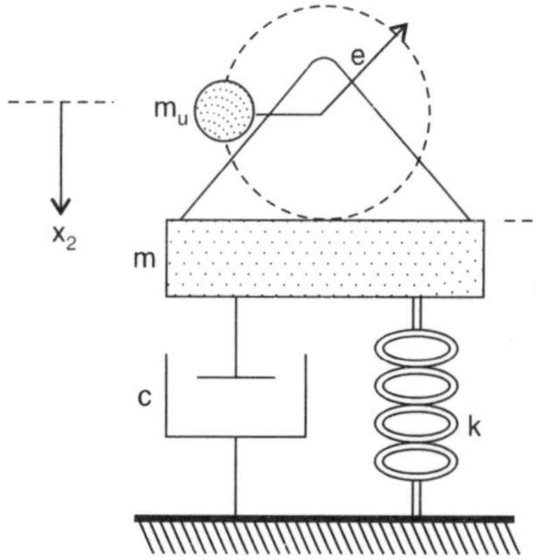

Fig. 1.17. Forced Vibrations from Rotating Eccentric Weight

When the machine is mounted on an elastic foundation, the motion of the foundation comes into play. As discussed earlier in section 1.7, displacement forced on an otherwise unforced system is equivalent to a system where a force is applied on that system. Therefore, a rotating machine with an eccentric mass in the rotor mounted on an elastic foundation behaves as if two distinct forcing functions are applied on it, one due to the eccentric mass whose frequency coincides with the angular speed of the rotor, and the other due to motion from the foundation. The frequency of the foundation motion will depend on the elastic stiffness of the foundation and the mass of the rotating machine. Damping characteristics of the foundation will control the amplitude of motion of the machinery from both sources of excitation, since the foundation will be in the direction of the load path traveling from the rotor to the foundation. Relative motion between the rotor and machine housing will be dependent on the material damping of the shaft and the bearings.

As an example of the above discussion, consider a centrifugal compressor weighing 3,000 lb. mounted on a simply supported elastic beam of negligible weight, of the type shown in Fig. 1.13. The elastic beam sags 9 in. under the weight of the compressor, and has a damping coefficient of 28 lb. for a velocity of 1 in. per second. The elastic beam is moved up and down at resonant speed with an amplitude of 1 in. Assume the center of the compressor to coincide with the center of the beam along the beam's length. The natural frequency of vibration is $(g/\delta_{st})^{1/2} = (386/9)^{1/2} = 6.6$ rad./second. At resonance the disturbing force is $\{(ka_o)^2 + (ca_o\omega)^2\}^{1/2}$. Here stiffness of the beam $k = 3000/9 = 333$ lb./in; $a_o = 1$ in.; damping $c = 28$ lb.-second/in; and natural frequency $\omega = \omega_n = 6.6$ radians/second. The disturbing force is then equal to 380 lbs. The amplitude of motion of the compressor then is disturbing force divided by the product of c and ω, and is found to be 2.07 in.

1.10 ROTATING AND RECIPROCATING UNBALANCE

An important group of vibration phenomena of practical importance in reciprocating machines is torsional oscillations in the crankshaft and in the shafting of driven machinery. The effect is caused by a combination of periodic accelerations of moving parts (piston, connecting rod, crank) and periodic variations in cylinder steam or gas pressure.

In a single cylinder vertical reciprocating machine the piston executes an alternating motion, and in the process experiences alternating vertical accelerations. When the piston accelerates downward there must be a downward force acting on it, and this force must have a reaction pushing upward against the stationary parts of the engine. Thus, the alternating acceleration of the piston is coupled with an alternating force on the cylinder frame, making itself felt as a vibration in the engine and its supports. In the lateral direction perpendicular to both the crankshaft and the piston rod, moving parts are also being accelerated, namely the crank pin and part of the connecting rod. The forces that cause these accelerations must have equal and opposite reactions on the frame of the machine. This last effect is the horizontal unbalance. In the crankshaft main axis no inertia forces appear, since all moving parts remain in planes perpendicular to the crankshaft.

These various inertia forces can cause moments. Consider a two cylinder vertical machine with the cranks set 180° apart. While one piston is accelerated downward the other piston is accelerated upward, and the two inertia forces form a couple tending to rock the machine about a lateral axis. Similarly, the horizontal or lateral inertia forces of the two cranks are equal and opposite, forming a couple tending to rock the machine about a vertical axis.

A rocking about the crankshaft axis can occur even in a single cylinder machine. If the piston is accelerated downward by a pull in the connecting rod, this pull exercises a torque about the crankshaft axis. Since the piston acceleration is alternating, this inertia torque is also alternating.

If the whole machine mounted on weak springs is considered as the mechanical system, the external torque is zero, and so any increase in clockwise angular momentum of the moving parts must be neutralized by an increase in counterclockwise angular momentum of the stationery parts of the machine. Taking merely the moving parts of the machine as the mechanical system, an increase in the clockwise angular momentum of the moving parts must be caused by a clockwise torque on these parts, which has a counterclockwise reaction torque on the frame. If the machine frame is mounted solidly on its foundation, this counter torque is transmitted to the foundation and may cause trouble. But if the machine is mounted on soft springs, reaction to the foundation cannot penetrate through these springs and the counter torque is absorbed as an inertia torque of the frame and cylinder block, and the machine block must vibrate.

In Fig. 1.18 inertia forces are excluded by assuming the machine to be turning at a slow constant speed ω, or the moving parts have negligible mass. The pressure force on the piston is P, and is varying with time, or with crank angle ωt. Besides acting on the piston, the gas pressure also presses upward against the cylinder head. Force P is transmitted through the piston rod (force 1) to the cross head. Neglecting friction, it is held there in equilibrium by forces 2 and 3. Of the forces acting on the cross head, 3 is a compression in the connecting rod, and 2 is a reaction pressure on the frame to the right, of magnitude $P \tan(\phi)$. Force 3 of magnitude $P/\cos(\phi)$ is transmitted through the connecting rod to the crank pin (force 4). Shifting this force parallel to itself to O and adding a torque $yP/\cos(\phi)$, the driving torque of the gas pressure is obtained. Force 5 is taken up by the main bearings at O and can be resolved into a vertical component 6 and a horizontal component 7. Triangles 1, 2, 3 and 5, 6, 7 are similar, so the magnitude of 6 is P and that of 7 is $P \tan(\phi)$.

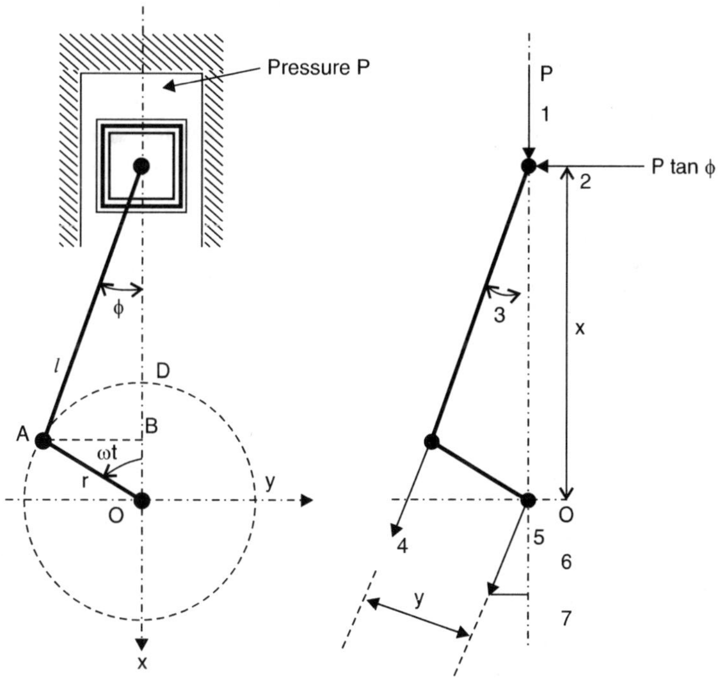

Fig. 1.18. Gas Pressure Load on a Single Cylinder Machine

The forces transmitted to the stationery parts of the machine are P upward on the cylinder head, $P \tan(\phi)$ to the right on the cross head guide, P downward on the main bearings at O and $P \tan(\phi)$ to the left on the main bearings at O. The total resultant force on the frame is zero, but there is a resultant torque $Px \tan(\phi)$. From Newton's law of action and reaction, this torque must be equal and opposite to the driving torque on the crankshaft, $yP/\cos(\phi)$. This is easily verified, because $y = x \sin(\phi)$. Thus, the gas pressure forces do not cause any resultant forces on the machine, producing only a torque about the main crankshaft axis.

In summary, no forces occur along the longitudinal crankshaft axis of an machine, while in the lateral and vertical directions only inertia forces appear. About the vertical and lateral axes only inertia torques are found, and in the longitudinal direction both an inertia torque and a cylinder gas pressure torque occur.

If the machine is assumed to be built of elastically non-deformable members, the problem is one of static balance only. The frame and other stationary parts generally fulfill this condition, but the crankshaft can be twisted significantly, making torsional vibrations possible. The subject may be divided into three categories:

(*a*) *Inertia Balance:* This refers to the balancing of the machine against vertical and lateral forces, and against moments about vertical and lateral axes.

(*b*) *Torque Reaction:* Under this heading the effects of torque due to inertia and gas pressure acting on the stationary parts about the longitudinal axis are analyzed.

(*c*) *Torsional Vibrations of Crankshaft:* Consequences of the longitudinal torque on the moving parts of the reciprocating machine are dealt with.

Effect *c* is of particular importance since many crankshafts have encountered failures due to the longitudinal torque.

1.11 VIBRATION ISOLATION

An unbalanced machine has to be installed in a structure where the vibrations are undesirable. Prime examples of this situation are gas turbine engines attached to an airplane wing and automotive engines. The problem is one of mounting the machine in such a manner that no vibrations will appear in the structure to which it is attached.

Probably the most popular solution consists of mounting the machine on springs. In Fig. 1.19 the machine is represented as a mass m with a force $P_o \sin(\omega t)$ acting on it. In figure *a* it is solidly attached to the structure, and in figure *b* it is mounted on springs with a combined stiffness k. For simplicity the machine is assumed to be rigid.

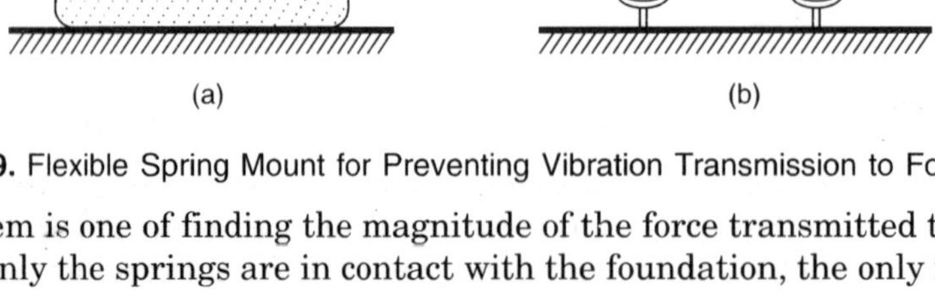

Fig. 1.19. Flexible Spring Mount for Preventing Vibration Transmission to Foundation

The problem is one of finding the magnitude of the force transmitted to the floor by the machine. Since only the springs are in contact with the foundation, the only force transmitted is the spring force, and that has the amplitude kx, damping being neglected here. If x_o is the displacement of the mass and $x_{st} = P_o/k$ is the static displacement, then:

$$\frac{x_o}{x_{st}} = \frac{x_o}{P_o/k} = \frac{kx_o}{P_o} = \frac{\text{spring force}}{\text{impressed force}}$$

$$= \frac{\text{transmitted force}}{\text{impressed force}} = \text{Transmissibility} \quad (1.38)$$

The ideal is to have this ratio zero, but practical considerations aim to make it small. If the spring constant $k = \infty$, as in figure a, the natural or resonant frequency is infinite. The operating frequency is then small compared to the natural frequency, ω/ω_n is nearly zero and the transmitted force equals the impressed force. Physically this is obvious, since the rigid floor does not permit mass m to move, and all the force P_o is transmitted to the foundation. It is necessary to design the support springs so as to make the natural frequency of the whole machine low compared with the disturbance frequency. Thus, the springs should be soft.

Other interesting observations can be made from (1.29). If ω is smaller than $\omega_n\sqrt{2} = \sqrt{(2k/m)}$, the springs make matters worse, since the transmissibility is greater than unity. If the natural frequency is one-fifth of the disturbing frequency, the transmissibility is 1/24, a good ratio.

So far the support has been considered without damping, which is the case in steel springs. If cork or rubber padding is used for the purpose, then damping is not negligible and must be accounted for. The system is then represented by Fig. 1.20. Now the displacement curve is not directly proportional to the amplitude of the transmissibility curve. The force is made up of spring force kx_o as well as the damping force $c\omega x_0$. As shown earlier the two forces are out of phase by 90°. Consequently, their vector sum is the total transmitted force. The amplitude x_0 is given by the following expression:

$$\text{Transmitted force} = P_o \frac{\sqrt{1+\left\{\frac{c\omega}{k}\right\}^2}}{\sqrt{\left\{1-\frac{\omega^2}{\omega_n{}^2}\right\}^2+\left\{2\frac{c\omega}{c_c\omega_n}\right\}^2}} \quad (1.39)$$

Transmissibility is defined as the ratio of transmitted force to impressed force. Thus, damping is advantageous only in the region $\omega/\omega_n < 1.41$, where spring mounting makes matters worse. For all values of ω/ω_n where spring mounting helps, the presence of damping makes the transmissibility worse. However, this statement is not as important as it may appear at first sight. For one thing the effect of damping is not great, and can be easily offset by making the springs weaker. Secondly, operation at resonance point $\omega/\omega_n = 1$ is never the intention, although it may sometimes unfortunately occur, and then the presence of damping is highly desirable. Thus, in spite of the dictum, some damping in the springs is generally advantageous.

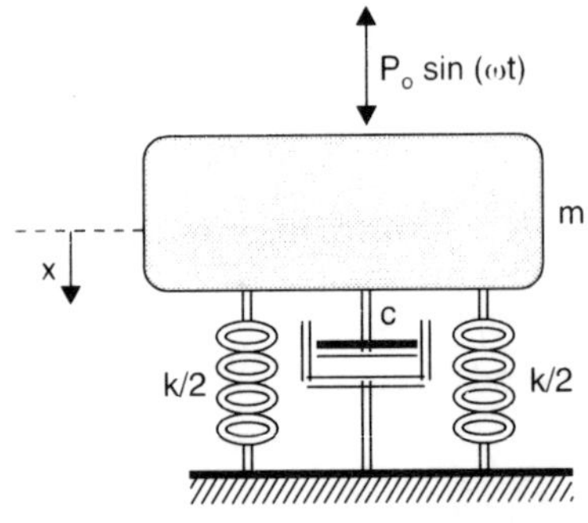

Fig. 1.20. A Spring Support with Damping

Practical cases of isolation by means of springs occur in many machines. Their main application lies in apparatus that is inherently unbalanced, or as in the case of most reciprocating machines, inherently has a non-uniform torque. The case of internal combustion engines used in automobiles is discussed in detail in the next section.

1.12 APPLICATION TO AUTOMOBILE ENGINES

Before discussing the vibration transmission of the automobile engine, a brief introduction to cylinder-pressure torque and its variation with crank angle, or time, will be useful. This topic will be discussed at length in later chapters. Internal combustion engines operate in a cycle that have four distinct regimes. During the intake phase fresh air is sucked into the cylinder through the inlet valves. In an engine operating on four-cycles this phase occurs when the piston is near 0° crank angle, close to the cylinder head, and lasts till the crank angle reaches approximately 180°. At this instant the piston reaches the lower dead center. At the end of the stroke the inlet valves close and the compression cycle begins, lasting from about 180° crank angle to 360° crank angle, when the piston is once again at the top dead center. The firing phase of the cycle then begins, with the spark ignition occurring just after the piston has reached top dead center. As the flame spreads throughout the cylinder and the crank turning, the gases expand and perform work on the piston by pushing it to the other end of the cylinder. At approximately 180° crank angle the exhaust valves open and the spent gases then exit the cylinder, at the end of which the piston reaches once again to the top dead center to commence the next cycle. In a four cycle single cylinder engine, at the four dead center positions during the two revolutions of a firing cycle, the torque is zero. Thus, the average torque delivered by the cylinder is only a small fraction of the maximum torque that occurs during the firing period. The fact that the torque is so irregular constitutes one of the primary disadvantages of the reciprocating engine as compared with a turbine where the torque curve as a function of time is a straight horizontal line.

According to Newton's law, every action is accompanied by an opposite reaction, and due to the torque generated by the cylinder gas pressure on the crankshaft the engine block would experience a counter rotating torque. With the engine mounted rigidly on the frame, these torque variations have reactions on the automobile's frame, and total such that the natural frequency of vibration is appreciably lower than $n/2$ times the running speed. Uncoupling of dynamic motion is accomplished by mounting the engine block on two journals, fore and aft, supported in bearings attached to the chassis, enabling the block to rotate about an axis nearly parallel to the torque axis and passing through the center of gravity, shown as axis AA in Fig. 1.21. With any other mount arrangement than the one described, the engine block would be free to rotate about axis AA. To prevent this, a cantilever leaf spring B between the engine block and the chassis is used, the stiffness of which is selected such as to make the natural frequency sufficiently low.

Besides cylinder pressure torque the engine also experiences horizontal and vertical inertia forces, which have to be reacted at mount points A and B. For this reason both bearings A as well as the chassis end of the leaf spring B are embedded in rubber. In the actual construction axis AA is not exactly parallel to the torque axis. This is the right procedure, because generally the torque axis is not the principal axis of inertia, and consequently does not coincide with the corresponding axis of rotation. Any rigid body has three principal axes of inertia. For instance, take an elongated solid rectangular steel bar, Fig. 1.22, and attach to it a weightless shaft passing through its center of gravity, but not coinciding with one of its principal axes of symmetry. Application of a sudden torque to the assembly will cause it to accelerate. Multiplying the acceleration with the masses of elements in the bar gives inertia forces, and multiplying the forces by their distance from the axis of rotation form a torque, which is equal and opposite to the impressed torque. Also these forces when multiplied by their distance from the axis perpendicular to the shaft have a torque about that axis. This will have reaction at the two bearings; the right hand bearing experiences a force out of the plane of the diagram, the left hand bearing force being pushed into the plane. In the absence of the bearings, under the influence of the torque the bar would not rotate about the torque axis, since forces are required at the bearings to make it do so. In general, then, a body under the influence of a torque will rotate about an axis not coinciding with the torque axis if that is not a principal axis.

The axis about which the automobile engine has to be suspended, therefore, should not be about the torque axis itself but the axis of rotation to which the torque axis belongs. Only when the torque axis is a principal axis do the two coincide. Several other design schemes of spring-supported automobiles are available, most of which are similar to that of Fig. 1.21. Some have one rubber support at the rear of the engine and two other rubber supports close together at the same height in the front. These two supports are virtually a combination of the single bearing A and the restoring spring B of Fig. 1.21.

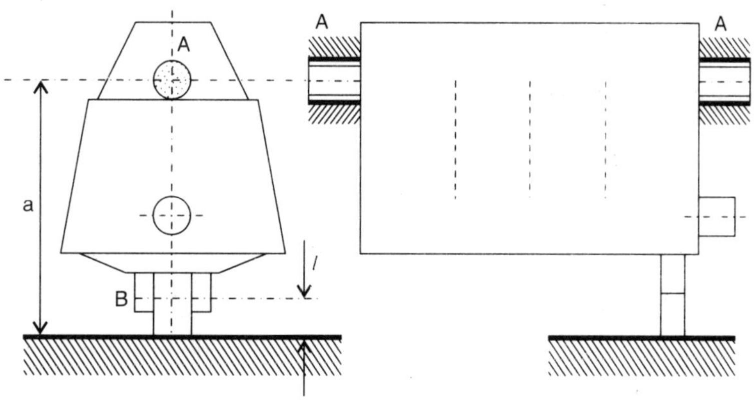

Fig. 1.21. Schematic Arrangement of Automobile Engine Mount

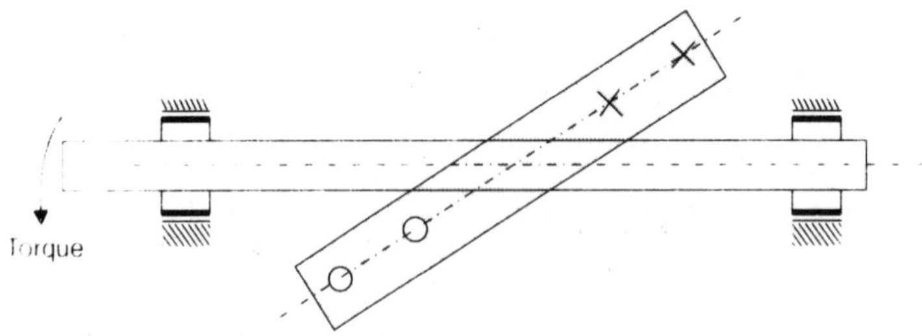

Fig. 1.22. Rotation of Bar about an Axis Different from its Principal Axis of Inertia

To illustrate the above discussion, consider a four cylinder automobile engine weighing 400 lb. supported in the same manner as in Fig. 1.21. The radius of gyration of the engine about axis *AA* is 6 in., distance a is 18 in. and length l of the cantilever is 4 in. The diameter of the rear wheels is 30 in., and in high gear the engine makes three revolutions per revolution of the rear wheels. The requirement is that the engine be in resonance at a speed corresponding to 3.5 miles per hour, or 61.6 in. per second, in high gear. Wheel circumference is $30\pi = 94.25$ in. At the critical speed the wheel makes 61.6/94.25 = 0.65 revolutions per second, so the engine runs at 3 × .65 = 1.95 r.p.s. Now, the engine torque curve goes through a full cycle for every firing, and since there are two firings per revolution in a four cylinder, four cycle engine, there will be 3.9 firings per second. For resonance, then, the frequency of the engine will be f_n = 3.9 cycles per second, or, $\omega_n^2 = 4\pi^2(3.9)^2 = 600$ rad^2/sec^2 $= k/I$. Here k is defined as the torque required in the cantilever per radian twist. The deflection at the end of the cantilever for a twist of φ radians is 18φ in. If k_1 be the linear stiffness of the cantilever in lbs/in., the spring force is $18k_1\varphi$ lbs acting on a moment arm of 18 in., so the torque is $18 \times 18k_1\varphi$ in.-lbs. Thus, $k = 324k_1$. Also, since the moment of inertia $I = 400 \times (6)^2/386 = 37.3$ in.-lb.-sec^2, then $\omega_n^2 = 600 = 324 \times k_1/37.3$ and $k_1 = 37.3 \times 600/324 = 69$ lbs per in.

Note that in this example if one cylinder was firing inadequately another periodicity is introduced in the torque curve for each two revolutions of the engine. Since the disturbance is four times as slow as the one discussed, it will be in resonance with the natural frequency of the engine at a speed of 4 × 3.5 = 14 miles per hour. Still another possibility is when the cylinders are not firing at the proper instant in time, when the overall dynamic characteristics will change.

1.13 EXAMPLE PROBLEMS

Example Problem 1.1: *A force 5 sin $(2\pi\, 60t)$ acts on a displacement of $(1/10)$ sin $(2\pi\, 60t - \pi/4)$. What is the work done during the first second? (Units: inch, pound, seconds).*

Solution: The force is $\pi/4$ radians (45°) out of phase with the displacement; so it can be resolved into two components, each of magnitude $5/\sqrt{2}$ lb. in phase and 90° out of phase with displacement. The component in phase with the displacement does no work. Work done by the 90° out of phase component per cycle is: $\pi P_o x_o = \pi \cdot (5/\sqrt{2}) \cdot (1/10) = 1.11$ in-lbs. In the first second there are 60 cycles, so work performed is 60 × 1.11 = **66.6 in-lbs.**

Example Problem 1.2: *In the above example what is the work during the first 1/1000 second?*

Solution: In the first 1/1000 second there will be 60 × 1/1000 = 0.06 cycle, or 0.06 × 360° = 21.6°. For part of the cycle, work is determined by integration.

$$W = \int P\,dx = \int P_o \sin(\omega t) \cdot x_o\omega \cdot \cos(\omega t - \varphi)\,dt$$

$$= 5(1/10)\int_0^{21.6} \sin(\omega t)\,.\,\cos(\omega t - 45^\circ)\,d(\omega t)$$

$$= \{(1/4)\cos(45^\circ)\,.\,\sin^2(\omega t) + (1/2)\sin(45^\circ)[\omega t/2 - (1/4)\sin(2\omega t)]\}\Big|_0^{21.6}$$

$$= (1/4) \times (0.707) \times (0.368)^2 + [21.6 \times 0.707/(4 \times 57.3)] - 0.707 \times 0.685/8$$

$$= \mathbf{+\ 0.030\ in\text{-}lb.}$$

Example Problem 1.3: In the system shown in Fig. 1.23 the mass weighs 2 lb., spring stiffness is 25 lb./in.; l = 8 in. and $a = b$ = 3 in. Also a damper is attached at the mid point of the beam, where the spring is fastened. The damper constant is 0.005 lb./in/sec. What is the damper critical damping?

Solution: The undamped natural frequency is $\omega_n = \sqrt{(k/m)}$. The equivalent spring constant is $k\,.\,a^2/l^2$, or 25 × (9/64) = 3.5 lb./in. Then:

$$\omega_n = \sqrt{(k/m)} = \sqrt{(3.5 \times 386/2)} = 26.0 \text{ radians/sec.}$$

The critical damping constant is:

$$c_c = 2m\omega_n = 2 \times (2/386) \times 26.0 = \mathbf{0.27\ lbs/in/sec.}$$

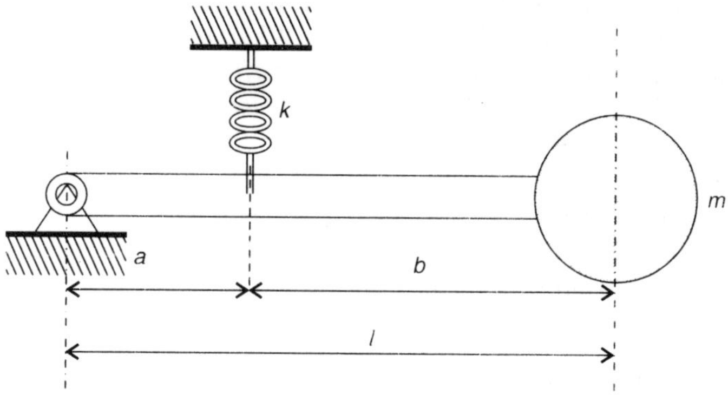

Fig. 1.23

Example Problem 1.4: Find the rate of decay of free vibration in the example problem # 1.3.

Solution: The rate of decay is:

$$\Delta x/x = \delta = 2\pi \times (c/c_c) = 2\pi \times (0.005/0.27) = \mathbf{0.116.}$$

Example Problem 1.5: A variable length cantilever beam consists of a strip of spring steel 0.20 in. wide and 0.020 in. thick. It carries a weight of 1/4 oz. at its free end. What should be the free length of the strip if it is required to have frequencies from 6 cycles/sec. to 60 cycles/sec?

Solution: The spring constant of a cantilever beam is $3EI/l^3$. The moment of inertia of the cross section is:

$$I = (1/12)bh^3 = .2 \times (.02)^3/12 = 1.33 \times 10^{-7} \text{ in}^4.$$

Bending constant $EI = 30 \times 10^6 \times 1.33 \times 10^{-7} = 4.0$ lbs-in², so spring constant $k = 12/l^3$ lbs/in. At the end, mass $m = 1/(4 \times 16 \times 386) = 4.05 \times 10^{-5}$ lbs-in^{-1}-sec^2.

Mass per inch length of strip $\mu = 0.2 \times 0.02 \times 0.28/386 = .29 \times 10^{-5}$ lbs-in^{-1}-sec^2. So total mass = $(4.05 + 0.29l) \times 10\text{-}5$ lbs-in^{-1}-sec^2.

Maximum length frequency is 6 cycles/sec, or, $\omega^2 = (2\pi \times 6)^2 = 1421$ (radians/sec)^2.

Natural frequency: $\omega^2 = k/m$, or $1421 = 12 \times 10^5/\{(4.05 + 0.29l)l^3\}$

By a method of trial and error, length of the beam is: **l = 5.33 in.**

Example Problem 1.6: *A flywheel is represented approximately by a solid steel disk of 4.5 in. diameter and 2 in. thickness. The shaft on which it is mounted has a torsional stiffness of 9.5 in.-lbs./radian. Find the natural frequency of vibration of the system.*

Solution: Weight of the flywheel is:

$$(\pi/4) \times (4.5)^2 \times 2.0 \times 0.28 = 8.9 \text{ lbs.}$$

and its moment of inertia is:

$$I = \frac{1}{2}(mr^2) = \frac{1}{2}(8.9/386)(2.252) = 0.0584 \text{ lbs-in-second}^2.$$

The natural frequency thus is:

$$\omega_n = \sqrt{(k/I)} = \sqrt{(9.5/0.059)} = 12.75 \text{ radians/second.}$$

or $$f_n = 12.75/2\pi = \mathbf{2.03\ cycles/second.}$$

Example Problem 1.7: *A stroboscope is a device for producing intermittent flashes of light with which rapid vibratory motions can be made to appear to stand still, or to move very slowly. The flashes of light are of extremely short duration. When a vibrating object illuminated with this light is adjusted to the same frequency as the vibration, the object will be seen in a certain position; then it will be dark, and the object is invisible while traveling through its cycle. When it returns to the first position after one cycle another flash of light occurs. Thus the object appears to stand still. Now consider a point located 4 in. from the axis of a machine rotating at 10,000 rpm. If it is required to see the point with a blurring of less than 1/32 in., what should be the duration of the light flashes?*

Solution: The point in question travels at:

$$10{,}000 \times 2\pi \times 4/60 = 4{,}189 \text{ in/sec.}$$

Flash time duration multiplied by blur size will give the velocity of the point, which is $4189 = t(1/32)$, so: $t = 1/(4189 \times 32) =$ **1/134,000 second.**

Example Problem 1.8: *A flywheel consists of a heavy rim of weight W = 475 lbs and mean radius R = 22.75 in. attached by four flexible prismatic spokes (Fig. 1.24). If the hub is held fixed, find the period of free rotational vibration of the rim about its central axis O. Neglect the mass of the spokes, and assume spoke bending stiffness B = 2,250 lbs/in.*

Solution: Consider the rim to have a small angle of rotation f from its equilibrium position as shown. Each spoke behaves as a beam built in at the hub and constrained to move with the rim at the other end. At the outer end of the spoke shear force Q and bending moment M act, whicn may be determined from formulas for beam stiffness:

$$Q = \frac{12B\Delta}{R^3} - \frac{6B\varphi}{R^2} = \frac{12 \times 2250\Delta}{22.75^3} - \frac{6 \times 2250\varphi}{22.75^2} = 2.293\text{D} - 26.08\varphi$$

$$M = \frac{6B\Delta}{R^2} - \frac{4B\varphi}{R} = \frac{6 \times 2250\Delta}{22.75^2} - \frac{4 \times 2250\varphi}{22.75} = 26.08\text{D} - 395.6\varphi$$

If the rim is assumed to be rigid, the tangent to the elastic line at the outer end of each spoke will be radial. Then deflection $\Delta = R\phi$, giving:

$$Q = 6B\phi/R^2 = 26.08\phi \quad \text{and} \quad M = 2B\phi/R = 197.8\phi$$

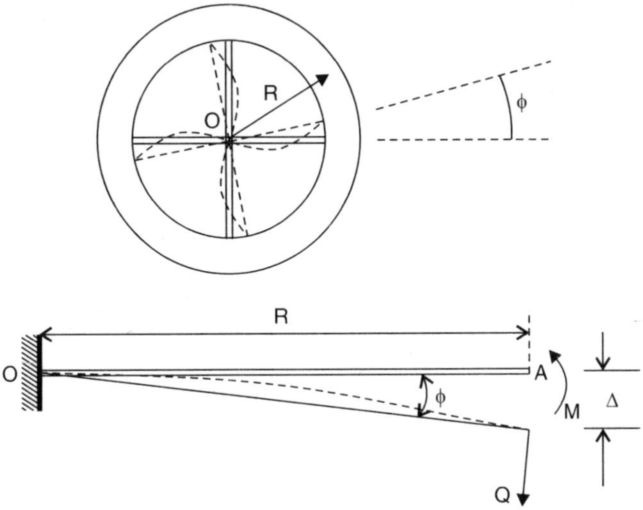

Fig. 1.24

Then the total moment on the rim will be:

$$M_t = 4QR - 4M = 16B\phi/R = 1582\phi$$

and the rotational spring constant is:

$$k_r = \frac{M_t}{\phi} = \frac{16B}{R} = 1582 \text{ in.-lbs/radian.}$$

Since the mass moment of inertia of the flywheel rim is:

$$I = WR^2/g = (475 \times 22.75^2)/386 = 636.9 \text{ in.-lbs/second}^2,$$

then the period of free natural vibration of the rim is:

$$\tau = 2\pi\sqrt{\frac{WR^3}{16gB}} = 2\pi\sqrt{\frac{JR}{16B}} = 2\pi\sqrt{\frac{636.9 \times 22.75}{16 \times 2250}} = \textbf{3.986 seconds.}$$

Example Problem 1.9: *Using Rayleigh's method for calculating the angular frequency of the fundamental mode of vibration of the beam with two masses, shown in Fig. 1.25. Assume* $W_1 = W_2 = W$, *flexural rigidity of the beam as EI and neglect beam mass.*

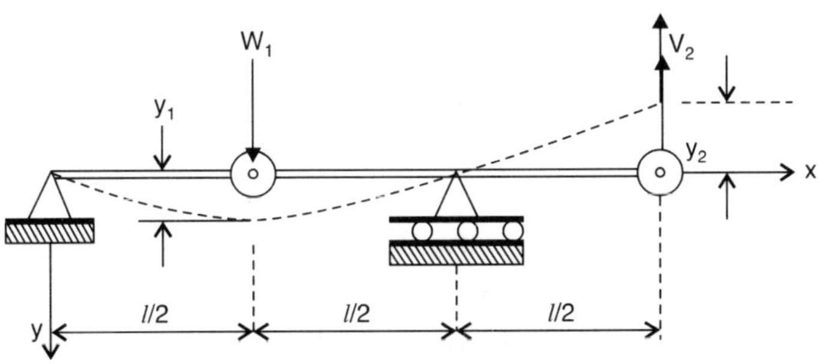

Fig. 1.25

Solution: From Rayleigh's method, for n masses on a beam, the formula for angular frequency is given by the expression:

$$p^2 = \frac{g\sum_{j=1}^{n} W_j y_j}{\sum_{j=1}^{n} W_j y_j^2}$$

It will be assumed that during vibration the beam maintains a shape similar to the static deflection curve due to the two weights acting in opposite directions. The corresponding static deflections are:

$$y_1 = \frac{W_1 l^3}{48EI} + \frac{W_2 l^3}{32EI} = \frac{5Wl^3}{96EI}$$

$$y_2 = \frac{W_1 l^3}{32EI} + \frac{W_2 l^3}{8EI} = \frac{5Wl^3}{96EI}$$

Substituting these values into the expression for angular frequency gives:

$$p = \sqrt{\mathbf{192}EIg/\mathbf{25W}l^3}.$$

Example Problem 1.10: *Figure 1.26 shows a mechanism used in the design of test machines to produce vibrations, and is capable of producing sine and cosine displacements. Crank r rotates at a constant angular velocity ω_r, and the projection of point P upon the x or y axes moves with simple harmonic motion. If the radius is 3.75 in., crank speed is 500 rpm and $\theta_r = \omega_r t = 65°$, determine the displacement, velocity and acceleration of point Q along the x axis.*

Solution: Displacement $OQ = x = r(\cos\theta_r) = 3.75\,[\cos(65°)]$

$= 1.585$ in.

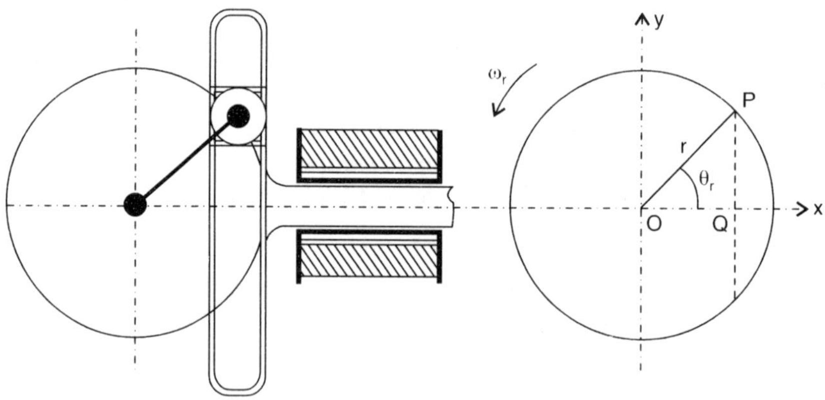

Fig. 1.26

Velocity: $v = dx/dt = r\omega_r \sin(\theta_r)$

$= 3.75 \times (2\pi \times 500/60) \times \sin(65°)$

= 177.95 in/second.

Acceleration: $a = r\omega_r^2 \cos(\theta_r)$

$= 3.75 \times (2\pi \times 500/60)^2 \times \cos(65°)$

$=$ **4344.87 in./second^2.**

Example Problem 1.11: *A variation of the slider and crank mechanism calls for increasing the size of the crank pin so that it is larger than the shaft to which it is attached, and at the same time offsetting the center of the crank pin from that of the shaft (Fig. 1.27). The enlarged crank pin, sometimes referred to as an eccentric, replaces the crank in the original mechanism. Point A is the center of the eccentric and O the center of the shaft. The motion with the equivalent crank length OA is identical with that of the slider and crank mechanism. Adequate lubrication between the eccentric and the rod may pose a problem and limit the amount of transmitted power.*

If AB is given as L = 34.5 in., crank length R = 4.25 in. and operating speed is 360 rpm, determine the displacement, velocity and acceleration of the slider when crank angle θ = 75°.

Solution: Expressions for displacement, velocity and acceleration can be readily derived with the aid of figure 1.28. Note that the projection of crank R along the vertical plane equals the projection of L. Now, displacement:

$$x = R + L - R[\cos(\theta)] - L[\sin(\phi)]$$

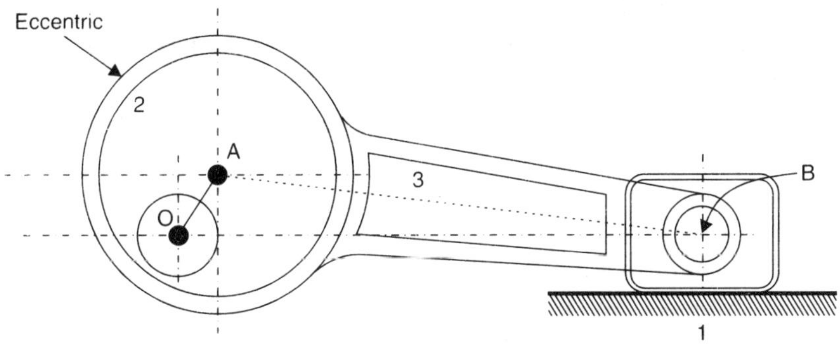

Fig. 1.27

$$= R[1 - \cos(\theta)] + L\left[1 - \sqrt{1 - \frac{R^2}{L^2}\sin^2(\theta)}\right]$$

Simplification of the expression under the radical sign can be obtained by using the first two terms in the binomial series, which will give sufficiently accurate results. Then:

$$x = R\,[1 - \cos(\theta)] + \frac{R^2}{2L}\sin^2(\theta)$$

$$= 4.25[1 - \cos(75°)] + 4.252 \times \sin 2(75°)/(2 \times 34.5)$$

$=$ **3.391 in.**

Expressions for velocity and acceleration of the ram are obtained by differentiating displacement with respect to time. Assume crank angular velocity to be constant.

Velocity: $$V = \frac{dx}{dt} = R\omega\left[\sin(\theta) + \frac{R}{2L}\sin(2\theta)\right]$$

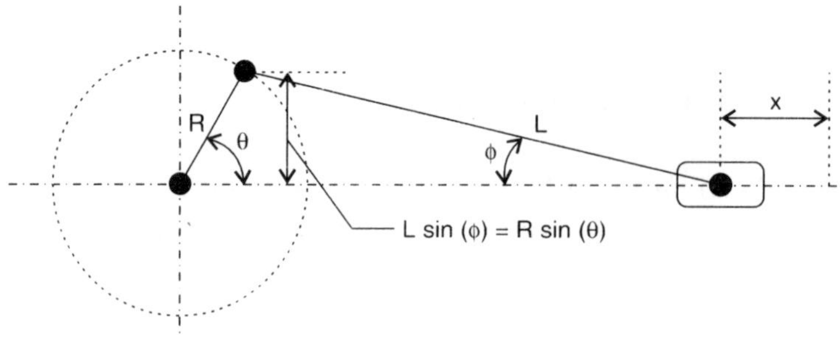

Fig. 1.28

$= 4.25 \times (2\pi \times 360/60) \times [\sin(75°) + 4.25 \times \sin(150°)/(2 \times 34.5)]$

$= \mathbf{159.71\ in/second.}$

Acceleration: $$A = \frac{d^2x}{dt^2} = R\omega^2\left[\cos(\theta) + \frac{R}{L}\cos(2\theta)\right]$$

$$A = 4.25 \times \{(2\pi \times 360/60)^2 \times [\cos(75°) + 4.25 \times \cos(150°)/(34.5)]\}$$

$= \mathbf{918.81\ in/second^2.}$

REFERENCES AND BIBLIOGRAPHY

Anwar, I., "Computerized Time Transient Torsional Analysis of Power Trains", *ASME Paper* No. 89-DET-74, 1989.

Choi, Y. S. and Noah, S. T., "Forced Periodic Vibration of Unsymmetric Piecewise—Linear Systems", *J. of Sound and Vibration,* 121(3):117-126,1988.

Crede, C. E., Ruzicka, J. E., "Theory of Vibration Isolation", Shock & Vibration Handbook, McGraw-Hill, New York, 1988.

Den Hartog, J. P., "Mechanical Vibrations", Dover Publications, Inc., New York, 1984.

Deutschman, A. D., Michels, W. J., Wilson, C. E., "Machine Design - Theory & Practice", MacMillan Publishing Co., New York, 1989.

Ehrich, F. F., Rotordynamic Response in Nonlinear Anisotropic Mounting Systems", *Proc. Of the 4th Intl. Conf. On Rotor Dynamics,* IFTOMM, 1-6, Chicago, September 7-9, 1994.

Ehrich, F. E., "Handbook of Rotordynamics", McGraw-Hill, New York, 1992.

Eshelman, R. L., "Torsional Vibrations in Machine Systems", Vibration, 3(2):3, 1987.

Gale, N. F., Widener, S. K., Sui, P. C., Zhang. H., "Analytical Engine Design Methods: A Review", SAE Paper # 950806, 1995.

Goldstein, H., "Classical Mechanics", Addison-Wesley Publishing Co. Inc., Mass, 1987.

Hildebrand, F. B., "Advanced Calculus for Applications", Prentice-Hall, New Jersey, 1983.

Jeon, H., Tsuda, K., "Theoretical Analysis of the Undamped, Coupled Torsional—Axial Vibration of Marine Diesel Engine Shaftings", Bulletin of the M. E. S. J., Vol. 4, No. 7, P41, 1989.

Law, B., "Prediction of Crankshaft and Flywheel Dynamics", Institute of Mechanical Engineers, Paper # C382/046, pages 487-498, Perkins Technology, 1989.

Mabie, H. H., Reinholtz, C. F., "Mechanism and Dynamics of Machinery", John Wiley & Sons, 1987.

McLachlan, N. W., "Ordinary Nonlinear Differential Equations", Oxford University Press, New York, 1956.

Meirovitch, L., "Analytical Methods in Vibration", Macmillan Publishing Co., Inc., New York, 1984.

Rao, S. S., "Mechanical Vibration", Addison-Wesley Publishing Co., Reading, Mass., 1990.

Rayleigh, "The Theory of Sound", Volume # I, Macmillan & Co. Ltd., London, 1894.

Roark, R. J., Young, W. C., "Formulas for Stress and Strain", 5th ed., McGraw-Hill, New York, 1985.

Robeson, G., "Cosworth - The Search for Power", Patrick Stephens Limited, London, 1991.

Schroen, B. G., "Die Dynamik der Verbrennungskraftmaschine", Springer-Verlag, Vienna, 1982.

Shemeld, D.E., "A History of Development in Rotordynamics, A Manufacturer's Perspective", CP2443, NASA, Washington, D. C., 1986.

Timoshenko, S., Young, D. H., Weaver, W., "Vibration Problems in Engineering", John Wiley & Sons, Inc., New York, 1984.

Wachel, J. C., Tison, J. D., "Vibrations in Reciprocating Machinery and Piping Systems", *Proceedings of the 23rd Turbomachinery and Piping Systems,* Texas A & M University, College Station, TX, 1994.

Wehrli, V. C., "Uber Kritische Drezahlen unter Pulsierender Torsion", *Ingenieur Arch,* 33:73-84, 1993.

Classification and Construction of Reciprocating Machines

2

2.1 INTRODUCTION

Reciprocating machines perform a plethora of services, come in every conceivable shape and size and are found in the home, on the road and on the high seas. They are used as internal combustion engines to transport people and materials, as nodding pumps at oil wells and as compressors for clean and moisture-free compressed air for underwater breathing. Compression ignition engines using diesel or other easily available hydrocarbon fuels provide large amounts of mechanical power for railroad locomotives, ships, factories and electric power generating plants. Petrochemical plants, off-shore oil platforms and utility services require reciprocating compressors, pumps and engines, all the while operating at exceptionally high levels of efficiency. Metal working reciprocating machines make parts that build other machines. Construction of bridges, hospitals, dams and buildings need bulldozers, excavators and cranes; all of them depend on engines, pumps, compressors and machines using the slider and crank mechanism, the heart of a reciprocating machine.

Two general types of machines are in common use for performing mechanical tasks. One is the reciprocating type where the pistons reciprocate (move back and forth or up and down) as it operates, the other is the rotary type. Within these two categories the machines may be further divided into two sub-categories. In the first one power generation is the prime goal, as in heat engines such as the internal combustion engine and gas turbine engine where thermal energy is used for the development of mechanical power. The other group of machines uses mechanical energy to perform other work, such as an electric motor driven centrifugal compressor or a reciprocating pump. In reciprocating compressors or pumps pressure is developed in the fluid because of the positive-displacement characteristics of the mechanism. As the piston pushes on the fluid contained inside the cylinder, it pressurizes the fluid while at the same time pushing it out through the discharge valves into the connecting pipe. On the other hand, centrifugal compressors operated by means of a shaft mounted fan or impeller are used to impart velocity to the working fluid. The velocity, or kinetic, energy is then converted into pressure, or potential, energy using a diffuser or volute type of mechanism. The latter components progressively increase in cross-sectional flow area, thereby reducing the flow velocity while also increasing the pressure of the pumped fluid.

An incredibly large variety of machines are in operation in today's world. Of necessity, then, it is impossible to focus on every single type of machine. What makes it even more interesting is the tremendously large selection of configurations available within each type of machine category. Internal combustion powered equipment includes automotive, off-highway vehicles, industrial machinery, marine and power generation sectors. The end use to which the equipment will be put to use plays a major role in the overall size and shape of the equipment.

A case in point is the internal combustion engine used in two applications, a lawn mower and a bulldozer. Even though the two engines operate on the same overall principles, cylinder size, power generated, and practically every other design parameter for the two engines will be vastly different. Thus, for all practical purposes both reciprocating machines may be identified to fall in their own specific category.

Returning to the earlier discussion on rotating and reciprocating machines, an important point comes up in the initial design stages of any equipment, which is whether to choose the design based on the rotating or reciprocating principles. A fundamental advantage of the reciprocating internal combustion engine over a steam engine is the absence of heat exchangers such as boilers and condensers. The absence of these components leads to mechanical simplification and also eliminates the loss inherent in the process of heat transfer through an exchanger of finite area. Gas turbines are used without heat exchangers, but require them for maximum efficiency in electric power generation facilities; hot exhaust gases are used to produce steam for steam turbines to generate additional power. In internal combustion engines the parts operate at temperatures well below the maximum cyclic temperature. This feature allows high maximum cyclic temperatures to be used, which results in high cyclic efficiencies.

The advantages of reciprocating internal combustion engine are of special importance in the field of land transportation, where small weight and bulk of the engine and fuel are essential factors. Considering the tremendous impact in the mode of life which the motor vehicle has, it may be safely said that the world as we know it cannot function without the internal combustion engine. The internal combustion *turbine* has relative mechanical simplicity, making it very attractive for aircraft propulsion. The absence of reciprocating parts gives freedom from vibration. It also has the significant advantage of lower weight to power ratio, making it the only power plant available for large military and commercial aircraft. Cooling the turbine blades helps to increase inlet gas temperature, which increases the efficiency and output from a given size unit. This requires sophisticated metals and manufacturing methods for the blades and elaborate pipes, sensors and valves to ensure adequate operational

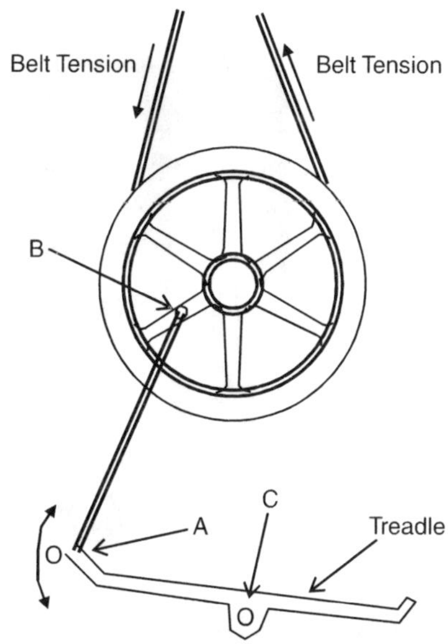

Fig. 2.1. Converting Reciprocating Motion to Rotating Motion

life of the components. Reciprocating engines for automobiles and small aircraft have the distinct advantage of cheaper materials and initial cost, which is perhaps the overriding factor for the average automobile and private aircraft owner. The combination of a reciprocating engine with an exhaust gas driven turbine driving a supercharger is in wide use.

An ordinary sewing machine belt wheel and treadle, illustrated in Fig. 2.1, offers a good example of the conversion of straight line reciprocating machine to rotary motion. One end of a connecting rod is fastened at *A* to the treadle. The other end of the connecting rod is fastened at *B*, a point located off-center on the belt wheel. When the treadle is pushed down with the toe, end *A* of the connecting rod moves downward and pulls the other end of the connecting rod, *B*, around for a part of the circle. If the operator continues to push the toe down, the belt wheel will stop at dead center. But if the operator pauses a moment the inertia of the wheel will carry it past the dead center position. The operator then pushes with the heel on the rear edge of the treadle, and the connecting rod end is pushed upward, causing point *B* on the belt wheel to move upward also. As *B* nears top dead center position, the operator pauses again to allow the wheel to be carried past the dead center position by its inertia. Repeating the process indefinitely, operator energy is exerted up and down in a reciprocating motion, but the belt wheel is given a rotating motion.

The basic parts of a typical reciprocating machine are the cylinder, piston, connecting rod and the crankshaft, as shown in Fig. 2.2. The cylinder has a smooth surface such that the piston can, with the aid of piston rings and a lubricant, create a seal so that no gases or fluids can escape between the piston and the cylinder walls. The piston is connected to the crankshaft by means of the connecting rod so that the rotation of the crankshaft causes the piston to move with a reciprocating motion up and down in the cylinder. The distance through which the piston travels is called the stroke. During each stroke the crankshaft rotates 180°. The limit of travel to which the piston moves into the cylinder is called the top dead center (or TDC), and the limit to which it moves in the opposite direction is called the bottom dead center (or BDC). For each revolution of the crankshaft there are two strokes of the piston, one up and one down if the cylinder is in a vertical position. It is important to understand the concept of top and

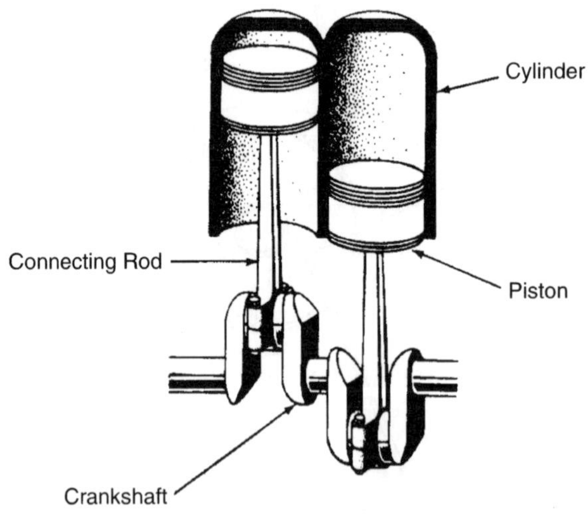

Fig. 2.2. Basic Parts of a Reciprocating Machine

bottom dead centers of a reciprocating machine, because these positions of the piston are used in setting the timing and determining valve overlap. Top dead center may be defined as the point the piston has reached when it is at its maximum distance from the centerline of the crankshaft. In a like manner bottom dead center may be defined as the position the piston has reached when it is at a minimum distance from the center line of the crankshaft. Fig. 2.5 illustrates the piston position at top and bottom dead centers.

2.2 INTERNAL COMBUSTION ENGINES

An internal combustion engine is so called because the fuel is burned inside the engine rather than externally, as in a steam turbine. Basically an internal combustion engine is a device for converting a source of energy to useful work. In heat engines the source of energy is the fuel burned to develop heat. The heat in turn is converted to power (scientifically described as the rate of doing work) by the engine. The reciprocating engine uses the heat to expand a combination of gases (air and the products of fuel combustion), thus creating a pressure against the piston inside the engine cylinder. The piston, connected to a crankshaft, causes the crankshaft to rotate, thus producing power. In a gas turbine engine the fuel is burned in a combustor, a separate chamber within the equipment, and the heat is used to expand the gases as it moves through the engine, with the result that the gas velocity is greatly increased. This kinetic energy of the gas is then imparted to the rotating turbine blades to produce shaft power. In an aircraft turbo-jet engine the turbine powers a fan to create a thrust to propel the aircraft, with the velocity of the exhaust gases from the turbine providing some additional thrust. In a turbo-shaft or turbo-prop engine a large portion of the energy is used to power the aircraft propellers or helicopter blades, with exhaust from the turbine often producing from 15 to 20 percent of the total thrust.

An engine cycle is a series of events that an internal combustion engine goes through while it operates and delivers power. In a four stroke, five event cycle they are intake, compression, ignition, combustion and exhaust. Since the events occur in a certain sequence and at precise intervals of time, they are said to be timed. Most piston engines operate on the four stroke, five event cycle principle developed by August Otto, and is named after him as the Otto cycle. Other cycles for heat engines are the Carnot cycle, the Diesel cycle and the Brayton cycle; they differ in the particular engine theories developed by the scientists whose name is associated with the cycle.

The four strokes of a four-stroke cycle engine are the intake stroke, the compression stroke, the power stroke and the exhaust stroke. In a four-stroke-cycle engine the crankshaft makes two revolutions for each complete cycle, so the ignition of the fuel and air mixture takes place only once in two complete revolutions of the crankshaft. Fig. 2.3 shows diagrams for port and valve arrangements in a two-stroke engine. During the intake stroke the piston starts at top dead center, the intake valve is open, the exhaust valve is closed, the piston moves downward, the fuel-air mixture is drawn into the cylinder, and at its end the intake valve closes. During the compression stroke both valves are closed, the piston moves toward the top dead center while compressing the working fluid and ignition takes place near the top of the stroke. In the power stroke both valves are closed, the ignited gases build up a lot of pressure and cause it to expand while forcing the piston toward the bottom dead center. The exhaust valve opens well before the bottom of the stroke. During the scavenge stroke the exhaust valve is open and the intake valve is closed, the piston moves toward top dead center, forcing the burned gases out through the open exhaust valve, and the intake valve opens near the top of the stroke.

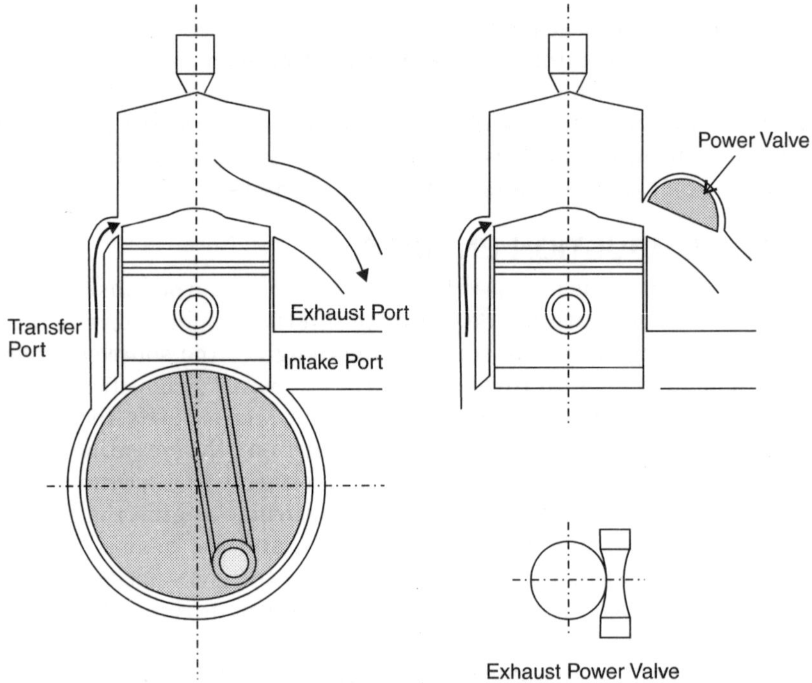

Fig. 2.3. Port and Valve Arrangements for Two-Stroke Internal Combustion Engine

The five-event sequence of intake, compression, ignition, power and exhaust is a cycle that must take place in the order given at the right crank angle position if the engine is to operate at all, and must be repeated continuously for the engine to operate. For example, if the gasoline supply is shut off, or the ignition switch is turned off, there will be no power. It should be emphasized here that during each event the crankshaft does not rotate through exactly 180°. The intake valve begins to open substantially before the piston reaches top dead center, and the exhaust valve closes after the piston reaches top dead center. This is called valve overlap, and is designed to take advantage of the inertia of the out-flowing gases to provide more complete scavenging and to allow the entering air-fuel mixture to flow into the combustion chamber at the right time; the technique helps to improve volumetric efficiency, which is discussed later in the chapter. Near the bottom center, likewise, the valve opening and closing time is designed with the twin objectives of optimum scavenging and cooling.

An engine cannot start until it is rotated to begin the sequence of operating events. A variety of starting systems are needed for this purpose. For hand tools and domestic use machines, a cord wrapped around a pulley provides enough force to start the sequence; a hand crank may also be used in many cases. For larger equipment electric inertia starters, combustion starters and direct cranking electric starters are used.

Although the four-stroke cycle is by far the most commonly used in most internal combustion engines, a few small engines such as in motor-cycles and ultra light aircraft operate on the two-stroke-cycle principle. The difference is in the number of strokes per operating cycle and the method of admitting the fuel-air mixture. Fig. 2.4 shows diagrams of some engines working on the two-cycle method. Like the four-cycle engine, the two-stroke cycle calls for a cylinder, piston, connecting rod, crankshaft and crankcase; however, the valve arrangement and fuel intake system are considerably different. The upward motion of the piston inside the cylinder creates low pressure in the crankcase. This reduced pressure causes the air-fuel mixture from the carburetor into the crankcase through a check valve. When the piston reaches top dead center the crankcase is filled with the fuel-air mixture and the inlet check valve is closed. The piston then moves downward in the cylinder and compresses the mixture in the crankcase. As the piston reaches the lowest point in its stroke the intake port is opened to permit the compressed fuel-air mixture from the crankcase to flow into the cylinder. This is the intake event. The piston then moves up in the cylinder, the intake port is closed, and the fuel-air mixture in the cylinder is compressed. At the top of the stroke a spark ignites the compressed fuel-air mixture. The power event next follows, and at the same time the fuel-air mixture in the crankcase is pressurized. A baffle on the top of the piston deflects the incoming mixture into one side of the cylinder as the exhaust gases rush out of the exhaust port on the other side.

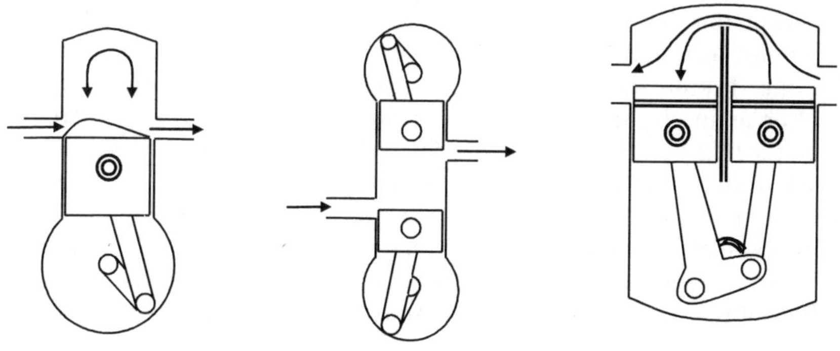

Fig. 2.4. Two-Stroke Engines. (*a*) Loop Scavenge, (*b*) Opposed Piston, (*c*) U-Cylinder

It is important to understand that in a two-stroke cycle two strokes of the piston are required for a complete cycle operation, so the crankshaft goes through one revolution during the cycle. Thus, all cylinders of a multi-cylinder two-stroke cycle engine fire at each revolution of the crankshaft. The two-stroke cycle, however, has significant drawbacks. Since the fuel-air mixture is drawn in while the exhaust gases are leaving the cylinder at the same time, some fuel-air charge escapes with the exiting gases, resulting in a loss of efficiency. Also the engine is more difficult to cool than a four-stroke cycle engine, chiefly because the cylinder fires at every revolution of the crankshaft. Another reason is that it is difficult to lubricate properly because the lubricant must be introduced with the fuel-air mixture through the carburetor.

The operating principle of a four-stroke cycle diesel engine superficially resembles that of a four-stroke cycle gasoline engine. The primary difference is that a diesel engine does not require electrical ignition. Also, the diesel engine operates on fuel oils that are heavier and cheaper than gasoline. On the intake stroke of the diesel engine only pure air is drawn into the cylinder. As the piston approaches the top of its stroke the fuel is injected into the cylinder under a high pressure in a finely atomized state. On the compression stroke the piston compresses the air to such an extent that the air temperature is high enough to ignite the fuel without the use of an electrical spark. The compressed hot air already inside the cylinder ignites the fuel. The fuel burns during the power stroke, and the waste gases escape during the exhaust stroke in the same manner as they do in a gasoline engine. On many diesel engines, particularly those used in automobiles, glow-plug igniters are installed to aid in starting the combustion of the fuel. Once the engine is running the igniters are not in operation. The compression ratio, to be discussed more fully later, of a diesel engine may be as high as 14:1 as compared with a maximum of 10 or 11:1 for conventional gasoline engines. It is common for a gasoline engine to have a compression ratio of about 7:1; however, certain high performance engines have a higher ratio.

2.3 DESIGN PARAMETERS FOR ENGINES

Perhaps the power output from any mechanical equipment is its most important parameter. Power is the rate of doing work. A certain amount of work is done when a weight is raised a certain distance. Power considers the time required to do that work, and thus depends on force, distance and time. James Watt, the inventor of the steam engine, found that an English work horse could work at the rate of 550 ft-lbs per second for a reasonable period of time. From this observation came the horsepower. In order to compute the power of an internal combustion engine, it is necessary to know various measurements of the engine, such as cylinder bore, piston stroke and piston displacement. The piston displacement of one piston is obtained by multiplying the cross sectional area of the cylinder bore by the total distance the piston moves during one stroke. Piston displacement is, thus, expressed in terms of cubic inches of displacement. For example, if the bore of a cylinder is 6 in. and the stroke is 8 in., the displacement is $\pi \times (6/2)^2 \times (8) = 226.2 \text{ in}^3$. The total displacement of an engine is the total volume displaced by all pistons during one revolution of the crankshaft. Other factors remaining the same, the greater the total displacement, the greater will be the horsepower developed by an engine. Displacement is one of the many factors in power plant design that is subject to compromise. If the cylinder bore is too large, fuel will be wasted and the intensity of heat and the restricted flow of the heat may be so great that the cylinder may not be cooled properly. If the piston stroke is too large excessive dynamic stresses and too much angularity of the connecting rods will be the undesirable consequences. Experience tells that an engine where the bore and stroke are equal (sometimes referred to as a 'square' engine) provides an optimum balance between the requirements of structural integrity, weight and performance. Increased engine displacement can be obtained by adding cylinders, thus also producing additional power output. Adding cylinders also offers the advantage of closer spacing of power impulses during cylinder firing, hence increasing engine operation smoothness.

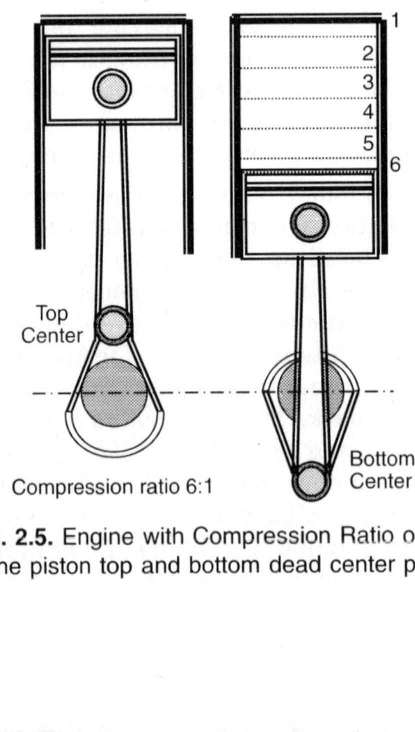

Fig. 2.5. Engine with Compression Ratio of 6:1
(Note the piston top and bottom dead center positions)

The compression ratio of a cylinder is the ratio of the volume of space in the cylinder when the piston is at the bottom of its stroke to the volume when the piston is at the top of its stroke. For example, if the volume of the space in the cylinder is 120 in^3 when the piston is at the bottom of its stroke and the volume is 20 in^3 when the piston is at the top of its stroke, the compression ratio is 120:20, or 6:1. Fig. 2.5 shows a piston and cylinder providing a compression ratio of 6:1. The compression ratio of any internal combustion engine controls the horsepower developed by it, and within reasonable limits the maximum horsepower increases as the compression ratio increases. It is found from experience that if the compression ratio is much greater than 10:1, premature ignition of the fuel-air mixture (called pre-ignition) or detonation may occur and cause overheating, loss of power and damage to the engine. Physically this can be explained that compression results in a temperature rise of the working fluid, and under proper circumstances the greater compression ratio could generate sufficient heat to ignite the fuel even before the spark from the igniter. If the compression ratio is as high as 10:1, it is necessary to use a fuel with higher anti-knock characteristics, either in the form of higher octane rating or higher performance number. The maximum compression ratio is also limited by the design limitations of the engine and the degree of supercharging.

An increase in the compression ratio of an engine may be accomplished by installing 'higher' pistons, by using longer connecting rods or by installing a crankshaft with a greater throw. In an engine with a removable cylinder head the combustion space in the head may be reduced by shaving the surface of the cylinder head where it mates with the top of the cylinder. Increasing the compression ratio of an engine has the advantage that it lowers the specific fuel consumption (defined as pounds of fuel burned per hour per horsepower) of the engine, and also increases its thermal efficiency. Thermal efficiency is the ratio of the heat converted to useful work to the heat value of the fuel consumed.

Indicated horsepower (*ihp*) is based on the theoretical amount of work done according to calculations made from the actual pressure recorded in the form of a diagram on an indicator card, as illustrated in Fig. 2.6. The indicator diagram shows pressure rising during the compression stroke and after ignition, and dropping as the gases expand during the power stroke. It emphasizes the fact that the force acting on the piston during the power stroke is not constant, because the fuel-air mixture burns almost instantaneously, with a resulting high pressure at the top of the stroke. Power computations for indicated horsepower are simplified by using the average pressure acting on the piston throughout the working stroke, often called the mean effective pressure, and is obtained from the indicator diagram. Sophisticated instrumentation with computers is utilized to determine the various engine measurements. Fig. 2.6, however, provides a graphic illustration of the process.

Indicated horsepower is obtained from heat energy converted to mechanical energy, and is the total horsepower developed by the engine. If the length of the stroke in inch is represented by L, the area of the piston in square inches by A, the indicated mean effective pressure in pounds per square inch by P, the number of working strokes per minute per cylinder by N and the number of cylinders by K, the indicated horsepower is calculated by the formula:

$$ihp = \frac{PLANK}{33{,}000} \tag{2.1}$$

Brake horsepower is the actual horsepower delivered to the drive train or device. Part of the total horsepower is needed to overcome friction in the engine and its accessories, and is called friction horsepower. Thus, brake horsepower is given by the difference between indicated horsepower and friction horsepower, and is the power available to perform work. Generally, in a properly maintained engine 10 to 15 percent of the indicated horsepower is lost to friction.

The brake horsepower of an engine can be measured by coupling the engine to a power absorbing device, such as an electric generator or a dynamometer which directly measures force. Large transport aircraft powered by internal combustion engines used a mechanism in the nose section of the engine which applies pressure to oil in a closed chamber in proportion to engine torque. Since the propeller reduction gear mates with a large stationary ring gear, the ring gear is used to develop an indication of engine torque. Variations in the engine torque is transmitted through the nose cone device to hydraulic pistons which are connected to a pressure gage in the cockpit. The gage may be calibrated to read brake mean effective pressure or brake horsepower.

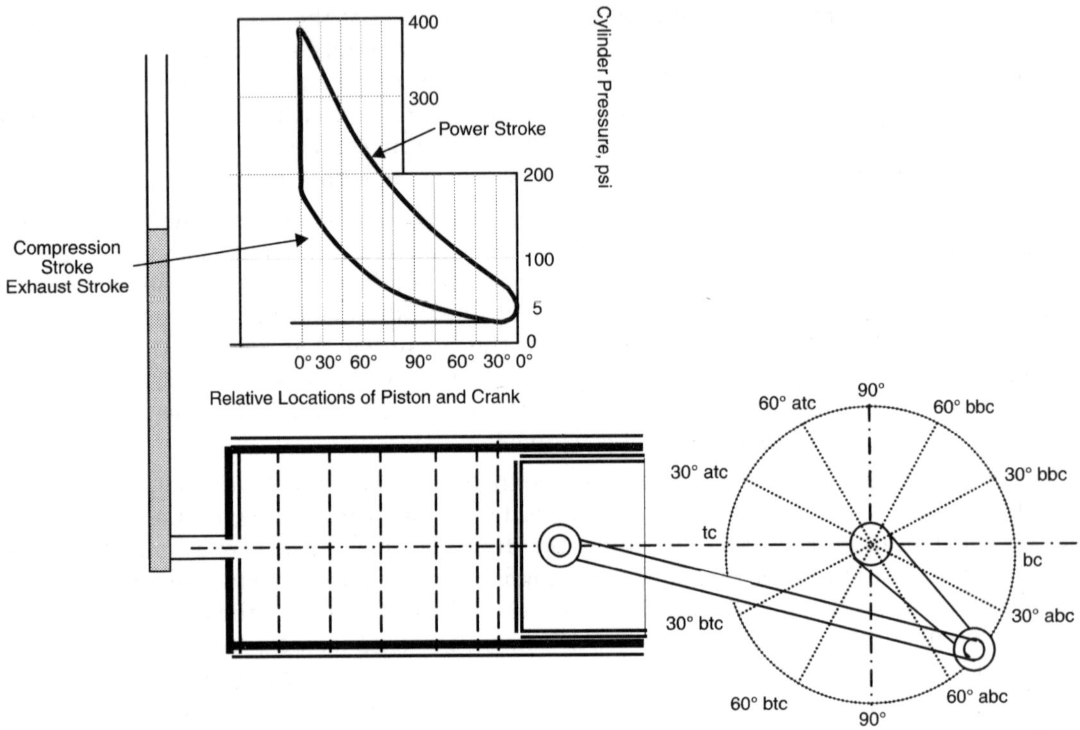

Fig. 2.6. Cylinder Pressure Indicator Diagram

The mechanical efficiency of an engine is measured by the ratio of the brake horsepower, or shaft horsepower, to the indicated horsepower, or power developed in the cylinders. Thermal efficiency is a measure of heat losses suffered in converting the heat energy of the fuel into mechanical work. In Fig. 2.7 the heat dissipated by the cooling system represents 25 percent, heat carried by the exhaust gases represent 40 percent, mechanical work on the piston to overcome friction and pumping losses represent 5 percent and useful work at the drive shaft represents 30 percent of the heat energy of the fuel.

The indicated thermal efficiency of an engine is given by the expression:

$$\frac{ihp \times 33{,}000}{\text{weight of fuel burned/min} \times \text{heat value (BTU)} \times 778} \tag{2.2}$$

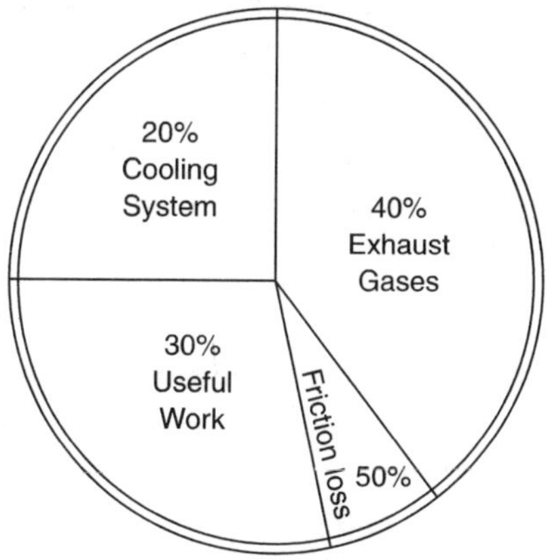

Fig. 2.7. Thermal Efficiency Chart

and brake thermal efficiency is given by:

$$\frac{bhp \times 33{,}000}{\text{weight of fuel burned/min} \times \text{heat value (BTU)} \times 778} \qquad (2.3)$$

As an example, consider an engine developing 104 bhp at 2600 rpm at full power, and burning 6.5 gallons per hour of gasoline. The heat value of the fuel is 19,000 to 20,000 BTU/lb. Since each gallon weighs about 6 lb., then 6 × 6.5/60 = 0.650 lbs/min. Hence, the brake thermal efficiency is (104 × 33,000)/(0.650 × 20,000 × 778) = 0.34, or 34 percent. Although a thermal efficiency of 34 percent may not appear high, it is in line with other types of engines. The thermal efficiency of many diesel engines is 35 percent if run at an output of one-half to three-fourths of full power, but when the output is increased to full power, the thermal efficiency of the diesel drops to less than one-half that of the usual carburetor type engine. This is because of incomplete combustion of thc fuel when a large amount of excess air is not available in the cylinder. Thermal efficiencies of 45 percent are available in some low-speed stationary or marine engines when operating under favorable conditions.

Volumetric efficiency is the ratio of the volume of fuel-air charge burned by the engine at atmospheric pressure and temperature to the piston displacement. Factors that tend to decrease the mass of air entering an engine have an adverse effect on volumetric efficiency. Typical factors are: improper valve timing, engine speed, carburetor air temperature, design of induction system and high combustion chamber temperature. A naturally aspirated (non-supercharged) engine has a volumetric efficiency of less than 100 percent due to resistance to airflow in bends, obstructions and surface roughness of the intake manifold. On the other hand, a supercharged engine is often operated above 100 percent volumetric efficiency because the supercharger compresses the air before it enters the cylinder.

2.4 AUTOMOBILE ENGINES

The reciprocating internal combustion engine is at the present time (2004) without any serious competitors in road transportation. The total power for these machines exceeds that of all other prime movers combined by at least on order of magnitude (ten times). Most of this installed power is in the form of spark ignition engines of less than 250 horsepower for motor vehicles. Road transportation depends entirely on reciprocating engines and accounts for about one-third of the total energy consumption of the world. The reasons for this predominance of reciprocating engines in automotive services are centered around their major characteristics as compared with those of other types. These are:

- Lighter weight (except for gas turbines)
- Smaller size (except for gas turbines)
- Lower cost (except in larger units)
- Better fuel economy
- Simplicity of operation and control
- Flexibility—ability to operate over a wide range of speed and load combinations.

  Possible future competitors of the conventional reciprocating internal combustion engine may be taken to include:

- Unconventional displacement engines
- Gas turbines

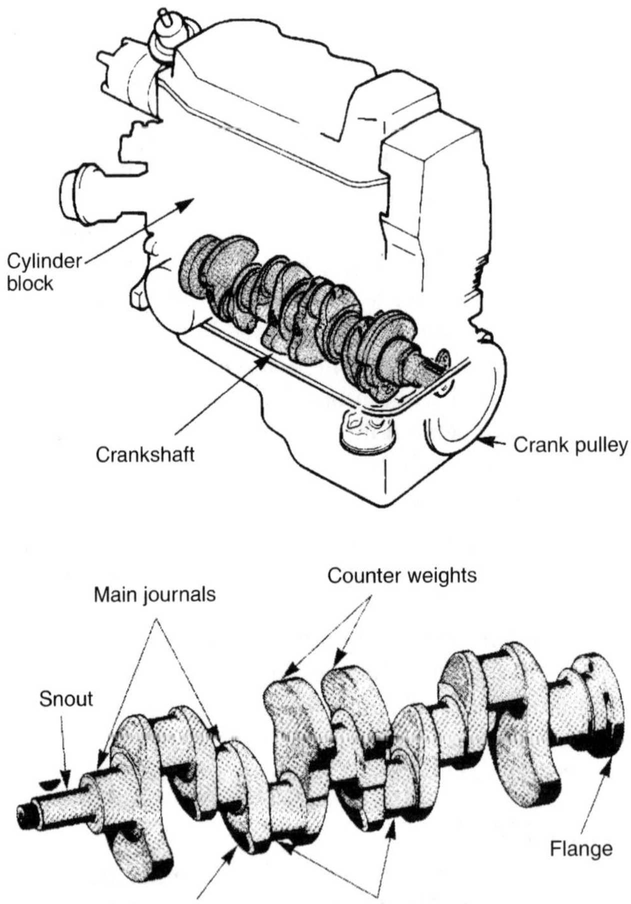

Fig. 2.8. Placement of Crankshaft in Cylinder Block of In-Line Automotive Engine (*upper*), Basic Parts of Typical Crankshaft (*lower*) (*Courtesy*: Ford, Peugeot)

about one-third of the total energy consumption of the world. The reasons for this predominance of reciprocating engines in automotive services are centered around their major characteristics as compared with those of other types. These are:

- Lighter weight (except for gas turbines)
- Smaller size (except for gas turbines)
- Lower cost (except in larger units)
- Better fuel economy
- Simplicity of operation and control
- Flexibility—ability to operate over a wide range of speed and load combinations.

Possible future competitors of the conventional reciprocating internal combustion engine may be taken to include:

- Unconventional displacement engines
- Gas turbines

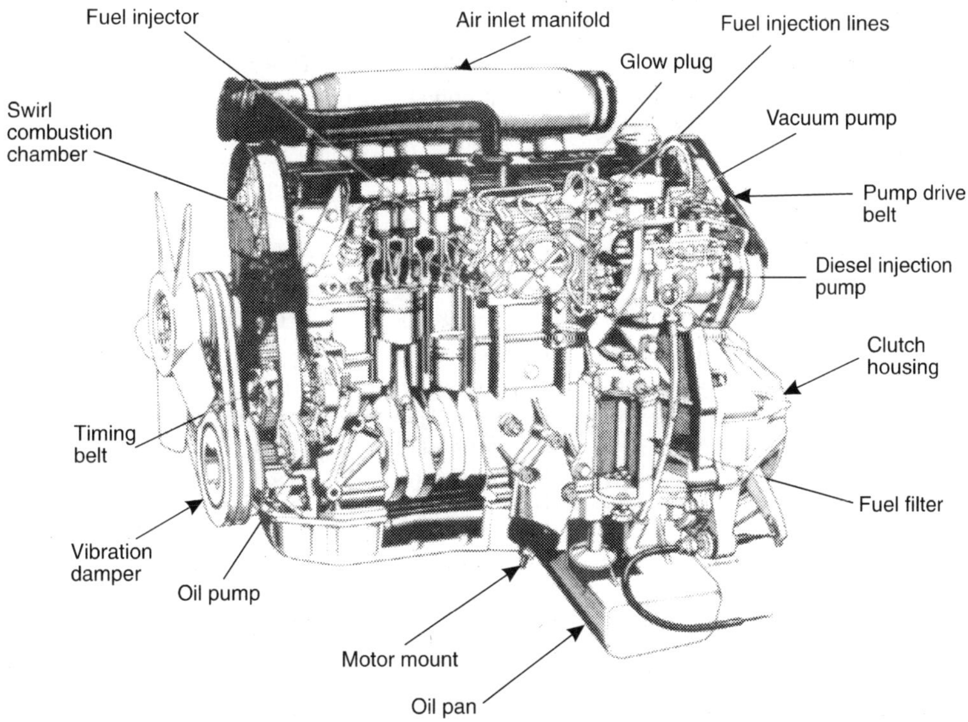

Fig. 2.9. Major Components of Typical Automotive Diesel Engine (*Courtesy*: General Motors)

The question pertaining to diesel versus spark ignition is decided by the type of fuel to be used. For example, only fuel suitable for diesel engines is available at dockside in quantities appropriate for large vessels, and in many parts of the world only gasoline is available at the roadside. Among other reasons, adverse effects of large cylinder size on detonation makes it difficult to design for spark ignition for the larger bore. For engines where the fuel supply is not restricted and bore is less than about 6 in., the chief reason for considering a diesel engine as an alternative to a spark ignition engine is fuel economy and fuel cost. Without the benefit of supercharging, the diesel engine is heavier, larger, more expensive to build, rougher and noisier than a spark ignition engine of equal power, while also requiring more sophisticated maintenance. For passenger automobiles use of diesel engines has been more common in Europe and Asia, chiefly due to high fuel prices. In United States diesel automobiles were rare until after the twin fuel crises of 1970s.

Table 2.1. Classification of Reciprocating Engines by Types of Service

Class	*Service*	*HP Range*	*Predominant type*		
			D or S.I.	*Cycle*	*Cooling*
Road Vehicles	Motorcycles, Scooters	1-100	S.I.	2, 4	A
	Ordinary Automobiles	20-150	S.I.	4	W
	Sport Automobiles	100-500	S.I.	4	W
	Light Commercial	50-200	S.I., D	4	W
	Heavy Commercial	150-500	D or S.I.	2, 4	W
Off-Road Vehicles	Light Vehicles	2-50	S.I.	2,4	A-W
	Agricultural	4-200	D or S.I.	2, 4	A-W
	Earth Moving	50-2000	D	2, 4	W
	Military	50-2500	D	2, 4	A-W
Railroad	Rail Cars	200-1000	D	2, 4	W
	Locomotive	500-4000	D	2, 4	W
Marine	Outboard	0.5-200	S.I.	2	W
	Inboard Motorboats	5-1000	D or S.I.	4	W
	Medium size Vessels	1000-4000	D	2, 4	W
	Ships over 2,000 tons	4,000-50,000	D	2, 4	W
Airborne Vehicles	Airplanes	65-3500	S.I.	4	A
	Helicopters	65-2000	S.I.	4	A
Home Use	Lawn Mowers	1-4	S.I.	2, 4	A
	Snow Blowers	3-6	S.I.	2, 4	A
	Light Tractors	3-10	S.I.	4	A
Stationary	Building Service	10-1000	D	2, 4	W
	Electric Power	20-30,000	D	2, 4	W
	Gas Pipe Line	1000-5000	S. I.	2, 4	W
Racing Specials	Boats, Vehicles	100-2000	S.I.	4	W
Toys	Model Airplanes, Autos	0.01-0.5	HW	2	A

S.I.-Spark Ignition D-Diesel HW-hot wire ignition of Carburated Mixture

The conventional spark ignition engine is supplied with a mixture of fuel oil and air, either from the carburetor or by fuel injection, which is quite homogeneous and essentially gaseous by the time ignition occurs. Combustion of the mixture ignited by the spark is characterized by the more or less rapid development of the flame that starts from the ignition point and spreads in a continuous manner outward from the ignition point and spreading to the end of the chamber without an abrupt change in its speed or shape. Ignition occurs at one or more fixed points and at a crank angle that is subject to accurate control. In diesel, or compression ignition, engines the physical aspects of the process of combustion is quite different. Here air, diluted by a small quantity of residual gas, is compressed through a volume ratio of from 12 to 20, and liquid fuel is sprayed into the cylinder near the top dead center position of the piston. Since both the pressure and temperature of the cylinder contents at the time of injection are quite high, the combustion process initiates.

Table 2.2. Engine Characteristics Emphasized by Type of Service

Service	*Critical*	*Important*	*Less Important*
Small engines for homes Outboard engines	Light weight, bulk Low first cost	Low noise Reliability	Fuel economy Long life
Passenger Automobiles	Low noise Low vibration Low first cost Reliability Flexibility Low maintenance	Fuel economy Weight Bulk	Long life
Engines for: Commercial Light Vehicles Marine Industrial	Reliability Fuel economy Low maintenance	Weight Bulk Low noise Low vibration Long life	First cost
Locomotive engines	Small bulk Fuel economy Low maintenance Reliability	Long life	First cost Weight Noise Vibration
Aircraft engines	Light weight Small bulk High take-off power Fuel economy Reliability	Low vibration Low maintenance	First cost Long life Noise
Racing engines	High output	Reliability	All other
Large engines (over 12 in. bore)	Fuel economy Long life Reliability	Low noise Low vibration Bulk	First cost Weight
Flexibility refers to smooth and efficient operation over a wide range of speed and load			

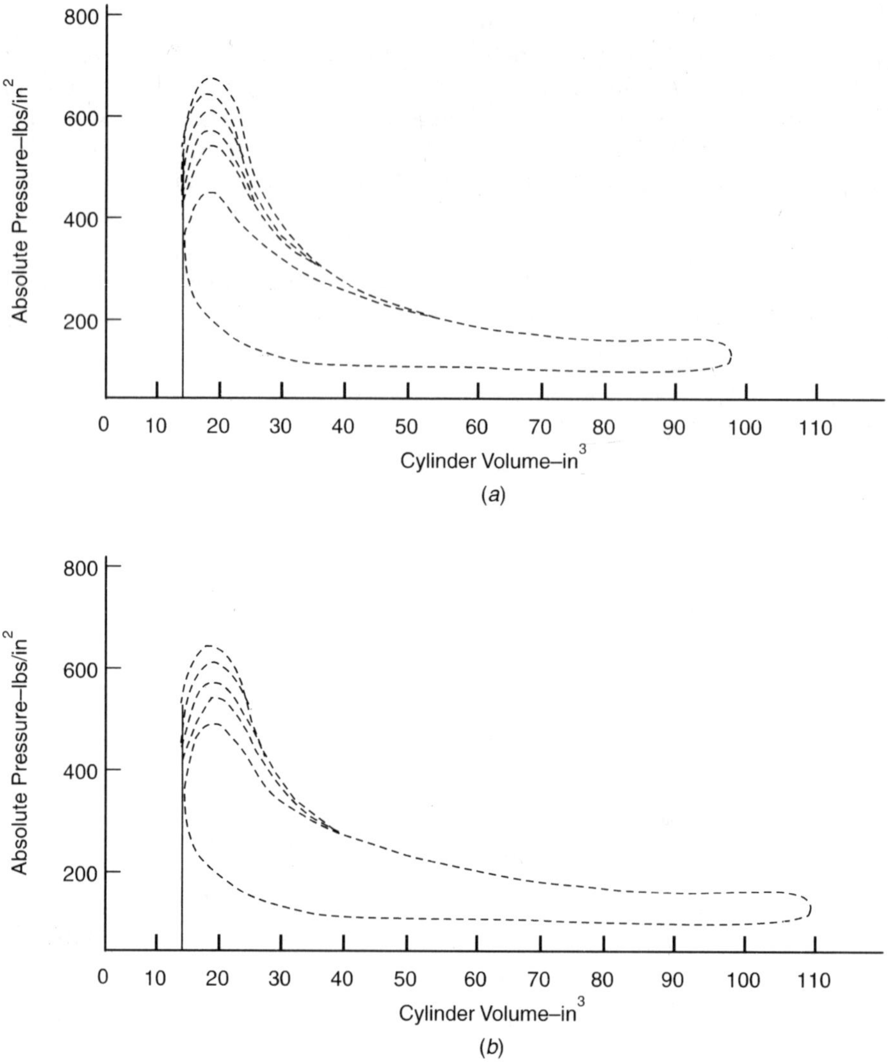

Fig. 2.10. Typical Cyclic Variation in a Spark Ignition Engine
(*a*) With a Lean Mixture, Peak Pressure Deviation = ± 28 %,
(*b*) With a Rich Mixture, Peak Pressure Deviation = ± 10 %

However, the chemical reaction starts slowly and a measurable pressure rise takes place only after a delay period. In a compression ignition engine the time and place where ignition occurs is not fixed and not so easily controlled as an ignition spark. The combustion rate is affected by the state and distribution of the fuel as well as by the pressure and temperature in the chamber.

One of the prominent characteristics of the spark ignition combustion process is a wide variation, from cycle to cycle, of the pressure versus crank angle diagram, as illustrated in figure 2.10. With spark timing adjusted for peak pressure at the same angle, it is found that the pressure variation in the pressure versus crank-angle curve is almost entirely in the early stages of combustion, during the first 10 per cent of flame travel. One reason may be incomplete mixing of fuel, air and residual gas, with consequent variations in conditions at the ignition point. Random turbulent motion in the cylinders in the presence of incomplete mixing may also contribute to the spark occurring in different regions at different time points, resulting in the pressure variations of figure 2.10. As will be seen later, repercussions of this phenomenon in the design and dynamic behavior of the crankshaft, connecting rod and piston are substantial.

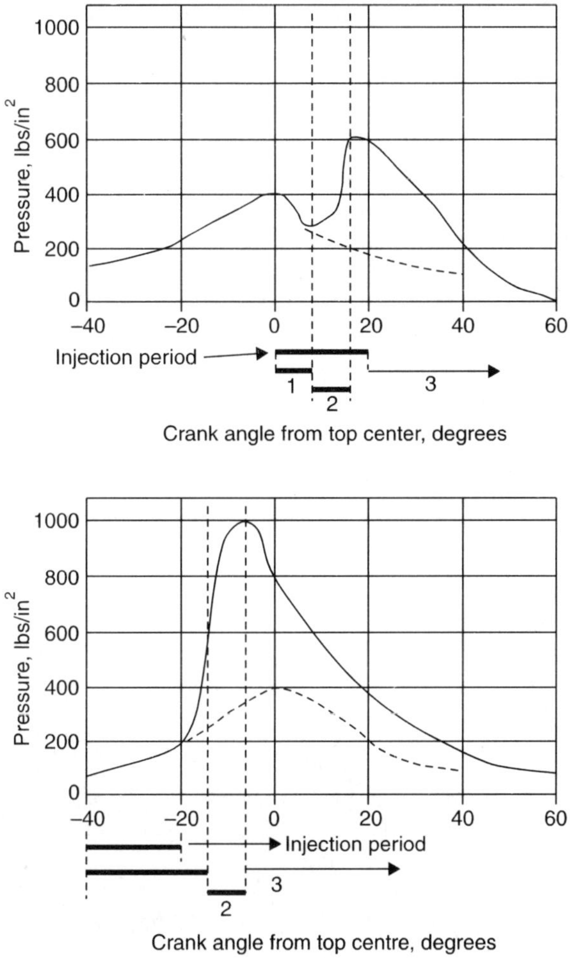

Fig. 2.11. Pressure-Crank Angle Diagrams of a Compression Ignition Engine Showing Three Stages of Combustion. Delay Period Shorter than Injection Period (*upper*), Delay Period Longer than Injection Period (*lower*)

The effects of the delay period in a compression ignition engine are illustrated in Fig. 2.11 in the form of pressure crank-angle diagrams. The delay period, period of rapid combustion and the final third stage of combustion are indicated on the diagrams. The effects of the delay period shorter than the injection period and one where the delay period is longer than the injection period are significant in terms of peak pressure developed in the cylinder.

In order to illustrate the discussion of various types of internal combustion engines and their detail design, cross sections of typical engines of different categories are shown in figures 2.12 through 2.21. Specifications for these engines are given in Tables 2.3 through 2.8 which give important design ratios for the engines illustrated and for a number of other engines. The tables were prepared for engines which have been in service ranging from a few to several years. Newer engine designs have not shown significant changes in the ratios except a few heavy duty diesel engines where crankshaft diameter to bore has increased slightly.

2.5 AIRCRAFT POWER PLANTS

The engine that emerged as the most practical and dependable for aircraft use from before World War I until after World War II was the conventional piston, or reciprocating, engine; today many light and medium-light aircraft use this engine. The gas turbine engine, developed since the 1950s, has advanced to the stage where it has proved to be a superior power plant for both military and commercial aircraft for which high power and high speed are important. The gas turbine engine can develop significantly more power for its weight than a piston engine. For example, a piston engine usually produces on an average less than 1 hp/lb of engine weight, whereas some turboprop engines can produce more than 4 hp/lb of engine weight. Since weight is a prime consideration in the design of any aircraft, the weight-power ratio of the engine is always an important factor in the selection of the aircraft power plant. The weight per horsepower depends in part on the size of the engine, but it depends also on the choice of metals and alloys, engine accessories, stress analysis factors, use of high performance fuels and the absence or presence of a supercharger to give better performance with lower fuel consumption. Use of a power recovery turbine by which a portion of the exhaust energy is returned to the crankshaft has proved to be a very effective way of controlling weight to power ratio and specific fuel consumption.

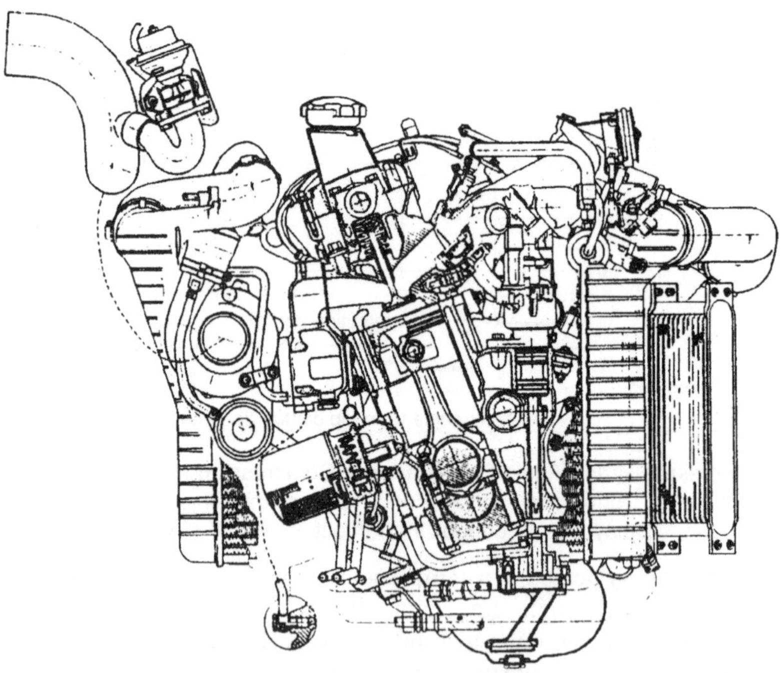

Fig. 2.12. Volvo B23FT 141 cub. in. (2.32 liter) supercharged and after cooled four cylinder automobile engine. Bore 3.78 in., stroke 3.15 in., compression ratio 8.7. Direct acting overhead camshaft with parallel valves. Combustion chamber with large squish areas. Fuel injection with mechanical plus electronic controls. Air cooled after cooler located in front of engine radiator. (*Courtesy: Volvo Car Corporation*)

Small engines are usually built for small airplanes which are operated by pilots with a limited budget. To keep down costs the manufacturer may select less expensive materials and use cheaper manufacturing processes. Large engines are normally built for airplanes operated by the airlines or by the government; hence, the original cost of the engine is not as important to the buyer as the efficiency of its performance over a long period of time. The use of techniques such as direct fuel injection and supercharging on small engines has greatly improved their efficiency, weight to power ratio and specific fuel consumption. In larger engines the use of lighter weight but more expensive materials lowers the weight; consequently, the ratio of horsepower to weight is greater than in the typical small engine.

In this section the design and construction of various types of piston engines for aircraft still in use will be examined. The engines are classified according to a variety of characteristics. The most common classification is according to cylinder arrangement with respect to the crankshaft as follows: (1) in-line upright, (2) in-line inverted, (3) V- type upright, (4) V-type inverted, (5) double V or fan type, (6) opposed or flat type, (7)

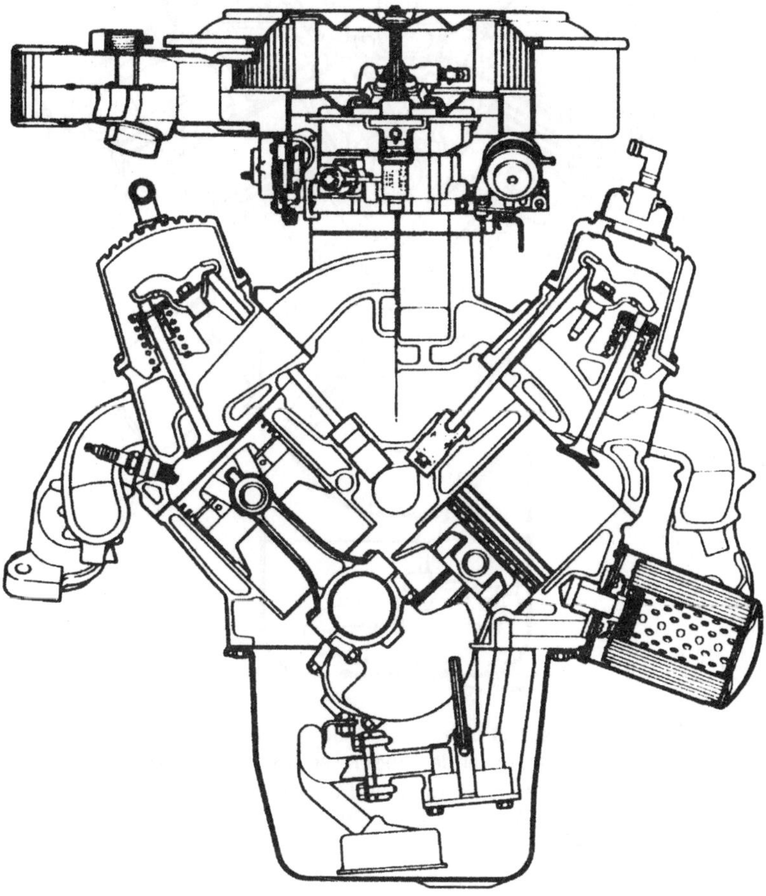

Fig. 2.13. Ford 90° V-8 automobile engine, 302 cub. in. (5 liters). Bore 4 in., stroke 3 in., compression ratio 8.4. With minor changes, this has been Ford's only engine for large cars. It is also offered with fuel injection and supercharging. Typical U.S. style, with push rods and squish style combustion chamber. (*Courtesy: Ford Motor Company*)

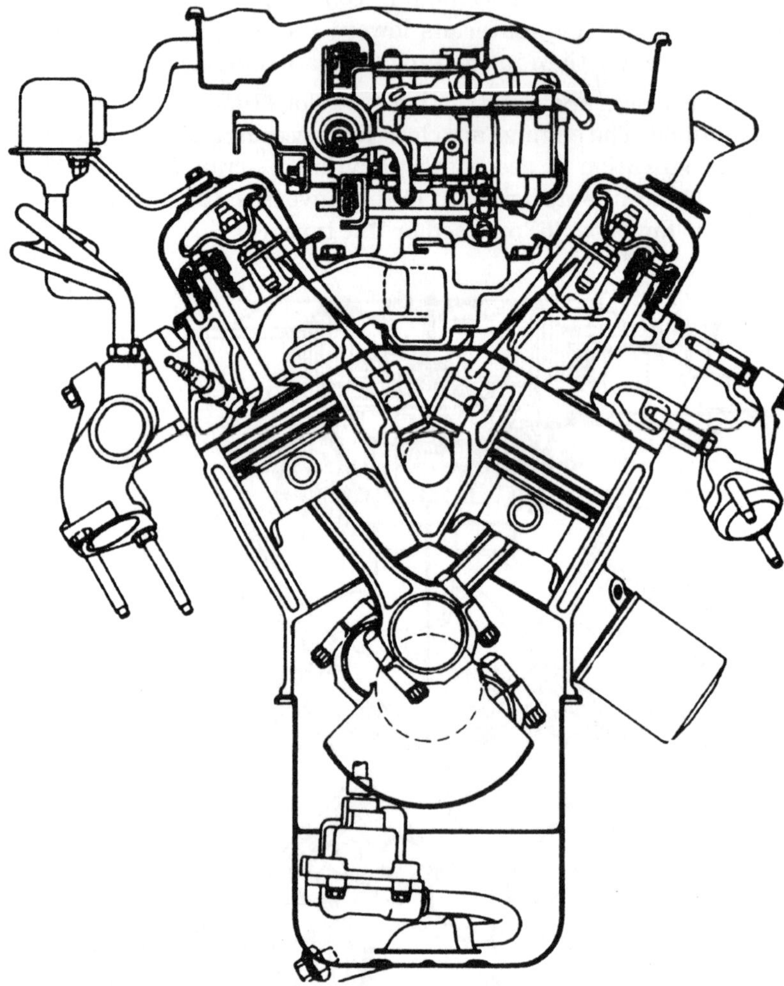

Fig. 2.14. Chevrolet 60° V-6 engine, 151 cub. in. (2.8 liters). Bore 3.5 in., stroke 2.99 in., compression ratio 8.5. Replaces In-line 6 cylinder engine so it can be placed transversely in front drive cars. Typical U.S. push rod valve gear and parallel valve combustion chamber with squish areas. (*Courtesy: General Motors Corp., Chevrolet Div.*)

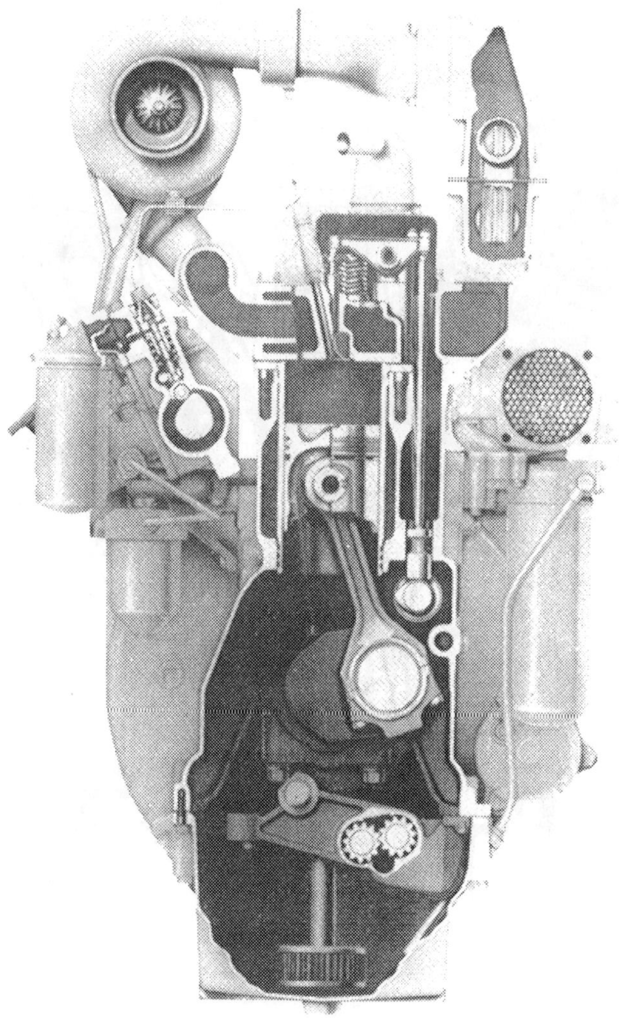

Fig. 2.15. Caterpillar 3306 six-cylinder truck engine, 638 cub. in. (10.5 liters). Bore 4.75 in., stroke 6.0 in. Heavy duty supercharged and after cooled engine. Open (DI) combustion chamber, water cooled after cooler. Sturdy, simple design. (*Courtesy: Caterpillar Tractor Company*)

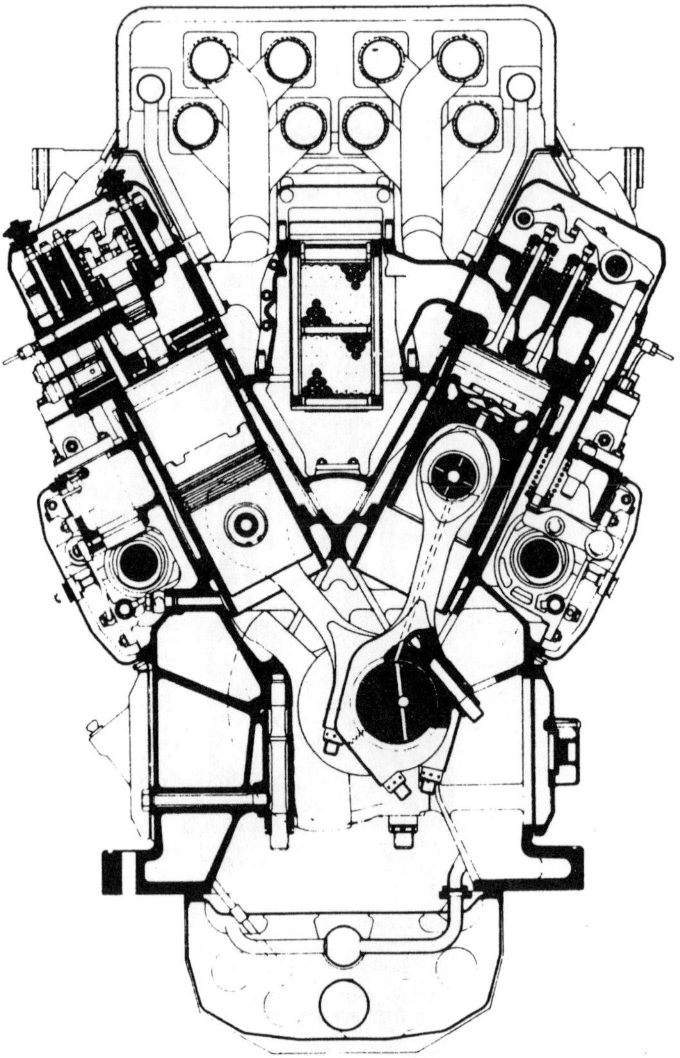

Fig. 2.16. Sulzer four cycle locomotive diesel engine, model ATV 25. *Special Features*: Supercharged and after cooled. Extremely rugged construction for high specific output. Welded steel framework, massive connecting rods split at angle to allow withdrawal through cylinders, maximum crank pin diameter, serrated rod junctions. Four valves per cylinder. Very high bmep. Multiple exhaust pipes indicate 'blow-down' type turbo-supercharger. (*Courtesy: Sulzer Brothers, Ltd.*)

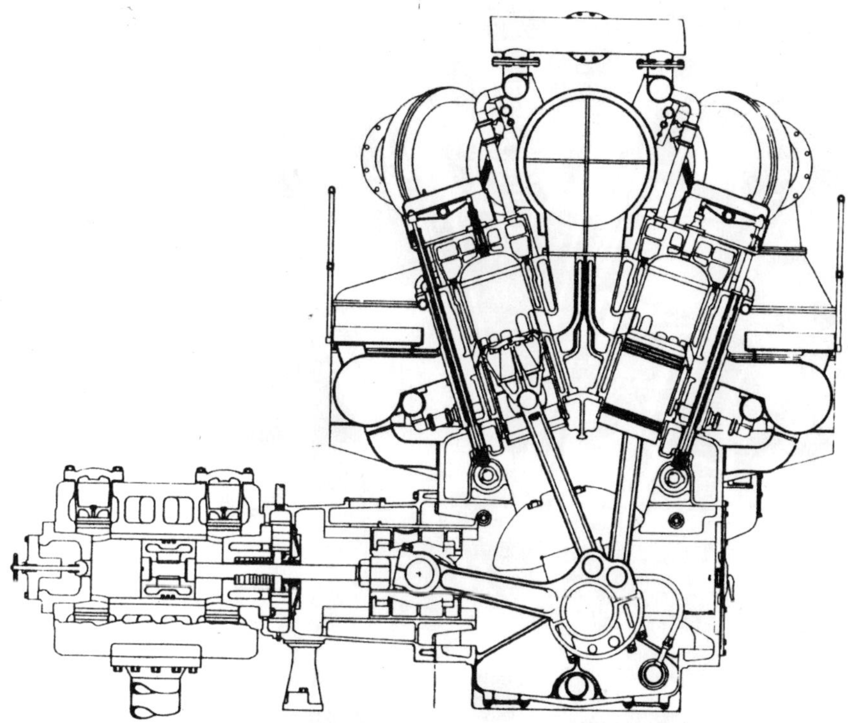

Fig. 2.17. Cooper-Bessemer V-250 2-Cycle Loop Scavenged Gas-Engine-Compressor *Special Features:* Scavenged and supercharged by turbo-compressors. After-cooled. Light structure suitable for low piston speed and maximum cylinder pressure. Crankshaft removable from side opening in crankcase. Cast iron framework. (*Courtesy: Cooper Bessemer Corporation*)

Fig. 2.18. Fairbanks-Morse opposed piston two cycle diesel engine, 1,000 bhp per cylinder. *Special Features:* Unusual arrangement for opposed piston engines reducing exhaust pistons to the category of piston type valves. Turbo-supercharged and after cooled. Small ratio exhaust/inlet port area. Exhaust crankshaft is connected to main crankshaft by a train of heavy spur

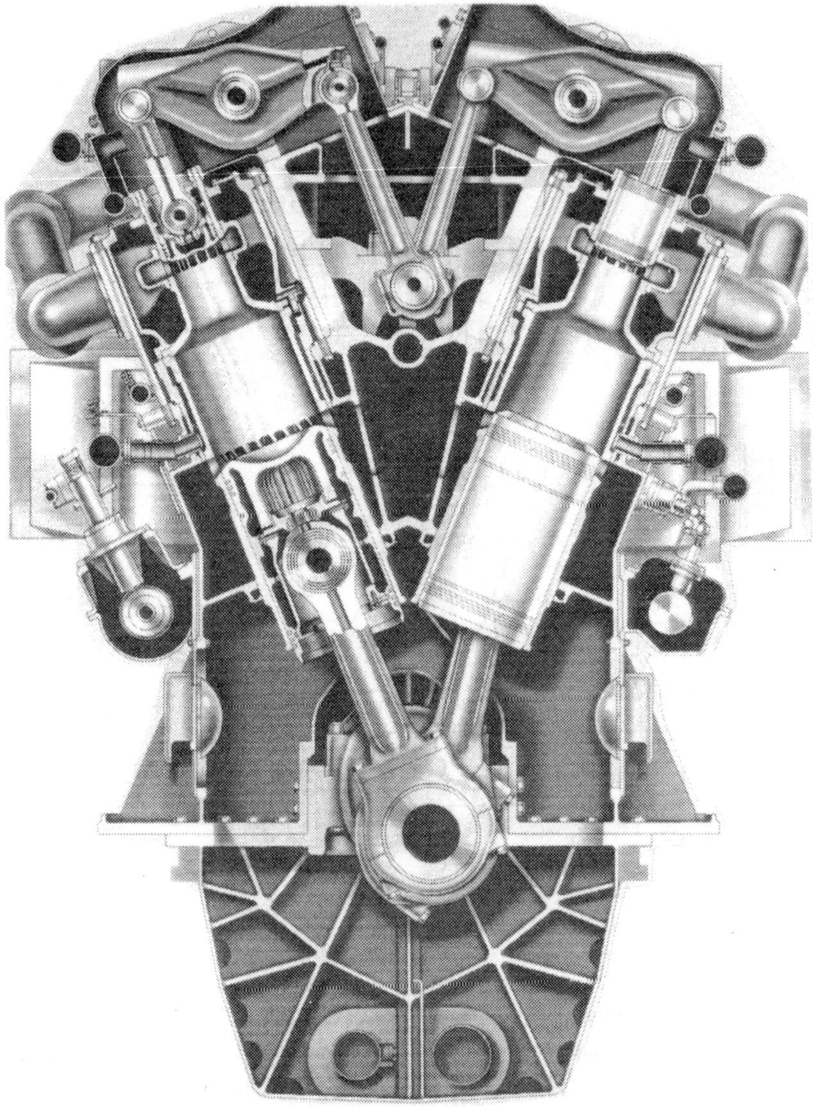

gears. Nodular iron crankshaft, welded steel frame. Available in In-line 6, V-8 and V-12 versions. (*Courtesy: Fairbanks-Morse, Inc.*)

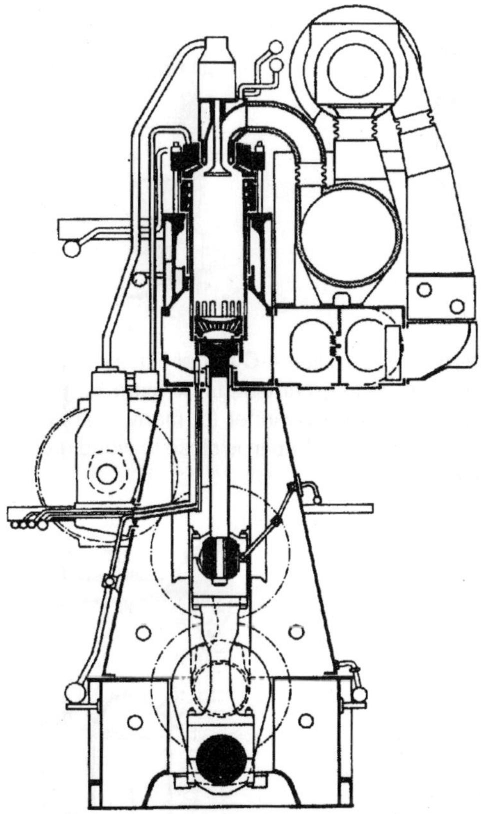

Fig. 2.19. Sulzer RTA84 two cycle marine diesel engine. Cross head type, 4,030 hp per cylinder. Bore 31.5 in., stroke 94.5 in. Available with 4 to 12 cylinders in-line, 16,120 to 48,360 bhp at 87 rpm. These giant engines (34 ft high) are designed for maximum fuel economy in long distance super tankers. Stroke is based on most economical propeller rpm and piston speed (1,370 ft/min); bore is limited in order to allow supercharging to 250 psi mep (equivalent to 500 psi in a four cycle engine). Fuel economy at rating 0.28 lb/bhp, 0.472 brake thermal efficiency. (*Courtesy: Sulzer Brothers, Ltd.*)

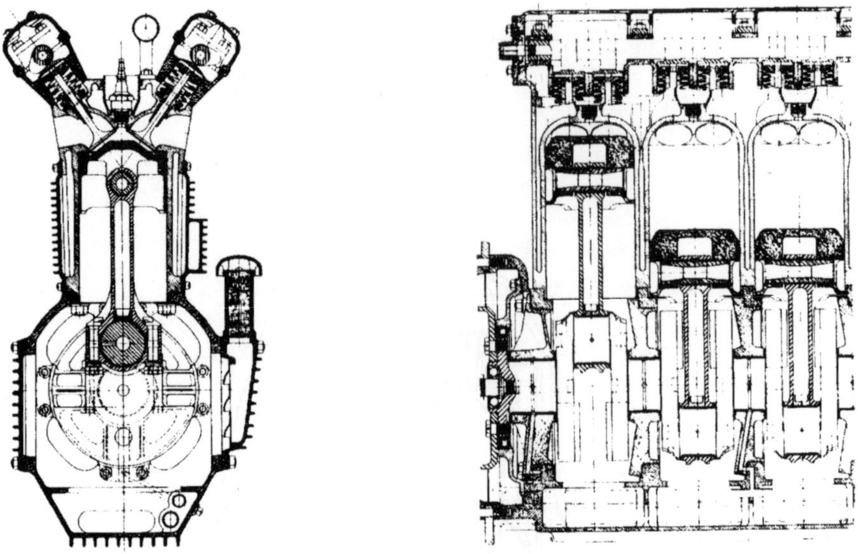

Fig. 2.20. Meyer-Drake-Offenhauser 4.16 liter 4-Cylinder Racing Engine Designed for maximum specific output. Overhead direct acting camshafts. 4 extra large valves per cylinder. Hemispheric heads. This engine was a steady winner for many years at Indianapolis Speedway. Specific output of 2.74 liter version is 11.6 bhp/in is close to world record. (*Courtesy: Drake Engineering & Sales Corporation*)

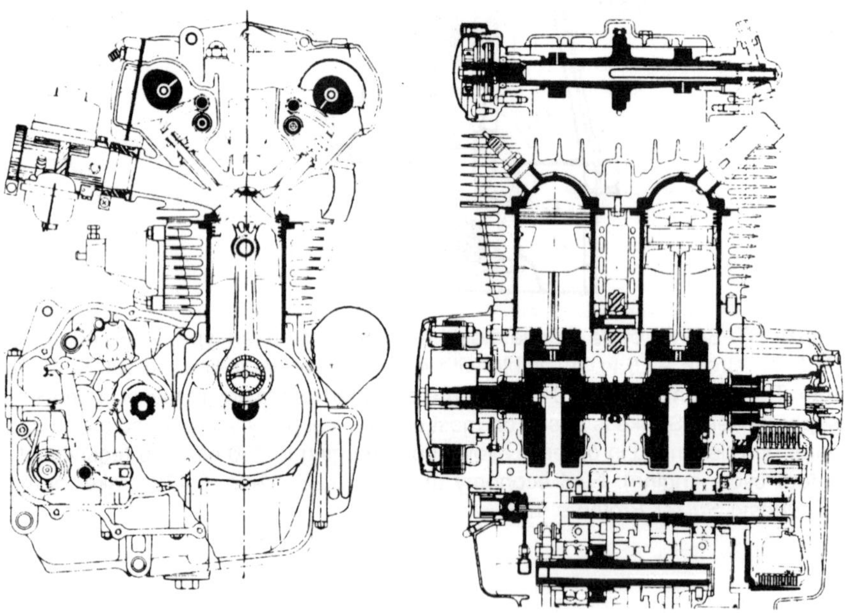

Fig. 2.21. Honda 450 4-Cycle Motorcycle Engine *Special Features:* High specific output, naturally aspirated. Domed cylinder head, excellent large valve and port design. 2 overhead camshafts. Valve seats in head, torsion bar valve springs. Built up crankshaft with roller bearings. Chain camshaft drive. (*Courtesy: Honda Motor Company*)

Table 2.3. Design Ratio for Automotive Diesel Engines

Machine Model	*Component Parameters*				
Commins V-265	Cylinder	Bore: 5.50	Stroke: 4.13	S/b: 0.75	
	Piston	l/b: 0.78	p/b: 0.32	h/b:	N_c: 2
Cylinder Arrangement	Valves	No.: 4	D_i/b: 0.34	D_e/b: 0.34	
V-8	Connecting Rod				
Displacement W: 2.65		L/R: 4.0	B/b:	w_r/p:	
x/b:	Crankshaft	P/b: 0.57	M/b: 0.64	N: 5	w_j/m:
4–Cycle		t/b:			
Caterpillar 61	Cylinder	Bore: 4.50	Stroke: 5.50	S/b: 1.22	
	Piston	l/b: 1.25	p/b: .34	h/b: .85	N_c: 2
Cylinder Arrangement	Valves	No.: 4	D_i/b: 0.41	D_e/b: 0.38	
V-8	Connecting Rod				
Displacement W: 4.00		L/R: 3.40	B/b:	w_j/m:	w_r/p: 0.43
x/b: 1.31	Crankshaft	P/b: 0.69	M/b: 0.80	N: 7	w_j/m: 0.31
4-Cycle		t/b: 0.24			
GMC Detroit	Cylinder	Bore: 4.25	Stroke: 5.00	S/b: 1.18	
Diesel 12-V-71	Piston	l/b: 1.40	p/b: .35	h/b: .9	N_c: 4
Cylinder Arrangement	Valves	No.: 4	D_i/b:	D_e/b: 0.28	
V-8	Connecting Rod				
Displacement W: 3.90		L/R: 4.1	B/b: 0.10	w_r/p: 0.36	
x/b: 1.26	Crankshaft	P/b: 0.65	M/b: 0.82	N: 7	
2-Cycle		w_j/m: 0.37	t/b: 0.19		
GMC Toro-Flow	Cylinder	Bore: 5.13	Stroke: 3.86	S/b: 0.75	
	Piston	l/b: 0.88	p/b: .31	h/b: .5	N_c: 2
Cylinder Arrangement	Valves	No.: 2	D_i/b: 0.49	D_e/b: 0.31	
V-8	Connecting Rod				
Displacement W: 2.00		L/R: 3.70	B/b: 0.08	w_r/p:	
x/b:	Crankshaft	P/b: 0.55	M/b: 0.74	N: 5	
4-Cycle		w_j/m:	t/b:		
Mack End-864	Cylinder	Bore: 5.0	Stroke: 5.50	S/b: 1.10	
	Piston	l/b: 1.10	p/b: .36	h/b: .67	N_c: 3
Cylinder Arrangement	Valves	No.: 2	D_i/b: 0.47	D_e/b: 0.39	
V-8	Connecting Rod				
Displacement W: 2.40		L/R: 3.80	B/b: 0.11	wr/p:	
x/b:	Crankshaft	P/b: 0.70	M/b:	N: 5	w_j/m:
4-Cycle		t/b:			

Table 2.3. (Continued) Design Ratio for Automotive Diesel Engines

International	Cylinder	Bore: 4.50	Stroke: 4.31	S/b: 0.98	
	Piston	l/b: 1.05	p/b: .34	h/b: .63	N_c: 2
Cylinder Arrangement	Valves	No.: 2	D_i/b: 0.45	D_e/b: 0.35	
V-8	Connecting Rod				
Displacement W: 2.40		L/R:4.0	B/b:	w_r/p: 0.4	
x/b: 1.25	Crankshaft	P/b: 0.58	M/b: 0.68	N: 5	w_j/m: 0.36
4-Cycle		t/b: 0.17			
Rolls Royce	Cylinder	Bore: 3.44	Stroke: 3.60	S/b: 1.05	
K-60	Piston	l/b: 1.12	p/b: .42	h/b: .73	N_c: 3
Cylinder Arrangement	Valves	No.:	Di/b:	D_e/b:	
OP-6	Connecting Rod				
Displacement W: 3.90		L/R: 3.40	B/b:0.08	w_r/p: 0.7	
x/b: 1.40	Crankshaft	P/b: 0.70	M/b: 0.80	N: 7	
2-Cycle		w_j/m: 0.45	t/b: 0.31		
GMC Detroit	Cylinder	Bore: 5.75	Stroke: 5.75	S/b: 1.00	
Diesel 16-V-149	Piston	l/b: 1.17	p/b: .43	h/b: .47	Nc: 4
Cylinder Arrangement	Valves	No.: 4	D_i/b:	D_e/b: 0.28	
V-16	Connecting Rod				
Displacement W: 4.20		L/R: 4.20	B/b: 0.1	w_r/p:	
x/b:	Crankshaft	P/b: 0.68	M/b: 0.85	N: 9	w_j/m: 0.67
2-Cycle		t/b: 0.28			

Table 2.4. Design Ratio for Locomotive Diesel Engines

Machine Model	*Component Parameters*				
ALCO 251-T	Cylinder	Bore: 9.00	Stroke: 10.50	S/b: 1.17	
	Piston	l/b: 1.23	p/b: .33	h/b: .75	N_c: 4
Cylinder Arrangement	Valves	No.: 4	D_i/b: 0.32	D_e/b:	
V-16	Connecting Rod				
Displacement W: 3.90		L/R: 3.75	B/b:	w_r/p:	
x/b:	Crankshaft	P/b: 0.67	M/b: 0.95	N: 9	
4-Cycle		w_j/m:	t/b:		
Cooper-Bessemer	Cylinder	Bore: 9.00	Stroke: 10.50	S/b: 1.17	
	Piston	l/b: 1.10	p/b: .35	h/b: .61	N_c: 4
Cylinder Arrangement	Valves	No.: 4	D_i/b: 0.32	D_e/b: .32	
V-16	Connecting Rod				
Displacement W: 3.70		L/R: 4.40	B/b: 0.13	w_r/p: .38	
x/b:	Crankshaft	P/b: 0.86	M/b: 0.92	N: 9	w_j/m: 0.55
4-Cycle		t/b: 0.33			
M. A. N.	Cylinder	Bore: 9.45	Stroke: 11.81	S/b: 1.25	
V8V 24/30	Piston	l/b: 1.37	p/b: .44	h/b: .78	N_c: 3
Cylinder Arrangement	Valves	No.: 4	D_i/b: 0.34	D_e/b: 0.31	
V-16	Connecting Rod				
Displacement W: 3.50		L/R: 4.10	B/b: 0.11	w_r/p: 0.48	
x/b: 1.95	Crankshaft	P/b: 0.79	M/b: 0.79	N: 9	w_j/m: 0.52
4-Cycle		t/b: 0.40			
Sulzer	Cylinder	Bore: 9.45	Stroke: 11.00	S/b: 1.16	
16LVA24	Piston	l/b: 1.40	p/b: .48	h/b: .92	N_c: 2
Cylinder Arrangement	Valves	No.: 4	D_i/b: 0.35	D_e/b: 0.35	
V-16	Connecting Rod				
Displacement W: 3.30		L/R: 4.10	B/b:	w_r/p:	
x/b:	Crankshaft	P/b: 0.85	M/b:	N: 9	w_j/m:
4-Cycle		t/b:			

Table 2.5. Design Ratio for Medium Size Industrial & Diesel Engines

Machine Model	*Component Parameters*				
Maybach-M.B.	Cylinder	Bore: 6.90	Stroke: 8.10	S/b: 1.17	
	Piston	l/b: 0.33	p/b: 0.42	h/b:	N_c: 3
Cylinder Arrangement	Valves	No.: 4	D_i/b: 0.30	D_e/b: 0.30	
V-12	Connecting Rod				
Displacement W: 2.30		L/R: 3.70	B/b:	w_r/p:	
x/b:	Crankshaft	P/b: 0.72	M/b:	N: 7	w_j/m:
4-Cycle		t/b:			
Poyaud	Cylinder	Bore: 5.92	Stroke: 7.09	S/b: 1.20	
820DB	Piston	l/b: 1.42	p/b: .38	h/b: .85	N_c: 4
Cylinder Arrangement	Valves	No.: 4	D_i/b: 0.32	D_e/b:	
V-8	Connecting Rod				
Displacement W: 4.00		L/R: 4.3	B/b:	w_r/p:	
x/b:	Crankshaft	P/b: 0.65	M/b: 0.79	N: 5	w_j/m:
4-Cycle		t/b:			
Caterpillar	Cylinder	Bore: 6.25	Stroke: 8.00	S/b: 1.28	
D-379	Piston	l/b: 1.32	p/b: .38	h/b: .85	N_c: 3
Cylinder Arrangement	Valves	No.: 2	D_i/b:	D_e/b:	
V-12	Connecting Rod				
Displacement W: 3.50		L/R: 4.2	B/b:	w_r/p: 0.45	
x/b:	Crankshaft	P/b: 0.80	M/b: 0.95	N: 7	
4-Cycle		w_j/m: 0.80	t/b: 0.24		
Nordberg	Cylinder	Bore: 13.5	Stroke: 16.5	S/b: 1.22	
VIB-115	Piston	l/b: 1.40	p/b: 0.41	h/b:	N_c: 4
Cylinder Arrangement	Valves	No.: 2	D_i/b: 0.40	D_e/b: 0.40	
V-16	Connecting Rod				
Displacement W: 3.40		L/R: 4.0	B/b: 0.10	w_r/p: 0.49	
x/b:	Crankshaft	P/b: 0.74	M/b: 0.96	N: 9	
4-Cycle		w_j/m: 0.40	t/b: 0.31		

Table 2.6. Design Ratio for Large Gas Engines

Machine Model	*Component Parameters*				
Cooper-Bessemer	Cylinder	Bore: 13.5	Stroke: 16.5	S/b: 1.22	
KSV-16SG	Piston	l/b: 0.38	p/b: 0.74	h/b:	N_c: 4
Cylinder Arrangement	Valves	No.: 4	D_i/b: 0.31	D_e/b: 0.31	
V-16	Connecting Rod				
Displacement W: 4.20		L/R: 4.90	B/b: 0.11	w_r/p: 0.74	
x/b:	Crankshaft	P/b: 0.85	M/b: 0.85	N: 10	w_j/m: 0.50
4-Cycle		t/b: 0.29			
Worthington	Cylinder	Bore: 16.0	Stroke: 18.0	S/b: 1.12	
Mainliner	Piston	l/b: 1.90	p/b: .41	h/b: 1.04	N_c: 4
Cylinder Arrangement	Valves	No.: 2	D_i/b:	D_e/b: 0.4	
V-18	Connecting Rod				
Displacement W: 4.50		L/R:	B/b: 0.86	w_r/p: 0.4	
x/b:	Crankshaft	P/b: 0.88	M/b: 0.88	N: 10	w_j/m: 0.63
2-Cycle		t/b: 0.34			
Ingersoll-Rand	Cylinder	Bore: 16.0	Stroke: 22.0	S/b: 1.37	
PKVT	Piston	l/b: 1.48	p/b: .44	h/b: .8	N_c: 4
Cylinder Arrangement	Valves	No.: 4	D_i/b: 0.30	D_e/b: 0.30	
V-16	Connecting Rod				
Displacement W: 2.70		L/R: 4.0	B/b: 0.08	w_r/p: 0.38	
x/b: 2.02	Crankshaft	P/b: 0.88	M/b: 0.88	N: 10	w_j/m: 0.38
2-Cycle		t/b: 0.38			
Nordberg	Cylinder	Bore: 14.0	Stroke: 16.0	S/b: 1.14	
Radial	Piston	l/b: 1.40	p/b: .41	h/b: .79	N_c: 4
Cylinder Arrangement	Valves	No.:	D_i/b:	D_e/b:	
Radial, 12 l. s.	Connecting Rod				
Displacement W: 3.90		L/R: 4.40	B/b: 0.07	w_r/p:	
x/b:	Crankshaft	P/b: 1.0	M/b: 0.85	N: 2	w_j/m: 0.58
2-Cycle		t/b: 0.36			

Table 2.7. Design Ratio for Large Diesel Engines

Machine Model	*Component Parameters*				
Enterprise	Cylinder	Bore: 17. 0	Stroke: 21.0	S/b: 1.23	
RV-16-3	Piston	l/b: 1.77	p/b: .38	h/b: .83	N_c: 5
Cylinder Arrangement	Valves	No.: 4	D_i/b:	D_e/b: .38	
V-16	Connecting Rod				
Displacement W: 3.40		L/R: 4.60	B/b:	w_r/p:	
x/b:	Crankshaft	P/b: 0.70	M/b:	N: 9	w_j/m:
4-Cycle		t/b:			
Fairbanks-Morse	Cylinder	Bore: 20.0	Stroke: 21.5	S/b: 1.07	
38A20	Piston	l/b: 1.75	p/b: .39	h/b: 1.0	N_c: 3
Cylinder Arrangement	Valves	No.:	D_i/b:	D_e/b:	
V-12 Opposed	Connecting Rod				
Displacement W: 2.90		L/R: 4.7	B/b:	w_r/p:	
x/b: 1.65	Crankshaft	P/b: 0.81	M/b:	N:	w_j/m:
2-Cycle		t/b:			
Fiat	Cylinder	Bore: 35.4	Stroke: 63.0	S/b: 1.78	
900-12S	Piston	l/b: 1.90	p/b: 0.5	h/b:	N_c: 6
Cylinder Arrangement	Valves	No.:	D_i/b:	D_e/b:	
I-12	Connecting Rod				
Displacement W:		L/R: 3.90	B/b:	w_r/p:	
x/b: 1.26	Crankshaft	P/b: 0.82	M/b: 0.82	N: 13	w_j/m:
2-Cycle		t/b:			
M. A. N.	Cylinder	Bore: 36.6	Stroke: 66.9	S/b: 1.83	
K293/170	Piston	l/b:1.90	p/b: .59	h/b:	N_c: 6
Cylinder Arrangement	Valves	No.:	D_i/b:	D_e/b:	
I-12 l. s.	Connecting Rod				
Displacement W: 3.40		L/R: 4.0	B/b: .135	w_r/p: .49	
x/b:	Crankshaft	P/b: 0.78	M/b: 0.78	N: 14	w_j/m: 0.59
2-Cycle		t/b: 0.43			

Table 2.7. Continued

Nordberg	Cylinder	Bore: 21.5	Stroke: 31.0	S/b: 1.44	
TS-2112-SC	Piston	l/b: 2.36	p/b: .56	h/b: 1.47	N_c: 5
Cylinder Arrangement	Valves	No.:	D_i/b:	D_e/b:	
I-12, l.s.	Connecting Rod				
Displacement W: 2.75		L/R: 5.0	B/b: .087	w_r/p: 0.4	
x/b:	Crankshaft	P/b: 0.69	M/b: 0.77	N: 13	w_j/m: .45
2-Cycle		t/b: .32			
Sulzer	Cylinder	Bore: 35.4	Stroke: 61.0	S/b: 1.72	
RD/RF 90	Piston	l/b: 1.10	p/b: .74	h/b:	N_c: 5
Cylinder Arrangement	Valves	No.:	D_i/b:	D_e/b:	
I-12, l.s.	Connecting Rod				
Displacement W: 3.00		L/R: 4.6	B/b:	w_r/p: .51	
x/b: 1.91	Crankshaft	P/b: 0.71	M/b: 0.71	N: 13	w_j/m: .72
2-Cycle		t/b: 0.41			
Stork-Werkspoor	Cylinder	Bore: 33.5	Stroke: 67.0	S/b: 1.74	
170-8	Piston	l/b: 0.50	p/b: 0.5	h/b:	N_c:
Cylinder Arrangement	Valves	No.: 4	D_i/b:	D_e/b: 0.3	
I-12 p.v.	Connecting Rod				
Displacement W: 3.00		L/R: 4.0	B/b:	w_r/p:	
x/b:	Crankshaft	P/b: 0.77	M/b: 0.77	N: 13	w_j/m:
2-Cycle		t/b:			
Burmeister and Wain	Cylinder	Bore: 33.0	Stroke: 77. 0	S/b: 2.12	
	Piston	l/b: 0.75	p/b: .61	h/b:	N_c: 6
Cylinder Arrangement	Valves	No.: 2	D_i/b:	D_e/b: 0.6	
I-12 p.v.	Connecting Rod				
Displacement W: 3.00		L/R: 4.0	B/b:	w_r/p:	
x/b:	Crankshaft	P/b: 0.66	M/b: 0.88	N:13	w_j/m:
2-Cycle		t/b:			

Table 2.8. Design Ratio for Super-Sport Car & Automobile Racing Engines

Machine Model	*Component Parameters*				
FERRARI	Cylinder	Bore: 2.87	Stroke: 2.32	S/b: 0.81	
6-90	Piston	l/b:	p/b:	h/b:	N_c:
Cylinder Arrangement	Valves	No.: 2	D_i/b: 0.57	D_e/b: 0.57	
I-6	Connecting Rod				
Displacement W: 2.90		L/R:	B/b:	w_r/p:	
x/b:	Crankshaft	P/b:	M/b:	N:	w_j/m:
4-Cycle		t/b:			
Ferrari	Cylinder	Bore: 2.99	Stroke: 2.69	S/b: 0.90	
Lancia	Piston	l/b: 0.78	p/b: 0.37	h/b: .51	N_c:
Cylinder Arrangement	Valves	No.: 2	D_i/b: 0.625	D_e/b: .625	
V-8	Connecting Rod				
Displacement W:		L/R: 3.90	B/b: 0.11	w_r/p:	
x/b:	Crankshaft	P/b: 0.72	M/b:	N: 5	w_j/m:
4-Cycle		t/b:			
Ford	Cylinder	Bore: 3.76	Stroke: 2.87	S/b: 0.76	
	Piston	l/b: .86	p/b: .27	h/b: .45	N_c: 2
Cylinder Arrangement	Valves	No.: 4	D_i/b: 0.39	D_e/b: 0.36	
V-8	Connecting Rod				
Displacement W: 1.57	L/R: 3.60	B/b: 0.12	w_r/p: 0.35		
x/b: 1.20	Crankshaft	P/b: 0.56	M/b: 0.61	N: 5	w_j/m: 0.53
4-Cycle		t/b: 0.18			
Coventry	Cylinder	Bore: 2.13	Stroke: 1.60	S/b: 0.75	
Climax	Piston	l/b:	p/b: 0.22	h/b:	N_c: 2
Cylinder Arrangement	Valves	No.: 2	D_i/b: .57	D_e/b: 0.55	
FLAT 16	Connecting Rod				
Displacement W:		L/R:	B/b:	w_r/p: 0.37	
x/b:	Crankshaft	P/b: 0.60	M/b: 0.60	N: 10	w_j/m: 0.50
4-Cycle		t/b: 0.22			

Table 2.8. Continued

Coventry-Climax	Cylinder	Bore: 3.00	Stroke: 2.68	S/b: 0.89	
FPE 2.5	Piston	l/b: 0.83	p/b: .29	h/b: .48	N_c: 2
Cylinder Arrangement	Valves	No.: 2	D_i/b: 0.57	D_e/b: 0.55	
V-8	Connecting Rod				
Displacement W: 2.25		L/R: 3.90	B/b: 0.12	wr/p: 0.35	
x/b: 1.40	Crankshaft	P/b: 0.71	M/b: 0.84	N: 5	w_j/m: 0.35
4-Cycle		t/b: 0.22			
Coventry-Climax	Cylinder	Bore: 3.20	Stroke: 2.88	S/b: 0.88	
FPF	Piston	l/b: 0.75	p/b: 0.3	h/b: .45	N_c: 2
Cylinder Arrangement	Valves	No.: 2	D_i/b: 0.53	D_e/b: .45	
I-4	Connecting Rod				
Displacement W: 2.80	L/R: 3.4	B/b: 0.10	w_r/p: .52		
x/b: 1.18	Crankshaft	P/b: 0.75	M/b: 0.88	N: 5	w_j/m: 0.50
4-Cycle		t/b: 0.28			
Drake	Cylinder	Bore: 4.28	Stroke: 4.38	S/b: 1.02	
Offenhauser	Piston	l/b: 0.67	p/b: .25	h/b: .34	N_c: 2
Cylinder Arrangement	Valves	No.: 4	D_i/b: 0.40	D_e/b: 0.40	
I-4	Connecting Rod				
Displacement W: 1.80		L/R: 3.50	B/b: 0.11	w_r/p: 0.60	
x/b:	Crankshaft	P/b: 0.50	M/b: 0.55	N: 5	w_j/m: 0.54
4-Cycle		t/b: 0.18			
GMC TORO-FLOW	Cylinder	Bore: 4.13	Stroke: 3.13	S/b: 0.76	
	Piston	l/b:	p/b:	h/b:	N_c:
Cylinder Arrangement	Valves	No.: 4	D_i/b: 0.37	D_e/b: 0.35	
V-8	Connecting Rod				
Displacement W: 2.00		L/R:	B/b:	w_r/p:	
x/b:	Crankshaft	P/b:	M/b:	N:	w_j/m:
4-Cycle		t/b:			

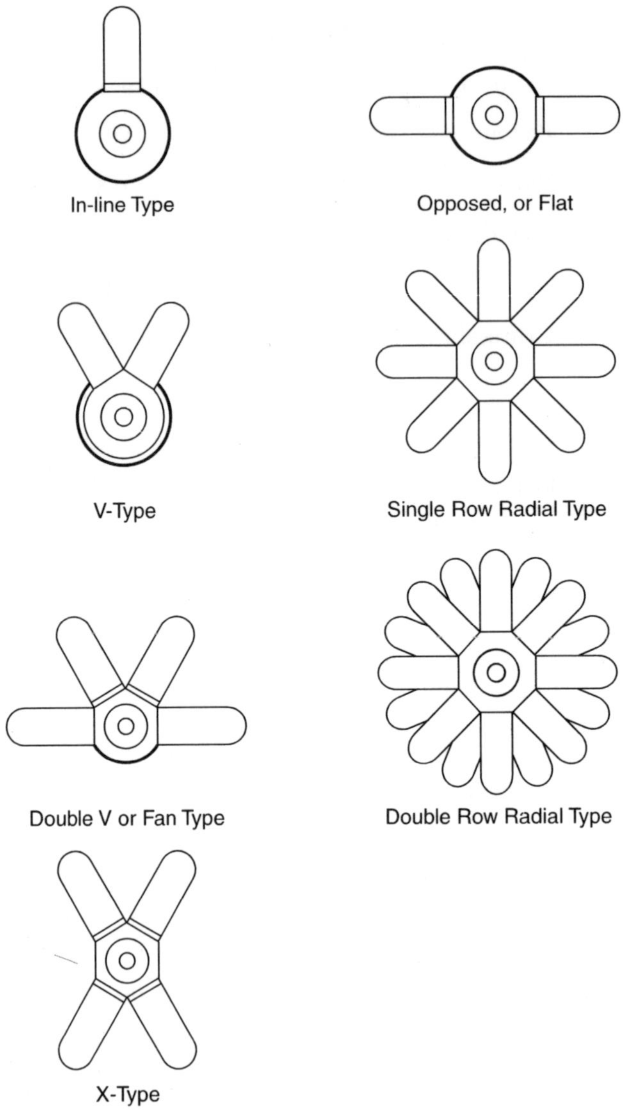

Fig. 2.22. Aircraft Engines Classified According to Cylinder Arrangement

X-type, (8) radial type, single row, (9) radial type, double row, (10) radial type, multiple row, or 'corncob'. Figure 2.22 illustrates some of these arrangements. The only piston engines used for aircraft propulsion in the United States are the opposed and radial types, together with some V-type and in-line engines.

The cylinders of in-line engines are arranged in a single row parallel with the crankshaft. The cylinders are either upright above the crankshaft or inverted, that is, below the crankshaft, the inverted version being more extensively used. The number of cylinders in an in-line engine is generally limited to six in order to facilitate cooling and to avoid excessive weight per horsepower. To obtain a better balance of firing impulses the number of cylinders is generally even. The crankshaft is located above the cylinders in an inverted engine. The engine is mostly air cooled; only a few liquid cooled engines are still in use. The in-line engine is generally confined to low and medium horsepower applications in small aircraft. The engine presents a small frontal area and is therefore adaptable to streamlining and resultant low drag nacelle configuration. With inverted cylinders pilot visibility is improved, while also helping to shorten the landing gear. However, the in-line engine has a greater weight to horsepower ratio than most other types. As aircraft engine size increases, it becomes difficult to cool the in-line type.

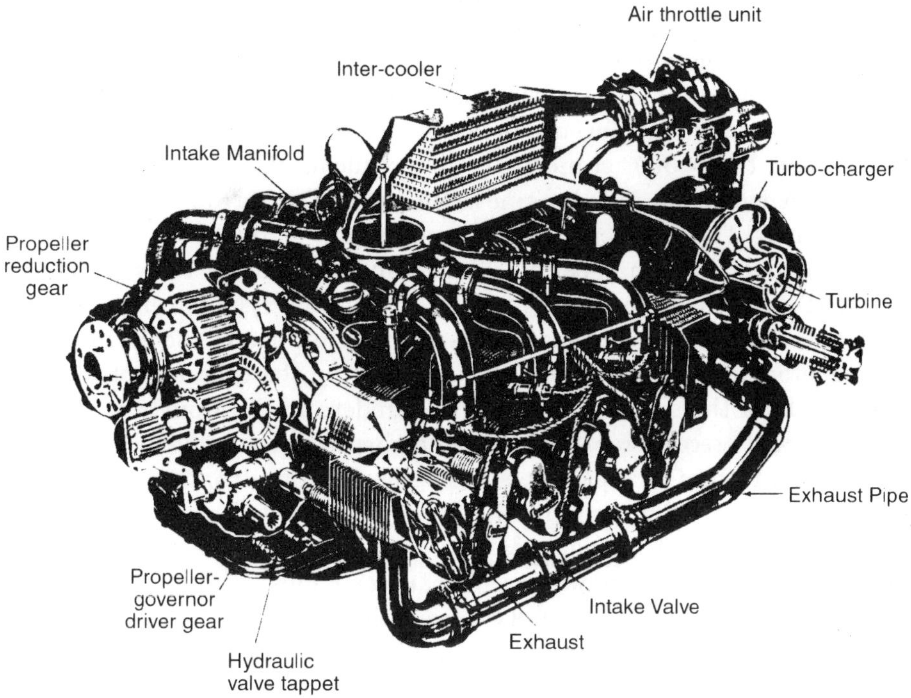

Fig. 2.23. Teledyne-Continental Six Cylinder Opposed Engine
(*Courtesy: Teledyne-Continental*)

The V-type engine has its cylinders arranged on the crankcase in two rows (or banks) forming the letter V. The angle between the banks is 90°, 60° or 45°. There is always an even number of cylinders in each row. Since the two banks of cylinders are opposite each other, two sets of connecting rods can operate on the same crank pin, thus helping to reduce weight per horsepower. The frontal area is only slightly larger than that of the in-line type; hence the engine cowling can be streamlined to reduce drag. As in in-line engines, if the cylinders are above the crankshaft the engine is known as the upright V-type. If the cylinders are below the crankshaft it is known as an inverted V-type. Better pilot visibility and shorter landing gear are possible if the engine is inverted.

The opposed type engine is the most popular for helicopters and light conventional aircraft, and is made in sizes delivering from less than 100 horsepower to more than 400 horsepower. Compared with other piston engine types opposed-type engines are most efficient, dependable and economical for light aircraft duty. Gas turbine engines are installed in some light aircraft, but their cost is prohibitive for the average private plane owner. The opposed-type engine is mounted with the cylinders and the crankshaft horizontal; in some helicopter installations, however, the crankshaft is vertical. The engine has a low weight to horsepower ratio, and because of its flat shape, it is very well adapted to streamlining and to horizontal installation in the nacelle. Figure 2.23 illustrates an opposed engine for general aircraft use.

The radial engine has been the workhorse of military and commercial aircraft since the 1920's. During World War II they were used in United States bombers and transport aircraft, as well as in most other categories of aircraft. They were developed to a peak of efficiency and dependability; even in today's jet age many of them remain operational throughout the world for all types of duty. A single row radial engine has an odd numbers of cylinders extending radially from the center line of the crankshaft. The number of cylinders is usually from five to nine. The cylinders are arranged evenly in the same circular plane, and all the pistons are connected to a single throw 360° crankshaft, thus reducing the number of working parts as well as the weight. A double row radial engine resembles two single row radial engines combined on a single crankshaft. The cylinders are arranged radially in two rows, and each row has an odd number of cylinders. Total number of cylinders in the engine is 14 or 18. Thus, the same effect is produced as having two seven cylinder or two nine cylinder engines joined on one crankshaft. A two-throw 180° crankshaft is used to permit the cylinders in each row to be alternately staggered on the common crank case. That is, the cylinders of the rear row are located directly behind the spaces between the cylinders in the front row. This allows the cylinders in both rows to receive ram air for cooling purposes. The radial engine has the lowest weight to horsepower ratio of all types of piston engine. The disadvantages of the radial engine are greater drag because of the large frontal area presented to the air, and it also has some cooling problems. Nevertheless, its dependability and efficiency have made it the most widely used type for large aircraft equipped with reciprocating engines. The largest and most powerful piston engine built and used in the United States is the 28-cylinder Pratt & Whitney R-4360. It was most extensively used at the end of World War II and afterward for both bombers and transport aircraft. With the development of gas turbine engines, the larger piston engines have been replaced by the more powerful and lighter weight turbo-prop and turbo-jet engines. The gas turbine engine has the additional advantage of having fewer moving parts, and so the maintenance cost is reduced. Time between overhauls is also increased.

Practically all aircraft piston engines in use today are air cooled, with few liquid cooled engines still in operation. Excessive heat is undesirable in any internal combustion engine for three principal reasons: (1) it adversely affects the combustion behavior of the fuel-air charge, (2) it weakens and shortens the life of engine parts, and (3) it impairs lubrication. If the temperature inside the cylinder is too high the fuel mixture will be preheated and combustion will occur before the proper time. Premature combustion causes detonation, 'knocking' and other undesirable conditions. It also aggravates the overheating problem, and possibly could result in failure of pistons and valves. The strength of engine parts depends on their heat treatment. Excessive heat weakens such parts and shortens their life. Also, the parts may undergo plastic deformation, and become elongated or warped to the extent that they freeze or lock together and stop engine operation. Excessive heat build-up can also 'crack' the lubricating oil, lowers its viscosity and deteriorates its lubricating properties.

In an air-cooled engine thin metal fins project from the outer surfaces of the walls and heads of the engine cylinders. When air flows over the fins it absorbs excess heat from the cylinders. Fins on the cylinder head are forged or cast as part of the head. Fins on the steel cylinder barrel are machined from the cylinder barrel forging. Deflector baffles, made of aluminum sheet, fastened around the cylinders direct the flow of air to obtain the maximum cooling effect. A cylinder with baffles is shown in figure 2.24. Engine operating temperature can be controlled by the movable cowl flaps located on the engine cowling. A large amount of heat generated in an engine cylinder is carried to the atmosphere with the exhaust gases, the amount varying from 40 to 50 percent. Proper valve timing adjustment is the critical factor in heat rejection through the exhaust, thereby maximizing heat transfer efficiency.

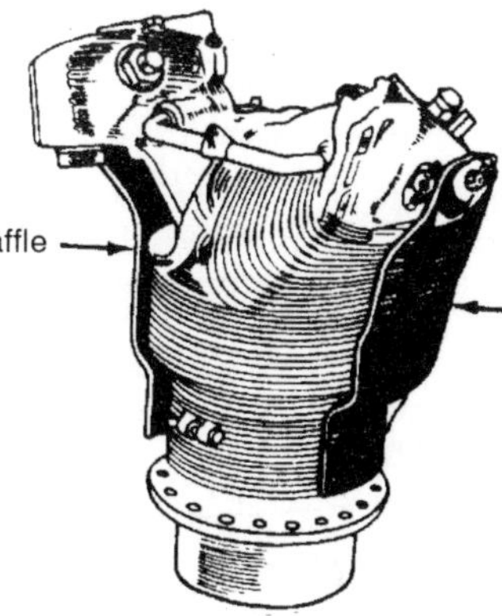

Fig. 2.24. Cylinder with Baffles for Cooling (*Courtesy: Teledyne-Continental*)

2.6 MARINE, INDUSTRIAL AND POWER GENERATION ENGINES

In the marine field for conventional boats and ships under 1,000 tons, and even higher for larger ships, the reciprocating internal combustion engine is practically the only propulsion prime mover of choice, and is much more widely used than steam engines, steam turbines or gas turbines. Gas turbines based on aircraft engine design, because of their lighter weight, are used to propel naval vessels and to generate electric power on off-shore oil production platforms. Reduced weight of aviation technology derived gas turbines comes from using thin walled casings, rotors and bearing support frames. But they face the twin problems associated with such designs, one of higher initial cost due to the more complex engineering and manufacturing, the other due to reduced operating life of the components. In terms of capital outlay, fuel efficiency, operating cost and component length of life reciprocating engines using pistons, connecting rod and crankshaft enjoy substantial advantages over gas turbines. The only service where the reciprocating internal combustion engine has been completely displaced is for transport and military airplanes, where the jet or the turbopropeller engine is the accepted norm. The basic reason for this situation is that gas turbines, as compared with reciprocating engines for the same service, are lighter in weight and enable the aircraft to attain much greater speeds and to fly at much higher altitudes. The poorer fuel economy for the compressor-burner-turbine engine on a horsepower basis has been offset by higher aircraft speed and higher altitude operation, so that the fuel consumption per ton-mile traveled is actually less than it was at the lower speeds and lower altitudes at which the reciprocating power was used at the end of World War II. Note that aircraft engines operate, except for short periods during take-off from the ground, at loads less than half maximum power, so the poor economy of gas turbines at light loads is not a serious drawback. In this service the higher initial cost of the power plant is also not a serious factor, considering the other gains made possible by its use. After a half a century of experience with aircraft gas turbine engines, it has been found that they are much more reliable and require less frequent overhaul than the high-strung reciprocating engines they replaced.

Ships use diesel engines of over 30,000 horsepower per unit. Except for very large installations (where gas turbines are sometimes used), the reciprocating engine is dominant for all classes of portable commercial power, including contractor's machinery, road building machinery, mining and oil well machinery and agricultural machinery.

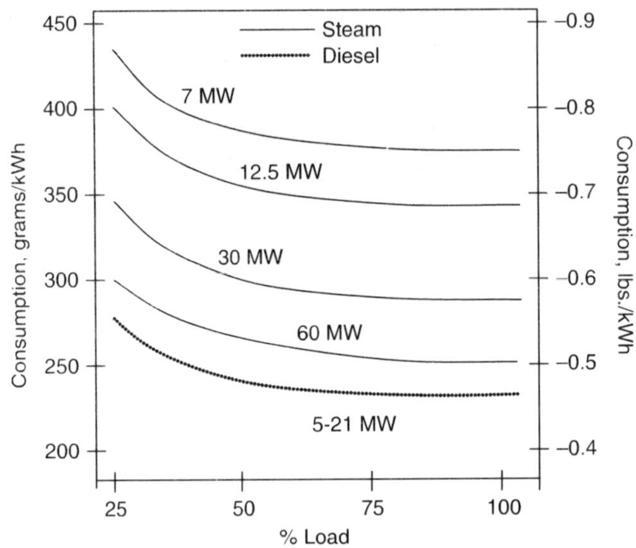

Fig. 2.25. Fuel Consumption for Power Stations

Figure 2.25 graphically illustrates the advantages of diesel over steam plants in regard to fuel economy.

Electric power stations of less than 10,000 kilowatts are almost universally diesel powered, and there are many much larger diesel-electric installations. However, gas turbines are sometimes used for peak and standby power as the capacity of existing stations is increased or new stations are established. Hospitals and other institutions typically rely on emergency power generating units powered by diesel reciprocating engines in case of a power failure. Koehler Corporation of Wisconsin offers electronically controlled internal combustion engine driven power generation units that kick in within one second of power loss from the electric utility's power lines; the back- up power is of more than passing interest to patients relying on kidney dialysis and other life saving devices.

In situations where diesel fuel may not be available for diesel powered engines, as in military operations, it is convenient to be able to use other common fuels, such as gasoline or aviation gas-turbine fuels. This multi-fuel capability is achieved principally by raising the compression ratio as required to fire the poorest (highest octane) fuel that may be expected to be used. For example, General Motors multi-fuel diesel engines have a compression ratio of about 21, whereas the standard model uses 17 to 18. The engine structure, of course, must be strong and stiff enough to withstand the resulting higher maximum pressure and rate of pressure rise. Compression ignition engines have also been run on various kinds of vegetable and animal oils. Satisfactory ignition, power and efficiency are obtainable with many such fuels, and many of them seem to have little trouble from excessive engine deposits or corrosion. Experiments on coal tar oils, such as creosote oil have been less successful because of the poor ignition quality of naphthenic compounds. Gaseous fuels are used in compression ignition engines either by introducing gas into the inlet manifold or through a special gas injecting system. Since most gaseous fuels have poor ignition quality, a pilot spray of fuel oil is used as an ignitor. The engine may then be treated as a spark ignition engine where the spark has been replaced by the pilot jet or as a compression ignition engine in which the combustion is augmented by the gaseous fuel. In this case the useful point of view depends on the relative energy contribution of the liquid and gaseous fuels. In some cases engines are so arranged that they can be easily changed over from gas to oil as the relative price of the two fuels fluctuates.

Gas burning engines have become of considerable importance for pumping gas through pipelines and for power generation in oil field regions where natural gas is plentiful, cannot be stored in substantial amounts, and so is cheap in price. Several attempts have been made to discover fuels that will appreciably increase engine output over that available with a normal fuel-air mixture operating under similar conditions. Acetylene has a significantly higher value of heat of combustion per standard cubic foot than that of normal fuels, but it has been found to be impractical both from economic and technical viewpoints. While hydrogen has the highest heat of combustion, its low specific weight results in low heat of combustion per unit volume of mixture. Ammonia has been of interest in the possibility of using anhydrous ammonia mixed with air under circumstances where ordinary fuel is not available, but only at great sacrifices in power and fuel economy. For racing purposes where there are no restrictions on the type or quantity of fuel, and where maximum power is a major requirement, methyl alcohol mixed with small amounts of other fuels is often used.

2.7 SUPERCHARGED ENGINES

Supercharging of spark ignition engines involves a compromise with efficiency, which can only be justified in a few cases, including:

1. Aircraft engines: Supercharging is used here to provide both high specific power output for take-off and to compensate for lower air density at operating altitudes. All but small engines for light aircraft are supercharged. The problem of detonation is solved by use of high-octane fuels. Automobile engines: With the advent of small, efficient turbo-superchargers and reliable electronic controls, many luxury and sports-type automobiles are supercharged. This usually lowers fuel economy in comparison with the same engine naturally aspirated.
2. The decision to use supercharging is more one of marketing than one of utility.
3. Racing engines: Here specific output has exaggerated its importance, and supercharging is used wherever it is allowed.
4. Large natural gas engines: In this case the saving in weight and bulk by means of supercharging is great, and the fuel used has a high resistance to detonation. Nearly all natural gas engines above 500 horsepower are supercharged. No limit on supercharging in diesel engines is imposed by combustion.

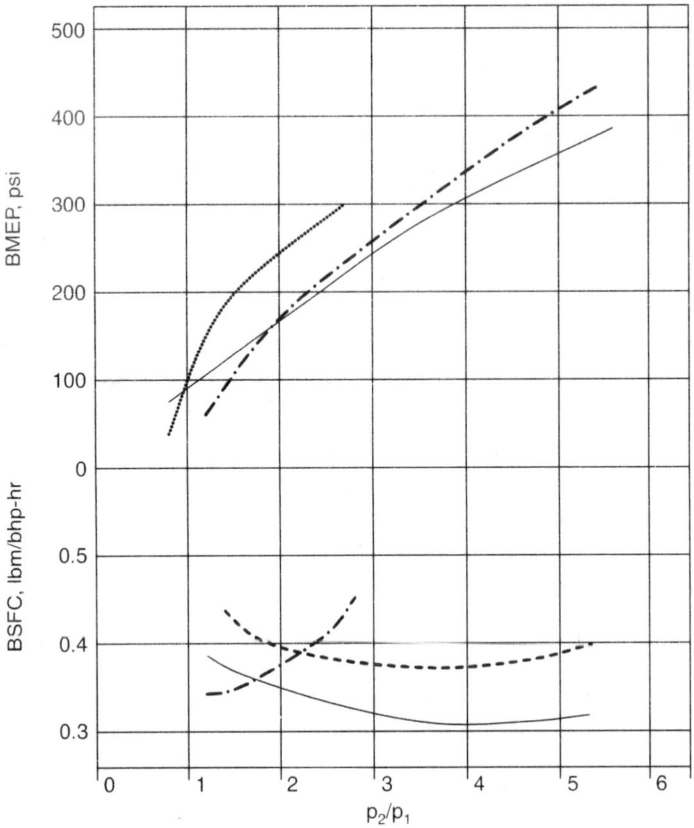

Fig. 2.26. Performance of three types of Turbo-Supercharged 4-Cycle Diesel Engines versus Compression Pressure Ratio

The decision whether or not to use supercharging, and if so how much, depends on a balance between the relative simplicity of the non-supercharged engine together with its generally lower mechanical and thermal stresses and the smaller size of a supercharged engine with the same rating. The initial cost may be in favor of either type, depending on the cost of the supercharging system and the reduction in cost due to decreased engine size. Diesel engines for trucks, buses, locomotives and medium and large marine engines are almost always supercharged. The same is true for stationary applications, such as power generation, except in the smallest sizes. The allowable amount of supercharging in the diesel engines depends on the question of reliability and durability, plus economic factors such as the cost of the supercharging equipment as a function of its capacity and pressure ratio.

Figure 2.26 compares some test results on three styles of four cycle diesel engines, where performance variation with compressor pressure ratio is shown. The curve for the engine variable compression ratio shows that gains in output can be made at a sacrifice in fuel economy.

2.8 PERFORMANCE FACTORS IN ENGINES

Earlier in the chapter engine power, mean effective pressure, rpm, displacement and other factors involved in the design and measurement of engine performance were discussed. These areas are explored in greater depth and are applied to actual operation of engines of all types.

Manifold pressure is the absolute pressure of the fuel air mixture immediately before it enters the intake port of the cylinder. Manifold pressure is of primary concern to the operator of a high performance spark ignition engine because it will be operating at a point near the maximum allowable pressure. Thus, when the danger of operating at an excessive pressure is real, a manifold pressure gage must be provided on the equipment to ensure operation within safe limits. Equipment operation at excessive manifold pressure will result in excessive cylinder pressures and temperature, which are likely to over stress the cylinders, pistons, piston pins, valves, connecting rods, bearings and crankshaft journals. Excessive pressure usually is accompanied by excessive temperature, and this leads to detonation, pre-ignition, and loss of power. Detonation usually results in engine damage if continued for more than a few moments. Damage may include piston failure by burning or cracking, failure of cylinder base studs, cracking of cylinder head and burning of valves.

Detonation is caused when the temperature and pressure of the compressed mixture in the combustion chamber reaches levels sufficient to cause instantaneous burning (explosion) of the mixture. Excessive temperatures are caused by high manifold pressure, high intake air temperature or an overheated engine. When detonation occurs the fuel air mixture may burn properly for a portion of its combustion and then explode as the pressure and temperature in the cylinder increases beyond the normal limit. Detonation will further increase the temperature of the cylinders and pistons, and may cause the head of a piston to melt. Extended detonation can also cause valves to burn.

Pre-ignition is caused when there is a hot spot in the engine that ignites the fuel air mixture before the spark plug fires. The hot spot may be red hot spark plug electrodes or carbon particles which have reached burning temperature and are glowing. Pre-ignition is detected by roughness of engine operation, power loss and high cylinder head temperature. To prevent pre-ignition and detonation the engine must operate with the proper fuel and within correct limits of manifold pressure and cylinder head temperature.

A mechanically reliable internal combustion engine must cost less to operate than the one that constantly needs overhauling or other maintenance. Its parts and accessories should serve well for a long period of time, and should be easily and inexpensively replaceable. Low fuel and oil consumption must also be considered, not only from the cost perspective, but as in aircraft engines, from the standpoint of reducing weight in flight. The engine should have the flexibility to run smoothly and perform in the desired manner at all speeds from idle to full power output and through all variations of atmospheric conditions.

Balance has several possible meanings in a discussion of fundamental requirements in an automobile or an aircraft, but the principal factor is freedom from vibrations. In a light airplane engine vibrations can reduce the life of certain parts or units and cut down the life of the airplane as a whole. In addition, vibration is fatiguing to the pilot and passenger, thus reducing the desirability of flying. Tubing, cables, conduits and accessories may be severely damaged or their function impaired by excessive engine vibrations.

One method of reducing engine vibrations is to design the engine with a large number of cylinders, thus reducing the effect of the pulsating torque delivered by the separate cylinders to the crankshaft. Rotating counterweights, dynamic dampeners, or other vibratory torque control mechanism may be used to balance the crankshaft and to reduce torsional vibrations. V engines of eight cylinders or more have excellent balance and relative freedom from vibrational problems. The V-8 engine fits very well into the space available in front-wheel drive automobiles, specially when the stroke to bore ratio is less than 1.0. This characteristic, together with its good balance and relatively simple design explains the great popularity of V-8 engines for higher powered automobiles and for many other applications where it proves to be a good compromise between compactness, high specific output and reasonably low cost. The V-12 engine is also an excellent type from all points of view, and is widely used for engines above 300 horsepower. V-engines of more than 12 cylinders are used for large power outputs when it is necessary to limit cylinder size, or in order to use a standard cylinder design for a wide range of power ratings. Two and four cylinder V-engines may have balance problems. They are used chiefly in small sizes where the resultant shape is appropriate, such as in motorcycles, and snowmobiles.

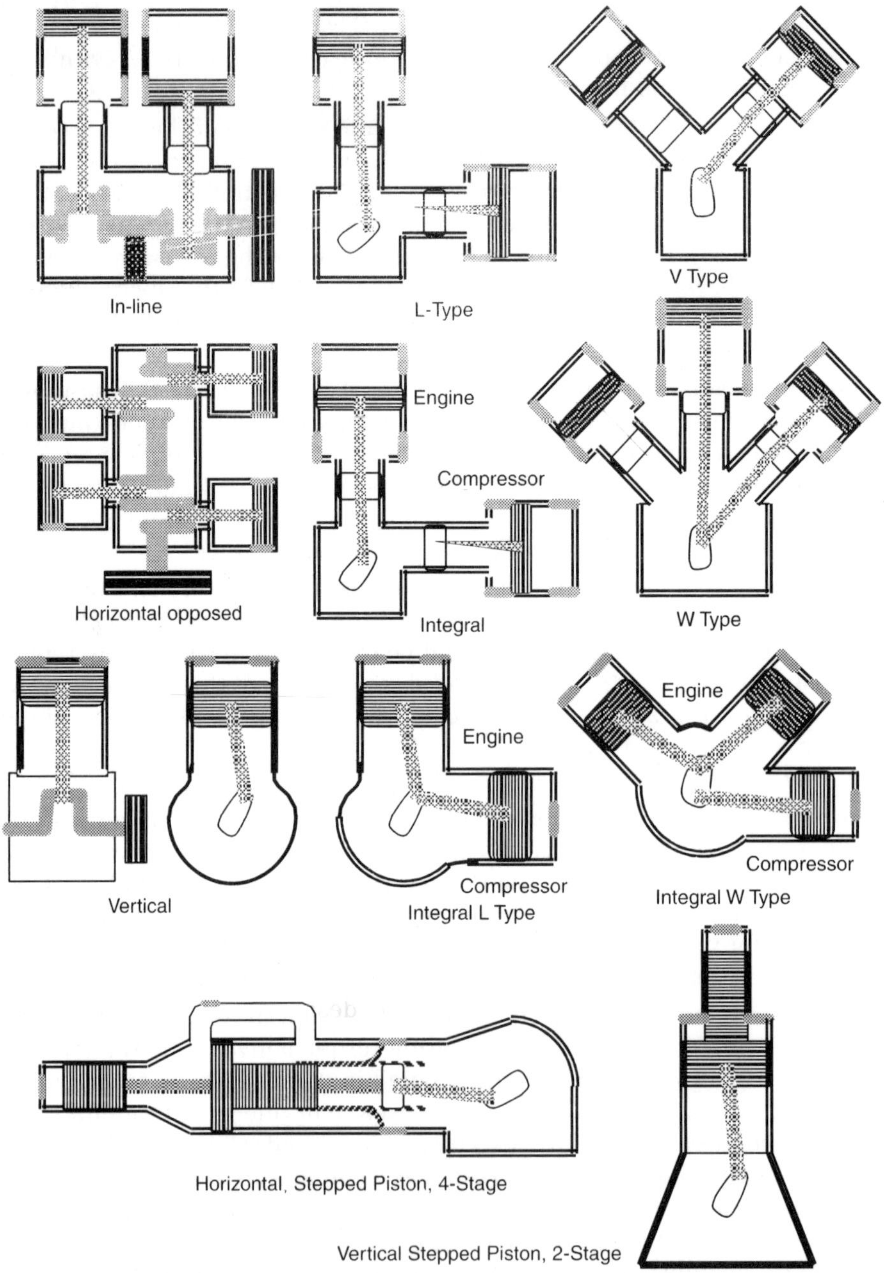

Fig. 2.27. Arrangement of Cylinders in Reciprocating Engines/Compressor Combination

2.9 AIR AND GAS COMPRESSORS

The reciprocating compressor is the patriarch of the compressor family. Higher efficiency and flexibility have given the reciprocating compressor a prominent role in new specialty process plants. A unique feature of this machine is the possibility of multiple services on one compressor frame. On a multi-stage frame each cylinder can be used for a separate gas service. Arrangement of the cylinders can be accomplished by using a large variety of configuration styles, as shown in Fig. 2.27. In choosing the correct compressor for a given installation, factors that need to be taken into account are: maximum, minimum and mean demand, the degree of freedom from moisture and oil (as might be required for underwater breathing or in a food processing plant), ambient conditions, methods of cooling available, purchase and running costs and environmental conditions. Reciprocating compressors are suitable for operating over a wide range of pressures, from a maximum of 150 psi for single stage machines to 1,000 psi for a two stage machine. Multi-stage compressors built for special applications are capable of delivering pressures in excess of 10,000 psi. Intermediate sizes comprise a variety of different configurations. The compressors may have one or more cylinders in-line, 'V', 'W', 'H' and 'L' configurations disposed about a common crankshaft. Variations include transferring the side thrust by employing a crosshead. Single acting compressors generally use the trunk (or automotive) type of piston. Differential or stepped pistons may also be used. A common arrangement uses three cylinders of a four-cylinder block to form the engine, with the fourth cylinder used to compress air or gas. This is a convenient way of balancing the forces with a single cylinder compressor. Figure 2.28 shows the cross section of a two-stage piston engine driven compressor.

For oil free applications special precautions have to be taken. Figure 2.29 shows a design to produce a high degree of cleanliness in the air. No lubrication is necessary in the upper cylinder because the pistons are fitted with PTFE rings; the combination of these rings, austenitic liners and a cross head guide to minimize piston side loads extends operating life without complicating maintenance. To prevent the oil from entering the cylinder, a spacer piece incorporating rubber seals is fitted between each cylinder and crankcase. Pure compressed air is produced without using further separation equipment.

An alternative to PTFE rings is to use a labyrinth seal on the piston. Fine grooves are cut in the cylinder wall and the piston skirt, and also on the piston rod.

Fig. 2.28. Two Stage Piston Engine Driven Compressor.

Reciprocating compressors operate at a fixed modulation in the basic configuration. One method of volume variation is by speed modulation. Another more common method is the use of clearance pockets, with or without valve unloading. With clearance pockets cylinder performance is modified. With valve unloading one or more inlet valves is physically open.

Figure 2.30 shows a typical idealized indicator card of a single stage compressor. At point 1 the piston is at the end of the stroke, with a certain amount of gas at discharge pressure trapped in the clearance volume existing between the piston and the head. As the piston begins its return stroke the clearance gas expands until the pressure in the cylinder is slightly lower than the pressure in the intake line. This condition occurs at point 2, and here the intake valves open to fill the cylinder with gas from the intake line.

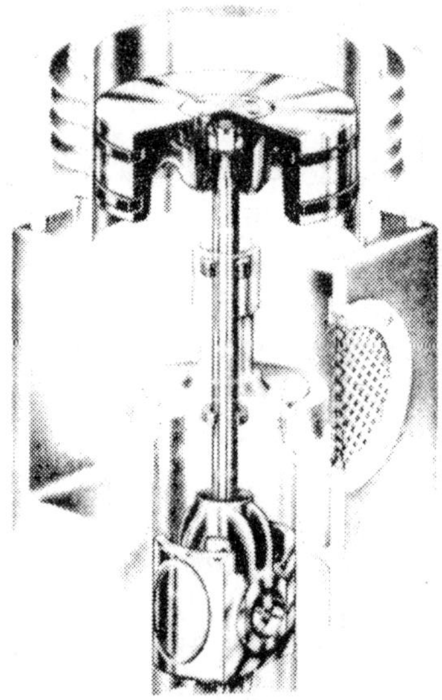

Fig. 2.29. Oil Free Air Compressor Using PTFE Piston Rings

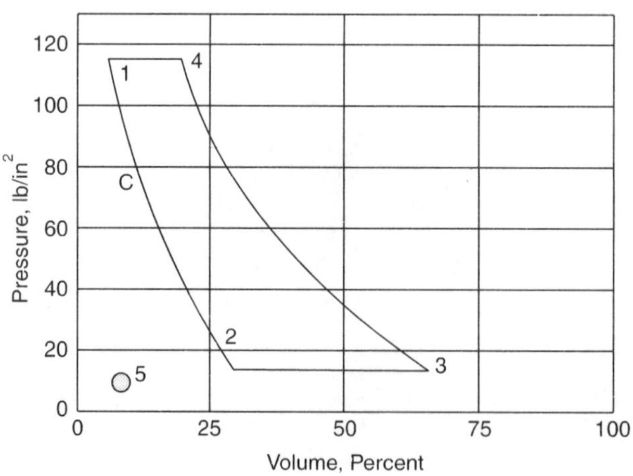

Fig. 2.30. Theoretical Indicator Card for Single Cylinder Compressor

This is represented by the line from 2 to 3 on the indicator card. At point 3 the piston is at the other end of the stroke, the inlet valve closes, and the gas is pressurized by the piston along line 3-4 on the return stroke. Even though the cylinder may be water cooled, in practice the compression process is fundamentally adiabatic due to the small amount of time in which it occurs. Further, the process is assumed to be perfectly adiabatic and reversible, or an isentropic process. In analyzing the performance of the compressor the compression curve 3-4 and the expansion curve 1-2 follow the general law pv^k = a constant, where the exponent k is equal to c_p/c_v, the specific heat values of the gas at constant pressure and volume. At point 4 the pressure in the cylinder is a little higher than that present in the discharge line. The discharge valves open to allow the piston to push the compressed air out of the cylinder into the discharge system along curve 4-1, when the cycle is repeated.

Some interesting points can be observed from the pressure volume diagram. The ratio of compression is discharge pressure divided by the intake pressure. Line 2-3 represents the actual capacity of the compressor. The entire stroke is represented by the travel of the piston from point 5 to point 3. The volumetric efficiency of the compressor is the ratio of the line 2-3 to line 5-3. Volumetric efficiency, however, is generally lower than that measured from the theoretical or actual indicator cards due to losses. A correction factor is needed to account for leakage past piston rings and valves. Gases such as hydrogen with a low molecular weight tend to be 'slippery', causing it to be more difficult to seal in the cylinder than air. The same is true in the case of non-lubricated compressors where the piston rings do not seal the gas as well as lubricated rings. There are other thermodynamic deviations from the ideal perfectly reversible adiabatic process, such as gas turbulence and heating of the incoming gas. These losses may be indicated on the idealized card as shown in Fig. 2.31, which shows an actual indicator card as might be taken on a machine in the field. The losses occurring within the cylinder may be summed up under the general term of compression efficiency. Fluid friction losses through the valves are the major factor behind reduced compression efficiency. Friction losses depend on flow velocity, valve configuration and specific gravity of the gas being compressed. In modern compressors compression the efficiency ranges from 85 to 93 percent.

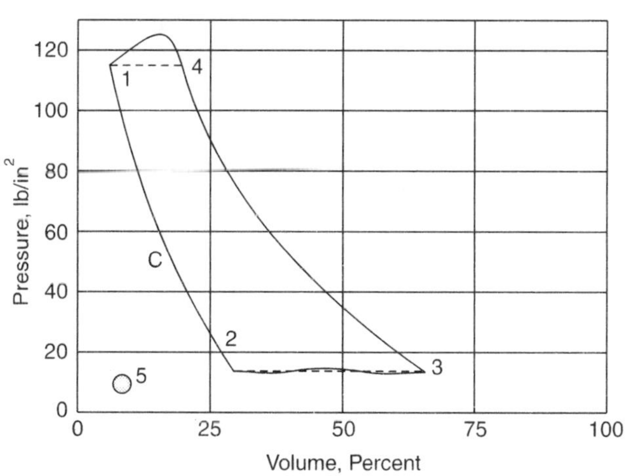

Fig. 2.31. Actual Indicator Card for Single Cylinder Compressor

Frictional losses in the piston rings, packing and bearings are accounted for by the mechanical efficiency of the compressor. Mechanical efficiency ranges from 88 to 93 per cent, and overall efficiency ranges between 75 and 85 percent when based on the ideal isentropic compression process.

Multi-staging of compression improves compression efficiency and reduces power requirements. Figure 2.32 illustrates two-stage compression by superposing the effect of two stages on a single stage card. The compression process in the first stage is carried only from point 3 to point 3a. At point 3a the gas is withdrawn from the cylinder and is cooled, reducing its volume back to point 3b. Compression in the second stage then occurs along the curve 3b-4a. The shaded area represents the saving in power achieved using two-stage compression is illustrated. In the case of compression of air from 0 to 100 psi, two-stage compression reduces power requirements by about 15 per cent compared to single stage compression. In the actual machine the effect of added inlet and discharge valve losses will offset some gains.

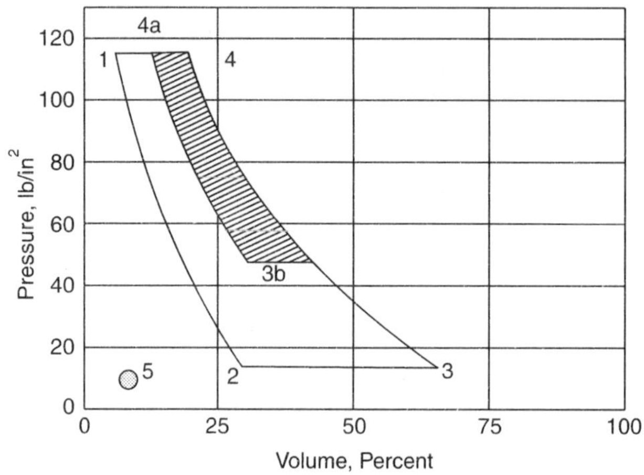

Fig. 2.32. Indicator Card for Two-Stage Compression

Multi-staging with cooling of gases between the stages also offers other advantages. Since the compression process is adiabatic the temperature of the gas continuously increases during compression. Cooling between the stages appreciably reduces the maximum gas temperatures in the cylinders, thereby easing difficulties with lubrication, carbon deposits and thermal stresses. Another advantage is the reduced pressure differential across each cylinder; this lightens the load and stresses imposed on valves, piston rings, piston rods and bearings, correspondingly increasing their life. This consideration of component load is often the prime reason for making a machine multi-staged. Certain applications in ammonia and propane refrigeration, for example, could well be handled with single stage compression except that the low isentropic exponent of the gases results in low volumetric efficiency, resulting in such large cylinders that piston rod loading becomes excessive for the frame to handle. One approach to this problem is to use two stages without inter-cooling, thus reducing piston rod load. To permit operation of the fixed-displacement reciprocating compressor at multiple operating points, one method is through use of clearance pockets. By using clearance pockets the cylinder capacity can be reduced. If the pocket is directly connected to the clearance region, the clearance volume can be increased. Increasing clearance reduces the capacity, but lowers the volumetric efficiency. Control of the pocket addition and removal is by means of a manual valve. If multiple pockets are used a step unloading system can be used, as shown in Fig. 2.33.

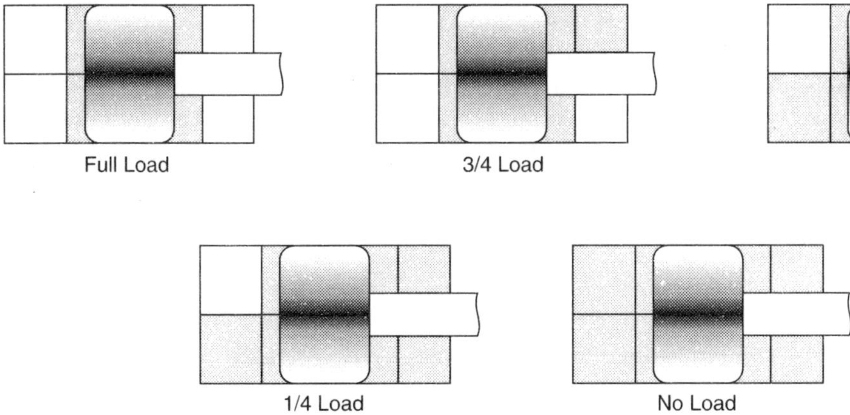

Fig. 2.33. A 5-Step Clearance Pocket Unloading Scheme

Fig. 2.34. Running Gear Elements of a Two-Stage Single Acting Compressor

The variable volume clearance pocket can provide an alternate unloading method. This device is normally attached to the outboard head. It consists of a piston-cylinder arrangement where the piston rod is threaded and attached to a hand wheel. Turning the hand wheel changes the clearance volume in an infinite number of steps up to the total pocket volume. On cylinders lacking physical space for pockets the same effect can be achieved by attaching external bottles and piping. Another capacity control method is the unloader. The plug type is used on all inlet valves for the unloaded end. The center of the valve is used for the unloader plug and port. The port type is used to replace one of the inlet valves on multiple inlet valve cylinders. It is normally used with low molecular weight applications. This unloader consists of a plug and port that uses up the total space of the valve it replaces. The plunger type of unloader is used on heavier gas applications where maximum unloaded flow area is needed. It operates by using plunger fingers to hold the valve plates open. Control of all the described unloaders is the same in that a piston operator is used. A few words of caution are in place when using the valve unloading method: a problem arises if there is loss of rod load reversal. Load reversal is needed to provide lubrication to some of the bearings. While the reversal problem is generally associated with unloading a double acting cylinder from one side, it should be checked for all unloaded cases, including pocket unloading. The second caution is about the duration of a totally unloaded case.

While the capacity is reduced to zero, gas in the outer end of the cylinder reciprocates as it follows piston movement. This movement of the gas will generate heat, and prolonged operation without cooling may cause overheating.

Regarding lubrication some machines employ a full pressure oil circulation system, where a positive displacement oil pump draws oil from the crankcase and delivers it under pressure to all sleeve bearings, and spray or drip lubrication to any anti-friction bearings. Other machines use a controlled splash lubrication system, as seen in Fig. 2.34. A portion of the crankshaft or connecting rod dips into the oil reservoir to produce a spray or mist.

An important group of non-lubricated reciprocating compressors are the labyrinth piston units made by Sulzer. The main design features are provided in Fig. 2.35. The distinguishing characteristic of the compressor is that, unlike the oil-free reciprocating compressors of traditional design with dry-running piston rings, they do not use piston rings or rider bands. Friction is absent in the cylinder and in the piston rod stuffing box because the labyrinth piston is provided with a number of grooves that generate a labyrinth-sealing action against the cylinder wall. The cylinder wall is likewise provided with grooves that are, however, finer than those of the piston. The piston moves within the cylinder with a clearance so that, even in the warm state, contact-free running is assured. Thermostats at the gas outlet detect any overheating that could lead to the piston scraping the cylinder wall. Thanks to this construction, lubrication of the cylinder and of the piston rod stuffing box is not necessary. Moreover, the suction and discharge valves are designed so that no lubrication is required. A gear pump driven from the crankshaft lubricates the driving mechanisms of the compressors. Depending on the requirement, compressors can be equipped additionally with an oil pump for preliminary lubrication (driven by an electric motor), an oil cooler, and a high-efficiency oil filter. The lubricating oil system supplies oil under pressure to the main bearings, the lower and upper connecting rod bearings, and the crosshead guide mechanisms. Splash oil lubrication is provided for the piston rod guide bearings.

When operating in the normal temperature range, the cylinder and crosshead guide mechanisms and the piston rod guide bearings are water-cooled. Where low suction temperatures apply, such as in refrigeration applications, cooling of these bearings can be dispensed with. The piston is guided from outside the compression space by the piston rod, which is located with relatively small clearance in the guide bearing, and by a precise guidance of the crosshead. Oil scrapers ensure separation between the oil-free portion and the oil-lubricated crank drive. The rod is provided with an oil slinger to prevent the thin film of oil remaining on the piston rod from creeping upward along it. The distance between the drive mechanism and the stuffing box is greater than the length of the piston stroke, so that the oil-wetted portion of the piston rod cannot penetrate into the oil-free stuffing box.

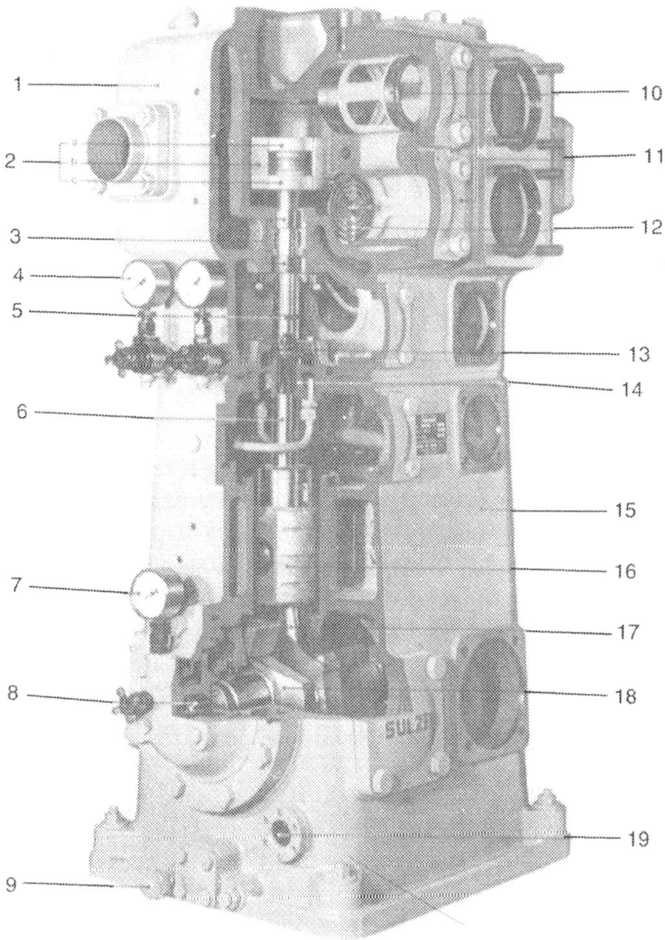

1. Cylinder block
2. Labyrinth piston
2a. Upper piston crown
2b. Piston skirt
2c. Lower piston crown
3. Piston rod gland
4. Pressure gauge with throttle valve
5. Piston rod shield
6. Piston rod
7. Oil pressure gauge
8. Geared oil pump
9. Oil discharge
10. Lantern
11. Valve cover
12. Plate valve
13. Oil scraper
14. Piston and guide bearing
15. Frame
16. Crosshead
17. Connecting rod
18. Crankshaft
19. Oil sight glass

Fig. 2.35. Labyrinth Piston Compressor (*Courtesy*: Sulzer-Burckhardt, Winterthur and Basel)

Intimately bound to the chemical industry, the technique of using large reciprocating compressors to obtain very high pressures (exceeding 1000 atmospheres) is generally associated with the manufacture of low-density polyethylene. A special characteristic feature of the high-pressure ethylene polymerization process is that a considerable difference in pressure is necessary between the inlet gas entering the reactor and the outlet of the recycle gas. The re-circulators, also called hyper compressors, work between the limits of 100 to 300 atmospheres on the suction side and 1500 to 3500 atmospheres on the delivery side for most of the process cycle, by far the most powerful machines in the production stream. The unit capacity may exceed 50 ton/h, and the corresponding power requirement is nearly 20,000 hp. The operating range of the secondary compressors takes place well above the critical point of ethylene, the thermodynamic behavior of the fluid lies somewhere between that of a gas and that of a liquid. This peculiar condition has two main effects: a small reduction of the specific volume with increasing pressure and a moderate rise of adiabatic temperature with increasing pressure. These thermodynamic conditions have a major influence the design of the compressors.

Compared with the conventional reciprocating compressor, the compression ratio is of little practical significance, but the important factor is the final compression temperature should not exceed the range of 80 to 1200C to avoid premature polymerization.

Sealing of the high-pressure compression chamber is a major problem. Labyrinth seals cannot function at very high pressures, so friction seals are currently used for moving and stationary seals. Metallic piston rings are used as moving seals in large high-pressure reciprocating type of compressor, and are made in three pieces: two sealing rings and an expander ring behind both of them to seal the gaps in the radial direction. The materials used are special grade cast iron, bronze or steel. The piston is of built-up design, comprising of a series of supporting and intermediate rings with a guide ring on top of them and a through bolt. All piston components are made of high tensile steel. Particular attention must be paid to the design and stress calculation of the central bolt, which is subjected to severe strain fluctuations because of fluctuation in the gas pressure during the suction and discharge cycles. The use of piston rings allows for a simple cylinder design, the main part of which is a liner that has been thermally shrunk to withstand the high variations of the internal pressure. The inner sleeve is generally made of a sintered material such as tungsten carbide. The use of the expensive material is justified because it has the twin beneficial characteristics of a hard surface and a high modulus of elasticity. Increased surface hardness reduces friction forces, while eliminating the danger of seizure. However, sintered materials have poor tensile strength, hence care must be taken to ensure the inner liner is always under compression even when the gas temperature rises.

Packed plungers for hyper compressors are the other answer to piston sealing. Metallic self-adjusting packing materials are widely used, held in place by supporting steel plates. The plates are thermally shrunk to resist large variations in internal pressure. Piston rings have also been used for pressures in the 2,000 to 3,000 lb/in^2 range. The choice between packing and rings depends largely on the process and the type of lubricant.

2.10 PUMPS

Reciprocating pumps have been in use for over two thousand years. Development of the very early power driven reciprocating pump, shown in Fig. 2.36, is attributed to Ctesibius of Alexandria, Egypt, in about 150 B.C., about 100 years after Archimedes invented his famous screw pump. Ctesibius's pump consisted of a water wheel driving a camshaft, which through levers caused a pair of pistons to alternately reciprocate. The Newcomen Pumping engine was one of the first applications of steam power to drive a pump, shown in Fig. 2.37. A conventional reciprocating water well pump was connected through a rocking beam to a steam driven piston. After steam from the boiler pushed the piston up, water condensed the steam and atmospheric pressure pushed the piston back down.

The general characteristic of a reciprocating piston pump is that it is a true positive displacement pump, with delivery directly proportional to swept volume and speed of reciprocation. The pump has an inherently high mechanical efficiency, with the overall efficiency ranging from 85% to 95%. Thus, nearly 90% of the power from the driver is converted into liquid power. The maximum design working pressure is established by the strength of the components subjected to the pressure. Maximum pump speed is limited by restraints on the piston speed, satisfactory valve operation and time for pump suction side to fill completely with the fluid charge.

Fig. 2.36. Ctesibius's Pump, circa 150 B. C. (*Courtesy:* Ingersoll Rand Co.)

The reciprocating pump is not a kinetic machine, as is a centrifugal pump, because it does not require velocity to achieve pressure. High pressures are achieved at low velocities, and therein lies one of the advantages of the reciprocating pump, permitting usage for pumping abrasive slurries and high viscosity liquids.

Vertical straight lines define the head-capacity (H-Q) characteristics. Therefore, a constant speed reciprocating power pump that pumps, for example, 100 gpm at 500 psig discharge pressure will pump very nearly 100 gpm at 3,000 psig discharge pressure. It is this feature that allows a controlled-volume pump to be used as a precise metering device. The H-Q characteristics are modified by slippage, which tends to increase with increasing head.

The type of valves employed on a reciprocating pump governs the pressure rating of the unit. Simple porting, sliding or rotary valves tend to have inherent pressure limitations. Higher-pressure rating can be obtained with seated valves, such as poppet valves. Multi-cylinder pumps are normally used for high pressures and large discharge flow, particularly for oil fluids. Speed rating for this type of pump for continuous running is around 1,500 rpm. They are usually in-line configuration of either horizontal or vertical layout. Larger units are designed for speeds of 100 to 600 rpm, and are generally of horizontal configuration with volumetric efficiency as high as 97%.

Reciprocating pumps work on the two-stroke principle. During the outgoing suction stroke the suction valve opens and the cylinder is filled with the fluid, Fig. 2.38(*a*). During the incoming delivery stroke the suction valve closes and the discharge valve opens to deliver the fluid, Fig. 2.38(*b*). The theoretical flow versus time curve for a single cylinder pump is shown in Fig. 2.38(*c*). Since the discharge flow remains zero during the suction stroke, the pulsating nature of the flow is exaggerated in a single cylinder pump. In the case of a double cylinder pump discharge flow is momentarily zero at each end of the stroke, but again with sinusoidal characteristics, as shown in Fig. 2.38. A simple method of smoothing the flow with a double acting pump is to introduce a cushion chamber into which a portion of the flow is forced under pressure and ejected when the pressure falls off. The effect is to provide a loop-over of sustained flow during the pause period between the end of one stroke and the beginning of next stroke. This helps to smooth out the pulsations as shown in Fig. 2.39. This procedure is commonly adapted in the case of single cylinder double acting (simplex) pumps.

Fig. 2.37. Newcomen Pumping Engine (*Courtesy:* Ingersoll Rand Co.)

Reciprocating pumps are usually classified by the following features:

- Orientation of the center line of pumping element (horizontal or vertical).
- Number of discharge strokes per cycle of each drive rod (single or double acting).
- Type of pumping element (piston, plunger or diaphragm) and adjustability of stroke length (fixed-stroke or variable stroke).

In a single acting pump the wetted part of the piston is only on one side and the other part exposed to the atmosphere, and has one set of suction and discharge valves. Figure 2.40 illustrates a double acting pump, with suction and discharge parts of the cycle performed at the same time on the two sides of the piston. A plunger pump can be made double acting by tying two coaxial plungers together with side rods and cross heads. The terms piston and plunger are often used interchangeably. A piston is a cylindrical disc mounted on a smaller diameter rod and fitted with sealing rings. A plunger is a smooth rod, similar to a piston rod. The sealing rings are stationary and the plunger slides through the rings. A diaphragm is a flexible tube or disc that serves to isolate the pumpage from the piston, plunger or atmosphere. The diaphragm may be connected directly to the reciprocating rod or hydraulically actuated, driven by the liquid that is oscillated by a piston or a plunger. Most larger industrial power pumps are manufactured with a fixed stroke.

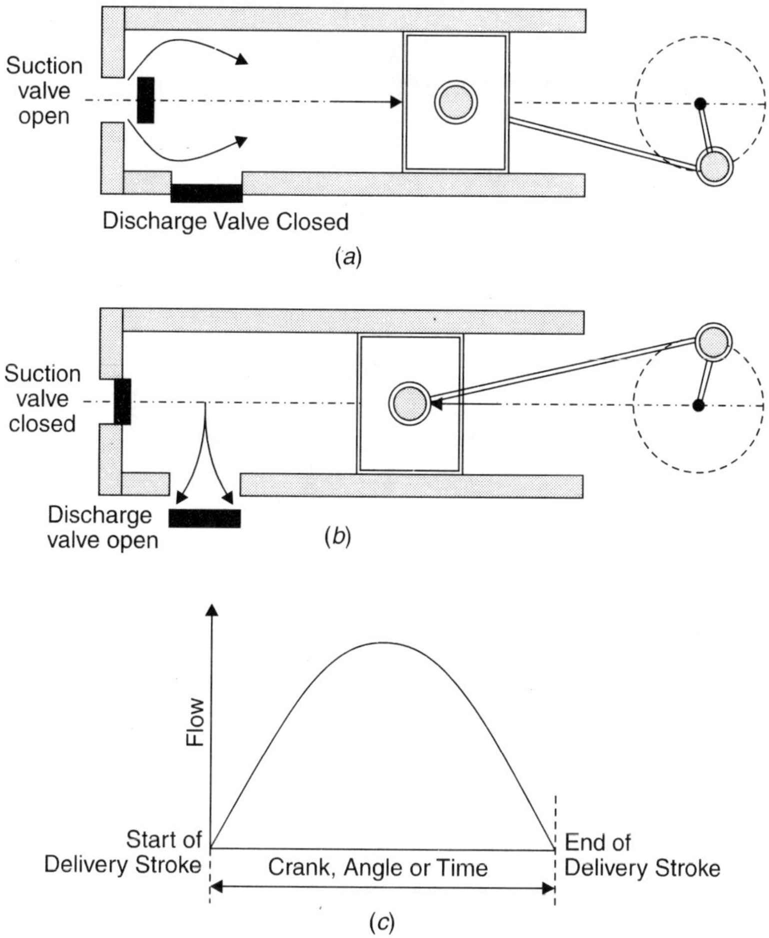

Fig. 2.38. (*a*) Suction, (*b*) Discharge, (*c*) Flow Curve for Single Cylinder Pump

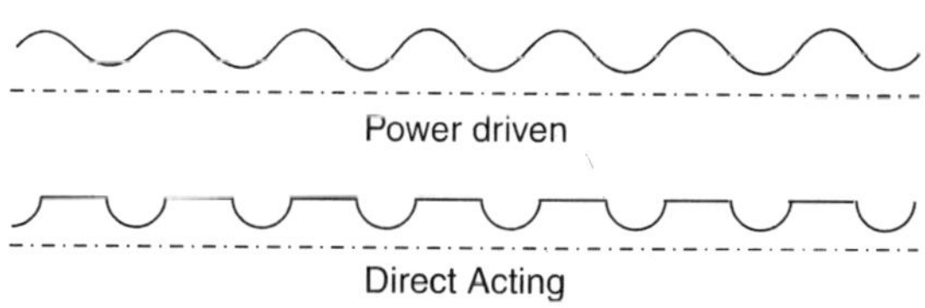

Fig. 2.39. Pressure Variation using a Cushion Chamber

Power pumps are manufactured in a variety of configurations. Vertical units are provided with power ends above or below the liquid ends. Some vertical plunger pumps push the plunger into the liquid cylinder, thus placing the connecting rods in compression and the tie studs connecting the power frame with the liquid end in tension. Other vertical pumps pull the plunger into the liquid cylinder. Horizontal plunger pumps generally push the plunger into the liquid cylinder. Valve assemblies are normally integrated with the fluid cylinder. In special applications, such as for slurries and for hot applications, the valves are mounted remotely. Figure 2.41 shows the power end of a horizontal plunger pump.

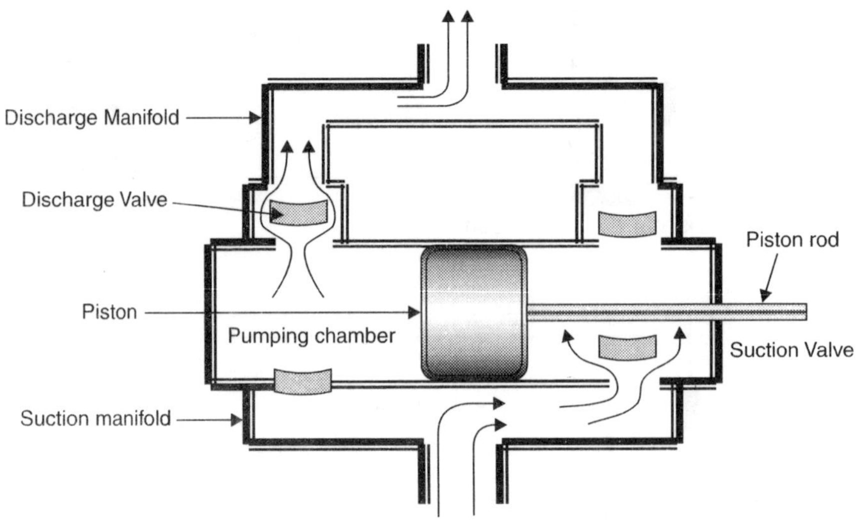

Fig. 2.40. Liquid End of Double Acting Piston Pump

Reciprocating pumps perform a large variety of services, from the small unit that provides pulsating jets of water for cleaning in confined spaces to pushing powdered coal slurry along a pipeline. In the drilling of oil and gas wells drilling mud is pumped by both piston and plunger pumps at pressures up to 5,000 psi. To lock the well casing into the drilled hole in the earth, a mixture of cement and water is pumped to the bottom of the well into the annular space between the casing and drilled hole; engine driven power pumps are used to pump this slurry into the well. During the drilling operation heavy hydraulically driven rams stand ready to slam against the drill pipe and seal the well if excessive pressure threatens to blow the well out. These rams are driven by hydraulic oil at pressures of 1,000 to 3,000 psi obtained from a motor driven reciprocating power pump. When the hydrocarbon formation pressure is insufficient to force the oil to the surface, reciprocating pumps are used to bring it to the surface. This is evidenced by the nodding pump units often seen at producing oil wells. Duplex piston power pumps are used to gather and transfer crude oil and hydrocarbon condensate. Power pumps are used to inject salt water into peripheral wells to maintain or increase formation pressure, the process sometimes referred to as 'water-flooding'. A slurry of glass beads in water is pumped into wells at a pressure of 15,000 psi to fracture rock formations for reaching multiple tight pockets holding the oil, and the service is performed by reciprocating power pumps. Reciprocating pumps are used to perform services in petroleum refineries requiring high pressure at low flow rates, such as circulation of glycol, amine and lean oil, for knock-out drum pump-out and for handling hot oil. In the chemical industry reciprocating pumps are used in fertilizer plants, detergent production and transporting plastic slurries at high pressure. Similar other services for the reciprocating pump can be listed in metals industry, nuclear power plants and pipeline services. They perform work ranging from water jet cleaning, water jet cutting, reverse osmosis, cryogenic services and food services. The tasks performed by the reciprocating pump are so varied and numerous that it suffices to say the world cannot function without it.

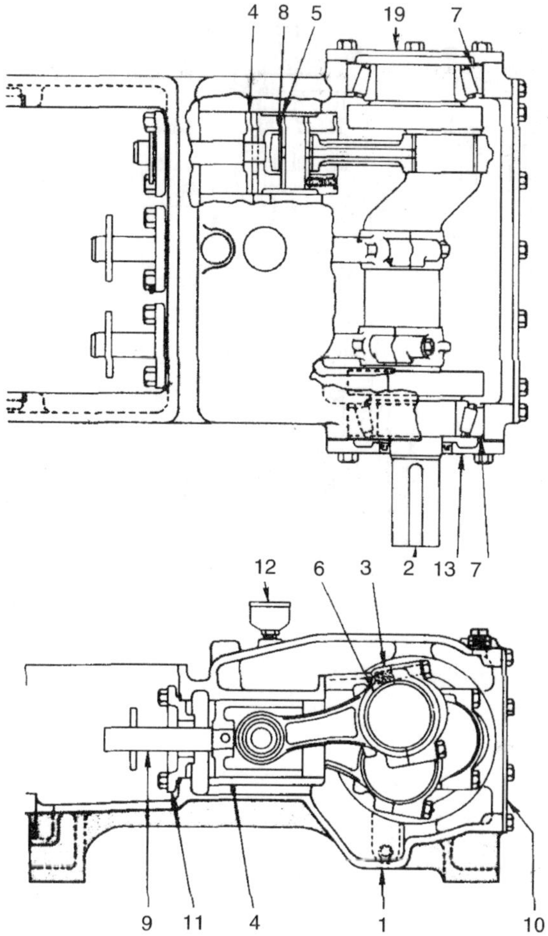

1. Power frame
2. Crankshaft
3. Connecting Rod
4. Power crosshead
5. Wrist pin
6. Crankpin bearing
7. Main crankshaft bearing
8. Wrist pin bearing
9. Crosshead extension
10. Crankcase cover
11. Wiper box
12. Breather
13. Crankshaft bearing

Fig. 2.41. Power End of Horizontal Plunger Pump (Courtesy: Union Pump Co.)

The primary disadvantage of reciprocating pumps results from the pulsating characteristics of the discharge flow. The basic cause of pressure pulsations is related both to the compressibility of the fluid and the geometry of the pump. The piston in any reciprocating pump has to travel some distance from the beginning of the discharge stroke before sufficient pressure is built up to open the discharge valve. This results in compression of the fluid in the cylinder. As soon as the valve opens, fluid suddenly escapes from the cylinder, at the same time relieving itself of the compression, and thus generating an instantaneous rise in pressure above the discharge pressure. In multi-cylinder pumps, as the number of cylinders working out of phase increases, pressure fluctuation of the discharge flow correspondingly reduces. In all cases, however, it is important, or even necessary, to incorporate some form of cushioning in the design. This is particularly true if the flow is carried through rigid pipes. The pulsating characteristics of the pump output is important for the piping system carrying the flow. This subject is explored in detail in the chapter on pressure pulsations from reciprocating pumps and compressors and its effects on the design and dynamic stresses in piping system.

Fig. 2.42 shows an illustration of a vertical triplex single-acting plunger pump.

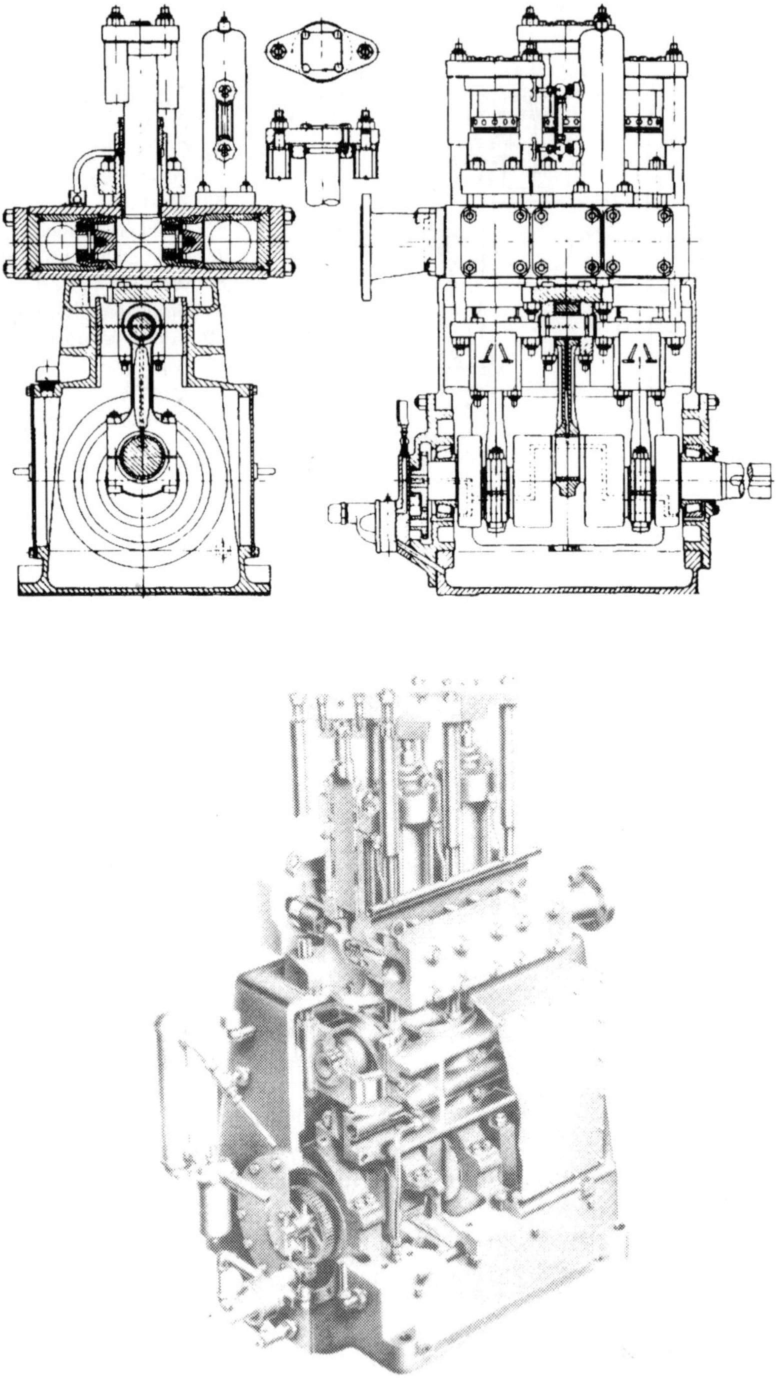

Fig. 2.12. Vertical Triplex Single-Acting Plunger Pump
(*Courtesy*: Worthington Pump Division, Dresser Industries)

2.11 VARIABLE STROKE MACHINES

The positive displacement nature of reciprocating pumps is put to good use for precise metering and proportioning duties. This type of service calls for delivering accurate volumes of fluid, with accuracy to within 0.2 per cent or better. Metering pumps also offer other advantages, such as the ability to carry out precise dosing consistently for long periods of time at high pressure. Another advantage is that accurate displacement of the pump is virtually unaffected by widely varying the discharge pressure. For additives and chemicals that are concentrated, metering pumps offer the advantage of eliminating spillage.

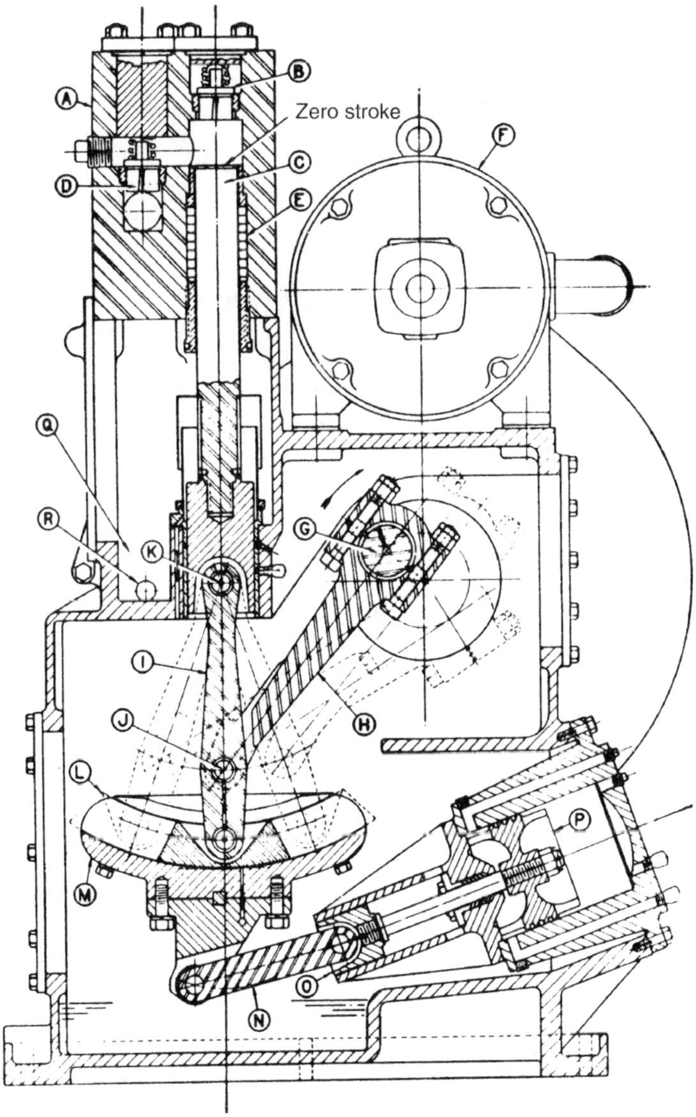

Fig. 2.43 *(a)*. Large Variable Stroke Power Pump, Adjusted for Zero Flow (Courtesy: Ingersoll Rand Co.)

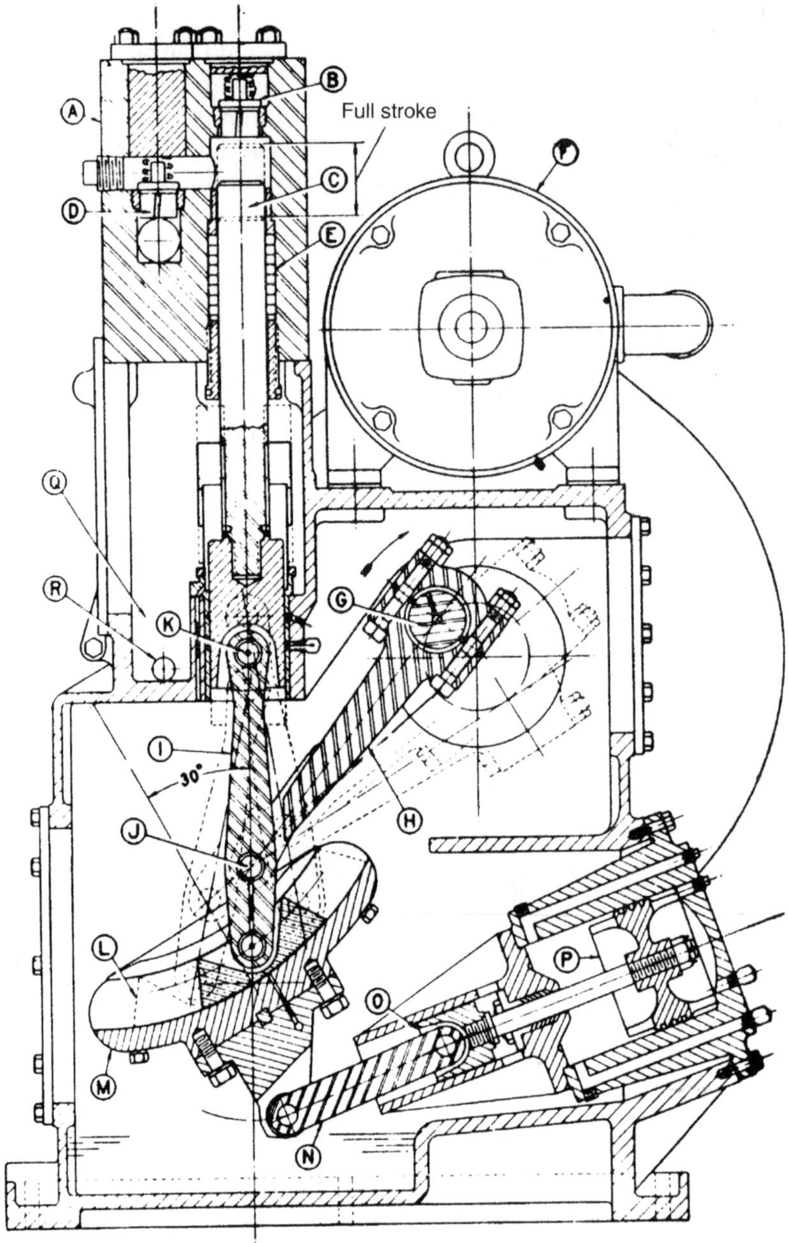

Fig. 2.43 (*b*). Large Variable Stroke Power Pump, Adjusted for Full Flow (Courtesy: Ingersoll Rand Co.)

Flow rate of the metering pump can be adjusted by varying stroke length. Since stroke length plays a primary role in the determination of inertia loads, balancing of inertia forces and moments as well as the dynamic characteristics of the crankshaft and its associated components, the variable stroke machine deserves a special mention. Figure 2.43 shows the arrangement for varying the stroke of a power reciprocating pump. In Fig. 2.43(*a*) the pump is adjusted to provide no flow, while Fig. 2.43(*b*) shows the mechanism set up for the other extreme of full stroke. Since the length of stroke controls the volume displaced by the piston, the flow rate can be adjusted in infinite steps through manipulation of the stroke control mechanism.

Variable compression ratio devices for diesel engines have been mentioned in order to improve upon thermodynamic efficiency at higher-level power output. Due to incomplete combustion of the air and fuel mixture at higher output the performance of the diesel engine goes down significantly compared with efficiency at mid level power setting, sometimes going down by as much as 50 per cent. One way to overcome this drawback is to use an increased compression ratio at full load operation. A variable compression ratio diesel engine was produced for military applications, where the height of the piston crown above the piston pin is hydraulically controlled. Increased air intake into the engine would permit greater quantity of fuel to be burnt, thus increasing the thermodynamic efficiency at higher power levels. A variable compression ratio could also be achieved by varying the stroke as described above for reciprocating pumps. However, the necessary increase in the complexity of the mechanical equipment and the chances of failure raise the question whether such an arrangement can be justified.

2.12 OTHER APPLICATIONS

Any discussion concerning reciprocating machinery always brings to mind the internal combustion engine, primarily due to the usage of the slider-crank mechanism in household automobiles. To a lesser extent, but in some ways performing more vital tasks, pumps and compressors also play an extensive role in all aspects of industrial use.

Besides these three categories there are innumerable applications of the piston, connecting rod and crankshaft mechanism. One important group deals with metal working machines. Hole punching in metal strips or sheets, plate bending machines, machines that shear metal plates and forming of complex shapes out of metal blanks are some common examples. Forging hammers used in foundries rely on the slider and crank mechanism. A number of these machines require the operator to activate a switch; then the sliding ram holding the metal-cutting or forming tool comes into contact with the work piece to perform the task. Here a pneumatically, hydraulically or electrically operated clutch may be provided in the machine. On the other hand if the task needs to be performed on a continuous basis another mechanism advances the metal strip by a pre-determined amount after the completion of each job execution, in which case operator assistance to actuate the clutch will be bypassed. In the continuous process the clutch automatically engages at some instant during the cycle for the job execution to proceed.

Similarities between the automobile engine's piston, connecting rod and crankshaft mechanism and a metal working machine's slider-crank mechanism exist. The piston in the engine takes the place of a reciprocating ram (to which the cutting or forming tool is attached) in the metal working machine. In a two-stroke cycle engine the piston performs useful work immediately after ignition for only a part of the cycle. In a similar manner the ram in metal working machines does useful work only during the cutting, or power, stroke. In both cases a flywheel is required to store mechanical energy and to maintain a more uniform rotational speed of the machine.

Metal shaping machines use a single point cutting tool to produce a flat surface. During the outward stroke the cutting tool chips away the metal along a straight line trajectory. On the return stroke a hinge arrangement on the tool holder permits the tool to idly float above the job surface. On the next cutting stroke a feed mechanism advances the tool by a preset distance in the direction perpendicular to the machine stroke. The ram carrying the tool holding attachment slides along guide ways, and its reciprocating motion is controlled by a crankshaft and a connecting rod. The cutting stroke on the machine can be adjusted in the same manner as the variable stroke mechanism shown in Fig. 2.43 (*a*) and (*b*).

Piston motors powered by compressed air operate at a running speed of 3,000 rpm or under, developing maximum power at 1,000 to 2,000 rpm. Most piston motors are multi-cylinder units, and may be designed as in-line, V, H, flat-four and radial variants. A radial configuration offers the possibility of weight reduction, and is frequently adopted for larger size motors. Odd or even number of cylinders may be used. The interesting part of this machine, however, is the power drive. The radial motor in figure 2.44 has all connecting rods mounted on a common crank. The V-4 piston motor has two cylinder banks, generally at 90° to each other. A drop forged two-throw crankshaft is used, with two connecting rods on each crank pin.

Since piston motors are positive displacement machines, air consumption is directly related to speed within the limitation of the air passages delivering the air. Power output is of the order of 5 to 10 kW per liter of cylinder volume. The motors may be provided with gears to increase speed and reduce torque. Reversal of direction of rotation is obtained by interchanging the inlet and secondary exhaust connections, which may be done through a control valve that may be located away from the motor. Axial piston motors can have four, five or six pistons, but are limited to smaller sizes, less than 2.5 kW. They have the advantage of low inertia and high starting torque, so they reach operating speed almost instantly.

Another variation of piston motors employs a free piston acting on an elliptic cam ring. This design does not use a connecting rod, hence the air supply need not be lubricated. This feature makes it useful for certain process industry applications where the presence of any lubricating oil is not acceptable. It is particularly suitable for low speed applications (up to 800 rpm), and can be driven through a reduction gear-box for slow speed process control type of work.

Figure 2.45 shows a 'wobble plate' engine. A number of stationary cylinders are equally spaced circumferentially around a central shaft. By appropriately proportioning the inertia of the piston and piston rods in relation to the inertia of the wobble plate, it is possible to balance the engine completely. Consider a design where the cylinder bores are equal, and piston and piston rods have a uniformly distributed mass around the axis of rotation. The wobble plate is a disk with its total weight W_{disk} uniformly distributed over its circular area of radius R_{disk}. The total weight of all the pistons and connecting rods is W_{pis}, supposedly concentrated on a circle of radius R_{pis} from the x-axis. If α is the angle the wobble plate makes with the y-axis, then the polar mass moment of inertia about the x-axis will be:

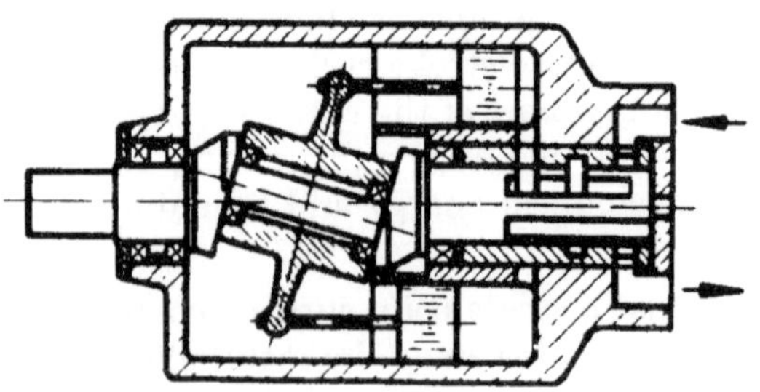

Fig. 2.44. Piston Motor (*Courtesy* : Atlas Copco)

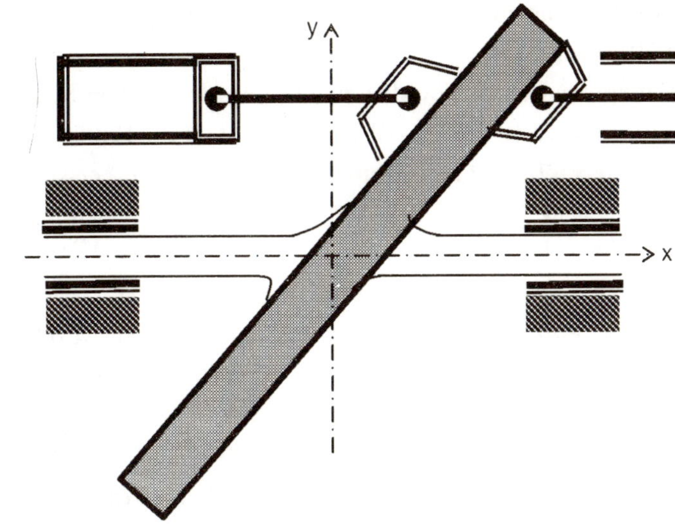

Fig. 2.45. Wobble-Plate Mechanism

$$I_{disk} = W_{disk} R_{disk}^2 [\cos^2(\alpha)]/2$$

The polar mass moment of inertia of the piston and piston rods will be:

$$I_{pis} = W_{pis} R_{pis}^2$$

Since the disk and the piston assemblies are rotating at the same angular velocity, by equating the kinetic energies of both the mathematical expressions for obtaining a proper balance will be $I_{disk} = I_{pis}$, or:

$$W_{disk} R_{disk}^2 [\cos^2(a)]/2 = W_{pis} R_{pis}^2$$

2.13 EXAMPLE PROBLEMS

Example Problem 2.1: Calculate the suction capacity, horsepower, discharge temperature and piston speed for the following double-acting single stage compressor:

Bore: 6 in.

Stroke: 12 in.

Speed: 300 rpm

Rod area: 2.5 in^2

Clearance: c = 12%

Gas: CO_2

Inlet pressure: 1720 lbs/ in^2

Discharge pressure: 3440 lbs/in^2

Inlet temperature: 115°F

Gas compressibility at inlet: 0.312

Gas compressibility at outlet: 0.575

Isentropic exponent of CO_2: 1.3

Reference temperature: 88°F

Reference pressure: 1073 lbs/ in^2

Slippage factor for high pressure differential: $L = 0.05$

Compressor efficiency: 79%

Solution: Rate of piston displacement is equal to piston area multiplied by stroke and speed.

Piston displacement $= \pi \times [2 \times (6)^2 - (2.5)] \times 12 \times 300/(4 \times 1728)$

$= 113.719$ ft^3/minute.

Piston speed $= \omega r = 2\pi \times 300 \times 12/(2 \times 12) = 942.5$ feet/minute

Calculate volumetric efficiency:

At inlet, $T_1 = 460 + 115 = 575°$ R

Reference temperature ratio at inlet $= T_r = 575/(460 + 88) = 1.049$

Reference pressure ratio at inlet $= P_r = 1720/1073 = 1.603$

Gas compressibility at inlet conditions $= Z_1 = 0.312$

At discharge, pressure ratio $r_p = 3440/1720 = 2.0$

Temperature $T_2 = 575[2^{(1.3-1)/1.3}] = 674.7°$ R

or $t_2 = 674.7 - 460 = 214.7°$F

Calculate discharge compressibility:

Reference temperature ratio at discharge $= T_r = 674.7/548 = 1.231$

Reference pressure ratio at discharge $= P_r = 3440/1073 = 3.206$

Gas compressibility at discharge conditions $= Z_2 = 0.575$

Compressibility ratio: $f = Z_2/Z_1 = 0.575/0.312 = 1.843$

Volumetric efficiency $E_v = 0.97 - [(1/f)r_p^{1/k} - 1]c - L$

$= 0.97 - [(1/1.843) + 2.0^{1/1.3} - 1]$

$\times 0.12 - 0.05 = 0.93$

Suction capacity: $Q = E_v \times$ (Displacement) $= 0.93(113.719) = 105.65$ ft^3/min.

Power required = HP $= \Delta p \times Q \times [r_p^{(k-1/k)} - 1] \times [k/(1-k)]/[33000 \times \text{efficiency}]$

HP $= [144 \times (3440 - 1720) \times (105.65) \times (2.3/1.3 - 1) \times (1.3/.3)]/[33000 \times 0.79]$

$=$ **754.46**

Example Problem 2.2: A freight ship is propelled by a 12-cylinder direct drive diesel engine of 30 in. bore giving 10,000 horsepower and weighing 120 lbs/bhp. The engine is 25 ft high, 8 ft wide and 45 ft long. It has been suggested that 8 cylinder engines of 12 in. bore be substituted, with gearing to the propeller shaft. The gearing is expected to weigh 20% as much as the engines and to add 15% to the floor space of the engine to which it is connected. Estimate the new dimensions and floor area required, assuming the smaller engines operate at the same stress level.

Solution: For the same output at the same piston speed and mean effective pressure, the number of 12 in. bore cylinders required will be $12 \times (30/12)^2 = 75$. Since there will be some losses in the gears, take 10 8-cylinder engines, or a total of 80 cylinders of 12 in bore. Total engine weight is:

$10{,}000 \times 120 \times (12/30) \times (80/75) =$ **512,000 lbs.**

This compares with 1,200,000 lbs for the single engine. To this must be added 20% for the gears, so the total weight of the engines and gears will be $512{,}000 \times 1.20 = 614{,}400$ lb. The height of the engines will be: $25 \times (12/30) = 10$ ft, which will allow two more useful decks over the engine room. The floor area covered by the engines plus gears will be: 15 × (80/75) = 16% greater than that of the single engine because floor area is proportional to the piston area with engines of similar design.

Example Problem 2.3: A diesel engine is rated at 2100 horsepower supercharged, with 12 cylinders 8.75 in. × 10.5 in. The engine weighs 39000 lbs. and is 5.33 ft wide and 16.1 ft long.

Compute the weight and floor area required for the same total output from another engine type for marine duty rated at 235 horsepower each, and each 3.04 ft wide, 5.7 ft long and weighing 2740 lbs.

Solution: The number of smaller engines required will be 2100/235 = 9. Total weight will be 9 × 2740 = 24660 lb. Floor area = 9 × 3.04 × 5.7 = 156 ft^2, compared with 5.33 × 16.1 = 85.8 ft^2 for the single engine. A single large engine of the *V* type occupies less floor area than the same power in 6 cylinder in-line engines. If the in-line engines were made in *V* type also, the floor area would be nearly the same as that occupied by the single large engine.

Example Problem 2.4: The large engine of example problem 2.3 has 12 cylinders, 8.75 in. × 10.5 in., and is rated at 2100 hp at 850 rpm. The smaller engines have 6 cylinders, 4.25 in. × 5.0 in., and are rated at 235 hp at 1800 rpm for commercial marine purposes. Which engine has the higher stress due to inertia forces and which has higher stress due to gas loads?

Solution: If the engines are of similar design (which they usually are), inertia stresses will be approximately proportional to the square of piston speed. For the larger engine piston speed:

$$V = \omega r = 2\pi \times 850 \times 10.5/(2 \times 12) = 2337 \text{ feet/minute}$$

and for the smaller engine:

$$V = \omega r = 2\pi \times 1800 \times 5.0/24 = 2356 \text{ feet/minute}$$

Inertia stresses, therefore, will not be substantially different. Gas loads will be nearly proportional to the indicated mean effective pressure (mep). If mechanical efficiency is assumed to be identical, then indicated mep will be equal to brake mep. For the larger engines piston area is:

$$\text{Area} = 12 \times (\pi/4) \times 8.752 = 720 \text{ in}^2. \text{ Therefore,}$$

$$\text{bmep} = (2100 \times 33000 \times 2)/(720 \times 2337) = 82.4 \text{ lbs/in}^2$$

and for the smaller engines:

$$\text{bmep} = (235 \times 33000 \times 2)/(85.17 \times 2356) = 77.3 \text{ lbs/in}^2$$

The larger engine will be carrying somewhat higher stress due to gas pressure. But the stress levels are quite close in the two engines.

Example Problem 2.5: Approximate the size of suction and discharge volume bottles for a single stage, single acting, lubricated, reciprocating compressor. The gas to be compressed is natural gas at the following conditions:

Cylinder bore	9 in.
Cylinder stroke	5 in.
Rod diameter	2.25 in.
Suction temperature	80° F
Discharge temperature	141° F
Suction pressure	514 lbs/in^2

Discharge pressure	831 lbs/in^2
Isentropic exponent	1.28
Specific gravity	0.60
Percent clearance	25.7%

Solution: Compression ratio is 831/514 = 1.617. Natural gas compressibility for specific gravity of 0.6 at both suction and discharge is about the same, 0.93. Then:

Suction volumetric efficiency = 0.93 – 0.257/(1.6171/1.28) – .03 = 0.723

Discharge volume efficiency = (0.823 × 1)/(1.6171/1.28) = 0.565

Then the total volume displaced per revolution is = (π/4) × 92 × 5 = 318.1 in^3

For the calculated volumetric efficiencies, from the volume bottle-sizing chart, the size multipliers are 13.5 for the suction and 10.4 for discharge sides. Then suction bottle volume = 13.5 × 318.1 = 4294 in^3, and discharge bottle volume = 10.4 × 318.1 = 3308 in^3.

To obtain bottle dimensions, use elliptical head with 2:1 ratio. Suction bottle diameter = 0.86 × 42941/3 = 13.98 in., and discharge bottle diameter = 0.86 × 33081/3 = 12.81 in. diameter. To calculate bottle length, suction bottle will be 2 × 13.98 = 27.96 in. long and discharge bottle will be 2 × 12.81 = 25.62 in.

Example Problem 2.6: A large, light weight reciprocating engine has 18 cylinders of about 6 in. bore and stroke. It is rated for maximum power of 3500 hp at bmep = 250 psi and piston speed = 3000 ft/min. Compression ration is 7.0 and weight is close to 1.0 lb. per maximum horsepower. The ratio of knock-limited indicated mep between 4 in. and 6 in. cylinders is 1.6. How much weight could be saved by using 4 × 4 in. cylinders at the same inertia stress level, supercharged to the same detonation level with the same fuel?

Solution: For the same inertia stress level piston speed will be identical. The weight per bhp will be proportional to the stroke and inversely proportional to mep. Therefore:

Weight per bhp of 4 in. cylinder = 1.0 × (4/6) × (1/1.6) = 0.417 lbs./hp

Comparison of the weight of the two engines using 6 × 6 in. and 4 × 4 in. cylinders are:

3500 × 1.0 = **3500 lbs. for 6 in. bore.**

3500 × 0.42 = **1458 lbs. for 4 in. bore.**

Example Problem 2.7: In problem # 2.6, how many cylinders are required for the smaller bore engine?

Solution: The number of cylinders will depend on the piston displacement, and is:

18 × (6/4)2/1.6 = **25 cylinders**

Example Problem 2.8: An 8-cylinder 4-stroke diesel engine has an indicated mean effective pressure of 145 lbs/in^2, 4.75 in. cylinder bore, 5.75 in. length of stroke and operates at 1250 rpm. If the engine experiences mechanical losses equivalent to 335 horsepower, determine the mechanical efficiency of the engine.

Solution: Area of the piston = π/4 × (4.75)^2 = 17.72 in^2. In a four stroke machine the number of power strokes per minute will be *N*/2 = 1250/2 = 625 strokes/min. Then from (2.1), the indicated horsepower is:

$$ihp = \text{PLANK}/33{,}000$$

Therefore, indicated horsepower is:

(145 × 5.75 × 17.72 × 625 × 8)/33,000 = 2,238 HP

Brake horsepower is then = (indicated horsepower) – (mechanical losses)

$$= 2{,}238 - 335 = 1{,}903$$

Mechanical efficiency of the engine $= bhp/ihp$

$$= 1903/2238 = 85.03\%.$$

Example Problem 2.9: A 4-stroke, 4-cylinder engine for an automobile develops 295 horsepower at 2975 rpm. Mechanical efficiency of this engine is estimated at 77%. If the bore to stroke ratio for this engine is defined at 0.92, determine the bore size. Assume indicated mean effective pressure is 135 lbs/in^2.

Solution: Let the cylinder bore be d in. Then the length of the stroke will be $d/0.92$ in. Piston area = $\pi/4 \times (d)^2$ in^2, and number of strokes is 2975/2 = 1487.5 strokes per minute. To determine indicated horsepower, apply (2.1):

$$ihp = [135 \times (d/0.92) \times \pi/4 \times (d)^2 \times 1487.5 \times 4]/33{,}000$$

$$= 20.78 \times d3$$

Also: $ihp = bhp/(\text{mechanical efficiency})$

$$= 295/0.77 = 383.12$$

Hence: $d^3 = 383.12/20.78 = 18.44$, and the cylinder bore $d = 2.64$ in.

Example Problem 2.10: Consider two lines of nearly similar diesel engines, one with cylinders of 4.25 in. bore, 5.0 in stroke, the other with 3.875 in. bore and 4.5 in. stroke. A 6-cylinder engine of the larger size gives 218 bhp at 2200 rpm with a brake specific fuel consumption (bsfc) of 0.412 lbs/bhp-hr. Estimate the power, rpm and bsfc of the smaller engine at the same piston speed.

Solution: At the same piston speed the brake mean effective pressure should be the same, and so power will be proportional to piston area. Therefore:

$$\text{Smaller engine } bhp = 218 \times (3.875/4.25)^2 = 181.2 \text{ bhp}$$

Engine speed will vary inversely with the stroke:

$$\text{rpm (smaller engine)} = 2200 \times (5.0/4.5) = 2444.$$

And since efficiency should not vary substantially within this small range of bore size, the bsfc of the smaller engine will also be 0.412 lbs/bhp-hr.

Example Problem 2.11: In the mechanism shown in Fig. 2.46 angular velocity of link 2 is constant, and the angular velocity of link 3 is to be determined. The coordinate system xy is fixed in link 3, as shown with its origin at point A_3. The system XY has its origin at point O_2.

Solution: When a point P is moving in a coordinate system xyz, which in turn is moving relative to a fixed coordinate system XYZ, the position of point P relative to the fixed system may be expressed as:

$$R_P = R_o + R$$

where R_P is the position vector of P in the fixed coordinate system, R_o is the position vector of the origin of the moving system and R is the position vector of the moving system's origin in the fixed coordinate system. The absolute velocity of point P relative to XYZ system, V_P may be obtained by differentiating the above equation with respect to time:

$$V_P = dR_P/dt = dR_o/dt + dR/dt$$

The velocity of P in the moving system xyz will be assumed to be V, while the velocity of the moving system origin in the fixed system will be assumed to be V_o. If vector R passing through a fixed base rotates about the base point with an angular velocity ω, then the velocity of the tip of vector R will be $\omega \times R$. Then the expression for VP is:

$$V_P = V_o + V + \omega \times R$$

The equation for velocity of point A_3 in Fig. 2.46 cannot be obtained directly with this equation, it is necessary to first find the velocity of A_2.

$$V_{A2} = V_o + V + \omega_3 \times R$$

where: $V_{A2} = (O_2A_2)\omega_2 = 5.25 \times 10 = 52.5$ in/sec, and is perpendicular to O_2A.

$V_o = V_{A3}$, direction perpendicular to O_3A_3, magnitude unknown.

V is directed parallel to O_3A_3, magnitude unknown.

$\omega_3 \times R = 0$, since $R = 0$.

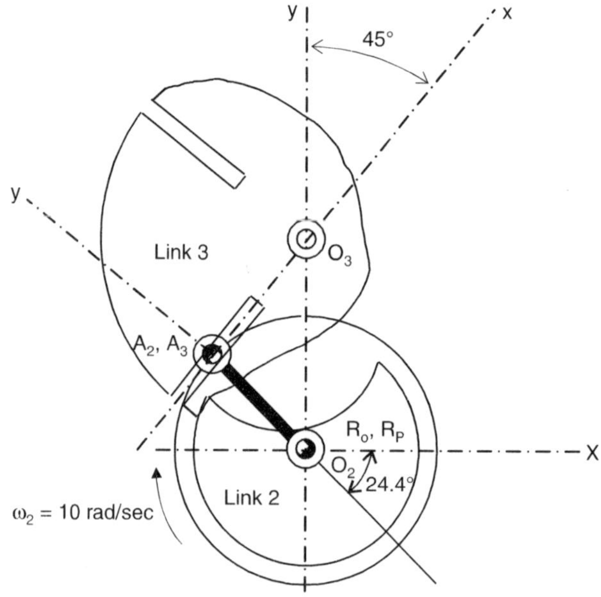

Fig. 2.46

V_{A2} will be solved with unit vectors i, j and k taken in the moving coordinate system, and ω_3 is calculated from the magnitude of V_{A3}.

$$V_{A2} = V_o + V + \omega_3 \times R$$
$$= VA_2(\cos 24.4i - \sin 24.4j)$$
$$= 52.5(0.9107i - 0.4131j)$$
$$= 47.81i - 21.69j$$
$$V_o = V_{A3} = -V_{A3}j$$
$$V = V_i$$
$$\omega_3 \times R = 0$$

Substituting in the expression for V_{A2},

$$47.81i - 21.69j = -VA_3j + Vi$$

Equating the i components:

$$V_i = 47.81i$$

so: $$V = 47.81 \text{ in/second.}$$

Equating the j components:

$$-V_{A3}j = -21.69j$$

$$V_{A3} = 21.69 \text{ in/second.}$$

Therefore: $$\omega_3 = \frac{V_{A3}}{O_3A_3} = \frac{21.69}{2.75}$$

$$= \mathbf{7.89\ radians/second\ counterclockwise.}$$

Example Problem 2.12: The angular velocity ω_2 of link 2 is to be determined in position IV for the slider and crank mechanism for a metal cutting machinery shown in Fig. 2.47. In the starting position I the angular velocity $\omega_2 = 0$. Torque T_2 on link 2 is constant at 10.2 N-m, acting counterclockwise. Force P on piston 4 varies uniformly from 716 N in position I to 160 N in position IV.

Solution: The energy method will be used here because of the ease with which it can be applied. Several procedures are available. One method is based on the distribution of kinetic energy in a mechanism, and according to the Energy Distribution Theorem, 'the percent of the total kinetic energy which a link of a mechanism contains will remain the same in any given position regardless of the speed'. It is tacitly assumed the mass or moments of inertia of the links will not change with speed, and that a linear relationship exists between the velocities of the various links in a given position. A convenient value is assumed for the velocity of the input link of the mechanism. A velocity analysis than establishes the velocity of the centers of gravity of all the links, as well as their kinetic energy. The ratio of the kinetic energy of any link to the total for all the links is expressed as:

$$\varepsilon = \frac{KE}{\Sigma KE}$$

This ratio is constant for any link in a particular position regardless of its velocity. Assume a value of 1 radian/second for the angular velocity ω'_2 of link 2 (prime denoting assumed value).

From a velocity analysis the velocities at other locations in the assembly in position IV are as follows. Kinetic energy of the links for the assumed condition may then be calculated:

$$KE_1' = 0$$

$$KE_2' = I_2(\omega_2)^2/2 = (0.00271)(1)^2/2$$

$$= 0.001355 \text{ N-m}$$

$$KE_3' = I_3(\omega_3)^2/2 + M_3(V_{g3})^2/2$$

$$= 0.01017 \times (0.049)^2/2 + 1.360 \times (0.0508)^2/2$$

$$= 0.001767 \text{ N-m}$$

$$KE_4' = M_4(V'_B)^2/2 = 0.9068 \times (0.0521)^2/2$$

$$= 0.001231 \text{ N-m}$$

Then, $$\Sigma KE' = 0.004353 \text{ N-m.}$$

From the energy distribution theorem, $\varepsilon_2 = \frac{KE_2'}{\Sigma KE'} = \frac{0.001355}{0.0043543} = 0.3112$

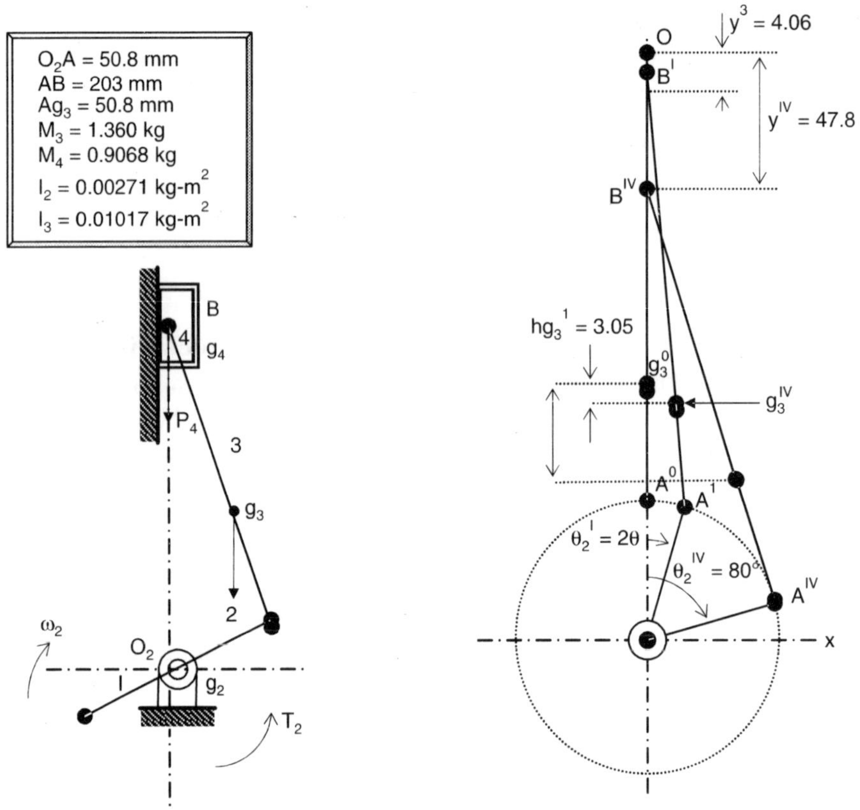

Fig. 2.47

The external work input to the mechanism by forces P_4 and W_3 and torque T_2 between positions I and IV is determined as follows:

$$Wk_{P4} = (P_4^{IV} + P_4^{I})(y^{IV} - y^{I})/2$$
$$= (160 + 716) \times (47.8 - 4.06)/(2 \times 1000)$$
$$= 19.16 \text{ N-m.}$$
$$Wk_{w3} = W_3(h_{g3}^{IV} - h_{g3}^{I})$$
$$= 1.360 \times 9.81 \times (42.7 - 3.05)/1000$$
$$= 0.5290 \text{ N-m}$$
$$Wk_{T2} = - T_2(\theta_2^{IV} - \theta_2^{I})$$
$$= - 10.2 \times (80 - 20) \times \pi/180$$
$$= - 10.68 \text{ N-m}$$

Hence: $\Sigma Wk = 9.009$ N-m

Since the change in kinetic energy must be equal to the work done, then:

$$\Sigma KE^{IV} - \Sigma KE^{I} = 9.009 \text{ N-m}$$

And since $\Sigma KE^{I} = 0$, then:

$$\Sigma KE^{IV} = 9.009 \text{ N-m}$$

The value of ω in position IV is obtained from:

$$KE_2^{IV} = I_2(\omega_2^{IV})^2/2$$

where:

$$KE_2^{IV} = \varepsilon_2^{IV}(\Sigma KE^{IV})$$
$$= 0.3112 \times 9.009$$
$$= 2.804 \text{ N-m}$$

Then:

$$\omega_2^{IV} = \left(\frac{2KE_2^{IV}}{I_2}\right)^{1/2} = \left(\frac{2 \times 2.804}{0.00271}\right)^{1/2}$$

$$= \mathbf{45.49 \text{ radian/second.}}$$

REFERENCES AND BIBLIOGRAPHY

API Standard 618, "Reciprocating Compressors for Petroleum, Chemical and Gas Industry Services", American Petroleum Institute, Washington, DC, 1995.

API Standard 671, "Special Purpose Couplings for Refinery Services", American Petroleum Institute, Washington, DC, Reaffirmed 1993.

Barber, A., "Pneumatic Handbook", Elsevier Science, Ltd., Oxford, UK, 1997.

Baxter, A. S., Rawlins, R. E., "The Design and Development of the MB190: A High Speed Diesel Engine Based on Medium Speed Principles", IDGTE Publication 428, February, 1985.

Bent, R. D., McKinley, J. L., "Aircraft Power Plants", McGraw Hill Book Co., New York, 1985.

Bloch, H. P. "A Practical Guide to Compressor Technology", McGraw-Hill, New York, 1995.

Brown, R. N., "Compressors - Selection and Sizing", Gulf Publishing Co., Houston, 1997.

Deutschman, A. D., Michels, W. J., Wilson, K., "Machine Design - Theory and Practice", Macmillan Publishing Co., New York, 1983.

Coplin, E. J., Davis, R. J., Brunson, D., "The Interaction of Engines and Finished Equipment Designed to Meet EEC Noise Legislation-The Engine Manufacturers Viewpoint", *Proceedings of I. Mech. Engineers International Conference-Small Internal Combustion Engines,* 1989-3, C372-029, 1989.

Haug, E. J., "Computer Aided Kinematics and Dynamics of Mechanical Systems, Vol. I: Basic Methods", Allyn & Bacon, Boston, 1989.

Havran, R. L., "Development of the Caterpillar G3412 Low Emission Engine with Electronic Ignition", ICE Vol. 22, *Heavy Duty Engines-A Look at the Future,* ASME, 1994.

Henshall, S. H., "Mirrlees MB430 Packs More Punch", Marine Engineering Review, April, 1985.

Henshaw, T. L., "Reciprocating Pumps", Van Nostrand Reinhold Co., New York, 1987.

Howe, G.: "Sound Proofing Diesels", Heavy Construction Magazine, September, 1984.

ISO 1940/1, "Mechanical Vibrations-Balance Quality Requirements of Rigid Rotors-Part I, Determination of Permissible Residual Unbalance", International Organization for Standardization, Geneva, Switzerland, 1986.

Karassik, I. J., Krutzsch, W. C., Fraser, W. H., Messina, J. P.: "Pump Handbook", McGraw Hill Book Co., New York, 1986.

Kochanowski, H. A., "Performance and Noise Emission of a New Single Cylinder Diesel Engine-With and Without Encapsulation", *Proceedings of I. Mech. Engineer. International Conference-Small Internal Combustion Engines,* 1989-3, C372-023, 1989.

Loomis, A. L., "Compressed Air & Gas Data", Ingersoll Rand Co., Woodcliff Lake, NJ, 1980.

Mead, G., "Quieter, Part by Part", Construction Plant and Equipment, October, 1983.

"New Directions in Automotive Engineering", Robert Bosch Corp., 2800 South 25th Ave., Broadview, IL 60153, 1993.

Miller, J. E., "The Reciprocating Pump: Theory, Design and Use", Krieger Publishing Co., Malabar, FL., 1995.

Nevinger, R. D., "Design and Development of the Caterpillar 3600 Spark-Ignited Engine Family", ASME-91-ICE-6, ASME, New York, 1991.

Poynton, W. A., "Designing a Range of Medium Speed Diesel Engines for High Reliability", *Inst. Of Mechanical Engineers., Seminar by Combustion Engines Group,* Sept. 1985.

Rollins, J. P., "Compress Air & Gas Handbook", Compressed Air and Gas Institute, New York, 1993.

SAE Publication # Sp-502, "Fuel and Combustion Effects on Particulate Emissions", 1982.

SAE Publication # PT-24, "Passenger Car Diesels", 1982.

Stevens, J. L., Tawil, N., "A New Range of Industrial Diesel Engines with Emphasis on Cost Reduction Techniques", *Proceedings of I. Mech. Engineers International Conference-Small Internal Combustion Engines,* 1989-3, C372-028, 1989.

Taylor, C. F., "The Internal Combustion Engine in Theory and Practice", Volumes 1 and 2, The M. I. T. Press, Cambridge, Mass, 1985.

Vetter, G., "Leak Free Pumps & Compressors", Elsevier Science, Ltd., Oxford, UK, 1995.

Vetter, G., "Reliability and Future Development of High Pressure Diaphragm Pumps for Process Service", *Proceedings of the 5th Pump Users Symposium,* Houston, 1988.

Vetter, G, Hering, L., "Oscillating Displacement Pumps", *Encyclopedia Fluid Mechanics,* Gulf Publishing Co., Houston, 1986.

Wachel, J. C., Tison, J. D., "Vibrations in Reciprocating Machinery and Piping Systems", *Proceedings of the 23rd Turbomachinery and Piping Systems*, Texas A & M University, College Station, TX 1994.

Warring, R. H., "Pumping Manual", Trade & Technical Press Ltd., Surrey, England, 1984.

The Crankshaft and Associated Components 3

3.1 INTRODUCTION

The detail design of machines so complex as internal combustion engines, reciprocating pumps and compressors and a host of other machines working on the slider and crank principle would require much more space than is available here, and could easily fill several volumes if covered in a thorough manner. This condition is attributable in part to the fact that the subject depends greatly on practical experience in a field subject to constant growth and change. In general, professionals active in this field do not have the time nor the inclination to write. Hence, most existing literature is written from an academic standpoint. A vast number of papers, manuscripts and publications tend to focus on rather narrow and special items of engineering design and evaluation. This is not surprising when the intent of major manufacturers is mass production, and where tiny errors in engineering and manufacturing could result in massive losses for the manufacturer. Engineering professionals, therefore, tend to specialize in their chosen area of expertise, which could be in combustion, fluid mechanics, structural integrity and materials technology, just to name a few among many. The objective of this and preceding chapters is to point out to the engineer working on the structural dynamic characteristics of any reciprocating machinery the differences in the design of the various machines and components, and their impact on the overall vibration and structural integrity of the machine. The intent is to emphasize only the major principles and to point out common mistakes and their corrections. The illustrations representing good practice are subject to improvement as knowledge, experience and development of new materials and techniques continue.

The detail design of reciprocating machines has developed over a period of some hundred years, during which time nearly all conceivable mechanical arrangements have been tried and the less practical and convenient ones discarded. To date the conventional crank, slider and connecting rod arrangement has invariably proved to be the best. Before proceeding to more detailed considerations, it may be said that all successful modern reciprocating machine designs are, in principle at least, copies of previous successful designs with, hopefully, as many improvements in detail as the experience and ingenuity of the designer can contribute. A thorough knowledge of good current practice is, therefore, essential to the creation of a new successful design. Another way of stating the problem is to say that it requires decisions based on previous experience that are combined with mathematical computer based solutions to come up with a design that will meet, sometimes even exceed, the established criteria for weight, bulk, cost per unit of power, reliability and durability of the machine at hand.

Once the major design decisions regarding the number of cylinders, their arrangement and size have been made based on performance characteristics, preliminary assembly drawings showing the relative position of all the major parts of the machine, usually referred to as 'layouts', will be made. In particular, the power section of the reciprocating machine, which is taken to include cylinders, pistons, connecting rods and crankshaft and their supporting structure are made available. Minor or even major changes in the preliminary design may be necessary as the detail design proceeds. A widely recognized practice is to initiate and develop the structural dynamics model of the machine from the very onset of the preliminary design and its development. Large amounts of useful data about the dynamic behavior of the machine system in its preliminary design phase can be obtained through such a program. In conjunction with weight, cost and performance, the dynamic analysis of the preliminary machinery design provides data regarding the suitability of bearing location, weight of components and even material properties. A number of assumptions about parameter values have to be made at this stage of the design concerning the component's physical dimensions, weight and stiffness characteristics; as more data about them becomes available from their computer based analytical and experimental tests, the overall system dynamic model can be upgraded.

3.2 GENERAL PROBLEMS IN DETAIL DESIGN

Important considerations in the design of any part always include:

- Structural integrity
- Method of manufacture
- Method of assembly and disassembly

In addition, the design may involve problems of:

- Thermal performance
  Efficiency
  Operating temperature
  Heat conductivity
  Thermal stresses
- Fluid flow performance
  Flow passage shape
  Leakage prevention
- Durability
  Wear
  Corrosion
  Accidental breakage
- Electrical and magnetic characteristics

Of all the design problems, that of securing a reliable structure to withstand the loads imposed by gas pressure, inertia forces and thermal expansion, without excessive weight, bulk or cost, is the most difficult and basically the most important. Without a sound, reliable and durable structure no amount of refinement of any other aspects of the design is of any value. Within the limits imposed by a particular service, that design is the most successful which fulfills its requirements with the smallest amount of material. The material in the perfect structure would be distributed so that all of it would be stressed to an equal fraction of its endurance limit. The degree to which this ideal can be approached depends on the skill of the designer, the amount of computer based finite element and experimental stress analysis available, and the necessary compromises with manufacturing costs, assembly and accessibility problems and non-structural functions to be performed by the same parts. In the case of reciprocating internal combustion engines this ideal was most nearly approached in the late models of large aircraft engines, where the need for high specific output was so acute as to justify high manufacturing costs for individual parts.

In nearly every case the geometry of a manufactured part is a compromise between a shape that is ideal from the functional point of view and a shape that facilitates manufacture. The designer therefore must have a basic knowledge of manufacturing methods and, where necessary, should consult with experts in production methods, specially when production quantity is large. In the case of automobile engines, for example, a saving of few pennies in either material or labor costs of a part can translate into millions of dollars per year for the car manufacturer due to the large volume of production.

For a design satisfactory with respect to assembly, the ideal shape from the functional point of view may again have to be compromised. Assembly and disassembly problems vary according to a number of considerations, however, removal and replacement of valves, cylinder sleeves, pistons, connecting rods and bearings in the field must be possible. Where cylinders and crank case are integral, as in a majority of automotive engines, there is the special problem of accessibility to the main and connecting rod bearings, and of the removal of the piston and connecting rod assembly. The latter operation is done either up through the cylinder bore or down past the crankshaft and out through the side cover or bottom cover of the crank case. The appropriate method will depend on the nature of the installation for which the machine is designed. In addition to the parts already mentioned auxiliaries, including ignition systems, spark plugs, injectors and injection pumps, oil and water pumps, filters, strainers, etc., must always be easily accessible in the field. This requirement often calls for compromise with other considerations.

Major problems regarding thermal performance concern the avoidance of excessive temperatures and thermal stresses in cylinders, pistons, valves and ports, and in the exhaust system including the exhaust turbine, if any. The solutions require use of correct material, minimizing the length of paths for heat flow from hot-side surface to the coolant, and minimizing or circumventing barriers to heat flow, such as bolted joints and gaskets. Questions of detail design to secure satisfactory thermal performance will be discussed further in connection with individual machine parts.

Problems pertaining to fluid flow are encountered in the inlet system, valves, ports, cooling system, exhaust system and fuel system. They will be covered in more detail in connection with the individual parts. The question of durability concerns chiefly problems of wear and corrosion. Electrical and magnetic characteristics apply to the detail design of ignition systems, starters and generators, and are not dealt with here.

3.3 COMPONENT MATERIALS

The number and types of materials at the disposal of the designer are so vast that the problem of making a choice is indeed formidable. Fortunately, the available materials are well classified and organized in publications issued by the ASM and the SAE, among others. This section is devoted to outlining a basic philosophy of material selection as related to the reciprocating machine design, in an attempt to establish a background for understanding the fundamental questions involved.

The choice of material for any part can depend on the following considerations:

1. General function: structural, bearing, sealing, heat conducting, space filling, etc., or combinations thereof.

2. Environment: loading, temperature and temperature range, exposure to corrosive environment or to abrasion and wear
3. Life expectancy
4. Space and weight limitations
5. Cost of the finished part and its maintenance and replacement
6. Special considerations such as appearance, customer prejudices, etc.

While every part must be designed to function satisfactorily in its operating environment for a suitable period of time, the relative importance of items 3 to 6 varies greatly with the type of service. For example, in aviation applications the importance of light weight and small space justifies higher cost and shorter life than might otherwise be desirable. In the case of passenger car engines low initial cost is so important that considerable sacrifice in life of the parts is often justified. On the other hand, large stationary and marine applications of reciprocating machinery are expected to have such

Table 3.1. Materials Typically Used in Reciprocating Machinery Parts

	Material				
Part	*Type*	*SAE No.*	*BHN*	*Remarks*	*Reasons*
Cylinder Heads	Gray cast iron, J431*a*	G3000 G3500 G4500	170-269	Usual	4, 7
	Cast Aluminum J465 Forged Aluminum, J454*c*	39, 322 A2218		Aircraft related Aircraft engines	1, 4 2
Cylinder Barrels	Gray cast iron, J431*a*	G3000 G3500 G4500	170-269	Usual	4, 7
	Steel	4130	300	Aircraft engines (nitrided)	2
	Cast Aluminum J465	34 39		Small engines (plated bore)	1, 4
Pistons	Sand cast aluminum, or Die cast aluminum	34, 39 309, 314, 321, 328		Engines with less than 10 in. bore	2, 3, 4
	Forged Aluminum, J454*c*	AA2018		Aircraft Diesel engines	1, 3
	Gray cast iron, J431*a*	G3000 G3500 G4500	170-269	Small engines Engines of bore more than 10 in.	4, 7

Table 3.1. Continued

Piston Pins	Steel	4140 4340	250-300	Usual	2
	Steel, case hardened	5015 4119 4320	670-720	Hard surfaced	1, 10
Piston	Special cast iron		150-250	Usual	5, 10
Rings	Steel chrome plated			Heavy duty	1, 10
Connecting Rods	Steel	1041 4130 4137	190-230	Small rods	1
		4340 9840	190-230	Large rods	2, 13
	m- or n-iron	J433/434	200-300	Small engines	4, 7
Bolts Studs	Steel	4137 4340	300-340	Highly stressed	2
Nuts		1137	200-250	Minor fastenings	1, 7
Crank-shafts	Steel	1046 4140 4340	230-275	Usual	1 2 2, 13
	Cast steel, J435*a*	0150 0175	311-363 363-415	Frequent	1, 4, 7
	n-iron, J-433	Gr. III Gr. IV	240-290 300-350	Rare	1, 4, 7
	m-iron, J434	53004	197-241	Rare	4, 7
Crank Cases	Gray iron, J431*a*	G3500 G4500	187-269	Automotive engines	4, 7
	Cast aluminum, J465	310, 311 312, 322		Aircraft, autom. engines	1, 4
	Forged aluminum, J454*c*	AA2218		Aircraft engine	2
	Welded steel, J410*b*	950C 945C	150	Large engines	1

Table 3.1. Continued

Main and Rod Bearings	Tin-based babbitt Lead-based babbitt	11, 12 13-15		Light duty non-automotive	5, 7, 10
	Lead-tin overlay Copper-lead Aluminum	19, 190 49,48,480 481 770-781		Heavy duty	5, 10, 11
Other Plain Bearings	Bronze	J459*a*		Lightly loaded	5, 10
	Aluminum	J461*a*, Table 10		Lightly loaded bearings, directly in casting material	5, 7, 10
	Cast iron	J431*a*	187-269		
	Sintered	J471*b*		Special purpose	5, 10
Anti-Friction Bearings	Steel	51100 52100	670-739	Small bearings	2, 10
	Steel, case hardened	5115 5120	670-739	Large bearings	2, 10
Cam-shafts	Special cast iron	G4000 *d*, *e*, *f*	241-321	Automotive engines	7, 10
	Steel, case hardened	4119 4317	650-700	Heavy duty	10
Cam Followers	Same as camshafts				10
Push rods	Steel tubing	.2 – .3 % carbon stl		Usual	1, 7
Rocker arms	Steel	4140, 4130		Rare	4, 7
	n-iron, J433	III or IV		Rare	4, 7
	Sintered steel	J471*b*		Rare	4, 7
	m-iron	J434		Rare	4, 7
Valves, Seats	Special steels	J775			8, 9, 10
Valve Springs	Alloy steel	4150 4350 6150	400-445	Shot peened	11

Table 3.1. Continued

Gears	Steel, *c. h.*	4119 4320 9317	455-525	Heavy duty	2, 10
	Steel	4150	375-425	Medium duty	1, 7
	Carbon steel		375-425	Light duty	7
	Bronze	J461*a*			
	Sintered	J471*b*			
Gear Cases	Cast iron	G2000 G3000	187-229	Usual	4, 7
	Cast aluminum, J452	33		Common	4, 1
	Cast magnesium, J465	50, 500, 502 504, 505		Aircraft	
Cylinder head & Spark plug gaskets	Copper or aluminum			Diesel, heavy duty and large gas engines	11
	Copper-asbestos			Automotive S.I., diesel engines	
Water Seals	Rubber O-rings	J14		Good sealing	12
	Cork Vellum Fiber	J90*a*		For flat surfaces	7
Low pressure gaskets	Cork Vellum Fiber	J90*a*		For flat surfaces	7

Numbers starting with J refer to sections in SAE Handbook. Other numbers are for specific composition.

Steel refers to forged or rolled type.

c. h.–case hardened.

Bearing surfaces of iron and steel shafts are commonly flame hardened to higher values.

Reason Legend

1. High strength/weight ratio
2. Very high strength/weight
3. High heat conductivity
4. Cast in intricate shapes
5. Good bearing properties
6. Best bearing properties
7. Low cost, adequate
8. High hot strength
9. Corrosion resistance
10. Wear resistance
11. Strength, resistance
12. Water tightness, durable
13. Good heat treatability

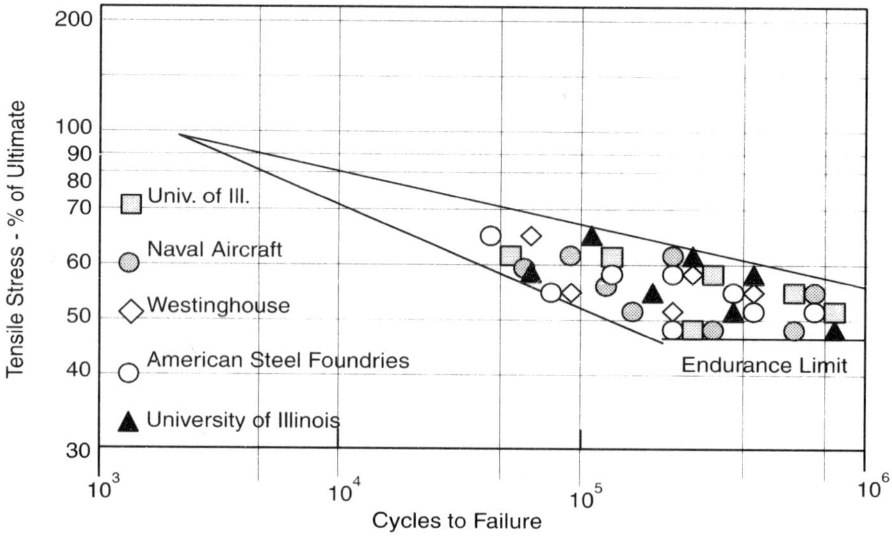

Fig. 3.1. S-N Diagram for Steels of Varying Tensile Strength (*ref.* Almen, 1983)

Extended service lives that space, weight and initial cost may be relatively great in order to attain these objectives. Materials commonly used for parts in reciprocating machinery are shown in table 3.1.

Materials whose essential role is to carry relatively high stresses are classified as structural. Heavily stressed materials include those which carry and transmit forces and torques developed by cylinder pressure and inertia of the moving parts in the power train. The success of structural materials is measured by their resistance to structural failure. In reciprocating machines the most prevalent structural failure is due to fatigue (failures due to wear, seizure, corrosion, oxidation, etc. are not considered structural.). Fatigue failure is fracture due to the repeated application of many cycles of varying stress. Although the physical and chemical causes are still not fully understood, fracture from fatigue is progressive in nature, starting as a crack beginning usually at the surface. The crack initiates at the point of maximum tensile or shear stress, usually a point of stress concentration, such as an original crack in the surface, a notch created by corrosion or in the manufacturing process, or by a point where the stress is high because of the shape of the part or the method of load application. The fatigue crack occurs at a maximum stress usually well below the elastic limit, and hence without any significant distortion of the surrounding material. The crack grows as the stress cycle is repeated, until a point is reached where the remaining material is stressed beyond its ultimate strength and fails suddenly, in tension or shear or a combination of both.

Comparative resistance to fatigue of materials is measured by laboratory tests such as the rotating beam, vibrating cantilever beam or the repeated torsion test. In these tests the load is usually reversed in a sinusoidal pattern, with average stress equal to zero. In homogeneous materials fatigue failure occurs in the region of maximum stress in the test specimen. The stress at this point is very much of principal interest. The results of such tests can be plotted as illustrated in Fig. 3.1. Each point on this plot represents a test to failure of a single specimen. The figure is referred to as an S-N plot because it records stress at failure against number of cycles to failure. Important characteristics are:

1. There is a general tendency for life to increase as stress is reduced.

2. In the case of the materials of Fig. 3.1 no failures appear to occur when stress is less than 48 per cent of the ultimate tensile strength. When environmental or operating conditions result in rough surfaces, the limit is reduced by about 25% of ultimate tensile strength.
3. Dispersion of test points increases with decreasing stress.

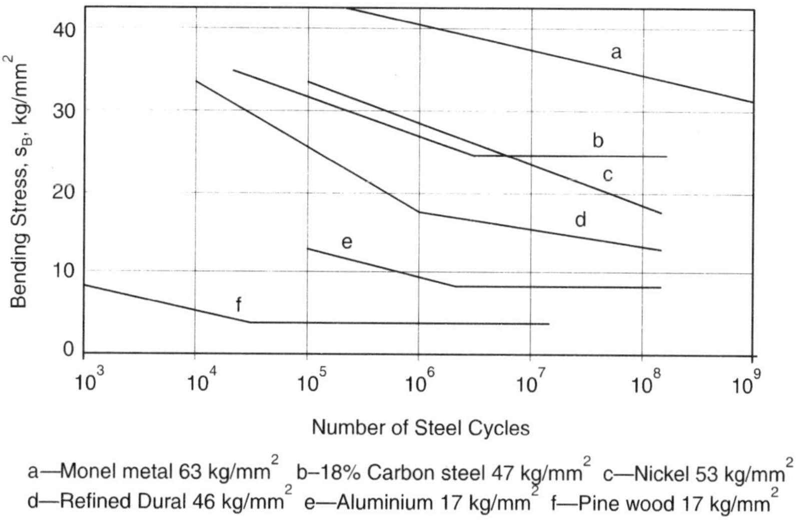

Fig. 3.2. S-N Diagrams (lower limit of test points) in Reversed Bending Versus Ultimate Tensile Strength (*ref.* Matthaes, 1984)

Figure 3.2 shows lower boundaries of S-N plots for various materials. Note that three of these materials (steel, pure aluminum and wood) show a definite endurance or fatigue limit, that is a stress beyond which failure does not occur even after 10^8 or 10^9 cycles. This type of curve is characteristic of most ferrous metals but is not typical for the common non-ferrous metals, including aluminum alloys. For materials that do not exhibit a definite fatigue limit of this type the nominal fatigue limit is taken as the no-failure stress at an arbitrary number of cycles, usually 10^8 cycles. Thus, according to Fig. 3.2 the fatigue limits for these materials are shown in table 3.2.

Table 3.2. Fatigue limit of Materials Shown in Figure 3.2

Material	*kg / mm²*	*10³ × lb / in²*
Monel metal	27	38.2
Low Carbon Steel	25	35.5
Nickel	19	27.0
Aluminum-Dural	12	17.0
Aluminum-pure	7.5	10.6
Wood-pine	4.0	5.7

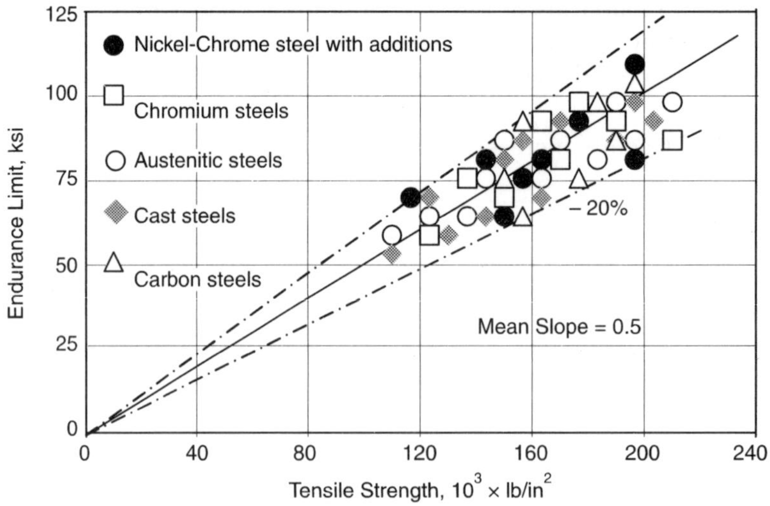

Fig. 3.3. Endurance Limit (Reversed Bending) versus Tensile Strength of Steel (*ref.* Matthaes, 1984)

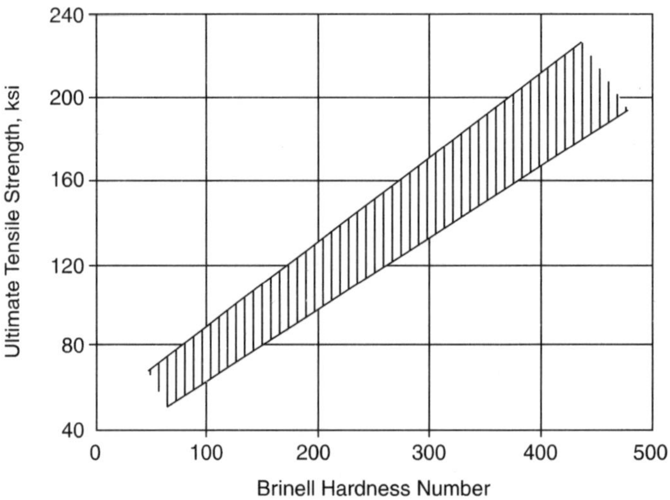

Fig. 3.4. Relation of Ultimate Tensile Strength to Brinell Hardness Number for Steel UTS ≅ 0.5 (BHN) ksi or 0.35 (BHN) kg/mm^2

Tests to obtain S-N curves require a considerable outlay in time and expense compared with simple one time tests such as elastic limit, ultimate strength and impact value. Much effort is expended on obtaining a correlation between endurance limit and one of the simpler tests. The only reasonably good correlation appears to be with ultimate tensile strength, hence it is customary to plot endurance limit versus ultimate tensile stress, as shown in Fig. 3.3 for steel.

In the case of steel the ultimate tensile strength (UTS) is nearly proportional to the Brinell hardness number (BHN), as shown in Fig. 3.4. Thus, for steel the BHN scale can be added to, or substituted for, the UTS scale. Figure 3.5 shows torsional shear endurance limits for steel. Note that the curves shown here are based on an arbitrary method of testing at room temperature through a sinusoidal stress pattern with average stress zero, and with highly polished specimens of arbitrary size and shape. It is therefore pertinent to inquire into the effects of different test conditions. Such an inquiry should include factors such as the effects of other values of frequency of test cycle, temperature, stress-cycle pattern, shape of specimen, surface finish and surface treatment, corrosion, size of specimen and direction of grain.

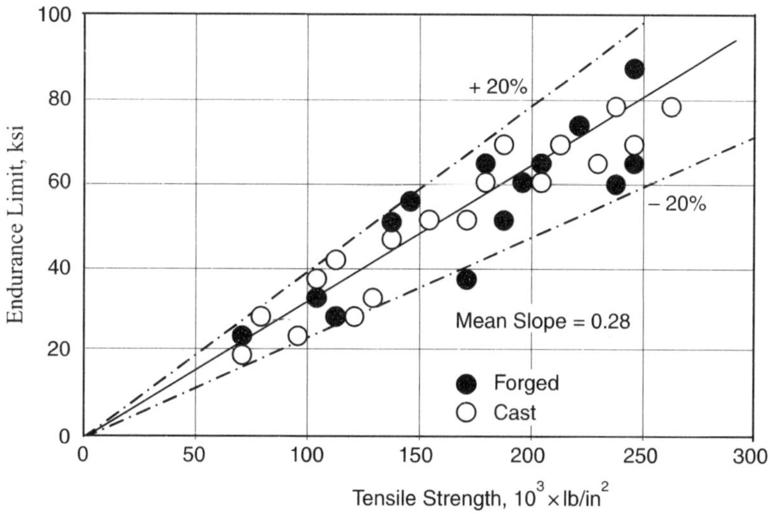

Fig. 3.5. Torsional Endurance Limit versus Tensile Strength for Steels (*ref.* Matthaes, 1984)

3.4 CYLINDER

Good cylinder design is one of the more important features of a good reciprocating machine design. Not only do the cylinders constitute a large fraction of the cost, weight and bulk, but they exert a major influence on performance, reliability, durability and maintenance.

Fig. 3.6. Fabricated Carbon or Stainless Steel Compressor Cylinder for Special Applications (*Courtesy*: Dresser-Rand Company, Painted Post, NY)

The cylinder barrel functions both as a structural member and a bearing surface for the pistons and rings. Cast iron is the most widely used material. Proper fillets are necessary at all changes of section, as for all other engine parts. Protection from rust on the outer surface of the barrel, where it comes in contact with cooling water, is essential. Failure of cylinder barrels because of weakening by corrosion is not uncommon where proper rust proofing is not provided. If long operating life is not required, as in passenger automobiles and small industrial engines, the cylinder barrel is cast integrally with the jacket structure. For long life in commercial and industrial use separate removable cylinder barrels are used. This construction method permits repair and replacement, while also allowing appropriate materials to be used, regardless of crankcase requirements. Cast iron and steel are the preferred materials.

Multi-cylinder units must have the cylinders designed for capacity, pressure, temperature and gas properties of a specific project. Available cylinder materials include cast iron, ductile or nodular iron and forged steel. Compressor manufacturers such as Dresser-Rand have built fabricated (welded) cylinders using carbon and stainless steel as the primary material (Fig. 3.6).

Where the cylinder barrel does not carry any axial load, stress due to gas pressure may not be the only design concern (Fig. 3.7). The requirements for maintaining circularity and for re-boring result in a thickness such that the pressure loading is not the decisive factor. When the barrel carries gas load from head to crankcase the design should avoid stress concentrations around the bolting flange. The design of aircraft cylinders incorporates the following special features to reduce stress concentration at the bolting flange: closely spaced studs, flange well fitted into the barrel, stud axis as close as possible to the bore, spherical sets for nuts to minimize uneven pressure distribution and to reduce bending in the bolts. If lightweight is not so important less costly methods are acceptable.

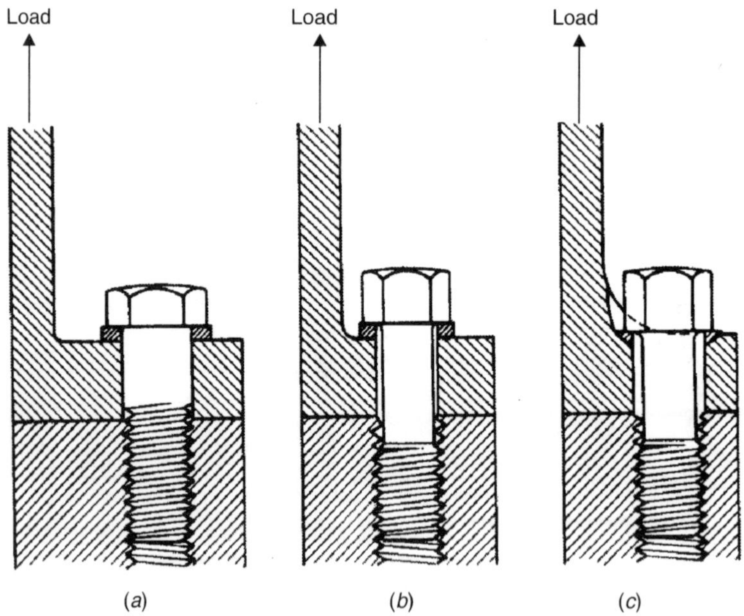

Fig. 3.7. Bolted Flange in Tension.
(*a*) Poor design - cap screw far from load, small fillets
(*b*) Better design - narrow-hex screw close to load, larger fillets
(*c*) Best design - maximum flange fillet, spherical washer and seat

Since the cylinder barrel is a heat flow path from the gases to the coolant, problems associated with this function increase with cylinder size. The following considerations must be checked:

1. At the head end, where heat flow is most rapid in engines, provide the shortest path from inner surface to the coolant.
2. Avoid barrel distortion. Distortion forces chiefly come from the head and jacket structure. Transmission from these components to the barrel should be minimized. With separate barrels the clamping and gasket arrangement should be such as to keep the barrel as free as possible from such forces.

Exhaust ports in two cycle engine cylinder barrels constitute a serious problem in bigger engines. The problem is acute in loop-scavenged cylinders, where the temperature difference between the intake and exhaust sides is large. Provision must be made to cool the bridges between the exhaust ports to reduce thermal distortion and stresses.

For most cylinder bores the wearing surface is cast iron. Chromium plating has been used for diesel engines. To minimize wear steel cylinders are sometimes nitrided. Final finish of the bore is usually by honing to a specified degree of surface roughness. The desirable degree of roughness depends on the material used for the bore, piston and rings, and to some extent, on design details, operating regime and fluid characteristics. Surfaces that are too rough cause rapid wear and those that are too smooth prevent fast break-in of piston rings and bore.

Cylinder heads for water cooled engines use cast iron; for lightweight or other circumstances aluminum may be selected. Use of long multiple aluminum head castings on iron cylinder blocks can cause differential expansion problems. The complex shape of engine cylinder heads, coupled with problems of structure, heat flow and fluid flow make it a more critical element in its design. The crux of the problem in a cylinder head with valves is to achieve an arrangement satisfactory from the point of view of valves and ports which will carry the gas loads and at the same time avoid excessive distortion and stress due to temperature gradients, while also avoiding excessively high cost or undue complexity. As in all parts subject to heat flow and accompanying thermal gradients, the problem becomes more difficult as the bore increases. For cylinders up to about 6 in. bore, well-designed iron or aluminum castings carrying gas and valve gear loads generally give no serious troubles from thermal stresses. As the bore increases from this point, the thickness necessary to carry the applied loads becomes large enough to introduce troublesome thermal stresses and distortion. Location of the areas of high thermal stress on a conventional head is shown in Fig. 3.7 (*a*). Possible methods of reducing these stresses are illustrated in Fig. 3.7 (*b*), (*c*). In general, the head stresses are minimized by using the following guidelines:

1. Careful detail design to reduce stress concentrations, including generous fillets.
2. Separation of functions of heat flow and load carrying.
3. Use of separate cylinder blocks on en-bloc cylinder assemblies.
4. Combined with 1 to 3, substitution of aluminum for iron for smaller cylinder heads.

Air-cooling is impractical for cylinders larger than 6 in. bore. Substitution of cooling fins for water jackets requires separate cylinders. Small low output engines sometimes use cast iron cylinders with integral cooling fins. Air cooled aircraft cylinder barrels are usually steel forgings, machined all over and screwed into the head, as in Fig. 3.8. Steel cylinders provide satisfactory service for aircraft engines where friction and wear considerations are significant, specially when they are nitrided.

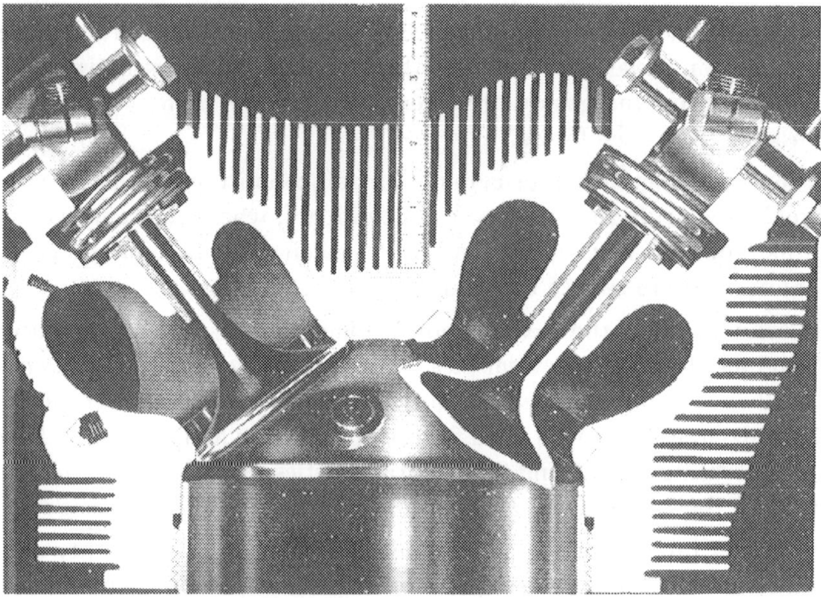

Fig. 3.8. Cylinder Head, Inlet and Exhaust Ports of Wright 'Cyclone' Aircraft Engine (*Courtesy:* Wright Aircraft Corporation)

3.5 PISTON

The piston combines problems related to structural strength, cooling, friction, lubrication, wear and noise suppression. Internal combustion engine pistons have a more acute problem of heat removal and cooling as opposed to most other reciprocating type machines. Also, a lighter piston reciprocating weight is crucial from the view point of crankshaft natural frequencies of vibration. Cast aluminum alloy is the most widely used material for pistons. Large aircraft engines and high output machines use aluminum forgings. Larger bore machines usually use cast iron, cast steel or a composite construction material of two different materials, for example, steel crown and iron skirt.

Figure 3.9 illustrates a typical piston assembly used in automotive engines. In non-supercharged engines of small bore the piston may be of light construction (Fig. 3.10). Noise suppression is critical in passenger cars and is usually solved by flexibility in piston skirt supports, so that the piston runs at all times with near zero skirt clearance. Another solution is to support the skirts by steel or Inconel plates inserted before pouring. The low expansion coefficient of these plates makes small cold clearance possible. Offsetting the piston pin also helps in noise suppression. Small-bore industrial machines require less noise suppression and may omit skirt flexibility in favor of lower cost and more rugged design. In diesel engines and gasoline supercharged spark ignition engines high maximum cylinder pressure requires more rugged design and avoidance of stress concentrations. Experimental and finite element stress analysis has shown likely areas of critical stress, principal among which is the upper side of the pin boss.

In small-bore (under 4 in.) non-supercharged engines pistons are usually cooled by conduction of heat from the head to the ring belt and skirt, from which heat then flows to the cylinder bore. For this method of cooling aluminum is helpful, with generous sections for heat conduction from the center of the piston head. Heat conduction to the cylinder walls is promoted by small skirt clearances. Rings run cooler if the distance from the top edge of piston to top ring is increased. For high performance and large bore spark ignition engines, and for most diesel engines, oil cooling of the piston may be necessary. The oil may be circulated by spraying from a fixed nozzle aimed at the inside of the piston, by supplying oil through the crankshaft and connecting rod to an enclosed space under the piston head, or through tubular passages in the piston head. Large cross-head engines often carry coolant for the piston through jointed or telescoping pipes. For oil cooled pistons the rate of oil circulation must be high enough to avoid breakdown of the oil into carbonaceous deposits.

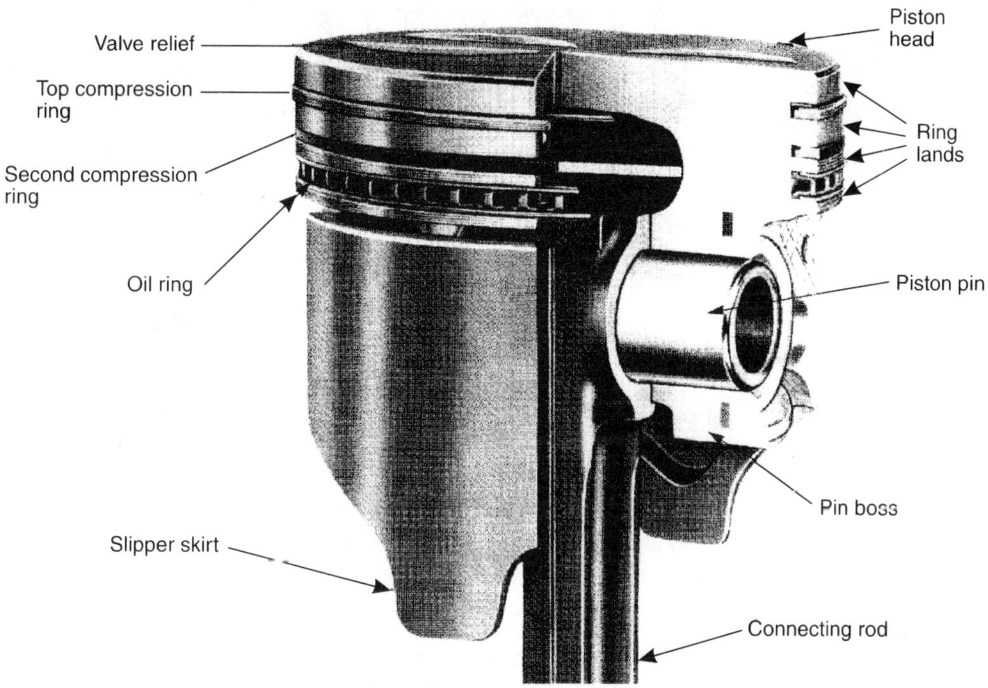

Fig. 3.9. Piston Assembly (*Courtsey*: Volvo)

Piston pins should be of the floating type, loose in both the pin and the rod, except where noise suppression is important. In the latter case the pin is usually clamped in the upper end of the rod. The end play of floating rings must be restricted by snap rings or by end plugs of a soft material (aluminum or bronze) finished to a spherical radius slightly less than the cylinder radius.

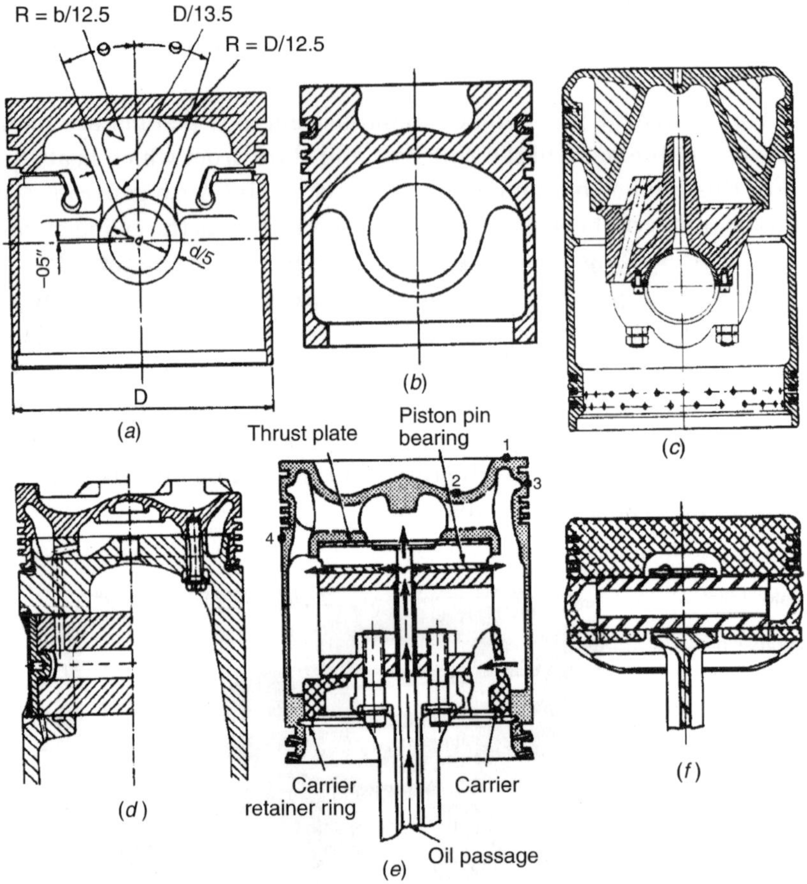

Fig. 3.10. Typical Piston Designs

(*a*) Passenger car type, cast aluminium alloy

(*b*) Automotive diesel type, cast aluminium alloy with cast iron ring belt

(*c*) Copper Bessemer V-250 2 cycle gas engine (Fig. 2.15), both piston parts cast iron

(*d*) M.A.N. 9.45 × 11.8 in. V8V24/30 highly supercharged diesel engine, piston with steel crown, aluminium alloy skirt

(*e*) GMC-Detroit Diesel 5.75 × 5.75 in. 2 cycle engine rotatable piston, cast steel

(*f*) Lycoming 5.375 × 3.875 in. aero engine, aluminium alloy piston

Piston pin stress is based on maximum gas pressure and inertia loads. Distribution of the load from the connecting rod and pin bosses to the pin surfaces is also a factor. Since the pin is a minor fraction of the piston weight, stiffness and strength of the pin should not be marginal. A rugged pin of larger diameter contributes to the reduction of critical stress in all three mating components, piston, pin and connecting rod. Use of a hollow piston pin satisfies the twin requirements of reduced overall reciprocating weight and increased stiffness characteristics, compared with a solid pin of equal weight.

The range of piston proportions in internal combustion engines is as follows:

Parameter	*Symbol*	*Max*	*Min*
Length/bore	l/b	2.36	0.64
Height from pin center to top of bearing surface	h/b	0.97	0.25
Pin diameter/bore	p/b	0.48	0.20
Number of compression rings	N_c	6	2

Airplane engines have achieved excellent reliability with piston lengths as short as 0.6 × bore. There would seem to be little reason for pistons to be longer than 1 × bore except where necessary to cover ports at top center in two cycle engines. Two cycle engines with trunk pistons require the piston to cover the ports at piston top center. In such cases minimum piston length is the stroke plus enough overlap to prevent leakage. Cross-head engines may not have this requirement. In trunk piston engines the ratio of pin diameter to bore runs from 0.20 to 0.48, and since no space and but little weight is saved by small pins, the recommended ratio is 0.35 or greater.

Where possible, it is generally desirable to locate the pin in the middle of the skirt bearing surface. However, successful engines have been produced with pin locations near the rings and also near the bottom of the skirt.

To minimize piston friction, use small length-to-bore ratio, minimum skirt bearing area, maximum skirt clearance, and small number of rings.

Figure 3.11 shows principles of good piston design for engines of automotive size or less. Areas of critical stress are greatly reduced in the preferred design. A common fault in piston design is a pin of too small a diameter; in such a case deflection of the pin due to bending increases stress concentration around the pin bosses. Another common mistake is to use ribs under the piston head. Such ribs act as stress raisers while rendering cooling of the head ineffective. Piston friction can be minimized by using smaller length to bore ratio, minimum skirt bearing area, maximum skirt clearance and fewer rings. However, the first three of these requirements are in conflict with conditions for minimum noise from piston 'slap', which is usually an important factor in engine mechanical noise. The above features have been used to their maximum effect in aircraft engines, where other sources of noise are so great that piston noise is concealed. On the other hand pistons for automobile engines must be quiet at the expense of greater piston friction.

As the specific output of engines is increased piston seizure is a common problem. It is due usually to one or more of the following causes:

- Inadequate clearance in the bore
- Inadequate cooling
- Distortion due to heating or loading

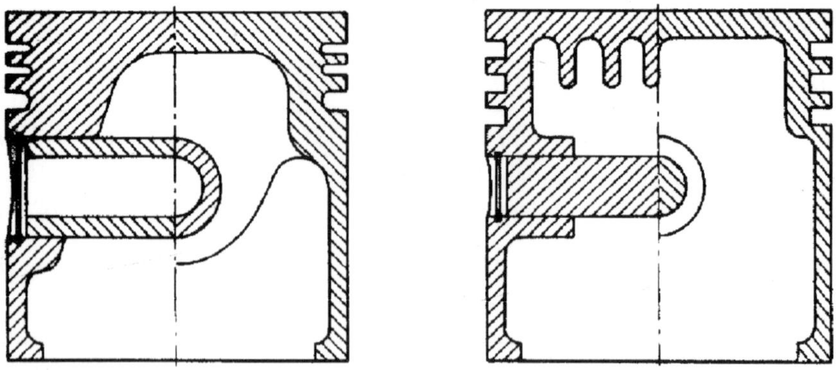

Acceptable Design

Remove fins
Narrow faced compression rings
Fillet ring grooves
Large hollow pin
Pin boss adequately braced to crown

Unacceptable Design

Fins create stress raisers
Rings have excessive width
Ring grooves very sharp
Pin too small
Pin boss to crown strength inadequate

Fig. 3.11. Principles of Good Piston Design for Heavy Duty

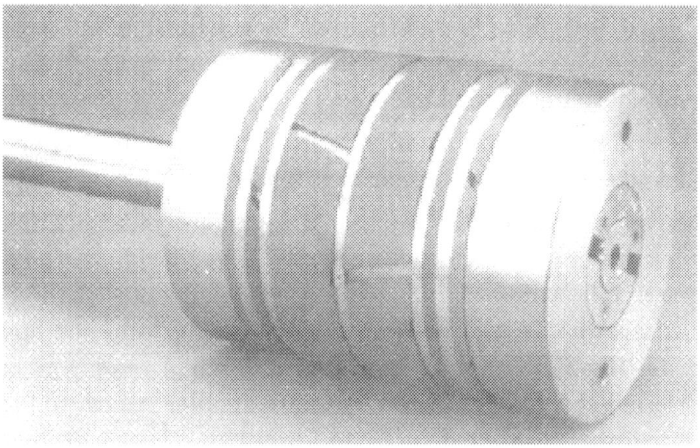

Fig. 3.12. Non-Lubricated Compressor Piston with PTFE Rider Bands for Contact Support (*Courtesy*: Dresser-Rand Company, Painted Post, NY)

- Heavy detonation, in spark engines
- Ring failure, followed by gas blow-by and piston burning.

Plating of piston skirts with tin is found sometimes to reduce the tendency toward piston seizure, especially with aluminum pistons.

Most process units for gas compression are equipped with Teflon (PTFE) piston rings. Two or three single-piece diagonal cut without expanders are used. High-pressure (over 300 atmosphere) applications may be provided with three-piece segmental bronze rings. Most non-lubricated compressors require PTFE rider rings. The rider rings support the weight of the piston and the rod, and may be of the split type located at the center of the piston (figure 3.12) or of the band type stretched on the surface of the piston.

3.6 PISTON RINGS

Piston rings have constituted a difficult design problem ever since the advent of the internal combustion engines. To achieve life and reliability commensurate with those of other wearing parts of the engine calls for attention to a host of engineering details. Their design has developed substantially from empirical basis, and the literature on the subject is extensive. Piston rings have two functions: to prevent gas leakage and to hold the flow of oil into the combustion chamber to the minimum necessary for adequate ring and piston lubrication. All rings take part in the control of oil flow, but there is usually at least one ring designed for this intent alone. Such rings are called oil control rings, while the others are compression rings. Most piston rings are made of gray cast iron because of its excellent wearing properties in all kinds of cylinder bores. Where ring leakage is a problem nodular iron or steel is used, usually with coated bearing surfaces. For maximum resistance to wear of both ring and cylinder bore rings are faced either with chromium plate or 'metallized molybdenum', a porous structure of molybdenum oxides. Oil control rings may be of iron or steel. For compression ring design the following items have been found to be desirable (Fig. 3.13).

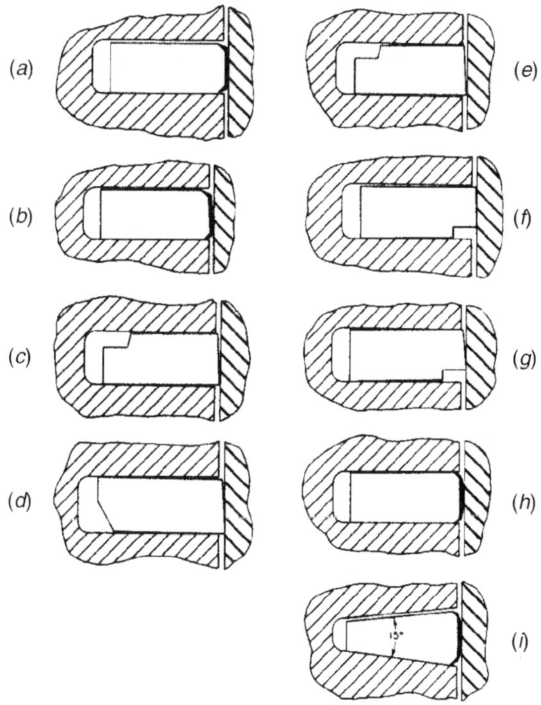

(*a*) Plain compression ring, chrome plated. (*b*) Taper faced compression ring, chrome-plated. (*c*) torsional twist compression ring. (*d*) reverse twist taper face compression ring. (*e*) torsional twist taper face compression ring. (*f*) scraper type of compression ring. (*g*) taper face compression ring. (*h*) barrel face compression ring. (*i*) wedge type compression ring. (*ref.* Binford, 1985)

Fig. 3.13. Current Practice in Compression Rings

1. Face width of the rings should be small. A ratio of face width to cylinder bore of 1/50 is typical. Small face width reduces ring mass and facilitates wear in.
2. Radial depth of the ring must be sufficient to give necessary spring pressure, but not so large as to cause rings to break during assembly.
3. Spring pressure of the ring against cylinder wall should be controlled to definite limits by the ring manufacturer.
4. Bearing surface of new rings should be slightly conical with larger diameter toward the crankcase (Fig. 3.13*b-g*). A taper may be created by machining off the upper inside corner of the ring, after the face has been cylindrically finished. The intent of this taper is both for oil control and faster break-in. As an alternative to the taper the ring may be barrel faced (Fig. 3.13*h, k*), either with or without chromium plating. A close fitting worn-in contact seems to be the essential requirement.
5. For heavy duty service and most diesel engines the face of the top of one or two rings is chromium plated. The plating is found to reduce both ring and bore wear significantly. Other coatings for this purpose include oxides of molybdenum, iron oxides and tin plating.
6. Where ring sticking, or seizure, in the groove is a problem, as in many diesel engines, its incidence may be reduced by wedge-shaping the grooves and rings (figure 3.13*k*). This shape apparently makes it more difficult for carbon particles to hold the ring tight in the groove.
7. Research data shows that gas pressure in ring grooves is about 50 per cent of the cylinder pressure behind the top ring, 20 per cent behind the second ring and zero behind the third ring. Thus, two good compression rings are all that may be necessary. Additional rings may be justified only for protection against leakage in the event of failure of the upper rings.
8. Radial clearance behind the rings must be large enough to prevent bottoming of the rings in their groove. Axial clearance must be controlled to reasonable limits. Giving the ring initial twist (Fig. 3.13 (*c*), (*d*)) helps reduce axial play of the ring in the groove.
9. In aluminum pistons service life of both piston and rings can be extended by mounting the rings in a belt of harder material, specially for diesel engines.

Compression ring failures occur through sticking, excessive wear, loss of spring tension or breakage. Clean combustion and adequate piston cooling are the best insurance against all of these failures, except possibly breakage. Breakage is often attributed to ring 'flutter', a form of ring vibration. It exists as a periodic motion of the ring out of contact with the cylinder surface. Small ring face width tends to reduce the incidence of ring breakage. Large axial clearance and high cylinder pressure tend to increase the dynamic bending stresses in the ring, leading to breakage. As a last resort to prevent ring breakage chromium plated steel rings may be used.

One or more oil control rings specifically designed to minimize oil flow to the combustion chamber is required in engines. Some successful configurations of the oil control ring are shown in Fig. 3.14. They are usually made of thin steel stampings and may or may not require a plated bearing surface against the cylinder bore. The ring is not loaded by gas pressure and must depend on its own spring action to ensure contact with the bore surface.

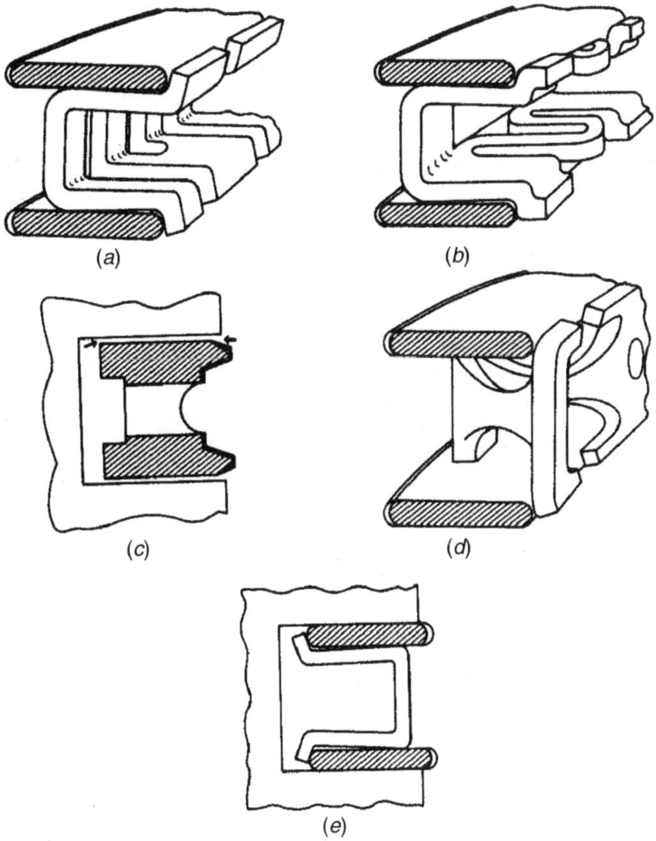

(*a*) Steel side sealing oil ring. (*b*) another type of steel oil ring. (*c*) cast iron oil ring (transient position). (*d*) special type of oil ring. (*e*) section of type *a* oil ring, showing side sealing feature.

Fig. 3.14. Oil Control Rings (*ref.* Binford, 1985)

3.7 CONNECTING ROD

Because of space restrictions and the necessity for minimum reciprocating mass the connecting rod is one of the more heavily stressed engine part. Most internal combustion engine rods are made from steel forging. Small engines may use forged aluminum or cast malleable iron.

The theoretical load on a connecting rod is readily calculated from the indicator diagram and inertia forces discussed in Chapter 5. The heaviest load comes at top center from peak cylinder pressure, resulting in a large column type of load on the rod. The total stress is a result of a combination of vibratory stress and non-symmetrical loading. Heavy non-symmetrical loading comes from the piston axis being out of line with the rod axis, creating bending moments in the connecting rod. Such bending moments cause greater stress than due to lateral inertia, or 'whip', of the connecting rod. This is one argument for orienting the H section of rod column 90° from the usual direction. This design often simplifies forging and machining. Near the top center inertia forces are opposite to gas pressure loads, thus reducing maximum gas load on the rod. In some cases pistons and connecting rods are made heavier than necessary in order to reduce crankpin bearing load.

In four-cycle engines the outward inertia load during the exhaust stroke is not offset by the gas pressure load, and may be large at high piston speeds. In radial engines, because of the large mass of the master-rod assembly, the inertia load on the crank pin and the rod can be very large, especially when the engine is running at high speed with low gas pressure in the cylinders. This problem became acute in airplane engines in combat during World War II, and was responsible for the development of bearings made of silver, lead and indium.

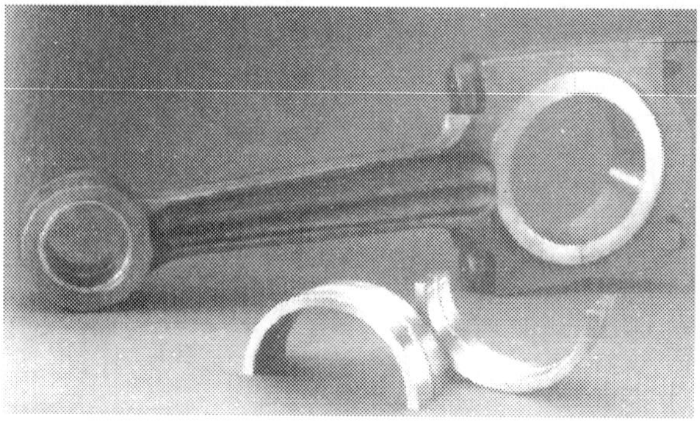

Fig. 3.15. Connecting Rod for Compressor (*Courtesy*: Dresser-Rand Company)

In two cycle engines the calculated outward inertia load never exceeds the gas load, and it has been assumed that there are no tension loads in the connecting rod or that load reversal does not occur in the piston pin. But since vibrating forces may be added to the inertia forces in such a manner that load reversal does take place in the connecting rod, this assumption is dangerous. In any case if compression pressure is lost, the full inertia load will fall on the connecting rod. Hence, it is necessary to allow for some load reversal in the case of two cycle engine connecting rods and their bearings.

Figure 3.15 shows a typical connecting rod for process reciprocating pump and compressors. The rods are generally made of forged steel and manufactured with a closed die to provide good grain flow throughout the piece. Forced lubrication oil passages are drilled through the length of the rod to permit flow of lubricant from the crankpin to the crosshead pin bushing. The replaceable crosshead bushings are made of bronze. Connecting rod bolts are special forgings, and larger sizes have rolled threads to provide maximum strength.

Torsional natural frequencies of vibration of the connecting rod should be checked in order to avoid resonance with engine operating speed. The length of the rod, torsional stiffness and polar mass moments of inertia of the piston (about its central axis) and portion of the crank pin will control the torsional frequencies of the connecting rod. Under proper circumstances the presence of torsional vibration frequencies in the operating speed regime provides adequate reasons for minimizing piston mass and making the connecting rod torsionally stiffer. Crankshaft torsional vibration influences load on the rod, but it is usually more critical to the crankshaft than to the rod.

Connecting rods can be classified as either single or multiple, the latter term applying to the case when more than one rod is combined to accept a single crank. Figure 3.16 illustrates good and bad practices in a single connecting rod design for heavy-duty service. The usual points of critical stress are indicated. Figure 3.17 shows the rod of the high output diesel locomotive engine of Fig. 2.16.

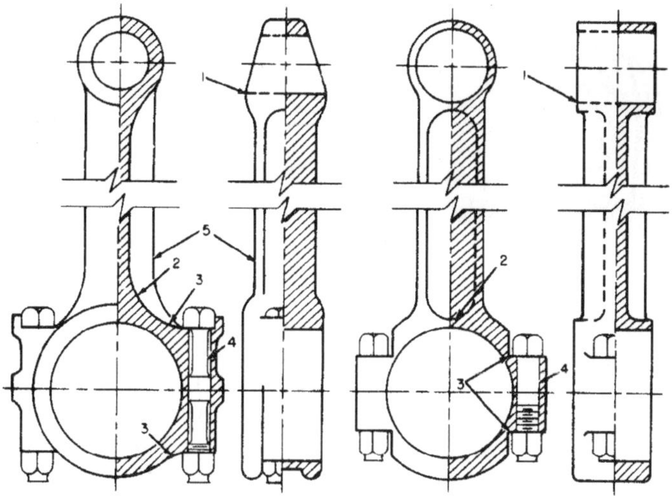

Acceptable Design

1. Thin edge avoided
2. Plenty of metal over bearing
3. No sharp grooves
4. Good bolt design
5. H section turned 90°

Unacceptable Design

1. Stress concentration at thin edge
2. Section too thin
3. Stress concentration at corners
4. Bolt improperly designed

Fig. 3.16. Principles of Good Connecting Rod Design

Successful ratios of connecting rod length to crank radius, *L/R*, range from 3.0 to 5.0. Within these limits this ratio is also affected by space considerations, including the relative importance of low engine height and the question of interference with the cylinder bore. The latter factor accounts for the fact that *L/R* varies in the same sense as the stroke/bore ratio.

To achieve proper torsional stiffness the rod is tubular, with a serrated parting surface turned at a steep angle to permit removal of the rod through the cylinder block. A stiff support for the crank pin bearing is also provided. Except in aircraft engines where engine length is at a premium, single rods placed side by side on a long crank pin is the simple and reliable system for V-engines. Where the crank pin length is short either the forked or plain system shown or an articulated system shown in Fig. 2.17 has proved successful. In the case of articulated rods the geometry of piston motion is different from that of the master rod and varies with the angle chosen between the planes intersecting the crank pin and the link pin and the centerline of the master rod, called the link-pin angle. With this angle the same as the *V* angle, the stroke will be the same as that of the master rod, but the top center will come at a crank angle not quite at crank alignment with the cylinder axis. With the angle such that the top center of the link-rod piston will coincide with the top center of the crank, the stroke will be longer than in the master rod cylinder. Either system gives satisfactory engine performance, provided the compression ratio is held constant by suitable positioning of the cylinders with respect to the crankshaft axis. The design of quality forked or articulated connecting rod requires great attention to the minimizing of stress concentration. It is best carried out with a combination of experimental methods and finite element stress analyses.

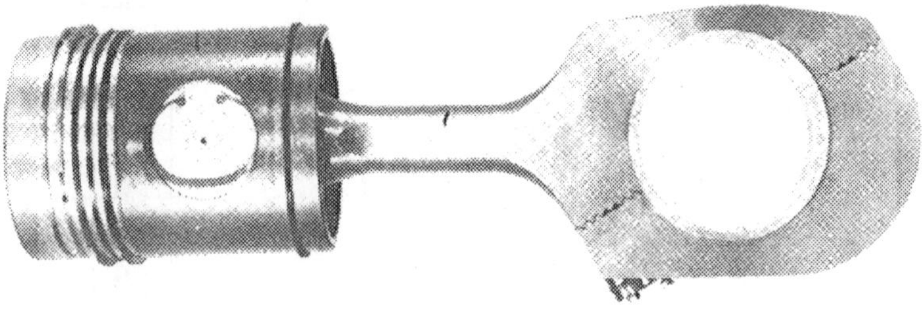

Fig. 3.17. Connecting rod of Sulzer Locomotive Engine of Fig. 2.16.
Tubular column with serrated parting surface turned at steep angle to facilitate removal of rod through cylinder with large crank pin diameter. Massive lower end of rod for bearing support.
(*Courtesy:* Sulzer Brothers, Ltd.)

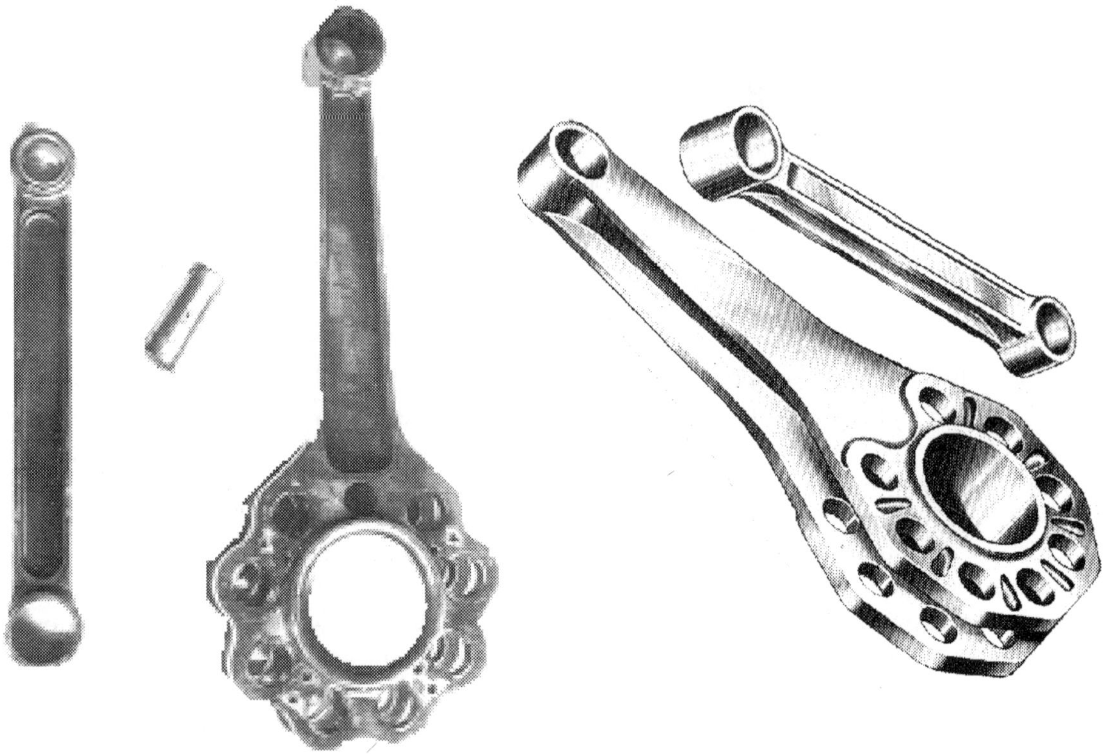

Fig. 3.18. Wright Radial Aircraft Engine Connecting Rod (left). Earlier design, complicated geometry, stress concentration at scallops, holes and transitions; (right) newer design with H section turned 90°, scallops, threads eliminated.

In the master rod system generally used in radial aircraft engines (Fig. 3.18), link-pin angles are equal to cylinder angles, and schedules of piston position versus crank angle for the link rods differ from each other and from that of the master rod. These differences are not great enough to affect cylinder performance measurably, provided the compression ratio in all cylinders is held the same by proper positioning of the link pins in the master rod. In spite of this some manufacturers have opted for more complex systems to obtain symmetrical piston motion. In any master and link rod system gas and inertia forces on the link-rod elements introduce bending moments on the master rod, and may be substantial. Bending strength of the connecting rod must be accommodated in its design. An improved design of the connecting rod system for a large aircraft engine is shown in Fig. 3.18 (*left*). The improvements arise from simpler construction, strength and ease of manufacture.

The sliding crosshead assembly shown in Fig. 3.19 is typically used for units over 150 kW in capacity. The components, made of cast steel, or cast or ductile iron are made to meet API 618 guidelines. The top and bottom shoes can be adjusted with shims, and are replaceable. Generally, the crossheads have a floating pin design, but larger units opt for a fixed pin version for long term and reliable operation.

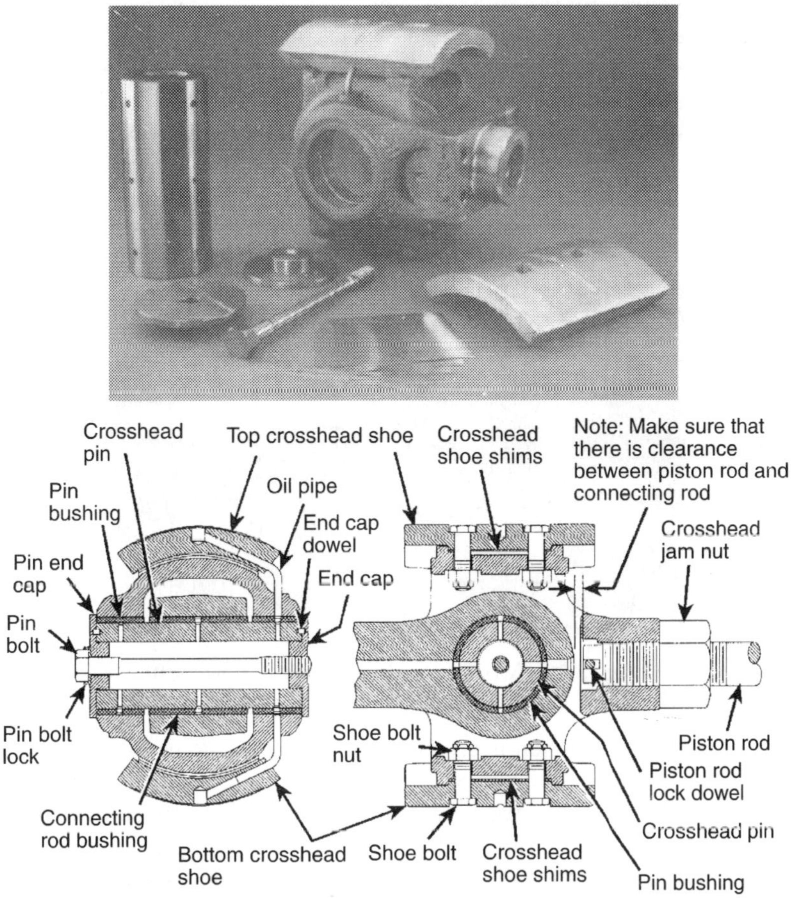

Fig. 3.19. Crosshead Components for Compressor
(*Courtesy*: Dresser-Rand Company, Painted Post, NY)

3.8 CRANKSHAFT

As in the case of other reciprocating machine parts, the crankshaft should be designed on the basis of previous successful practice, combined with modifications imposed by overall machine requirements. Advances in newer materials, manufacturing methods and engineering technologies should also be incorporated in order to obtain better performance from the machine.

Crankshaft loads at crank pin and main journal bearings are obtained from indicator diagrams and calculated inertia forces using rigid body assumptions. As the design of all other associated parts in the machine system are better defined, knowledge of component and system dynamics and thermal gradients provide greater flow of information about the loads in greater detail. An application of the theory of similitude and computer-based advanced experimental and analytical calculation methods provides the final design and the product.

Crankshafts are made from steel forgings or cast in steel, nodular iron, malleable iron or gray iron. With a given design the strength of the shaft lies approximately in that order. Where high specific output is required forged steel is the preferred material. Gray iron is used only for small low-cost engines. Modern techniques in metallurgy and foundry practice have made it possible to produce shafts of the best cast materials that are very near to the steel shafts in structural load carrying capabilities. Near-net shape of the casting also means reduced machining to obtain the finished crankshaft. Forged steel shafts are generally of 0.3 to 0.4 per cent carbon or alloy steel, treated to 250 to 275 hardness, giving tensile strength of 125 to 150 ksi. Physical properties of cast materials are provided in Table 3.3. The object of using cast shafts in place of steel forgings is to reduce the expense of manufacture, or to achieve geometric configuration not possible with the forged shaft. For example, counter-weights can be cast integrally. Cast crankshafts are found most often in passenger automobile engines and other small engines, but some large engines also use cast shafts.

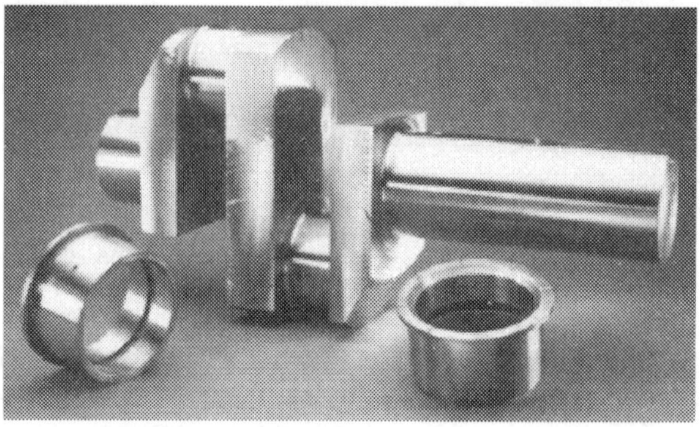

Fig. 3.20. Two Throw Compressor Crankshaft
(*Courtesy*: Dresser-Rand Company, Painted Post, NY)

Qualitatively speaking, the loads on the crankshaft result in bending, shear and torsion stresses throughout its entire length. Even if the loads are accurately known, the geometry of the crankshaft makes the task of stress calculations complex. In spite of these difficulties, much has been done toward rationalization of crankshaft design, largely by finite element methods combined with experimental techniques. Significant useful information about the crankshaft design can be obtained from crankshaft models subjected to a variety of simulated loads under laboratory conditions. Figure 3.16*a* shows the substantial effect of fillet radius on the maximum stress in a straight shaft and a simple crank. Given reasonably well chosen ratios of major dimensions, fillet radius is the most important factor in crankshaft strength. The larger the fillet radius the better, provided adequate space is available for the required bearing length. Where bearing length is at a premium, under cut fillets, figure 3.16b, are far better than inadequate conventional fillets. Fillets with non-circular contours are slightly better than fillets of fixed radius, but are practicable mostly in large crankshafts. With the same overall dimensions and fillet radius, the crank of figure 3.16a is weaker than the stepped shaft, showing that the stress between the pin and the cheek is not uniformly distributed around the end of the pin, but tends to concentrate near the inside corner. Figure 3.16b illustrates this point and shows how better distribution of the forces between pin and journal can be obtained, with consequent lowering of the stress concentration at the fillets. As an introduction to the principles of crankshaft design a reading of 'The Structural Durability of Crankshafts' (Sulzer Technical Review # 2) is recommended.

Table 3.3. General properties of cast metals

Material Characteristic	*Gray Iron*	*Malleable Iron*	*Nodular Iron*	*Cast Steel*
U. T. S., ksi	20–50	50–90	60–160	60–175
Elastic Modulus, 10^6 psi	15–20	23	25	28–30
B. H. N.	170–269	163–285	160–360	131–415
Endurance Ratio	0.35–0.50	0.50–0.60	0.40–0.60	0.40
Elongation, %	0	2–18	0–19	6–32
Specific Gravity	7.01–7.315	7.40	7.1–7.3	7.7
Heat Conductivity BTU/ft-°F-hr	22–35	23	23	26
Coefficient of Expansion in/in/°F x 10^6	6.0	6.6	6.6	6.5
Relative Machinability	very good	good	good	fair to good
U. T. S., ksi	17–45	18–40	30–110	
Elastic Modulus, 106 psi	10	—	10–12	
B. H. N.	40–125	very soft	wide range	
Endurance Ratio	0.30	0.40	0.40	
Elongation, %	0–12	1–7	8–22	
Specific Gravity	2.6–2.9	1.80	7.8–8.9	
Heat Conductivity BTU/ft-°F-hr	60–95	77–90	14–226	
Coefficient of Expansion in/in/°F x 10^6	12.1–13.8	16	9.0–9.8	
Relative Machinability	excellent	excellent	good to excellent	

3.9 CRANKSHAFT DESIGN

Designing the crankshaft is far from an exact science. It must depend on an intelligent application of judgment based on previous successful practice, finite element evaluation of stress and dynamic characteristics of the crankshaft and its attached components and experimental test results.

In subsequent discussions, the following symbols are used to designate crankshaft dimensions, having the dimension of length.

B = diameter of bore in pin or journal

M = diameter of main journals

O = overlap between pin and journal

P = diameter of crank pin

R = crank radius (1/2 stroke)

r = fillet radius

t = axial thickness of crank cheek

w_c = transverse width of crank cheek

w_j = axial length of journal

w_p = axial length of crank pin

When journal and crank pin diameters are sufficiently large in relation to the crank radius, circles representing the journal and pin will overlap when viewed along the crankshaft axis. The amount of overlap will be defined as:

$$O = (P + M)/2 - R \qquad ...(3.1)$$

When this number is positive it gives the radial distance by which the two circles overlap. A negative value indicates the minimum distance between the two non-overlapping circles. Figure 3.21(*c*) shows the effect of overlap on the strength of a crank in bending. As overlap increases, the strength of the shaft, as measured by endurance tests in bending, increases. This trend is due to the fact that with increasing overlap more and more of the load is carried directly from pin to main journal through the overlap area, thereby reducing stress in the fillets.

Figure 3.22 shows results of endurance tests on single cranks. The length of the vertical lines represents reversed torsional loads that these cranks could withstand for several million cycles without failure. The design ratios held constant for these series of tests are as follows:

$R/M = 1.10$, $P/M = 1.0$, $t/M = 0.45$, $O/M = -0.12$,

$r/M = 0.1$ (approximately), $w_p/P = 1.5$ approximately.

For the bored shaft: $B/M = B/P = 0.76$.

The following observations are obtained from Fig. 3.22:

1. The solid crank *a* is by far the weakest of the group, with large improvements in strength obtained by boring the pin and main bearings with straight holes *g* and still further gains by more complicated boring *b*, *c*. These trends can be explained on the basis of Fig. 3.21(*b*).
2. Beveling crank *I* to shape *h* had no effect, but this bevel was used with a relatively thick cheek, $t/M = 0.45$. With thinner cheeks the effect of beveling can be detrimental, as will be explained subsequently.
3. Increasing the cheek contour as viewed along the journal axis had favorable results with the straight bore crank (*I* to *j*) but was less effective with the two-diameter bore (*c* to *d*). Further increases in contour, *d* to *f* and *k* to *l*, had little effect.

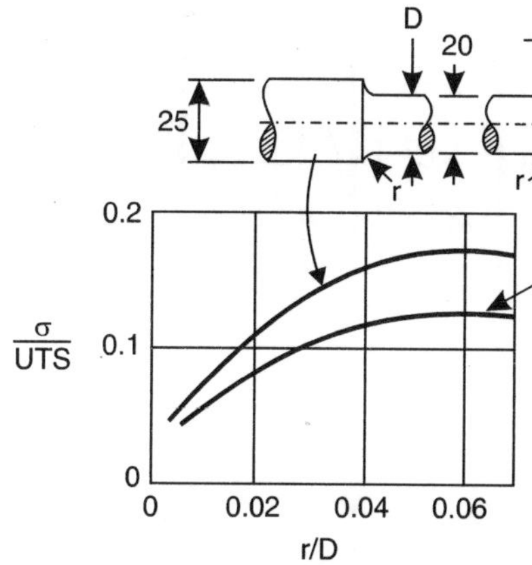

(a)

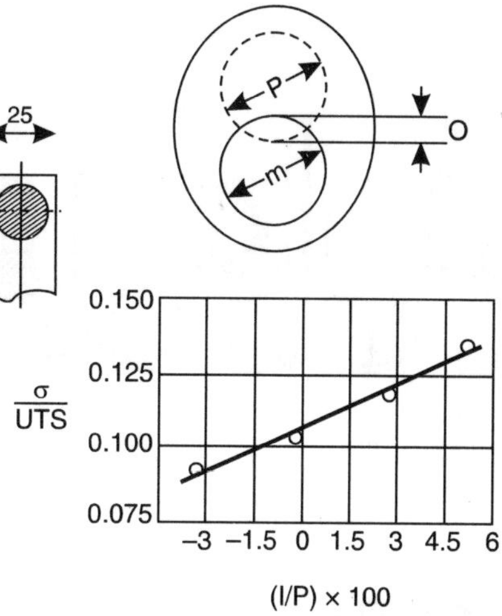

(c)

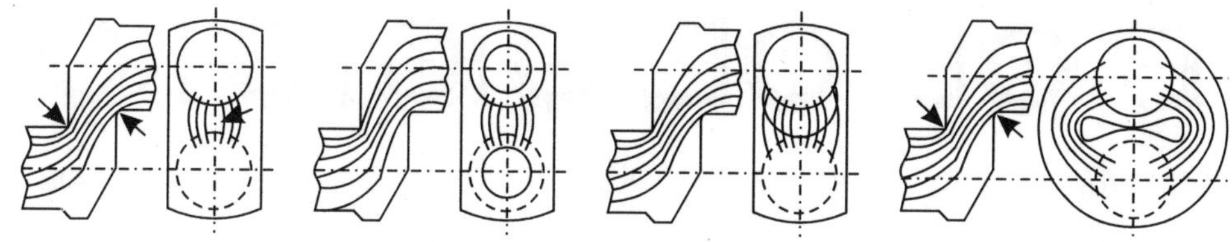

Normal Crankshaft Marked stress risers	Bored Crankshaft Some improvement in force flux distribution	Klose's Solution Reduced stress concentration	Fremont's Solution Divided and diverse force flux

(b)

Fig. 3.21. Crank Stress Patterns (*ref.* Matinaglia, 1982). (*a*) Endurance limit nominal stress in torsion (σ) versus fillet radius. Numbers are dimensions of test pieces in cm; (*b*) 'flux' of force from pin to journal. Stress is smaller as flux lines are farther apart. (*c*) Effect of overlap on crank endurance strength in bending. The dimensions held constant, in millimeters are as follows: Crank throw 73; pin length 52; pin bore 40; journal length 37; journal bore 50; cheek thickness 22; cheek width 126. The dimensions varied in millimeters are as follows:

O/P	*Pin diameter*	*Journal diameter*
0.055	68	85
0.028	66.4	83.4
0.000	64.4	81.4
– 0.032	62.4	79.4

σ = nominal endurance limit O = overlap P = pin diameter

The results illustrated in Fig. 3.21 and 3.22 show the relative loads that the various crank designs can carry without experiencing fatigue failure. Actual stresses were not measured.

Figure 3.23 show stresses at various points measured by means of short gauge length strain gauges on crank models subjected to arbitrary moments of bending and torsion. The following design ratios were held constant (cheeks beveled as shown in Fig. 3.23(*b*):

$$P/M = 0.8,\ t/M = 0.27,\ r/M = 0.05,$$

$$w_p/P = 0.79, \; w_m/M = 0.57, \; w_c/M = 1.18$$

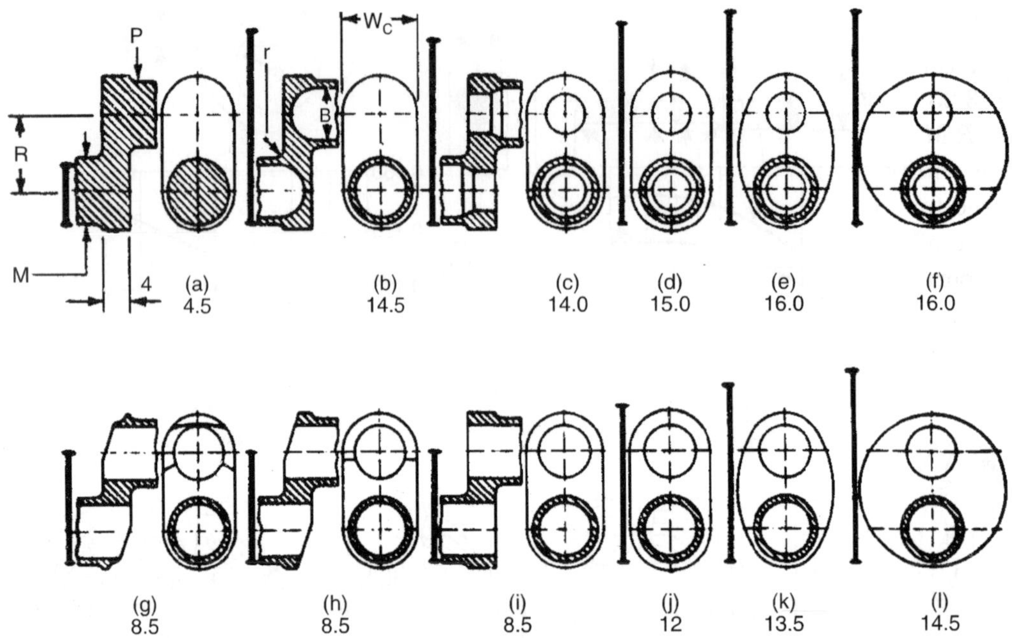

Fig. 3.22. Influence of Form on Relative Strength of Crankshafts Made of Same Material. Strength Measured by Reversed Torsion Tests on Models. Numbers and Vertical Lines Indicate Relative Strength of Cranks (*ref.* Lurenbaum, 1977).

Bending moment was applied as shown at *e* in the figure. Torsion was applied to a complete six throw shaft supported by rollers on main journals. The results given in figure 3.23 apply to number 2 crank, which was found to have higher stresses than others. The variables investigated were the diameters of the bore in crank pin and main journal, and the length of the crank throw. Changing the throw varied the overlap between journal and crank pin from O/M = 0.20 to O/M = – 0.04. The short-throw crank with O/M = 0.20 can be taken as typical value for this parameter. The conclusions that can be drawn from this study are as follows (letters refer to Fig. 3.23):

1. The maximum stress in a crank can be several times higher than the stress values using the simple beam theory. Fillet stresses up to five times the nominal torsional stress and three times the nominal bending stress in the pin are shown.
2. Reducing the crank throw from R/M = 0.97 to R/M = 0.70, which increases the overlap from O/M = – 0.04 to O/M = 0.20, decreased the stresses, except with large values of bores in pin and journal (see diagrams *g* through *m*).
3. Boring of pins and journals to half their outside diameters gives an appreciable reduction in fillet stress. Larger boring tends to increase fillet stresses (*j, k, l*) and to introduce large stresses at the cheek edges.
4. The bevel shown in Fig. *b* and used in all the tests of Fig. 3.23 was probably detrimental to strength, at least in the case of the bored shaft, and the stresses at section *K*(figure *f*) might have been greatly reduced. The same may be said for traverse *I* in *m*.

The stresses shown at the edge of the bored hole could have been reduced by rounding these edges to a radius similar to the fillet radius. Sharp corners invite stress concentration and should be avoided in all machine parts.

In comparing the results of Figs. 3.22 and 3.23 it is well to consider the long throw shaft of Fig. 3.23 (*R*/*M* = 0.97, *O*/*M* = – 0.04) which had a negative overlap near that of Fig. 3.22 shafts (*R*/*M* = 1.10, *O*/*M* = – 0.12). The effect of boring can be demonstrated by comparing crank *a* with cranks *h* and *i*, Fig. 3.22. In this case boring to 76 per cent outside diameter increased torsional strength by 90 per cent. In the long-throw crank the strength in bending was little affected (Fig. 3.23*g, k*), while the strength in torsion was actually decreased by a similar boring operation (*f, l, m*). One reason for the smaller effect of boring may be due to the smaller cheek thickness, *t*/*m* = 0.27, while in figure 3.23 it was 0.45. The heavier cheek probably gave higher fillet stress concentration in the solid shaft, which was relieved by the boring.

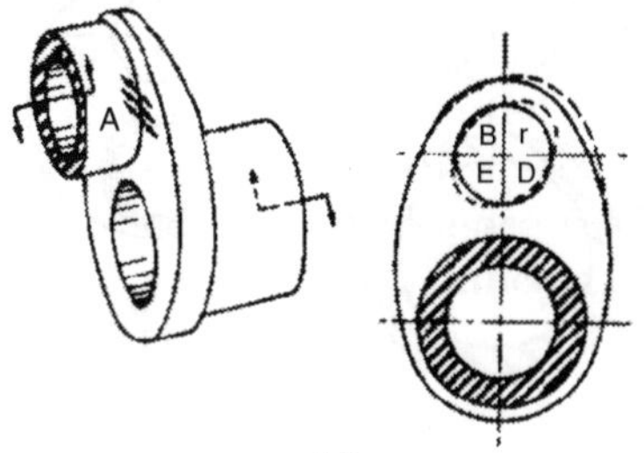

(*a*)
Location of critical stress, distortion
Hollow crank pin subject to torsion

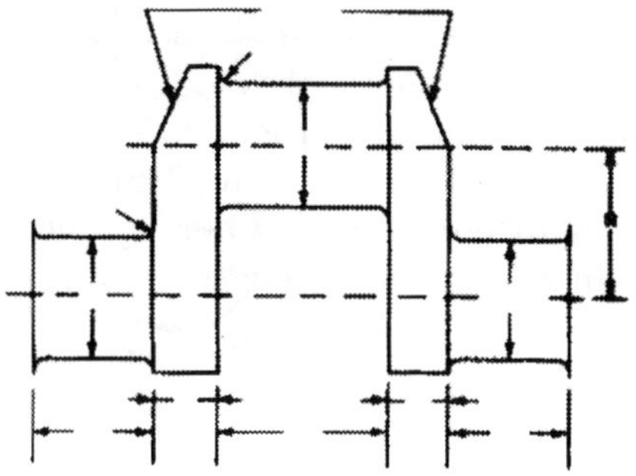

(*b*)

Crank Throw Dimensions

M = 3.75 in. (95 mm), P/M = 0.8, t/M = 0.27, r/M = 0.05, w_c/M = 1.18, w_p/0.79, w_j/M = .57
R/M = 0.7, 0.8, 0.9, corresponding O/M values 0.20, 0.10, – 0.04

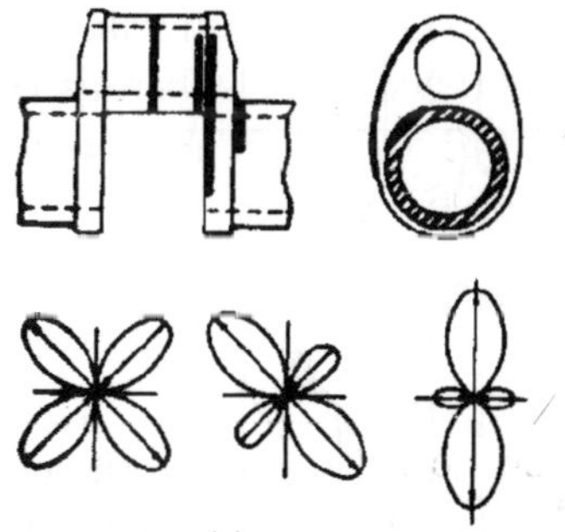

(*c*)

Section G Shear — Section H, L Normal & shear — Section I, K Tension

(*d*)
Location of critical bending stress

(*e*)
Method of applying bending moment

Fig. 3.23. Measured Stresses in Model Crankshafts (ref. Gold and van de Griff, 1984)

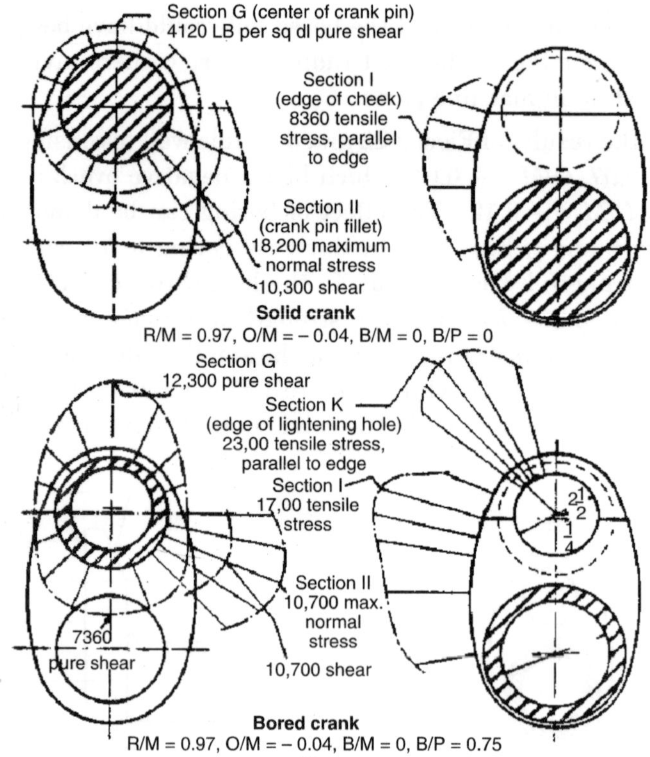

(*f*)

Stresses in solid and bored crank in torsion

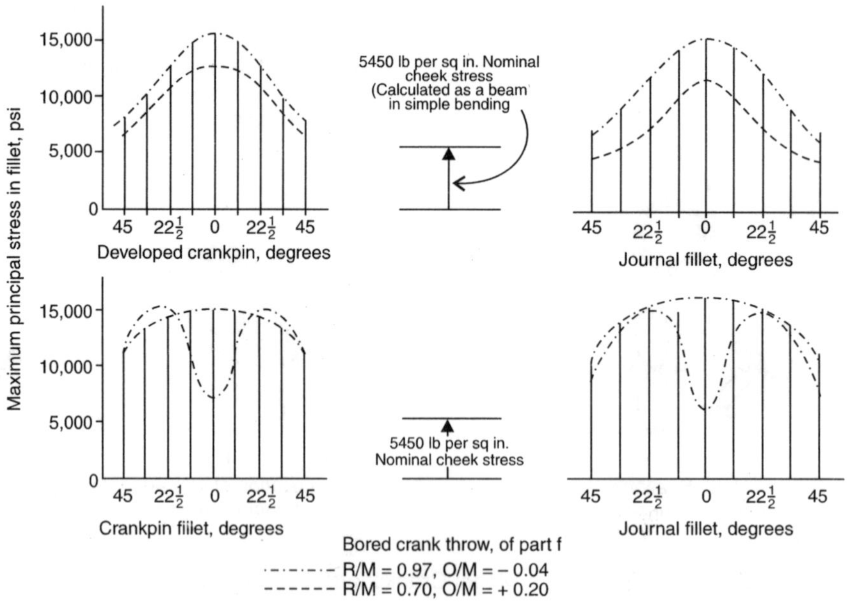

(*g*)

Bending stresses in the solid and bored cranks of part f, with bending moment applied as shown in part e

Fig. 3.23. Continued

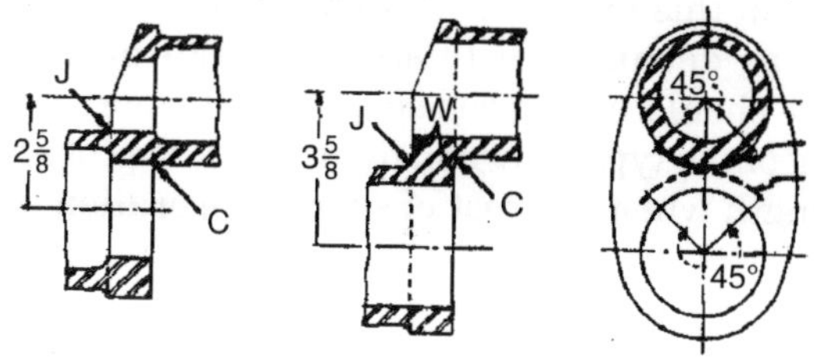

(h)

Location of bending stresses shown in part g.
J = journal fillet, C = crank pin fillet

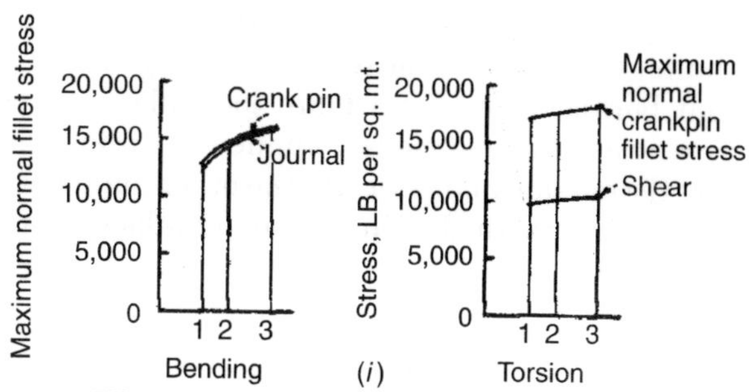

(i)

Effect of varying crank radius of solid crank throw
1. R/M = 0.70, O/M = + 0.20
2. R/M = 0.80, O/M = + 0.10
3. R/M = 0.97, O/M = − 0.04

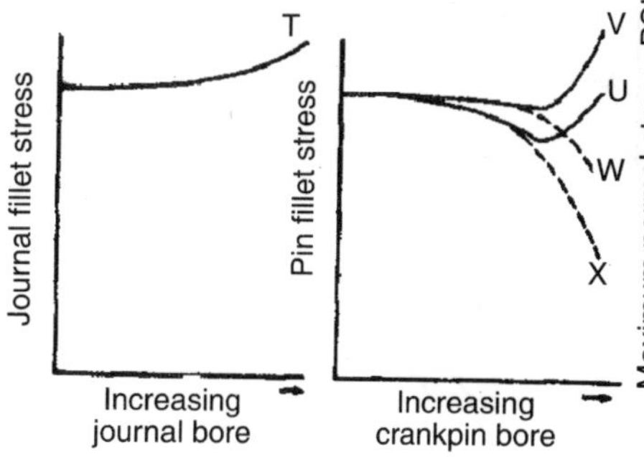

(j)

Effect of lightening holes on journal-fillet stress concentration in bending

T maximum stress, crank pin bore constant
U maximum stress, journal bore constant
V maximum stress, increasing both journal and pin bores
W and X are stresses of point M, part d.
Maximum pin stress occurs at regions NN of part d.

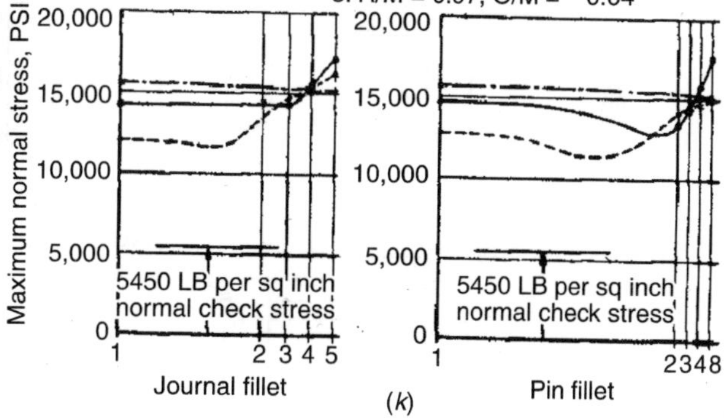

(k)

Effect of boring journal and crankpin on the maximum fillet stresses in bending.

Station	*B/M*	*B/P*
1	0	0
2	0.67	0.50
3	0.70	0.58
4	0.74	0.67
5	0.77	0.75

-·-·-·-·- R/M = 0.97, O/M = − 0.04
——— R/M = 0.80, O/M = + 0.10
- - - - - - R/M = 0.70, O/M = + 0.20

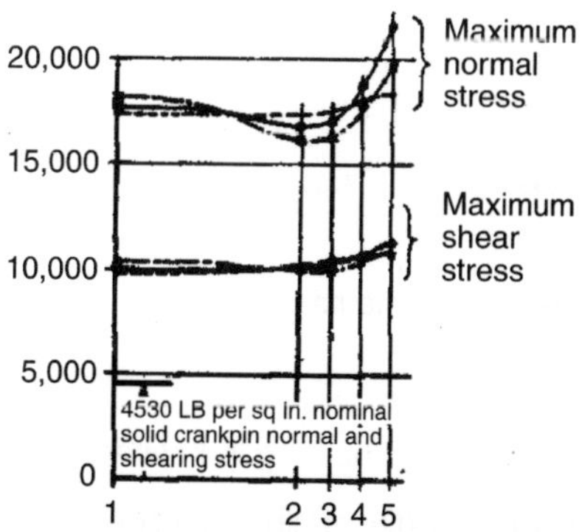

(l)

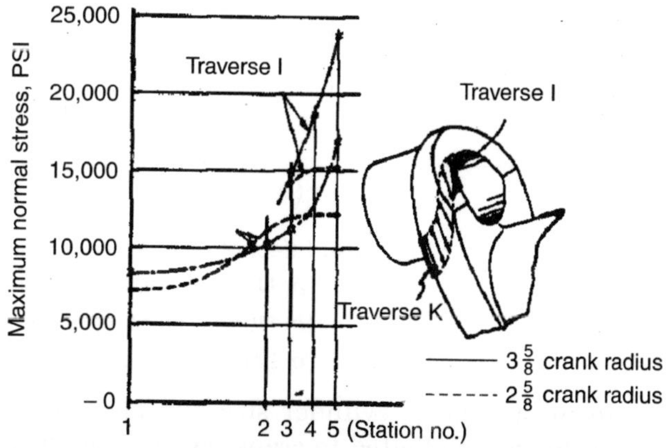

(m)

Fig. 3.23. Measured Stresses in Model Crankshafts (*ref.* Gadd and van Degrift, 1984)

Other considerations in this connection are that all large aircraft engines use bored shafts, which are subjected to very large loads. From this fact alone it seems safe to conclude that boring of pins and main journals need not be detrimental to strength, and probably beneficial when judiciously applied. A uniform bore should not exceed substantially more than half the outside diameter. On the other hand boring of the type shown in figure 3.22*c* is probably an effective method of increasing crankshaft strength when the cost of such an operation can be justified.

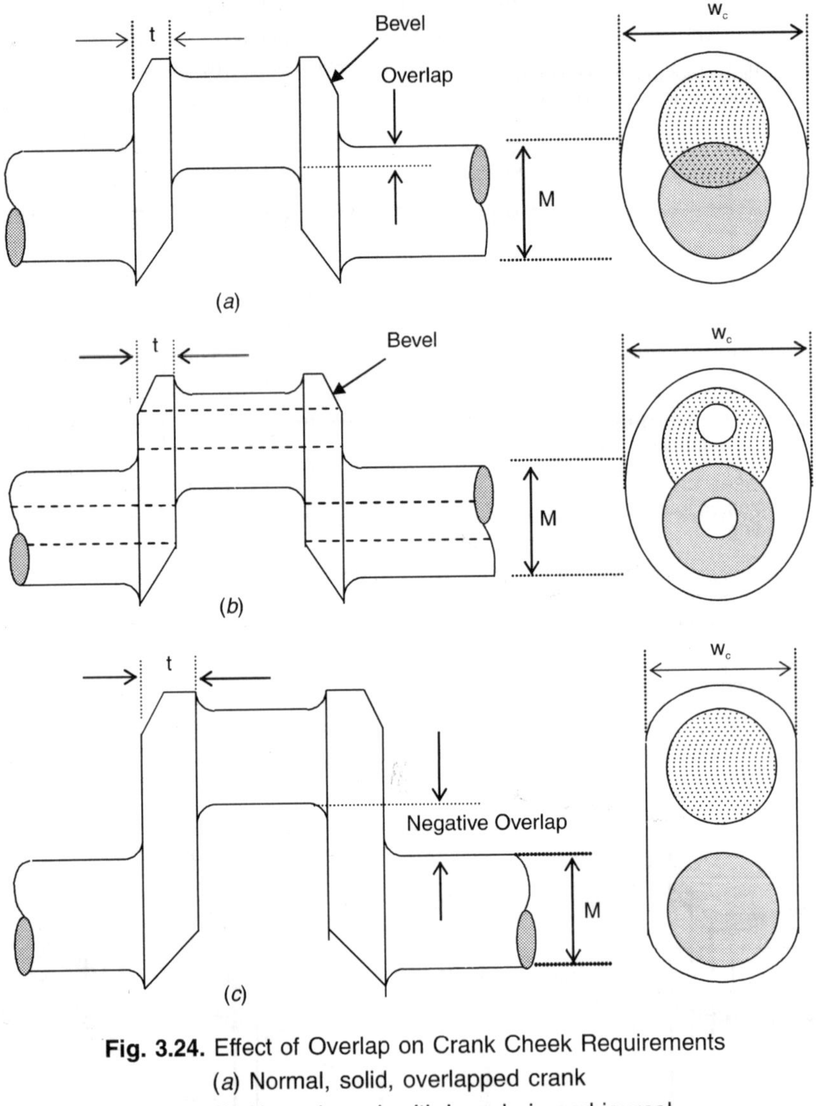

Fig. 3.24. Effect of Overlap on Crank Cheek Requirements
(*a*) Normal, solid, overlapped crank
(*b*) Normal crank with bored pin and journal
(*c*) Long stroke crank, or two cranks with no bearing in between

In most engines cylinder size and spacing determine length of the crankshaft, and whatever length is allotted to crank cheek is at the expense of the length for crank pins and main journals. Where crank pin and journal lengths tend to be short, longitudinal thickness *t* in figure 3.24 must be held to a minimum. Short stroke engines ($S/b < 1.0$) always have a crankshaft with considerable overlap between crank pin and main journal as shown at *a* in Fig. 3.24. Since test results show a beneficial effect of overlap, it would appear that much of the bending and torsional load is transmitted through the overlap area. This observation, together with good design practice, indicates that in well-overlapped solid shafts ratio t/M can be small. With hollow pins or journals the cheek should be designed to distribute the loads around the end periphery of these parts.

For engines with large stroke/bore ratio there is usually a negative overlap. These conditions also apply to cheeks that connect cranks without a main journal between them. The cheek dimensions should still be large in this case as shown in Fig. 3.24 (*c*). Figures 3.25 (*a*) and *b* show interesting methods of reducing fillet stresses in cast crankshafts, especially where the cheek connects two crank pins without a main journal between them. The designs endeavor to transmit the loads from the pin to the journal or from pin to pin in such a way as to spread the load around the pin or the journal end circumference rather than to allow it to concentrate at the fillet in the region of closest approach.

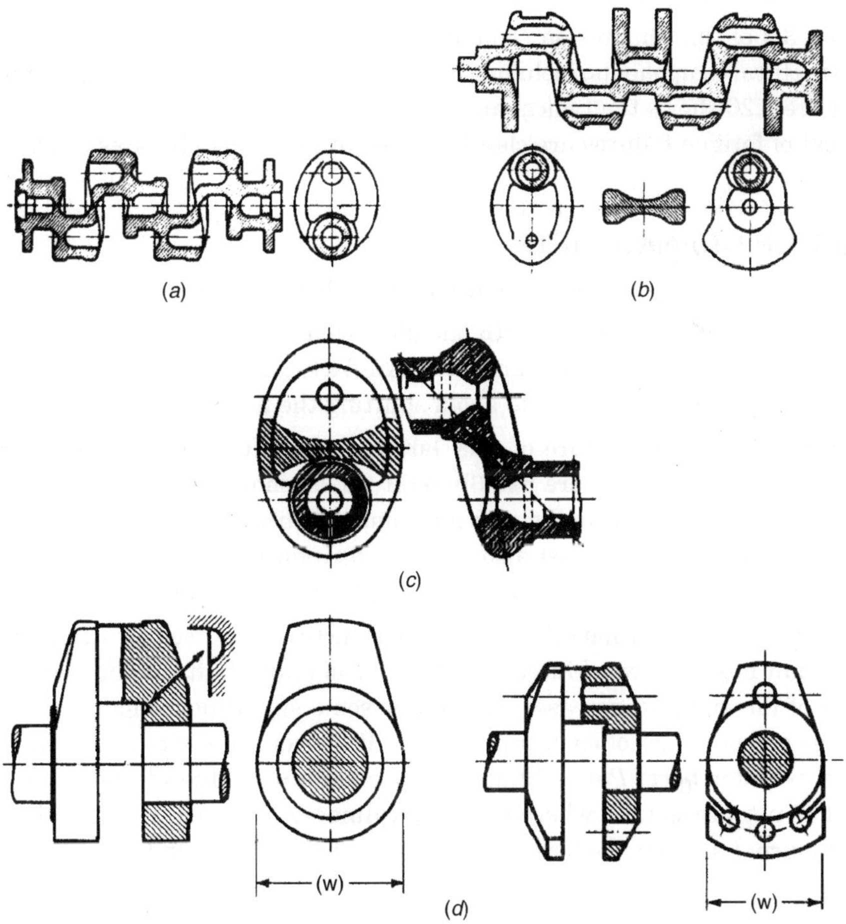

Fig. 3.25. *a, b, c.* Sulzer Designs for Cast Crankshafts, *d.* Built up Shaft for Large Engine (*ref.* Matinaglia, 1982)

Counterweights may be necessary to improve engine balance or to relieve main journal bearing load. An advantage of cast shafts is that counterweights can be included as an integral part of the casting. It is difficult to form adequate counterweights on forged shafts, and welding or bolting is used where such weights are considered essential.

The most common mistakes in built up crankshaft designs involve unnecessary stress concentrations due to oil holes passing close to the fillets, screw threads in stressed areas, sharp corners (either concave or convex) and inadequate fillets. Drastic beveling, specially with bored pins and journals, must also be considered an error. Bearings and journals must always be finished to a relatively smooth surface. Crankshafts for high specific output, including aircraft engines, are generally finished all over to a very smooth surface for improved endurance. In order to reduce wear crank pins and journals are often hardened by locally heat treating the surfaces, either by flame or by electric induction. When this process includes fillets some strengthening of the shaft results. Aircraft and other high cost engines use nitrided pins and journals. Rolling also helps to improve fillet strength. If stresses are not high crank cheeks may be allowed to remain in the as-forged or as-cast condition.

Section 3.3 of this chapter on component materials gave a good description on fatigue failures of structural components, including circumstances under which failures may be expected to occur. Figure 3.26 shows the typical appearance of a fatigue failure in a crankshaft. 'Beach marks' typical of fatigue failures are clearly visible in the crank cheek as a result of reversed cyclic loading.

3.10 BUILT UP CRANKSHAFTS

Shafts built up from separate pieces are used for the following reasons:

(*a*) To accommodate assembly with one piece connecting rods. Many small single crank engines use this arrangement to accommodate anti-friction bearings. Built up shafts have also been used for radial aircraft engines to avoid splitting the master rod.

(*b*) To facilitate manufacture of very large crankshafts. For large engines of the type shown in Fig. 2.19, crankshafts are usually built up, as shown in Fig. 3.25 (*d*), by shrinking heavy crank cheeks over individual crank pins and/or journals. The ratio t/M runs as high as 0.58 for this type. The width w must also be large, as indicate in Fig. 3.25 (*d*).

The principal dimensions of a crankshaft in relation to cylinder bore and stroke are limited by practical considerations of strength and structure. The need for strength and stiffness in the crankshaft has led toward increasing the size of crankshafts as the mean pressure and piston speed of the engines increases. Except for some six cylinder engines, the shafts have main bearings on both sides of each crank throw. If the connecting rod needs to be removable through the cylinder bore, P/b is limited to a maximum value of about 0.6. Where this requirement is not present, or where separate cylinders are used, ratios as high as 1.0 are found. Main journal diameter to bore ratio M/b runs between 0.6 and 1.0.

Fig. 3.26. Fatigue Failure of Automotive Crankshaft
A. Start of fracture in fillet;
B. 'Beach-marks' typical of crack progress under intermittent operation;
C. Line of final sudden failure.
(*Courtesy: Lessel and Associates, Inc.*)

It is remarkable that the range of these ratios is about the same for all classes of engines, and that it is nearly independent of the number and arrangement of cylinders. It is also remarkable that these design ratios are not dependent on engine specific output. As a general rule, pin and journal diameters should be chosen on the high side for engines of high specific output, particularly when cast crankshafts are to be used. From the foregoing discussion, the following recommendations pertaining to the crankshaft design are submitted:

1. Crank pin diameter should be at least 0.6 time the bore. Crank pin length should be enough to accommodate connecting rod bearings of not less than 0.3 times crank pin diameter.
2. Main journal diameter should be larger than crank pin diameter. Main journal length can be as short as 0.3 times the journal diameter when centrifugal crank loads are counterbalanced.
3. Overlap of pins and journals improves crankshaft strength, provided fillets are adequate.
4. Judicious boring of pins and journals tends to reduce stresses.
5. With bored shafts cheeks should be designed to spread the load over the ends of pins and journals. Excessive beveling should be avoided.
6. Fillet radii should be as large as possible (at least 0.05 times the journal diameter), and all sharp corners should be well rounded off. Fillets should have a smooth finish, and may be rolled if possible.
7. The shaft should be as free as possible from holes and keyways, particularly from threaded holes near areas of high stress. All holes should be chamfered at their openings.
8. Where dimensions are not unduly restrictive, cast crankshafts are a satisfactory alternative to forged steel crankshafts.

3.11 CRANK PIN AND JOURNAL BEARINGS

Bearings in the power train are critical elements of any reciprocating machine. Most machines use plain journal bearings. The use of 'anti-friction' ball, roller or needle bearings will be discussed later.

Table 3.4. Bearing & Bushing Material Properties

<table>
<tr><th>Material</th><th>SAE No.</th><th>Nominal Composition</th></tr>
<tr><td>Tin-Base Alloys</td><td>11</td><td>Sn 87.5% Sb 6.75% Cu 5.75%</td></tr>
<tr><td></td><td>12</td><td>Sn 89.0% Sb 7.50% Cu 3.50%</td></tr>
<tr><td colspan="3">Method of Manufacture: Cast on steel, bronze or brass backs, or directly in housing.
Characteristics: Soft, corrosion resistant, moderately fatigue resistant.
Applications: Main & connecting rod bearings, motor bushings. Operate with hard or soft journal.</td></tr>
<tr><td>Lead-Base Alloys</td><td>13</td><td>Sn 6% Sb 10% As-Pb remaining</td></tr>
<tr><td></td><td>14</td><td>Sn 10% Sb 15% As-Pb remaining</td></tr>
<tr><td></td><td>15</td><td>Sn 1% Sb 15% As 1%, Pb remaining</td></tr>
<tr><td colspan="3">Method of Manufacture: SAE 15 cast on steel, others on steel bronze or brass, or directly in housing.
Characteristics: Soft, corrosion resistant, moderately fatigue resistant.
Applications: Main & connecting rod bearings, motor bushings. Operate with hard or soft journal.</td></tr>
<tr><td>Lead-Tin Overlays</td><td>19</td><td>Sn 10% Pb 90%</td></tr>
<tr><td></td><td>190</td><td>Sn 7% Pb 93%</td></tr>
<tr><td colspan="3">Method of Manufacture: Electro-deposited as a thin layer on copper-lead or silver bearing face.
Characteristics: Soft, corrosion resistant, runs against soft shaft throughout coating life.
Applications: Heavy duty high speed main & connecting rod bearings.</td></tr>
<tr><td>Copper-Lead Overlays</td><td>49</td><td>Cu 76% Pb 24%</td></tr>
<tr><td></td><td>48</td><td>Cu 70% Pb 30%</td></tr>
<tr><td></td><td>480</td><td>Cu 65% Pb 35%</td></tr>
<tr><td></td><td>481</td><td>Cu 60% Pb 40%</td></tr>
<tr><td colspan="3">Method of Manufacture: SAE No. 49, 48-Cast or sintered on steel back.
SAE No. 480, 481-Cast on steel back.
Characteristics: Moderately hard; subject to acid attack. Fatigue resistance good to fair. Listed in order of decreasing hardness and fatigue resistance.
Applications: Main & connecting rod bearings. Higher lead alloys may be used unplated against soft shaft, lower lead alloys used against hard shaft, or with overlay against soft shaft.</td></tr>
</table>

Table 3.4. Bearing & Bushing Material Properties *Continued*

Material	*SAE No.*	*Nominal Composition*			
Aluminum-Base Alloys	770	Sn 6.25%	Ni 1%	Cu 1% Al-remaining	
	780	Sn 6%	Ni .5%	Cu 1% Al-remaining Si 1.5%	
	781	Si 4%	Cd 1%	Al-remaining	
Method of Manufacture: 770-Cast in permanent molds, work hardened. 780-Bonded to steel back 781-Bonded to steel back; also as casting or wrought strip. *Characteristics*: Hard, corrosion resistant to oil, extremely fatigue resistant. *Applications*: Main & connecting rod bearings, with suitable overlay.					
Other Copper Based Alloys	795	Cu 90%	Pb 0%	Sn .5%	Zn 9.5%
	791	Cu 88%	Pb 4%	Sn 4%	Zn 4 %
	793	Cu 84%	Pb 8%	Sn 4%	Zn 4%
	798	Cu 84%	Pb 8%	Sn 4%	Zn 4%
	792	Cu 80%	Pb 10%	Sn 10%	Zn 0%
	797	Cu 80%	Pb 10%	Sn 10%	Zn 0 %
	794	Cu 73.5%	Pb 23%	Sn 3.5%	Zn 0%
	799	Cu 73.5%	Pb 23%	Sn 3.5%	Zn 0%
Method of Manufacture: SAE No. 795, 791 - Wrought solid bronze. SAE No. 793, 798–Cast, sintered on steel back. SAE No. 792, 797–Cast, sintered on steel back. SAE No. 794, 799–Cast, sintered on steel back. *Characteristics*: SAE 795–Hard, strong, fatigue resistance good. SAE 791, 793,798–Good shock, load capacity, high temperature resistant. Hard shaft desirable. SAE 792, 797–Maximum shock, load carrying capacity; hard, fatigue, corrosion resistant. SAE 794,799–Improved surface action for higher speeds, less corrosion resistant. *Applications*: SAE 795–Oscillating motion bearings: tie rods, brake shafts. SAE 791–Moderate to high loads. Piston pin bushings, thrust washers. SAE 793, 798–Medium to high loading. Transmission, chassis bushings. SAE 792, 797–Heavy loads, oscillating or rotating motion. Piston pin, steering knuckle, differential axles, thrust washers, wear plates. SAE 794, 799–Intermediate load application for oscillating and rotating motion, rocker arm bushings, transmissions.					

Power train journal bearings consist of an iron or steel journal supported in a bearing of a material usually softer than the journal. Table 3.4 shows materials generally used. Desirable characteristics of the bearing material include:

1. Adequate endurance strength.
2. Deformability, or the ability to deform plastically at points of high pressure, and to conform to the shape of the journal.
3. Embeddability, or the ability to allow particles of foreign material to embed in the bearing material rather than to circulate in the oil film and score or wear the journal and bearing surfaces.
4. Resistance to scoring and seizure, or attaching bearing material to the shaft.
5. Resistance to corrosion, usually from elements in the lubricant.

The motion of the connecting rod calls for an oscillating journal type of bearing; this is also true of link pin bearings of articulated connecting rods. 'Slipper' type of bearings in radial engines also falls in this category. Since oscillating bearings operate at relatively low average surface speeds the problem of wear due to abrasive material is less serious than in rotating bearings. Also, the work done to overcome friction, and hence the heat generated, is less serious. Hence, high endurance strength takes precedence over embeddability for the oscillating bearings. A good bearing bronze, such as SAE 791, or a bearing grade aluminum is satisfactory for these bearings. With aluminum pistons the piston pin can bear directly into the aluminum, and some piston pins bear directly in cast iron pistons. Hard surface pins are desirable. With proper lubrication oscillating bearings can withstand high maximum unit pressure, as evidenced by the fact that their projected area is usually half that of the crank pin bearing carrying nearly the same maximum load.

Crankshaft bearings are among the most critical elements in engine design. Earlier designs and also present low specific output applications use tin base or lead base 'babbitt' material. Such material has all the desirable characteristics of a bearing, except high endurance strength. It may be cast directly into the connecting rod and finished after casting, or cast into removable steel or bronze bushings. It has been superseded mostly for reasons of strength alone, as it is otherwise a nearly ideal bearing material. Materials similar to 'babbitt' called lead-tin overlays are used as a very thin coating on stronger bearing materials, such as copper-lead, silver or aluminum. This overlay acts as a protection against seizure and as an absorber of foreign material smaller than the thickness of the overlay. Copper-lead bearings consist of a spongy matrix of copper with lead particles distributed in the interstices of the spongy structure. This combination was developed in order to combine structural strength of copper with the desirable bearing properties of lead. This material ranks next to 'babbitt' in resistance to scoring and seizure, and has good endurance strength. It is widely used for crank pin and crankshaft bearings. Its principal weakness is susceptibility of lead to corrosion by acid constituents in the lubricant, but can be overcome by using oils with suitable anti-corrosive additives. The use of aluminum has increased rapidly because it ranks high in strength and seizure, but low in deformability and embeddability. Modern methods of high precision manufacture and effective oil filtration have reduced the need for a high degree of these two characteristics. For heavy duty the bearing should consist of a thin layer of aluminum on a steel shell, preferably with a very thin lead-tin overlay.

The load on power section bearings can be plotted against crank angle from the indicator diagram and the inertia forces for any assumed condition of engine speed and load. The loads vary with time in both direction and magnitude. Such calculations give results well within the accuracy of estimates of what a given bearing will carry. Figure 3.27 illustrates graphically an example of a crank pin bearing for a particular engine. The subject of load variation on journal bearing friction has been extensively explored theoretically but does not give satisfactory

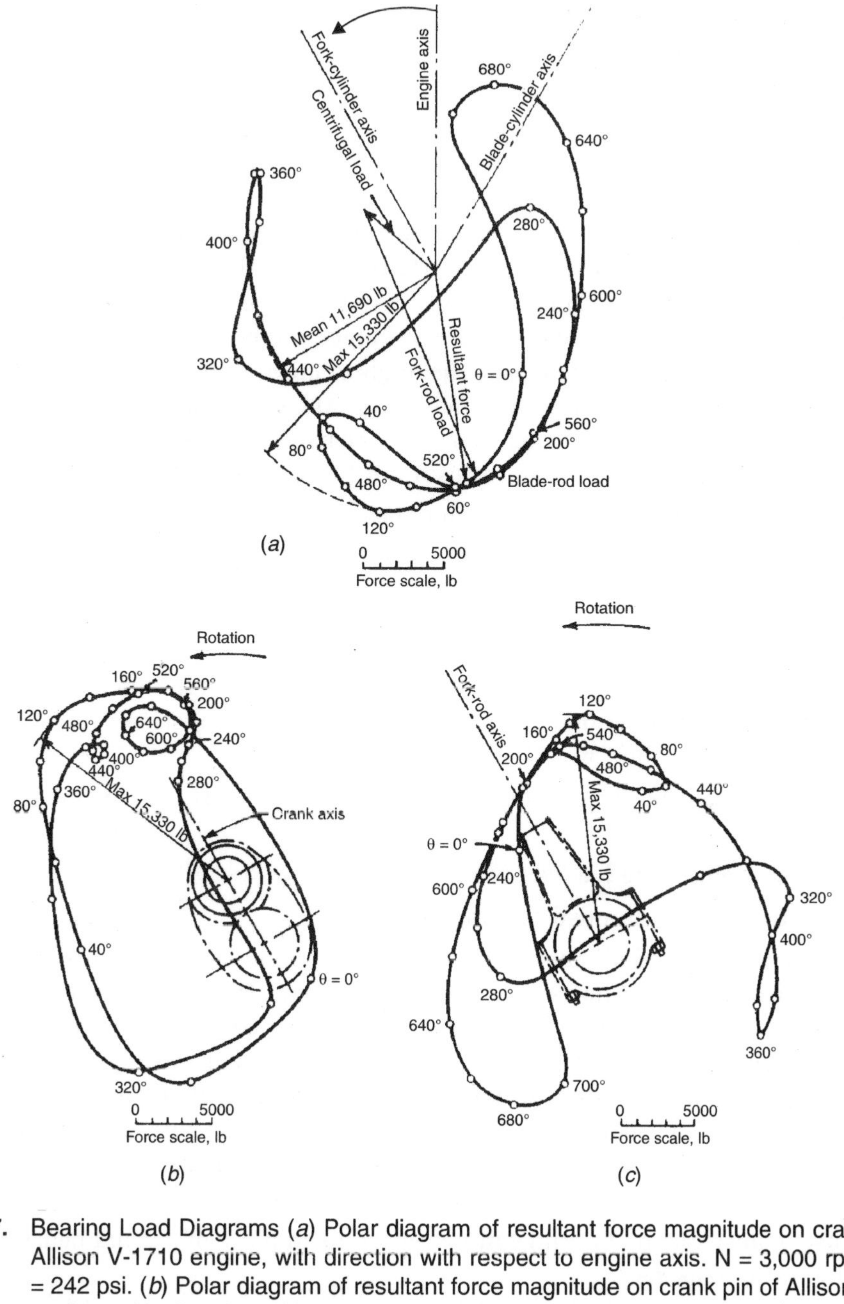

Fig. 3.27. Bearing Load Diagrams (*a*) Polar diagram of resultant force magnitude on crank pin of Allison V-1710 engine, with direction with respect to engine axis. N = 3,000 rpm, IMEP = 242 psi. (*b*) Polar diagram of resultant force magnitude on crank pin of Allison V-1710 engine, with direction with respect to crank axis. N = 3,000 rpm, IMEP = 242 psi. (*c*) Polar diagram of resultant force magnitude on crank pin of Allison V-1710 engine, with direction with respect to fork rod axis. N = 3,000 rpm, IMEP = 242 psi. (*Courtesy: NACA*)

explanation of bearing behavior. For example, theory indicates that when loads vary sinusoidally at a frequency equal to ½ the journal speed the oil film will rupture, with resultant high friction and wear. However, in four stroke internal combustion engines a large component of the load varies at 1/2 journal rpm, yet very heavy loads are carried. Another important aspect is that, in general, the supporting structure is more flexible, and distortions are relatively greater.

The load that may be carried by a journal bearing depends on the characteristics of the bearing as also that of the lubricant. Theoretically, the oil film will not fail if $\mu N/p$ (viscosity × rpm/pressure) does not fall below its critical value. In practice bearing failures are attributable to overloading of the oil film. Empirical formulas for the limit load capacity of bearings that include maximum pressure and surface velocity are widely used by designers. The maximum pressure involves strength, and surface velocity influences the rate of heat generation of the bearing. From considerations of unit pressure it would appear that the longer a bearing the more load it can carry. On the other hand, the longer the bearing the greater the chance of distortion and axial misalignment, as also the angle of misalignment. Thus, it is better to keep length to diameter ratio for crank pin and main journal bearings to less than 1.0, even if space considerations permit a longer bearing. For main bearings with non-counterweighted cranks at nearly the same angle on each side, ratios up to 1.0 may be necessary.

The usual recommendation for crankshaft bearing clearance has been in the neighborhood of 0.001 to 0.002 times the journal diameter. Small clearances are objectionable both because they do not allow enough for distortion and because they have less safety against wear and scoring by foreign particles. Objections to a large clearance include a reduction in the value of the Sommerfeld number and hence a lower factor of safety against oil film breakdown. However, most bearing failures are due to reasons other than direct oil film failure. Since manufacturing tolerances are more or less independent of size, small bearings may have to have larger clearance/diameter ratios than would otherwise be used.

Structural bearing failures (not due to corrosion or wear) is usually due to one or both of the following causes:

- Oil starvation to the loaded area
- Distortion of the intended bearing geometry

The principles of good oil supply are two, introduction of oil at points of minimum load and avoidance of oil grooves or holes in heavily loaded areas. Distortion of the bearing geometry may act to cut down oil supply to loaded areas or to concentrate the load over a smaller area of the bearing. A particularly harmful type of distortion is from misalignment of the axes of bearing and journal, usually due to bending of the crankshaft or engine frame, or both. Maintenance of the cylindrical shape of the bearings, together with proper axial alignment, should be a paramount consideration in the design of connecting rods, crankshafts and engine frames.

The use of anti-friction bearings in power trains in place of journal bearings may be for any of the following reasons:

1. Lubrication: Ball and roller bearings do not require flooding with oil. They are adequately lubricated by a thin mist of oil in the crankcase or by oil mixed with fuel in two cycle engines with crank case compression.

2. Space: For a given load capacity ball and roller bearings can be axially shorter than journal bearings, but are considerably larger in outside diameter.
3. Friction: Use of anti-friction bearings reduces starting friction, and may also have a favorable effect on overall running friction.

Disadvantages of ball and roller bearings include the problem of assembly on a crankshaft, and the fact that bearing failure is likely to be more sudden and disastrous than in journal bearings. Ball bearings are specially useful for 'thrust' loads and are so used in aircraft engines. Roller bearings lend themselves to direct use against hardened cylindrical shafts, and so are used on crankshafts in preference to ball bearings. The problem of assembling roller bearings on a crankshaft may be solved by one of the following methods:

1. For crank pin bearings, since there is little side load, the rod may be split and the rollers allowed to bear directly in the rod material.
2. The crankshaft may be separable to allow assembly.
3. Rollers may be assembled piece by piece through a suitable slot, with the cage added as a final operation.

For single crank machines, main bearings may be directly assembled over the end of the crankshaft.

Needle bearings are roller bearings where diameter to length ratio of the rollers is small, as the name suggests. They require more lubrication than conventional ball or roller bearings. They are used in piston pin ends of connecting rods, mostly for small two cycle engines. They also serve as crank pin bearings for some small engines.

3.12 LUBRICATION

Lubrication is an important parameter in reciprocating machinery performance, its function being to dissipate frictional heat, reduce wear on sliding surfaces, reduce internal leakage and protect parts from corrosion. It can also act to flush away wear products, contaminants in the air and moisture. Modern air compressors for processing (food preparation, for example) dispense with oil in the compression chamber so as to avoid having to separate the oil carried over with the air, thus avoiding the risk of contamination at source. Lubrication of parts away from the compression chamber is necessary. The table below gives some factors affecting lubrication in the cylinders of reciprocating machines.

The choice of lubricating system and the lubricant itself are important, because lubrication requirements differ with different types of machinery, as well as in different sections of the machine. Most modern equipment rely on pressure rather than splash lubrication, except for small single cylinder units. Problems are seldom encountered if the manufacturer's recommendations regarding lubricant type, setting on the pressure relief valve on the oil pump and maintenance schedule for oil and filter changes are observed.

Experience shows that certain lubrication systems are best for specific types of machines, although alternative choices are available for the designer. The most common systems are:

Cause	*Effect*
Deposits on valve seats	Increase in discharge temperature
Deposits in discharge passages	Increase in discharge temperature, pressure
High discharge temperature	Increased oil oxidation and deposits
Contamination from hard particles	Abraded cylinder surfaces, interference with valve and piston seating
Excess oil feed	Deposit formation
Moisture, condensation from	Displaced oil film on metallic surfaces cylinder over cooling
Rust	Increased wear, oil oxidation
Water droplets entrained in suction gas between stages	Excess wash off lubricant in cylinder, excessive wear, rust
Oil droplets entrained in suction gas between stages	Oil exposed to oxidizing conditions, causing rapid deposit build-up
High cylinder pressure on pistonrings	Increased rubbing pressure, ruptured oil film

- splash lubrication
- flood lubrication (gravity)
- force feed lubrication
- injection lubrication

The simplest arrangement is splash lubrication, shown in Fig. 3.28, where a projection on the connecting rod dips into oil held in the crankcase, which acts as an oil reservoir and must be closed, with the exception of a small breather hole. Lubricant is dispersed mainly in the form of an oil mist, reaching the bearings, sliding surfaces and cylinders.

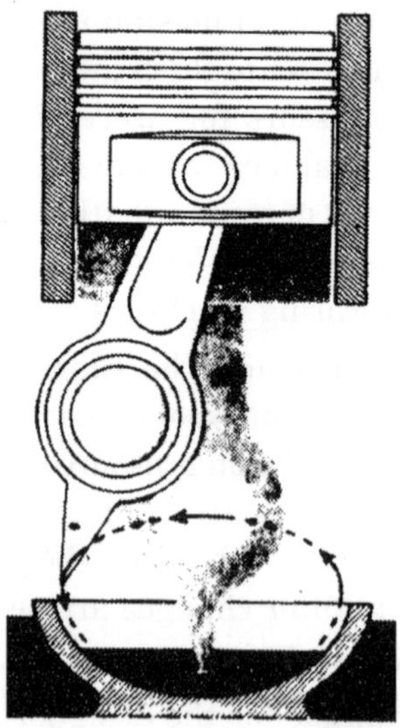

Fig. 3.28. Splash Lubrication with Controlled Feed (*Courtesy:* Atlas Copco)

For best control the splash oil should be taken from a trough, or by using some other means to keep the crank case oil at a constant level. Water condensate or other fluid at the bottom of the crankcase should be drawn off at specified intervals, either manually or automatically. The method has several disadvantages, although relatively simple:

- the lubricant cannot be filtered, and its life is limited
- lubrication at start up may be marginal
- bearing film is thin and is not supplied under pressure
- the level of oil cannot be guaranteed unless maintenance schedule is followed strictly, and the machine is kept properly leveled at all times.

Flood lubrication is widely used on horizontal reciprocating machinery for lubricating the bearings and cross head. The oil flow is under the pressure of gravity after being elevated by a disc, oil ring chain or collar on the crankshaft which lifts it to a trough with distribution tubes, or to the upper part of the journals. This provides better control of the flow, but the pressure is low and generally not sufficient for modern higher speed machines.

Fig. 3.29. Six Cylinder W-type Single Acting Compressor with Force Feed Lubrication (*Courtesy:* Atlas Copco)

Force feed lubrication using a circulatory system is normally the preferred system; it is used widely in compressor system lubrication. Feed is taken from an oil reservoir by a plunger or gear pump (Fig. 3.29). The figure also shows different lubrication flows for a typical reciprocating machine. An externally mounted gear pump may be used with a pressure relief valve to control oil pressure. Oil is drawn into the pump through a submerged strainer, then passes through a pressure filter with a disposable element. Oil is then fed to the main bearings, through the connecting rod to the small end bearing, then drains back into the crank case. Actual circulation may be by gravity when the oil is pumped to an overhead reservoir, or may be directly fed from the pump under pressure through pipes to the different lubrication points. Oil pressure is maintained at a constant value by a spring-loaded over flow valve. Piston type lubricators are used for lubrication of cylinders of machines with cross heads. Figure 3.30 shows a typical oil circulation system that includes provision for cooling the oil.

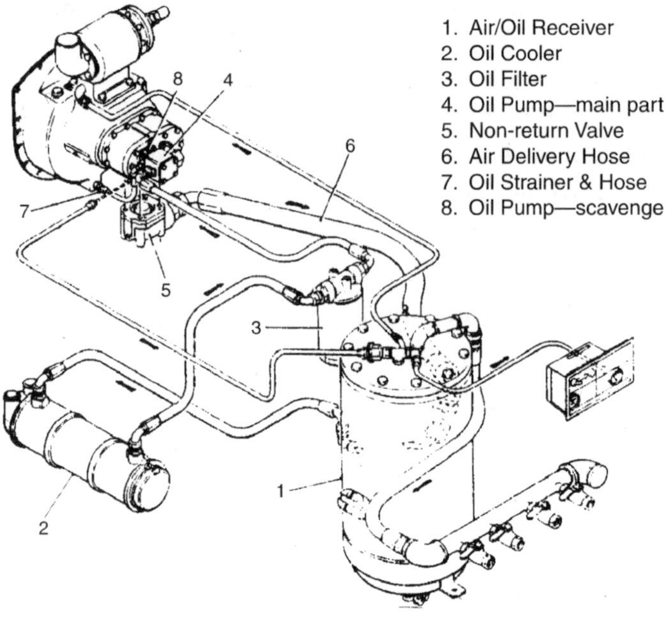

Fig. 3.30. Lubrication System (*Courtesy:* Atlas Copco)

Sight feed drip oilers are occasionally used to lubricate small parts. They are similar to cylinder lubricators and can be controlled within close limits if maintained properly. Siphon wick oilers may be used when undue variation in oil level in the container is not permitted. Grease lubrication may be used at certain points, for example, to lubricate rolling bearings. In this case the lubricating points are fitted with grease nipples or screw-down oil cups.

Internal compressor fires and explosions sometimes occur in reciprocating compressors. The phenomenon is not completely understood. It appears to be associated with the use of an incorrect oil, running the machine with low oil level or not keeping strictly with right maintenance procedures. Carbon deposit build up may serve as a starting point for fires. Presence of aluminum or magnesium in the system can contribute to the danger. Cleaning and strict attention to the manufacturer's recommendations usually alleviates the problem.

3.13 CRANKCASE

Materials for the crank case, or the machine frame as it is called in many reciprocating machinery applications, include cast iron, cast aluminum, forged aluminum and forged steel, usually welded in the latter case. Which type is appropriate depends principally on considerations of machine type and manufacturing cost versus the importance of weight saving. For large diesel and gas engines the alternate materials for crank case structures are cast iron and welded steel. For stationary purposes where weight is not of paramount importance the tendency is to use cast iron. Welded steel crank cases are generally used in high output locomotive engines and for very large marine engines.

General principles to be followed in the design of cast crank cases include the following:

1. Thick sections cool more slowly than thin sections, and abrupt changes in section thickness should be scrupulously avoided in order to minimize shrinkage strains and cracks.

2. Avoid large plane sections. Curve or dish all large thin sections.
3. The custom of adding ribs or thin webs to improve strength usually has the opposite effect. For a given moment the stress in a beam in bending is proportional to c/I where c is the maximum distance from neutral axis to the outer fibers of the beam and I is its section modulus. Adding widely spaced ribs has little effect on I but increases c by the depth of the rib. Even if the ribs are closely spaced, ribbing could increase the stress. A finite element type of stress evaluation is necessary to determine any beneficial advantages of the ribs if the size and location of the ribs are judiciously selected. As a rule, however, ribs do not help to reduce overturning moment loads; in the absence of adequate fillets where the rib meets the base structure, stress concentration may be high where a fatigue failure may initiate.

The heaviest structural load on the crankcase is mostly the tensile load from cylinder head to crankshaft main bearings, caused by maximum cylinder pressure. In small en-bloc machines this load is usually taken through the casting itself. In larger highly stressed in-line engines the tensile load is often taken up by long bolts extending from near the top of the cylinder barrels to the bottom of the frame. Alternatively, this load may be taken up by long studs from cylinder head to crank case. In V engines, in addition to the tensile load, there are also large side loads on the main bearings when cylinder pressure is high. In small gasoline engines the main block casting absorbs this load. However, when cylinder pressures are very high, as in diesel engines, the crank case may need reinforcement by horizontal bolts or studs at the level of the main bearings. An interesting case of reinforcement by two bolts and a stud from cylinder head to cylinder head is used in some opposed cylinder engines.

An important factor in crank case design is the question of accessibility of moving parts for inspection, repair and replacement. The solution of this problem depends on several factors, including cylinder arrangement, machine mounting and surroundings, as also methods of repair and maintenance.

En-bloc machines generally require that pistons and connecting rods be removable through the cylinder bores. Accessibility to crankshaft and main bearings in automotive engines is generally provided for by a removable cover at the bottom of the engine and by main bearing caps that are removable from below. Stationary and marine engines are usually not accessible from below, and with such engines the usual practice is to provide main bearing caps bolted on top of the crankshaft and accessible through side plates. Examples are shown in figures 2.18 and 2.19. In the engine and compressor unit of figure 2.17 the crankshaft is removable from the side. In all these machines the main bearing shells are usually replaceable by turning the lower bearing shell around to the top or side, without removing the shaft.

Military engines from tanks and other vehicles are usually removed as a whole from the vehicle for repairs requiring major parts. This system is also generally used for locomotive engines and fleets of trucks and busses.

3.14 VALVES

The poppet valve is used extensively in four-cycle engines, replacing sleeve and piston valves because of their high cost, friction and their adverse effect on oil consumption. Figure 3.31 is presented to show the different parts in a typical valve gear assembly. The same disadvantages apply to a great number of rotary and slide valves, perhaps to a larger extent. The advantages of poppet valves over other types include:

- It provides larger valve flow area to piston area.
- When properly designed it has excellent flow coefficients.
- The manufacturing cost of a poppet valve system is lower.
- Frictional losses are lower, and requires less lubrication.

Its main drawback is that cooling of exhaust valves can be a problem. To achieve satisfactory results the valve design must address this issue, and must provide for gas flow, structural strength, lubrication and wear. It must also have provision for replacement and repair. For four cycle engines the inlet valve Mach index is critical, and is defined as:

$$Z = \frac{A_p}{A_i} \frac{s}{C_i a} \tag{3.2}$$

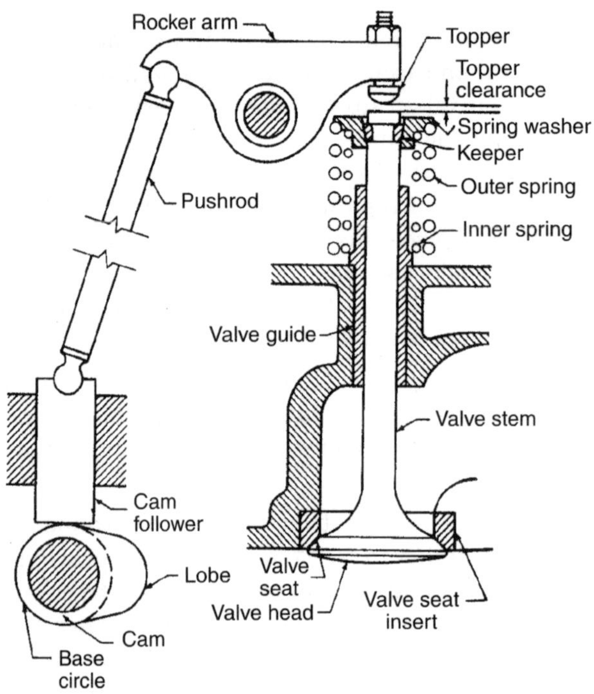

Fig. 3.31. Poppet Valve Assembly (*ref.* Taylor, 1985)

where A_p = piston area

A_i = inlet valve nominal area, $\pi D^2/4$ times number of valves

D = valve outside diameter

s = mean piston speed

a = sound velocity of inlet gas

C_i = mean valve flow coefficient based on valve area A_i, measured experimentally

Exhaust valve flow capacity can be less than the inlet valve capacity. Typically the ratio of exhaust valve to inlet valve flow capacity is 0.7 to 0.75. Valve capacity is also dependent on cylinder head design. Valve flow capacity with flat cylinder heads can be adequate for all but exceptionally high piston speeds. The limiting speed is calculated for Z = 0.6, sound velocity a

of 1200 ft/second and a value for $C_i = 0.35$. The assumed flow coefficient can be achieved with a lift-diameter ratio of 0.25. Wedge shaped cylinder heads have the same valve area limitations as those of two-valve flat heads, provided squish area requirements do not encroach upon the area available for valves. Where maximum valve capacity is required, as in racing engines, the cylinder head can be domed. The disadvantages of doming include more complex valve gear geometry and higher manufacturing cost.

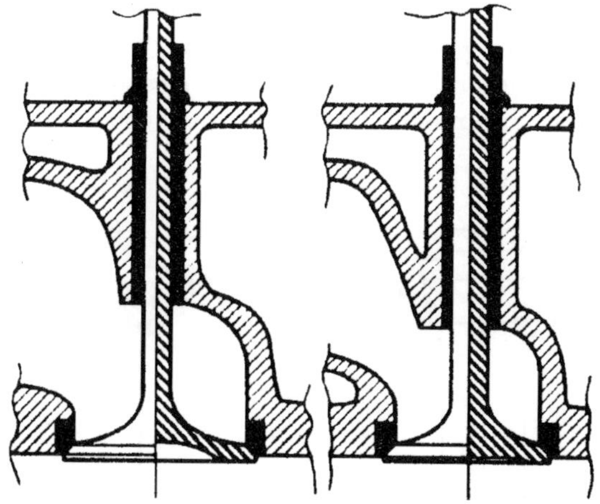

(*a*) Inlet valve

1. Emphasis on high flow capacity
2. Diameter = 1.15 × exhaust valve dia.
3. Small stem diameter (.15 × D_1)
4. Free length of stem long
5. Port has generous flow area

(*b*) Exhaust valve

1. Emphasis on cooling
2. Large stem diameter ($D_c/4$)
3. Free length of stem short
4. Coolant passages all around seat and guide

Fig. 3.32. Well Designed Poppet Valves (*ref.* Cherrie, 1983)

Lifting the valve more than one quarter of its diameter gives diminishing returns in flow capacity. Since valve gear load increases as the square of lift, a reasonable compromise between maximum flow capacity and valve gear load is required. There may also be a problem of interference between valves and piston at top dead center. With multiple valves the valve area to piston area ratio is larger than for two valves in a flat cylinder head. In addition to improved flow capacity, compelling reasons for using multiple valves include reduction of temperature and thermal stresses and reduced valve gear stresses at a given piston speed. In most cases where it is advisable to use multiple valves, specially for large cylinders, the four valve design is the best choice.

Cooling of exhaust valves has been a difficult problem ever since the advent of the internal combustion engine. The heat receiving surface is much larger than the heat rejection surface, which consists only of the stem and seat. Also, both the mean gas temperature and the gas velocity relative to the surface are higher than at any other point in the engine. The solution is to use improved valve materials that can withstand high temperatures without failure due to breakage, warping, burning or excessive wear. Most engines use austenitic steels. However, even with the best of steels the question of cooling and heat flow must receive careful attention during the design phase. When the valve is closed most of the heat flows to the coolant through the valve seat. But when it is open, the only path for the heat to flow to the engine coolant is through the stem and valve guide, obviously a poor one.

Typical failures of exhaust valves start with leakage due to poor seating. This may be due to warping of the valve, seat, or valve guide support structure, or due to accumulation of deposits on the seat. Severe leakage results in over heating the valve and ultimately in complete failure. Failure may be by burning and/or breakage of the stem just above the head. Failure may also be due to high seating velocity because of excessive engine speed, improperly designed valve gear or excessive valve clearance. Another possible point of failure is at the grooves for the keeper, that is, at the anchorage of the spring support. These grooves should be of minimum depth, carefully filleted and smoothly finished. As in the case of threads, rolling of grooves is to be preferred over machining. Figures 3.32 and 3.33 show good and poor valve stem and spring anchorage designs.

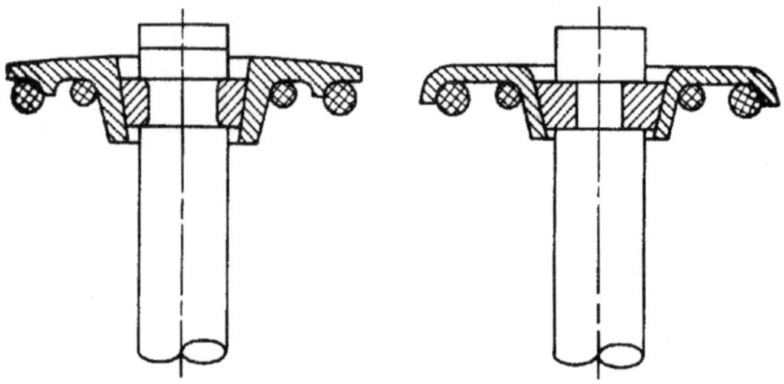

Good practice	Poor practice
Maximum neck diameter	Neck diameter too small
Well filleted neck	No fillets in neck
Rolled groove	Machine finished groove
Sturdy washer	Washer too thin
Springs centered on inside	Spring centered on outside
Hard button on top	Soft top

Fig. 3.33. Good and Poor Practice in Valve Spring Retainer Design (*ref.* Cherrie, 1983)

The practice of enclosing the valves and valve gears and to circulate oil over valve spring ends is extensive. A difficult problem is to provide enough lubrication for the valve stems without involving extensive oil consumption. An oil seal ring at the top of the valve guide minimizes oil loss through the guides.

3.15 EXAMPLE PROBLEMS

Example Problem 3.1: An automobile engine has the following characteristics: 8 cylinders, *V* configuration, 4 in. bore, 3.5 in. stroke, maximum rating 250 horsepower at 5,000 rpm, compression ratio 7.2, weight 550 lbs, overall length 38 in., width 30 in., height 30 in. By using the same detail design in a 12 cylinder *V* engine with the same maximum output at the same stress level, compute the weight and dimensions. Assume the *V* angle remains at 90°.

Solution: At the same stress level the 12-cylinder engine will operate at the same mep and piston speed, and therefore will have the same piston area. Bore = $4\sqrt{(8/12)}$ = 3.266 in. Weight is approximately proportional to displacement, which is proportional to piston area times the stroke. Also, bore diameter is proportional to stroke length.

$$\text{Weight} = 550 \times 3.266/4.0 = \mathbf{449\ lbs.}$$

Width and height will be proportional to the bore:

$$\text{Width} = (30 \text{ in.}) \times 3.26/4.0 = \mathbf{24.5 \text{ in.}}$$

$$\text{Height} = (30 \text{ in.}) \times 3.26/4.0 = \mathbf{24.5 \text{ in.}}$$

Length will be proportional to number of cylinders times bore:

$$\text{Length} = (38 \text{ in.}) \times (12/8) \times (3.26/4.0) = \mathbf{46.5 \text{ in.}}$$

Example Problem 3.2: The above example was solved on the assumption of no change in compression ratio. With the smaller bore the compression ratio could be changed from 7.2 to 7.9, with a corresponding increase in indicated efficiency from a fuel air cycle value of 0.34 to 0.35. Thus, with the same fuel the 12-cylinder engine could give greater power by a factor of 0.35/0.34 = 1.03, or its bore could be reduced by $1/\sqrt{1.03} = 0.985$ to $0.985 \times 3.26 = 3.21$ in. The resultant weights and dimensions would be:

$$\text{Width} = (30 \text{ in.}) \times 3.21/4.0 = \mathbf{24.1 \text{ in.}}$$

$$\text{Height} = (30 \text{ in.}) \times 3.21/4.0 = \mathbf{24.1 \text{ in.}}$$

$$\text{Length} = (38 \text{ in.}) \times (12/8) \times (3.21/4.0) = \mathbf{45.74 \text{ in.}}$$

$$\text{Weight} = (550 \text{ lbs.}) \times 3.21/4.0 = \mathbf{441.4 \text{ lbs.}}$$

Example Problem 3.3: Estimate the weight, length and relative fuel economy of 6-cylinder and 4-cylinder engines to give the same output as the 8-cylinder engine of example 3.1, using the same fuel.

Solution: If the engines run at the same compression ratio and piston speed the cylinder bores required will be:

$$4\sqrt{(8/6)} = \mathbf{4.62 \text{ in.}} \text{ for 6 cylinder engine}$$

$$4\sqrt{(8/4)} = \mathbf{5.66 \text{ in.}} \text{ for 4 cylinder engine}$$

Compression ratio will be limited by engine knock considerations. For these bore sizes, the compression ratios will be assumed to be 6.7 and 5.8, and the corresponding fuel air cycle efficiencies, 0.33 and 0.315. To equal the power of the 8-cylinder engine, the bores will have to be enlarged so that the product of piston area and efficiency is the same as that of the 8-cylinder machine.

For the 6-cylinder engine:

$$4.62\sqrt{(0.35/0.33)} = 4.76 \text{ in.}$$

Knock limited compression ratio $r = 6.6$ at this bore.

Fuel air efficiency at this compression ratio is unaffected, remaining at 0.33.

For the 4-cylinder engine:

$$5.65\sqrt{(0.35/0.315)} = 5.96 \text{ in.}$$

Knock limited compression ratio $r = 5.6$.

Fuel air efficiency at this compression ratio is 0.31, so a further enlargement of the bore is required.

$5.95\sqrt{(0.315/0.31)} = 6.0$ in., at which bore the knock limited compression ratio.

$r = 5.6$ and fuel air cycle efficiency is 0.31 as required.

Example Problem 3.4: In figure # 3.34 combination K and M represents a machine with force $P_o \sin(\omega t)$ acting on it, resulting in unacceptable vibrations in the machine. To remedy the situation a dynamic absorber in the form of a comparatively smaller vibratory system k, m is attached to the main mass. $Mg = 10$ lbs, $mg = 1$ lbs, $P_o = 10$ lbs, and $K = 102$ lbs/in. For the first attempt $k = 10.2$ lbs/in. Find the maximum deflection amplitude of the larger mass.

Solution: In the case of a dynamic vibration absorber of the type under consideration the deflection of the main mass x_1 in terms of $x_{st} = P_0/K$ is given by:

$$x_1/x_{st} = \sqrt{\{1 + (2m/M)\}}$$

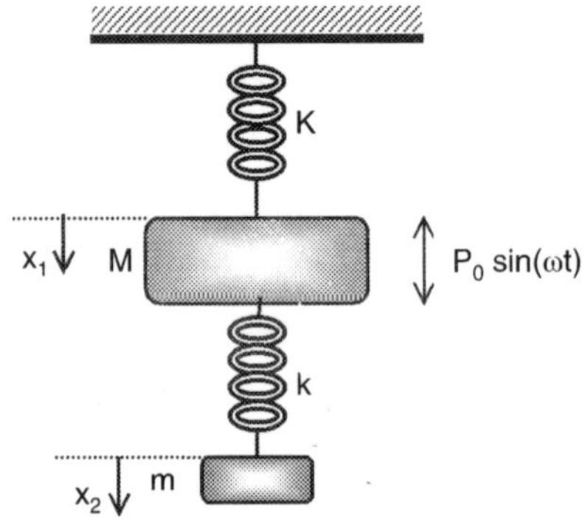

Fig. 3.34. Dynamic Vibration Absorber

Here $x_{st} = P_0/K = 10/102 = 0.98$ in. Then:

$$x_1 = 0.098 \times \sqrt{(1 + 0.2)} = \mathbf{0.107\ in.}$$

Example Problem 3.5: The crankshaft speed of the slider crank mechanism of a single cylinder 4-stroke internal combustion engine shown in figure 3.35 is 3,000 rpm. Using the data given, determine the loads on the mechanism when the crank angle $\theta_2 = 60°$, and calculate the forces transmitted to bearings located at the piston wrist pin, crank pin and the main bearings. Also calculate the crankshaft torque.

Solution: For crank angle $\theta_2 = 60°$ the engine is going through the power stroke, when the gas pressure is 200 psi. Corresponding gas load on the piston (pressure × cylinder bore area) $P =$ 1410 lbs. Inertia force F_{04}(mass M_4 multiplied by acceleration A_b) also acts on the piston, equaling 383 lbs. Inertia force F_{03} of the connecting rod has the magnitude M_3A_{g3} (= 1230 lbs) arising from acceleration of the mass center of the rod, with a line of action offset by e_3 because of the angular acceleration α_3. Component weights are assumed small compared with operating loads.

In Fig. 3.36 the force analysis of the mechanism is shown, in which P, F_{04} and F_{03} are the known loads on the linkage. Superposition of forces is used to determine the unknown forces. In Fig. 3.36 forces in the linkage due to loads P and F_{04} on the piston are determined, and in Fig. 3.37 forces due to F_{03} on the connecting rod are determined. Finally, the resultant forces are determined by superposition as shown in Fig. 3.38. Friction forces are not included, since for well lubricated bearings and cylinder walls they are negligible.

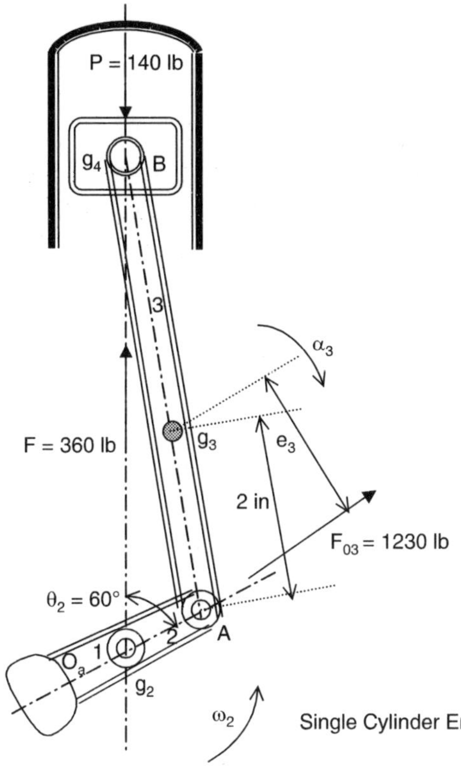

Engine data

Crankshaft speed ; 3,000 rpm
Stroke : 4 inch
Crank Length : 2 inch
Piston weight : 2 lbs.
Piston area : 7.05 inch^2
Connecting rod weight : 3 lbs.
Connecting rod length : 8 inch
Connecting rod moment of inertia I_3 : 0.0075 lb-ft-sec^2

Single Cylinder Engine Slider Crank Mechanism

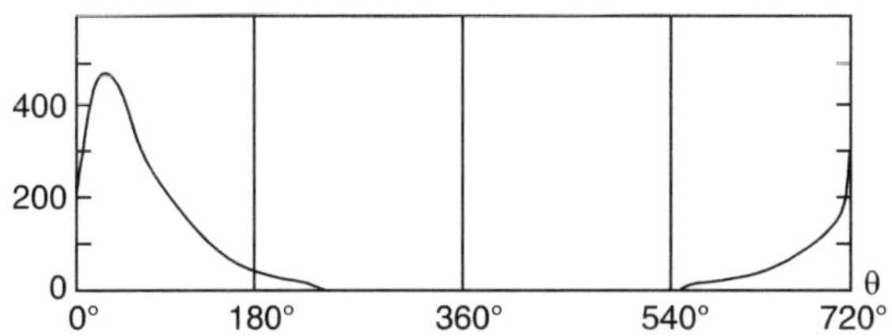

Gas Pressure/Crank Angle Diagram

Gas force on piston

$P = pA_p = 200 \times 7.05$
$= 1410$ lbs.

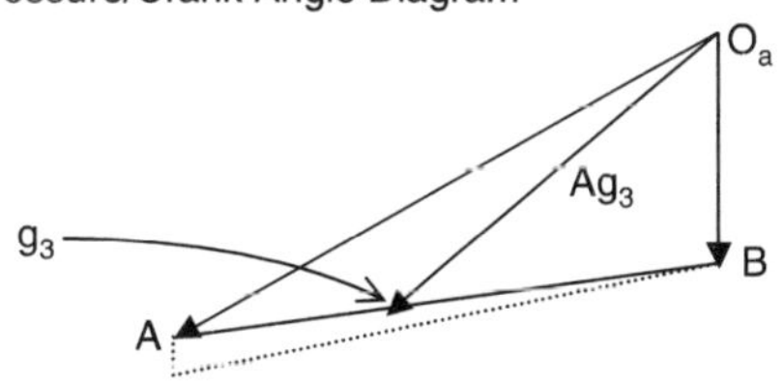

Acceleration Polygon for connecting Rod

Fig. 3.35

Acceleration and Inertia Force Calculations

(From Acceleration Polygon of Fig. 3.30)

$$O_aA = A_A = (2\pi \times 3000/60)^2 \times 2/12 = 16449 \text{ ft/sec}^2$$

$$O_aB = A_B = 16449 \times [\cos 60 + (2/8) \cos 120] = 6168 \text{ ft/sec}^2$$

$$O_a g_3 = A_{g3} = 13200 \text{ ft/sec}^2$$
$$\alpha_3 = 21900 \text{ rad/sec}^2$$
$$F_{04} = M_4 A_B = W_4 A_B / g = 2 \times 6168/32.2 = 383 \text{ lbs.}$$
$$F_{03} = M_3 A_{g3} = W_3 A_{g3}/g = 3 \times 13200/32.2 = 1230 \text{ lbs.}$$
$$e_3 = I_3 \alpha_3 / F_{03} = (0.0075)(21900)(12)/1230 = 1.60 \text{ in.}$$

In Fig. 3.36 F_4 is the resultant of collinear forces P and F_{04}. Beginning with the free body diagram of the piston three concurrent forces are shown at B. Direction of connecting rod force F_{34}' on the piston is known because only two forces act on link 3. Direction of cylinder wall force F_{14}' on the side of the piston is normal to the wall when friction is not present, and the line of action of F_{14}' is through the point of concurrence at B. Beginning with the known force F_4, the equilibrium force polygon is constructed to determine the magnitude and direction of forces F_{34}' and F_{14}'. The two collinear forces F_{43}' and F_{23}' on the free body of the connecting rod are equal in magnitude to F_{34}' of the polygon. Also, the two parallel but non-collinear forces F_{32}' and F_{12}' on the free body of the crank are equal in magnitude to F_{34}'. Thus, all unknown forces are obtainable from one force polygon. As shown on the free body of the crankshaft, torque T_s' at O_2 is in counterclockwise equilibrium to the couple formed by F_{32}' and F_{12}'. From Fig. 3.37 the known force on link is F_{03} of the connecting rod. By isolating the connecting rod as a free body as shown, three concurrent forces are noted. The direction of the piston force F_{43}' on the rod is known since only two forces act on the piston, one of which is normal to the piston side. Direction of the crank force F_{23}'' on the rod is through point k of concurrence, determined from the intersection of lines of action of F_{03} and F_{43}''. Construction of the equilibrium force polygon determines the magnitude and sense of F_{43}'' and F_{23}''. The two non-collinear forces F_{32}'' and F_{12}'' on the free body a large force, because the moment arm very nearly acts through crank pin center at A.

The above process uses a graphical approach. The polygon of Fig. 3.37 involving forces F_4, F_{14}' and F_{34}' could also be calculated analytically because the direction of the forces can be written in unit form. The polygon of forces F_{03}, F_{23}'' and F_{43}'' in Fig. 3.37 is more easily solved graphically, however, because the direction of forces F_{03} and F_{23}'' cannot be readily expressed in unit vectors without further calculations to determine angles.

The resultant forces obtained by superposition are shown in Fig. 3.38. The free body of the connecting rod shows the resultant forces acting at the pin-connected ends of the rod. At the upper end of the rod the resultant force F_{43} is transmitted through the wrist pin bearing. F_{43} is the vector sum of F_{43}' and F_{43}''. Similarly, resultant force F_{23} transmitted through the crank pin bearing at A is the vector sum of F_{23}' and F_{23}''. Note that the connecting rod is in equilibrium under the action of these resultant forces and the inertia force F_{03}, with the three forces intersecting at a common point k'. The resultant force through the main bearings is F_{12}, which is identical to force F_{23} through the crank pin bearing. Crankshaft torque T_s is the algebraic sum of T_s' and T_s''.

Example Problem 3.6: Dynamically equivalent system of two masses is widely used for the connecting rod of reciprocating machinery, primarily because of the advantages of simplification in analysis. Figure 3.39 shows a typical connecting rod for which the center of gravity, weight and moment of inertia are provided. One equivalent mass M_B is located at the wrist pin bearing

B, from which the inertia force may be calculated if the piston acceleration is available. The location of the second mass M_P is shown, and its value is determined from the expression:

$$l_P l_Q = \frac{I}{M}$$

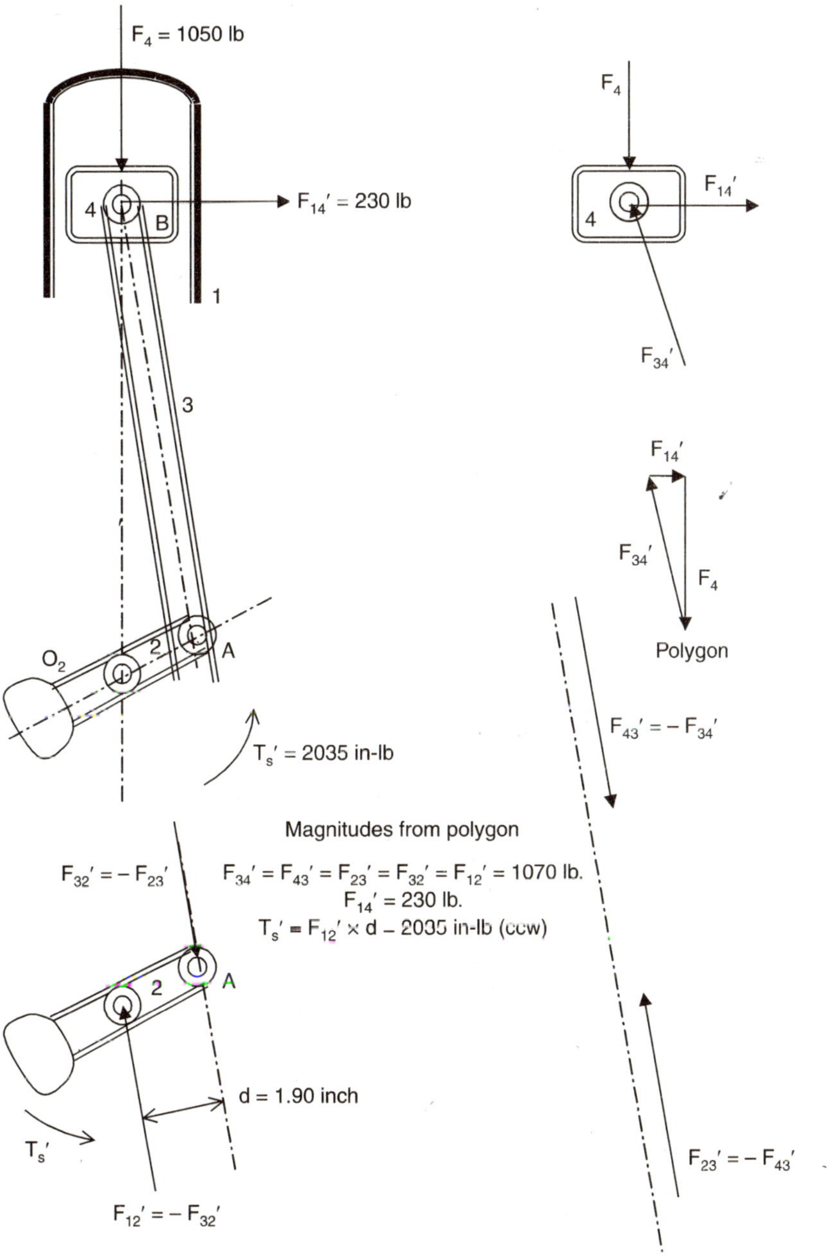

Fig. 3.36

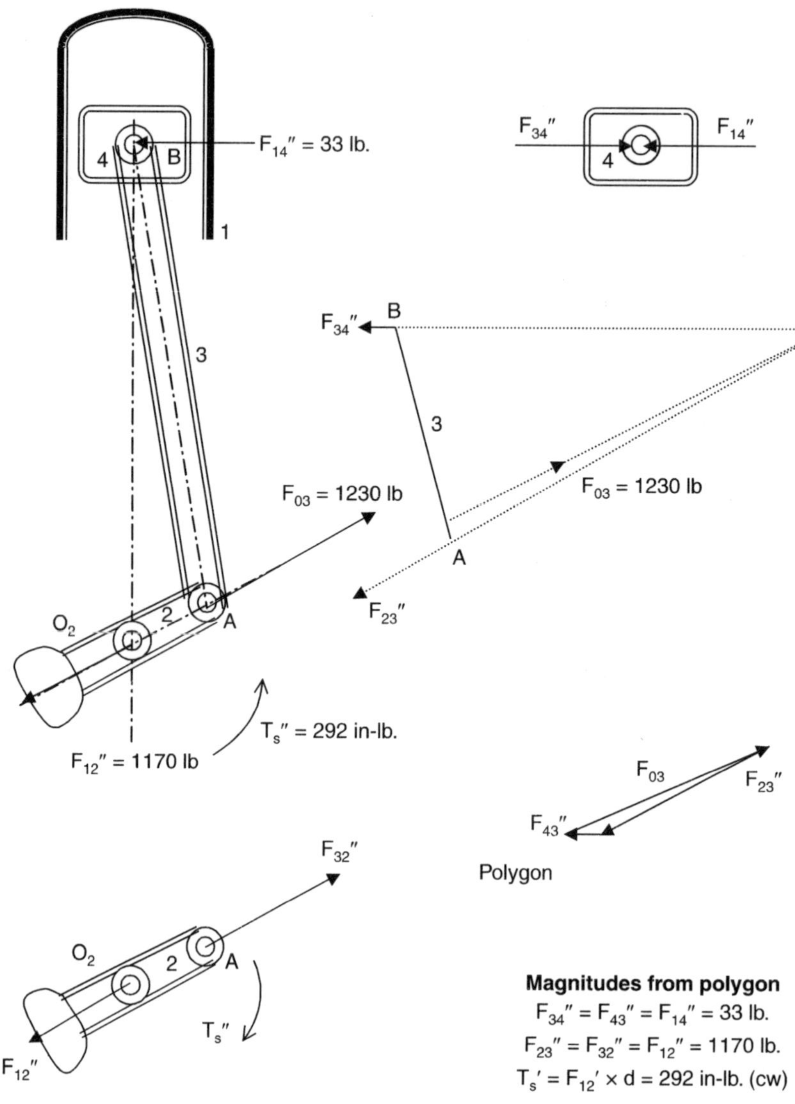

Fig. 3.37

where M and I are the mass and moment of inertia of the rod. Because of the shape of the connecting rod the center of gravity is near the crank pin center A, as shown. Since P and A are close to each other, it may be assumed that $l_P = l_A$ with little error in the solution. Thus, with the second mass at the crank pin center, the acceleration of the second mass can be readily calculated for constant crankshaft speed.

$$W = 8.65 \text{ lb.}$$

$$M = W/g = 8.65/386.0 = 0.02241 \text{ lbs-second}^2\text{/in.}$$

$$I = 4.3545 \text{ lbs-second}^2\text{-in.}$$

Solution:

$$l_P = I/Ml_B = 4.3545/[0.02241 \times 24.72]$$

$$= 7.860 \text{ in.}$$

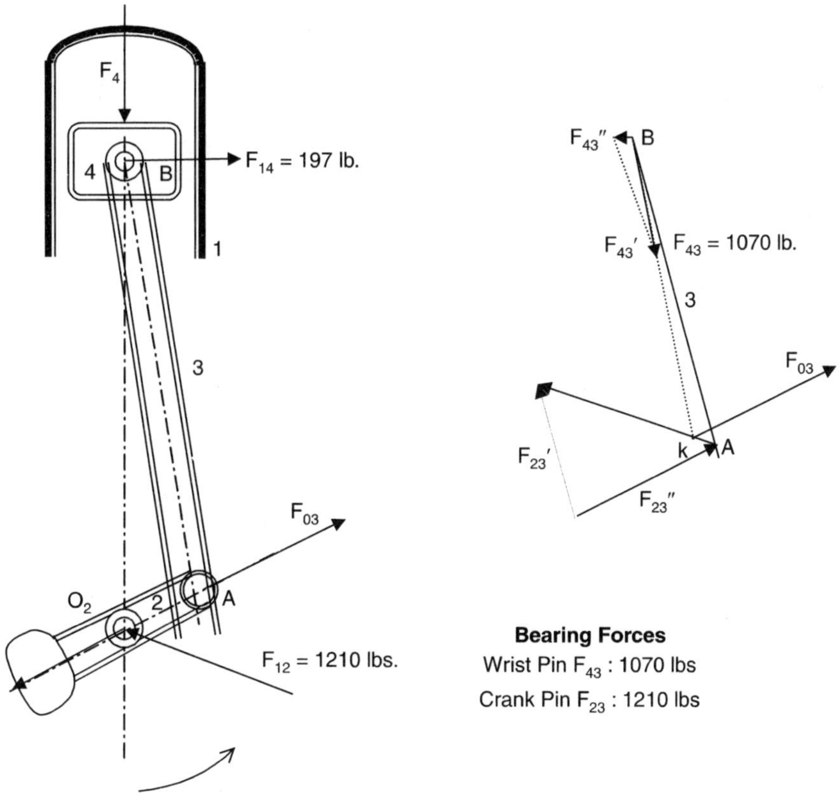

Fig. 3.38

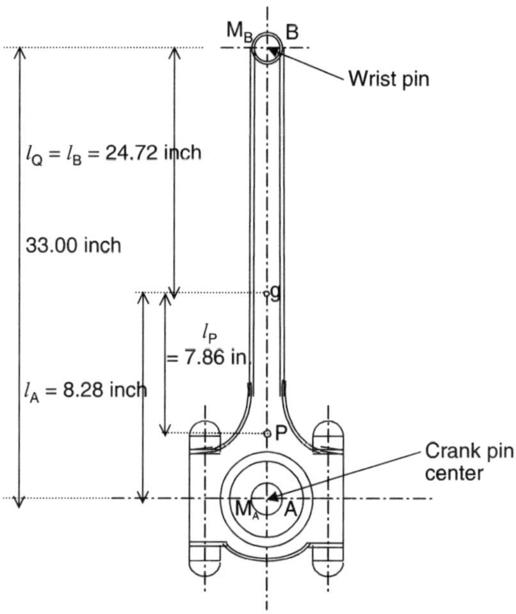

Fig. 3.39

$$M_B = M[l_P/(l_P + l_B)]$$
$$= [0.02241 \times 7.86]/[7.86 + 24.72] = 0.00540 \text{ lb-sec}^2\text{/in.}$$
$$M_P = M - M_B$$
$$= 0.02241 - 0.00540 = 0.0170 \text{ lbs-sec2/in.}$$
$$M_A = M - M_B$$
$$= 0.02241 - 0.005406 = 0.0170 \text{ lb-sec}^2\text{/in.}$$
$$I = Ml_A l_B$$
$$= 0.02241 \times 8.28 \times 24.72$$
$$= \mathbf{4.5869 \text{ lbs-sec}^2\text{-in.}}$$

The approximation leads to a somewhat larger moment of inertia of the equivalent system than the true moment of inertia of the rod. This is because $Ml_A l_B$ is greater than $Ml_P l_B$.

Since M_A is a rotating mass about the crankshaft axis with constant centrifugal force when angular speed is constant, counterweights attached to the crankshaft may be sized so as to counterbalance M_A of the connecting rod as well as the mass of the crank.

REFERENCES AND BIBLIOGRAPHY

API Standard 618, "Reciprocating Compressors for Petroleum, Chemical and Gas Industry Services", American Petroleum Institute, Washington, DC, 1995.

API Standard 671, "Special Purpose Couplings for Refinery Services", American Petroleum Institute, Washington, DC, Reaffirmed 1993.

Almen, G., Black, K., "Residual Stress and Fatigue of Metals", Mc-Graw Hill, New York, 1983.

Barber, A., "Pneumatic Handbook", Elsevier Science, Ltd., Oxford, UK, 1997.

Baxter, A. S., Rawlins, R. E., "The Design and Development of the MB190: A High Speed Diesel Engine Based on Medium Speed Principles", IDGTE Publication 428, February, 1985.

Bent, R. D., McKinley, J. L., "Aircraft Power Plants", McGraw Hill Book Co., New York, 1985.

Binford, P., "Piston Ring Designs-Have They Changed", SAE Paper # 850483, 1985.

Bloch, H. P. "A Practical Guide to Compressor Technology", McGraw-Hill, New York, 1995.

Brown, R. N., "Compressors - Selection and Sizing", Gulf Publishing Co., Houston, 1997.

Cherrie, B. J., "Factors Influencing Valve Temperatures in Passenger Car Engines", SAE Paper # 830484, May, 1984.

Coplin, E. J., Davis, R. J., Brunson, D., "The Interaction of Engines and Finished Equipment Designed to Meet EEC Noise Legislation-The Engine Manufacturers Viewpoint", *Proceedings of I. Mech. Engineers International Conference - Small Internal Combustion Engines,* 1989-3, C372-029, 1989.

Gadd, J., van Degrift, N., "Analysis and Applications to Study of Crankshaft Stress", *J. Applied Mech.* (ASME, 9), 1984.

Haug, E. J., "Computer Aided Kinematics and Dynamics of Mechanical Systems, Vol. I: Basic Methods", Allyn & Bacon, Boston, 1989.

Havran, R. L., "Development of the Caterpillar G3412 Low Emission Engine with Electronic Ignition", *ICE Vol. 22, Heavy Duty Engines-A Look at the Future,* ASME, 1994.

Henshall, S. H., "Mirrlees MB430 Packs More Punch", Marine Engineering Review, April, 1985.

Henshaw, T. L., "Reciprocating Pumps", Van Nostrand Reinhold Co., New York, 1987.

Howe, G., "Sound Proofing Diesels", Heavy Construction Magazine, September, 1984.

ISO 1940/1, "Mechanical Vibrations - Balance Quality Requirements of Rigid Rotors - Part I, Determination of Permissible Residual Unbalance", International Organization for Standardization, Geneva, Switzerland, 1986.

Karassik, I. J., Krutzsch, W. C., Fraser, W. H., Messina, J. P., "Pump Handbook", McGraw Hill Book Co., New York, 1986.

Kochanowski, H. A., "Performance and Noise Emission of a New Single Cylinder Diesel Engine-with and without Encapsulation", *Proceedings of I. Mech. Engineer. International Conference-Small Internal Combustion Engines,* 1989-3, C372-023, 1989.

Loomis, A. L., "Compressed Air & Gas Data", Ingersoll Rand Co., Woodcliff Lake, NJ, 1989.

Lurenbaum, K., "Einfluss von Formgebung und Werkstoff auf die Gestalfestigkeit geschmiedeter und gegossener Flugmotoren-Kurbelwellen", Jahrbuch 1977 der deutschen Luftfahrtforschung, II/129.

Mabie, H. H., Reinholtz, C. F., "Mechanisms and Dynamics of Machinery", John Wiley & Sons, New York, 1987.

Matinaglia, E., "The Structural Durability of Crankshafts", Sulzer Technical Review No. 2, 1982.

Matthaes, P., "Fatigue Strength of Airplane and Engine Materials", NACA Tech. Memo # 743.

Mead, G., "Quieter, Part by Part", Construction Plant and Equipment, October, 1983.

Miller, J. E., "The Reciprocating Pump: Theory, Design and Use", Krieger Publishing Co., Malabar, FL., 1995.

Nevinger, R. D., "Design and Development of the Caterpillar 3600 Spark-Ignited Engine Family", ASME-91-ICE-6, ASME, New York, 1991.

Poynton, W. A., "Designing a Range of Medium Speed Diesel Engines for High Reliability", *Inst. Of Mechanical Engineers., Seminar by Combustion Engines Group,* Sept. 1985.

Rollins, J. P., "Compress Air & Gas Handbook", Compressed Air and Gas Institute, New York, 1983.

Stevens, J. L., Tawil, N., "A New Range of Industrial Diesel Engines with Emphasis on Cost Reduction Techniques", *Proceedings of I. Mech. Engineers International Conference - Small Internal Combustion Engines,* 1989-3, C372-028.

Taylor, C. F., "The Internal Combustion Engine in Theory and Practice", Volumes 1 and 2, The M. I. T. Press, Cambridge, Mass, 1985.

Vetter, G., "Leak Free Pumps & Compressors", Elsevier Science, Ltd., Oxford, UK, 1995.

Vetter, G., "Reliability and Future Development of High Pressure Diaphragm Pumps for Process Service", *Proceedings of the 5th Pump Users Symposium,* Houston, 1988.

Vetter, G, Hering, L., "Oscillating Displacement Pumps", *Encyclopedia Fluid Mechanics,* Gulf Publishing Co., Houston, 1986.

Wachel, J. C., Tison, J. D., "Vibrations in Reciprocating Machinery and Piping Systems", *Proceedings of the 23rd Turbomachinery and Piping Systems,* Texas A & M University, College Station, TX 1994.

Warring, R. H., "Pumping Manual", Trade & Technical Press Ltd., Surrey, England, 1984.

4 Crank Mechanism Dynamics

4.1 INTRODUCTION

A unique feature of all reciprocating machinery is the peculiar motion of the connecting rod linking the piston pin with the crank pin on the crankshaft. Looking at the kinematic motion of the system, the piston executes a purely translational motion, while the crank pin moves in a completely circular path. This can only be accomplished when one end of the connecting rod also moves along a linear path and the other end of the rod travels along a circular path. In other words, the connecting rod has a swinging type of motion to satisfy the paths of both the ends. In large compressors, pumps and steam and gas engines the forces perpendicular to the line of motion of the piston are substantial, and the piston skirt and cylinder walls are not capable of withstanding the transverse direction loads. A piston rod, cross head and guide bar is then used to absorb these side loads. The cross head is then connected with the connecting rod through the cross head pin.

Since the piston moves in a straight line and the crank pin travels along a circular path, it is relatively easy to calculate the inertia forces acting on them. Once the expression for displacement is obtained, differentiating it with respect to time twice will yield acceleration, from which the inertia forces are obtained. In the connecting rod, on the other hand, besides the two end points all other points travel in an elliptical path, and this requires a considerable amount of algebraic calculations to determine the displacement, acceleration, inertia forces and subsequent integration. Fortunately, however, this is not necessary. If the connecting rod is replaced by another structure having the same mass and center of gravity location so that the path traveled by the center of gravity is not changed, then the total inertia force of the rod is equal to that of the new structure. This follows directly from Newton's law which states that the component of the inertia force of a body in a certain direction equals the product of its mass and acceleration of the center of gravity in that direction.

With the aid of this relationship the problem can be easily solved by distributing the connecting rod mass at the reciprocating and rotating ends of the rod, so that the center of gravity location and the total weight remain unaffected. This distribution of masses of the connecting rod is the same as the one obtained by placing the two ends of the connecting rod horizontally above the center of two weight measurement scales (Fig. 4.1). Note that this procedure will leave the total weight and center of gravity as in the original connecting rod, but it will not represent the inertia characteristics of the rod. This implies that the inertia forces calculated using this method will be correct, but the moments due to these forces, or the inertia couples, will not be exact. One way to obtain greater numerical accuracy is to use a finite element representation of the connecting rod, which will be discussed later in the chapter on advanced methods.

Mathematical expressions for displacement, velocity and acceleration for both ends of the connecting rod will be developed in order to calculate inertia calculate inertia and gas pressure torque loads acting on it, the latter coming from the piston. The loads from the connecting rod will then traverse into the crankshaft and on to the drive train. In a single cylinder reciprocating machine only one connecting rod will be transmitting the loads to the crankshaft, as opposed to a multi-cylinder machine where more than one reciprocating mass and connecting rod act on it at a different phase angle. At first sight multi-cylinder reciprocating systems may appear very complex, but a systematic approach using phase angle relationships between individual firing systems can be developed. Once again, multi-cylinder machines also come in a variety of different configurations, the most common being in-line and V-types. In many machines, piston and connecting rod size is identical for all cylinders, which means that the reciprocating mass values for all cylinders are the same. In the case of two stage V-type compressors with an intercooler, however, the cylinder sizes are different, resulting in a more elaborate mathematical model to calculate inertia and air pressure loads acting on the system components through a full cycle. A large variety of other configurations of reciprocating machines exist, each of which can be mathematically simulated and analyzed using similar guidelines that will be developed next.

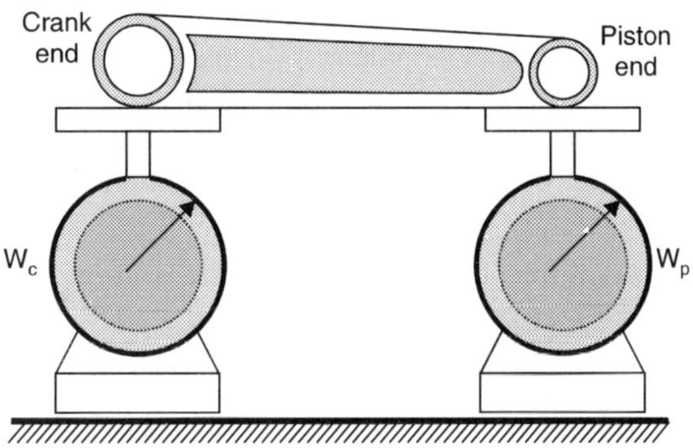

Fig. 4.1. Connecting Rod Mass Distribution

4.2 PISTON DYNAMIC MOTION

Let Fig. 4.2 represent a piston and crank, and let:

x_p = downward displacement of piston from top

ωt = crank angle from top dead center

r = crank radius

l = length of connecting rod.

Assume that ω is constant, so the crankshaft is rotating at uniform speed. The first objective is to calculate the position of the piston in terms of the angle ωt. As the crank turns through this angle, distance x_p is equal to length DB plus a term due to the fact that the connecting rod has assumed a slanting position as given by the angle ϕ. Distance DB is given by $r\{1 - \cos(\omega t)\}$. The correction factor is given by $AC - BC$, which is equal to $l\{1 - \cos(\phi)\}$. Angle ϕ can be given in terms of ωt by the fact that $AB = l \sin(\phi) = r \sin(\omega t)$, or:

$$\sin(\varphi) = \frac{r}{l} \sin(\omega t) \tag{4.1}$$

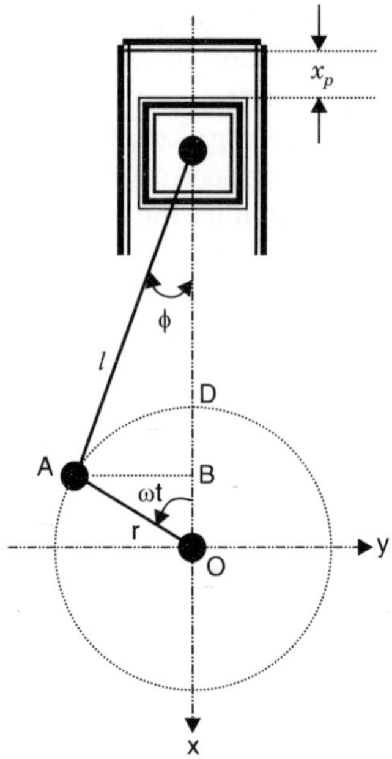

Fig. 4.2. Crank Mechanism

and consequently, $$\cos(\varphi) = \sqrt{1 - \frac{r^2}{l^2}\sin^2(\omega t)} \tag{4.2}$$

Thus, the exact expression for piston displacement x_p in terms of crank angle ωt is given by the expression:

$$x_p = r\{1 - \cos(\omega t)\} + l\left\{1 - \sqrt{1 - \frac{r^2}{l^2}\sin^2(\omega t)}\right\} \tag{4.3}$$

Due to the square root this expression is not convenient for further calculation. It can be simplified by noting that for most reciprocating machines the ratio r/l is of the order of 1/4, and $(r/l)^2$ is 1/16, so the second term under the square root sign is small in comparison to unity. Expanding into a power series and retaining only the first term, equation (4.3) then becomes:

$$x_p \approx r\{1 - \cos(\omega t)\} + \frac{r^2}{2l}\sin^2 \omega t \tag{4.4}$$

Further simplification is obtained through the use of trigonometric identities for cos (2ωt) and $\sin^2(\omega t)$, from which the piston displacement becomes:

$$x_p = \left\{r + \frac{r^2}{4l}\right\} - r\left\{\cos(\omega t) + \frac{r}{4l}\cos(2\omega t)\right\} \tag{4.5}$$

The velocity and acceleration of the piston follow from differentiation of the displacement:

$$dx_p/dt = r\omega\,[\sin(\omega t) + (r/2l)\cos(\omega t)] \tag{4.6}$$

$$d^2x_p/dt^2 = r\omega^2\,[\cos(\omega t) + (r/l)\cos(2\omega t)]$$

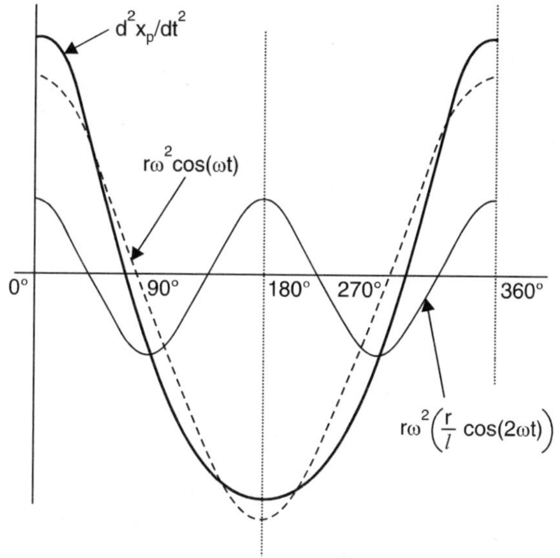

Fig. 4.3. Piston Acceleration as a Function of Crank Angle for $l/r = 4$

After multiplication by the reciprocating masses, these expressions will yield the vertical momentum and the vertical inertia force. They consist of two terms, one varying with the same frequency as the crank rotation and known as the 'primary' term, and the other varying at double frequency and is known as the 'secondary' term. If the connecting rod is infinitely long the secondary term becomes zero and the piston executes a harmonic motion. With a short connecting rod the motion, and particularly the acceleration, deviates substantially from a sinusoid. Figure 4.3 shows graphically piston acceleration, or inertia force, of a reciprocating machine in which $l/r = 4$.

4.3 CRANK ROTATING MOTION

The inertia force due to the mass of the rotating crank is the same as the resultant of all the small inertia forces on the various small parts of the crank. The problem is simplified by concentrating the entire rotating crank mass at its center of gravity. Next the mass is shifted from the center of gravity to the crank pin A, but in the process it is diminished inversely proportional to the distance from the center of the shaft, so the inertia force, which in this case is centripetal in nature, remains unchanged. In addition, there is the mass of the rotating portion of the connecting rod acting at the crank pin, mentioned earlier. The whole crank is thus replaced by a single mass m_c at the crank pin, and the vertical displacement can be found directly from Fig. 4.2:

$$x_c = r\{1 - \cos(\omega t)\} \tag{4.7}$$

So the vertical components of velocity and acceleration become:

$$dx_c/dt = r\omega \sin(\omega t) \tag{4.8}$$
$$d^2x_c/dt^2 = r\omega^2 \cos(\omega t)$$

The horizontal components are:

$$y_c = -r\sin(\omega t)$$
$$dy_c/dt = -r\omega\cos(\omega t) \tag{4.9}$$
$$d^2y_c/dt^2 = -r\omega^2\sin(\omega t)$$

The momentum (or inertia force) is obtained from velocity (or acceleration), by multiplying the expression with the rotating crank mass m_c.

Next consider the connecting rod. Having divided the connecting rod mass into a part moving with the piston (reciprocating) and a part moving with the crank pin (rotating), the reciprocating and rotating masses can be represented by m_{rec} and m_{rot}. Thus, m_{rec} is the total mass of the piston and a portion of the connecting rod, and m_{rot} represents the total equivalent rotating mass of the crank pin and the other part of the connecting rod. For a cylinder oriented such that the piston executes a path along the vertical direction, the vertical inertia force for all moving parts is:

$$X = m_{rec}\, d^2x_p / dt^2 + m_{rot}\, d^2x_c / dt^2$$

$$X = (m_{rec} + m_{rot})\, r\omega^2 \cos(\omega t) + m_{rec}\, (r^2/l)\, \omega^2 \cos(2\omega t) \tag{4.10}$$

and the horizontal inertia force is:

$$Y = m_{rot}\, d^2y_c/dt^2 = m_{rot}\, r\omega^2 \sin(\omega t) \tag{4.11}$$

Putting it in words, the vertical component of the inertia force consists of two parts, a 'primary part' equal to the inertia action of the combined reciprocating and rotating masses as if they were moving up and down harmonically with crankshaft frequency and amplitude r, and a 'secondary part' equal to the inertia action of a mass $m_{rec}(r/4l)$ moving up and down with twice the frequency of the crankshaft with the same amplitude r. The horizontal, or lateral, force has only one primary part due to the rotating mass.

The torque due to these inertia forces about the longitudinal axis O also needs to be calculated. Recall that to determine the vertical and horizontal forces the connecting rod when replaced by two masses at the piston and crank pin gives exact results, but for the inertia torques the result thus obtained is no longer exact. It will be correct to an acceptable degree of approximation. Thus, the piston rod and crank mechanism is replaced by a reciprocating mass m_{rec} and a mass m_{rot} rotating uniformly about O, so that it has no torque about O. The inertia torque is caused wholly by the reciprocating mass m_{rec}, and its magnitude can be deduced from Fig. 1.18 (*b*), where the torque equals the downward piston force multiplied by $x \tan(\phi)$. Note that in Fig. 1.18 (*b*) gas pressure force is used instead of the inertia force due to the reciprocating mass of the piston, but the argument remains the same. Multiplying the reciprocating mass by the expression for piston acceleration, and ignoring all terms proportional to second or higher powers of r/l (involving an error of the same order as that incurred in the calculation of piston acceleration), the inertia torque becomes:

$$M = \frac{1}{2} m_{rec}\, \omega^2 r^2 \left\{ \frac{r}{2l} \sin(\omega t) - \sin(2\omega t) - \frac{3r}{2l} \sin 3(\omega t) \right\} \tag{4.12}$$

This equation for the inertia torque, acting on the shaft in the direction of its rotation, and on the frame about O in the opposite direction, is accurate for the usual reciprocating machine where the connecting rod design calls for two substantial bearings at its ends joined by a relatively light stem.

4.4 RADIAL AIRCRAFT ENGINE

A reciprocating aircraft engine typically has a number of cylinders arranged with their centerlines oriented in the radial direction. The connecting rod arrangement differs from typical reciprocating machines in that a master connecting rod has a crank end carrying not only the crank pin bearing but $n - 1$ other bearings to which the n -1 connecting rods are attached. For this case the connecting rod cannot be replaced by two concentrated masses at the ends. The exact connecting rod analysis is of interest.

Figure 4.4 shows the crank rotating in a counterclockwise direction at uniform speed ω, moving with it the connecting rod. Since the piston inertia force is already calculated, it is considered massless. Moreover the piston will be assumed to move without friction, so the reaction force of the cylinder wall on the rod must be F_4. Further, forces F_1 and F_2 represent forces exerted by the crank pin on the rod, which travels in a prescribed manner under the influence of the three forces. Since the connecting rod moves in a plane, its motion is governed by the three equations of Newton:

In the x-direction: $F_1 = m\, d^2x_G/dt^2$

In the y-direction: $F_2 + F_3 = m\, d^2y_G/dt^2$

Moments about center of gravity: $-F_1\, a \sin(\varphi) + F_2 a \cos(\varphi) - F_3\, b \cos(\varphi) = I_G d^2\varphi/dt^2$

The geometry of the motion is known; in particular the center of gravity moves thus:

$$x_G = x_p + (x_c - x_p)(b/l) = (x_p a/l) + (x_c b/l) \tag{4.13}$$

and

$$y_G = y_c b/l \tag{4.14}$$

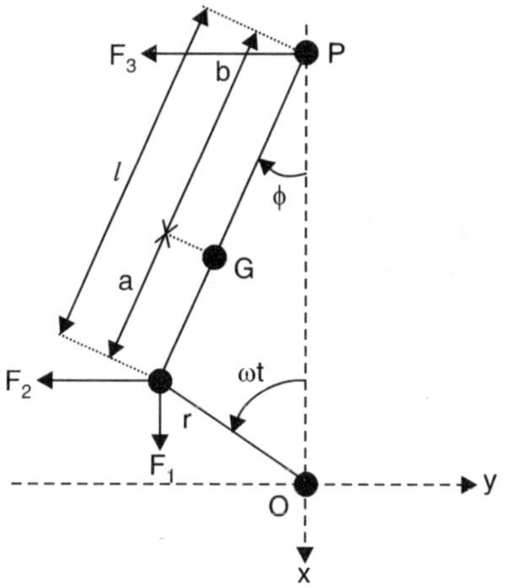

Fig. 4.4. Forces on a Crank

where the subscripts c and p denote crank and piston, and a and b are distances to the center of gravity G. Accelerations of the piston and of the crank pin in both the directions are given by equations (4.6), (4.8) and (4.9). Also the expression for ϕ and its functions are known. Thus, the Newton equations can be solved for all three unknown forces. To calculate the torque in the clockwise direction exerted on the shaft by the inertia of the rod, all terms containing powers of r/l higher than 2 may be neglected without loss of generality, leading to the following equation:

$$M_{\text{shaft}} = \frac{1}{2} m_{\text{rec}} \omega^2 r^2 \left\{ \frac{r}{2l} \sin(\omega t) - \left[1 + \frac{ab - k^2}{al} \right] \sin(2\omega t) - \frac{2r}{2l} \sin(3\omega t) \right\} \tag{4.15}$$

in which k is the radius of gyration of the rod, defined by $I_G = mk^2$. Except for neglecting of the higher powers of r/l, the result is exact, providing the expression for the inertia torque on the shaft of a connecting rod consisting of two concentrated masses ma/l and mb/l at distances b and a from the center of gravity. Such a rod has a radius of gyration $k^2 = ab$.

In order to observe how (4.15) differs numerically from (4.12), two cases of rods that have no concentrations are considered. First take the uniform rod with $a = b = l/2$ and $k^2 = l^2/12$. In this case the double frequency term of (4.15) is a third greater than the term in the approximate formula (4.12). Next consider a rod with $m_{rec} = 0$, $(b = l)$, having its center of gravity at the crank pin and a certain dimension around it, which is a rough picture of the master connecting rod of a radial aircraft engine. Assuming $k^2 = l^2/10$, the middle term in (4.15) which is the same as that in (4.12) if only m_{rec} is replaced by $m/10$. Note, however, that the sign is reversed.

The aircraft master rod of actual practice is a combination of both cases just discussed, and the increase in moment due to the uniform rod effect more or less balances the decrease in moment due to the large moment of inertia of the crank end. Thus, even for an unusual connecting rod as that of a radial reciprocating engine, the approximate result from (4.12) is reasonably accurate.

The torque acting on the frame of the engine about the shaft center O (Fig. 4.4) is found by multiplying the force F_3 by its moment arm and solving for F_3 from the Newton equation. Neglecting higher powers of r/l, the result is:

$$M_{\text{frame}} = \frac{1}{2} m_{\text{rec}} \omega^2 r^2 \left\{ \left[\frac{(r^2 + 8l^2)(k^2 - ab)}{4ral^2} + \frac{r}{2l} \right] \sin(\omega t) - \left[\frac{ab - k^2}{al} + 1 \right] \right.$$

$$\left. \sin(2\omega t) - \left[\frac{3r(k^2 - ab)}{4al^2} + \frac{3r}{2l} \right] \sin(3\omega t) \right\} \tag{4.16}$$

Again, for the connecting rod with two concentrated ends ($k^2 = ab$) this result reduces to (4.12). For the general connecting rod the inertia torques on the shaft and the frame are not equal but differ by the moment of the inertia forces of the various rod points about O. Only when the rod degenerates into concentrated masses is this moment zero, since both inertia forces are along the center line and along a radius, both passing through O.

4.5 VARIABLE ELASTICITY CHARACTERISTICS OF RECIPROCATING SYSTEMS

All of the discussion thus far considered is limited to that described by linear differential equations. Physically speaking, all masses and inertias are constant, all spring forces are proportional to relative deflections and all damping forces are proportional to the velocity. In practice these assumptions may not remain valid for a number of reasons. Manufacturing techniques used to make and assemble the components into a complete reciprocating system, inherent geometric features, operational characteristics and environmental conditions often times impose a number of factors which render the linear theory assumptions invalid. Deviations from the classical theory can be categorized into two groups, one where the differential equations of motion are linear but the coefficients are functions of time, the other where the equations are non-linear. In a non-linear case the displacement x or its derivatives do not appear in the first power. Two important cases directly affecting the dynamic behavior of the rotating system of reciprocating machines will be explored at length, one arising from variable elastic properties of the components, and in the next section, from the variable inertia characteristics of the piston and crank mechanism.

A typical linear equation describing the motion of a system where the spring rate varies with time can be expressed as:

$$md^2x/dt^2 + cdx/dt + f(t)x = 0 \tag{4.17}$$

Assume that two different solutions of this equation are known:

$$x = \varphi_1(t) \quad \text{and} \quad x = \varphi_2(t) \tag{4.18}$$

Then $C_1\varphi_1(t)$ is also a solution, and

$$x = C_1\varphi_1(t) + C_2\varphi_2(t) \tag{4.19}$$

is the general solution. The principle of superposition holds for the solutions of linear differential equations with variable coefficients, but it is not true for non-linear equations. In the latter case the general solution can still be written in a form containing two arbitrary constants C_1 and C_2, since the solution process is one of double integration. For linear equations with time varying coefficients a knowledge of two particular solutions leads to the general solution in the form of (4.19). Except for a few cases, however, this is not true for non-linear equations.

Examples of cases where systems display variable elasticity will be discussed next physically, with an explanation of their dynamic behavior. Consider a disk mounted on the middle of a vertical shaft running in two bearings B of which only the upper one is shown in Fig. 4.5. The shaft cross section is not completely circular, but is of such a nature that two principal directions can be distinguished, one of maximum and one of minimum stiffness, as, for example, in a rectangular section. Assume the shaft to have two circular spots A, A close to the disk. The spots can slide without friction in two straight guides, confining the motion of the disk in the plane perpendicular to shaft axis. The disk on the flexible shaft is a vibrating system of a single degree of freedom. While the shaft is rotating the spring constant varies with time, from a maximum of $k + \Delta k$ to a minimum of $k - \Delta k$, twice during each revolution. The equation of motion is:

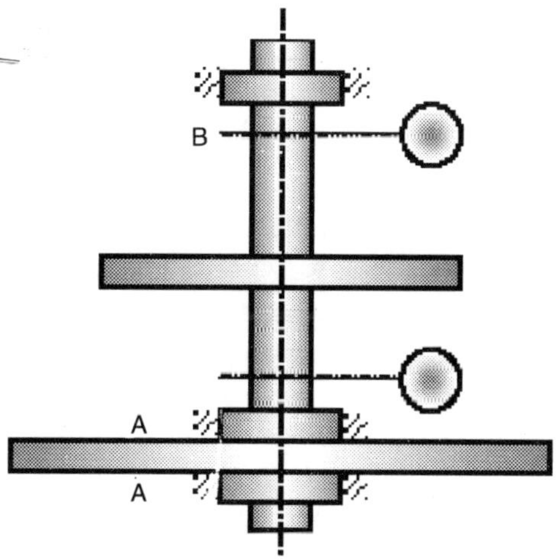

Fig. 4.5. Disk Mounted on a Shaft with Non-uniform Elasticity

$$md^2x/dt^2 + \{k + \Delta k \sin(\omega_k t)\}x = 0 \tag{4.20}$$

where ω_k is twice the running speed of rotation of the shaft, subscript k being used to denote variation in elasticity k.

If this shaft is placed horizontal with guides A vertical so that the vibration of the disk is restricted to the vertical direction, weight W of the disk acts as an additional force, and the equation of motion changes to :

$$md^2x/dt^2 + \{k + \Delta k \sin(\omega_k t)\}x = W \tag{4.21}$$

If the elasticity were constant there would no significant difference between (4.20) and (4.21), because (4.21) could be transformed into (4.20) merely by taking another origin at a location representing the static deflection of the shaft at the disk. With variable elasticity, however, this is not so. Taking a new variable $y = x + C$, where C is a constant to be determined, (4.21) becomes:

$$md^2y/dt^2 + \{k + \Delta k \sin(\omega_k t)\}y = W + kC + C\Delta k \sin(\omega_k t) \tag{4.22}$$

If the variation in the elasticity were zero, C could be chosen to equal $-W/k$, and transform (4.22) back to (4.20). If W were to be zero, it is interesting to note that in (4.20) by a mere shift of the origin of x can be given a right-hand member which can be classed as an extraneous exciting force of frequency ω_k.

Since (4.20) and (4.21) are definitely different, assume that the variations in k are small, 10 per cent or less of k. Then the elastic force is principally that of k, and the motion of the disk is nearly harmonic with the frequency $\omega_n = \sqrt{(k/m)}$.

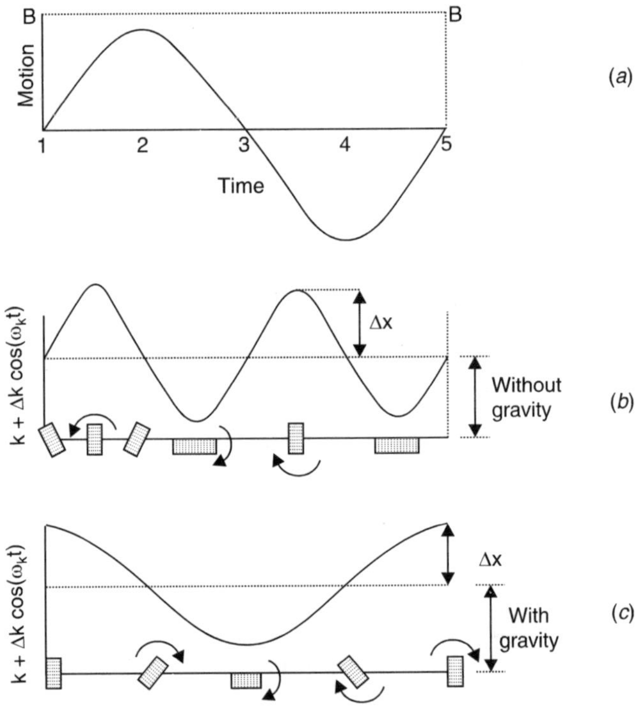

Fig. 4.6. Instability in a Flat Shaft at Half and Full Critical Speed

The curves of Fig. 4.6 (a) and (b) illustrate the motion x of the disk with a frequency ω_n and a variation in shaft stiffness taking place at twice that frequency. The diagrams pertain to the case of no gravity (vertical shaft), so OA is the equilibrium line where there is no bending in the shaft. The elastic force is therefore the product of the ordinates of figures 4.6 (a) and 4.6 (b), measured from OA. With the phase relation shown in the figure the disk is moving away from the center position (1-2 and 3-4) when the spring force is smaller than its average value, and the disk is moving toward the center when the spring force is greater than its average value (2-3 and 4-5). Thus the spring force is small when it opposes the motion and large while it is helping the motion. Hence, over a full cycle the spring force does work on the system and the vibration, once initiated, builds up, resulting in instability.

When the shaft is oriented in the horizontal direction, with gravity the spring force is the product of k and the amplitude of Fig. 4.6 (a), but here the ordinate is not measured from base OA, it is measured from base BB distant from OA by the static deflection of the shaft due to W. This represents another variation of the stiffness k. In Fig. 4.6 (c) $2\omega = \omega_n$, so the shaft is running at half critical speed. This is so because the spring force is small (2 to 4) while the disk is going away from its equilibrium position BB and large (1 to 2 and 4 to 5) while it is coming toward BB. The physical analysis leads to the following conclusions:

- In the vertical shaft with flats any small vibrations that may exist at the natural frequency $\omega_n = \sqrt{(k/m)}$ will be increased to large amplitudes if the shaft runs at its full critical speed ($\omega_k = 2\omega_n$).
- For the case of the horizontal shaft with flats the same type of instability exists at the full critical speed as well as at half the critical speed.

In this example $f(t)$ is a periodic function of time of the form sin ($\omega_k t$). The resulting differential equation of motion is known as Mathieu's equation, and its general solution containing two arbitrary integration constants does not exist. Rather than the exact shape of the motion, the question of interest generally is whether the system is stable or unstable. A stable solution implies a system in which an initial disturbance dies down with time as in a damped vibration, while an unstable motion is one where the amplitude increases with time.

If the ripple time function $f(t)$ attached to the variable spring constant Δk has the frequency ω_k, the motion, although not periodic, will show some regularities after each interval $T = 2\pi/\omega_k$. For any general $f(t)$ the differential equation of motion:

$$md^2x/dt^2 + \{k + \Delta k\, f(t)\}x = 0 \tag{4.23}$$

cannot be solved. The particular case of a 'rectangular ripple' Δk on the spring constant k is a comparatively simple solution, shown in Fig. 4.7. In most cases the ripple is more sinusoidal than rectangular, but the general behavior of the system is much the same as that of a system with a harmonic ripple on the spring force. In such a case the motion in one half of the cycle will depend on the modified natural frequency:

$$p_1 = \sqrt{\omega_n^{\,2} + \frac{\Delta k}{m}} \tag{4.24}$$

and for the other half of the cycle:

$$p_2 = \sqrt{\omega_n^{\,2} - \frac{\Delta k}{m}} \tag{4.25}$$

The stability of the system will depend on two variables, p_1/ω_k and p_2/ω_k, or more significantly, ω_n/ω_k (the ratio of 'free' and 'elasticity' frequencies) and $\Delta k/k$ (stiffness variation ratio).

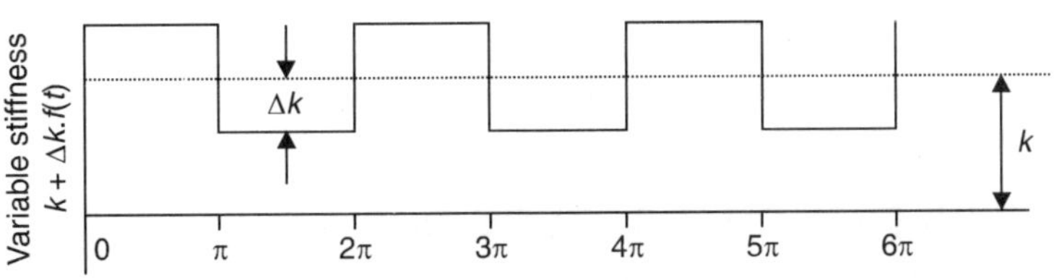

Fig. 4.7. Variable Stiffness Coefficient

The result is shown graphically in Fig. 4.8. Shaded regions indicate stability of motion, while non-shaded regions denote unstable regime. The numbers ½, 1, 1½, etc., inscribed in the regions of instability indicate the number of vibrations of the system during one ω_k-period of the variation in stiffness. In interpreting the results the 'rectangular ripple' assumption should be remembered, so the results are only approximate for most actual cases where the variation is nearly harmonic. However, the approximation is a good one. Also, no damping has been considered. The stiffness ratio variation $\Delta k/k$ and the average natural frequency $\omega_n = \sqrt{(k/m)}$ are constant, so the only variable in the system is the frequency of variation in elasticity ω_k. The line for $\Delta k/k = 0.4$ is drawn and marked *OA*. A slow variation in ω_k corresponds to a point on that line far from the origin *O*, while points close to the origin have a small value of $(\omega_n/\omega_k)^2$ and therefore a large ω_k. Most of the points on *OA* are in the stable region where vibrations will not be a problem, but there are theoretically an infinite number of narrow regions of instability, occurring at ω_n/ω_k = ½, 1, 1½, 2, 2½, etc.

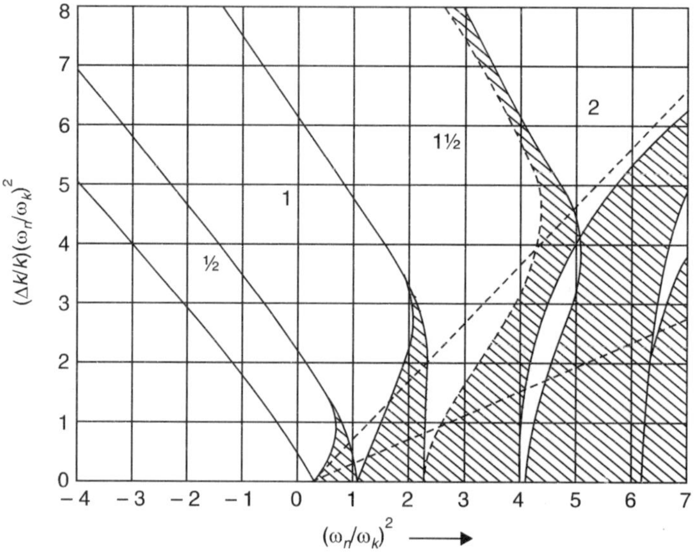

Fig. 4.8. Diagram Determining Stability of a System with Variable Elasticity. Shaded Regions are Stable, Blank Regions are Unstable (*ref.* Van der Pol and Strutt, 1979)

Experiments with different types of shaft cross sections provide interesting results. A shaft with two flat sides will pass through a great number of critical speeds. However, only the highest two of the critical speeds proved to be of importance in the experiment, one occurring at half the usual primary critical speed and the other at that speed itself. A shaft with a keyway cut in it gave rise to critical speeds of one half the normal, but by cutting two additional dummy keyways symmetrically placed so that the keyways are 120° from each other, makes the stiffness uniform in all directions. In another example of a two-pole turbo-generator rotor slots are cut for electric windings, the solid parts forming the pole faces. Here the non-uniform elastic characteristics of the rotor around the full circumference cannot be avoided, so a two-pole rotor will always be rough at half its critical speed.

4.6 VARIABLE INERTIA CHARACTERISTICS

The shafting of all reciprocating systems, with all its cranks, pistons, connecting rods and flywheels is a complicated structure. In order to systematically analyze it for dynamics the inertia characteristics need to be established. First consider the equivalent moment of inertia of the mechanism. The moment of inertia I_{rot} of the purely rotating parts offers no problem, but it is not evident what should be done with the reciprocating weight. In Fig. 4.9(*a*) and (*b*) the piston is shown in two positions. Consider the crankshaft when it is not rotating but executing small torsional oscillations. In Fig. 4.9(*a*) when the piston is at either dead center this takes place without any motion of the piston, but in Fig. 4.9(*b*) the motion and acceleration of the piston is nearly equal to that of the crank pin. The equivalent inertia at the top or bottom dead center is zero, but when the crank angle is at 90° or 270° it is $m_{rec}r^2$. Thus, when the crankshaft is turning the total moment of inertia of the piston, connecting rod and crank mechanism varies between I_{rot} and $I_{rot} + m_{rec}r^2$, with an average value of $I_{rot} + \frac{1}{2}(m_{rec}r^2)$. In the calculation of torsional natural frequencies of vibration this parameter plays a major role, and common practice is to replace the system with variable inertia by one of constant inertia I, where:

$$I = I_{rot} + \frac{1}{2}(m_{rec}r^2) \tag{4.26}$$

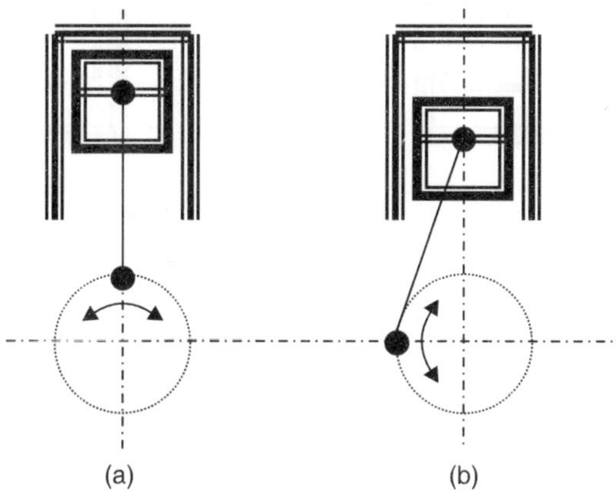

Fig. 4.9. Variation of Equivalent Moment of Inertia of a Piston with its Position

In most mechanical engineering problems it is usually the elasticity factor that varies, but the piston, connecting rod and crank mechanism represents a special case where the inertia varies with the crank angle, and thus with time. To understand the implications on the dynamic stability of the system due to this time varying factor the same mathematical basis as that of variable elasticity may be employed, provided damping is not included. The equation of motion may be written as:

$$m(t)d^2x/dt^2 + kx = 0, \quad \text{or} \quad d^2x/dt^2 + kx/m(t) = 0 \tag{4.27}$$

This equation describes a system of unit, or constant, mass and of variable elasticity, and the mathematical treatment will be analogous to that discussed in the previous section. The physical behavior of such a system can be observed using the one shown in Fig. 4.10 where a piston and crank mechanism are coupled through a flexible shaft k to a flywheel of very large inertia. The flywheel is rotating at constant speed. This system is a torsional one of a single degree of freedom with constant elasticity k and a variable moment of inertia due to the reciprocating masses.

Since the system is mathematically equivalent to the one with variable elasticity and constant mass, Fig. 4.7 applies. According to this figure the system will experience critical speeds when the average natural frequency $\omega_n = \sqrt{(k/I)}$ is ½, 1, 1½, 2, 2½, ... times the frequency ω_k of mass variation. Since the inertia value changes twice in one shaft revolution, the frequency of mass variation is twice the shaft speed. Thus, the critical speeds should appear for ω_n = 1, 2, 3, ... times the rotor speed.

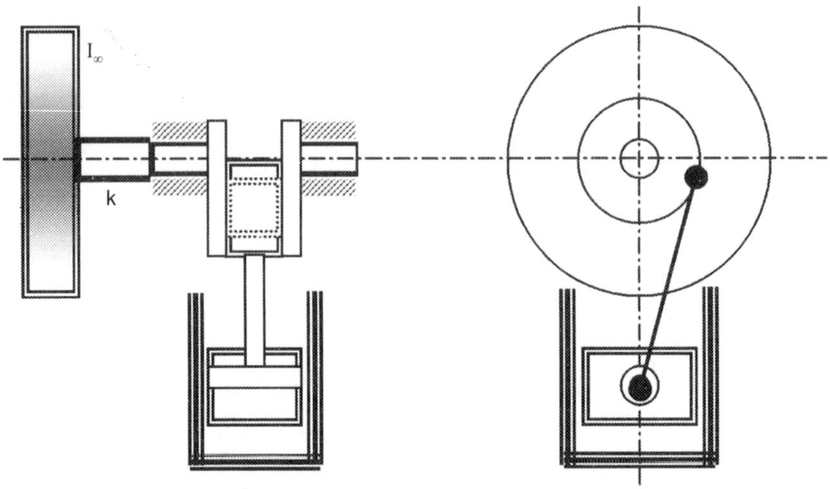

Fig. 4.10. Classical Reciprocating Mechanisms with Periodically Varying Inertia

4.7 ELECTRIC LOCOMOTIVE WITH SIDE ROD

Electric locomotives do not come under the classic definition of reciprocating machines. However, the side rod transferring power from the motor to the locomotive main wheels through a pinion and gear set executes reciprocating type motion. It is discussed in this section since technically it represents perhaps one of the more important case of a system with variable elasticity. Violent torsional vibrations have been detected in the drive train in several speed ranges. They are caused by a periodic pulsation in the spring constant, which can be understood from Fig. 4.11 representing a simpler construction of this type. An electric motor is mounted on the frame and coupled to a driving axle by one connecting rod on each side of the locomotive. To avoid a dead center in the system the two rods are offset by 90°. During normal operating conditions the wheels are locked to the rails by friction, but the motor is capable of rotating slightly against the flexibility of the two side rods. When a side rod is at one of its dead centers it does not prevent the motor from turning through a small angle, so the motor's contribution to the stiffness is absent. When it is 90° from its dead center, it constitutes a very stiff spring since it has to be elongated, or the crank pins have to be bent, to allow a small rotation of the motor. The spring constant of one side rod, therefore, varies between a maximum and practically zero and performs two full variation cycles for each revolution of the wheel. The variation in the flexibility of the combination of two side rods is less pronounced and shows four cycles per revolution.

Curves 1 and 2, of Fig. 4.12 show the torque at the motor necessary for one unit of twisting angle, if only side rod 1 or 2 is attached. Curve 3 is the sum of curves 1 and 2, giving the resultant k for the whole system.

The torsional oscillations of the motor on its side rod springs will take place superposed on the overall rotation of the motor. The phenomenon is represented by (4.20) where ω_k is four times as large as the angular velocity of the wheels. Serious vibrations may thus be expected when $\alpha_n = \frac{1}{2}.4.2\pi = 4\pi$ radians per second.

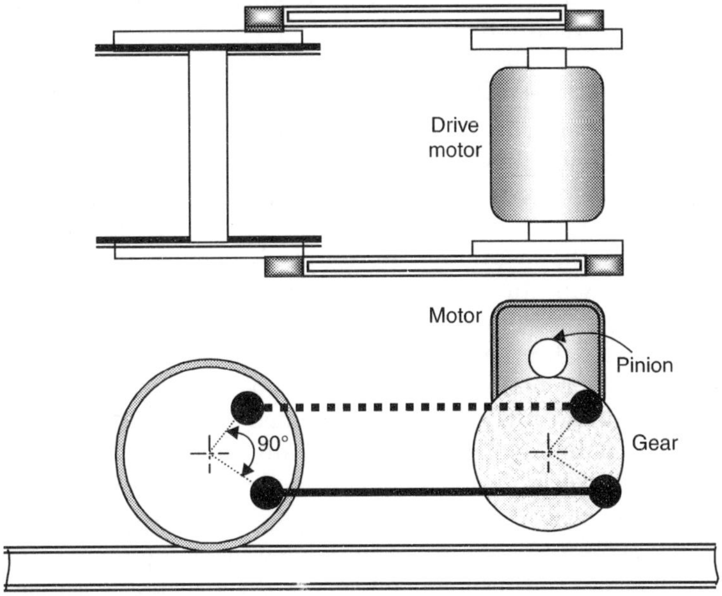

Fig. 4.11. Torsional Vibration in Electric Locomotive with Side Rod

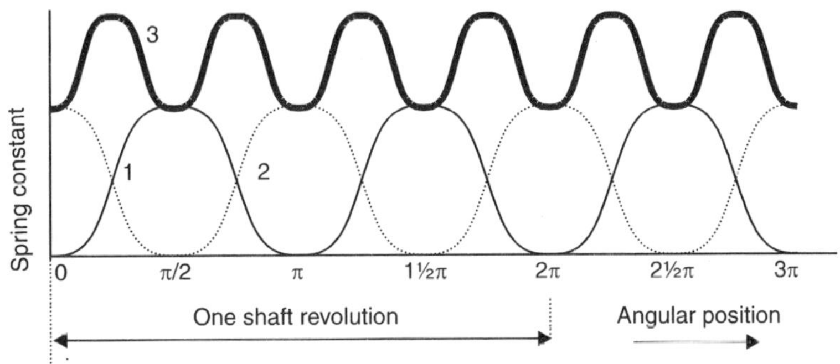

Fig. 4.12. Torsional Spring Constants of Locomotive Side Rod as Function of Angular Position

From Fig. 4.12 the dynamic behavior of the system can be deduced. When the electric locomotive is starting slowly and increasing its speed gradually, the variation in the side rod elasticity frequency (which increases four times as fast as the rotation of the wheels) equals twice the natural frequency of torsional vibration. Along *OA* in Fig. 4.7 this means a motion from infinity to the point where $(\omega_n/\omega_k)^2 = 1/4$, and it is seen that an infinite number of critical speeds has been passed.

In section 4.6 it was noted that the two speeds corresponding to $\omega_n/\omega_k = 1$ and $\omega_n/\omega_k = 1/2$ are the most significant and that the other critical speeds are of lesser importance. Nevertheless it is not possible to avoid these low speed instabilities by changes in design. To avoid problems arising from these vibrations torsionally flexible couplings with springs between the motor gear and its crank or between the driving wheel and its crank have been successfully used. These couplings act in two ways. First the natural frequency ω_n is shifted to a low value so that all critical regions lie below a low value, say 20 miles per hour. At this low speed the energy input level is not high. Furthermore, the springs, specially of the leaf type, have some internal friction in them and thus introduce damping.

4.8 SELF EXCITED VIBRATIONS

Besides free and forced vibrations, disturbances have been observed in reciprocating machinery that belong to a class of their own, and may be characterized as self excited vibrations. In a self excited vibration the alternating force that sustains the oscillating motion is created or controlled by the motion itself; when the motion stops the alternating force disappears. In a forced vibration, on the other hand, the sustaining alternating force exists independently of the motion and persists even when the vibratory motion is stopped. To recognize the difference, consider two applications. In a single cylinder steam engine the piston executes reciprocating motion, which may be considered a 'vibration'. The force maintaining this vibration comes from the steam, pushing alternately on both sides of the piston. Next consider an unbalanced disk mounted on a flexible shaft and running in two bearings on both sides of the disk. The center of the disk vibrates, the motion of the disk being maintained by the centrifugal force of the unbalance, pushing the disk away from the undeflected shaft axis.

The steam engine is a case of self-excited vibration, while the disk executes an ordinary forced vibration. If the piston is prevented from moving by clamping the flywheel, the valves also do not operate and no alternating steam force acts on the piston. The disk can be prevented from vibrating by bringing the two bearings close to the disk on either side. But the unbalance remains rotating, so the centrifugal force also remains.

Another way to look at the situation is by defining a self excited vibration as a free vibration with negative damping. A regular positive viscous damping force is proportional to the velocity of vibration and directed opposite to it. A negative damping force is also proportional to the velocity but acts in the same direction as the velocity. Instead of reducing the amplitudes of free vibration the negative damping will increase them. Note that even with this definition the damping force, whether positive or negative, vanishes when the motion stops.

Examine the differential equation of motion of a single degree of freedom with negative damping:

$$md^2x/dt^2 + cdx/dt + kx = 0 \tag{4.28}$$

The equation differs only in the sign of c, hence its solution can be written as:

$$x = e^{+\frac{c}{2m}t} \{A \cos(qt) + B \sin(qt)\} \tag{4.29}$$

which is a vibration with exponentially increasing amplitude, graphically shown in Fig. 4.13.

A system with positive damping may be considered as dynamically stable, where as one with negative damping is dynamically unstable. A mechanical system is statically stable if a displacement from the equilibrium position sets up a force or couple tending to drive the system back to the equilibrium position. It is statically unstable if the force thus set up tends to increase the displacement. Hence, static instability implies a negative spring constant k, or more generally, a negative value of one of the natural frequencies ω^2.

Note that dynamic stability always presupposes static stability (Fig. 4.13(*c*)), but the converse is not true. A statically stable system may yet be dynamically unstable (Fig. 4.13(*b*). About the frequency of self-excited vibrations, in most cases the negative force is very small in comparison to the elastic and inertia forces of the motion. If the damping force were zero the frequency would be the natural frequency. The damping force, whether positive or negative, lowers the natural frequency, but the difference is negligible for most practical cases, so the natural frequency is the frequency of the self excited vibrating system. In the steam engine example damping force is much larger than the spring force, which is totally absent, so the frequency of vibration differs appreciably from the natural frequency.

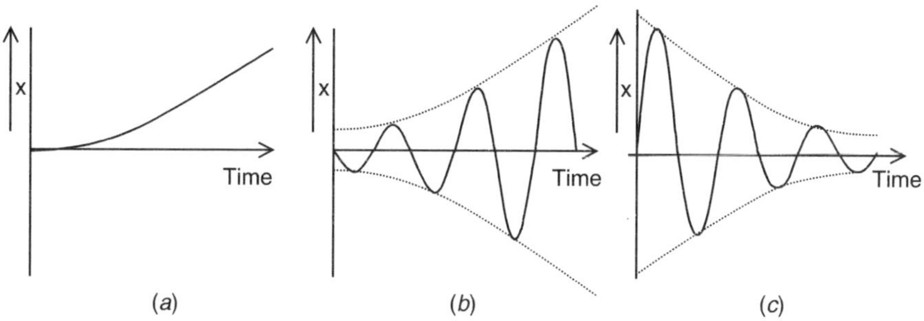

Fig. 4.13. The Free Motion of a System in Different States of Stability
(*a*) Statically Unstable (*b*) Statically Stable but Dynamically Unstable
(*c*) Statically and Dynamically Stable

Looking at energy considerations, with positive damping the damper does negative work, being always opposed to the velocity. The damping energy is taken from the vibrating system and converted into heat by the damper, with successive vibrations having less energy and kinetic energy absorbed by the damper. In the case of negative damping, the damping force which is a driving force, does positive work on the system. This work is converted into additional kinetic energy of the increased vibration. The self excited vibration cannot exist without an external source of energy, which in the case of the steam engine is the steam from the boiler. The energy source does not need to have the alternating frequency of motion; it could be a steady wind or water under pressure or the steady torque of an engine. But there are cases where the source is alternating of a frequency much higher than that of the vibration it excites. Because more energy is put in during each cycle, the vibration could become infinitely large with time, which is contrary to observation. In most systems the self exciting and positive damping mechanisms occur together and separately. For a linear system the positive damping energy follows a parabolic curve since the dissipation is $\pi c \omega x_o^2$. In all practical cases the input and damper dissipating curves intersect, where energy equilibrium exists. The motion thus executed is then an undamped steady state free vibration.

4.9 STABILITY CRITERION

For single degree of freedom systems a simple physical reasoning suffices to show the negativity of the damping constant. With greater degrees of freedom the physical conception is helpful but does not give a full interpretation of the process. With a linear self-excited system the amplitude will become infinitely large in time, since more energy must be put into the system during each cycle, which does not occur in practice. When self excitation and damping are present, the energy per cycle follows a parabolic curve, because the dissipation per cycle is $\pi c \omega x_o^2$. A mathematical approach is necessary, calling for setting up of differential equations of motion. For small vibrations the equations are linear and of second order. Their solution is found by assuming:

$$x_1 = x_{1\,max}\, e^{st}$$
$$x_2 = x_{2\,max}\, e^{st}$$
$$\cdots\cdots\cdots\cdots$$
$$x_n = x_{n\,max}\, e^{st} \tag{4.30}$$

where s is a complex number the real part of which determines the damping and the imaginary part the natural frequency. Substituting into the differential equations of free vibrations transforms the equations into a set of n equations of the complex unknowns $x_{1\,max}$, ... $x_{n max}$. A process of algebraic elimination is then performed with these results such that only one equation is obtained which does not contain any of these variables. This equation is the 'frequency equation' of the degree $2n$ in s. The roots of the equation are usually complex as shown below, and occur in conjugate pairs:

$$s_1 = p_1 + jq_1$$
$$s_2 = p_1 - jq_1$$
$$s_3 = p_2 + jq_2$$
$$s_4 = p_2 - jq_2 \tag{4.31}$$

The solution of the first differential equation is:

$$x_1 = C_1 e^{s_1 t} + C_2 e^{s_2 t} + C_3 e^{s_3 t} \tag{4.32}$$

The terms can be combined by pairs as follows:

$$C_1 e^{s_1 t} + C_2 e^{s_2 t} + e^{p_1 t}\,\{A \sin (q_1 t) + B \cos (q_1 t)\} \tag{4.33}$$

so that the real part of s determines the rate of damping and the imaginary part the natural frequency. If the real parts of all the values of s are negative the system is dynamically stable; but if the real part of any of the values of s is positive then the system is dynamically unstable.

Thus, the stability can be determined by an examination of the signs of the real parts of the solutions of the frequency equation. It is not necessary to solve the equation, because certain rules exist by which from an inspection of the coefficients of the equation a conclusion about the stability characteristics of the system can be drawn. These rules, prepared by Routh in 1877, are complicated for frequency equations of the higher order, but for practical cases using third or fourth degree equations, they are simple. Consider first the cubic equation:

$$s^3 + A_2 s^2 + A_1 s + A_0 = 0 \tag{4.34}$$

which occurs in two degrees of freedom systems. One of the roots of a cubic equation must always be real, and the other two are either real or complex conjugate. If the roots of the cubic equation are s_1, s_2, s_3, then separating the real and imaginary parts provides the following results:

$$s_1 = p_1$$
$$s_2 = p_2 + jq_2$$
$$s_3 = p_2 - jq_2 \tag{4.35}$$

Substituting into (4.34) and going through algebraic elimination yields:

$$A_2 = -(p_1 + 2p_2)$$
$$A_1 = 2p_1 p_2 + p_2^2 + q_2^2$$
$$A_0 = -p_1(p_2^2 + q_2^2) \tag{4.36}$$

The criterion of stability is that both p_1 and p_2 be negative. This means that all coefficients A_2, A_1 and A_0 must be positive, because if any one of them were negative, (4.36) requires that either p_1, p_2 or both p_1 and p_2 must be positive. This requirement can be proved to hold for higher degree equations as well. Hence, a frequency equation of any degree with one or more negative coefficients determines an unstable motion.

Even if the coefficients A_2, A_1 and A_0 are all positive, the third equation of (4.36) requires that p_1 be negative. On the boundary between stability and instability, p_2 must pass from a positive to a negative value through zero. By making $p_2 = 0$ in (4.36), eliminating p_1 and q_2 and substituting a trial solution yields the complete criterion for stability of the cubic (4.34), namely, that all coefficients A are positive and that:

$$A_1A_2 > A_0 \tag{4.37}$$

Next consider the quartic equation:

$$s^4 + A_3s^3 + A_2s^2 + A_1s + A_0 = 0 \tag{4.38}$$

for which the procedure is similar. Here the quartic can be resolved into two complex conjugates. Following a similar method, the complete criterion for stability of the quartic (4.38) is that all coefficients A are positive and that:

$$A_1A_2A_3 > A_1^2 + A_3^2A_0 \tag{4.39}$$

Systems with three degrees of freedom generally have a sextic for their frequency equation and in degenerated cases a quintic. In such cases there are three real parts of the roots of s, and besides the requirement that all coefficients A are positive, there are two other requirements which are quite lengthy. For more information refer to the original work of Routh. Applications of this theory are made in the hunting of speed control governors, discussed in section 4.14.

4.10 INTERNAL SHAFT HYSTERESIS

Hysteresis is a deviation from Hooke's stress-strain law for the shaft metal, and appears in many materials subjected to alternating stresses. In the diagram 4.14(*a*) Hooke's law would appear as a straight line, and a fiber of a vibrating shaft experiencing alternately tension and compression should move up and down the line between P_1 and P_3. Actually the stress-strain law is represented by a narrow elliptic figure running through in a clockwise direction, so narrow that it cannot be readily distinguished from a straight line.

Consider a vertical shaft rotating in two bearings with a central disk as shown in Fig. 4.14(*b*). During the whirling motion the center of the shaft S describes a circle about point B on the bearing center line. Point B is the normal or equilibrium position of S when no whirl exists. Figure 4.14(*c*) shows a cross section of the middle of the shaft, with the dotted circle as the path of S during the whirl. Assume that the rotation of the shaft and the whirl are both clockwise as shown. Line AA is the neutral line of strain, which due to the deviation from Hooke's law, does not coincide with the neutral line of stress.

In order to understand this, consider the point P_1 in Fig. 4.14(*c*) that will be taken as a reference mark on the shaft. As the shaft rotates the mark travels from Q_1, P_2, P_3, etc. Meanwhile the shaft whirls, whereby S and line AA run around the dotted circle. The speed of rotation and the speed of whirl are wholly independent of each other. In case the speed of rotation equals the speed of whirl, mark P_1 will always be in the elongation of line BS, and so P_1 will always be the fiber of maximum elongation. If the rotation is faster than the whirl, P_1 will gain on S and consecutively reach P_2 (of no elongation), P_3 (of maximum shortening), etc. On the other hand if the rotation is slower than the whirl, P_1 will go the other way (losing on S) and go through the sequence P_1, P_4, P_3, P_2, etc.

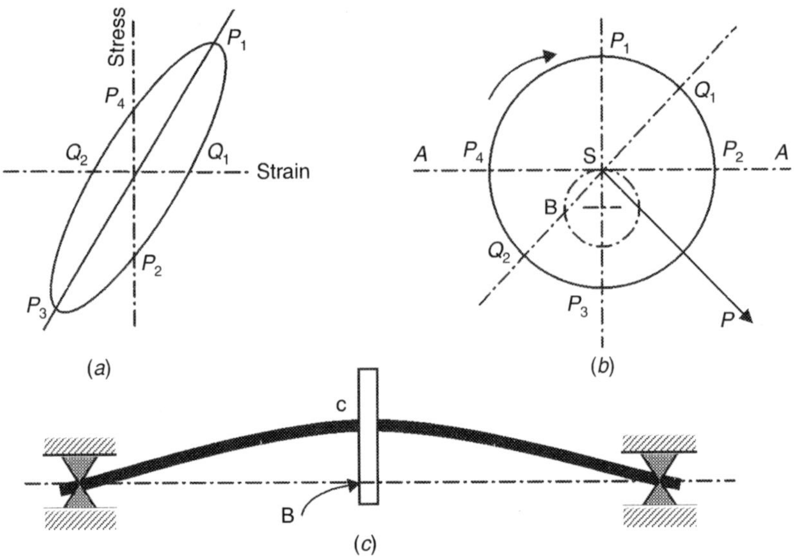

Fig. 4.14. Shaft Whirl Arising from Internal Hysteresis

First, investigate a rotation faster than the whirl. The state of elongation of the shaft fibers P_1, P_2, P_3, P_4 of Fig. 4.14(*c*) is indicated by the same letters in Fig. 4.14(*a*), where Q_1 of no stress lies between P_1 and P_2. Point Q_1 is now drawn in Fig. 4.14(*c*) and the same is done with Q_2 between P_3 and P_4. Thus line Q_1Q_2 is the line of no stress, and all fibers above Q_1Q_2 have tensile stress while those below Q_1Q_2 have compressive stress. The stress system described sets up an elastic force $\overline{P}$ as shown. This elastic force has not only a component toward B (the usual elastic force) but also a small component to the right, tending to drive the shaft around in its path of whirl. Thus, there is a self excited whirl. Note that if the rotation is slower than the whirl, the inclination of Q_1Q_2 reverses and the elastic force has a damping instead of a driving component.

The whirling motion is determined primarily by the elastic force of the shaft toward the center B combined with the inertia forces of the disk, thus taking place with the natural frequency. Internal hysteresis of the shaft material acts as damping on the whirl below the critical speed; above that speed a self excited whirl may build up.

Internal hysteresis in the shaft material may be small, but a more pronounced hysteresis loop is found in cases where actual slipping occurs, as in loose shrink fits or mechanical joints between rotating components. Slippage at bolted flange joints in rotors is noticeably higher in older machines that have accumulated a large number of operating cycles. Bolt holes may get elongated in the radial or tangential directions, and bolts may get bent. Thus, a shrunk disk will probably develop a whirl at the natural frequency above the critical speed.

4.11 AIRPLANE AND SHIP PROPELLERS

From vibration considerations an aircraft or a ship engine propeller blade is a complicated system. Its natural frequencies are different for various running speeds, because the centrifugal force due to its mass and speed tends to force the vibrating beam back to its middle position, thus acting like a spring force. The propeller blade can be idealized as a twisted cantilever beam, but the mass pr unit length and bending stiffness *EI* vary along the length. Each frequency can be calculated by Rayleigh's method by assuming some probable shape of deformation and determining the potential and kinetic energies. The potential energy will consist of two parts, one due to bending the other due to the centrifugal effect. As is true with Rayleigh's method, the natural frequency thus found is a little too large. To avoid the complex calculations in this manner, Southwell's theorem may be applied:

- If in an elastic system the spring forces can be divided into two parts so that the total potential energy can be divided into two partial potential energies, then the natural frequency ω of that system can be calculated approximately from:

$$\omega^2 = \omega_1^{\,2} + \omega_2^{\,2} \tag{4.40}$$

  where ω_1 and ω_2 are the exact natural frequencies of the (modified) system where one of the spring effects is absent. The value of ω thus found is a little too low.

Applied to a propeller blade, the theorem says that a good approximation for natural frequency ω when rotating can be found from (4.40) from the exact natural frequency at standstill (ω_1) and the exact natural frequency of a chain without bending stiffness of the same mass distribution as the blade and rotating at full speed (ω_2). The standstill frequency ω_1 can be determined by experimental methods on the actual propeller. The chain frequency ω_2, which expresses the effect of rotation, can be calculated without much difficulty. In the case of a central hub of negligible dimensions compared with the blade length, the chain frequency is the same as the rotating speed, or:

$$(\omega_2) = \Omega \tag{4.41}$$

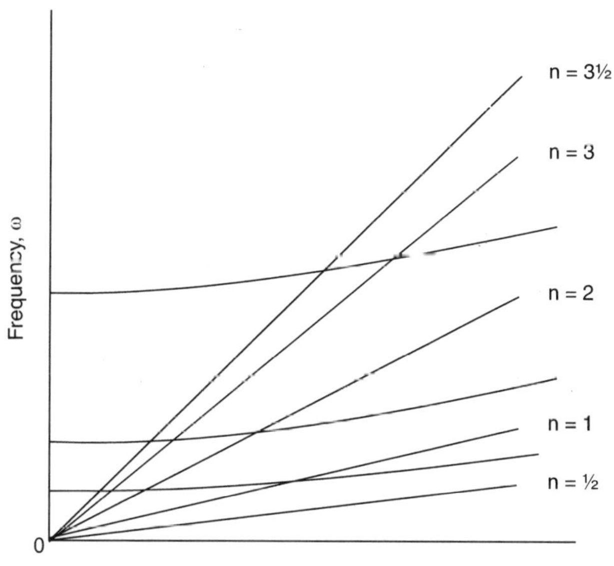

Fig. 4.15. Exciting Frequency Straight Lines and Natural Frequency Hyperbolas

Regardless of the mass distribution along the length of the chain, its natural frequency when rotating about one of its ends as a center is equal to the angular velocity of its rotation. Physically this can be explained as follows. The flat side of the propeller blade lies practically in the plane of rotation. Elements in the blade will move nearly perpendicular to this plane, that is, parallel to the axis of rotation. But the elements will be brought back to the straight, or equilibrium, position due to their own centrifugal force effect. The centrifugal force field may be considered as a spring force. The frequency of vibration of the elements will be the same as the rotational frequency setting up the centrifugal force, which is Ω. This effect is known as the centrifugal stiffening effect.

The principal source of excitation of blade vibration is found in torsional impulses on the crankshaft of the engine. An approximation for the first natural frequency of vibration of the rotating propeller blade is :

$$\omega^2 = \omega_1{}^2 + \Omega^2 \tag{4.42}$$

For higher modes the result can be expressed generally by:

$$\omega_{rot}{}^2 = \omega_{non\text{-}rot}{}^2 + a\Omega^2 \tag{4.43}$$

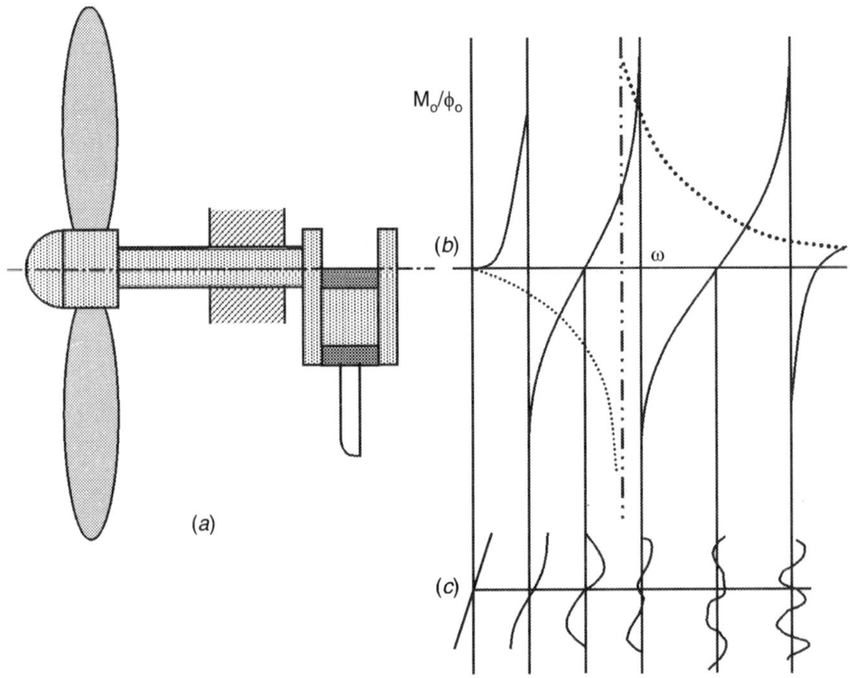

Fig. 4.16. (*a*) Propeller Blade and Crankshaft, (*b*) Mechanical Impedance: Full line for Blade, Dashed Line for Shaft, (*c*) Blade Mode Shapes at Various Frequencies

where Ω is the speed rotation and a is a numerical factor. This factor has a value of about 1.5 for the first mode, 6.0 for mode 2 and 12.0 for mode 3. A common manner in which (4.43) is plotted is shown in Fig. 4.15, in which two families of curves are given. The first set are hyperbolas showing the relation between the natural frequencies of a blade in its various modes and the speed of rotation as given by (4.43). The other set is a star of straight lines passing through the origin, expressing the relation between the exciting frequency and speed. These straight lines have slopes equal to the order of vibration, that is, to the number of oscillations per revolution. For four cycle internal combustion engines commonly used on aircraft, the orders occurring are either integer or half integer numbers, as shown in the figure. Any intersection of one of the straight lines of exciting frequency with one of the curves of natural frequency indicates a possible condition of resonance in torsional vibration. The determination of the natural frequencies of the non-rotating blade is not as simple as it may appear at first sight. This is because the bending frequency of the blade is coupled together with the torsional oscillation of the engine crankshaft. Thus, the frequency of vibration of the blade is different for different engines to which it is attached. But it is still desirable to have a criterion by which to determine characteristics of the propeller independent of the engine. This is possible from the reasoning used with Fig. 4.16(*a*), (*b*), (*c*).

Consider the shaft cut at the hub of the propeller. The amplitude of torque transmitted through the cut is $M_o \sin(\omega t)$, and the amplitude of torsional oscillation is $\phi_o \sin(\omega t)$. In the propeller blade there is a definite ratio between M_o and the frequency of ϕ_o. The ratio M_o/ϕ_o is called the mechanical impedance of the propeller, and is plotted in Fig. 4.16(*b*). The full lines are for the free blade, the dashed lines for the shaft. Mode shapes belonging to various frequencies are shown in Fig. 4.16(*c*). The actual condition of the hub lies between that of the completely free and that of completely clamped conditions. The natural frequencies of the combined system are the intersections between the dotted and full curve of Fig. 4.16(*b*). These frequencies are used on the ordinate axis of Fig. 4.15, and then Fig. 4.15 determines the critical speeds of the system caused by a purely torsional excitation.

Vibrations can take place in propeller blades due to other causes. When the engine is out of balance, the center of the propeller hub may move back and forth laterally, a motion independent of torsion and associated with the displacement of the engine. Besides unbalance, this may also be caused by torque variation. If there is clearance in the main bearings of the engine, or if the crankshaft is flexible, periodic thrust variations on the crank pin due to the firing cylinders may cause the crankshaft to deform and move within its bearing clearance so that engine center of gravity is displaced. A lateral motion of the center of the propeller hub at a frequency ω does not, however, cause stress variations in the propeller of the same frequency, but of the frequency $\omega + \Omega_{prop}$ or $\omega - \Omega_{prop}$. This is because the lateral motion of the blade root will cause bending vibrations in the blade of the same frequencies as the root displacement.

Still another possible source of excitation of bending vibrations is from aerodynamic forces in airplane engines. The airplane propeller is mounted in front of a wing and consequently each blade comes close to the wing twice during one revolution. The velocity of the air stream close to the wing is different from that at some distance from it, so the aerodynamic forces acting on the propeller blade will pass through a periodic change twice per revolution, causing bending vibrations in the blade.

The excitation of a ship propeller falls into two classes, torsional and linear. When an individual blade passes by the hull of the ship or the 'bossing' that holds the propeller tube in place, it is in a region of flow different from that in the more or less open water. Hence, the hydrodynamic forces are different, and so these forces experience variation with the blade frequency, that is, the frequency of revolution of the propeller multiplied by its number of blades. The torque variation caused by this effect is more serious when the 'bossings' are close to the propeller.

Another effect of the propeller is found on their reactions on the ship's hull and the 'bossings'. Hydrodynamic forces are dependent on the clearance between the 'bossings' and the propeller and between the hull and the tip clearance of the propeller. These hull forces are responsible for vibrations usually observed on the afterdecks of ships. If the propeller forces are of the same frequency as that of the entire ship as a 'free-free' beam, oscillations of considerable magnitude can be set up. One way to cure the problem is to replace the propeller by that of another configuration, using different number of blades.

4.12 BEARING OIL FILM

A self excited vibration known as oil whip is caused by certain properties of the oil film in generously lubricated sleeve bearings. To understand this phenomenon, it is necessary to know that in a horizontal shaft rotating in a counterclockwise direction in a bearing does not seek a central position but is deflected somewhat to the right (Fig. 4.17). Since journal load or weight *W* is acting downward, resultant *P* of the oil pressures is equal and opposite to *W*, making an angle α with line *OA* connecting the centers of the bearing and the journal.

Consider a vertical guide bearing with a shaft in it. If there are no lateral loads acting on the shaft, it will seek a central position in the bearing. If the shaft starts whirling for any reason, it will occupy an eccentric position in the bearing. Also, if the oil pressures are the same as shown in Fig. 4.17 (where *W* is now replaced by a centrifugal force), there is no equilibrium between P and the centrifugal loading, but there is a small resultant force tending to drive the journal around in the bearing in a counterclockwise direction. Thus the oil pressure distribution will encourage, or self excite, a whirl in the direction of rotation, but will damp a counter-rotating whirl if one were to set in.

Conditions under which oil film pressures during whirling will be the same as in a steady state case for a horizontal bearing with gravity loading need to be understood. Take two extreme cases, those in which the ratio of angular velocity of whirl and the angular velocity of rotation is either very small or very large. In the first case the shaft makes say 100 revolutions while the whirl moves forward 5°. In such a slow drift there is no effect on the pressure distribution, so for a slow whirl a succession of steady states occurs and the oil whirl develops. In the second case the journal center whirls around while the journal itself rotates very little. No oil film develops at all and the shaft merely vibrates in a bath of oil, which effectively damps the motion. The ratio of $\omega_{rotation}/\omega_{whirl}$ at which the damping passes from a positive to a negative value can only be determined by experiment. It has been found experimentally that if whirl frequency is equal to or smaller than half the rotation frequency, oil whip develops. This constitutes a serious problem in vertically mounted high speed machines with oil lubricated bearings, which may be difficult to overcome. Section 4.15 deals with dry friction due to rub between the rotor and the stator (or a bearing), where it will be shown that the direction of whirl is opposite to that of shaft rotation.

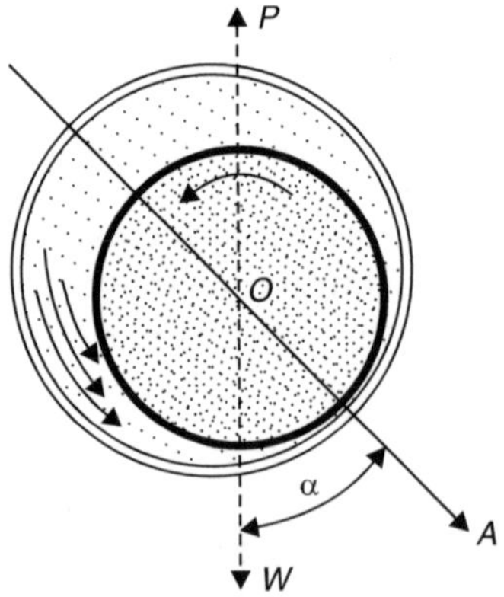

Fig. 4.17. Oil Film Lubrication in Bearing

For horizontal bearings with a certain loading oil whip also appears for speeds above twice the critical frequency. During the whirl oil pressures will not have a purely radial direction, it has a tangential component as well. The tangential component may be driving the whirl during part of the cycle and retarding during another part. For excitation it is necessary that the total work done by the tangential force component on the motion during a whole whirling cycle is positive, so that the average value of the tangential force component is positive, or driving.

4.13 FUEL INJECTION VALVE

A common construction for a liquid fuel injection valve and nozzle for diesel engines is shown in Fig. 4.18. Chamber V is permanently filled with liquid fuel oil and is connected to the fuel pump through the passage B. The normal position of valve A is on its seat N. At the instant the engine piston is ready to start on its firing stroke the fuel pump pushes a certain amount of fuel into V, where the pressure rises significantly. Since the valve stem has a greater diameter above than below, the pressure tends to push the valve up. As soon as the pressure is large enough to overcome the force in spring S the stem will go up and the liquid is forced through nozzle N into the cylinder head. At the end of the pump stroke the pressure in V falls and spring S closes the valve again.

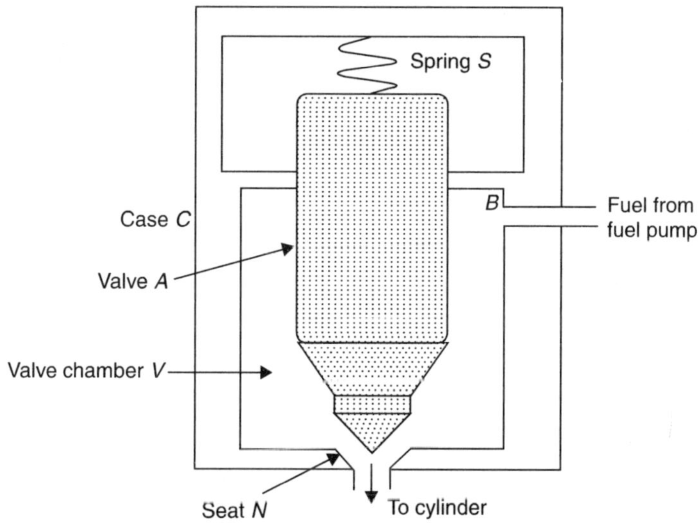

Fig. 4.18. Diesel Engine Fuel Injection Valve

Self excited vibrations of the valve have been observed, as shown in Fig. 4.19, where the upward displacement x of the valve is plotted against time. The shaded box represents the interval during which the fuel pump is delivering and flowing through passage B. Case c is for positive damping, case b is no damping and case a uses negative damping. The physical action may be understood as follows. During vibration part of the valve stem retreats from chamber V, oil flows in at B and out at N, all of which affects the pressure in V. If the average pressure is greater during the upward stroke of the valve stem than during its downward stroke, there is a feeding of energy into the vibration. If this energy is greater than the friction loss in the gland, the vibration is self excited.

In the absence of gland damping the system is unstable, which can be seen physically as follows. Consider only the period during which the fuel pump is operating, and assume that the fuel oil is flowing in at a constant rate through passage B. Oil outflow through the nozzle is varying, depending on the position of the valve stem. Assume the valve stem is vibrating about some average position. In this average position the outflow through the nozzle equals the inflow through B; when the stem is higher than the average position the outflow is greater than the inflow, and lesser when the stem is lower than the average position. Pressure in chamber V depends on the amount of oil in it; the more oil, the greater pressure. Consider the valve stem going up from its average position. During the next two quarter cycles of vibration the outflow exceeds the inflow and the pressure diminishes. Hence, when the valve stem is going down from its neutral position the pressure is at a minimum. In the same manner when the stem is in the middle of its upward stroke the pressure is a maximum. Thus, the pressure does work on the vibration.

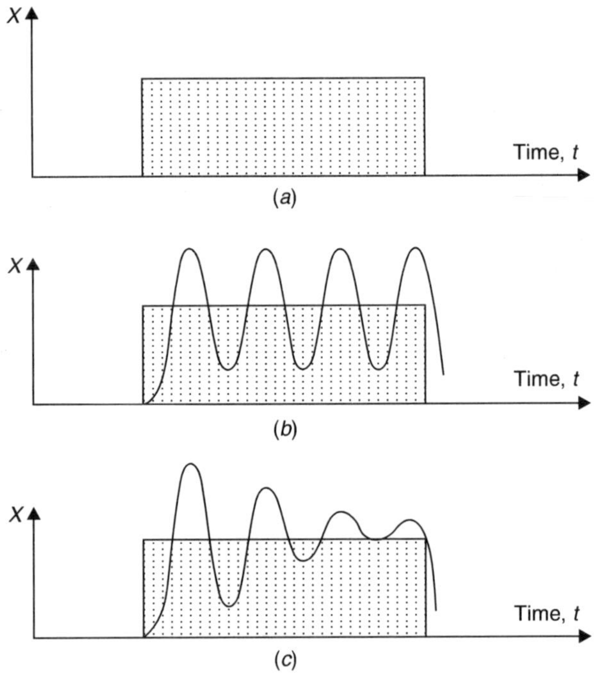

Fig. 4.19. Mechanical Vibrations of: (*a*) Unstable, (*b*) Neutral, (*c*) Stable Valve System

In the discussion it has not been mentioned that due to its motion the valve stem changes the volume in chamber V, thus causing pressure variations. The total pressure caused by the fuel pump is so great that these variations are not considered to affect the outflow, which is determined by the nozzle opening only. Also, these pressure variations are in phase with valve displacement, and thus act as an oil spring and not as damping.

4.14 HUNTING OF SPEED CONTROL GOVERNOR

Interesting self excited vibration phenomena are observed in constant speed reciprocating steam engines and other machinery operating in conjunction with an inertia governor of the direct acting type. By this is meant that the speed sensitive part of the governor, namely, the fly balls, is in direct mechanical connection with the steam supply throttle valve. Figure 4.20 shows the system schematically. When the speed of the engine increases due to some reason, the fly balls lift the sleeve b of the governor higher, thereby reducing the flow of the steam flow through valve c. Less steam is admitted and engine speed falls. Since there is inertia in the system the speed will decrease below normal, resulting in the governor opening the valve more than normal. Hence, engine speed oscillations occur, which may be damped or self-excited, depending on the circumstances. The unstable case has occurred repeatedly, and technically it is referred to as 'hunting'. In order to understand the phenomenon it is convenient to start with the differential equations. The governor is an ordinary vibrating system consisting of a mass, a spring and a dashpot, giving three terms in the equation.

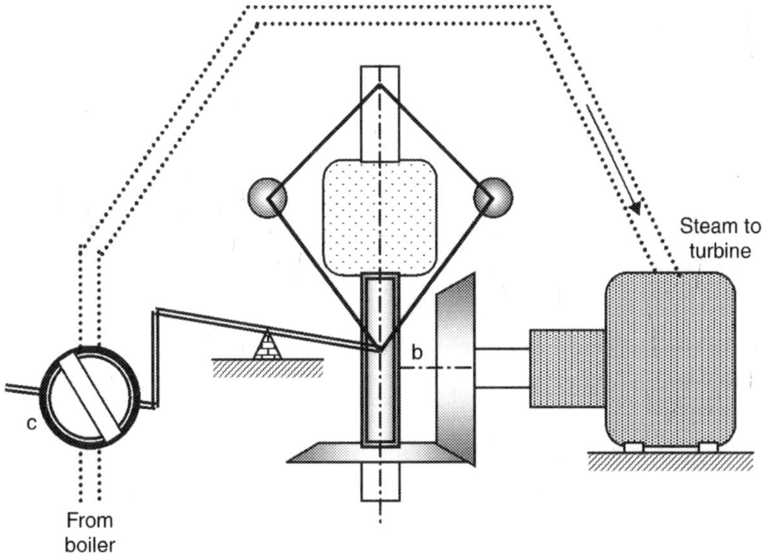

Fig. 4.20. Watt Governor Directly Coupled to Throttle Valve of Steam Engine

Also, the governor is connected to the engine in such a manner that when engine speed increases an additional force on the governor mass ensues, caused by the centrifugal force of the fly balls. This gives the equation:

$$md^2x/dt^2 + cdx/dt + kx = C_1 d\phi/dt \tag{4.44}$$

where x = upward displacement of governor speed measured from the normal position at a certain load

m = equivalent mass of governor sleeve

c = damping coefficient at governor sleeve

k = stiffness of governor spring

$d\phi/dt$ = difference between instantaneous engine speed and average speed at a certain load

C_1 = increase in upward force on governor sleeve from centrifugal action caused by increase in engine speed of 1 radian/second.

The two coordinates x and $d\phi/dt$ are based on normal engine running condition, when both parameter values are zero. x is positive when the governor sleeve is higher than normal; similarly, $d\phi/dt$ is negative when engine speed is slower than normal. The damping torque from the engine crankshaft will be neglected in this investigation. The engine is coupled to the governor in a sense that when the governor sleeve is lower than normal (negative x), the throttle is open wider than usual, so that an extra driving torque is exerted on the engine. Its equation of motion is given by the expression:

$$I\frac{d\phi}{dt} = C_2 x \tag{4.45}$$

where I = equivalent moment of inertia of the engine

C_2 = increase in steam torque of engine caused by a lowering of governor sleeve by 1 in.

Equations (4.44) and (4.45) represent the free vibrations of the engine-governor system, with no periodic force present. The solution of the differential equation may be expressed in the form:

$$e^{st} \tag{4.46}$$

where $s = p + jq$ is a complex number, resulting in the 'complex frequency'. Substitution into (4.44) and (4.45) gives a set of algebraic homogeneous equations which have a solution only if:

$$s^3 + \frac{c}{m}s^2 + \frac{k}{m}s + \frac{C_1C_2}{mI} = 0 \tag{4.47}$$

Equation (4.47) is the frequency equation of the set (4.44) and (4.45). The equation is a cubic instead of a quartic since the spring constant is absent in the engine system. Of the two criteria for stability shown in section 4.9, the first one requiring all coefficients to be positive is satisfied. The other criterion, expressed by (4.37), will then become:

$$c > (mC_1C_2/kI) \tag{4.48}$$

If damping in the governor dashpot is greater than the value indicated by this equation, the system will come to rest after a sudden change in the load. For any damping less than this the system is inoperative.

In case the engine is rigidly coupled to an electric generator feeding a large network, the problem becomes more complicated. There is an 'engine spring', since the network tends to keep the generator in a definite angular position. Any deviation from that synchronous position is opposed by a torque caused by the magnetic spring in the air gap of the generator. An electric damper winding is used in the generator to alleviate the hunting trouble. It consists of a short-circuited copper winding in the pole faces of the rotating part of the generator. As long as the generator runs at constant (synchronous) speed no current flows in this winding, and consequently, does not impede the motion. With changes in speed currents are induced in the winding, which together with the magnetic field in the air gap produce a torque proportional to the deviation of the angular velocity from the synchronous. This torque is braking if the engine temporarily runs too fast and driving when running too slow.

4.15 INSTABILITY DUE TO FRICTION

In a number of cases friction, instead of providing positive damping, gives rise to negative damping. A classic example is that of the violin string being excited by a bow.

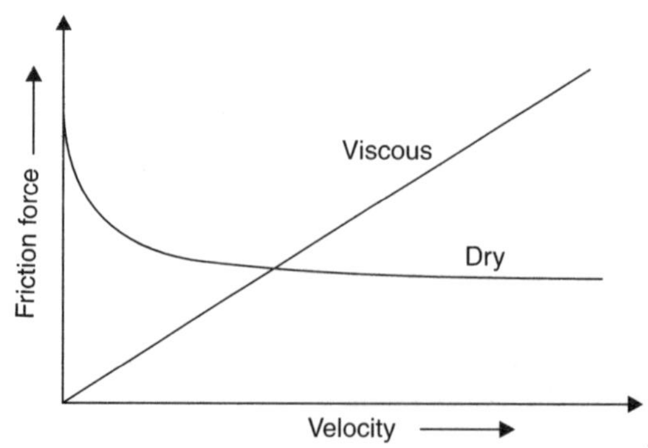

Fig. 4.21. Damping Forces with Positively and Negatively Sloping Characteristics

The string is a vibrating system, and a steady pull of the bow provides the required source of non-alternating energy. The friction between the string and the bow has the characteristic of being greater for small slipping velocities than for larger ones, which is completely opposite to that of viscous friction (Fig. 4.21). Note that the large friction force acts in the direction of motion of the string, whereas the small friction force acts against the motion of the string. Since the string executes a harmonic motion, work done by the friction force during one half stroke is $2Fx_o$ where F is an average value of the friction force and x_o the amplitude of vibration. Since F is greater during the forward stroke (when friction does positive work on the string) than during the receding stroke (when negative work is done), the total work done by friction over a full cycle is positive, and so the vibration builds up.

In mechanical engineering certain vibrations, usually referred to as 'chatter', can be explained in the same manner. The cutting tool on a lathe may chatter and also the driving wheels of a locomotive. When starting a heavy train these drivers sometimes slip on the rails. Though slipping usually takes place in a uniform manner, chattering slip is sometimes observed. Besides the major slipping rotation the wheels execute torsional oscillations, causing large alternating stresses in the crank pins and side rods. A negative slope of the friction-velocity characteristic between the wheels and the rails is essential for this phenomenon. Other examples of this phenomenon may be observed, such as the door that binds and screeches when opened, and the piece of chalk held perpendicular to the board while writing. Another case is that of rubbing the rim of a water glass with a wet finger, causing them to sing.

Torsional vibration of this type occurs in ship's propellers when turning at low (creeping) speeds. The shaft is usually supported by one or two outboard bearings of the hard rubber type, which are water lubricated. At low speeds a water film does not form and the bearings are dry, causing a torsional vibration of the shaft at one of its natural frequencies. The propeller blades have their own natural frequencies that are not too far removed, and act as loud speakers, making this starting squeal detected at great distances under water.

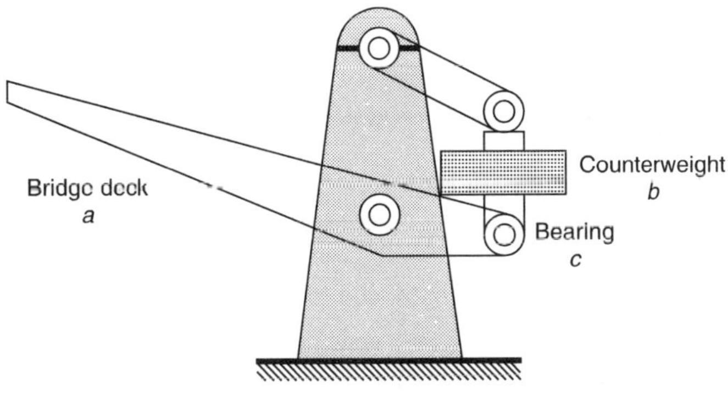

Fig. 4.22. Drawbridge

A striking example of self-excited vibrations caused by dry friction is shown in Fig. 4.22, which represents a large drawbridge. Bridge deck a is counter balanced by a large concrete counter weight b, which together with the guiding links and the supporting tower forms a parallelogram. After nearly a year's operation one of the towers of the bridge broke. On inspection the failure proved to be unmistakably caused by fatigue. Experiments with the other half of the bridge, still standing, showed that violent vibrations of the whole structure took place at a very low frequency when the deck was raised or lowered, the frequency being about six cycles during the time period of raising the deck. The explanation was found in bearing c, carrying the large load of counter weight b. Whatever grease happened to be in this bearing at the beginning of the new bridge was squeezed out, and the bearing was found to be entirely dry. Dry friction chatter thus caused was sufficiently violent to cause the failure. The obvious remedy in this case is proper grease cups kept in proper order and inspected daily.

An important case of dry friction excitation occurring in rotating machines is 'shaft whipping' caused by a loose guide or by a poorly lubricated bearing with excessive clearance. In Fig. 4.23 circle A represents the inside of a bearing or a stator, and B the cross section of the vertical shaft running in it. The shaft is rotating clockwise and is temporarily deflected from its equilibrium position so it strikes A at left.

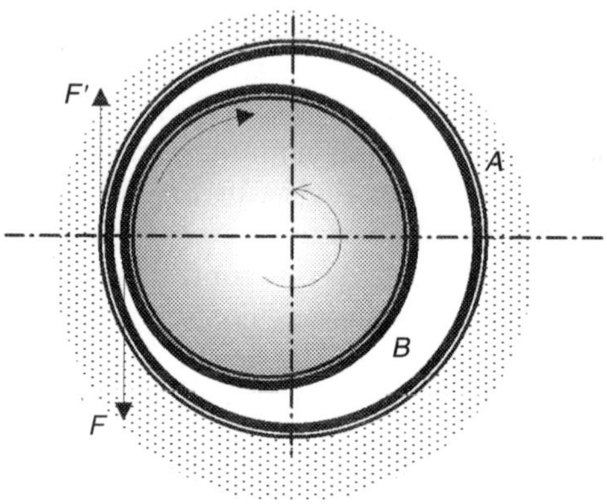

Fig. 4.23. Shaft Whirl Caused by Dry Friction

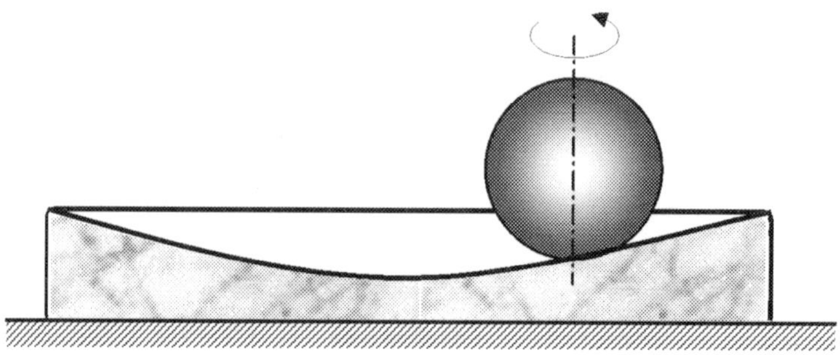

Fig. 4.24. Ball Whirling Around due to Friction at Point of Contact

Due to its rotation the shaft sets up friction forces F and F', of which F acts on the shaft and $F' = -F$ acts on the stator, or the bearing. Force F can be replaced by a parallel force of equal magnitude through the center of the shaft B and a couple Fr, where r is the offset between F and F'. The couple acts as a brake on the shaft which is driven at uniform speed, so the only effect of the couple is to require additional driving torque and is inconsequential. The force F through the center of the shaft drives it in a direction tangent to the circle A. The direction of F changes with the position of the shaft B in A, so the shaft will be driven around as indicated by the dotted circle. Notice that the shaft is driven around the clearance in a direction opposite to that of its own rotation. If the shaft rotates in the center of the bearing or stator without touching it, the shaft is stable. As soon as it strikes the stator for any reason, the shaft is set into a violent whirling vibration.

This effect is present in many modifications. A simple model for demonstrating it is as follows. Consider a shallow conical cup and a steel ball (Fig. 4.24). The ball is given a spin between the fingers at the bottom of the cup. This position is an unstable one for the rotating ball, because due to a small accidental displacement from the center of the cup, the point of contact with the cup no longer coincides with the vertical axis of rotation. There will be a slip and a friction force perpendicular to the paper causing the ball to drive around in a circle. The direction of rolling of the ball will be opposite to the direction of the spin.

The phenomenon is not restricted to cylindrical guides or bearings but is also observed on thrust bearings. Figure 4.25 shows the schematic representation of a thrust bearing and a shaft, the equilibrium position of which is central and vertical. Assume that this elastic system is capable of a natural mode of motion whereby the shaft center line whirls around the vertical with an eccentricity δ and an inclination α. Collar disk center A describes a circle of radius δ and the shaft a cone of apex angle 2α. This mode of motion will be self excited by friction, because during the vibration the collar rests on the bearing on one side only. This causes a tangential friction force on that side, pushing point A around the center line in a direction opposite to that of the rotation of the collar disk. A clear way to prevent this type of disturbance is to make the bearing support flexible, so that in spite of the angular deviation the pressure on all parts of the bearing remains uniform.

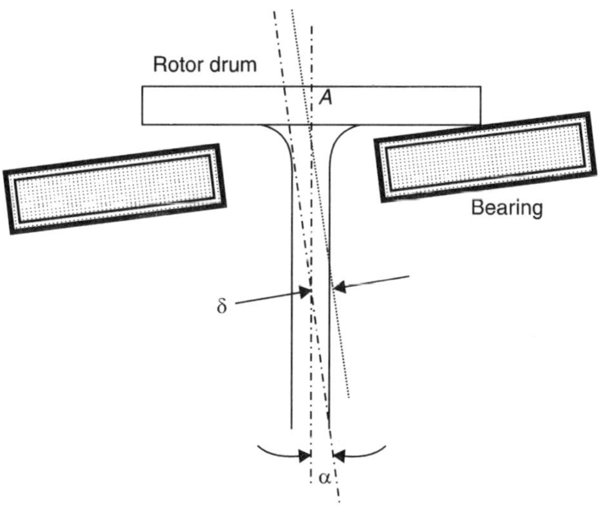

Fig. 4.25. Whirl due to Dry Friction in a Thrust Bearing

4.16 EXAMPLE PROBLEMS

Example Problem 4.1: In the system shown in Fig. 4.26 the following numerical data is given:

Weight of piston	= 6.0 lbs.
Weight of connecting rod	= 3.0 lbs.
Total engine weight	= m = 500 lbs.
Engine speed	= 600 rpm
Crank radius	= 8 in.

Length of connecting rod = 24 in.

Spring stiffness = k =11500 lbs/in.

Neglecting damping, find the maximum displacement of the engine from its equilibrium position during steady state forced vibrations of the system. Assume that one-third of the connecting rod weight (1lb) is reciprocating, so total reciprocating weight is 7.0 lbs and total rotating weight is 2.0 lb. Assume also that the crankshaft is perfectly balanced.

Solution: The natural frequency of vibration of the total reciprocating machine system will be calculated first.

$$\omega_n = \sqrt{\frac{k}{m}} = \sqrt{\frac{(11500)(386)}{500}} = 94.3 \text{ sec}^{-1}$$

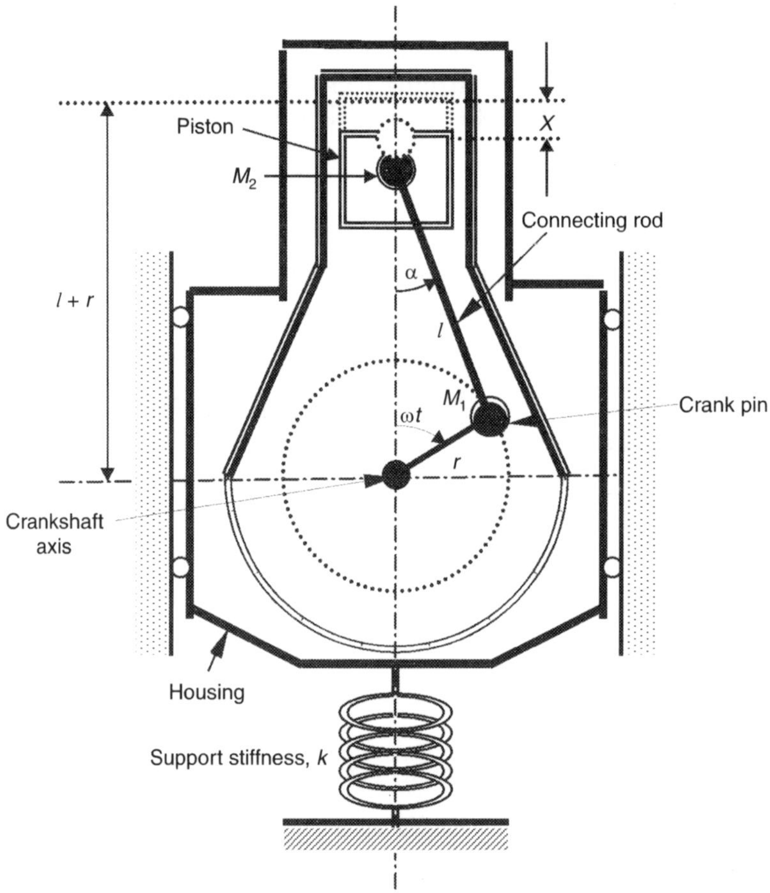

Fig. 4.26

Also, the operating speed is:

$$\omega = \frac{(600)(2\pi)}{60} = 62.83 \text{ sec}^{-1}$$

Thus: $$\frac{\omega}{\omega_n} = \frac{62.83}{94.3} = \frac{2}{3}$$

The primary disturbing force corresponding to cos (ωt) will be working below resonance, while that proportional to cos ($2\omega t$) will be working above resonance. Neglecting higher frequency components of the disturbing force, the primary vertical direction inertia force for all moving parts is:

$$X_1 = (m_{rec} + m_{rot})r\omega^2$$

$$= \left\{\frac{7+2}{386}\right\}(400\pi^2)(8) = 736 \text{ lb.}$$

The secondary vertical inertia force is:

$$X_2 = m_{rec}\frac{r^2}{l}\omega^2$$

$$= \frac{7}{386}\frac{8^2}{24}(400\pi^2) = 191 \text{ lb}$$

The maximum undamped forced vibration produced separately by the two disturbing forces is:

$$x_1 = \frac{X_1}{k}\left\{\frac{1}{1-\omega^2/\omega_n^2}\right\} = \frac{736}{11500}\left\{\frac{1}{1-4/9}\right\} = 0.115 \text{ in.}$$

$$x_2 = \frac{X_2}{k}\left\{\frac{1}{1-4\omega^2/\omega_n^2}\right\} = \frac{191}{11500}\left\{\frac{1}{1-16/9}\right\} = 0.0214 \text{ in.}$$

The maximum displacement occurs at $\omega t = \pi$; then:

$$(x_1 + x_2)_{max} = 0.115 + 0.0214 = \mathbf{0.136 \text{ in.}}$$

Example Problem 4.2: A single degree of freedom system is subjected to a disturbing force $F(t)$ that varies with time according to the diagram shown in Fig. 4.27. Neglecting effects of damping, find the steady state forced vibrations that will be produced if the mass m and the spring constant k are such that $\omega/\omega_n = 0.9$.

Solution: A periodic disturbing force of any kind can be represented in the form of a Fourier (or trigonometric) series as follows:

$$F(t) = a_0 + a_1 \cos(\omega t) + a_2 \cos(2\omega t) + \ldots$$

$$+ b_1 \sin(\omega t) + b_2 \sin(2\omega t) + \ldots \quad (4.49)$$

$$= a_0 + \sum_{i=1}^{i=n}\{a_i \cos(i\omega t) + b_i \sin(i\omega t)\}$$

The period of disturbing force is $T = 2\pi/\omega$. Quantities a_0, a_i, and b_i are constants in the Fourier trigonometric series, and are to be determined from the expressions given below:

$$a_0 = \frac{1}{T}\int_0^T F(t)\,dt$$

$$a_i = \frac{2}{T}\int_0^T F(t)\cos(i\omega t)\,dt \quad (4.50)$$

$$b_i = \frac{2}{T}\int_0^T F(t)\sin(i\omega t)\,dt$$

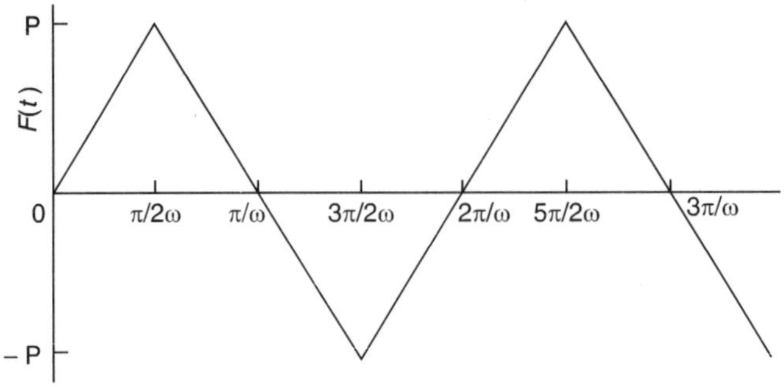

Fig. 4.27

The coefficients can be calculated if $F(t)$ is known analytically. Here it is a straight line in the period T, and so can be readily expressed mathematically. If $F(t)$ is given graphically and analytical expression is available, some approximate numerical method for calculating the integrals must be used. Note that the integral of $F(t)$ in the period $T = 2\pi/\omega$ is the area under the saw-tooth diagram in Fig. 4.27 between the ordinates $t = 0$ and $t = T = 2\pi/\omega$. This area is zero, hence $a_0 = 0$. Considering the next equation for a_i, each ordinate of the diagram in Fig. 4.27 must be multiplied by $\cos(i\omega t)$ and then integrated from $t = 0$ and $t = 2\pi/\omega$. From the antisymmetry of $F(t)$ with respect to $t = \pi/\omega$, together with the symmetry of $\cos(i\omega t)$ with respect to the same time, it can be concluded again that the integral for a_i vanishes, and a_i is also zero.

Finally, in the equation for b_i, each ordinate of the diagram in Fig. 4.27 must be multiplied by $\sin(i\omega t)$ and then integrated from $t = 0$ and $t = 2\pi/\omega$. In this case $F(t)$ from $t = 0$ to $t = \pi/\omega$ is symmetric about $t = \pi/2\omega$; while from $t = \pi/\omega$ to $t = 2\pi/\omega$, it is symmetric about $t = 3\pi/2\omega$. However, when i is an even integer, corresponding parts of $\sin(i\omega t)$ are antisymmetric about $t = \pi/2\omega$ and $t = 3\pi/2\omega$ respectively. Thus, for $i = 2, 4, 6, \ldots$, it can be concluded that $b_i = 0$. When i is an odd integer both $F(t)$ and $\sin(i\omega t)$ are antisymmetric about the ordinate $t = \pi/\omega$. Then:

$$b_i = \frac{\omega}{\pi}\int_0^{2\pi/\omega} F(t)\sin(i\omega t)\,dt = \frac{4\omega}{\pi}\int_0^{\pi/2\omega} F(t)\sin(i\omega t)\,dt \tag{4.51}$$

Referring to Fig. 4.27, in the interval $t = 0$ to $t = \pi/2\omega$

$$F(t) = \frac{2P\omega t}{\pi} \tag{4.52}$$

Substituting into equation (4.25), then:

$$b_i = \frac{8P}{i^2\pi^2}\int_0^{i\pi/2} u\sin(u)du$$

$$= \frac{8P}{i^2\pi^2}\sin\frac{i\pi}{2} = \frac{8P}{i^2\pi^2}(-1)^{(i-1)/2} \tag{4.53}$$

where $i = 1, 3, 5, 7, \ldots$

Substituting the expressions for $a_0 = 0$, $a_i = 0$ and for b_i, the trigonometric series then becomes:

$$F(t) = \frac{8P}{\pi^2}\left\{\sin(\omega t) - \frac{1}{3^2}\sin(3\omega t) + \frac{1}{5^2}\sin(5\omega t) - \ldots\right\} \tag{4.54}$$

Therefore, to represent the saw-tooth diagram in Fig. 4.27 by a trigonometric series, the sine curves with an odd number of full waves in the interval $t = 0$ and $t = 2\pi/\omega$ must be superimposed. Furthermore, the series (4.27) converges rapidly, so only the first term is of practical importance. That is, the saw-tooth disturbing force produces approximately the same effect as a sinusoidal disturbing force of slightly smaller amplitude given by:

$$F(t) = \frac{8P}{\pi^2} \sin(\omega t) \tag{4.55}$$

To judge the insignificance of the second term, note that for $\omega/\omega_n = 0.9$ the magnification factor is:

$$\frac{1}{1-(3\omega/\omega_n)^2} = -0.159$$

Thus, the amplitude of forced vibration produced by the second term is only $0.159/3^2 = 0.0177$ times that produced statically by the force $8P/\pi^2$, while the magnification factor for the first term is:

$$\mathbf{\frac{1}{1-\omega^2/\omega_n^{\,2}} = 5.26}$$

It can therefore be concluded that using the approximate expression (4.55) results in an error of less than 0.4%, and approximate solution for response is (k = spring constant):

$$\mathbf{x = \frac{8P(5.26)}{\pi^2 k} \sin(\omega t)}$$

Example Problem 4.3: Figure 4.28 shows a hollow round shaft supported by self-aligning bearings at the ends. A V-belt pulley is mounted on the shaft, exerting a radial load of 445 lbs. on the shaft. A gear is also mounted on it, and it results in a radial load of 137.5 lbs. The two loads act in the same direction. Bending stress in the shaft is not to exceed 10,000 psi. Assume ratio of shaft outer and inner diameters = 2.0, and ignore weight of the shaft. Determine inner and outer diameters of the shaft.

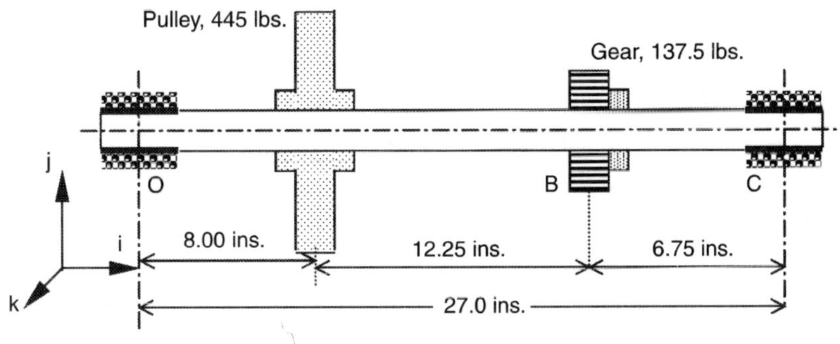

Fig. 4.28

Solution: Self-aligning bearings permit rotation of the shaft, hence the shaft may be considered simply supported, with concentrated load reactions. Load diagram is shown in Fig. 4.29. Taking moments about C,

$$\Sigma M_C = -27R_1 + 19(445) + 6.75(137.5) = 0$$

Hence, $R_1 = 347.5$ lbs.

Next, consider overturning moments about O:

$$\Sigma M_O = 27.0R_2 - 8(445) - 20.25(137.5) = 0$$
$$R_2 = 235.0 \text{ lbs.}$$

As an alternative, the problem may also be solved using vector algebra. Unit vectors **i**, **j**, **k** are associated with axes x, y, z. Vectors from the origin O to load points A, B and C will be:

$$\mathbf{r}_A = 8.0i \quad \mathbf{r}_B = 20.25\mathbf{i} \quad \mathbf{r}_C = 27.0\mathbf{i}$$

and the force vectors are:

$$\mathbf{R}_1 = R_1\mathbf{j} \quad \mathbf{F}_A = -445\mathbf{j} \quad \mathbf{F}_B = -137.5\mathbf{j} \quad \mathbf{R}_2 = R_2\mathbf{j}$$

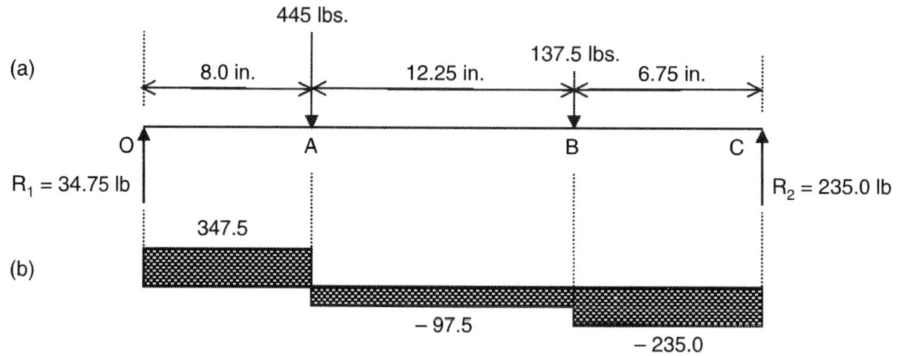

Fig. 4.29

Summation of moments about O will yield the vector identity:

$$\mathbf{r}_A \times \mathbf{F}_A + \mathbf{r}_B \times B + \mathbf{r}_C \times \mathbf{R}_2 = 0 \tag{4.56}$$

The vector cross products are:

$$\begin{vmatrix} i & j & k \\ 8 & 0 & 0 \\ 0 & -445 & 0 \end{vmatrix} + \begin{vmatrix} i & j & k \\ 20.25 & 0 & 0 \\ 0 & -137.5 & 0 \end{vmatrix} + \begin{vmatrix} i & j & k \\ 27.0 & 0 & 0 \\ 0 & R_2 & 0 \end{vmatrix} = 0$$

or

$$-3560\mathbf{k} - 2784.4\mathbf{k} + 27\mathbf{R}_2\mathbf{k} = 0$$

Solving the algebraic equation gives reaction $R_2 = 235.0$ lbs., or $\mathbf{R}_2 = 235.0\mathbf{j}$. Then:

$$\mathbf{R_1} = -\mathbf{F_A} - \mathbf{F_B} - \mathbf{R_2} = -(-445\mathbf{j}) - (-137.5\mathbf{j}) - 235.0\mathbf{j} = 347.5\mathbf{j} \text{ lbs.}$$

Figure 4.29(b) shows the shear force diagram. Maximum bending moment occurs where shear force changes sign at A, and has the value:

$$M_A = 347.5 \times 8 = 2780 \text{ in-lbs.}$$

Section modulus may be computed from the equation:

$$\frac{I}{c} = \frac{I}{d_o} = \frac{\pi(d_o^3 - d_i^3)}{32} = 0.0982(d_o^3 - d_i^3)$$

$$\text{Bending stress: } \sigma = \frac{M}{I/c} = \frac{2780.0}{.0982(d_o^3 - d_i^3)}$$

Substituting $\sigma = 10{,}000$ psi, and using $d_o = 2d_i$ gives $d_i = 0.7396$ in. and $d_o = 1.479$ in. Use 1.5 in. shaft outer diameter, 0.75 in. inner diameter.

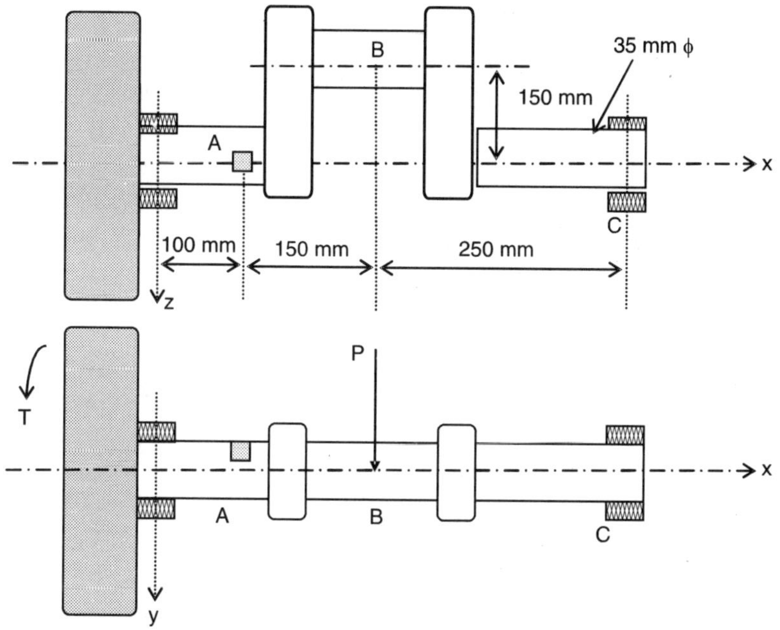

Fig. 4.30

Example Problem 4.4: A portion of the kinetic energy stored in the flywheel of a one-cylinder reciprocating compressor's drive mechanism shown in Fig. 4.30 is used to produce some of the piston force *P*. The flywheel delivers 1,200 N-m torque to the crankshaft, all of which develops the piston force. A stress element is located at the top surface of the crankshaft journal at *A*, located 100 mm from the flywheel end bearing. Element sides are parallel to the *x* and *z* axes. Determine the state of stress at the element *A*.

Solution: Since load vector *P* is offset from *A* along axes *x* and *z*, one overturning moment component will be generated. Point *A* lies along the load-carrying path from the flywheel to the piston, hence the torque load will create shear stress. The moment due to crank offset of 150 mm will result along the shaft axis *x*, which opposes the torque *T*. Thus, at point *A* total torque loading is:

$$T = (1200 \times 1000) = 1200000 \text{ N-mm.}$$

(*a*) Torsional shear stress: $\tau_{xy} = \dfrac{T.R}{J} = \dfrac{(1200000)(17.5)}{\dfrac{\pi}{32}(35)^4}$

$= 142.5 \text{ N/mm}^2$

(*b*) Overturning moment at *A*: $M_A = P/2 \times 100$

$= [1200 \times 1000/150]/2\} \times 100$

$= 4 \times 10^5$ N-mm.

(*c*) Bending stress at *A*: $\sigma_A = (M_A c/I)$

$= (400000 \times 17.5)/(\pi \times 35^4/64)$

$= 95.0 \text{ N/mm}^2$

(*d*) Maximum principal stress at *A*:

$$\sigma_{max} = \frac{\sigma_x + \sigma_y}{2} + \sqrt{\left[\frac{\sigma_x - \sigma_y}{2}\right]^2 + \tau_{xy}{}^2}$$

$$= \frac{95.0}{2} + \sqrt{\left[\frac{95.0}{2}\right]^2 + 142.5^2}$$

$$= 197.71 \text{ N/mm}^2$$

(*e*) Minimum principal stress at *A*:

$$\sigma_{min} = \frac{\sigma_x + \sigma_y}{2} - \sqrt{\left[\frac{\sigma_x - \sigma_y}{2}\right]^2 + \tau_{xy}{}^2}$$

$$= \frac{95.0}{2} - \sqrt{\left[\frac{95.0}{2}\right]^2 + 142.5^2}$$

$$= -\ 102.71 \text{ N/mm}^2$$

(*e*) Principal shear stress at *A*:

(*f*)
$$\tau_{max} = +/- \sqrt{\left[\frac{\sigma_x - \sigma_y}{2}\right]^2 + \tau_{xy}{}^2}$$

$$= +/- \sqrt{\left[\frac{95.0}{2}\right]^2 + 142.5^2}$$

$$= +/-\ 150.2 \text{ N/mm}^2$$

(*g*) Direction of principle stress at *A*: The angle 2ϕ from the *x*-axis counter clockwise to σ_A is:

$$\tan(2\phi) = -\frac{\sigma_A}{2\tau_{xy}} = -\frac{95.0}{2 \times 142.5} = -\ 0.3333$$

$$2\phi = -\ 18.4°$$

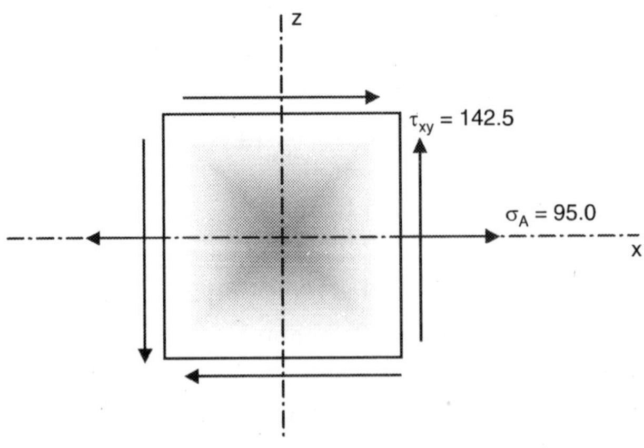

Fig. 4.31. Stress Components at Element *A*

Figure 4.31 shows the state of stress at A. Normal and shear stresses are shown on the element surfaces. Figure 4.32 provides representation of the principal stresses.

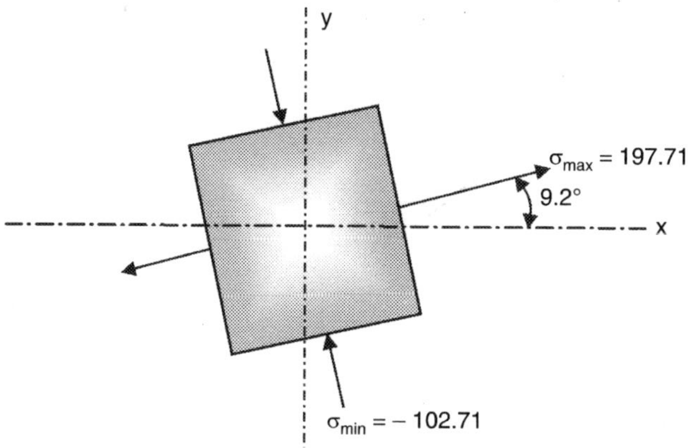

Fig. 4.32. Principal Stresses

Example Problem 4.5: Slider crank mechanisms may be analyzed by the loop closure equations and complex numbers. The mechanism of Fig. 4.33 has the crank rotating at a constant angular velocity ω_2. If the position of point B relative to the fixed point O is given by the vector r_B, determine the velocity and acceleration of B.

Solution: From Fig. 4.33(*b*) two independent vector equations may be written for point B: $r_B = r_1$ and $r_B = r_2 + r_3$, and so $r_1 = r_2 + r_3$. Using complex notation, then:

$$r_1 e^{i\theta 1} = r_2 e^{i\theta 2} + r_3 e^{i\theta 3}$$

where r_2 and r_3 are the length of links 2 and 3. Angle θ_1 of vector r_1 will be taken to be 0, hence $e^{i\theta 1} = 1$. Two successive differentiation of this expression will yield velocity and acceleration of point B, as shown below:

$$d(r_1 e^{i\theta 1})/dt = V_B = r_2\omega_2(e^{i\theta 2}) + r_3\omega_3(e^{i\theta 3})$$

$$d^2(r_1 e^{i\theta 1})/dt^2 = A_B = r_2(i\alpha_2 - \omega_2^{\,2})e^{i\theta 2} + r_3\,(i\alpha_3 - \omega_3^{\,2})e^{i\theta 3}$$

Note that these expressions require angular velocity and acceleration of links 2 and 3. In the expressions the known quantities are: r_2, r_3, θ_2, ω_2 and α_2, while the unknown quantities are: r_1, θ_3, ω_3, α_3 and the velocity and acceleration of r_1 . r_1 and θ_1 may be found by separately equating the real and imaginary parts:

$$r_1 = r_2[\cos(\theta_2) + i\sin(\theta_2)] + r_3[\cos(\theta_3) + i\sin(\theta_3)]$$

$$r_1 = r_2\cos(\theta_2) + r_3\cos(\theta_3)$$

$$0 = r_2\sin(\theta_2) + r_3\sin(\theta_3)$$

Hence:
$$\theta_3 = \sin^{-1}\left(\frac{-r_2}{r_3}\sin(\theta_2)\right)$$

From this expression r_1 may also be determined. In a similar manner, the unknowns for velocity of r_1 and ω_3 may be obtained from the equation for V_B by separately equating the real and imaginary parts.

$$d(r_1 e^{i\theta 1})/dt = r_2\omega_2[i\cos(\theta_2) - \sin(\theta_2)] + r_3\omega_3[i\cos(\theta_3) - \sin(\theta_3)]$$

$$d(r_1 e^{i\theta 1})/dt = -r_2\omega_2\sin(\theta_2) - r_2\omega_3\sin(\theta_3)$$

$$0 = r_2\omega_2 \cos(\theta_2) + r_3\omega_3 \cos(\theta_3)$$

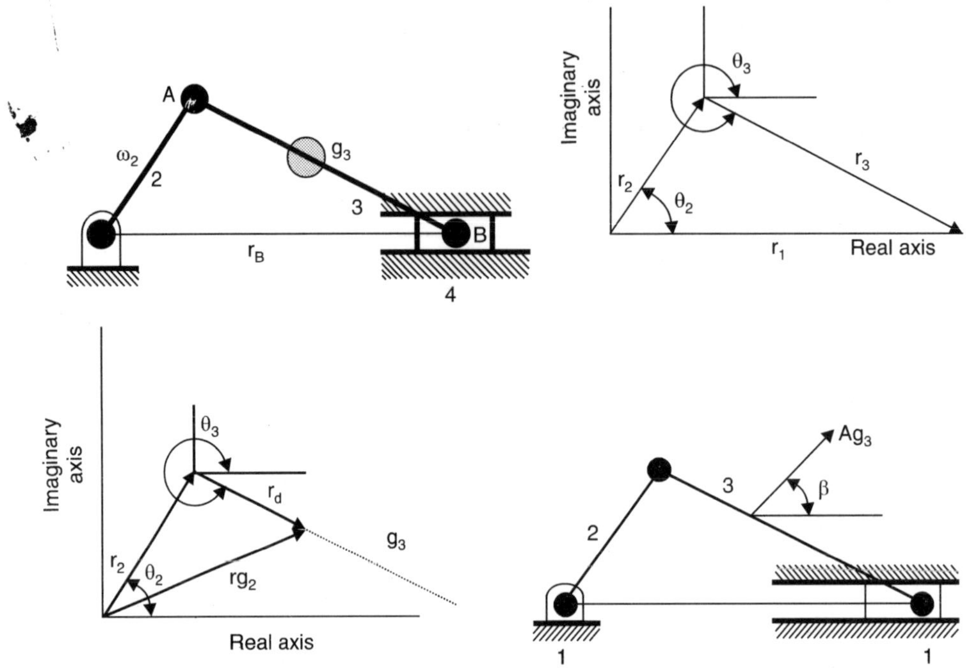

Fig. 4.33

Hence:
$$\omega_3 = -\,\omega_2 \left(\frac{r_2 \cos(\theta_2)}{r_3 \cos(\theta_3)}\right)$$

The remaining unknowns, acceleration of r_1 and α_3, may be determined from the real and imaginary parts of the expression for acceleration of r_1.

$$d^2(r_1 e^{i\theta 1})/dt^2 = -\,r_2[\omega_2 \cos(\theta_2) + \alpha_2 \sin(\theta_2)] - r_3[\omega_3 \cos(\theta_3) + \alpha_3 \sin(\theta_3)]$$

$$0 = r_2[\alpha_2 \cos(\theta_2) - \omega_2 \sin(\theta_2)] + r_3[\alpha_3 \cos(\theta_3) - \omega_3 \sin(\theta_3)]$$

Then angular acceleration α_3:

$$\alpha_3 = \frac{r_2(\omega_2{}^2 \sin(\theta_2) - \alpha_2 \cos(\theta_2)]}{r_3 \cos(\theta_3)} + \frac{\omega_3{}^2 \sin(\theta_3)}{\cos(\theta_3)}$$

From this expression the acceleration of r_1 may also be calculated. When angular velocity of link 2 is constant, its acceleration α_2 will be zero, and this will simplify some of the expressions shown above.

Some of the kinematic quantities of interest to the engineering professional, such as the velocity and acceleration of the sliding point *B* and angular velocity and acceleration of the connecting rod ω_3 and α_3 may be calculated from the above equations for all values of θ_2 of the crank. Angular velocity of the crank ω_2 and its *L*/*R* ratio must also be known.

The velocity and acceleration of other locations within the mechanism may also be of interest. For example, the acceleration of the mass centers of individual links are important since they are related to the forces acting on the links. In considering the acceleration of the mass center of link 3, the following equation results from the vector addition:

$$r_{g3} = r_2 + r_d = r_2 e^{i\theta 2} + r_d e^{i\theta 3}$$

$$dr_{g3}/dt = V_{g3} = r_2\omega_2(ie^{i\theta 2}) + r_d\omega_3(ie^{i\theta 3})$$

$$d^2r_{g3}/dt^2 = A_{g3} = r_2(i\alpha_2 - \omega_2^{\,2})e^{i\theta 2} + r_d(i\alpha_3 - \omega_3^{\,2})e^{i\theta 3}$$

If angular speed of the crank is constant, α is zero, and then $A_{g3} = a_{g3} + ib_{g3}$. The magnitude of A_{g3} is:

$$A_{g3} = \sqrt{a_{g3}^{\,2} + b_{g3}^{\,2}}$$

and the angle β that A_{g3} makes with the real axis may be determined from:

$$\tan(\beta) = b_{g3}/a_{g3}.$$

REFERENCES AND BIBLIOGRAPHY

Brown, R. D., Hart, J. A., "A Novel Form of Damper for Turbomachinery", CP2443, NASA, Washington, D. C., June, 1986.

Childs, D. W., "Force and Moment Rotordynamic Coefficients for Pump-Impeller Shroud Surfaces", *Proceedings of Fourth Workshop on Rotordynamic Instability Problems in High Performance Turbomachinery,* NASA CP, Texas A & M University, College Station, Texas, 1986.

Childs, D. W., "Identification and Avoidance of Instabilities in High Performance Turbomachinery", Von Karman Institute for Fluid Dynamics, lecture, Rhode Saint Genese, Belgium, 1987.

Choi, Y. S., Noah, S. T., "Non-Linear Steady State Response of a Rotor Support System", Journal of Vibration, Acoustics, Stress and Reliability in Design, 169(3), pages 255-263, 1987.

Crede, C. E., Ruzicka, J. E., "Theory of Vibration Isolation", Shock and Vibration Handbook, Mc Graw Hill, New York, 1988.

Den Hartog, J. P., "Mechanical Vibrations", Dover Publications, Inc., New York, 1984.

Ehrich, F. F., "Self Excited Vibrations", Shock and Vibration Handbook, Mc Graw Hill, New York, 1987.

Eshelman, R. L., Lewis, F. M., "Torsional Vibration in Reciprocating and Rotating Machines", Shock and Vibration Handbook, Mc Graw Hill, New York, 1987.

Fowlie, D. W., Miles, D. D., "Vibration Problems with High Pressure Compressors", 85-PET-28, ASME, New York, 1985.

Hildebrand, F. B., "Advanced Calculus for Applications", Prentice-Hall, New Jersey, 1986.

McLachlan, N. W., "Ordinary Nonlinear Differential Equations", Oxford University Press, New York, 1956.

Mabie, H. H., Reinholtz, C. F., "Mechanisms and Dynamics of Machinery", John Wiley & Sons, New York, 1987.

Meirovitch, L., "Analytical Methods in Vibration", Macmillan Publishing Co., Inc., New York, 1987.

Muszynska, A., Bentley, D., "Rotor Active Anti-Swirl Control", Rotating Machinery Dynamics, Volume 2, pages 215 - 222, ASME, New York, 1987.

Myklestad, N. O., "Vibration Analysis", Mc-Graw Hill, New York, 1944.

Porter, B. G., "Harmonic Coefficients of Engine Torque Curves", Journal of Applied Mechanics, ASME, 10, Page A-33, March, 1943.

Shemeld, D. E., "A History of Development in Rotor Dynamics - A Manufacturer's Perspective", CP2443, NASA, Washington, D. C., 1986.

Taylor, C. F., "The Internal Combustion Engine in Theory and Practice", Volume 2, The M. I. T. Press, Cambridge, MA., 1997.

Timoshenko, S., Young, D. H. and Weaver, W., "Vibration Problems in Engineering", John Wiley & Sons, Inc., New York, 1974.

Balancing of Periodic Inertia and Gas Pressure Loads

5

5.1 INTRODUCTION

Two groups of vibration phenomena are peculiar to reciprocating machinery, one of them transmitted to the foundation by the machine as a whole, the other arising from torsional oscillations in the crankshaft and in the shafting of coupled machines. Each of these two effects is caused by a combination of periodic accelerations of the moving parts (pistons, rods and cranks) and the periodic variations in the cylinder gas or fluid pressure. Consider a single cylinder reciprocating machine. The piston executes an alternating motion, thus experiencing alternating accelerations along the cylinder axis. When the piston is accelerated inboard (toward the crankshaft), there must be an inboard force acting on it, and this force must have a reaction pushing outboard (toward the cylinder head) against the stationary parts of the machine. Thus an alternating acceleration of the piston is coupled with an alternating force on the cylinder frame, which makes itself felt as a vibration in the reciprocating machine and in its supports. In the lateral direction perpendicular to both the crankshaft and the connecting rod, moving parts are also being accelerated, specifically the crank pin and a portion of the connecting rod. The forces that cause these accelerations must have equal and opposite reactions on the frame of the machinery. In the direction of the crankshaft no inertia forces appear, since all moving parts remain in planes perpendicular to the crankshaft.

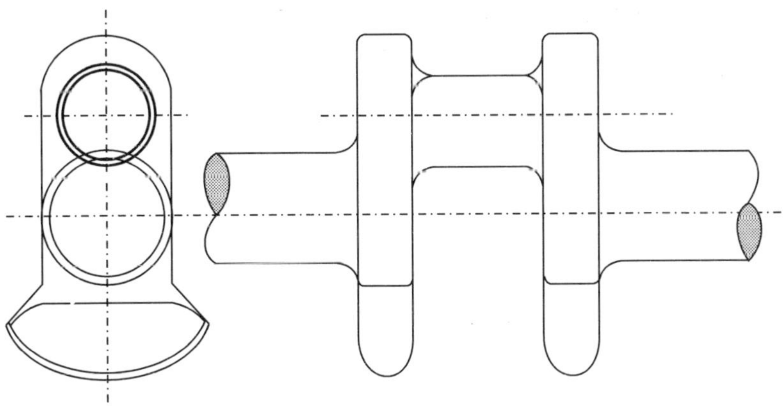

Fig. 5.1. Cranks with Counter Balance Weights

The unbalance or inertia forces on a single cylinder machine are given by (4.10) and (4.11). In these equations the reciprocating mass $m$rec is always positive, but the rotating mass $m$rot can be made zero or even negative by 'counter-balancing' the crank as shown in Fig. 5.1. Hence, it is possible to reduce the inertia force perpendicular to cylinder axis (in the plane of crank motion) to zero, but the unbalance force along the cylinder axis always remains, leading to the conclusion that a single cylinder reciprocating machine is inherently unbalanced.

Next take the case of a two-cylinder machine with the cranks opposed to each other, or at 180° from each other. Forces perpendicular to the cylinder axes are also in opposition, thus canceling each other. There is, however, a moment about the axes of the cylinders. Since both pistons move against each other, the same is true for the primary forces along the cylinder axis. However, the secondary forces along the cylinder axis are in the same direction and add. To understand this it is convenient to visualize all the forces as the projection, along the direction perpendicular to the cylinder axis, of rotating vectors. The method can best be understood using the case of a four-cylinder machine.

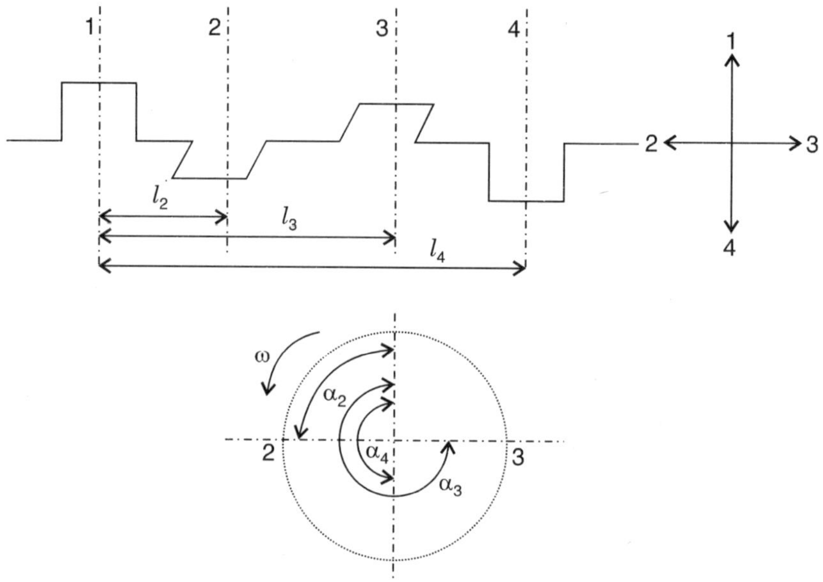

Fig. 5.2. Primary Froces in a 4-Cylinder Reciprocating Machine

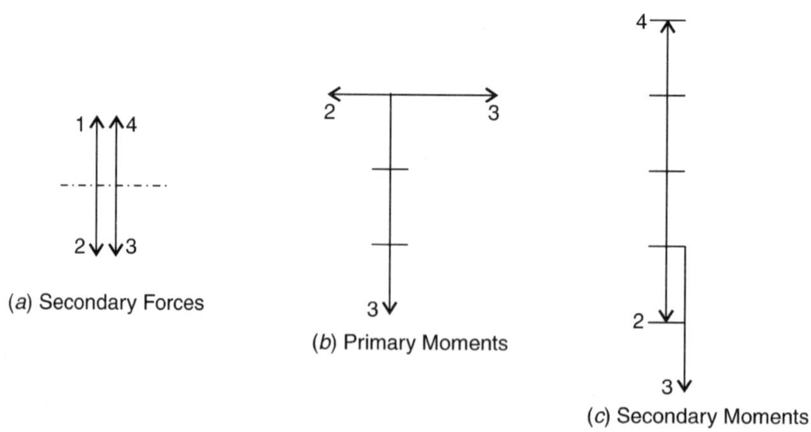

Fig. 5.3. Crankshaft Loads for Four-Cylinder Machine

In Fig. 5.2 the first crank is shown in the vertical position, corresponding to a maximum value of the primary vertical inertia force. The distance and angle between the nth crank and the first crank are l_n and α_n. The second crank is α_2 radians ahead of the first one, hence its vertical primary inertia force has passed through the maximum value α_2/ω second earlier. If the rotating vector for the primary vertical inertia force of the first cylinder is in the vertical direction, that due to the second cylinder is in the position α_2 and for the nth cylinder in position α_n. The same is true for the primary horizontal inertia force. The crank diagram of Fig. 5.2*b*, regarded as a vector diagram (Fig. 5.2*c*), represents the primary inertia forces in the machine. In this example of a 4-cylinder machine the primary inertia forces are balanced.

The secondary forces rotate twice as fast as the crankshaft. The secondary force diagram appears as a star with angles $2\alpha_2$, $2\alpha_3$, .. $2\alpha_n$ between the vectors. Fig. 5.3*a* shows the diagram for the example of Fig. 5.2.

Similar reasoning holds for the moments of these forces about a lateral axis. The moment of the nth inertia force about the center of the first crankshaft is that force multiplied by the moment arm l_n. The plane in which such a moment operates is defined by the direction of the force and the centerline of the crankshaft. Thus, the moment can also be represented by a vector in the same direction as the inertia force, its length being multiplied by the appropriate moment arm l_n.

5.2 GAS PRESSURE FORCES

Under the assumption that the components shown in Fig. 5.4 are all rigid bodies, the forces due to gas pressure in the cylinder can be taken as a force $-P$ on the piston, which is a variable with time, or with the crank angle ωt. The gas pressure not only pushes the piston downward but it also presses upward against the cylinder head, shown as force P. The piston force is transmitted through the connecting rod to the crank pin, and is balanced by an equal and opposite force supplied by the frame to the main bearings. The resultant force on the machine frame is zero, but there is a torque on the frame due to the couple $-Pr$ Sin (θ), which balances the useful torque from the shaft and which must be absorbed by elements outside the machine frame. This torque will be analyzed in section 5.7.

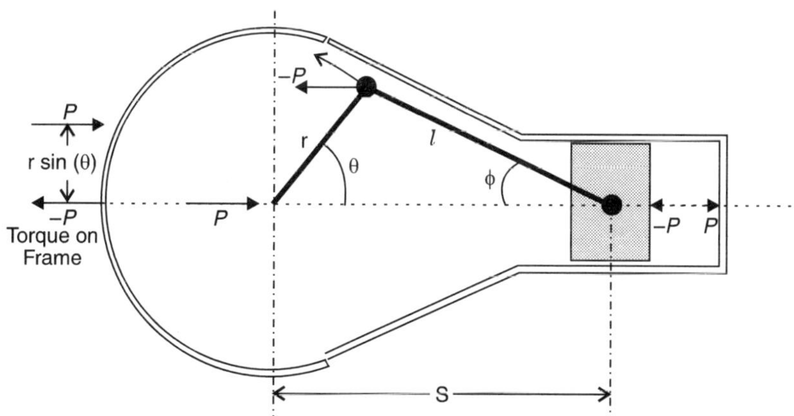

Fig. 5.4. Gas Pressure Force and Torque

5.3 NON-HARMONIC PERIODIC FUNCTIONS

If a function $f(t)$ is a curve taken from a pressure-volume indicator diagram (or from an actual vibration record), a mathematical expression for it is not available. In order to determine the instantaneous forces acting in the cylinder due to gas pressure, the harmonic analysis method is most suitable, a review of which is provided next.

A 'periodic' motion has the property of repeating itself in all details after a certain interval of time, the 'period' of motion. All harmonic motions are periodic, but not every periodic motion is harmonic. For example, Fig. 5.5 represents the motion:

$$x = a \sin(\omega t) + \frac{a}{2} \sin(2\omega t) \tag{5.1}$$

which is the superposition of two sine waves of different frequencies. It is periodic but not harmonic. Mathematical theory shows that any periodic curve $f(t)$ of frequency ω can be split up into a series of sine curves of frequencies ω, 2ω, 3ω, 4ω, etc. or:

$$f(t) = A_0 + A_1 \sin(\omega t + \varphi_1) + A_2 \sin(2\omega t + \varphi_2) + A_3 \sin(3\omega t + \varphi_3) + \ldots. \tag{5.2}$$

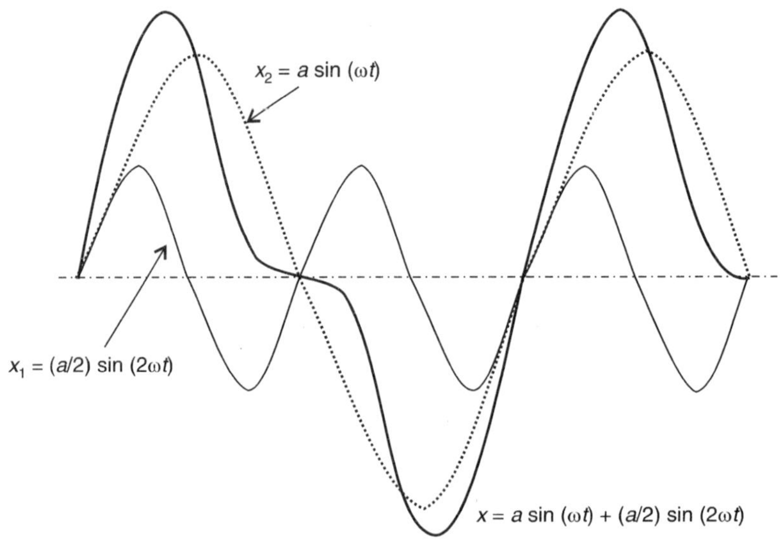

Fig. 5.5. Sum of Two Harmonic Functions of Different Frequencies

provided that $f(t)$ repeats itself after each interval $T = 2\pi/\omega$. The amplitudes of the waves $A_1, A_2, \ldots$. and their phase angles $\varphi_1, \varphi_2, \ldots$ can be determined analytically when $f(t)$ is given. Series (5.2) is the Fourier series representing function $f(t)$. Since:

$$\sin(n\omega t + \varphi_n) = \sin(n\omega t) \cos(\varphi_n) + \cos(n\omega t) \sin(\varphi_n) \tag{5.3}$$

the series can also be written as:

$$f(t) = a_1 \sin(\omega t) + a_2 \sin(2\omega t) + \ldots\ a_n \sin(n\omega t) + \ldots + b_0 + b_1 \cos(\omega t) + b_2 \cos(2\omega t) + \ldots + b_n \cos(n\omega t) + \ldots \tag{5.4}$$

The constant term b_0 represents the 'average' height of the curve $f(t)$ during a cycle. For a curve which is as much above the zero line during a cycle as it is below, term b_0 is zero. The amplitudes $a_1, a_2, \ldots a_n, \ldots b_1, b_2, \ldots b_n$ can be determined from the following expressions:

$$a_n = \frac{\omega}{\pi} \int_0^{2\pi/\omega} f(t) \sin(n\omega t)\, dt \tag{5.5}$$

$$b_n = \frac{\omega}{\pi}\int_0^{2\pi/\omega} f(t)\cos(n\omega t)\,dt$$

The relations between a_n, b_n and the quantities A_n , φ_n are as follows:

$$A_n^2 = a_n^2 + b_n^2, \quad \text{and} \quad \tan\varphi_n = b_n/a_n \tag{5.6}$$

With the aid of (5.5) it is possible to find a_n and b_n for any given periodic curve. The branch of mathematics concerning this type of problems is known as harmonic analysis. Thus, if a pressure-volume type of indicator diagram is available from the cylinder of an engine it is possible to determine the harmonic coefficients, from which the mathematical representation of the curves may be obtained. Note, however, that the evaluation of the integrals can be done only for a few simple shapes of $f(t)$. Also, for an indicator diagram the mathematical expression for it is not available. The integrals can be obtained either graphically or numerically by means of a machine known as a harmonic analyzer.

The harmonic analyzer operates on the same principles as Watt's steam engine indicator. The indicator traces a closed curve of which the ordinate is the steam pressure (or piston force) and the abscissa is the piston displacement. The area of the closed curve is the work done by the piston force per cycle. Equations (5.5) state that the coefficients are ω/π times the work done per cycle by the force $f(t)$ on a certain displacement of which the velocity is expressed by $\sin(n\omega t)$.

The mechanism of the harmonic analyzer measures the area of the closed curve. Electronic harmonic analyzers give an extremely rapid analysis of the total harmonics $A_n = (a_n^2 + b_n^2)^{1/2}$, without giving information on the phase angles. Other analyzers require the original curve to be in the form of an electrical voltage varying with time, such as results from a vibration pick-up. This voltage, after proper amplification and going through band-pass filters to suppress all frequencies except those in a narrow band, is fed into an electrical network to be Fourier analyzed.

5.4 UNBALANCED FORCES ON SINGLE CYLINDER MACHINES

Inertia forces and moments arising from the mass properties of the piston, connecting rod and crank pin were determined in section 4.2, Piston Dynamic Motion and in section 4.3, Crank Rotating Motion. Equations (4.1) through (4.6) established the displacement, velocity and acceleration of the reciprocating masses, while equations (4.7) through (4.9) calculate the same parameters at the rotating end of the connecting rod. Inertia forces and torques along the piston line of motion and perpendicular to it were next determined by equations (4.10) through (4.12). In the analysis it was assumed that the reciprocating machine is oriented such that the piston axis of motion is vertical, giving rise to vertical force X and horizontal force Y. The selection of orientation of the coordinate axes was purely for convenience, since it is easier to understand the procedure. In chapters 2 and 3, however, it was seen that for overwhelming number of cases the piston line of motion does not lie in the horizontal or vertical direction, but is skewed in many different directions. This is particularly true of V, W, T and other multi-cylinder configurations of reciprocating machines. For better definition, F_p will be used henceforth to describe forces along piston axis of motion, F_{cp} describes the force acting on the crank pin directed radially outward, F_a describes the reaction on the machine frame due to F_p, and F_{ct} gives the equal and opposite reaction to F_{cp} (see Fig. 5.6).

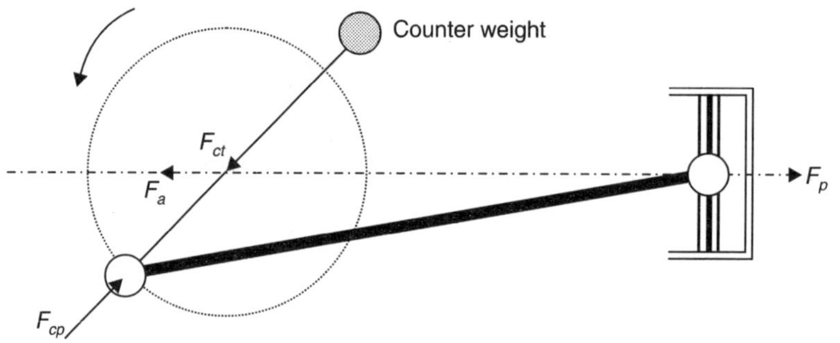

Fig. 5.6. Counter Weight for Rotating Mass

From this analysis it can be seen that a single crank and connecting rod mechanism operating at a constant rotational speed produces forces as follows:

1. A force F_p along the cylinder axis, acting on the piston to produce acceleration of the piston assembly (piston, piston rings, piston pin and equivalent mass of the connecting rod). For double acting reciprocating machines with a cross head and piston rod, their mass is also included.
2. A force F_{cp}, acting on the crank pin and lower end of the connecting rod, directed radially toward the center of the crankshaft, producing centripetal acceleration of the parts revolving with the crank pin. The force F_{ct} is balanced by suitable counter weights on the crankshaft, as show in Fig. 5.6. If this force is balanced, the only unbalanced force acting on the machine is F_a. If the machine is mounted rigidly, the force acting on the machine is F_a, and the reaction on the supports is $-F_a$.

The unbalanced inertia force F_a acts along the cylinder axis only. It is due to the inertia of the reciprocating assembly and has no side component. It may be expressed as follows:

$$F_a = -F_p \cong Z\{\cos(\theta) + (r/l)\cos(2\theta)\} \tag{5.7}$$

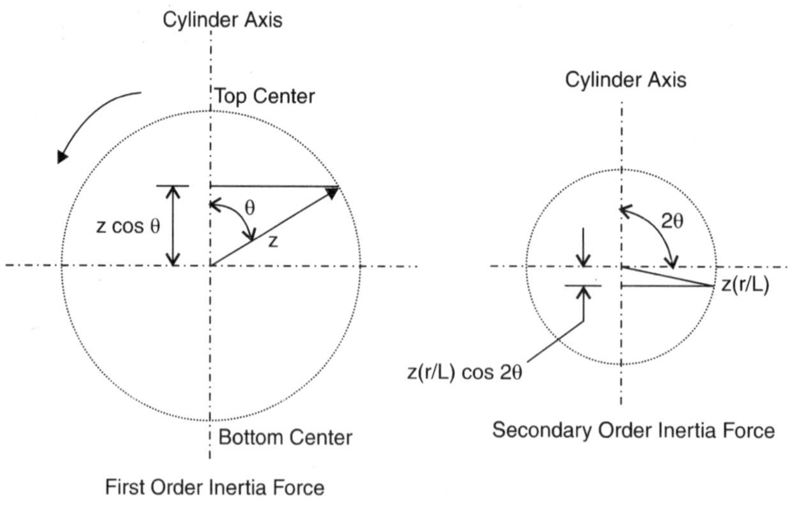

Fig. 5.7. First and Second Order Vectors

where $Z = m_{rec}\omega^2 r$ and $\theta = \omega t$, used in (4.10). The coefficient of θ will be referred to as *order* of the harmonic. Thus, the first term in (5.7) represents the first order of the force, varying periodically once per shaft revolution, while the second term gives the second order force, varying twice per shaft revolution. The total force is the sum of two sinusoidal forces of different amplitudes and frequencies, both having maximum amplitude at $\theta = 0$. Figure 5.7 shows the first and second order inertia force vectors for a single cylinder engine. The first order vector has its projection on the cylinder axis, and is the first order inertia force $Z \cos(\theta)$. The second order has a length represented by $Z(r/l)$ and rotates at twice crank speed in the crank direction. Its projection on the cylinder axis is $Z(r/l)\text{Cos}(2\theta)$, which is a second inertia force. In both cases only the axial component of the vector represents the force in question.

5.5 FORCES ON MULTI-CYLINDER IN-LINE MACHINES

In these reciprocating machines the unbalanced reaction forces F_a from individual pistons add algebraically, since their lines of action are parallel. Thus the resultant unbalanced reaction force on the machine frame is:

$$\Sigma F_a = F_{a1} + F_{a2} + F_{a3} + \ldots \tag{5.8}$$

where F_{a1}, F_{a2}, ... represent individual reactions from the cylinders. These individual forces may be obtained from (5.7) by using appropriate values of θ for each cylinder. Ordinarily the stroke, reciprocating mass, and lengths of crank and connecting rod will be the same for individual cylinders. As noted in chapters 2 and 3, however, compressors, pumps and internal combustion engines do not always have identical designs for all the cylinders, particularly if an inter cooler is used between the stages for improved efficiency. In the latter case, appropriate values for component parameters need to be used for all of the cylinders in (5.7) to obtain F_{a1}, F_{a2}, F_{a3}, ... Then:

$$\Sigma F_a = Z\{[\cos(\theta_1) + \cos(\theta_2) + \cos(\theta_3) + \ldots] \\ + r/l[\cos(2\theta_1) + \cos(2\theta_2) + \cos(2\theta_3).\,.]\} \tag{5.9}$$

The angles θ_1, θ_2, θ_3, ... for all the cylinders may be expressed in terms of θ_1. If the cranks are evenly spaced, and α is the smallest angle between two cranks, then:

$$\alpha = 2\pi/N_{cp} \quad \text{when} \quad N_{cp} = \text{total number of crank positions.} \tag{5.10}$$

The force F_{ct} mentioned in the case of single cylinder machines is usually balanced by the arrangement of cranks in multi-cylinder machines, specially if the cylinders and corresponding reciprocating and rotating masses are identical n design. If this form of balance cannot be achieved, suitable counterweights may need to be attached to the crankshaft.

Example problem 5.3 illustrates the case of a 3-cylinder reciprocating machine with cranks evenly spaced. The example illustrates the general rule that the resultant forces for any order will be zero (star diagram) except for those orders that are an even multiple of the number of separate crank positions. When *n* is an even multiple, the vectors coincide. Where each crank is at a different angle, the resultant unbalance force is N_{cp} times that of a single cylinder. Thus, in the 3-cylinder example problem # 5.3, the lowest order which adds is the sixth.

Applying the rule to in-line reciprocating machines with evenly spaced cranks gives the following tabulation:

Number of Cylinders	*Cycle*	N_{cp}	*Orders That Add*
2	4	1	1, 2, 4, 6, ...
2	2	2	2, 4, 6, ...
3	2 or 4	3	6, 12, 18, ...
4	4	2	2, 4, 6, ...
4	2	4	4, 8, 12, ...
5	2 or 4	5	10, 20, 30, ...
6	4	3	6, 12, 18, ...
6	2	6	6, 12, 18, ...

In 4-stroke internal combustion engines with even number of cylinders, for evenly spaced firing there must be two cranks at the same angle θ, so that N_{cp} will be $N_c/2$, where N_c is the number of cylinders. In the case of two cylinders on each crank all unbalanced forces are equal to twice those of one cylinder.

Although the forces due to acceleration of the pistons in an in-line machine are in the same plane, they do not have the same line of action, giving rise to unbalanced moments in the plane of the cylinders. When the forces are balanced, it is convenient to take moments about the intersection of the crankshaft center line and the axis of one cylinder. For a 3-cylinder machine, if the distance between the cylinder axes is uniformly equal to d, take the moments about an axis normal to the axis of the first cylinder. Then the resultant first order moment may be calculated from the expression:

$$M_1 = Z[d \cos(\theta_2) + 2d \cos(\theta_3)] \tag{5.11}$$

Similarly the second order moments will be:

$$M_{11} = Z(r/l)[d \cos(2\theta_1 + 2\alpha) + 2d \cos(2\theta_1 + 4\alpha)] \tag{5.12}$$

Example problem 5.6 provides the method for determining the moments thus created in a 3-cylinder case.

In order to eliminate all moments, in-line machines (and as will be seen later, in V configurations as well) can be made with even number of cylinders and with cranks arranged symmetrically about a plane perpendicular to the crankshaft axis. Fig. 5.8*a* and 5.9*a* illustrate symmetrical shafts. The unbalanced force at each crank is balanced by an equal and opposite force on another crank at 180°, so placed that the moment arms around the plane of symmetry are equal. Conventional 4-cycle 4, 6 and 8 cylinder internal combustion engines have symmetrical crankshafts. For two cycle engines, however, it is undesirable to have two cylinders fire at the same time, so the symmetrical crank arrangement is seldom used here.

An anti-symmetrical crankshaft is one in which the crank of the last cylinder is 180° from the first cylinder, that from the next to last cylinder is 180° from the second cylinder, and so on, as shown in Fig. 5.8*b* and 5.9*b*. With anti-symmetrical crankshafts moments about the center of the engine due to secondary forces is zero. In cases where the primary moment can be offset by counter weights, all important moments are thus eliminated, if the rotating masses are suitably counter weighted.

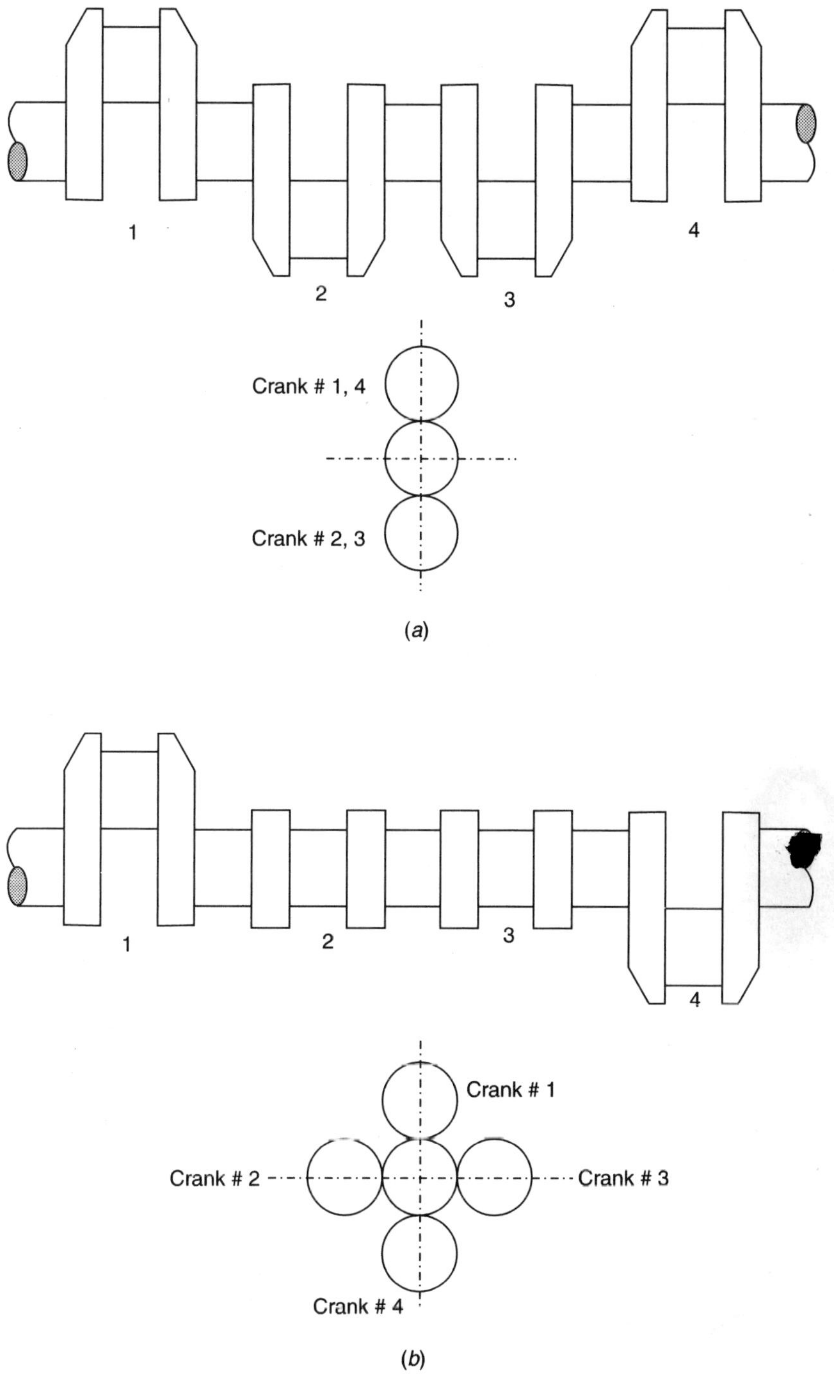

Fig. 5.8. (*a*) Symmetrical and (*b*) Anti-symmetrical 4-Throw Crankshafts

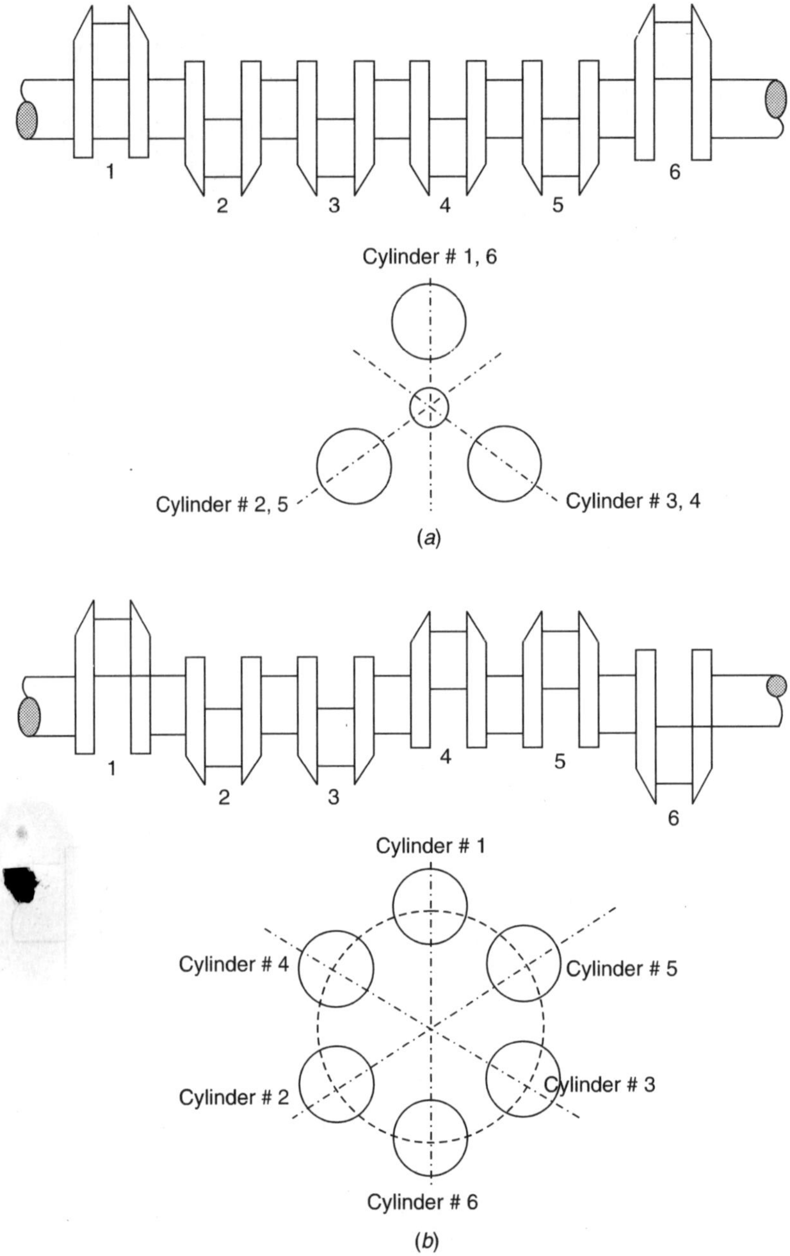

Fig. 5.9. (*a*) Symmetrical and (*b*) Anti-symmetrical 6-Throw Crankshafts

5.6 V MACHINES WITH SIMPLE CONNECTING ROD

In many V machines the forces due to piston acceleration and resulting moments are balanced for each bank of cylinders. This is the case in four stroke 12- and 16-cylinder internal combustion engines. In many cases, specially in 90° *V* machines, it is possible to obtain excellent balance, even though individual banks of cylinders may not themselves be balanced. To analyze the situation, consider a unit consisting of cylinders that act on the same crank (Fig. 5.10). If *YOY* is the axis of symmetry and *XOX* is perpendicular to it through the crankshaft center, the unbalanced forces from each cylinder may be projected on the *XY* axes and added to find the unbalanced forces of the unit.

With the *V* angle β, let the crank angle for the right bank cylinder be θ; that for the left bank cylinder will be 2π – β + θ, which is trigonometrically equivalent to θ – β. For the right cylinder inertia force on the frame opposite to the piston force is:

$$F_{aR} = Z[\cos(\theta) + (r/l)\cos(2\theta)] \quad (5.13)$$

and for the left cylinder:

$$F_{aL} = Z[\cos(\theta - \beta) + (r/l)\cos 2(\theta - \beta)] \quad (5.14)$$

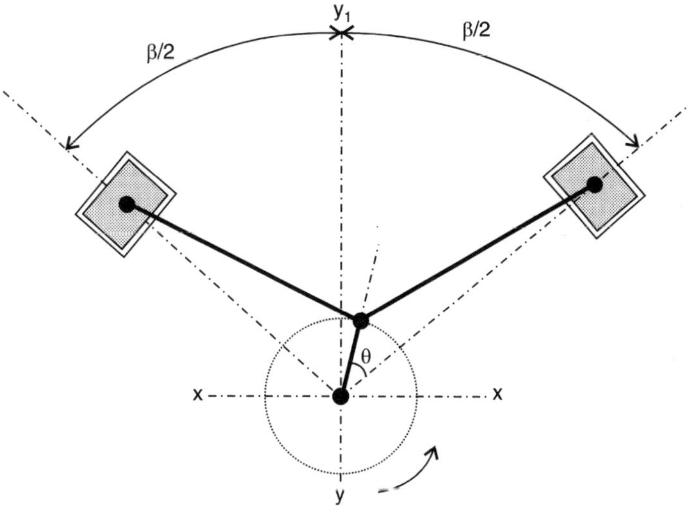

Fig. 5.10. 90° V Balance Diagram

Projecting these forces on the X and Y axes and simplifying the results after using trigonometric identities yields the following first order terms:

$$F_{aY1} = 2Z\{\cos^2(\beta/2)[\cos(\theta - \beta/2)]\} \quad (5.15)$$

$$F_{aX1} = -2Z\{\sin^2(\beta/2)[\mathrm{Sin}(\theta - \beta/2)]\} \quad (5.16)$$

For a 90° V machine, β = π/2, so that:

$$F_{aY1} = Z[\cos(\theta - \beta/2)] \quad (5.17)$$

$$F_{aX1} = -Z[\sin(\theta - \beta/2)] \quad (5.18)$$

These forces are the same as would be produced by a mass m_{rec} concentrated at the crank pin. Hence, they may be eliminated by a suitable counter weight opposite the crank. Because of this possibility the cranks of a 90° V reciprocating machine may be arranged without regard to primary forces. The second order terms after simplification in the same manner as the first order terms are:

$$F_{aY11} = (2Zr/l)[\cos(\beta/2)]\{\cos(\beta)[\cos 2(\theta - \beta/2)]\} \quad (5.19)$$

$$F_{aX11} = -(2Zr/l)[\sin(\beta/2)]\{\sin(\beta)[\sin 2(\theta - \beta/2)]\} \quad (5.20)$$

For a 90° V angle, $F_{aY11} = 0$ and $F_{aX11} = -\sqrt{(2)(Zr/l)\sin 2(\theta - \beta/2)}$. The second order balance is therefore in a horizontal plane only.

Conventional V-8 internal combustion engines use the anti-symmetrical crankshaft (Fig. 5.8*b*) with four crank positions ($\alpha = 90°$). This arrangement of crank throws balances the primary and secondary forces in each bank and, therefore, in the engine. The anti-symmetrical shaft eliminates the secondary moments. The primary moments which the anti-symmetrical crankshaft would ordinarily produce are eliminated by cancelling the primary forces of each pair of V cylinders by means of counter weights opposite the cranks. Note that the 90° V-8 arrangement results in the simultaneous arrival of two pistons at top center, thus the form is adapted for use in 4-stroke engines. The 12-cylinder 90° V engine with an anti-symmetrical shaft (cranks at 60°) is similarly free from primary and secondary unbalanced forces and moments and is adaptable to 2-stroke engines, since each of the pistons arrives at top center 30° after the previous piston.

5.7 TORQUE ON CRANKSHAFT DUE TO GAS PRESSURE

Moments around the crankshaft axis are classified as machine torque. The instantaneous torque T_p acting on the crank due to gas pressure may be found by noting that, without friction, work done on the piston is the same as work done on the crankshaft:

$$pdV = pA_p dS = T_p d\theta \tag{5.21}$$

where p is the pressure on the piston, A_p the piston area, S the length of stroke and V the cylinder volume. Re-arranging the terms:

$$T_p = pA_p(dS/d\theta) \tag{5.22}$$

Since: $dS/d\theta = [dS/dt]/[d\theta/dt] = [dS/dt]/\Omega$

Therefore, $$T_p = pA_p[dS/dt]/\Omega \tag{5.23}$$

Using $\theta = \omega t$ and $\Omega = \omega$ in (4.6) gives the values of $[dS/dt]/\Omega$:

$$[dS/dt]/\Omega = r[\sin(\theta) + 2a_2 \sin(2\theta) + 4a_4 \sin(4\theta) + \ldots\} \tag{5.24}$$

Values of p must be found from an indicator diagram. Multiplication of the instantaneous pressure, piston area and the ratio of piston velocity to crank velocity gives T_p. A typical plot of pressure versus crank angle and the resultant torque versus crank angle diagram are shown in Fig. 5.11 for a 4-cycle internal combustion engine. The instantaneous torque for each cylinder may be analyzed and expressed as a Fourier series as follows:

$$T_p/T_m = 1 + U_{1/2}\sin(\theta/2) + U_1\sin(\theta) + U_{3/2}\sin(3\theta/2) + U_2 \ldots$$
$$+ V_{1/2}\cos(\theta/2) + V_1\cos(\theta) + V_{3/2}\cos(3\theta/2) + V2 \ldots \tag{5.25}$$

where T_m is the mean torque, U_n and V_n are cosine and sine coefficients for gas torque and θ is the crank angle measured from the firing top center. For a 4-cycle engine the expression for work done is:

$$T_m(4\pi) = \text{work done in two revolutions}$$

and so: $$T_m = (\text{mep } x \text{ piston displacement})/4\pi$$

Values of the coefficients U_n and V_n for a 4-cycle spark ignition engine are plotted in Fig. 5.12. In this case the indicated mean effective pressure is 155, ratio r/l is 0.291 and compression ratio is 6.

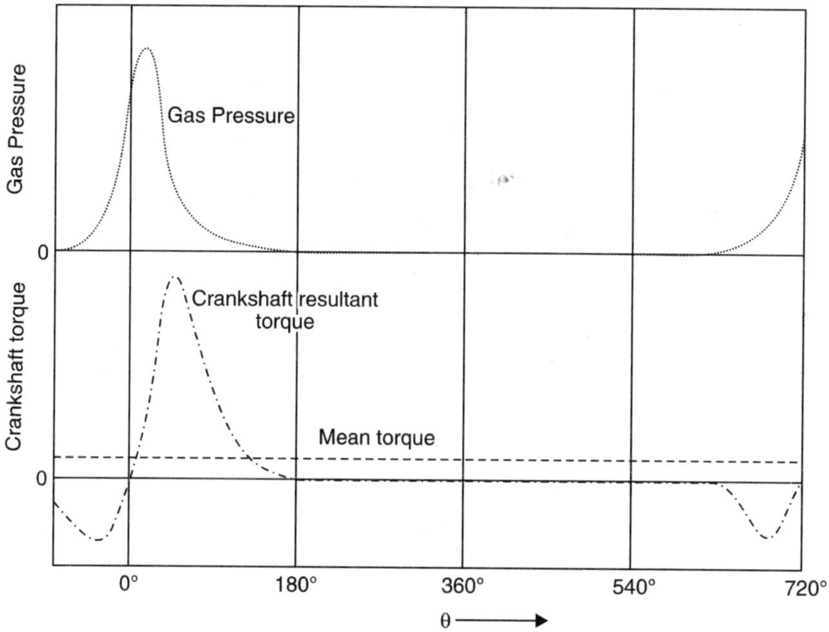

Fig. 5.11. Typical Gas Pressure versus Crank Angle Curve for 4-Cycle Engine, with Resultant Torque on Crankshaft

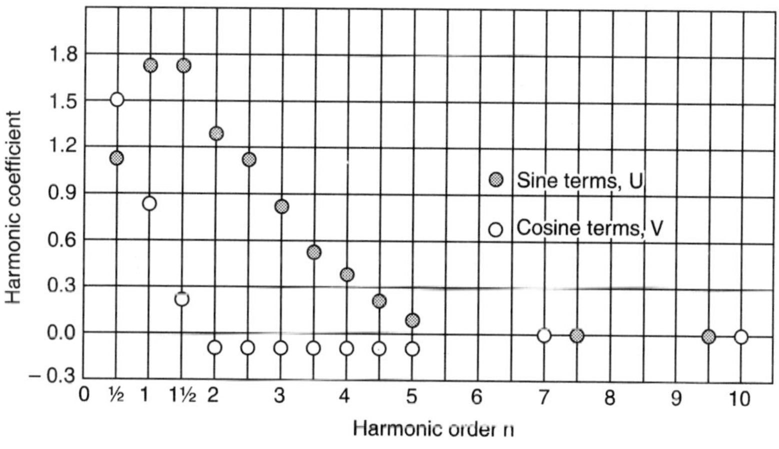

Fig. 5.12. Harmonic Coefficients of Gas Pressure Torque for a 4-Cycle Engine, with Compression Ratio of 6.0 (*ref.* Taylor and Morris)

5.8 TORQUE ON CRANKSHAFT DUE TO INERTIA OF MOVING PARTS

The instantaneous torque T_t may be found by noting that the change in kinetic energy of the piston must be equal to the work done by the crankshaft on the connecting rod; hence:

$$d\{m_{rec}\,[dS/dt]^2/2 = -\,T_t\,d\theta \tag{5.26}$$

Dividing by dt, and since $d\theta/dt = \Omega$, the expression for torque is:

$$T_t = -\,m_{rec}[dS/dt][d^2S/dt^2]/\Omega \tag{5.27}$$

Since S is a Fourier series in terms of θ, and S/Ω is also a Fourier series as shown in (5.24), the torque will be the product of two series. By combining terms, and using trigonometric identities, it can be shown that:

$$T_t = m_{rec}\,\Omega^2 r^2 \{t_1 [\sin(\theta_1) + \sin(\theta_2) + \sin(\theta_3) + ...] + t_2[\sin(2\theta_1) + \sin(2\theta_2) + \sin(2\theta_3) + ...] + t_3 [\sin(3\theta_1) + \sin(3\theta_2) + \sin(3\theta_3) + ...] + ...\} \quad (5.28)$$

where $t_n = f(a_1, a_2, a_3, ...)$.

Since t_n is a function of $a(dS/dt)$, it must also be a function of r/l. Fig. 5.13 is a plot of t_1, t_2, t_3 and t_4 versus r/l. Other terms are too small. Note that $\theta_1, \theta_2, \theta_3$, ... are the angles measured from the top center position of each cylinder under consideration. The torque applied by the inertia of the rotating parts to the shaft depends only upon θ for individual cylinders; it is independent of the angle between cylinder axes.

Next consider the angular acceleration of the connecting rod. Forces and torques acting on a connecting rod that is equivalent to two masses concentrated at its ends has already been considered. The motion of center of gravity of such a fictitious rod is the same as that of the actual rod, and the forces acting are therefore the same. However, the moment of inertia of the fictitious connecting rod is not in general equal to that of the actual connecting rod.

If the actual moment of inertia of the connecting rod is I_L, the torque that must act on the crankshaft to impart angular acceleration to the rod can be established in a process similar to that used to find the series representing the acceleration of the piston. This expression for the torque on the crankshaft is:

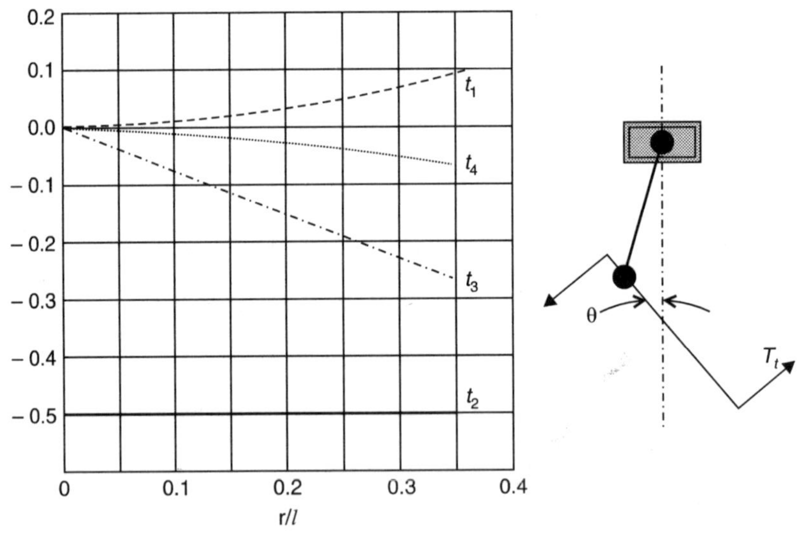

$$T_t = M_{rec}\Omega^2 r^2 \sum_{n=1}^{\infty} t_n \sin(n\theta)$$

where T_t = crankshaft torque from acceleration of reciprocating parts

$-T_t$ = reaction engine torque from acceleration of reciprocating parts

Fig. 5.13. First, Second, Third and Fourth Order Inertia Torque Coefficients

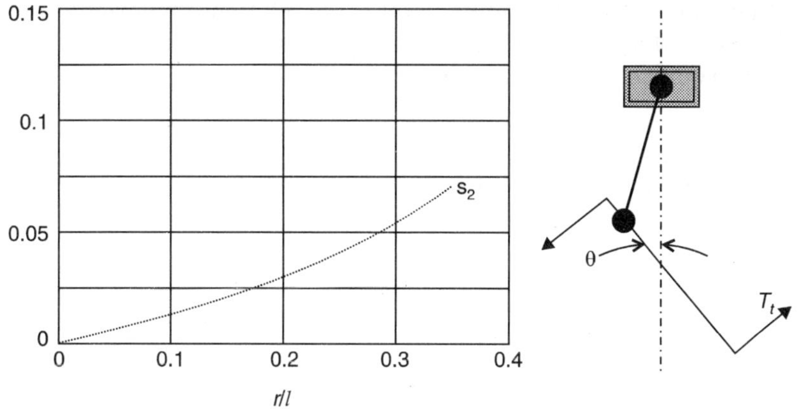

Fig. 5.14. Coefficient for Correction of Inertia Torque on Crankshaft for True Moment of Inertia of Connecting Rod

$$I_L\Omega^2\{s_2 \sin(2\theta) + s_4 \sin(4\theta) + ...\} \tag{5.29}$$

where
$$s_2 = +\left[\frac{1}{2}\left(\frac{r}{l}\right)^2 - \frac{1}{32}\left(\frac{r}{l}\right)^6 + ...\right]$$

$$s_4 = -\left[\frac{1}{4}\left(\frac{r}{l}\right)^4 + \frac{1}{8}\left(\frac{r}{l}\right)^6 + ...\right]$$

A plot of s_2 is shown in Fig. 5.14. Quantity s_4 is too small of importance. However, a correction torque from forces due to differences in the actual and the fictitious rod needs to be applied, which is:

$$hjm_L\Omega^2\{s_2 \sin(2\theta) + s_4\sin(4\theta) + ...\} \tag{5.30}$$

where h and j are distances from the center of gravity of the rod to its centers at the small and large ends and m_L is the mass of the connecting rod. The torque on the crankshaft due to the inertia of the moving parts is then:

$$T_t + T_L = m_{\text{rec}}\, \Omega^2 r^2 \sum_{1}^{\infty} t_n \sin(n\theta) + (I_L - hjm_L)\Omega^2 \sum_{1}^{\infty} s_{2n} \sin(2n\theta) \tag{5.31}$$

5.9 SUMMATION OF TORQUES ON CRANKSHAFT

From (5.25) it is evident that if all the coefficients U and V are zero, the torque on the crankshaft is equal to the mean torque and is a constant quantity. This condition is desirable in the interest of smooth machine operation, and is approached as the coefficients cancel out with increasing number of cylinders. Since the value of the coefficients is small above the eighth order, it is particularly desirable that orders below eight cancel in multi-cylinder machines.

The process of summing the torques in a multi-cylinder machine is simple, since all torques act about the same axis and are applied to the same body, and thus add directly. In the expressions for gas pressure torque the angle α between the first order vectors is equal to the number of crank degrees between successive arrivals at firing at top center for an internal combustion engine. Two examples of torque vector addition are shown in Fig. 5.15. The examples illustrate the general principle that with evenly spaced firing torque vectors for 4-cycle engines add when n is a multiple of half the number of cylinders ($Nc/2$), and cancel for other values of n. In the case of 2-cycle engines the vectors add when n is a multiple of the number of cylinders and cancel for other values. Table 5.1 summarizes these relations for various number of cylinders with even firing. Note that cylinder arrangement does not make any difference as long as firing intervals are equal.

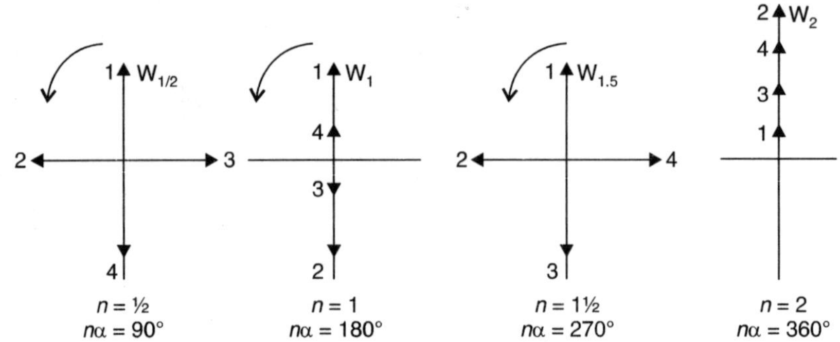

(*a*) Fourth order torque components—Four multiples torque from one cylinder four cylinder, four cycle engine (α = 720/4 = 180°)

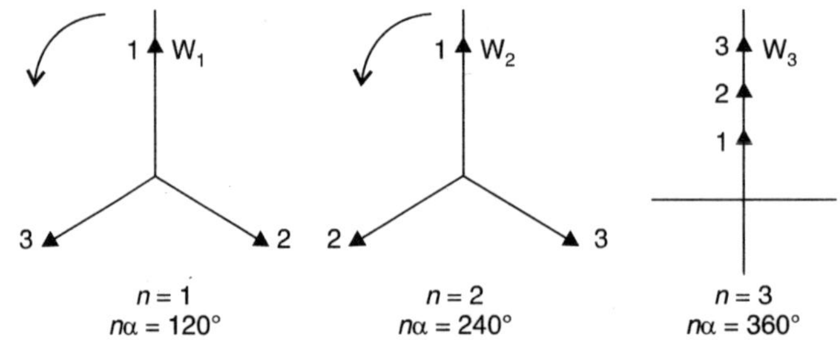

(*b*) Third order torque components—Three multiples torque from one cylinder three cylinder, two cycle engine (α = 360/3 = 120°)

Fig. 5.15. Vector Diagrams for Torque Output for Two Engines With Cyclic Uniformity and Equal Firing Intervals; $\alpha = 720/Nc$, if even for 4-cycle engines; for others $\alpha = 360/Nc$

Table 5.1. Orders of Gas Pressure Torque Appearing in Multi-Cylinder Engines with Even Firing Intervals and Uniform Pressure versus Crank Angle Diagrams

Number of Cylinders	*Cycle*	*Orders Appearing, n*
1	2	1, 2, 3, 4, etc.
1	4	½, 1, 1½, 2, 2½, 3, etc.
2	2	2, 4, 6, 8, etc.
2	4	1, 2, 3, 4, etc.
3	2	3, 6, 9, 12, etc.
3	4	1½, 3, 4½, 6, 7½, etc.
4	2	4, 8, 12, 16, etc.
4	4	2, 4, 6, 8, etc.
5	2	5, 10, 15, 20, etc.
5	4	2½, 5, 7½, 10, etc.
6	2	6, 12, 18, 24, etc.
6	4	3, 6, 9, 12, etc.

Note: The coefficient of each torque order appears n times the value for one cylinder.

It is to be noted that the preceding discussion assumed theoretically that indicator diagrams for all cylinders of a reciprocating engine are identical. In practice the diagrams vary to an appreciable extent between strokes of the same cylinder and also between the cylinders during a crankshaft revolution for a number of reasons. Due to this non-uniformity, all engines will show some degree of torque at all harmonics of the cyclic variation as well as of the crank speed. But since the differences are small, the magnitude of torque variations in orders that theoretically cancel will be small compared with those that theoretically add in the lower (< 8) orders.

5.10 EFFECTS OF ARTICULATED CONNECTING ROD

Figure 5.16 shows the geometry of a master articulated connecting rod system. When the hinge pin angle γ is equal to the angle between the cylinder axes β, the articulated piston has the same length of stroke as that of the master rod piston.

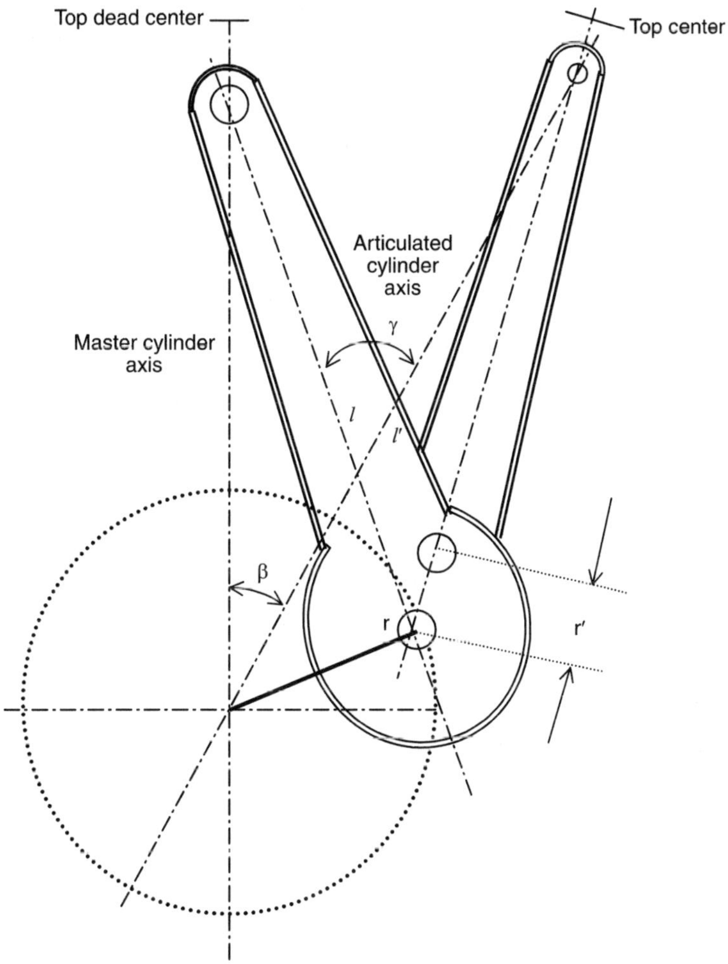

Fig. 5.16. Geometry of Master and Articulated Connecting Rod System

Analysis of the articulated rod motion is a difficult task because of its configuration. With $\gamma = \beta$, the acceleration of the articulated piston may be expressed as follows:

$$a = -\Omega^2\{[-r\cos(\theta) + 4A_2\cos(2\theta) + 16A_4\cos(4\theta)\ldots] + [4B_2\sin(2\theta) + 16B_4\sin(4\theta)\ldots]\} \quad (5.32)$$

where each A and B is equal to r multiplied by a function of r/l, r/l', r'/l' and angles α and β. Note that the cos (θ) term is the same as for the plain, or master, or master connecting rod. As in the case of the plain rod, only even harmonics appear after the first, but the appearance of the sine term indicates that the acceleration is not exactly in phase with that of the master piston. This is due to the difference in the arrival time of the piston at top center. For typical values such as $r/l = 0.25$, $r/l' = 0.308$, $r'/l' = 0.21$, and $\gamma = \beta$, crank positions for top and bottom centers are shown below:

(atc, bbc refer to crank position)

β	*0°*	*45°*	*60°*	*90°*	*120°*	*150°*	*180°*
Degrees, atc	0	0.53	1.09	2.66	3.23	1.99	0
Degrees, bbc	0	0.8	1.73	4.35	6.36	6.88	0

This table is for the case where crank position is from the master cylinder toward the articulated cylinder. In general, the departures shown have no effect on performance with best valve and ignition timing for the master cylinder. For the cited case values of A/r and B/r (where A and B are articulated rod acceleration cosine and sine terms shown in equation 5.32) are as follows:

β	*0°*	*45°*	*60°*	*90°*	*120°*	*150°*	*180°*
A_1/r	– 1	– 1	– 1	– 1	– 1	– 1	– 1
A_2/r	– 0.625	– 0.565	– 0.585	– 0.631	– 0.853	– 1.11	– 1.225
A_4/r	0.00003	0.00003	0.00006	0.00006	0.00009	0.00009	0.00009
B_2/r	0.0	– 0.0021	– 0.0041	– 0.0098	– 0.0127	– 0.0092	0
B_4/r	0.0	0.00003	0.00003	0.00003	0.00003	0.00003	0

It is evident that the first order term is the same as for the master rod $\theta = 0$. Only the second order A and B terms are large enough to be significant. From engine balance considerations, the following general characteristics of the articulated rod system can be observed:

1. Only the first and second order terms are important, as with the plain rod system.
2. Second order articulated rod forces are not fully in phase with the crank.
3. Master rods behave qualitatively like plain rods.
4. Connecting rod masses and moments of inertia will be different for both master and articulated rods as compared with the corresponding plain rod assembly.

The V-8 reciprocating machine with master and articulated rods will require first order counter weights different from those of the plain rod machine, and will have a small second order unbalance, because the second order forces of the articulated rods are not fully in phase with those of the master rods.

Overturning moments due to unbalanced forces in articulated rod systems can be handled in the same way as with plain rods. Again, only first and second order terms are usually important. Second order moments will have small sine components and will be somewhat out of phase with 2θ.

The expression for inertia torque from articulated rod systems is a Fourier series containing sines and cosines, except cos (2θ), for all orders of θ. Orders above 2 are small. For ordinary designs the overall result is not substantially different from that of a plain rod system of the same total mass.

Gas pressure torque from the master cylinder of an articulated rod system is the same as that for a plain rod system. The articulated rod has a lever arm around the crank pin which varies with the crank angle, resulting in a somewhat different torque versus crank angle curve. As with plain rods all half orders from the ½ up to the nth come into play for 4-stroke engines, and all integer orders with 2-stroke engines. In multi-cylinder engines, orders cancel and add as in the case of plain rods, but with small different values at most crank angles.

The differences in the gas pressure torque versus crank angle torque curves between articulated and plain rods have very little practical importance. It should be noted, however, that the torque on the master rod produced by the articulated rods is very important from the point of view of master rod stresses, master rod design and side forces on the master rod piston.

5.11 NON-UNIFORM FIRING OF ENGINE CYLINDERS

In practice, a good many engines are made with non-uniform firing intervals. The reason for such an arrangement may be for mechanical simplicity (for example, to reduce the number of cranks), for engine compactness (for example, small V angle) or to avoid a serious vibration resonance at critical frequency (see examples in table 5.3). For such cases all normal orders are present in the torque diagram, even assuming cyclic uniformity.

Gas forces do not contribute to the vertical direction shaking, but they do produce a shaking couple, as do the inertia forces of the reciprocating masses. Since output torque and shaking couple are equal in magnitude (and opposed in direction), output torque diagrams show the variations of the shaking couple in the engine cycle due both to gas force and to inertia forces of the reciprocating masses. The resultant shaking force on an engine block may be reduced to zero in some instances by combining several slider cranks to form a multi-cylinder engine in which the individual shaking forces balance one another. The corresponding resultant shaking couple, however, cannot be eliminated.

5.12 BALANCE OF TYPICAL MACHINES

Table 5.2 provides balance characteristics for 4-cycle internal combustion engines with various cylinder configurations, and table 5.3 lists similar data for 2-cycle engines. Both tables assume evenly spaced firing, except as noted.

5.13 EXAMPLE PROBLEMS

Example Problem 5.1: The curve c of Fig. 5.17 shows approximately the damping force caused on a body in harmonic motion. Shifting the origin of the coordinates by a quarter cycle to the left, the curve can be expressed mathematically as:

$$f(\omega t) = \sin^2 (\omega t) \quad \text{when } 0 < \omega t < \pi$$

$$f(\omega t) = -\sin^2 (\omega t) \quad \text{when } \pi < \omega t < 2\pi$$

Find the amplitudes of the various harmonics of this curve.

Table 5.2. Four Cycle Engines with Even Cylinder Spacing and Even Firing Intervals

INERTIA BALANCE & FIRING INTERVALS						
Arrangement of Crankshaft and Cylinders	*No. of Cyl.*	*Primary*	*Shaking*	*Secondary*	*Shaking*	*Firing*
		Force	*Moment*	*Force*	*Moment*	*Interval*
FOUR CYCLE IN-LINE ENGINES						
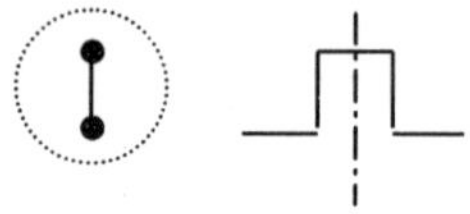	1	$F_c = 0$ $V = Z\cos\theta$ $H = 0$	$M = 0$	$F_c = 0, H = 0$ $V = \lambda Z \cos 2\theta$	$M = 0$	720°
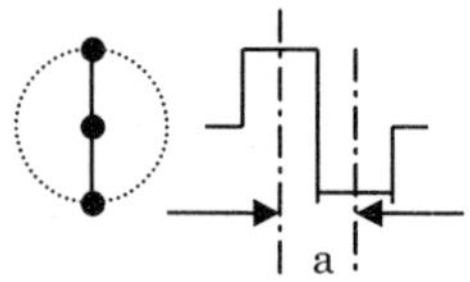	2	$F_c = 0$ $V = 0$ $H = 0$	$M_v = aZ \cos\theta$	$F_c = 0, H = 0$ $V = 2\lambda Z \cos 2\theta$	$M = 0$	180° 540°
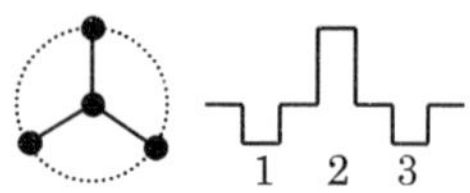	3	$F_c = 0$ $V = 0$ $H = 0$	$Mv = aZ\sqrt{3} \sin\theta$	$F_c = 0$ $V = 0$ $H = 0$	$Mv = a\sqrt{3}\lambda Z \sin 2\theta$	240°
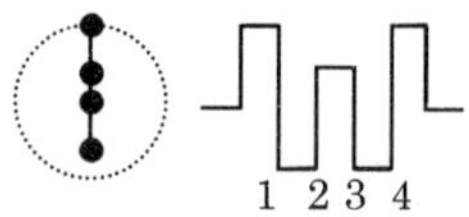	4	$F_c = 0$ $V = 0$ $H = 0$	$M = 0$	$F_c = 0, H = 0$ $V = 4\lambda Z \cos 2\theta$	$M = 0$	180°
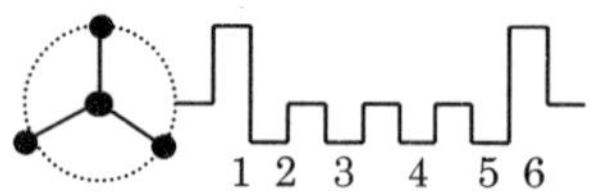	6	$F_c = 0$ $V = 0$ $H = 0$	$M = 0$	$F_c = 0$ $V = 0$ $H = 0$	$M = 0$	120°
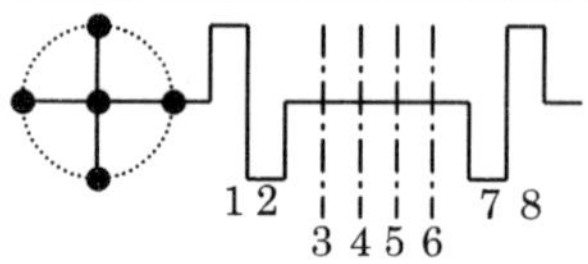	8	$F_c = 0$ $V = 0$ $H = 0$	$M = 0$	$F_c = 0$ $V = 0$ $H = 0$	$M = 0$	90°
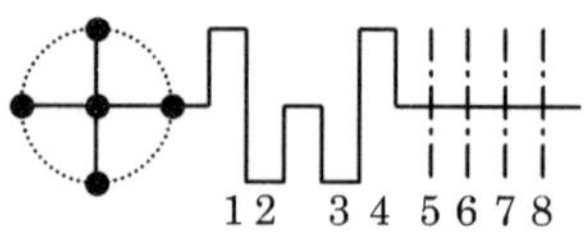	8	$F_c = 0$ $V = 0$ $H = 0$	$M = 0$	$F_c = 0$ $V = 0$ $H = 0$	$Mv = 4a\lambda Z \cos 2\theta$	90°
FOUR CYCLE IN-LINE ENGINES						
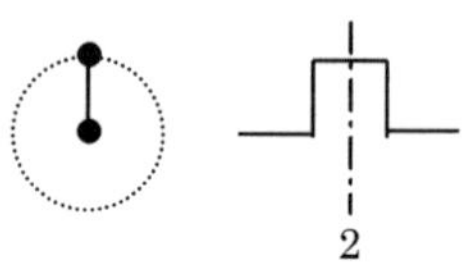	2	$F_c = 0$ $V = 2Z \cos 2\theta$ $H = 0$	$M = 0$	$F_c = 0$ $V = 0$ $H = 0$	$M = 0$	180° 540°

(Contd.)...

Table 5.2. Continued

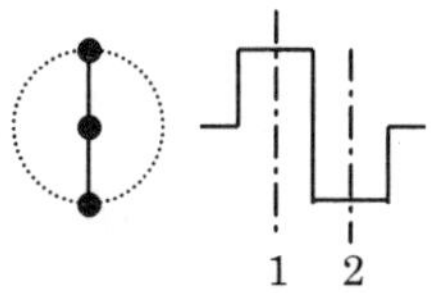	2	$F_c = 0$ $V = 0$ $H = 0$	$Mv = aZ$ $\cos\theta$	$F_c = 0$ $V = 0$ $H = 0$	$Mv = 2\lambda Z$ $\cos 2\theta$	360°
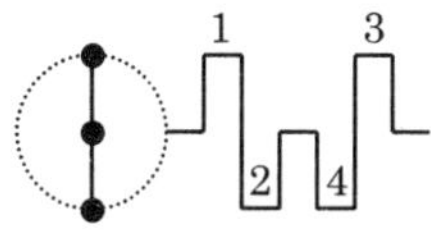	4	$F_c = 0$ $V = 0$ $H = 0$	$Mv = aZ$ $\sin\theta$	$F_c = 0$ $V = 0$ $H = 0$	$Mv = 2a\lambda Z$ $\cos 2\theta$	180°
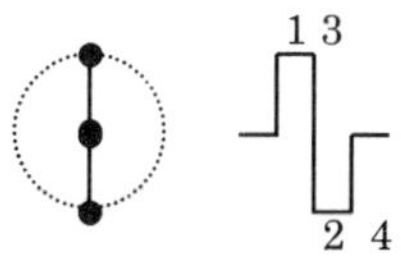	4	$F_c = 0$ $V = 0$ $H = 0$	$M = 0$	$F_c = 0$ $V = 0$ $H = 0$	$M = 0$	180°
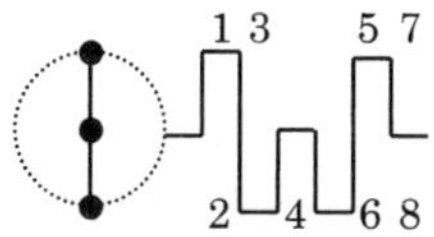	8	$F_c = 0$ $V = 0$ $H = 0$	$M = 0$	$F_c = 0$ $V = 0$ $H = 0$	$M = 0$	180° 2 cylin-ders simul-taneous
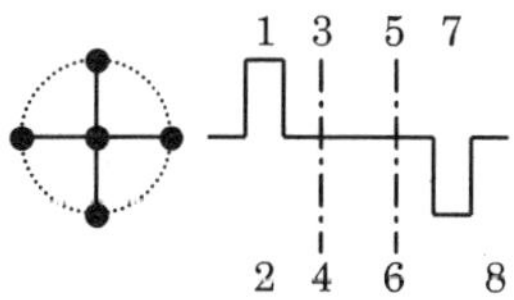	8	$F_c = 0$ $V = 0$ $H = 0$	$Mv = 2(2a$ $+ b)Z\cos\theta$ $- 2b\,Z\sin\theta$	$F_c = 0$ $V = 0$ $H = 0$	$M = 0$	90°
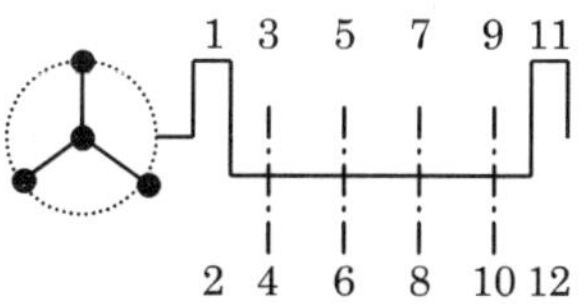	12	$F_c = 0$ $V = 0$ $H = 0$	$M = 0$	$F_c = 0$ $V = 0$ $H = 0$	$Mv = 0$	90°
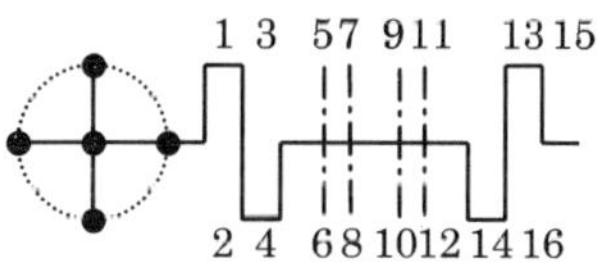	16	$F_c = 0$ $V = 0$ $H = 0$	$M = 0$	$F_c = 0$ $H = 0$ $V = 0$	$M = 0$	90º 2 Cylin-ders Simul-taneous
FOUR CYCLE V ENGINES						
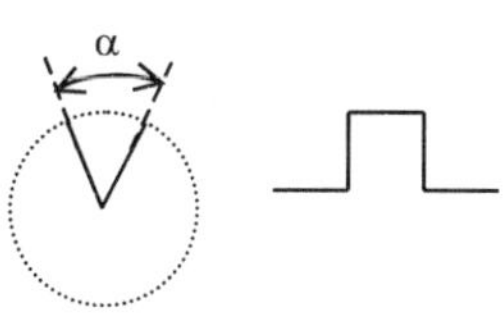	2	$F_c = 0$ $V =$ $2Z\cos\theta$ $\cos^2(\alpha/2)$ $H =$ $2Z\sin\theta$ $\sin^2(\alpha/2)$	$M = 0$	$F_c = 0$ $V = 2\lambda Z$ $\cos\theta\cos\alpha$ $\cos(\alpha/2)$ $H = 2\lambda Z$ $\sin\theta\sin\alpha$ $\sin^2(\alpha/2)$	$M = 0$	

(Contd.)...

Table 5.2. Continued

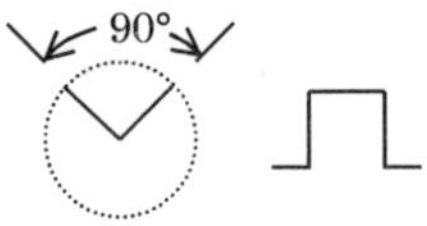	2	$F_c = Z$ $V = 0$ $H = 0$	$M = 0$	$F_c = 0$ $V = 0$ $H = \sqrt{2}\,\lambda Z$ $\cos 2\theta$	$M = 0$	450° 270°
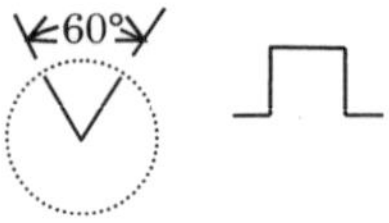	2	$F_c = Z/2$ $V = Z \cos \theta$ $H = 0$	$M = 0$	$F_c = 0.5\sqrt{3}\,\lambda Z$ $H = 0$ $V = 0$	$M = 0$	420° 300°
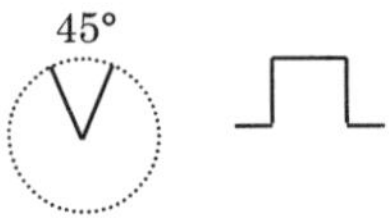	2	$F_c =$ $.293Z$ $V = \sqrt{2}\,Z$ $\cos \theta$ $H = 0$	$M = 0$	$F_c = .541$ λZ $V = .765$ $\lambda Z \cos 2\theta$ $H = 0$	$M = 0$	405° 315°
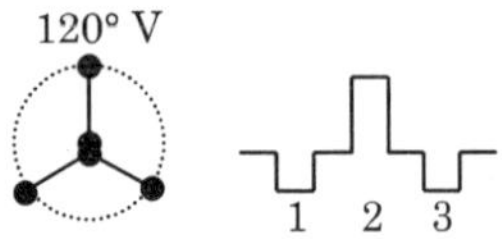	6	$F_c = 0$ $V = 0$ $H = 0$	$M_v = 0.5a$ $\sqrt{3}\,Z \sin \theta$ $M_h = 1.5a$ $\sqrt{3}\,Z \cos \theta$	$F_c = 0$ $V = 0$ $H = 0$	$M_v = 0.5a$ $\sqrt{3}\,Z \sin 2\theta$ $M_h = 1.5a$ $\sqrt{3}\,Z \cos 2\theta$	120°
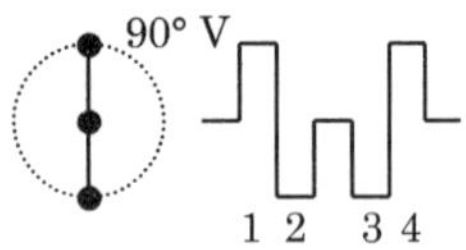	8	$F_c = Z$ $V = 0$ $H = 0$	$M = 0$	$F_c = 0$ $V = 0$ $H = 4\sqrt{2}\,\lambda Z$ $\cos 2\theta$	$M = 0$	90°
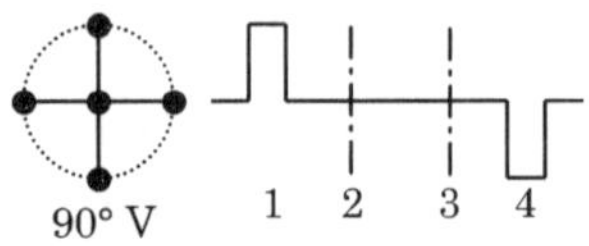	8	$F_c = Z$* $V = 0$ $H = 0$	$M = 0$	$F_c = 0$ $H = 0$ $V = 0$	$M = 0$	90°
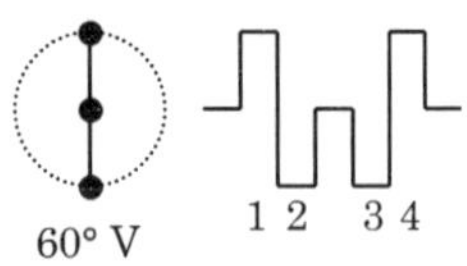	8	$F_c = 0$ $V = 0$ $H = 0$	$M = 0$	$F_c =$ $2\sqrt{3}\,\lambda Z$ $H = 0$ $V = 0$	$M = 0$	60° 120° 60° 120° Etc.
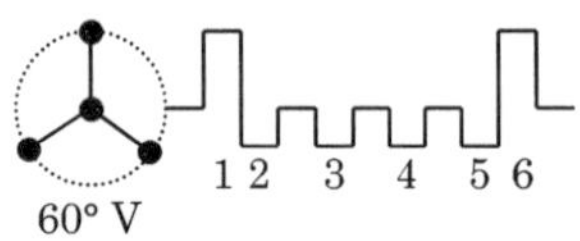	12	$F_c = 0$ $V = 0$ $H = 0$	$M = 0$	$F_c = 0$ $V = 0$ $H = 0$	$Mv =$	60°

*Opposite each crank

(Contd.)...

Table 5.2. Continued

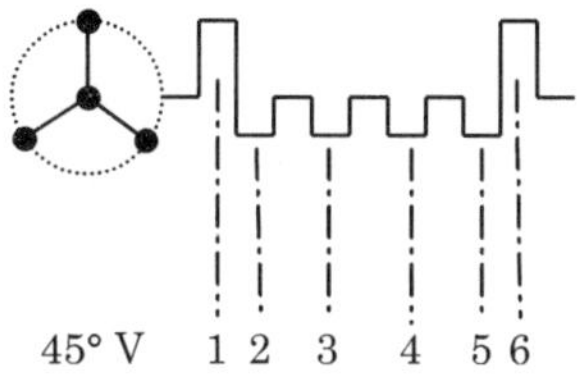	12	$F_c = 0$ $V = 0$ $H = 0$	$M = 0$	$F_c = 0$ $H = 0$ $V = 0$	$M = 0$	45° 75° 45° 75° Etc.
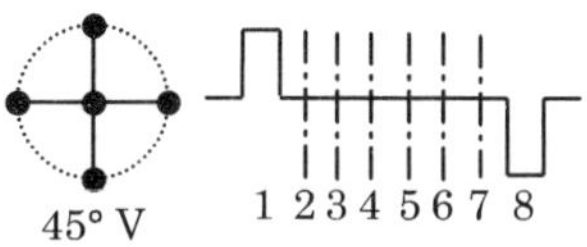	16	$F_c = 0$ $V = 0$ $H = 0$	$M = 0$	$F_c = 0$ $V = 0$ $H = 0$	$M = 0$	45°
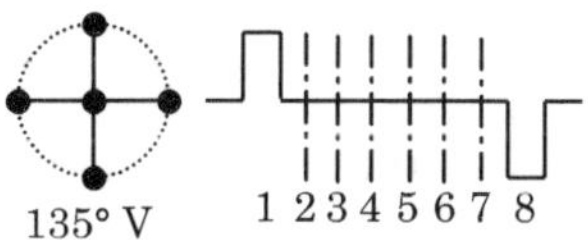	16	$F_c = 0$ $V = 0$ H = 0	$M = 0$	$F_c = 0$ $V = 0$ H = 0	$M = 0$	45°
RADIAL ENGINES, COUNTERWEIGHTED, WITH ALL CONNECTING RODS OPERATING AROUND CRANK PIN CENTER						
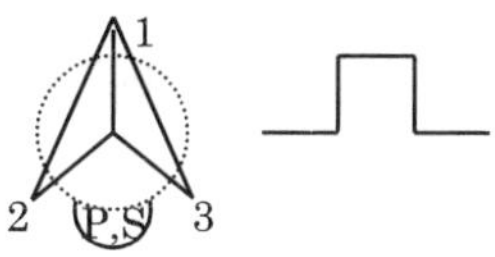	3	$F_c = 1.5Z$ $V = 0$ $H = 0$	M = 0	$F_c = 1.5\lambda Z$ $H = 0$ $V = 0$	$M = 0$	240°
RADIAL ENGINES OF THIS TYPE WITH MORE THAN 3 CYLINDERS IN EACH ROW HAVE NO PRIMARY OR SECONDARY UNBALANCED FORCES OR MOMENTS. FOUR-CYCLE RADIALS MUST HAVE AN ODD NUMBER OF CYLINDERS IN EACH ROW FOR EVEN FIRING AT INTERVALS OF 720°/N : RADIAL ENGINES WITH MASTER AND LINK ROD SYSTEMS HAVE A SMALL SECOND ORDER UNBALANCE IN EACH ROW.						
FOUR CYCLE RADIAL ENGINES; WITH LINK RODS						
Single Row Single Crank n Cylinders, $n > 3$ l = Distance from crank pin to piston wrist pin Must meet following conditions: r = constant, $L = r + l$ Angle between wrist pins same as angle between cylinders.	n	$F_c = Z.n/2$ $V = 0$ $H = 0$	$M = 0$	Force on same side as crank pin when Master at TDC $F_c = r\lambda Z/l$ $V = 0$ $H = 0$	$M = 0$	720°/n

(Contd.)...

Table 5.2. Continued

INERTIA BALANCE & FIRING INTERVALS						
Arrangement of Crankshaft and Cylinders	*No. of Cyl.*	*Primary Force*	*Shaking Moment*	*Secondary Force*	*Shaking Moment*	*Firing Interval*
FOUR CYCLE W ENGINES						
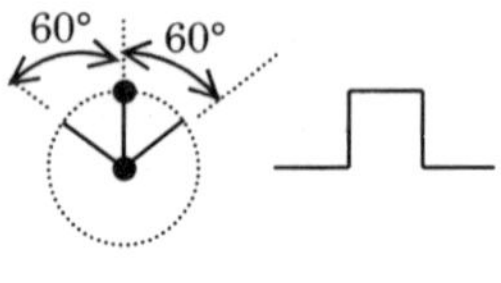	3	$F_c = 1.5\ Z$ $V = 0$ $H = 0$	M = 0	$F_c = .5\lambda Z$ $V = 0$ $H = \lambda Z$ $\sin 2\theta$	$M = 0$	120° 300° 300°
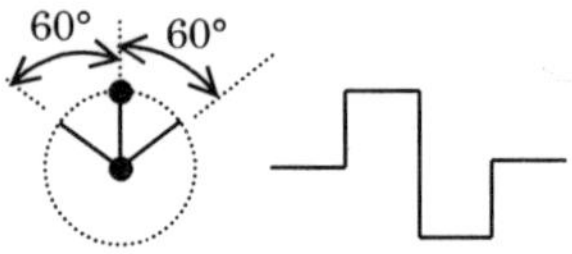	6	$F_c = 1.5Z$* $V = 0$ $H = 0$	$M = 0$	$F_c = \lambda Z$ $V = 0$ $H = 2\lambda Z$ $\sin 2\theta$	$M = 0$	120°
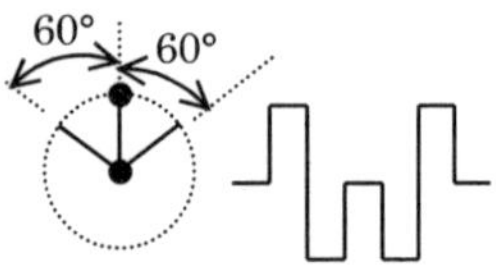	12	$F_c = 0$ $V = 0$ $H = 0$	$M = 0$	$F_c = 2\lambda Z$ $V = 0$ $H = 4\lambda Z$ $\cos 2\theta$	$M = 0$	60°
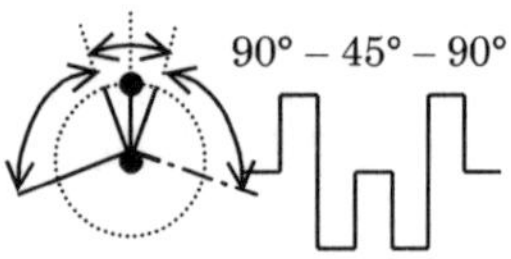	16	$F_c = 0$ $V = 0$ $H = 0$	$M = 0$	$F_c = 0$ $V = 0$ $H = 0$	$M = 0$	45°

$Z = M_{rec} R\Omega^2$

F_c = rotating force balanced by counter weights

H = residual horizontal unbalance inertia after balancing

V = residual vertical unbalance inertia after balancing

θ = crank angle measured from clockwise indicated position

a = distance as shown

b = distance as shown

c = distance as shown

M_v = resultant moment in vertical plane, after balancing

M_h = resultant moment in horizontal plane, after balancing

$\lambda = R/L$

---

*Opposite each crank

Table 5.3. Two Cycle Engines with Even Cylinder Spacing and Even Firing, Except as Noted

No. of Cyl.	*In-Line Engines*		*Unbalanced Force Orders**		*Unbalanced Moments Order**		
	Crank Diagram	*Firing Order*	*1st* Z	*2nd* ZR/L	*1st Rotate* dM_R	*1st Vert.* dZ	*2nd* dZR/L
1	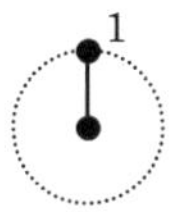	1	1.0	1.0	0	0	0
2		1-2	0	2.0	1.0	1.0	0
3	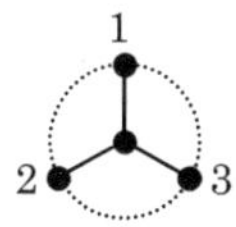	1-2-3	0	0	$\sqrt{3}$	$\sqrt{3}$	$\sqrt{3}$
4	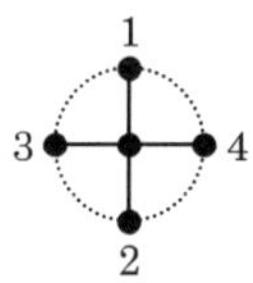	1-3-2-4	0	0	$\sqrt{2}$	$\sqrt{2}$	4
5	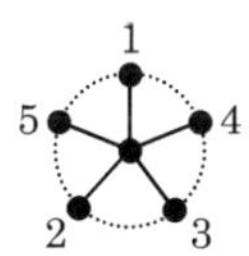	1 - 5 - 2 - 3 - 4	0	0	0.449	0.449	4.98
6	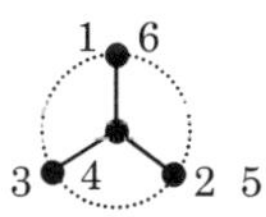	1, 6-3, 4-2, 5 Double Firing	0	0	0	0	0
6	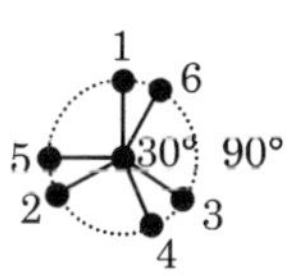	1-5-2-4-3- 6 Uneven Firing	0	0	0.89	0.89	1.74
6	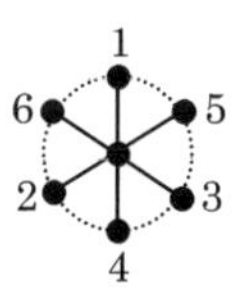	1-6-2-4-3-5	0	0	0	0	3.464

(Contd.)...

* Multiply Z and related quantities by numbers provided.

d = distance between adjacent cylinder axes.

Table 5.3. Continued

6	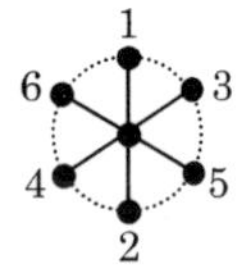	1-6-4-2-5-3	0	0	2	2	6.928
6	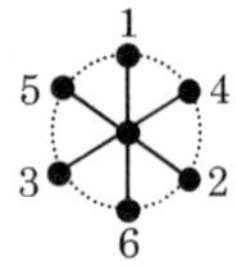	1-5-3-6-2-4	0	0	3.46	3.46	0
7	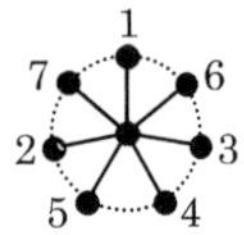	1-7-2-5-4-3-6	0	0	0.267	0.267	1
7	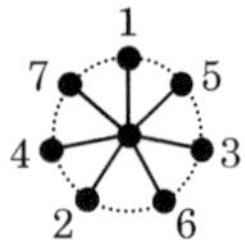	1-7-4-2-6-3-5	0	0	0.9	0.9	5.53
8	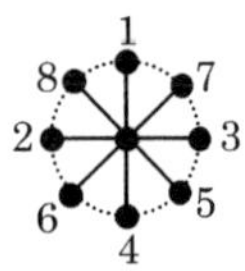	1-8-2-6-4-5-3-7 X = 36 50 O = 53 10	0	0	0	0	0
8	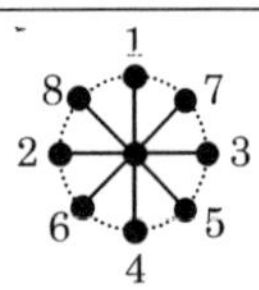	1-8-2-6-4-5-3-7	0	0	0.448	0.448	0
9	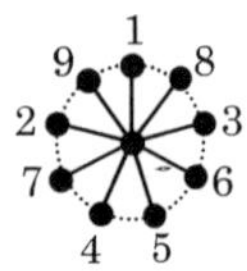	1-9-2-7-4-5-6-3-8	0	0	0.194	0.194	0.548
9	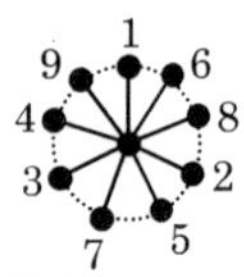	1-9-4-3-7-5-2-8-6	0	0	0.92	0.92	1.28
10	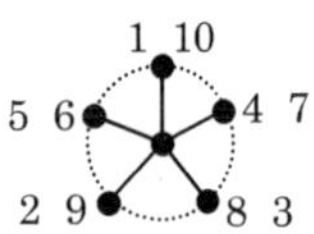	1, 10-5, 6-2, 9-3, 8-4, 7 Double Firing	0	0	0	0	0
10	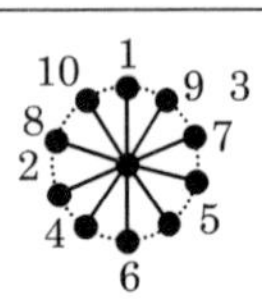	1-10-2-8-4-6-5 -7-3-9	0	0	0	0	0.896

(Contd.)...

11	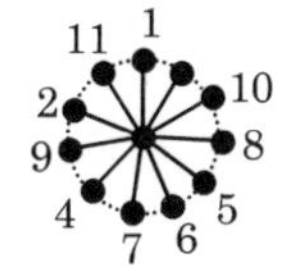	1-11-2-9-4-7-6 -5-8-3-10	0	0	0.153	0.153	0.382
12	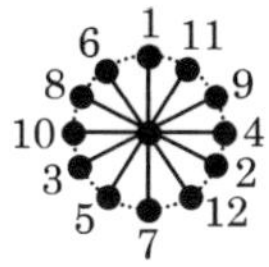	1-6-8-10-3-5-7 -12-2-4-9-11	0	0	0	0	0

All 2 cycle V engines with even firing and with less than 12 cylinders have some degree of unbalance regarding either primary or secondary moments, or both.

Solution: This curve is anti-symmetric, because the values of the function are equal and opposite on both sides of the origin. Sine waves are anti-symmetric and cosine waves are

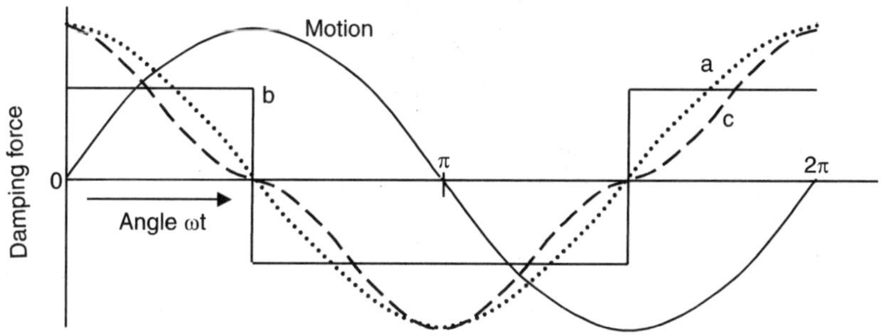

Fig. 5.17

symmetric. Anti-symmetric curves cannot have cosine terms, hence all b_n are zero. For the sine terms:

$$a_n = \frac{\omega}{\pi}\int_0^{2\pi/\omega} f(\omega t)\sin(n\omega t)\,dt$$

$$= \frac{1}{\pi}\left[\int_0^{\pi} \sin^2(\omega t)\sin(n\omega t)\,d(\omega t) - \int_{\pi}^{2\pi} \sin^2(\omega t)\sin(n\omega t)\,d(\omega t)\right] \qquad ...(5.33)$$

$$= \frac{4}{\pi}\frac{\cos(n\pi)-1}{n(n^2-4)}$$

For even values of n, an is zero, while for odd values of n the harmonic exists. In particular, for $n = 1$ the fundamental harmonic:

$$a_1 = 8/(3\pi) = 0.85$$

Thus, the amplitude of the fundamental harmonic is 85% of the maximum amplitude of the curve itself, and the Fourier series is:

$$f(\omega t) = \frac{8}{3\pi}\left[\sin(\omega t) - \frac{1}{5}\sin(3\omega t) - \frac{1}{35}\sin(5\omega t) - \frac{1}{63}\sin(7\omega t) - ....\right] \qquad (5.34)$$

Example Problem 5.2: A triple expansion steam engine has pistons of which the weights are in ratios of 1:1½:2. If it is desired to balance this engine for primary forces, how should the crank angles be prescribed?

Solution: The vectors in the star diagram of Fig. 5.18 have lengths in the required proportion, with the vector of 2 units pointing vertically down. Equilibrium requires that the other two vectors be arranged so that their horizontal components balance, and that the sum of their vertical components equals 2 units. Then:

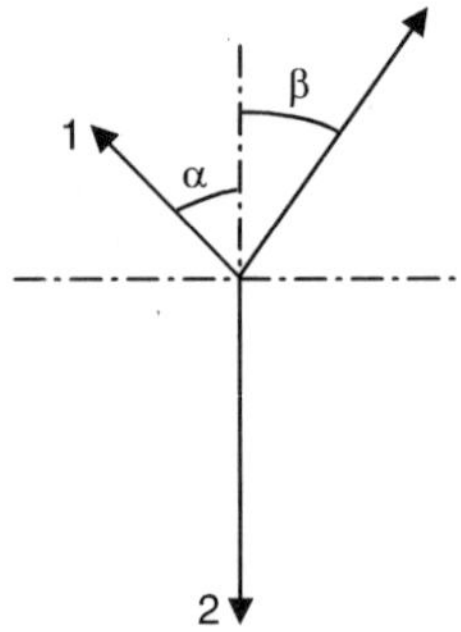

Fig. 5.18

$$1 \times \sin(\alpha) = 1.5 \times \sin(\beta)$$
$$1 \times \cos(\alpha) + 1.5 \times \cos(\beta) = 2$$

From the first equation:

$$\cos(\alpha) = \sqrt{1 - \sin^2(\alpha)} = \sqrt{1 - 2.25\sin^2(\beta)}$$

Using substitution and simplifying, $6 \times \cos(\beta) = 5.25$, and $\beta = 29°$.

Also, $\cos(\alpha) = 2 - 3 \times 0.875/2 = 0.6875$, and $\alpha =$ **46.6°.**

Example Problem 5.3: A 3-cylinder in-line reciprocating compressor, shown in Fig. 5.19, has the following design values:

Piston assembly weight	55 lbs.
Bore	6.25 in.
Stroke	5.75 in.
Connecting rod length	8.25 in.
Operating speed	950 rpm

The cranks are evenly spaced. Calculate the first and second order inertia forces.

Solution: With cranks evenly spaced, $N_{cp} = 3$, and $\alpha = 2\pi/3$ or 120°. Then $\theta_2 = (\theta_1 + \alpha)$, $\theta_3 = (\theta_1 + 2\alpha)$, $2\theta_2 = 2(\theta_1 + \alpha)$ and $2\theta_3 = 2(\theta_1 + 2\alpha)$. From equation (5.9) the first and second order inertia forces are:

$$\Sigma F_a = Z\{[\cos(\theta_1) + \cos(\theta_1 + \alpha) + \cos(\theta_1 + 2\alpha)]$$
$$+ r/l[\cos(2\theta_1) + \cos(2\theta_1 + 2\alpha) + \cos(2\theta_1 + 4\alpha)..]\}$$

Here $Z = m_{rec}\,\omega^2 r = (55/386) \times (2\pi \times 950/60)^2 \times 5.75/2 = 4054$ lbs.

For $t = 1$ second, $\theta_1 = \omega t = (950/3600) \times 360° = 95.0°$.

So, the first order inertia forces are:

$$4054 \times [-0.0872 - 0.8192 + 0.9063] = 0 \text{ lbs.}$$

The second order inertia forces are:

$$Z(r/l)[\cos(2\theta_1) + \cos(2\theta_2) + \cos(2\theta_3)..]$$
$$= 4054 \times (5.75/2) \times (1/8.25) \times (-0.9848 + 0.3420 + 0.6428)$$
$$= 0 \text{ lbs.}$$

The first and second order force components are shown as separate vectors in Fig. 5.19.

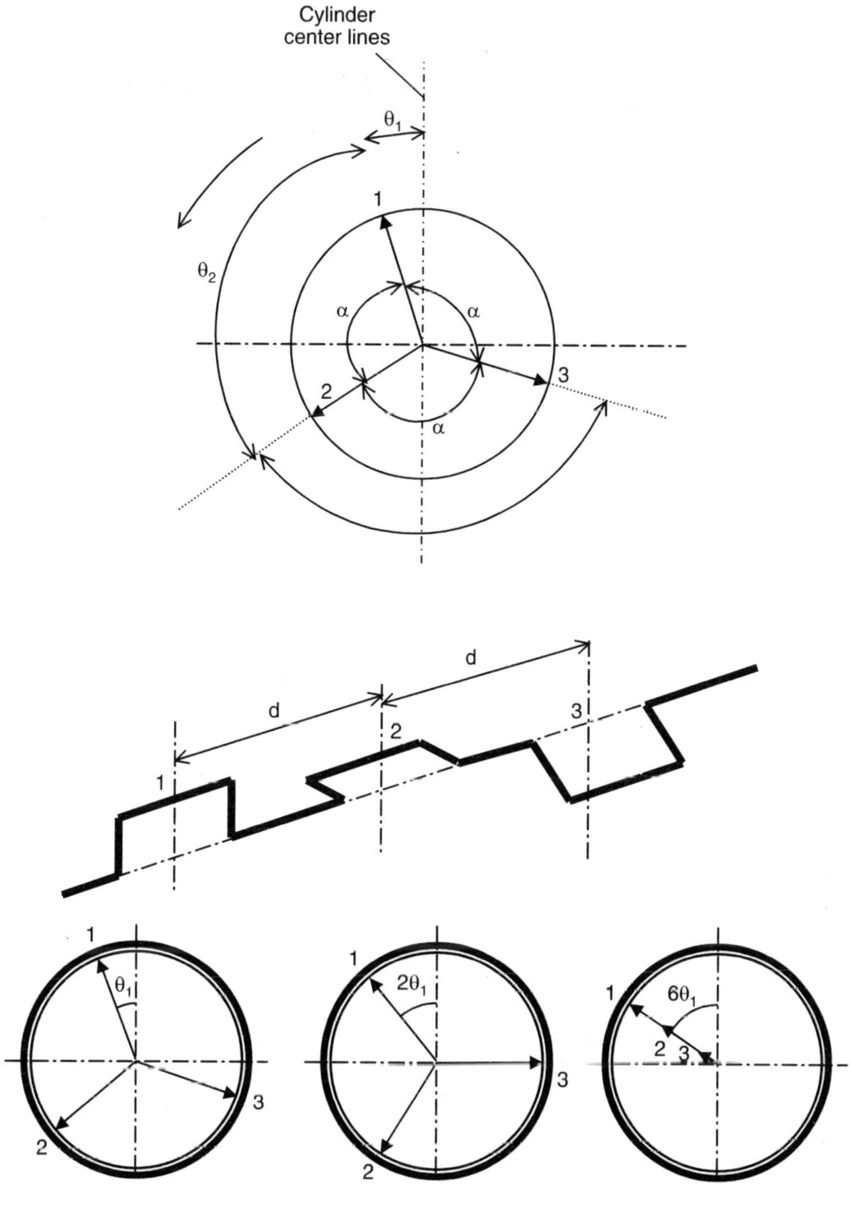

First order, Fa_1
$n\alpha = 120°$
Crank spacing = 120°

Second order, Fa_{11}
$n\alpha = 240°$
Crank spacing = 240°

Third order, Fa_{111}
$n\alpha = 720°$
Crank coincidental

Fig. 5.19. Inertia Forces for 3-Cylinders In-Line, with Equally Spaced Cranks

Example Problem 5.4: Fig. 5.20 shows the arrangement of cranks in a 6-cylinder in-line reciprocating machine, with the cranks spaced 120° to each other. All slider and crank parts may be assumed to be identical in size, shape and weight of 7.83 lbs., radius R = 4.0 ins. and speed N = 1200 rpm. Determine the inertia force on each crank.

Solution: The inertia force of a reciprocating mass is:

$$F = MR\omega^2 \left(\cos(\theta) + \frac{R}{L}\cos(2\theta)\right)$$

where M is the combined mass for a single cylinder, R is the crank length, L is the length of the connecting rod, ω is the crank angular velocity and θ is the rank angle from top dead center.

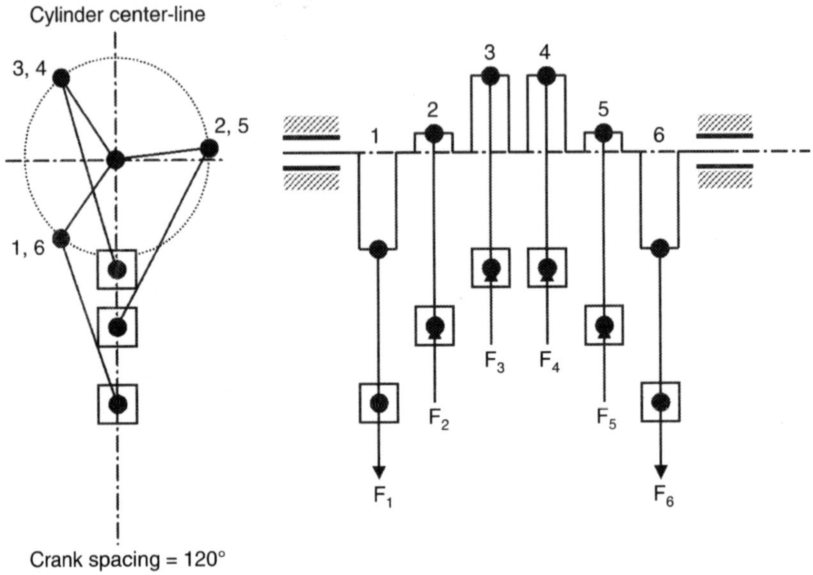

Fig. 5.20. Six Cylinder In Line Machine

Cylinder No.	θ *degrees*	*cos* (θ)	*cos* (2θ)	*[R/L] cos* (2θ)	*cos* (θ) + *[R/L] cos*(2θ)	*F lbs.*
1	20	0.940	0.766	0.191	1.131	1450
2	260	– 0.174	– 0.940	– 0.235	– 0.409	– 525
3	140	– 0.766	0.174	0.044	– 0.722	– 925
4	140	– 0.766	0.174	0.044	– 0.722	– 925
5	260	– 0.174	– 0.940	– 0.235	– 0.409	– 525
6	20	0.940	0.766	0.191	1.131	1450
		$\Sigma = 0$	$\Sigma = 0$	$\Sigma = 0$	$\Sigma = 0$	$\Sigma = 0$

The table shown above provides the arrangement of cranks and the inertia forces for $\theta = 20°$. The resultant of the six inertia forces is zero, but it can be shown that this resultant force value is valid for any value of θ. Thus, no shaking force is transmitted to the main bearings supporting the crankshaft or the engine block. This situation is well known for six cylinder in-line machines. The single, two, three, four and five cylinder reciprocating machines of the in line type cannot be inherently balanced against shaking by the reciprocating masses in a similar manner.

Example Problem 5.5: Derive mathematical expressions for the primary and secondary inertia forces and moments of a 3-cylinder in-line reciprocating machine (Fig. 5.21) for any crank angle θ_1 of the first cylinder. Assume all reciprocating and rotating masses are identical, as also the R/L ratio, for all cylinders. Also obtain expression for the total force on the crankshaft.

Solution: θ_1 of the first crank locates the clockwise position of the crankshaft in the engine cycle, ϕ_2 and ϕ_3 are the fixed angles of cranks 2 and 3, respectively, measured from crank # 1. Three cylinders only are shown in the figure, but any number of cylinders may be considered. The inertia force F of any cylinder at crank angular location θ is:

$$F = MR\omega^2\left(\cos(\theta) + \frac{R}{L}\cos(2\theta)\right)$$

$$= MR\omega^2 \cos(\theta) + M\frac{R^2}{L}\omega^2 \cos(2\theta)$$

The first term is the first harmonic, or the primary force F_p, and the second term is the second harmonic or secondary force F_s. The summation of the inertia forces of a multi-cylinder engine is the resultant shaking force S:

$$S = \Sigma F = \Sigma F_p + \Sigma F_s$$

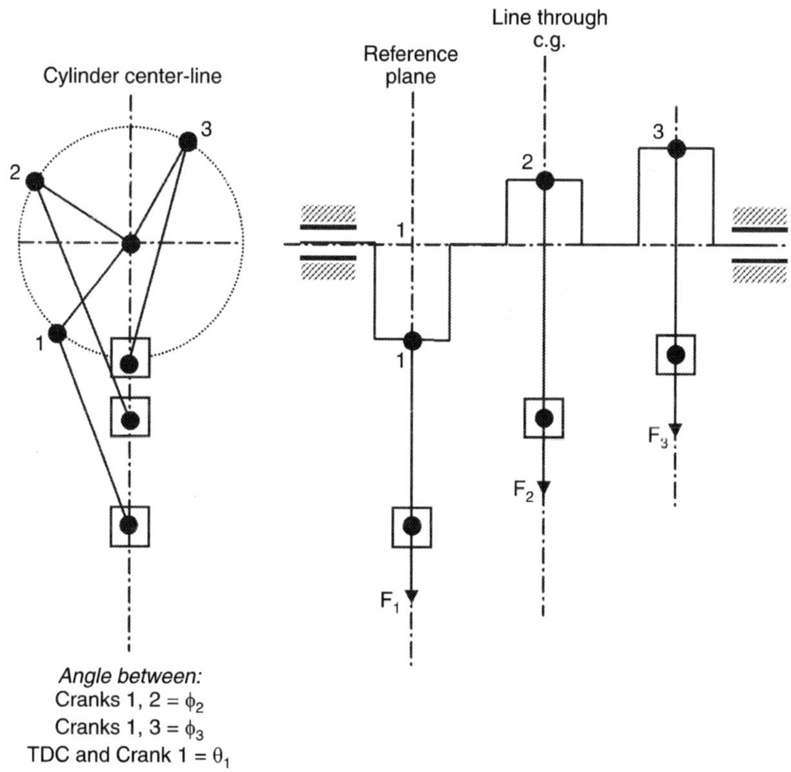

Fig. 5.21

Balancing of primary forces in some configurations of multi-cylinder reciprocating may also be the case. The equation of unbalance of primary forces is developed as follows, where $\theta = \theta_1 + \phi$.

$$\Sigma F_p = \Sigma MR\omega^2 \cos(\theta)$$
$$= MR\omega^2 \Sigma \cos(\theta)$$
$$= MR\omega^2 \Sigma [\cos(\theta_1)\cos(\phi) - \sin(\theta_1)\sin(\phi)]$$

Now $\cos(\theta_1)$ and $\sin(\theta_1)$ are constant for all cylinders. Hence:

$$\Sigma F_p = MR\omega^2 [\cos(\theta_1)\Sigma\cos(\phi) - \sin(\theta_1)\Sigma\sin(\phi)]$$

In a similar manner the expression for secondary forces may be determined from the expression:

$$\Sigma F_s = M[R^2/L]\omega^2 \{\cos(2\theta_1)\Sigma\cos(2\phi) - \sin(2\theta_1)\Sigma\sin(2\phi)\}$$

Thus, for any arrangement of cylinders where the angles ϕ are known, then $\Sigma\cos(\phi)$, $\Sigma\sin(\phi)$, $\Sigma\cos(2\phi)$ and $\Sigma\sin(2\phi)$ may be evaluated to determine the unbalance forces as a function of θ_1 only. For complete balance of the shaking forces, the following conditions must be met:

$$\Sigma\cos(\phi) = 0$$
$$\Sigma\sin(\phi) = 0$$
$$\Sigma\cos(2\phi) = 0$$
$$\Sigma\sin(2\phi) = 0$$

Multi-cylinder reciprocating machine configurations must also contend with another source of unbalanced load. In Fig. 5.21 if the schematic diagram is viewed from the side, the line of action of the resultant shaking force in the axial plane may not lie on a line of symmetry between the main bearings. Also, this line of action of the resultant force S can shift axially in the axial plane as a function of θ_1. In such a situation the engine will oscillate in an end over end rotational mode. The line of action of S may be determined from the principle of moments in terms of a primary moment Mp and a secondary moment M_s, where the moments are taken with respect to a reference plane at the first cylinder. If a is the distance from the reference plane to the line of action of the inertia force of any cylinder, then:

$$M_p = \Sigma F_p a = MR\omega^2 \Sigma\{a\cos(\theta)\}$$
$$= MR\omega^2[\cos(\theta_1)\Sigma a\cos(\phi) - \sin(\theta_1)\Sigma a\sin(\phi)]$$

and:

$$M_s = M[R^2/L]\omega^2\{\cos(2\theta_1)\Sigma a\cos(2\phi) - \sin(2\theta_1)\Sigma a\sin(2\phi)\}$$

The total moment load is: $M = M_p + M_s$.

Distances a_R of the line of action of the shaking force S may be obtained from the resultant moment M about the reference plane as follows:

$$aR = M/S$$

In some cases where the shaking force is balanced, the resultant moment M may still be present due to the presence of an end over end axial shaking couple.

Example Problem 5.6: Determine the unbalance shaking force S and the axial shaking couple M due to the reciprocating masses of a 4-cylinder machine shown in Fig. 5.22. The cranks are separated 180° from each other.

Solution: The fixed angles ϕ are shown in the Fig. The following summations are required for obtaining constant terms in the equations of unbalance:

$$\Sigma\cos(\phi) = 1 - 1 - 1 + 1 = 0$$
$$\Sigma\sin(\phi) = 0 + 0 + 0 + 0 = 0$$
$$\Sigma\cos(2\phi) = 1 + 1 + 1 + 1 = 4$$
$$\Sigma\sin(2\phi) = 0 + 0 + 0 + 0 = 0$$

Then the primary forces are: $\Sigma F_p = MR\omega^2 [\cos(\theta_1) \times 0 - \sin(\theta_1) \times 0] = 0$.

The secondary forces are:

$$\Sigma F_s = M[R^2/L]\omega^2 \{\cos(2\theta_1) \times 4 - \sin(2\theta_1) \times 0\}$$
$$= 4M[R^2/L]\omega^2 \cos(2\theta_1)$$

Hence: $$S = \Sigma F_p + \Sigma F_s$$
$$= 4M[R^2/L]\omega^2 \cos(2\theta_1)$$

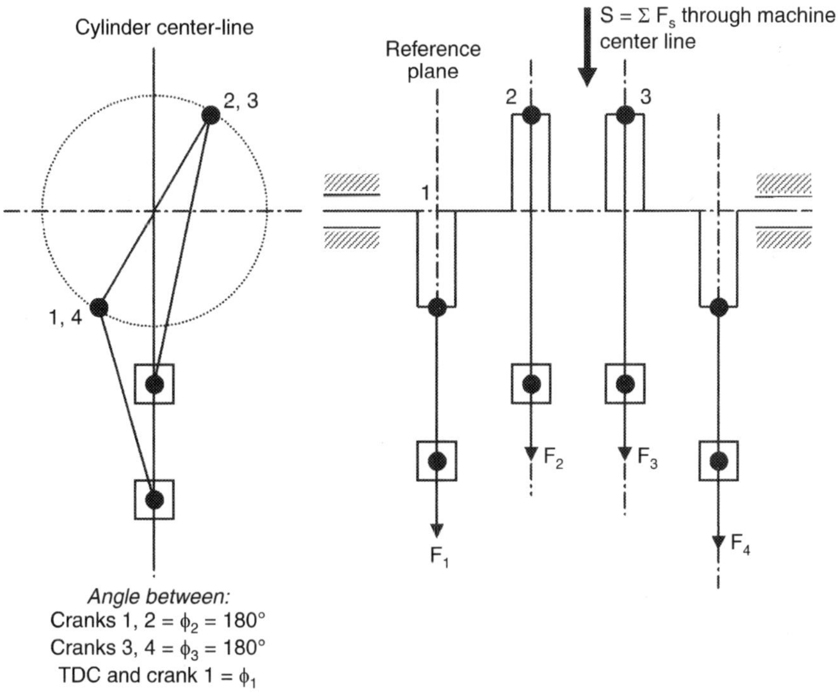

Fig. 5.22

This expression for the total shaking force is plotted as a function of θ_1 in Fig. 5.23.

In the calculations for moments the following summations are required:

$$\Sigma a \cos(\phi) = a(1 \times 0 - 1 \times 4 - 1 \times 8 + 1 \times 12) = 0$$
$$\Sigma a \sin(\phi) = a(0 \times 0 + 0 \times 4 + 0 \times 8 + 0 \times 12) = 0$$
$$\Sigma a \cos(2\phi) = a(1 \times 0 + 1 \times 4 + 1 \times 8 + 1 \times 2) = 24a$$
$$\Sigma a \sin(2\phi) = a(0 \times 0 + 0 \times 4 + 0 \times 8 + 0 \times 12) = 0$$

Hence, the primary moments are zero. Secondary moment exists about the reference plane, given by the expression:

$$C = C_p + C_s = 24aM[R^2/L]\omega^2 \cos(2\theta_1)$$

The line of action of the shaking force S is determined as follows:

$$aR = C/S$$
$$= \{24M[R^2/L]\omega^2 \cos(2\theta_1)\}/\{4M[R^2/L]\omega^2 \cos(2\theta_1)\} = 6.0$$

The only unbalance is a shaking force due to the secondary forces which tend to cause an up and down vibration of the machine. A common device in four cylinder internal combustion engines is the Lanchester balancer, shown in Fig. 5.24. It consists of two meshed gears with eccentric masses. The pitch point of the meshing gears is directly below the centerline of the engine, so the resultant inertia force of the rotating masses counterbalances the shaking force S. A cross helical gear on the crankshaft drives the balancing gears at twice the crankshaft speed in order that the balancing forces be of the same circular frequency as the unbalanced secondary forces. When unbalance is only from the primary shaking forces the gears rotate at crankshaft speed.

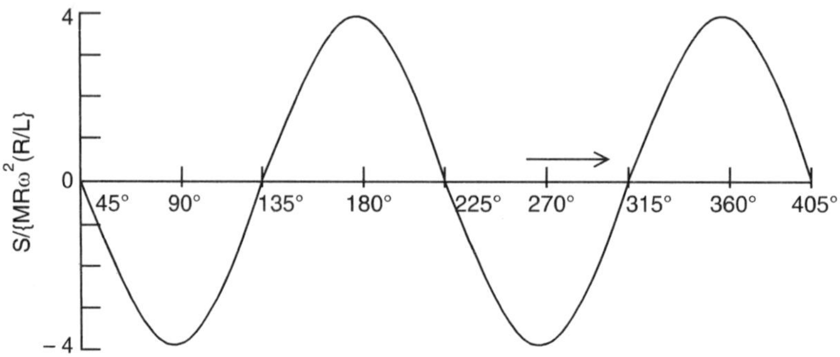

Fig. 5.23

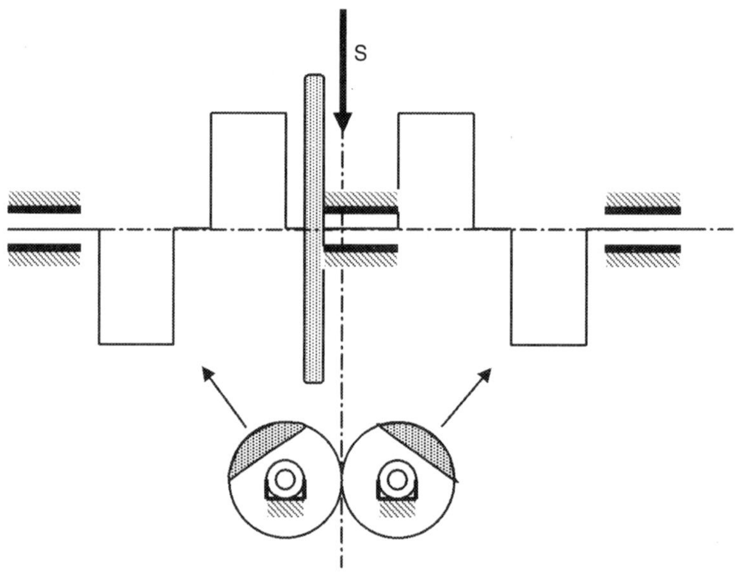

Fig. 5.24. Lanchester Balancer

Example Problem 5.7: Analyze the balance arising from reciprocating masses in the 5-cylinder machine depicted in Fig. 5.25.

Solution: The fixed angles ϕ are:

$$\phi_1 = 0°;\ \phi_2 = 216°;\ \phi_3 = 144°;\ \phi_4 = 72°;\ \phi_5 = 288°$$

$$\Sigma \cos(\phi) = \cos 0° + \cos 216° + \cos 144° + \cos 72° + \cos 288°$$

$$= 0$$

$$\Sigma \sin(\phi) = \sin 0° + \sin 216° + \sin 144° + \sin 72° + \sin 288°$$
$$= 0$$
$$\Sigma \cos(2\phi) = \cos 0° + \cos 432° + \cos 288° + \cos 144° + \cos 576°$$
$$= 0$$
$$\Sigma \sin(2\phi) = \sin 0° + \sin 432° + \sin 288° + \sin 144° + \sin 576°$$
$$= 0$$

Unbalanced shaking forces are:

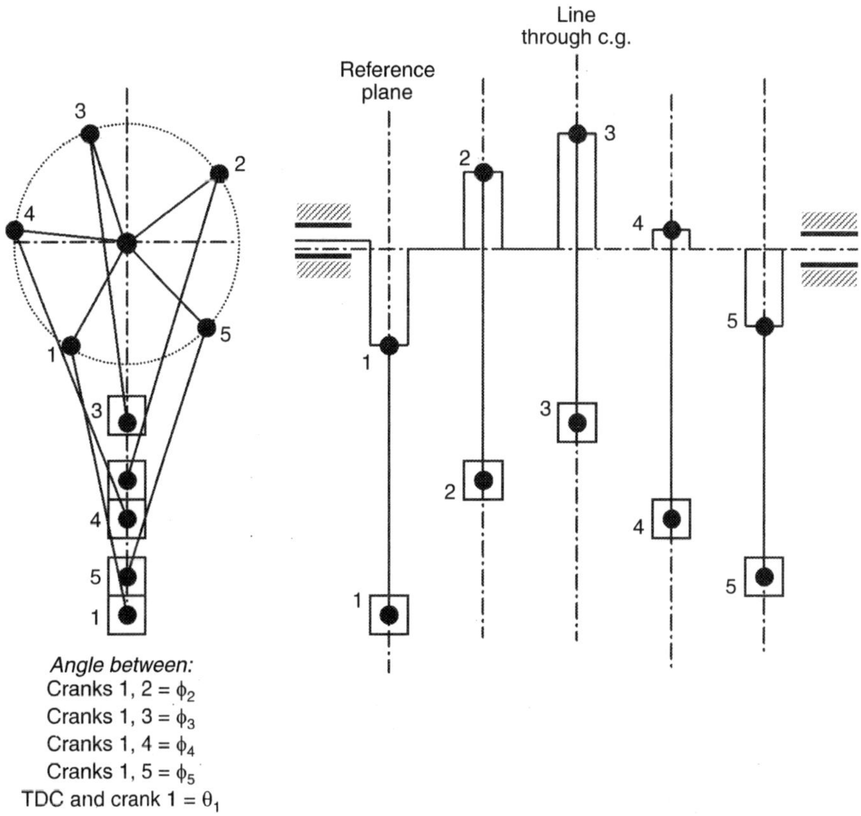

Fig. 5.25

$$\Sigma F_p = MR\omega^2 [\cos(\theta_1)\, \Sigma \cos(\phi) - \sin(\theta_1)\, \Sigma \sin(\phi)]$$
$$\Sigma F_s = M[R^2/L]\omega^2 \{\cos(2\theta_1)\, \Sigma \cos(2\phi) - \sin(2\theta_1)\, \Sigma \sin(2\phi)\}$$

Hence both the primary and secondary forces are balanced. The unbalanced shaking moments are:

$$M_p = MR\omega^2[\cos(\theta_1)\, \Sigma a \cos(\phi) - \sin(\theta_1)\, \Sigma a \sin(\phi)]$$

and: $M_s = M[R^2/L]\omega^2 \{\cos(2\theta_1)\, \Sigma a \cos(2\phi) - \sin(2\theta_1)\, \Sigma a \sin(2\phi)\}$

Now: $\Sigma a \cos(\phi) = 0 \times 1 + a(-0.809) + 2a(-0.809) + 3a(0.309) + 4a(0.309)$

$$= -0.264a$$

$$\Sigma a \sin(\phi) = 0 \times 1 + a(-0.5878) + 2a(0.5878) + 3a(0.9511) + 4a(-0.9511)$$
$$= -0.3633a$$

$$\Sigma a \cos(2\phi) = 0 \times 1 + a(0.309) + 2a(0.309) + 3a(-0.8090) + 4a(-0.8090)$$
$$= -4.736a$$

$$\Sigma a \sin(2\phi) = 0 \times 1 + a(0.9511) + 2a(-0.9511) + 3a(0.5878) + 4a(-0.5878)$$
$$= -1.5389a$$

Hence: $M_p = MR\omega^2[\cos(\theta_1)(-0.264a) - \sin(\theta_1)(-0.3633a)]$

$$= MaR\omega^2[-0.264 \cos(\theta_1) + 0.3633 \sin(\theta_1)]$$

and: $M_s = M[R^2/L]\omega^2\{\cos(2\theta_1)(-4.7360a) - \sin(2\theta_1)(-1.5389a)\}$

$$= Ma[R^2/L]\omega^2\{-4.736 \cos(2\theta_1) + 1.5389 \sin(2\theta_1)\}$$

REFERENCES AND BIBLIOGRAPHY

Carter, P.C., Forshaw, S., "Torsiograph Observations on a Merlin II Engine with Five Pitch Settings of Propeller Blades", *ARC Tech. Rep. R + M*, HM Stationery Office, London, 1983.

Cataford, G. F., Lancee, R. P., "Oil Free Compression on a Natural Gas Pipeline", ASME paper # 86-GT-293, New York, June, 1986.

Colby, H. C., "Torsional Vibration - A Case History", *Trans. SAE, 92*, 186, 1984.

Crede, C. E., Ruzcika, J. E., "Theory of Vibration Isolation", Shock & Vibration Handbook, McGraw-Hill, New York, 1988.

Cretchlow, B. G., Bean, P. R., "Crankshaft Bending Vibration", *SAE Quarterly Transactions 1,* July, 1987.

Den Hartog, J. P., "Mechanical Vibrations", Dover Publications, Inc., New York, 1985.

Dunavesky, M., "Analysis of Distortion of Cylinders and Conformability of Piston Rings", Tribology Transactions, Vol. 33, pages 33-40, 1990.

Gale, N. F., Widener, S. K., Sui, P. C., Zhang. H., "Analytical Engine Design Methods: A Review", SAE Paper # 950806, 1995.

Geiger, F. G., "Determination of Crankshaft Stresses in Critical Regions with Damping", Auto. Aviation Ind., 92, April, 1985.

Heath, A. R., "The Measurement and Prediction of Flywheel Whirl", Institute of Mechanical Engineers, Auto Congress, 1987.

Hodgetts, D., "Vibration of A Crankshaft", Institute of Mechanical Engineers, Paper # C99/81, 1981.

Holzer, "Die Berechnung der Dreschwingungen", Springer-Verlag, Berlin, 1921.

Jeon, H., Tsuda, K., "Theoretical Analysis of the Undamped, Coupled Torsional - Axial Vibration of Marine Diesel Engine Shaftings", Bulletin of the M. E. S. J., Vol. 4, No. 7, P41, 1989.

Kimball, J., "Optimum Ratio Between Crank Arm and Connecting Rod for Smooth Operation of Reciprocating Engines", Journal Franklin Institute, *246*, 409, 1978.

Law, B., "Prediction of Main Bearing and Crankshaft Loading in Reciprocating Engines", CIMAC Paper # 6358C/424A, 1983.

Law, B., "Prediction of Crankshaft and Flywheel Dynamics", Institute of Mechanical Engineers, Paper # C382/046, pages 487-498, Perkins Technology, 1989.

Lewis, F. M., *Trans. ASME,* 78:APM 377, 1975.

Maas, H., "Calculation of Crankshaft Plain Bearings", 9th CIMAC Conference, Paper # A-22, Stockholm, 1981.

Maas, H., "Calculation of Crankshaft Plain Bearings", 11th CIMAC Conference, Pages 43-68, Vol. 1, Barcelona, 1985.

Mabie, H. H., Reinholtz, C. F., "Mechanisms and Dynamics of Machinery", John Wiley & Sons, New York, 1987.

Malychevitch, D. P., "Torsional Vibration of Diesel Engine Crankshafts", Auto Ind., Part 1 & 2, March, 1989.

Myklestad, N. O., "Fundamentals of Vibration Analysis", McGraw-Hill, New York, 1956.

Nestorides, E. J., "A Handbook of Torsional Vibration", Cambridge University Press, 1958.

Oldenburg, H. G., "Frequency Response", Macmillan & Co., New York, 1986.

Porter, F, "Evaluation of Effects of Torsional Vibration", Society of Automotive Engineers, War Engineering Board; also *Trans.* ASME, 65:A-33, 1943.

Porter, F., "Harmonic Coefficients of Engine Torque Curves", Journal of Applied Mechanics, 10(1):A33-A48, March, 1963.

Rao, S. S., "Mechanical Vibration", Addison-Wesley Publishing Co., Reading, Mass., 1990.

Roark, R. J., Young, W. C., "Formulas for Stress and Strain", 5th ed., McGraw-Hill, New York, 1985.

Saito, J., Odajima, M., Harada, T., "Isuzu New V8-V12 PE-Series Diesel Engines", SAE Paper # 950525, 1995.

Schroen, B. G., "Die Dynamik der Verbrennungskraftmaschine", Springer-Verlag, Vienna, 1982.

Spaetgens, S. G., "Holzer Method for Forced Damped Torsional Vibrations", *Transactions ASME 72*, March, 1950.

Taylor, C. F., "The Internal Combustion Engine in Theory and Practice - Vol. 2: Combustion, Fuels, Materials, Design", The MIT Press, Cambridge, Massachusetts, 1985.

Taylor, C. F., Morris, T., "Harmonic Analysis of Engine Torque", *J. Aero Sci.*, 3, 129, 1966.

Taylor, E. S., "Balancing the Four Cylinder Aircraft Engine", Western Flying *12*, July, 1982.

Thomas, H. R., Hoersch, V. A., "Stresses due to the Pressure of Onc Elastic Solid upon Another", University of Illinois Engineering Experimental Sta. Bulletin # 212, 1990.

Timoshenko, S., Young, D. H., Weaver, W., "Vibration Problems in Engineering", John Wiley & Sons, New York, 1984.

Wakabayashi, K., Honda Y., Kodama, T., Iwamoto, S., "The Characteristics of Bending Vibration Stress Coupled with Torsional Vibration of Automotive Diesel Engine Shaftings", SAE Paper # 950543, 1995.

Wakabayashi, K., Iwamoto, S., Matsumoto, N., "Analysis of Vibrations of Reciprocating Engine Shaftings by the Transfer Matrix Method - The Second Report, Calculation and Comparison with Measured Data of Torsional Vibration", Bulletin of the M.E.S.J., Vol. 8, No. 1, pages 40-50, 1980.

Wakabayashi, K., Seki, T., Iwamoto, S., Ito, H., "Analysis of Vibrations of Reciprocating Engine Shaftings by the Transfer Matrix Method - The First Report, Analysis of Forced Vibration of a Crankshaft", Bulletin of the M.E.S.J., Vol. 8, No. 1, pages 27-39, 1980.

Wilson, W. K., "Practical Solutions of Torsional Vibration Problems", John Wiley & Sons, Inc., New York, 1982.

Yeo, S. D., Koo T. K., Chang, S. H., "Torsional Vibration Characteristics of a Crankshaft with a Rubber Damper", SAE Paper # 891232, 1989.

Zdanowich, G., Wilson, T. S., *Proc. Inst. Mech. Engrs. (London),* 1943:182, 1960.

6 Torsional Vibrations

6.1 INTRODUCTION

Because of their construction features reciprocating machines are subjected to unbalance arising from reciprocating components and from the rotational motion of the crankshaft. Torsional vibration is an oscillatory angular motion causing twisting in the shaft of a rotating system; the oscillatory motion is superimposed on the steady rotational motion of the shaft. Even though the vibration cannot be detected without special measuring equipment, its amplitude can be destructive. As an example, gear sets used to alter speeds of power transmission systems transmit the vibration to the casing. Similarly, slider crank mechanisms in engines, compressors and pumps convert torque to radial force that is discernible to human perception but is not measurable due to the insensitivity of the test equipment and background noise. If gear boxes or reciprocating machines are part of a drive train, excess noise and vibration can indicate trouble. Standards and measurement methods dealing with acceptable magnitudes are required to ensure trouble free operation of such a system.

Motion by itself is rarely a concern with torsional vibration unless it affects the operating function of the system. It is stresses that affect the structural integrity and life of components, and thus determine the allowable magnitude of the torsional vibration. Torsional vibratory motions can produce stress reversals that cause metal fatigue. Components tolerate less stress reversals than steady stress. In addition, stress concentration factors associated with machine components decrease the effectiveness of load bearing materials.

In modeling the system for torsional analysis, customary practice is to calculate the natural frequencies of vibration of the rotor as a body free in space, not connected to the supporting structure. For reciprocating machinery this assumption is a very good approximation. While it makes for simple and accurate analysis, it implies that such systems have a level of damping that is extremely low and essentially limited to internal material damping of the crankshaft. The torsional damping constant of the bearings-the proportionality constant between bearing torque and rotational speed - is negligibly small for both rolling element and journal bearings. Thus, unlike lateral vibrations of a rotor system where significant damping is derived from the bearings and support structure, the simple torsional system is one with very little inherent damping.

Large amplitudes of torsional vibration can be present in the rotor of a machine with little or no vibration induced in the stationary structure by the torsional vibration. If harmonic exciting torques are present in the system, either in the driving element or in the reactive torque of the driven element, then the possibility of a destructive failure of the shafting elements of the rotating system due to torsional fatigue exists.

Systems with gearing tooth contact force between mating gear elements impose lateral forces on the bearings and moment reactions on the gear case. If the system is in a state of torsional vibration, then cyclic force reactions are superimposed on these steady forces. Some vibration can be induced in the stationary structure by the cyclic forces, but in general, it is masked by the background machinery noise. However, if torsional vibration amplitude is large enough to unload the gear teeth, then clattering noise ensues as a result of the backlash in the gearing, providing a warning of the presence of torsional vibrations. This large amplitude vibration, if allowed to continue, can damage tooth contact surfaces. Under such circumstances the cyclic component of the load on the bearings provides some damping as a result of squeeze film action.

When the frequency of any harmonic component of the exciting torque is equal to one of the natural torsional frequencies of the rotor, then a condition of resonance exists. The speed at which this coincidence of frequencies occurs is called a 'critical speed'. For each natural frequency a series of critical speeds correspond to the various harmonics of the exciting torque. It is important that a dynamic analysis of the system be made and that the forced response of the system be evaluated for these critical speeds in or near the operating range of speed. High levels of response in the operating range of speed should be avoided if at all possible; otherwise special damping devices are required to limit the resonant response. Initially considering a simple two inertia system connected by a shaft with torsional flexibility will highlight the basic concepts in torsional analysis. Traditional calculation procedures as they relate to reciprocating machinery will next be reviewed, each system characterized by a different source of torsional excitation.

6.2 SIMPLE TWO INERTIA SYSTEM

The simple two-inertia system is a convenient starting point for making estimates, particularly where the driving and driven elements of a machinery system may each be approximately represented by a lumped inertia. Figure 6.1 shows two disks with polar mass moments of inertia J_1 and J_2 connected by a massless shaft with torsional stiffness K_{12}. Angular rotations of the disks are denoted as θ_1 and θ_2. A sinusoidal exciting torque equal to T_1 sin (ωt) is applied to disk 1. The exact solution will be based on governing differential equations of vibration. Three different conditions of damping are considered:

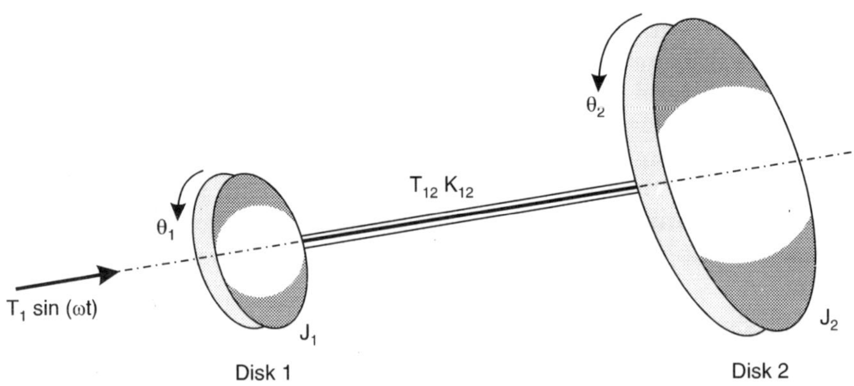

Fig. 6.1. A Simple Two Inertia System

1. A viscous damping torque proportional to the rate of change of the angle of twist in the shaft connecting the disks
2. A viscous damping torque proportional to the angular velocity of disk 1
3. A viscous damping torque proportional to the angular velocity of disk 2.

With the damping torque equal to the viscous damping coefficient B_{12} times the difference in angular velocities of the two disks ($d\psi/dt = d\theta_1/dt - d\theta_2/dt$) a steady state solution yields a relationship for the amplification factor of the angle of twist. In terms of the undamped natural frequency:

$$\omega_n = \sqrt{\left[\frac{K_{12}}{J_1 J_2/(J_1 + J_2)}\right]} \tag{6.1}$$

and the damping ratio:

$$\xi = \frac{B_{12}}{2\omega_n J_1 J_2/(J_1 + J_2)} \tag{6.2}$$

while the amplitude of twist angle is given by:

$$\frac{\psi}{\psi_{st}} = \frac{1}{\sqrt{\{[1 - (\omega/\omega_n)^2]^2 + [2\xi(\omega/\omega_n)]^2\}}} \tag{6.3}$$

where ω is the exciting frequency and ψ_{st} is the shaft static angle of twist due to an exciting torque T_1 applied statically to disk 1. The amplitude of torque T_{12} in the connecting shaft at resonance is:

$$| T_{12} | = \frac{T_1[J_2/(J_1 + J_2)]\pi}{\delta} \tag{6.4}$$

where the damping ratio is a single function of δ (the logarithmic decrement), $\xi = \delta/2\pi$. A representative value for the logarithmic decrement of a rotor steel might be 0.01, giving a value of the damping ratio equal to 0.0016 and an extremely high amplification factor of 314 at resonance. This damping is too low to be of practical significance. It may be necessary to add special damping provisions in the machinery if a condition of torsional resonance exists.

For the case in which the damper acts on the same disk where the sinusoidal exciting torque is applied, the viscous damping torque on disk 1 is equal to $B_1(d\theta_1/dt)$. Here, the simplicity of the solution for damping condition 1 is not possible. The general solution for angular displacement at each disk must be derived separately from the solution of the differential equations. A more manageable perspective can be obtained by examination of the response at resonance ($\omega/\omega_n = 1$), which will yield:

$$\theta_1 = -\frac{T_1}{B_1\omega_n}(\cos \omega_n t) \tag{6.5}$$

$$\theta_2 = \frac{T_1}{B_1\omega_n}\frac{J_1}{J_2}\cos(\omega_n t) \tag{6.6}$$

Shaft torque is then:

$$| T_{12} | = \frac{T_1}{B_1}\left[\frac{K_{12}(J_1 + J_2)J_1}{J_2}\right]^{1/2} \tag{6.7}$$

For the case in which the damper acts on the disk remote from where the sinusoidal exciting torque is applied, the viscous damping torque on disk 2 is equal to $B_2(d\theta_2/dt)$. The angular amplitudes at resonant condition ($\omega/\omega_n = 1$) is obtained from the expression:

$$\theta_1 = \frac{-T_1}{K_{12}}\frac{J_2}{J_1}\left\{\sin(\omega_n t) + \frac{J_2}{J_1}\frac{K_{12}}{B_2\omega_n}\cos(\omega_n t)\right\} \tag{6.8}$$

$$\theta_2 = \frac{T_1}{K_{12}} \frac{J_2}{J_1} \frac{K_{12}}{B_2\omega_n} \cos(\omega_n t) \tag{6.9}$$

and the amplitude of shaft torque is found to be:

$$|T_{12}| = T_1 \frac{J_2}{J_1}\left[1 + \frac{K_{12}}{B_2^2}(J_2 + J_1)\left(\frac{J_2}{J_1}\right)\right]^{1/2} \tag{6.10}$$

6.3 ENERGY BALANCE METHOD

The energy balance method is often used to calculate the response of a system to a harmonic exciting torque when the frequency of the exciting torque is equal to an undamped natural frequency of the system. The energy input from the exciting torque and the energy dissipated by the damping are evaluated on the basis of the undamped mode shape of the system. Thus, the method is particularly useful in the case of systems where there is little damping due to shaft twisting.

For the system of Fig. 6.1 the natural frequency is given by (6.1), and the mode shape is given by the expression:

$$\theta_2 = -\theta_1(J_1/J_2) \tag{6.11}$$

For a harmonic exciting torque of amplitude T_1 and frequency ω_n acting on disk 1, energy input per cycle E_i for the component of torque in phase with the harmonic angular velocity of amplitude $\omega_n\theta_1$ is given by:

$$E_i = \pi T_1\theta_1 \tag{6.12}$$

With the damping torque equal to the viscous damping coefficient B_{12} times the difference in angular velocities of the two disks ($d\psi/dt = d\theta_1/dt - d\theta_2/dt$, damping condition #1), the amplitude of viscous damping torque acting on disk 1 is equal to $\omega_n B_{12}(\theta_1 - \theta_2)$, which is equal and opposite to the torque acting on disk 1, and is also in phase with it. The energy dissipated per cycle is:

$$E_d = \pi\omega_n B_{12}(\theta_2 - \theta_1)^2 \tag{6.13}$$

Equating the input energy with the dissipating energy, and expressing the amplitude θ_2 in terms of amplitude θ_1 gives:

$$E_i = \pi T_1\theta_1 = E_d = \pi\omega_n\theta_1^2 B_{12}[(J_1 + J_2)/J_2]^2 \tag{6.14}$$

From this the angular amplitudes of each inertia is derived:

$$\theta_1 = \left[\frac{T_1}{\omega_n B_{12}}\right]\left[\frac{J_2}{J_1 + J_2}\right]^2 \tag{6.15}$$

$$\theta_2 = -\left[\frac{T_1}{\omega_n B_{12}}\right]\frac{J_1 J_2}{(J_1 + J_2)^2} \tag{6.16}$$

and the amplitude of shaft torque is:

$$|T_{12}| = \frac{T_1[J_2/(J_1 + J_2)]\pi}{\delta} \tag{6.17}$$

which is identical to (6.4).

If hysteretic damping is assumed instead of viscous damping, the energy dissipated per cycle E_d, for low values of damping, may be taken as twice the logarithmic decrement times the total vibrational energy of the system. This concept is useful for systems of greater complexity. The total vibrational energy of the system can be expressed as a summation of kinetic energies for the inertial elements in the system at zero amplitude, when the kinetic energy is a maximum and the potential energy is zero.

$$E_d = \delta\omega_n^2[J_1\theta_1^2 + J_2\theta_2^2] \tag{6.18}$$

and the amplitude of vibratory shaft torque is:

$$T_{12} = T_1(\pi/\delta)[J_2/(J_1 + J_2)] \tag{6.19}$$

Due to the relation between δ and ξ, this result is effectively identical to (6.17) and also (6.4).

With a damper located at disk 1 the viscous damping torque on disk 1 is equal to $\omega_n B_1\theta_1$.The energy balance then becomes:

$$E_i = \pi T_1\theta_1 = E_d = \pi\omega_n B_1\theta_1^2 \tag{6.20}$$

From this energy balance, together with (6.11), the angular amplitudes of the inertias are the same as (6.5) and (6.6), and the amplitude of the shaft torque is the same as (6.7).

With the damper located at the remote inertia end from torque application location, the amplitude of viscous damping torque on disk 2 is equal to $\omega B_2\theta_2$, and the energy balance equation becomes:

$$E_i = \pi T_1\theta_1 = E_d = \pi\omega_n B_2\theta_2^2 \tag{6.21}$$

From this energy balance together with (6.11), the angular amplitudes of the inertia are given by the expressions:

$$\theta_1 = [T_1/(\omega_n B_2)](J_2/J_1)^2 \tag{6.22}$$

$$\theta_2 = -[T_1/(\omega_n B_2)](J_2/J_1) \tag{6.23}$$

and the amplitude of shaft torque is:

$$|T_{12}| = \frac{T_1J_2}{B_2J_1}[K_{12}(J_2 + J_1)(J_2/J_1)]^{1/2} \tag{6.24}$$

The amplitudes given by (6.22), (6.23) and (6.24) are not in precise agreement with the amplitudes given by (6.8), (6.9) and (6.10) of the exact solution. Instead, they represent limiting values of the exact solution amplitudes as the damping B_2 approaches zero. This is consistent with the previously mentioned limitation of the energy balance method, and is applicable only to systems with low damping.

6.4 MATRIX METHOD

The common approach for analyzing the dynamics of rotating systems is to use the state vector matrix. The method is particularly well suited for 'chain-like' structures such as multi-cylinder crankshafts having crank pin and main journal bearings, cheeks and counter weights at a number of locations. The first use of this method is attributed to Holzer (1921) in the area of torsional vibrations. Other refinements have made this a reliable and widely used analytical procedure for reciprocating machinery manufacturers and rotor dynamics experts throughout the world.

The primary advantage of this method is that it does not require the storage and manipulation of large system arrays. Matrix algebra is used for the analysis of torsional vibration problems. The procedure is also applicable for calculation of natural frequencies and critical speeds of other eigen value problems.

Two basic calculation procedures are available, the fundamental difference between them lying in the assumed independent variable. In any eigen value problem a unique mode shape of the system is associated with each natural frequency. In the matrix iteration method, first described by Stodola, the mode shape is the independent variable. A mode shape is assumed and improved by successive iterations until the desired accuracy is obtained; its associated natural frequency is then calculated.

On the other hand in Holzer's transfer matrix method a frequency is assumed, and the mode shape of the system is calculated. If the mode shape fits the boundary conditions, the assumed frequency is a natural frequency, and a critical speed is thus obtained. Determining the correct natural frequencies amounts to a controlled trial and error process. Some of the essential boundary conditions arising from the geometry and natural boundary conditions from force considerations are assumed. For example, in the torsional system of Fig. 6.2 the natural boundary conditions would be zero torque at both ends. The torque at station number 1 is made zero, and the torsional vibration amplitude is set at unity. Then T_4 is calculated as a function of ω and plotted to find the natural frequencies. The calculations proceed according to the system transfer matrices, or functions, the elements of which reflect the dynamic characteristics of the system.

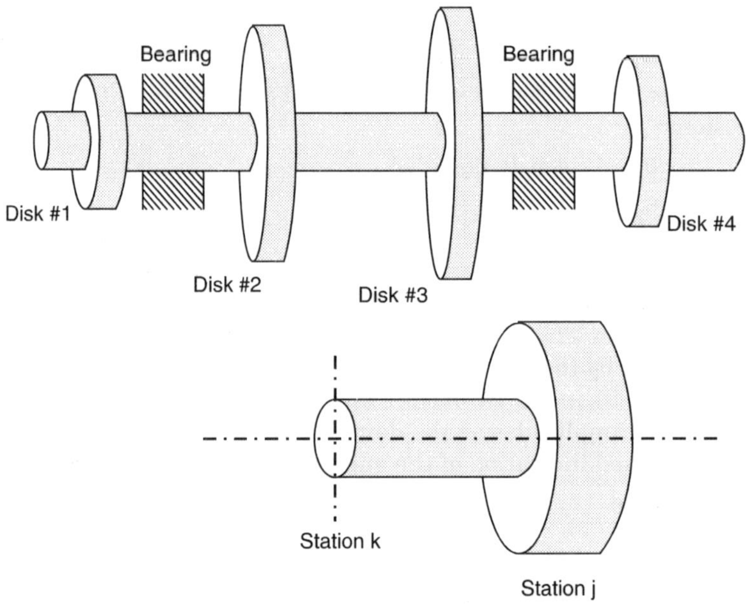

Fig. 6.2. Typical Torsional Vibration Model

No accuracy is lost with the transfer matrix method because of coupling of the mode shapes. Accuracy is lost with the matrix iteration method, however, because each frequency calculation is independent of the others. A disadvantage of the transfer matrix method is the large number of points that must be calculated to obtain the T_4 versus ω curve, but with digital computers the problem is readily overcome.

A typical station at location 4 is shown in Fig. 6.2. This general station and the transfer matrix equation shown below are used in the execution of the Holzer type of calculations for a given angular velocity ω across the full model.

$$\begin{bmatrix} \theta \\ T \end{bmatrix}_n = \begin{bmatrix} 1 & 1/k \\ -\omega^2 J & -(\omega^2 J/k)+1 \end{bmatrix}_n \begin{bmatrix} \theta \\ T \end{bmatrix}_{n-1} \tag{6.25}$$

natural frequencies. The calculations proceed according to the system transfer matrices, where
θ = torsional motion, radians

T = torque, in.-lbs.

J = station polar mass moment of inertia, in-lbs.-second^2

k = torsional stiffness of shaft section between stations, in-lbs./radian.

Because of the complex geometric configuration of a crankshaft, with all its cheeks and offset journals, stiffness k deserves special mention. In the case of a hollow circular shaft with outer diameter D, inner diameter d and length l, the stiffness k is given by:

$$k = GI/l = (\pi/32)G(D^4 - d^4)/l \tag{6.26}$$

where G is the shaft material shear modulus of elasticity. Since most shaft cross sections are circular, this expression will see extensive usage in all rotor dynamics calculations. Stiffness as well as inertia characteristics of the more irregularly defined elements of a typical crankshaft is covered at length in section 6.6.

The stiffness and polar moment of inertia of each station are entered into equations (6.25) to determine the transfer effect of each element of the model. Thus, the calculations begin with station number 1 to include its stiffness and inertia, which results in the output torque T_1 and rotation θ_1 for the given input values. At station number 1 these values generally are set at 0 and 1. The equations will next be applied at station number 2, yielding T_2 and θ_2. The process is repeated to obtain values T_n and θ_n at the other end of the rotor.

6.5 FINITE ELEMENT METHOD

The finite element method is a numerical procedure to calculate the natural frequencies of vibration, mode shapes and forced response of a discretely modeled rotor system. Finite element computer programs based on this procedure perform three useful functions:

- They incorporate modeling objects that transform a conceptual model such as a crankshaft into a set of matrix equations of the form expressed by (6.25).
- They incorporate a rich set of rules for performing a wide range of analyses, including transient and frequency response.
- They contain a rich set of data processing operations, including random response and response spectrum analyses.

The rotating system is composed of an assemblage of discrete smaller finite elements that are continuous structural members. The displacements and rotations are forced to be compatible, and force and torque balance is required at the joints, commonly referred to as nodes.

Figure 6.3 shows a uniform torsional element in local coordinates. The x-axis is taken along the axis of centroid. Physical properties of the element are density (ρ), area (A), shear modulus of elasticity (G), length (l) and polar area moment of inertia (I). $T(t)$ are the torsional forcing functions.

The torsional displacement within the element can be expressed in terms of rotations $\theta_1(t)$ and $\theta_2(t)$ at the joints as:

$$\theta(x, t) = U_1(x)\,\theta_1(t) + U_2(x)\,\theta_2(t) \tag{6.27}$$

where $U_1(x)$ and $U_2(x)$ are called 'shape functions'. Since $\theta(0, t) = \theta_1(t)$ and $\theta(l, t) = \theta_2(t)$, the shape functions must satisfy the following boundary conditions:

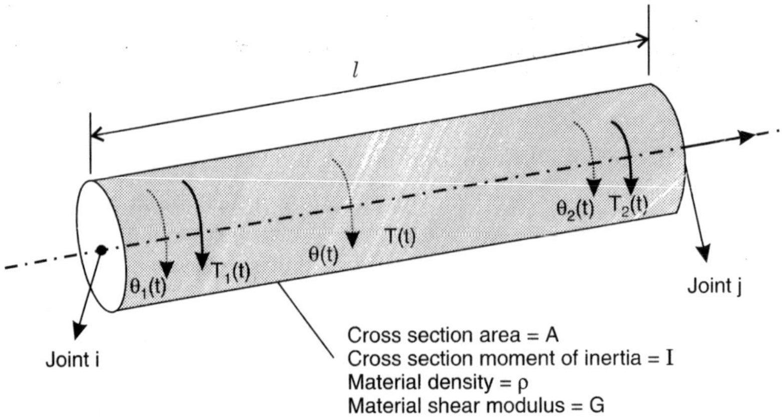

Fig. 6.3. Finite Element Model for Torsional Vibration

$$U_1(0) = 1 \quad U_1(l) = 0$$
$$U_2(0) = 0 \quad U_2(l) = 1$$

The shape function for the torsional element is assumed to be a polynomial with two constants of the form:

$$U_i(x) = a_i + b_i x \quad \text{where } i = 1, 2 \tag{6.28}$$

Selection of the shape function is left to the engineering analyst's judgement, and is a part of the art and science required for performing an accurate modeling of the finite element model of the rotating configuration at hand.

Thus, with four known boundary conditions the values of a_i and b_i can be determined from (6.28):

$$U_1(x) = 1 - x/l \quad U_2(x) = x/l$$

Then from (6.27):

$$\theta(x, t) = \left(1 - \frac{x}{l}\right)\theta_1(t) + \frac{x}{l}\,\theta_2(t) \tag{6.29}$$

Kinetic energy, strain energy and virtual work are used to formulate the finite element mass and stiffness matrices and force vectors. The quantities are then used to form the equations of motion. The derived matrices are:

$$\{J\} = \frac{\rho I l}{6}\begin{bmatrix} 2 & 1 \\ 1 & 2 \end{bmatrix}$$

$$\{K\} = \frac{GI}{l}\begin{bmatrix} 1 & -1 \\ -1 & 1 \end{bmatrix}$$

$$\bar{T} = \begin{Bmatrix} T_1(t) \\ T_2(t) \end{Bmatrix} = \begin{Bmatrix} \int_0^l T(x,t)\left(1 - \frac{x}{l}\right) dx \\ \int_0^l T(x,t)\left(\frac{x}{l}\right) dx \end{Bmatrix} \tag{6.30}$$

where: $\{J\}$ = mass matrix

$\{K\}$ = stiffness matrix

$\overline{T}$ = torque vector

ρ = shaft material density

I = polar area moment of inertia

G = shaft material shear modulus

l = length of element

For modeling convenience the elements thus defined are in local coordinates. Since the system as a whole must be analyzed as a complete unit, the elements must be transformed into one global coordinate system. Figure 6.4 shows the local element within a global coordinate system. The mass and stiffness matrices and the joint force vector of each element must be expressed in the global coordinate system to find the vibration response of thc complete system.

Using transformation matrices, mass and stiffness matrices and force vectors are used to set up the system equation of motion for a single element in global coordinates, given by the expression:

$$[J]_e \{d^2\, \Theta(t)/dt^2\} + [K]_e \{\Theta(t)\} = \{T_e(t)\} \tag{6.31}$$

The complete system is an assemblage of a number of finite elements required to adequately model and analyze. The joint displacements of the elements in the global coordinate system are labeled as $\Theta_1(t)$, $\Theta_2(t)$, .. $\Theta_m(t)$, or can be expressed as a column vector:

$$\{\Theta(t)\} = \begin{Bmatrix} \Theta_1(t) \\ \Theta_2(t) \\ \vdots \\ \Theta_m(t) \end{Bmatrix}$$

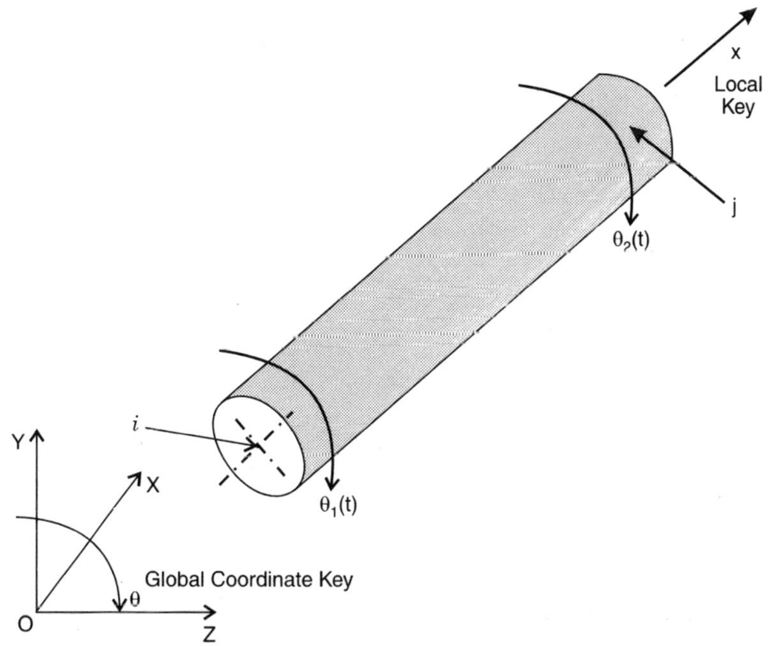

Fig. 6.4. Local and Global Joint Displacements of an Element

Using global joint displacements, mass and stiffness matrices and force vectors, the equations of motion are developed:

$$[J]_{n\times n} \{ d^2\Theta(t)/dt^2\}_{n\times 1} + [K]_{n\times n} \{\Theta\}_{n\times 1} = \{T\}_{n\times 1} \tag{6.32}$$

where n denotes the number of joint displacements in the system. In the final steps prior to solution appropriate boundary conditions and constraints are introduced into the global model.

The equations of motion for free vibration are solved for the eigen values (natural frequencies) using the matrix iteration method. Modal analysis is used to solve the forced torsional response. The finite element method is available in commercially available computer programs for the personal computer. The analyst is required to select the information for the joints (nodal location, materials, shape functions, geometry, torque and constraints) to model the system for computation of natural frequencies, mode shapes and torsional response. Similar to any other analytical effort, experience in the form of capabilities of the computer program and the rotating crankshaft system enable the engineer to obtain reasonably accurate results.

6.6 CRANKSHAFT STIFFNESS AND INERTIA CHARACTERISTICS

The stiffness of the circular sections of the crankshafts may be determined without any difficulty. Table 6.1 shows some common formulas for calculating the torsional stiffness of a hollow circular shaft, a tapered circular shaft and a combination of two geared shafts. In the case of the geared shaft set the stiffness is referred to the rotational speed of shaft no. 1.

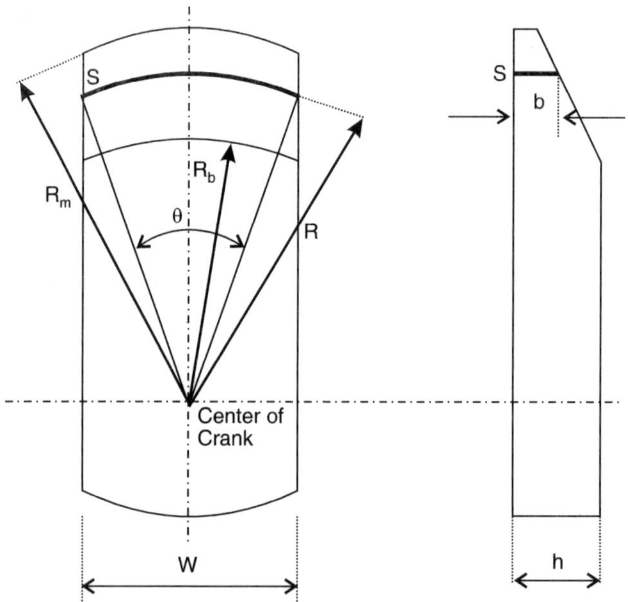

Fig. 6.5. Crank Web

The crank web stiffness, on the other hand, is easily the most uncertain element in the torsional vibration calculations. A common practice is to measure the stiffness of either the full crankshaft, or sections of it, experimentally by twisting the shaft with a known torque. Another possibility is to deduce the stiffness from observed values of the critical speeds in a running engine. Semi-empirical formulas for stiffness calculation are also available. Some recommended equations are given below. Refer to Figs. 6.5 and 6.6 for definition of dimensions;

l_e is the length of a solid shaft of diameter D_s equal in torsional stiffness to the section of crankshaft between crank centers.

Wilson's formula:

$$\frac{l_e}{D_s^4} = \frac{b + 0.4d_s}{D_s^4 - d_s^4} + \frac{a + 0.4D_c}{D_c^4 - d_c^4} + \frac{r - 0.2(D_s + D_c)}{hW^3} \tag{6.33}$$

Ziamanenko's formula:

$$\frac{l_e}{D_s^4} = \frac{b + 0.6hD_s/b}{D_s^4 - d_s^4} + \frac{0.8a + 0.2(W/r)D_s}{D_c^4 - d_c^4} + \frac{r^{3/2}}{hW^3 D_c^{1/2}} \tag{6.34}$$

Constant's formula:

$$\frac{l_e}{D_s^4} = \frac{1}{\alpha_1\alpha_2\alpha_3\alpha_4}\left[\frac{b}{D_s^4 - d_s^4} + \frac{a}{D_c^4 - d_c^4} + \frac{0.94}{hW^3}\right] \tag{6.35}$$

where α_1, α_2, α_3, and α_4 are modifying factors defined as follows:

$$\alpha_1 = 1 - \frac{0.0825}{\sqrt{\frac{W_s - d_s}{2W_s} + \frac{W_c - d_c}{2W_c} - 0.32}} \tag{6.36}$$

Assume $\alpha_1 = 0.9$ if the shaft is solid. Factor α_2 is a web thickness modification determined as follows: If $4h/l$ is greater than 2/3, then $\alpha_2 = 1.666 - 4h/l$. If $4h/l < 2/3$, assume $\alpha_2 = 1$.

Factor α_3 is a modification factor for web chamfering, determined as follows: If the webs are chamfered, estimate α_3 by comparison with the cuts on Fig. 6.6. For cuts *AB* and *A′B′* use $\alpha_3 = 1.00$; for cut *CD* alone $\alpha_3 = 0.965$; for cuts *CD* and *C′D′* $\alpha_3 = 0.930$; cut *EF* alone $\alpha_3 = 0.950$; for cuts *EF* and *E′F′* $\alpha_3 = 0.900$; if ends are square $\alpha_3 = 1.010$. Factor α_4 is a modification for the bearing support, and is given by:

$$\alpha_4 = \frac{Al^3w}{D_c^4 - d_c^4} + B \tag{6.37}$$

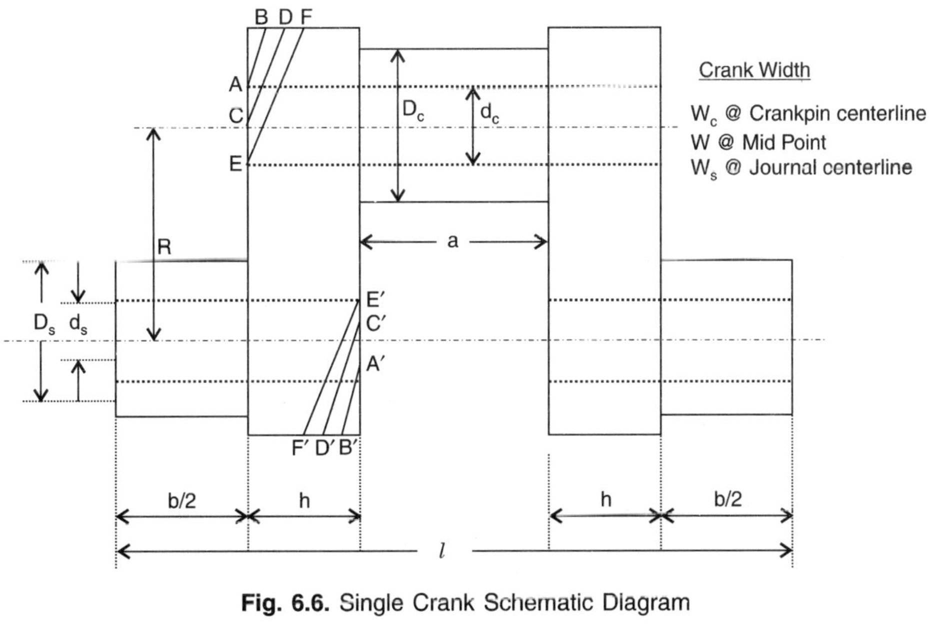

Fig. 6.6. Single Crank Schematic Diagram

For marine and large stationary engine crankshafts A = 0.0029 and B = 0.91. For automotive and aircraft engine crankshafts A = 0.010 and B = 0.84. If α_4 as given by (6.37) is less than 1.0, assume a value of 1.0. Constant's formula (6.35) is recommended for shafts with large bores and heavy chamfers.

Besides crank webs, crankshaft designs include elements such as changes in cross section, collars, shrunk and keyed armatures and other design features. An assessment of the stiffness can only come from experience on similar other design of the elements. For a fillet radius equal to 10% of the smaller diameter the stiffness can be estimated by assuming that the smaller shaft is lengthened, and the larger shaft shortened, by a length λ obtained from the curve of Fig. 6.7. This may also be applied to flanges where D is the bolt diameter. Stiffening effects from collars may be ignored.

Stiffness of shrunk and keyed parts is difficult to estimate because the stiffening effect depends on how tight the shrink fit remains during the machine's operating condition. The interference fit present at the mating surfaces of the two components during assembly will change due to inertia effects, distortion due to operating loads and thermal conditions, if existing at the location. Preferred approach is to ignore the stiffening effects of the interference fit between, say, an armature and the shaft. In that case only the inertia effects from the mass of the armature may be included at the proper location on the shaft. Some armature designs call for spiders and flywheels with significant flexibility in the arms; the treatment of these is discussed at length in the next section.

Numerous types of commercially available elastic couplings are available. The torsional spring constant and inertia value of the coupling is of interest in a torsional vibration analysis of the rotating system, and are obtainable from the manufacturers.

Table 6.1. Torsional Stiffness Equations of Common Rotating Elements

Series Spring Connection	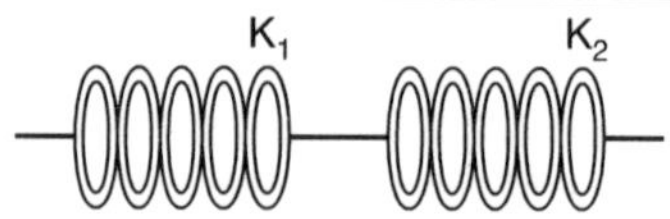	$K = 1/[1/K_1 + 1/K_2]$
Parallel Spring Connection	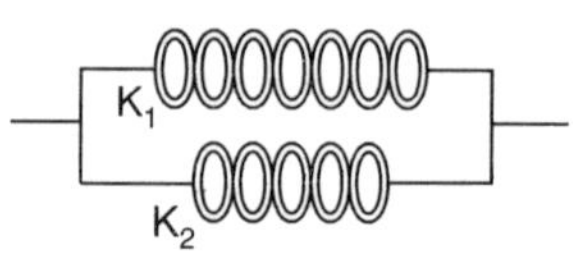	$K = K_1 + K_2$
Uniform Hollow Shaft	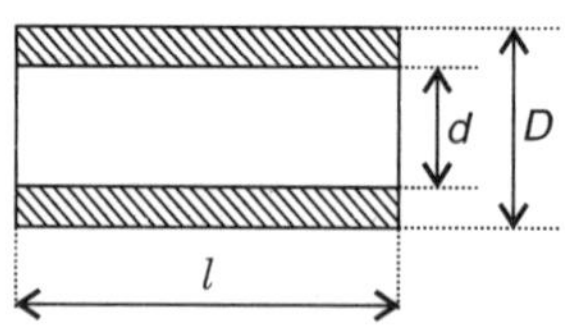	$K = \pi G(D^4 - d^4)/(32l)$
Circular Tapered Shaft	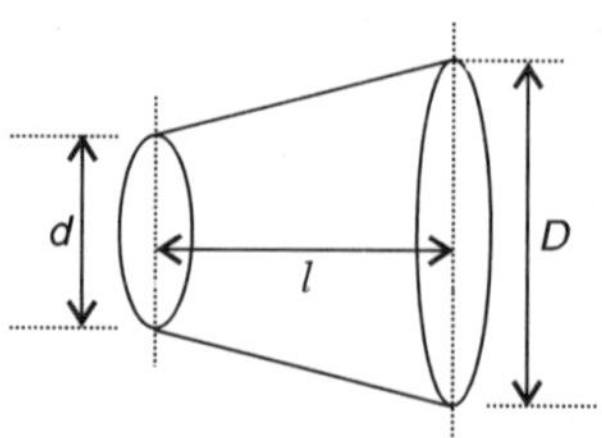	$K = \dfrac{3\pi}{32} \dfrac{d^4}{l(n + n^2 + n^3)}$ $n = d/D$

Pair of Geared Shafts (With reference to shaft # 1)	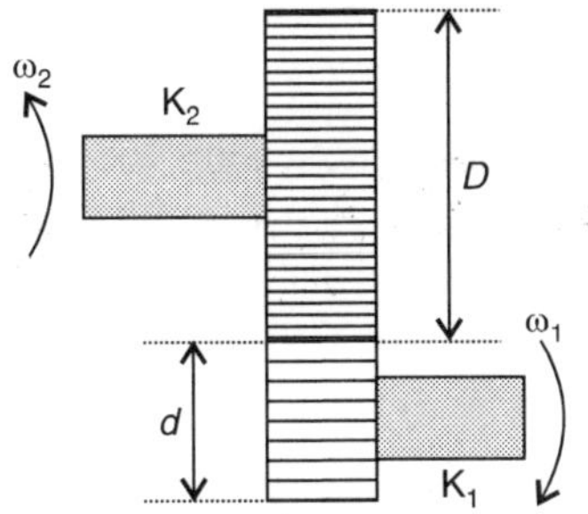	$K = [K_1K_2]/\{n^2[K_1 + K_2]\}$ $n = D/d = \omega_1/\omega_2$
K = Torsional Stiffness	ω = Shaft speed	G = Material Shear Modulus

Interestingly, if the presence of a natural frequency at a speed point is objectionable due to large vibration amplitudes, selection of another coupling with different flexibility and inertia characteristics may be used to shift it to another speed.

The polar moment of inertia required in torsional vibration analysis is relatively easy to determine. The expression is $J = \int r^2\, dm$, where r is the instantaneous radius and dm the differential mass. For a circular disc or cylinder rotating about a perpendicular axis the polar moment of inertia is given by:

$$J = \pi(D^4 - d^4)L\rho/(32g) \tag{6.38}$$

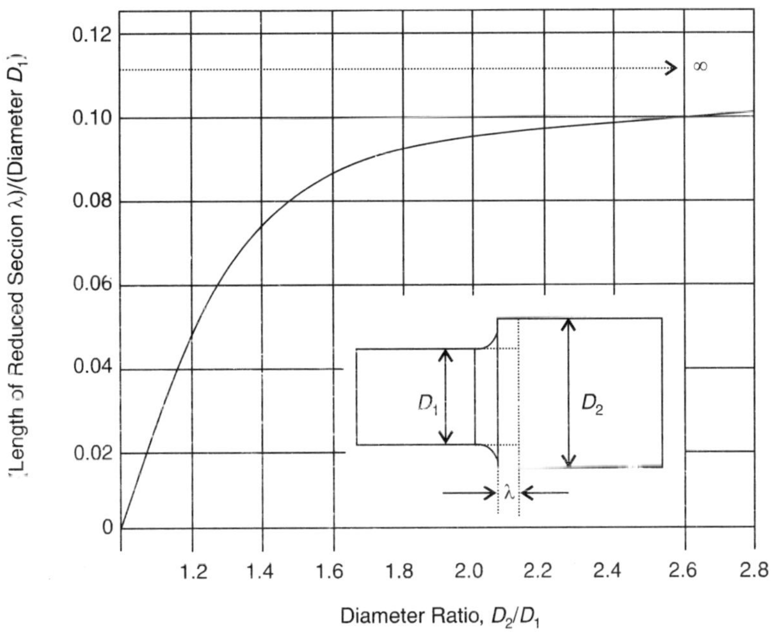

Fig. 6.7. Reduction in Stiffness due to Change in Shaft Diameter (*ref.* Porter, 1983)

where: J = polar moment of inertia, lbs.-in/second^2

ρ = material density, lbs./in^3

D = outer diameter of disc or cylinder, in.

d = inner diameter of disc or cylinder, in.

L = axial length of disc or cylinder, in.

g = acceleration due to gravity, 386.4 in./second^2

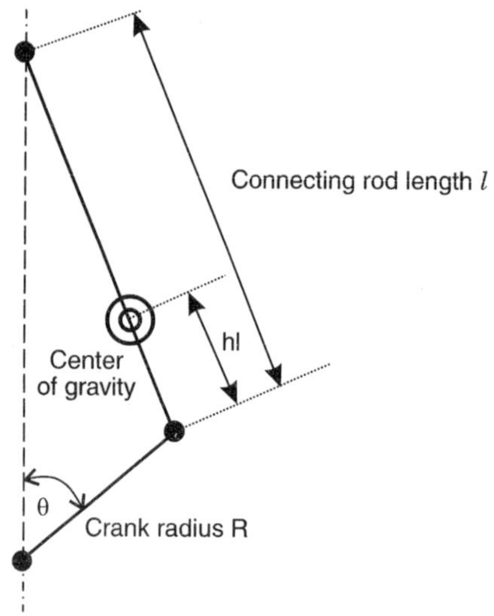

Fig. 6.8. Crank and Connecting Rod

The piston and connecting rod of a reciprocating machine represents a variable mass problem having a complicated solution. Figure 6.8 shows the mechanism schematically. The exact solution shows that the effect of the piston and the connecting rod can be closely approximated by representing them as a concentrated rotor of polar inertia defined by:

$$J = \left\{ \frac{W_p}{2} + W_c \left(1 - \frac{h}{2}\right) \right\} \frac{R^2}{g} \tag{6.39}$$

where W_p = weight of piston, piston pin and cooling fluid, lbs.

W_c = weight of connecting rod, lbs.

h = fraction of rod length from crank pin to center of gravity

R = crank radius, in.

The polar inertia of crank webs may be determined by using the following procedure. Consider the web to be intercepted by a series of concentric circles of radius R. The polar inertia of the crank is:

$$J = \left\{ \frac{\rho}{g} \int_{R_b}^{R_m} R^2 S \, dR \right\} + J_b \tag{6.40}$$

where: ρ = weight density of web material, lbs./in^3

R_m = maximum radius of web, in.

R_b = radius of base cylinder, in., see Fig. 6.5

J = polar inertia of portion of crank web within base cylinder

$S = bR\theta/57.3$, with θ measured in degrees

The integral in the above expression is the area of the R^2S curve between the two radii. Substituting for S in (6.40) yields:

$$J = \left\{ \frac{\rho}{57.3g} \int_{R_b}^{R_m} bR^3\theta \,.\, dR \right\} + J_b \text{ lbs.-in-second}^2 \tag{6.41}$$

A similar procedure is used to determine the polar inertia of propellers and other irregular parts. For propellers, hydraulic couplings and impellers an allowance must be made for the inertia of the fluid rotating with the component. For marine propellers this is generally 26% of the propeller inertia.

For complex rotating elements such as couplings and small flywheels it may be easier to determine the polar moment of inertia experimentally than to calculate it. In one procedure the component is suspended from three equally spaced vertical wires, as shown in Fig. 6.9. The component is then set into torsional motion and the period of vibration measured. The polar moment of inertia is then obtained from the following expression:

$$J = Wr^2\tau^2/[(2\pi)^2 l] \tag{6.42}$$

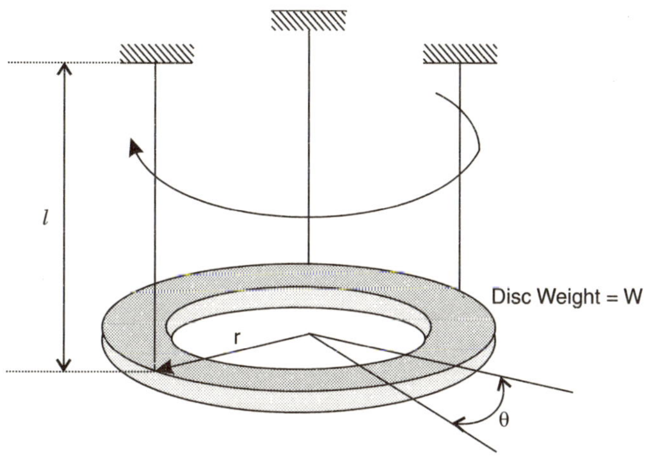

Fig. 6.9. Measurement of Polar Moment of Inertia Using Wires to Measure Torsional Oscillations

where: τ = period of vibration, second/cycle

W = weight of element, lbs.

l = length of cables, in.

r = radius of cables, in.

6.7 GEARED AND BRANCHED SYSTEMS

Natural frequencies of vibration of a geared system may be calculated by replacing the system in which the speed of the diver unit is a multiple of n, where n is the ratio of the speeds of the driver unit and the driven unit. The inertia and elastic constants on the driven side of the unit are multiplied by $1/n^2$, then calculate the system natural frequencies as if the gears do not exist. If the system calls for a damping device on the driven side, the damper constant of the device must also be multiplied by $1/n^2$. Torques and deflections thus obtained on the driven side of the substitute system, when multiplied by n and $1/n$ respectively, are equal to those in the actual geared system. Alternatively, the driven side may be used as the reference, in which case the driver constants are multiplied by the factor n^2.

When two or more drivers are geared to a common load, hydraulic or electrical type of couplings may be placed between the drivers and the gear set. The couplings serve a dual purpose role, one of isolating the gears from driver induced vibrations, the other as a clutch for connection and disconnection of load between the components. In many such coupling designs the vibration isolation may be complete enough to permit termination of the driver end of the unit at the coupling. The damping effect upon the vibration in the driver end of the system is small and may be disregarded in the amplitude calculations.

In many applications calling for two identical driver units hydraulic or electrical couplings are not employed. Then the modes of vibration are of two types:

- Opposed phase modes, in which the drivers vibrate against each other, with a node at the gear. The modes are calculated for a single branch in the usual manner, with the model terminating at the gear. The condition for a natural frequency is that shaft twist amplitude be set to zero at the gear.
- In phase modes, in which both drivers vibrate in the same direction against the driven machinery. To calculate these frequencies the inertia and stiffness constants of the driver side of one branch are doubled; the calculations are then performed for only one driver. The condition for a natural frequency is zero residual torque at the end.

If the two identical drivers rotating in the same direction are so phased that the same cranks are vertical simultaneously, all orders of the opposite phase modes will be eliminated. The two drivers can be phased such that certain like-phase modes can be eliminated. For example, if the number 1 cranks in the two branches are placed at an angle of 45° with respect to each other, the fourth, twelfth, twentieth, etc., orders, but not others, will be eliminated. If the drivers are connected with clutches, these phasing possibilities cannot be utilized.

In the general case of non-identical branches the calculation scheme proceeds as follows: Reduce the system to a 1:1 gear ratio. Call the two branches a and b. Perform the sequence calculation for the a branch, with initial twist angle equal to unity, and for the b branch the initial amplitude set to an unknown value x. At the junction equate the amplitudes and find x. With this numerical value of x substituted, the torques in the two branches and the torque of the gear are added; the sequence calculation is then continued up to the last mass.

The branch may consist of a single member elastically connected to the system. Examples of such a branch are a flywheel with flexible spokes or an armature with a flexible spider. Assume J to be the moment of inertia of the flywheel's rim and k the elastic constant of the connection. Then the flexibly mounted flywheel is equivalent to a rigid flywheel of an equivalent moment of inertia given by the expression:

$$J' = \frac{J}{1 - J\omega^2/k} \tag{6.43}$$

6.8 CRITICAL SPEEDS AND MODES

The crankshaft of a reciprocating machine and all rotating components attached to it comprise a torsional elastic system. Such a system has several modes of free torsional oscillation. Each mode is characterized by a natural frequency and by a pattern of relative amplitudes of all parts of the system when it is oscillating at its natural frequency. The harmonic components of the driving torque provide the excitation for vibration of the modes. If the frequency of any harmonic component of the torque is equal to, or close to, the frequency of any mode of vibration, a condition of resonance prevails, and the machine is said to be running at a critical speed. Operation of the machinery system at such critical speeds can be dangerous, and in an extreme condition result in fracture of the shafting.

The number of complete oscillations of the elastic system per unit revolution of the shaft is called the order of a critical speed. An order of a critical speed that corresponds to a harmonic component of the torque from the reciprocating machine is called a major order. A critical speed can also be excited by the harmonic component of the torque curve of a single cylinder. The fundamental period of torque from a single cylinder in a four-cycle internal combustion engine is 720°; critical speeds in this engine will then be of ½, 1, 1½, 2, 2½, etc., order. In a two-cycle engine only critical speeds of 1, 2, 3, etc. order will prevail. All critical speeds except those of the major orders are called minor critical speeds. Note, however, that this term does not necessarily mean they are not important.

When a reciprocating machine's crankshaft and other machinery's attached rotors have many masses, the Holzer method can be used to calculate the natural frequencies. A frequency is assumed; then starting at one end of the system a balance of torques and rotations is obtained in a step by step procedure. The final external torque required to achieve balance is called the residual torque. If the residual torque is zero, the assumed frequency is one of the natural frequencies of the system. The mode associated with this natural frequency is identified by the number of changes of the sign of the amplitudes at various mass locations. Figure 6.10 gives a plot of residual torque versus frequency, from which the natural frequencies of the system within the frequency range of interest can be obtained.

In many multi-cylinder reciprocating machines the polar inertia representing the cranks and pistons is equal, as are also the stiffness constants of the crankshaft between these locations. Advantage of this similarity in the stiffness and inertia properties can be taken to reduce the amount of computational effort. Assume J to be one of the equal polar moments of inertia and k one of the equal spring constants. Divide all the polar moments of inertia by J and all shaft stiffness values by k. The modified system is dynamically similar to the original system, referred to as the unity system. Then if ω_n is the angular natural frequency of the actual system and ω_s that for the unity system,

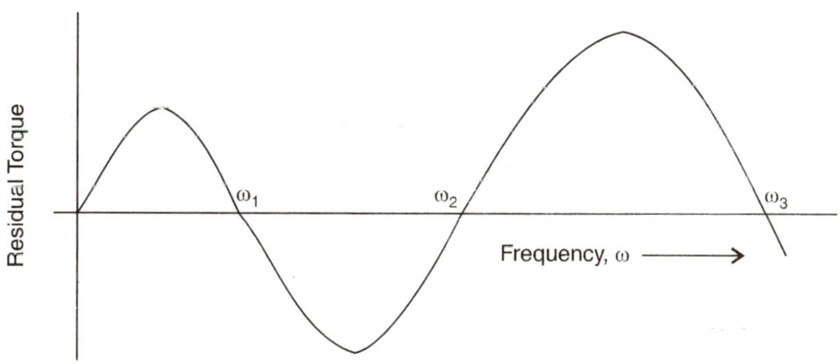

Fig. 6.10. Typical Curve of Residual Torque as a Function of Assumed Frequency ω

$$\omega_n^2 = \omega_s^2 \frac{k}{J} \tag{6.44}$$

Before starting the sequence calculation it is necessary to know the approximate frequency of the first mode. For a reciprocating machine with small inertias in the presence of large ones the following procedure is useful, giving generally satisfactory results. Add all the polar inertias within the reciprocating machinery portion of the system. Assume 40% of this amount is placed at the crank farthest from the flywheel or attached machine's rotor (number 1 crank). Find the stiffness with reference to the flywheel, or driven or driving machine's rotor, to the number 1 crank. If a large mass such as a flywheel is not attached, the system reduces to a two-mass system, and equation (6.1) gives the approximate lowest natural frequency of the system. For complicated systems with many masses and stiffness coefficients of differing values, general rules of this type are not available.

The method of Holzer for calculation of torsional natural frequencies of vibration is widely used and recognized in reciprocating machinery design, mainly due to its relative simplicity and accuracy of results. The method is best understood by means of an example. In section 6.14 example problems 6.1 and 6.2 look at an eight-cylinder engine to illustrate the procedure.

A dynamic analysis of a reciprocating machine involves several steps. Natural frequencies of modes that are likely to play an important role in the torsional vibration of the system must be calculated. The calculation may be limited to the two lowest modes, but additional modes may be necessary if the frequencies of the forces causing the vibrations are in the higher range. Vibration amplitudes and stresses around the operating speed range and at the critical speeds must be determined. Based on the results a study of remedial measures to mitigate the vibrations may also be necessary.

6.9 RESONANT RESPONSE

The torsional vibration response of a reciprocating machine's rotating system is determined by the magnitude, points of application and phase angle relations of the exciting torques produced by gas pressure and inertia, and by the magnitude and point of application of the damping torques. Damping may be due to a number of sources, including pumping action in the bearings, hysteresis in the shafting and between fitted parts, energy absorbed by the machine's frame and by the foundation. In some cases, such as a marine propeller or a pump's impeller, damping from the fluid on the propeller, or the impeller, predominates. When a reciprocating machine is fitted with a damper, the effects of damping will be substantial.

Analytical procedures are available for calculating the torsional vibration amplitudes in the transfer matrix method and in the modal analysis, which can be implemented in the finite element or the lumped parameter models. The Holzer table calculation sequence can be readily extended for this purpose. The initial amplitude is treated as an algebraic unknown θ. At each cylinder where an exciting torque may be present, the torque is added. Initially, assume that there is no torque from damping. Then the residual torque at the last rotor station is of the form $a\theta + b$, where a and b are numerical constants. Since torque at the last station is zero, $\theta = -b/a$. The amplitude and torque at any point of the system are found by substituting this numerical value of θ at the proper point in the calculation sequence. At frequencies removed from resonance damping has little effect and can be neglected.

Damping may be introduced into the system in the form of an exciting torque equal to the imaginary quantity $-jc\omega\theta$, where c is the damping constant and θ the amplitude at the point of application of the damper. Relative damping between two inertia points can be treated as a spring of stiffness constant equal to the imaginary quantity $+jc\omega$.

For major critical speeds the exciting torques are in phase, and are real numbers. For minor critical speeds the exciting torques are out of phase; these are entered as complex numbers of amplitude and phase as determined from the phase diagram (see section 6.14) for the critical speed of the order under consideration. With damping and out of phase exciting torques introduced, a and b in the equation $a + b\theta = 0$ are complex numbers, and θ must be entered as a complex number in the tabulation in order to determine the angle and torque at any point. These angles and torques are then of the form $r + js$, where r and s are numerical constants and the amplitudes are $\sqrt{(r^2 + s^2)}$.

In the forced vibration calculations the amplitude of the exciting torque is also assumed to be unity. It is advisable to start the sequence calculation at the flywheel end of the system, particularly when the intent of the analysis is to design the damper. This way introduction of the exciting and damping torques is deferred further down the calculation sequence.

Classical modal analysis of vibrating systems can be used to obtain forced response of multi-station rotor systems in torsional motion. Natural frequencies and mode shapes of the system are found using the transfer matrix method. The response of the rotor to periodic phenomena (not necessarily a harmonic or a shaft frequency) is determined as a linear weighted combination of the mode shapes of the system. Damping is to be entered in modal form; damping forces are thus a function of the different modal velocities. Formation of equivalent viscous damping constants that are some percentage of critical damping is required; the critical damping factor is obtained from the system modal inertia.

The modal analysis technique can be used for a torsional distributed mass model of a reciprocating machine using modal damping. Non-synchronous speed excitations are permitted. Shaft sections of the modeled rotor have distributed mass properties and lumped end masses (including rotary inertia). A transfer matrix analysis is performed to obtain a finite number of natural frequencies. The number required depends on the range of forcing frequencies used in the problem. The natural frequencies are substituted back into the transfer matrices to obtain the mode shapes. A function consisting of a weighted average of the mode shapes is formed to determine the torsional response.

Example problems # 6.3 through 6.7 provide the procedure for determining the torsional response using Holzer's sequence of calculations. The problems present a widely accepted and used method to calculate exciting torque, maximum torque, shaft twist angle, shaft stress and the effect on vibration amplitude due to damping introduced at a location in the system.

6.10 DYNAMIC VIBRATION ABSORBERS

In order to improve upon a machine or its component experiencing obnoxious vibrations when subjected to a steady alternating force of constant frequency, especially when the frequency is close to resonance, dynamic vibration absorbers have been applied quite successfully. The technique was invented by Frahm in 1909, and does not include any damping device. In Fig. 6.11 let the combination K, M be the schematic representation of a machine, with force $P_o \sin(\omega t)$ acting on it. The vibration absorber consists of a comparatively small vibratory system k, m attached to the main mass M. The natural frequency $\sqrt{(k/m)}$ of the attached absorber is chosen to be equal to the frequency ω of the disturbing force. It can be mathematically proved that the main mass M does not vibrate at all, and that the smaller system k, m vibrates in such a way that its spring force at all instants is equal and opposite to $P_o \sin(\omega t)$. Because of this no force acts on mass M, and so it will not vibrate at all.

To prove this concept, start with the equations of motion:

$$Md^2x_1/dt^2 + (K + k)x_1 - kx_2 = P_o \sin(\omega t) \qquad (6.45)$$

$$md^2x_2/dt^2 + k(x_2 - x_1) = 0$$

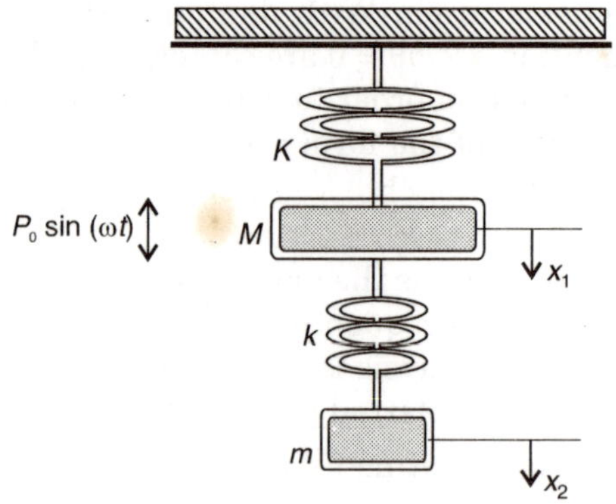

Fig. 6.11. Dynamic Vibration Absorber

The forced vibration of the system will be of the form:

$$x_1 = a_1 \sin(\omega t)$$
$$x_2 = a_2 \sin(\omega t) \tag{6.46}$$

After differentiating twice the sine function is still a sine function. Substituting into equations (6.45) will produce algebraic entities, which are easier for calculating the solution:

$$a_1(-M\omega^2 + K + k) - ka_2 = P_o$$
$$-ka_1 + a_2(-m\omega^2 + k) = 0 \tag{6.47}$$

For simplification, the dimensionless form may be introduced using the following symbols:

$x_{st} = P_o/K$ = static deflection of main spring

$\omega_a^2 = k/m$ = natural frequency of absorber

$\Omega_n^2 = K/M$ = natural frequency of main system

$\mu = m/M$ = mass ratio = absorber mass/main mass

Then the amplitudes of vibration will be:

$$\frac{a_1}{x_{st}} = \frac{1-\dfrac{\omega^2}{\omega_a^2}}{\left(1-\dfrac{\omega^2}{\omega_a^2}\right)\left(1+\dfrac{k}{K}-\dfrac{\omega^2}{\Omega_n^2}\right)-\dfrac{k}{K}} \tag{6.48a}$$

$$\frac{a_2}{x_{st}} = \frac{1}{\left(1-\dfrac{\omega^2}{\omega_a^2}\right)\left(1+\dfrac{k}{K}-\dfrac{\omega^2}{\Omega_n^2}\right)-\dfrac{k}{K}} \tag{6.48b}$$

The first equation reveals the contention that the amplitude of the main mass is zero if $\omega = \omega_a$, and this occurs when the frequency of the forcing function is the same as the natural frequency of the absorber. In the second equation, if $\omega = \omega_a$, the equation reduces to:

$$a_2 = -\frac{K}{k}x_{st} = -\frac{P_O}{k} \tag{6.49}$$

With the main mass standing still the damper mass has the motion $[-P_o/k] \cdot \sin(\omega t)$, and the force in the damper varies as $-P_o \cdot \sin(\omega t)$, which is equal and opposite to the external force.

These relations are good for any value of ω/Ω_n. However, adding the absorber is useful only if the original system is in resonance, or near it. Hence, consider the case where:

$$\omega_a = \Omega_n \quad \text{or} \quad k/m = K/M \quad \text{or} \quad k/K = m/M$$

The ratio $\mu = m/M$ then defines the size of the damper when compared with the size of the main machine. For this special case (6.48) becomes:

$$\frac{x_1}{x_{st}} = \frac{1-\dfrac{\omega^2}{\omega_a^2}}{\left(1-\dfrac{\omega^2}{\omega_a^2}\right)\left(1+\mu-\dfrac{\omega^2}{\omega_a^2}\right)-\mu}\sin(\omega t)$$

$$\frac{x_2}{x_{st}} = \frac{1}{\left(1-\dfrac{\omega^2}{\omega_a^2}\right)\left(1+\mu-\dfrac{\omega^2}{\omega_a^2}\right)-\mu}\sin(\omega t) \quad (6.50)$$

The denominators in the two equations are equal, which is due to a physical reason. When multiplied out and equated to zero, the denominator becomes a quadratic in $(\omega/\omega_a)^2$ having two roots. Thus, for two values of the external frequency ω both denominators become zero, and consequently x_1 and x_2 become infinitely large. These two frequencies are the natural frequencies of the system. As a result the compression and extension of the damper spring, as well as the force in it, are appropriately balanced.

For a torsional system such as the crankshaft of a reciprocating machine the Frahm dynamic vibration absorber takes the shape of flywheel A that can rotate freely on the shaft on bearings B and is held to it by mechanical springs k only, as shown in Fig. 6.12(a). Since the torsional impulses, for instance, in an internal combustion engine, are harmonics of the firing

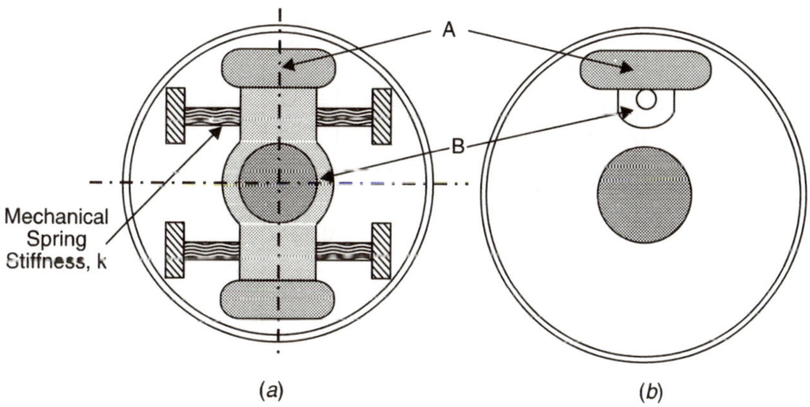

Fig. 6.12. Torsional Dynamic Vibration Absorber (a) With Mechanical Springs (b) With Centrifugal Springs

frequency (which is dependent on the engine speed), the absorber will work for one engine speed only. Note that there will also be two neighboring speeds at which the shaft will resonate. In order to overcome this situation the system is modified by replacing the mechanical springs by the 'centrifugal spring' of Fig. 6.12(b). The pendulum in the centrifugal field acts in the same manner as an ordinary gravity pendulum, where the field g is replaced by the centrifugal field $\omega^2 r$. Since the frequency of a gravity pendulum is $\sqrt{(g/l)}$, the frequency of a centrifugal pendulum becomes $\omega\sqrt{(r/l)}$, hence it is proportional to the machine's shaft speed. Thus, a centrifugal pendulum will act as a Frahm dynamic vibration absorber at all speeds. Further details of this device are discussed in the next section.

It is worth noting that several reciprocating machines operate at a fixed speed. For example, a diesel engine coupled to a power generator needs to generate alternating current electricity with a reasonably constant frequency, which calls for the diesel prime mover to operate at a constant shaft speed. The dynamic absorber with mechanical springs should be able to take care of the vibrations arising from rotor unbalance.

In the case of damped vibration absorbers, a dashpot c is arranged parallel to the damper spring k, between the masses m and M. Following a similar scheme of mathematical arguments, it is instructive to observe and analyze the expression for the amplitude of motion of the main mass:

$$\frac{x_1^{\,2}}{P_o^{\,2}} = \frac{(k - m\omega^2)^2 + \omega^2 c^2}{[(-M\omega^2 + K)(-m\omega^2 + k) - m\omega^2 k]^2 + \omega^2 c^2(-M\omega^2 + K - m\omega^2)^2} \tag{6.51}$$

For particular cases this equation reduces to known results as previously obtained.

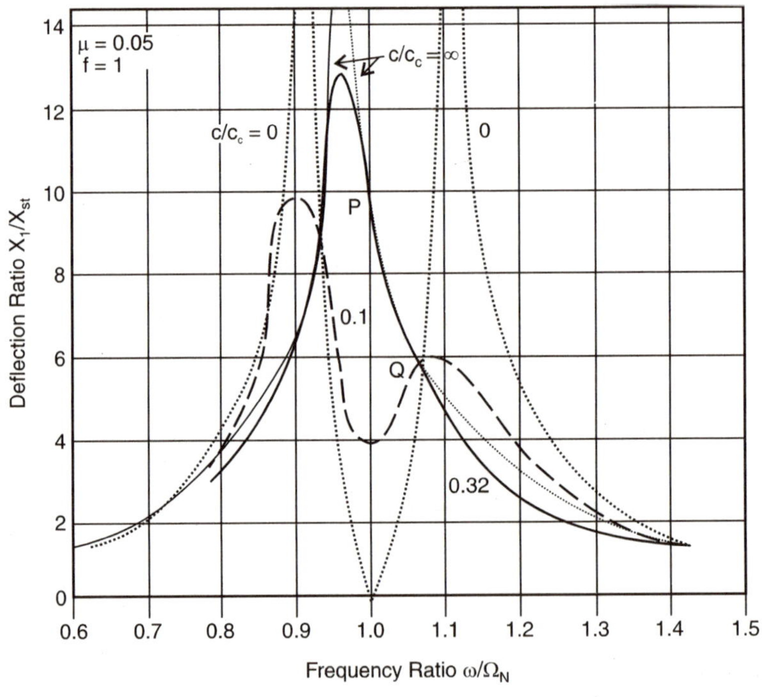

Fig. 6.13. Amplitudes of Main Mass of Fig. 6.11 for Various Absorber Damping Values

Figure 6.13 shows a plot of x_1/x_{st} as a function of frequency ratio $\mathbf{g} = \omega/\Omega_n$, $\mathbf{f} = \omega_a/\Omega_n = 1$, $\mu = 1/20$, for various values of c/c_c. The plots describe the behavior of a system where the main mass is 20 times the dynamic absorber mass and the frequency of the damper is equal to the frequency of the main system.

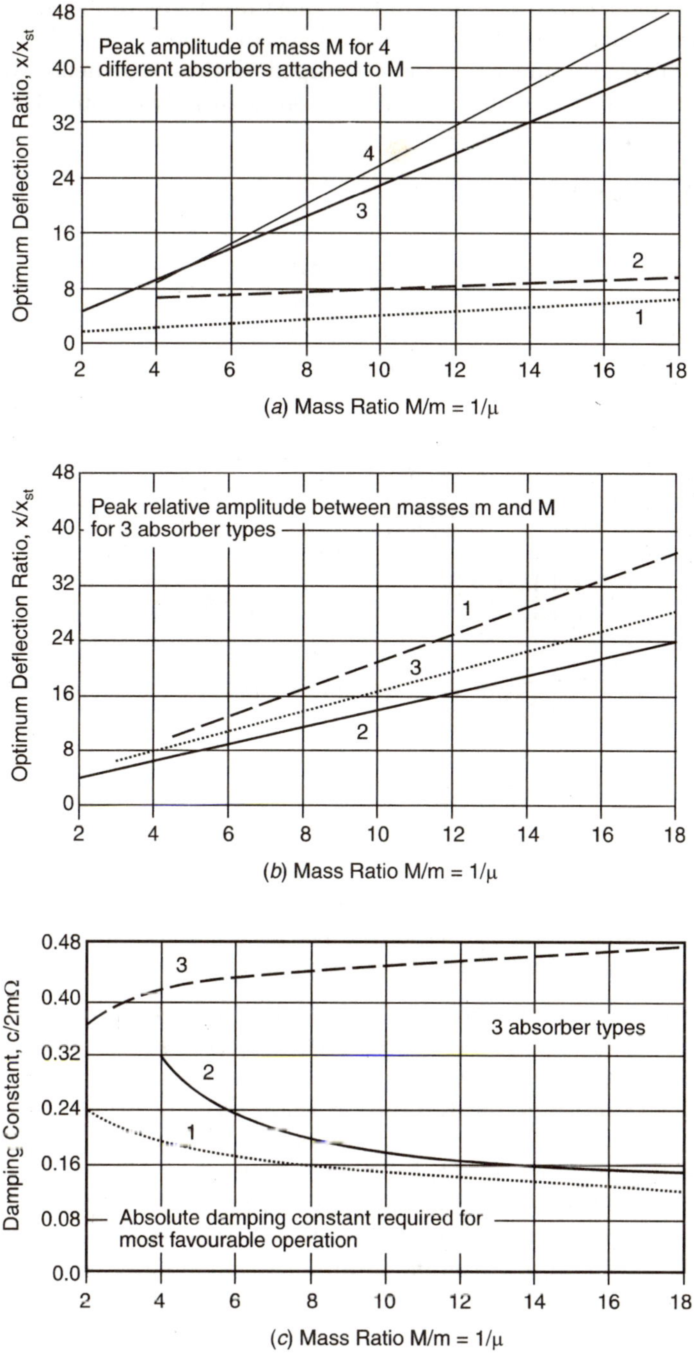

Fig. 6.14. Vibration Characteristics for Dynamic Vibration Absorbers

Some other cases commonly encountered in actual machines are shown in Fig. 6.14. Figure 6.14(*a*) provides peak amplitudes of the main mass as a function of μ for various absorbers attached to it. Figure 6.14(*b*) gives peak relative amplitudes between masses *m* and *M*, while Fig. 6.14(*c*) determines damping constants required for the most favorable operation of the absorber. In all three plots curve 1 represents the most favorably tuned and damped absorber; curve 2 represents the most favorably damped absorber tuned to the frequency of the main system; curve 3 considers the most favorably damped viscous Lanchester damper (discussed at length in the next section); and curve 4 for the most favorably damped Coulomb Lanchester damper (also discussed in the next section).

6.11 TORSIONAL DAMPERS

The Lanchester damper provides an effective method of mitigating torsional vibrations in reciprocating machines. It consists of two disks a (Fig. 6.15(a)), which rotate freely on the shaft bearings b. Between the two parts is a hub disk h solidly keyed to the machine's crankshaft (or to the attached machine's shaft). Hub h carries a brake lining c on its faces against which disks a can be pressed by screwing down the springs s.

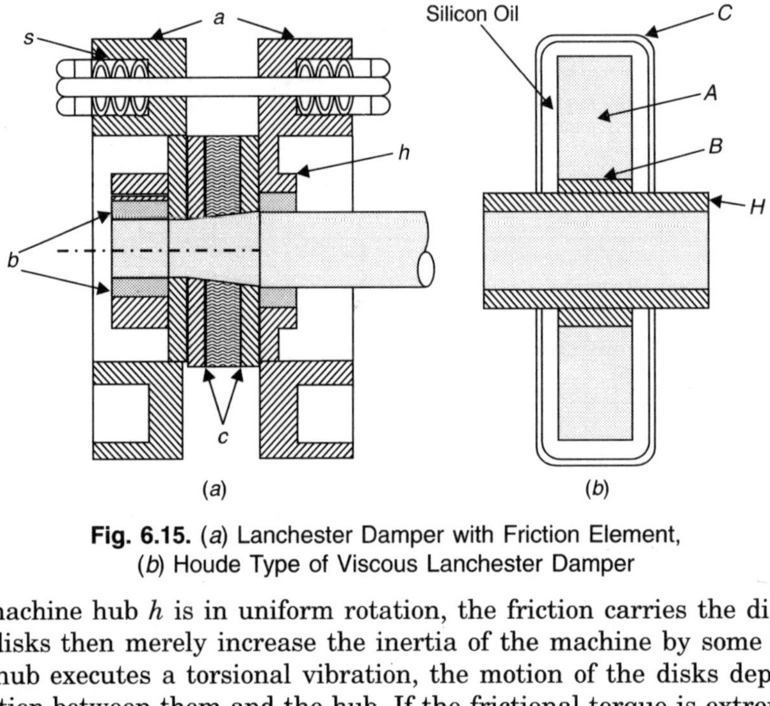

Fig. 6.15. (*a*) Lanchester Damper with Friction Element, (*b*) Houde Type of Viscous Lanchester Damper

If the machine hub h is in uniform rotation, the friction carries the disks a with the shaft, so the disks then merely increase the inertia of the machine by some percentage. If, however, the hub executes a torsional vibration, the motion of the disks depends upon the amount of friction between them and the hub. If the frictional torque is extremely small the disks rotate uniformly, and there is a relative slip between the hub and the disks that is equal to the amplitude of the hub's motion. Since the friction torque is nearly zero, little work is converted into heat. But if the friction torque is very large, the disks lock on the hub and follow its motion. There is then no relative slip and hence no energy dissipation. Between these two extremes there is both slip and a friction torque, so that energy is destroyed. There is an optimum value of the friction torque at which the energy dissipated is a maximum, as shown in Fig. 6.16.

The earlier version of the Lanchester damper depends on Coulomb, or dry, friction between the steel and asbestos-type of brake lining. The coefficient of friction of such an assembly is dependent on the degree of contamination with oil, and since the damper is located at the end of the crankshaft, this is difficult to avoid. In order to be more certain of the coefficient of friction, the dry friction can be replaced by viscous friction. Figure 6.15(b) shows the Houde type of viscous Lanchester damper. H is a hub, splined inside, to facilitate assembly of the damper on a shaft. A represents a flywheel, which, through a bronze ring B, can rotate freely about the hub. C is the assembly housing, welded to the hub. The clearance between the flywheel and the housing is small and accurately held to a dimension, and completely filled with a fluid permanently sealed in the space. The fluid is a type of silicone oil selected because of its advantage that its viscosity does not change substantially due to temperature change. A rotating motion of A relative to C and H is associated with a viscous-friction torque, so the damper substantially acts like the dry friction type of Fig. 6.16.

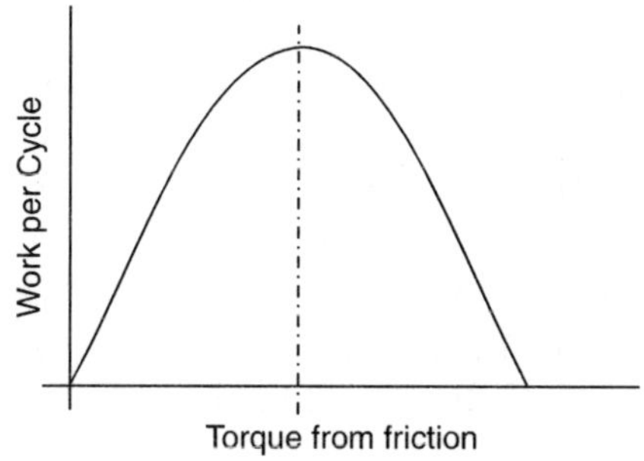

Fig. 6.16. Energy Dissipation in Lanchester Damper

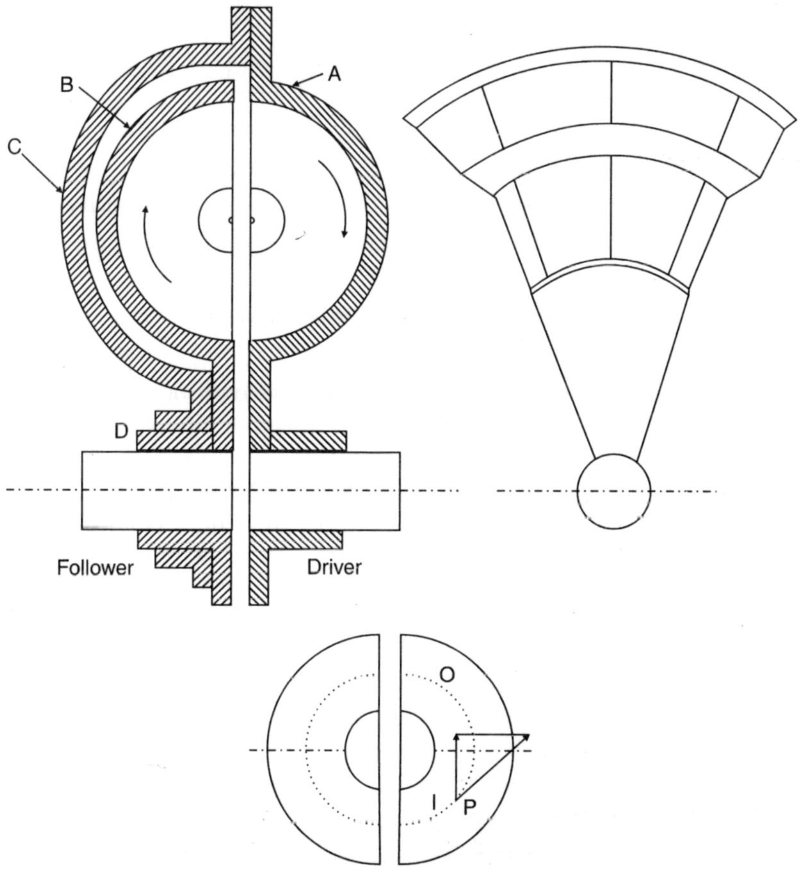

Fig. 6.17. 'Fluid Flywheel' to Reduce Torsional Vibrations

Another device useful for reducing torsional vibrations is in the form of a hydraulic coupling, sometimes referred to as a 'fluid flywheel' (Fig. 6.17). It consists of piece A in the shape of a half doughnut keyed to the driver shaft. A similar piece B is attached to the follower shaft. Cover C is solidly attached to A on the driver shaft, and can turn relative to the follower shaft. At D there is a hydraulic seal with no friction. The interior of the doughnut is filled with a fluid, thin oil or water, and the sole purpose of C is to hold the fluid in place. The doughnut-shaped space is sub-divided into a large number of open compartments by many thin vanes, each in the form of a semi-circle arranged in a radial plane. By Newton's law of action and reaction the torques on the driver and the follower must be equal and opposite. Since the device does not operate at peak efficiency, the speed of the follower is less than that of the driver, the speed ratio being the same as the efficiency, between 97 and 99%. The fluid in the coupling is under the influence of centrifugal force, which is greater in the driver than in the follower due to the speed difference. Thus, a circulation is set up, moving the fluid outward in the driver and inward in the follower. This circulation transmits torque between the two shafts.

The paddle type of damper is illustrated in Fig. 6.18, and utilizes machine-lubricating oil supplied through the crankshaft. It has the damping constant:

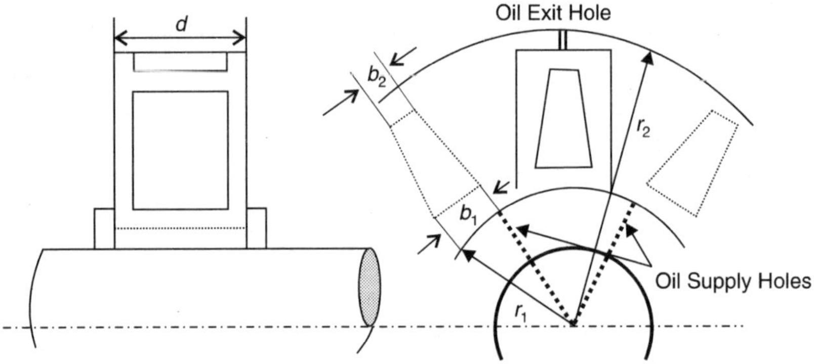

Fig. 6.18. Paddle Type of Torsional Damper

$$c = \frac{3\nu d^2 (r_2{}^2 - r_1{}^2)^2 n}{h^3 \left[\dfrac{d}{b_1} + \dfrac{d}{b_3} + \dfrac{4(r_2 - r_1)}{b_1 + b_2} \right]} \text{ lbs.-second/in.} \tag{6.52}$$

where n is the number of paddles, ν is fluid viscosity, b_1, b_2 are paddle tangential width at inner and outer radii, r_1, r_2 are paddle inner and outer radii and d is paddle axial length.

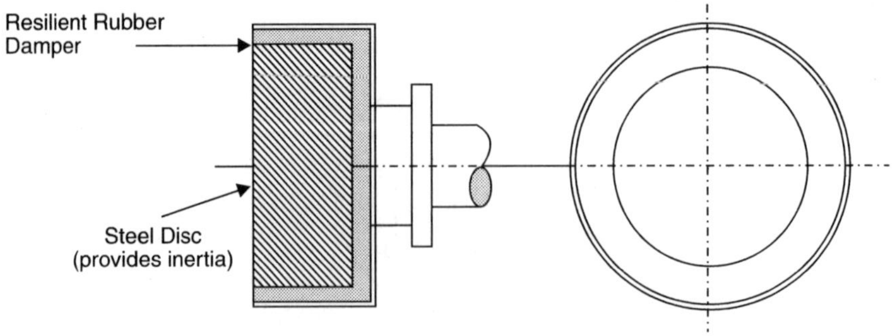

Fig. 6.19. Bonded Resilient Damper

The effectiveness of these dampers can be improved by connecting the flywheel to the engine by a spring of appropriate stiffness, in addition to the fluid friction. In one form, Fig. 6.19, the connection is in the form of rubber bonded between the flywheel and the shaft. The rubber acts as a spring as well as an energy-absorbing member by hysteresis.

6.12 PERMISSIBLE AMPLITUDES

Failure caused by torsional vibrations in the crankshafts of reciprocating machinery invariably initiates in the form of fatigue cracks that start at points of stress concentration. Typical locations are at fillets where there is a change in shaft diameters, at oil holes in a crankshaft and at the ends of keyway slots. At shaft oil holes the crack begins on a line at 45° to the shaft axis, then grows in a spiral pattern until failure occurs. The stress at the edge of the oil hole is theoretically 4 times the mean shear stress in the shaft, and failure may be expected if this concentrated stress exceeds the fatigue limit of the material. The problem of estimating the stress required to result in failure is further complicated by the presence of steady stress from the mean driving torque and the variable bending stresses. In practice, the severity of a critical speed is judged by the maximum nominal torsional stress from the equation:

$$\tau = 16T_m/(\pi d^3) \tag{6.53}$$

where T_m is the maximum torque amplitude and d is the crank pin diameter. The calculated nominal stress must be modified to include the effects of stress increase due to manufacturing procedures, operational and environmental considerations, then compared with the fatigue strength of the material.

Military standards issued by the U.S. Department of Navy state that the limit of acceptable nominal torsional stress within operating range is:

τ = (Ultimate strength)/25 for steel

τ = (Ultimate strength)/6 for cast iron

If the full-scale shaft has been fatigue tested (which is a difficult test), then:

τ = (Torsional fatigue limit)/2 for both materials.

For critical speeds inside the operating range, passed through during machine start up and shut down, the nominal torsional stress must not exceed 1¾ times the above values. Crankshaft steels having ultimate tensile strength between 75,000 and 115,000 lbs./in^2 usually have a torsional stress limit between 3,000 and 4,600 lbs./in^2.

For gear drives the vibratory torque across the gears at any operating speed must not be greater than 75% of the driving torque at the same speed, or 25% of the full load torque, whichever value is smaller.

The American Petroleum Institute considers the following sources of torsional excitation significant: problems arising from gears, including unbalance, pitch line run-out and eccentricity; start up conditions resulting from inertial impedances; and torsional transients from synchronous and induction electric motors. Torsional natural frequencies of the machine train must be at least 10% above or below any possible excitation frequency within the specified operating speed range. Torsional critical speeds at integer multiples of operating speed (for example, pump vane passing frequency) should be avoided, or should be shown to have no adverse effect if excitation frequencies exist. Non-synchronous torsional excitation frequencies must be considered in the evaluation. Identification of torsional excitations is the mutual responsibility of the purchaser and the vendor.

When torsional resonances are calculated to fall within the ± 10% margin, and the purchaser and vendor have agreed that all efforts to remove the natural frequency from the limiting frequency range have been exhausted, a stress analysis should be performed to demonstrate lack of adverse effect on any portion of the machine system.

In the case of synchronous motor driven units, the vendor is required to perform a transient torsional vibration analysis, with acceptance criteria previously agreed upon by the purchaser and the vendor.

6.13 SHIFTING OF CRITICAL SPEEDS

Among other procedures, shifting critical speed by making appropriate changes in mass and elasticity has been successfully implemented in new machine designs. If the stiffness of all the shafting in a system is increased by ratio a, then all the frequencies will increase by the same ratio a, provided that there in no increase in the inertia accompanying it. In crankshafts the diameters are made as large as practical to ensure bearing pressures are maintained within limits, hence it is rarely possible to stiffen the main and crank pin journals. If bearing diameters are increased the increase in the critical speeds will be much smaller than indicated by the ratio a, because the corresponding increase in inertia will also be substantial. Changes in the stiffness of a system made near a nodal point will have the maximum effect. Also, changes in inertia near a loop will have the greatest effect; those near a node will have little impact.

Elastic couplings have been used very effectively to place certain critical speeds below the operating speed, where they are passed through only in starting and stopping the machine. This leaves a clear range above the critical speed. The procedure must be used with caution. For example, in an eight-cylinder four-cycle internal combustion engine, fourth order torsional vibrations are so violent that it may be dangerous to pass through them. If the acceleration through the critical speed is rapid, some reduction in the amplitude may be attained, but with a controlled rate of engine acceleration the reduction in the amplitudes may not be large enough. The rate of deceleration when the machine is stopped is sometimes more important from this consideration. Mechanical clutches are sometimes used to disconnect the driven machinery until the driver has reached a speed above the dangerous critical speeds.

Elastic couplings take many forms. Helical coils arranged tangentially are an example; flat leaf springs arranged longitudinally or radially are other examples. Smaller diameter shaft sections of high tensile strength and shafts with grooves machined in them serve a similar purpose. Other configurations call for use of disks made of some resilient material between flanged coupling halves.

In internal combustion engines the choice of firing order and crank arrangement can be effectively used to reduce vibration amplitudes at certain minor critical speeds. Proper selection of crank arrangement and firing sequence can reduce the total shaft angle of twist and maximum shaft torque. But considerations of balance, bearing pressures and internal bending moments restrict the choice. Also, an arrangement that decreases the amplitude at one order of critical speed invariably increases the vibration amplitude at others. Tables 6.2 and 6.3 provide torsional vibration characteristics for eight-cylinder two- and four-cycle engines for crank arrangements and firing sequence with the most desirable properties. The values of Σb are calculated by assuming $\beta = 1$ for the cylinder furthest from the flywheel, $\beta = 1/n$ for the cylinder next to the flywheel (where n is the number of cylinders), and a linear variation in between. The arrangements assume equal crank angle locations and firing order. Reverse arrangements (mirror images) produce the same effects.

Table 6.2. Torsional Vibration Characteristics for Eight-Cylinder, Four-Cycle Engine with 90° Crank Spacing

Firing Order	*Σβ of Orders**				
	½, 7½, 8½, 3½, 4½	*1½, 5½, 6½*	*2, 6, 10*	*1, 3, 5, 8, 7*	*4, 8*
1-6-2-5-8-3-7-4	0.745	1.440	0	0	4.5
1-6-2-4-8-3-7-5	0.686	1.480	0	0	4.5
1-3-2-5-8-6-7-4	1.480	0.686	0	0	4.5
6-3-5-7-8-6-4-2	1.740	0.176	0	0	4.5
1-7-4-3-8-2-5-6	0.176	1.740	0	0	4.5

*0 values in the Σβ column represent very small values of actual β distribution.

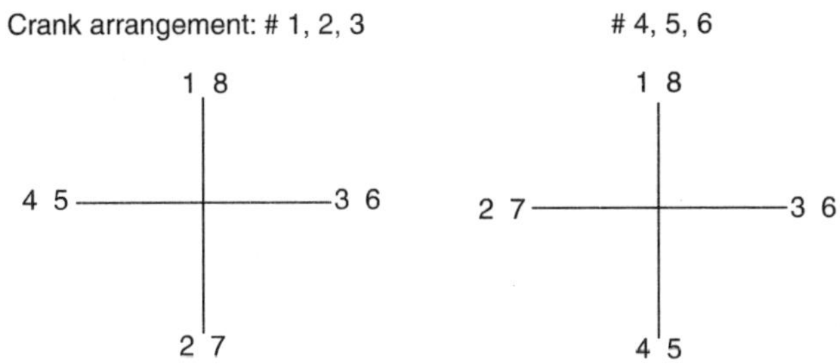

In *V*-type engines it is possible to choose the angle of the *V* that will cancel certain critical speeds. If ϕ is the *V* angle between cylinder banks and r the order number of the critical, the general equation is:

$$r\phi = 180°, 540°, 1080°, \text{etc.} \tag{6.54}$$

Thus, in an eight-cylinder engine the eighth order is canceled at angles of 22½°, 67½°, 112½°, etc. In four-cycle engines ϕ is taken as the actual bank angle if the second bank cylinders fire directly after the first. If the second bank cylinders omit a revolution before firing, ϕ is replaced by $(360 + \phi)°$, the cancellation equation becomes:

$$r \times (360 + \phi)° = m \times 180° \tag{6.55}$$

Table 6.3. Torsional Vibration Characteristics for Eight-Cylinder, Two-Cycle Engine with 45° Crank Spacing

Firing Order	*Σβ of Orders*				
	1, 7, 9	*2, 6, 10*	*3, 5, 11*	*4, 12*	*8, 16*
1-8-2-6-4-5-3-7	0.056	0	0.79	2.0	4.5
1-7-4-3-8-2-5-6	0.175	0	1.61	0	4.5
1-6-5-2-7-4-3-8	0.112	0	1.58	0.5	4.5

where m = 1, 3, 5, etc. Thus, to cancel a 4½ order critical the bank angle will be $\phi = 180°/4.5 = 40°$ for direct firing, and $\phi = 11 \times 180°/4.5 - 360° = 80°$ for the 360° delayed firing.

If a reciprocating machine can be arranged with approximately equal flywheel (or other rotors) at each end so that the node of a particular mode is at the center of the machine, $\Sigma\beta$ will cancel for the major orders of that mode. This procedure also needs to be used with caution, since the double flywheel arrangement may reduce the natural frequency in such a manner that the lower order minor critical speeds of large amplitudes take the place of canceled major critical speeds.

In marine installations damping arising from the hydraulic action on the propeller can be used to reduce torque amplitude in the shaft below a desired level by making the flywheel moment of inertia J of sufficient magnitude. The ratio of the propeller amplitude to the engine amplitude increases as the flywheel becomes larger, hence increasing the effectiveness of the propeller as a damper.

6.14 EXAMPLE PROBLEMS

Example Problem 6.1: Figure 6.20 shows the actual mass and elastic distribution of an eight-cylinder, four-cycle engine. Four of the cranks are fitted with counterweights, so the crank inertias are not identical. Dividing the moment of inertia by 74.2 and the stiffness constants by 342×10^6, the constants for the unity system are obtained.

The sum of the inertia constants through station 11 is 11.812. The stiffness of the shaft to station 10 is:

$$\frac{1}{7 + 1/4.003 + 1/2.20 + /1.331 + 1/5.341} = 0.1157$$

The two-rotor system can be represented by a two degree of freedom model. The approximate value of ω_s^2 for the unity system of sequence calculations is determined from the expression:

$$\omega_n^2 = \omega_s^2 \times (k/J) \tag{6.56}$$

$$\omega_s^2 = (0.1157 \times 46.80)/(4.725 \times 42.08) = 0.02724 \text{ (radians/second)}^2$$

The Holzer sequence of calculation is started with $\omega_s^2 = 0.02724$. The calculations are shown in Table 6.4. Rotor station numbers and the corresponding inertia constants are placed on alternate lines. β values opposite rotor station numbers define the first-mode relative amplitude curve, and torques between station numbers are shaft torques between corresponding rotor station numbers. Stiffness constants are placed on intermediate lines 1-2, 2-3, etc., in the last column. For the trial value of $\omega_s^2 = 0.02724$, $J\omega_s^2$ is computed next. An amplitude value of 1.0 is taken for β at station 1. In the torque column on line 1, $T = 1.0 \times 0.02724 = 0.02724$. The torque in the shaft section between stations 1 and 2 is equal to this inertia reaction. This torque divided by stiffness, 1.00 in section 1-2, is the relative deflection of stations 1-2, and is placed in the β column of line 1-2. Subtracting the relative deflection 0.02724 from the deflection at rotor station 1 will give deflection value 0.973 at station 2. This deflection multiplied by 0.0199 ($J\omega_s^2$ for station 2) gives inertia reaction 0.0194 of station 2. The inertia torque of station 2 is added to the torque in section 1-2 to give the torque in section 2-3. Thus, the sequence calculation is executed up to the last station 12. The final figure in the torque column gives the residual torque. Note that all additions and subtractions are done algebraically.

The residual torque has the following physical significance: If a simple harmonic torque of amplitude T_r and angular frequency ω is applied at station 10, then all rotors will oscillate with the amplitudes as calculated in the β column, and torque values will be as given in the T column. If the Holzer calculations are made over a sufficiently large range of ω_s to cover all the natural frequencies, T_r will follow the general shape of the curve shown in Fig. 6.10. The number of intersections the curve makes with the x-axis, excluding the origin, is equal to the number of rotor stations minus 1. The location of the trial values of ω_s^2 on this diagram can be determined from the sign of T_r and the number of changes in the sign of the relative amplitudes β of the rotor stations in the sequence table. In this example there is no sign change in β, and T_r takes a value greater than zero. Hence, the assumed value of ω_s^2 lies below the first mode.

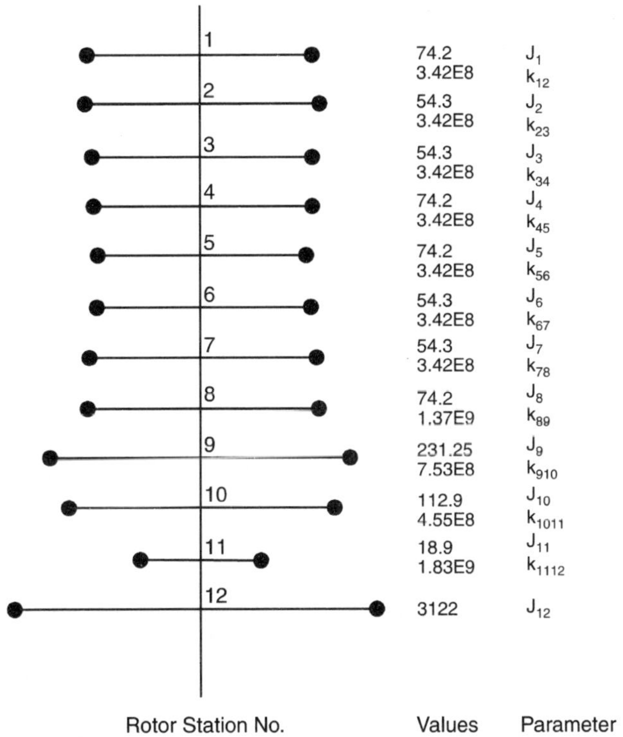

Fig. 6.20. Moments of Inertia and Stiffness of an 8-Cylinder Engine Driving a Centrifugal Compressor

To find the first mode assume a new trial value of $\omega_s^2 = 0.15$. This leads to $T_r = -0.1944$. Interpolating for a zero between $\omega_s^2 = 0.02724$ and 0.15 yields 0.087. Table 6.5 shows the sequence for this value, and represents the first mode of torsional vibration. The frequency error is less than one-thousandth, a much higher degree of accuracy than is justified by the magnitude of the rotor's geometric data. One sign change occurs in β, and T_r is negative (nearly zero), hence the selected value for ω_s^2 lies close to the first mode.

Table 6.4. Holzer Calculations for First Mode
(ω_s^2 = 0.02724)

Rotor Station #	*J*	$J\omega_s^2$	*Deflection* β	*Torque* *T*	*Stiffness* *k*
1	1.000	0.0272	1.0000	0.0272	
			0.0272	0.0272	1.000
2	0.732	0.0199	0.9728	0.0194	
			0.0466	0.0466	1.000
3	0.732	0.0199	0.9261	0.0185	
			0.0651	0.0651	1.000
4	1.000	0.0272	0.8611	0.0235	
			0.0885	0.0885	1.000
5	1.000	0.0272	0.7725	0.0210	
			0.1096	0.1096	1.000
6	0.732	0.0199	0.6629	0.0132	
			0.1228	0.1228	1.000
7	0.732	0.0199	0.5402	0.0108	
			0.1336	0.1336	1.000
8	1.000	0.0272	0.4066	0.0111	
			0.0361	0.1446	4.003
9	3.117	0.0849	0.3705	0.0314	
			0.0800	0.1761	2.200
10	1.522	0.0414	0.2904	0.0120	
			0.1414	0.1881	1.331
11	0.247	0.0067	0.1491	0.0010	
			0.0354	0.1891	5.341
12	42.075	1.1460	0.1137	0.1302	
		Residual Torque =		0.1881	

The second mode frequency corresponding to the condition in which there are two nodes in the engine shaft is approximated by:

$$\omega_s^2 = 22/n^2$$

where n is the number of machine cylinders. In this example, $\omega_s^2 = 22/8^2 = 0.344$. Through Holzer's calculations, this value is low; a value that will make T_r nearly zero is

Table 6.5. Holzer Calculations for First Mode ($\omega_s^2 = 0.0870$) (Second Iteration)

Rotor Station #	J	$J\omega_s^2$	*Deflection* β	*Torque* T	*Stiffness* k
1	1.000	0.0870	1.0000	0.0870	
			0.0870	0.0870	1.000
2	0.732	0.0637	0.9130	0.0581	
			0.1451	0.1451	1.000
3	0.732	0.0637	0.7679	0.0489	
			0.1940	0.1940	1.000
4	1.000	0.0870	0.5739	0.0499	
			0.2439	0.2439	1.000
5	1.000	0.0870	0.3299	0.0287	
			0.2726	0.2726	1.000
6	0.732	0.0637	0.0573	0.0036	
			0.2763	0.2763	1.000
7	0.732	0.0637	– 0.2190	– 0.0139	
			0.2623	0.2623	1.000
8	1.000	0.0870	– 0.4814	– 0.0419	
			0.0551	0.2205	4.003
9	3.117	0.2711	– 0.5364	– 0.1455	
			0.0341	0.0750	2.200
10	1.522	0.1324	– 0.5705	– 0.0755	
			– 0.0004	– 0.0005	1.331
11	0.247	0.0215	– 0.5702	– 0.0122	
			– 0.0024	– 0.0127	5.341
12	42.075	3.6606	– 0.5678	– 2.0784	
		Residual Torque =		– 0.0005	

ω_s^2 = 0.714. The torsional natural frequencies of the actual system is obtained from (6.56):

$$\omega_n^2 = (342.0 \times 10^6 \times \omega_s^2)/74.2 = 4.609 \times 10^6 \times \omega_s^2$$

Then the first two torsional natural frequencies of the eight-cylinder reciprocating engine are:

First: $\omega_s^2 = 0.0870$; $\omega_n^2 = 401{,}000$; f = **100.8 Hz, or 6,047 rpm.**

Second: $\omega_s^2 = 0.7140$; $\omega_n^2 = 3{,}292{,}000$; f = **288.8 Hz, or 17,326 rpm.**

Example Problem 6.2: A six-cylinder four-cycle engine with 120° crank spacing has three equally set firing impulses per revolution. What are the critical speeds?

Solution: Major orders for a six-cylinder engine are 3, 6, 9, 12, etc. Consider the engine force balance at frequencies of the major order times shaft speed. The critical speeds occur at $(60 \times f_n)/q$ rpm where f_n is the natural frequency of one of the modes in Hz, and q is the order of the critical speed. Although many critical speeds exist in the operating range of an engine, only a few are likely to be important. Their effects will result in higher maximum torque and stress in the crankshaft.

Major orders of torsional critical speed representing the number of complete oscillations of the elastic system per unit of revolution of the engine crankshaft are obtained from the phase diagram, shown in table 6.6 for the eight-cylinder engine of problem 6.1. In example problem 6.1, the first two natural frequencies have been calculated. Applying the phase diagram method, the critical speeds, shown in rpm, are calculated to be:

Order	*4*	*4½*	*5*	*5½*	*6*	*6½*	*7*	*7½*	*8*	*12*
1—noded	1512	1344	1210	1100	1008	930	864	806	756	504
2—noded	4332	3851	3466	3151	2888	2666	2475	2310	2166	1444

Example Problem 6.3: The engine in example problem # 6.1 has cylinder bore of 9 in. and stroke of 12 in. Harmonic coefficients A_r are obtained from a harmonic analysis of an actual indicator card, shown in table 6.7. Determine the exciting torque for various orders in this engine.

Solution: The exciting torque is expressed by:

$$T_e = (\pi/4) \times 9^2 \times (12/2) \times (100 \times A_r) = 38170A_r$$

Table 6.8 provides the exciting torque values.

Example Problem 6.4: Using Lewis formula and an empirical factor $\Re = 50$ for the engine in example problem # 6.3, calculate the maximum torque.

Solution: For a great many engine applications the damping factor is not known. Experimental tests have provided empirical equations to determine the response to torsional excitation. The formulas may yield a variation of 30% or more compared with test results, when applied to different engines. Better agreement cannot be expected even for identical engines, where amplitudes may differ by a ratio of 2 to 1. This is not surprising, when differences due to length of service, bearing clearances, engine mount, engine condition and operating conditions are considered.

The Lewis formula has been known to provide useful results:

$$T_m = \Re \times T_e \times \Sigma\beta$$

The empirical factor $\mathfrak{R}$ in the equation includes the effects of the uncertainties in geometry and operating conditions. In larger engines of 20 in. bore and more $\mathfrak{R}$ may be assumed to be in 50 – 60 range. For bore size of 8 in. it may be 40 – 50, and for 4 in. bore it may be around 35. Experience with other engines of a similar class and make generally provides a good indication for the value of $\mathfrak{R}$.

For the assumed value of $\mathfrak{R}$ = 50 table 6.8 provides values of the maximum torque.

Table 6.6. Phase Diagrams and Orders for 8-Cylinder Internal Combustion Engines

Phase Diagrams (Firing Sequence)	*Orders*
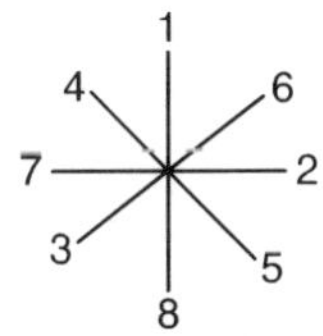	½, 4½, 8½, ... 3½, 5½, 7½, ... (Mirror Image)
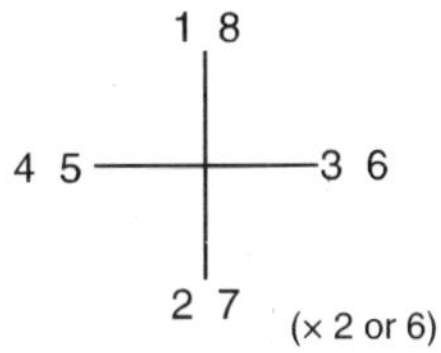	1, 5, 9, ... 3, 7, 11, ... (Mirror Image)
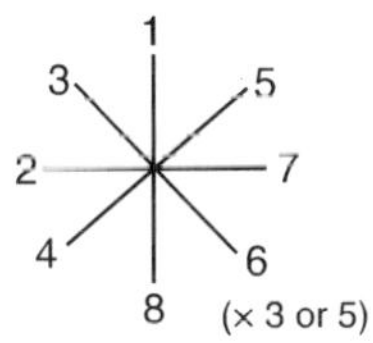	1½, 5½, 9½, ... 2½, 6½, 10½, ... (Mirror Image)
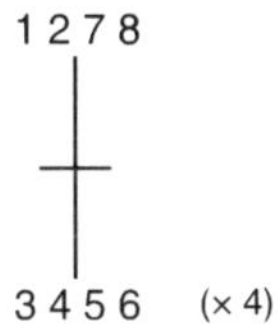	2, 6, 10, ...
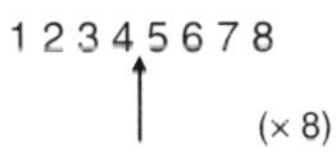	4, 8, 12

Example Problem 6.5: If the crank pin diameter is 6½ in., determine the nominal torsional stress in the engine of example problem # 6.1, using harmonic coefficients provided in Table 6.8.

Solution: Torsional shear stress is given by:

$$\tau = T_m \times r/J, \text{ where } J = [\pi \times (6.5)^4]/32 = 175.25 \text{ in}^4$$
$$= 0.0185 \times T_m \text{ lbs/in}^2$$

Table 6.7. Normalized Harmonic Coefficients of Gas Pressure Torque
Single Cylinder Four Cycle Diesel Engine

Order of Harmonic, r	A_r	a_r/b_o	b_r/b_o
0	1	0	1
½	2.148	1.357	1.665
1	2.322	2.139	0.903
1½	2.230	2.221	0.195
2	1.915	1.906	– 0.186
2½	1.569	1.537	– 0.315
3	1.290	1.235	– 0.373
3½	1.053		
4	0.827		
4½	0.630		
5	0.485		
5½	0.378		
6	0.285		
6½	0.202		
7	0.142		
7½	0.105		
8	0.075		
8½	0.046		
9	0.025		
9½	0.017		
10	0.013		
10½	0.008		
11	0.011		

$A_r = (1/b_o)[a_r^2 + b_r^2]^{1/2}$, a_r = Sine Coefficient, b_r = Cosine Coefficient (*ref:* Cylinder Gas Pressure Harmonic Coefficient Tables by Porter, Set # *P*2, Mean Indicated Pressure = 140 lbs/in²)

Shear stress in the crank pin is tabulated in table 6.8.

Table 6.8. Response Amplitudes for First Mode of Example Problems 6.1–6.5

Order	*RPM*	$\Sigma\beta$	A_r	T_e	T_m	*Stress*	θ_1
4	1512	3.642	0.295	11260	2050490	38027	0.0217
4.5	1344	0.453	0.235	8970	203171	3768	0.00215
5	1210	0.118	0.185	7062	41663	773	0.00044
5.5	1100	1.085	0.15	5726	310611	5760	0.00328
6	1008	0.351	0.12	4580	80387	1491	0.00085
6.5	930	1.077	0.088	3359	180882	3354	0.00191
7	864	0.102	0.65	2481	12653	235	0.00013
7.5	806	0.455	0.55	2099	47761	886	0.00050
8	756	3.642	0.039	1489	271082	5027	0.00287
12	504	3.642	0.14	534	97311	1805	0.00103
4	950	—	0.295	11260	192000	3561	0.00203

Example Problem 6.6: Determine the amplitude of the fourth-order vibration at the maximum speed of 950 rpm in the engine of example problem # 6.1.

Solution: The frequency in the unity system corresponding to fourth order vibration at 950 rpm in the actual system is given by:

$$\omega_s^2 = [2\pi \times 950 \times 4/60]^2 \times [69.4/(342.0 \times 10^6)] = 0.03213$$

Hence: $\omega_s = 0.1793$. Table 6.9 provides the calculations for $\omega_s = 0.1793$. Assuming no damping the residual torque in the actual system is:

$$T_r = 0.5992 \times \theta + 1.919 = 0$$

from which $\theta = -3.203$. Maximum torque occurs in section 10 – 11, and may be determined from:

$$T_{10-11} = 1.3565 \times \theta = -4.344$$

In section 6 – 7 the torque is:

$$T_{6-7} = 1.1598 \times \theta + 7.2677 = 3.553$$

Fourth-order torque T_e in the actual engine is 11,260 lbs.-in (table 6.8); in the actual engine fourth order torque in section 10 – 11 is:

$$T = 4.344 \times 11{,}260 = 48{,}913 \text{ lbs.-in.}$$

At cylinder # 1 the amplitude of the unity system is:

$$-7.3587 \times \theta - 43.231 = -19.66$$

So in the actual system it is:

$$[-19.66 \times 11{,}260/342.0 \times 10^6] = -0.000647 \text{ radians.}$$

Example Problem 6.7: If a concentrated damping of constant c = 1470 lbs.-in/radian acts at crank # 1, determine the amplitude of rotation.

Solution: In the unity system the corresponding damping constant is:

$$c = 1470/[74.2 \times 342.0 \times 10^6]^{1/2} = 0.00923$$

Table 6.9. Holzer Calculations for Forced Vibrations of Engine Problem 6.6 at $N = 950$ rpm ($\omega_s^2 = 0.0321$)

Rotor			*Deflection*		*Torque*		*Stiffness*
Station #	J	$J\omega_s^2$	$\theta \times$ *Inertia*	*Exciting*	$\theta \times$ *Inertia*	*Exciting*	k
12	42.075	1.3506	1.0000	—	1.3506	—	
			0.2529	—	1.3506	—	5.341
11	0.247	0.0079	0.7471	—	0.0059	—	
			1.0195	—	1.3565	—	1.331
10	1.522	0.0488	– 0.2724	—	– 0.0133	—	
			0.6105	1	1.3432	1	2.200
9	3.117	0.1000	– 0.8829	– 1	– 0.0883	– 0.10004	
			0.3135	1.899957	1.2549	7.60538	4.003
8	1.000	0.0321	– 1.1964	– 2.899958	– 0.0384	– 0.09308	
			1.2165	7.512294	1.2165	7.51229	1.000
7	0.732	0.0235	– 2.4129	– 10.41225	– 0.0567	– 0.24459	
			1.1598	7.267700	1.1598	7.2677	1.000
6	0.732	0.0235	– 3.5727	– 17.67995	– 0.0839	– 0.41532	
			1.0759	6.852381	1.0759	6.85238	1.000
5	1.000	0.0321	– 4.6486	– 24.53233	– 0.1492	– 0.78748	
			0.9267	6.064893	0.9267	6.06489	1.000
4	1.000	0.0321	– 5.5753	– 30.59723	– 0.1790	– 0.98217	
			0.7477	5.082722	0.7477	5.08272	1.000
3	0.732	0.0235	– 6.3230	– 35.67995	– 0.1485	– 0.83815	
			0.5992	4.244565	0.5992	4.24456	1.000
2	0.732	0.0235	– 6.9222	– 39.92451	– 0.1626	– 0.93786	
			0.4366	3.3067	0.4366	3.30671	1.000
1	1.000	0.0321	– 7.3587	– 43.23121	– 0.2362	– 1.38772	
			Residual Torque =		0.5992	1.91897	

At crank # 1 the torque $-jc_s\omega_s\theta$ is added. This torque is:

$$T = -j \times 0.00923 \times 0.175 \times (-7.3587 \times \theta - 43.231)$$
$$= j(0.0119 \times \theta + 0.0698)$$

The resulting residual torque is given by:

$$T_r = 0.5992 \times \theta + 1.919 + j(0.0119 \times \theta + 0.0698) = 0$$

Solving for θ gives: $\theta = [-1.919 - 0.0698j]/[0.5992 + 0.0119j]$

Then $|\theta| = 3.204$. Thus, at this speed the effect of external damping is negligible.

Example Problem 6.8: Figure # 6.11 shows a spring-mass combination K and M to represent a machine with force $P_o \sin(\omega t)$ acting on it, resulting in unacceptable vibrations in the machine. To remedy the situation a dynamic absorber in the form of a comparatively smaller vibratory system k, m is attached to the main mass. If Mg = 10 lbs., mg = 1 lbs., P_o = 10 lbs., K = 102 lbs/in. and k = 10.2 lbs/in.:

(*a*) what is the best damping coefficient across the absorber?

(*b*) what is the amplitude of the main mass?

(*c*) what is the maximum stress in the absorber spring?

Solution: The answers can be found in Fig. 6.15 (*a*), (*b*), (*c*).

(*a*) From Fig. 6.14(*c*): $c/2m\Omega_n = 0.205$, so $c = 0.41 \times 20 \times \pi/386 =$ **0.067 in-lbs/sec.**

(*b*) Figure 6.14(*a*) gives $x/x_{st} = 7.2$, then $x = 7.2/[1/102] =$ **0.0706 in.**

(*c*) Figure 6.14(*b*) gives $x_{rel}/x_{st} = 12.8$, so $x_{rel} = 12.8/102 = 0.1255$ in. The force kx_{rel} is $10.2 \times .126 =$ **1.28 lbs.**

Example Problem 6.9: If the requirement $k/K = m/M$ is dropped, for what value of k is the best overall effect obtained?

Solution: The most favorable tuning follows from $\omega_a/\Omega_n = 1/[1 + m/M] = 10/11$, so $(\omega_a/\Omega_n)^2 = 100/121$. Since m, M, and K are defined, $(\omega_a/\Omega_n)^2$ is proportional to k, and the new absorber spring $k = 10.2 \times (100/121) =$ **8.43 lbs/in.**

REFERENCES AND BIBLIOGRAPHY

American Petroleum Institute, "Centrifugal Compressors for General Refinery Service", API STD 617, Fifth ed., Washington, D.C., 1988.

Anwar, I., "Computerized Time Transient Torsional Analysis of Power Trains", ASME paper No. 89-DET-74, 1989.

Cataford, G. F., Lancee, R. P., "Oil Free Compression on a Natural Gas Pipeline", ASME paper # 86-GT-293, ASME, New York, June, 1986.

Constant, H., *Brit. Aero. Res. Comm. R and M,* no. 1201, 1928.

Crede, C. E., Ruzcicka, J. E., "Theory of Vibration Isolation", Shock & Vibration Handbook, McGraw-Hill, New York, 1988.

Den Hartog, J. P., "Mechanical Vibrations", Dover Publications, Inc., New York, 1985.

Ehrich, F. E., "Handbook of Rotordynamics", McGraw-Hill, New York, 1992.

Eshelman, R. L., "Torsional Response of Internal Combustion Engines", *Trans. ASME,* 96(2):441, 1984.

Eshelman, R. L., "Torsional Vibrations in Machine Systems", Vibration, 3(2):3, 1987.

Eshelman, R. L., Eubanks, R. A., "Effects of Axial Torque on Rotor Response: An Experimental Investigation", ASME paper # 80-WA/DE-14, , New York, December, 1980.

Holzer, "Die Berechnung der Dreschwingungen", Springer-Verlag, Berlin, 1921.

Jery, B., Acosta, A., Brennen, C., Caughey, T., "Hydrodynamic Impeller Stiffness, Damping and Inertia in the Rotordynamics of Centrifugal Pumps", CP2338, NASA, pages 137-160, May, 1984.

Lewis, F. M., *Trans. ASME,* 78:APM 377, 1955.

Myklestad, N. O., "Fundamentals of Vibration Analysis", McGraw-Hill, New York, 1956.

Nestorides, E. J., "A Handbook of Torsional Vibration", Cambridge University Press, 1958.

Porter, F, "Evaluation of Effects of Torsional Vibration", Society of Automotive Engineers, War Engineering Board; also *Trans.* ASME, 65:A-33, 1943.

Porter, F., "Harmonic Coefficients of Engine Torque Curves", Journal of Applied Mechanics, 10(1):A33-A48, March, 1983.

Rao, S. S., "Mechanical Vibration", Addison-Wesley Publishing Co., Reading, Mass., 1990.

Roark, R. J., Young, W. C., "Formulas for Stress and Strain", 5th ed., McGraw-Hill, New York, 1985.

Shemeld, D.E., "A History of Development in Rotordynamics, A Manufacturer's Perspective", CP2443, NASA, Washington, D. C., 1986.

Sohre, J. S., "Transient Torsional Criticals of Synchronous Motor-Driven, High Speed Compressor Units", ASME Paper No. 86-FE-22, June, 1985.

Taylor, C. F., "The Internal Combustion Engine in Theory and Practice - Vol. 2: Combustion, Fuels, Materials, Design", The MIT Press, Cambridge, Massachusetts, 1985.

Timoshenko, S., Young, D. H., Weaver, W., "Vibration Problems in Engineering", John Wiley & Sons, New York, 1984.

U. S. Navy Department: "Military Standard Mechanical Vibrations of Mechanical Equipment", MIL-STD-167 (SHIPS).

Wehrli, V. C., "Uber Kritische Drezahlen unter Pulsierender Torsion", *Ingenieur Arch,*33:73-84, 1993.

Zdanowich and Wilson, T. S., *Proc. Inst. Mech. Engrs. (London),* 1943:182, 1940.

7 Flexural Vibrations

7.1 INTRODUCTION

Failures in the crankshaft of reciprocating machinery generally result from fatigue due to torsional stress as it fluctuates between extreme values. In addition to torsion, fibers on the surface of a rotating shaft are subjected to bending loads, undergoing both tension and compression for each revolution of the shaft. If the crankshaft of a reciprocating pump is coupled to the shaft of an electric motor operating at 1,740 rpm, the shaft is stressed in tension and in compression 1,740 times. If the shaft is also axially loaded (caused, for example, by a helical or worm gear), the shaft has an axial stress superimposed upon the bending component. These and other loads occurring in the shaft produce stresses often referred to as repeated, alternating or fluctuating stresses.

Bending in the shaft may arise from a number of operational factors, some of which arise internally in the machine and others from outside the machine. Dynamic balance of the machine's rotating assembly is an internal machine factor that causes bending in the shaft. As a result of manufacturing procedures and limitations, tolerances have to be provided on machined components. Thus, a minimum diameter shaft journal may be paired with a maximum diameter bearing, causing a larger than normal radial clearance in the bearing. In contrast with torsional vibrations, bearing and support stiffness plays a significant role in the shaft's lateral vibrations. Since the hydro-dynamic stiffness of the bearing's lubricant film is reduced as a consequence of the larger bearing clearance, the shaft's natural frequency of vibration also is reduced, causing the machine to operate at or close to the natural frequency. Rotor unbalance arises from many sources attributable to manufacturing and field operation. With the appropriate amount and location of unbalance in the rotor, substantial bending moments are caused when operation takes place near a critical speed, resulting in flexural shaft vibrations along its lateral axes.

Shaft misalignment is another factor responsible for flexural vibrations. At the coupling between an internal combustion engine and an electric power generator thermal conditions may throw out the rotor from an aligned position during cold condition. Radial and angular misalignment of the rotor can radically alter the flexural vibration signature of a rotating assembly, with consequent bending moments in the shaft. This also has the adverse effect of putting unusual wear pattern in the bearings.

For flexural vibrations the most commonly encountered stimulus is rotor unbalance and shaft bow, resulting in whirling or whipping of the shaft. Unstable operation of the machine in the form of chatter (or stick-slip) and large deflections and loads are other undesirable consequences. To deal with such vibrations appropriate design strategies are required. For instance, placing the system's critical speeds out of the steady state operating range will minimize the degree of super-critical whirling and whipping phenomena. If for practical reasons it is not possible to raise the critical speed, or operation above the fundamental mode of vibration cannot be avoided, introduction of damping to reduce critical response peaks and to cancel the destabilizing forces in the operating range will be required. Minimizing the excitation through better balancing and aligning procedures will go a long way towards the stated objective of smoother operation of the machine.

An easy to use and accurate computational tool for prediction of the vibration characteristics of any equipment design is an indispensable requirement in the design process. The design process is too complex to permit automation in a direct sequence from specification to definition of the finished design. In spite of this drawback, a direct sequence can be approximated by a speedy iterative process that starts with posing design alternatives, then analytically evaluating them, followed by selecting and quantitatively refining the design that satisfies design requirements. Such analytical tools are made available in the form of computer codes, discussed later in this and subsequent chapters. Analytical methods are useful for predicting the rotor dynamic behavior of the preliminary design configuration, to assess its impact and provide quantitative data for design refinement. The mathematical model developed provides valuable insight during testing of a new machine; when unexpected data from accelerometers during operations indicates trouble, analytical models are studied to provide understanding and guidance in remedial design, fault detection and correction. The analytical model may range from an idealized version of a single mode to refined and detail finite element model of the total system to provide precise information in the form of natural frequencies, mode shapes, deflections and loads throughout the assembly.

Mathematical models for determining the vibration characteristics of a rotating machine simulate the kinematics of important degrees of freedom. They permit the determination of the response of the system due to the interaction of the geometric configuration of the rotor, bearings and support structure, with the dynamic loads acting on the system. The analytical model is used to analyze free vibrations, forced vibrations and self excited vibrations.

For each cycle or revolution of the crankshaft the most careful analysis may reveal that the maximum dynamic stresses are below the ultimate strength of the material, and quite frequently even below the yield strength; even then a fatigue failure is encountered. The primary distinguishing characteristic of fatigue failures is that the stresses have been repeated a very large number of times, and hence the failure is called a fatigue failure. A fatigue failure begins with a small crack, which may be small enough to prevent detection by the naked eye or even to locate in a Magnaflux or x-ray inspection. The crack will develop at a point of discontinuity in the material such as a keyway, a change in the cross section or a hole. Less obvious points at which fatigue failures are likely to start are inspection or stamp marks, internal cracks or irregularities caused by machining. Once a crack has developed, the stress concentration effect is increased and the crack propagation progresses at a faster pace. As the stressed area decreases in size, the stress increases in magnitude until the remaining area fails abruptly. The fatigue failure is characterized by two distinct areas of failure, the first due to progressive development of the crack, the second due to sudden fracture. This form of failure is in contrast to machine parts failing from a static loading; in this case the parts develop large deflection because the stress has exceeded the yield strength. Thus, many static failures are visible ones and give advance warning. Fatigue failures give no warning; they are sudden, total and hence dangerous.

It is relatively easier to design against a static failure since load conditions are better defined and understood. Fatigue is a more complicated phenomenon, and requires a far more intensive design engineering effort to come up with a successful product.

7.2 DYNAMICS OF SHAFT MOTION

Transverse, or lateral direction, vibrations of a shaft are analytically different from torsional and axial vibrations in several respects:

- The deformation of the rotor is due to a combination of bending and shearing, and is inherently more complex than simple stretching or twisting.
- Shaft bending is not confined to a single plane, and motion must be looked at in two orthogonal planes. Within each plane it is necessary to characterize the motion of each rotor station by two independent parameters, displacement and angular rotation.
- Gyroscopic and other asymmetric effects cause the motion of the shaft in the horizontal and vertical direction to be coupled.
- The location and dynamic characteristics of the bearings play a substantial role in transverse direction vibrations. This calls for development of detailed models of the bearings and their support structure.

Despite these complications it is possible to develop equations of motion for transverse vibration of the rotor, which are analogous to torsional vibrations. To establish some important concepts consider a simplified model of a high-speed rotor that retains many of the essential characteristics of the complex system in its response to unbalance. A disk is carried at the center of a flexible but massless shaft supported in two bearings (Fig. 7.1). Two types of bearing supports will be considered, rigid and flexible. In the model with rigid bearings the only damping c will come from the fluid medium surrounding the disk. Two types of unbalance must be considered separately, one due to eccentricity in the disk where the center of the disk is offset from the elastic axis of the shaft by a radial distance q, the other due to shaft bow, s.

In the eccentric disk case the whirl radius at disk $r$wd is calculated from the steady state solution of the differential equations for response to rotating unbalance (ref. Myklestad):

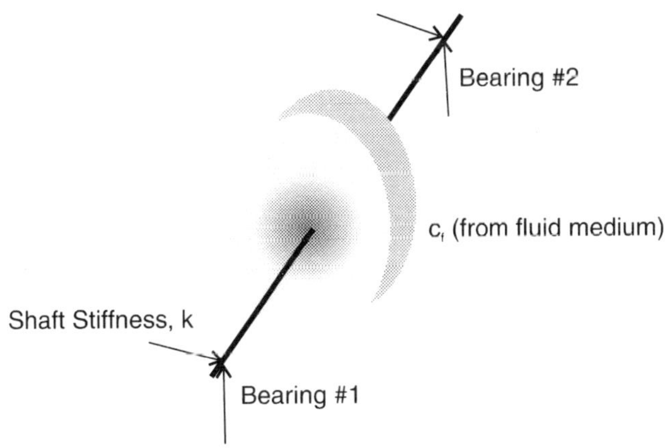

Fig. 7.1. Shaft Supported in Two Bearings

$$r_{wd} = \frac{q(\omega/\omega_m)^2}{\{[1-(\omega/\omega_{ns})^2]^2 + (2\xi_f\ \omega/\omega_{ns})^2\}^{1/2}} \tag{7.1}$$

where ω is the rotational speed, $\omega_{ns} = \sqrt{(k/m)}$ is the undamped natural frequency, m is the disk mass, k is the rotor stiffness and $\xi_f = c_f/(2m\omega_{ns})$ is the damping ratio. To understand the nature of critical speed consider a situation where the rotor has no unbalance, or $q = 0$, and no damping. The inertia force $m\omega^2 r_{wd}$ will equal shaft elastic spring force kr_{wd}. At the critical speed this force balance is satisfied for all values of whirl radius. There exists a state of indefinite unbalance, which will be easily upset by the slightest amount of unbalance or damping. Thus, a critical speed is defined as one where the inertia forces and elastic restoring forces are in balance for an undamped system.

For the case of the bowed shaft, assume mass eccentricity q to be zero. The rotor would be in perfect balance except for a permanent bow in the elastic axis of the shaft, which could be due to thermal distortion during operation, manufacturing deviation or physical damage. Thus, at standstill the elastic center is assumed to be displaced from the rotation axis by a radial distance s, the amount of shaft bow. A steady state solution of the differential equations yields:

$$r_{wd} = \frac{s}{\{[1-(\omega/\omega_{ns})^2]^2 + (2\xi_f\omega/\omega_{ns})^2\}^{1/2}} \tag{7.2}$$

Here, too, an analogy with planar beam vibration exists. If a radial sinusoidal force is applied to the stationary disk the resulting planar beam vibration is governed by the same equations that apply to the whirling motion of the bowed rotor. However, the analogy is limited by factors such as gyroscopic stiffening associated with the mass moment of inertia of the rotor, internal damping in the rotor and certain system parameters (as in fluid film bearings) which are dependent on shaft rotation.

When the shaft is supported on flexible bearing supports both the stator and the rotor are treated as flexible elements. Damping in the system, cb, will be associated with the bearing support only. If the rotor is infinitely stiff, the undamped critical speed is given by $\omega_{nb} = \sqrt{(K_s/m)}$, K_s is support stiffness and the new damping ratio $\xi_b = c_b/(2m\omega_{nb})$. If ω_{nb} is substituted for ω_{ns} and ξ_b for ξ_f, then (7.1) and (7.2) for the flexible rotor on rigid bearing supports may be used to represent the response to unbalance of the rigid rotor on flexible supports.

A steady state solution of the differential equations of motion for the full system with a flexible rotor and flexible supports yield the following response at the bearing and at the disk to unbalance:

$$r_{wb} = \frac{q(\omega/\omega_{nb})^2}{[c_1^2 + (c_3\, c_2)^2]^{1/2}} \tag{7.3}$$

$$\phi_{wb} = \arctan\left(\frac{-c_3\, c_2}{c_1}\right) \tag{7.4}$$

$$r_{wd} = r_{wb}\,[(1+\kappa)^2 + (c_3\,\kappa)^2]^{1/2} \tag{7.5}$$

$$\phi_{wd} = \arctan\left[\frac{-c_3}{(1+\kappa)c_1 + c_3^2\kappa c_2}\right] \tag{7.6}$$

where $c_1 = 1 - (1+\kappa)(\omega/\omega_{nb})^2$

$c_2 = 1 - \kappa(\omega/\omega_{nb})^2$

$c_3 = 2(\omega/\omega_{nb})\,\xi_b$

κ = stiffness ratio of support and rotor = K_s/k

The phase angle is the relative angle between the displacement vector and the unbalance vector q. The speed at which the whirl amplitude peaks is referred to as the damped, or actual, critical speed. A common design practice is to plot the response at the disk and at the bearing as functions of the ratio of the actual speed to the undamped critical speed. Fig. 7.2 shows a plot for a given set of parameter values. It is of special interest to obtain values of maximum response and the corresponding speed at which it occurs. Two significant observations can be made here. For a given damping level shaft response increases substantially with an increase in the stiffness ratio κ. Thus, as the bearing support stiffness increases (or shaft stiffness decreases), the rotor becomes more responsive to unbalance. For a given stiffness ratio, shaft response minimizes at some intermediate value of damping. As damping is reduced to negligible levels, the shaft response reaches infinity. Also, as damping becomes very large, shaft response becomes very large because of rigidity in the bearings.

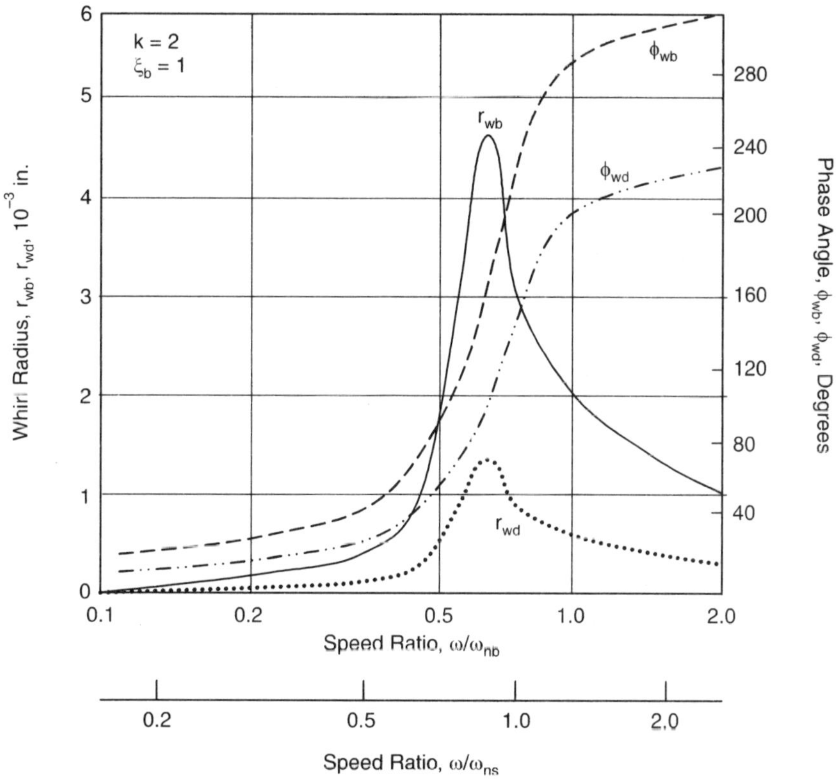

Fig. 7.2. Shaft Supported in Flexible Bearings-Response to Unbalance

7.3 SUPPORT FLEXIBILITY AND CRITICAL SPEEDS

In the discussion so far the effect of mass distribution along the length of the rotor is absent, and so its usefulness for long rotors is limited. In addition to the fundamental, or first, critical speed, with distributed mass there will be an infinite number of critical speeds. Each critical speed has a characteristic mode shape. To understand this, consider a rotor of uniform cross section. Linn and Prohl calculated the effect of bearing flexibility for the first four critical speeds of a uniform undamped rotor. The solution is derived from the differential equation of the elastic curve of the whirling shaft. The first critical speed on rigid bearings is given by:

$$\omega_1 = \pi^2 \left(\frac{gEI}{WL^3} \right)^{1/2} \tag{7.7}$$

where g = acceleration due to gravity

E = shaft material modulus of elasticity

I = moment of inertia of shaft cross section

W = weight of rotor

L = Length of rotor.

Fig. 7.3 shows the impact of bearing flexibility on the first four critical speeds of a uniform rotor. A parameter representing support flexibility is plotted along the abscissa and the ratio of critical speed to first critical speed on rigid bearings along the ordinate. For a given value of rigid bearing first critical speed, the first and second critical speeds continuously decrease with increasing bearing flexibility. This is accompanied by a gradual transformation of the mode shapes for these critical speeds. The numerical value between the first two critical speeds is equal to $\sqrt{3}$ for large values of bearing flexibility, the ratio obtained from theoretical considerations for rigid body modes. When bearing flexibility approaches infinity, these rigid body critical speeds approach zero. The third and fourth critical speed ratios initially decline rapidly with increasing bearing flexibility, then level out asymptotically to definite values. The third and fourth modes are referred to as free-free modes. This arises from the fact that conditions in the bearings at both ends have little impact on them.

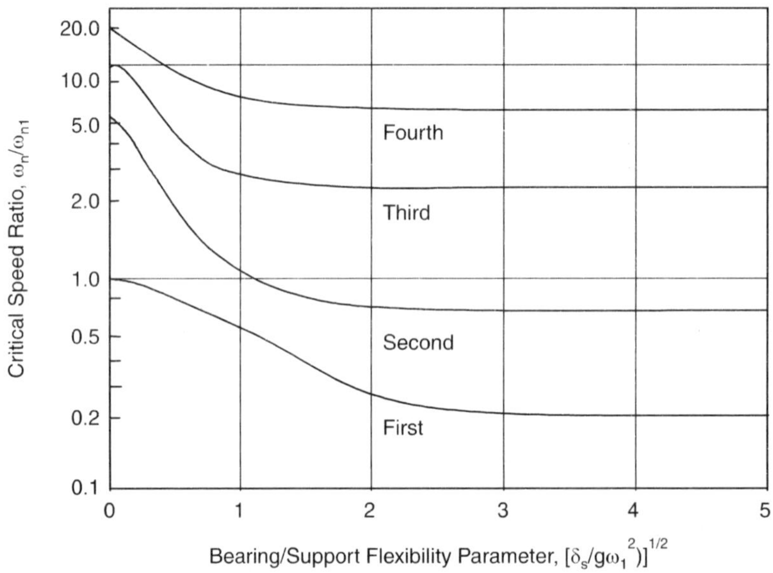

Fig. 7.3. Critical Speed as a Function of Bearing Flexibility - Uniform Rotor

Location of the bearings along the length of a rotor will have a major effect on shaft response to unbalance. For instance, if the bearings were placed closer to each other and away from the ends to the nodal points of the first mode of the free-free rotor, as shown in Fig. 7.4, supporting structure damping will not be effective. Excessive response to unbalance will be experienced when operating near the first critical speed.

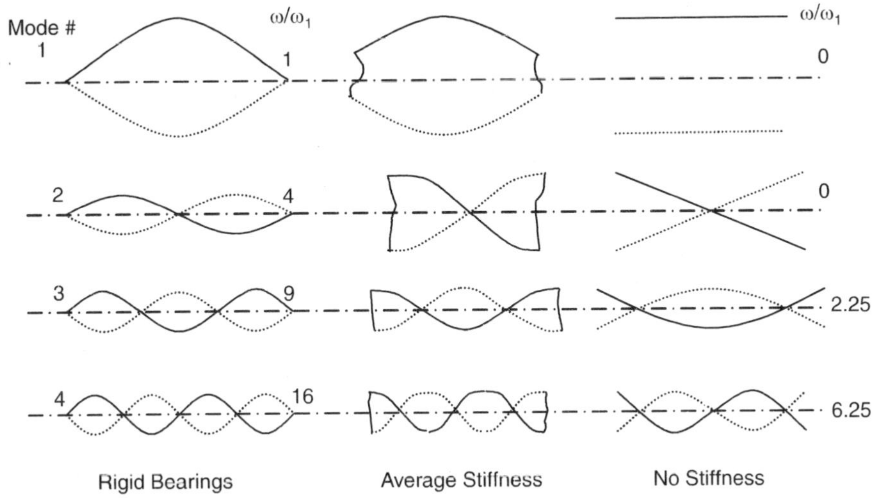

Fig. 7.4. Mode Shapes of a Uniform Rotor as Function of Bearing Stiffness

7.4 CRITICAL SPEED POSITIONING

Positioning of critical speeds in the operating speed range of a machinery plays a major role in the design of rotors. A simple check of the undamped natural frequencies of vibration using proven support stiffness assumptions suffices in many cases. New machines under development and new applications call for a more detailed evaluation using appropriate computational methods. Evaluation of damped response to rotor unbalance is of special importance. Rotor instability as a source of excitation should also be explored.

Design of the rotor and bearing system must be carefully examined for forced synchronous vibrations resulting from residual unbalance. Residual unbalance may be due to a lack of coincidence of axis of the mass centers with the elastic axis of the shaft. It could also be the result of a bow in the elastic axis, or due to a combination of the two factors. Rotor bow often results from a temporary lack of symmetry in the temperature distribution in the rotor. Gas turbine rotors used in power generating plants have been known to end up with a permanent bow at startup or shutdown times if the machine is not permitted to soak the temperatures at the proper speed level and time period. The soak cycle is required since the thickness of the shaft and length between bearing center lines is large. A similar effect is encountered if splines or couplings are not properly fabricated and assembled. Although differences exist in the response curves away from critical speeds for the two types of unbalance, the importance of positioning of the critical speed in the operating speed range of the machine cannot be over stated.

The discussion will be illustrated for a single shaft two bearing gas turbine shown in Fig. 7.5. All major components of the system will be assumed to be axially symmetric. For proper positioning of critical speeds in the operating speed range of the machine, stiffness and inertia characteristics of both the rotor and the bearings and its supports at the right axial location must be known. The machine stator is generally more rigid than the rotor, and is not included.

The modes of vibration of the system may be divided into two parts, one where the rotor remains mostly rigid, the other where the rotor undergoes flexure and whirls about the unbent axis.

There are two rigid body modes of the machine. In the first or fundamental mode the rotor supports move in phase with one another, and the shaft axis describes a cylindrical motion. In the second rigid body mode at a higher frequency the supports move out of phase with one another, with a node near the midpoint of the rotor. The rotor center line describes two cones with their apex at the node. Usually these frequencies occur well below the frequency of the shaft flexural mode. If the stator mass is very large compared to the rotor mass, low rigid body mode frequencies will prevail, even with stiff mounts. When the rotor is operated at a speed at which these rigid body modes are excited, the rotor and stator will move in phase with little relative motion between them, assuming that the bearings are adequately stiff. With limited motion between rotor and stator, internal structural integrity of the machine will not be affected. There may, however, be undesirable transmission of vibrations to the surroundings.

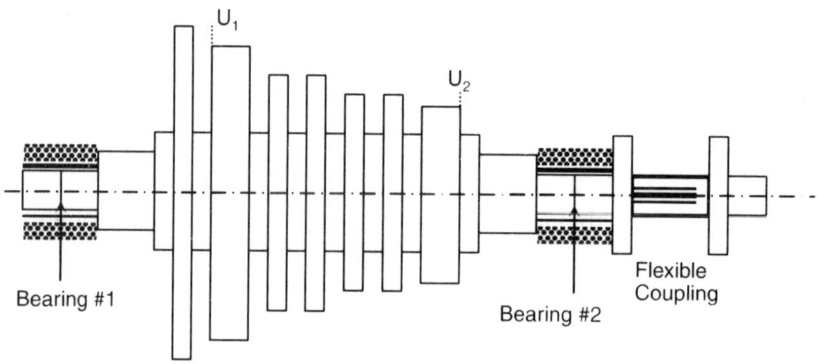

Fig. 7.5. Gas Turbine Rotor

From design considerations the rotor's flexural modes are of greater interest, and sometimes of concern. At each flexural mode the rotor whirls with a characteristic deflection shape, and the rotor and stator can vibrate out of phase. Thus, the potential for rotor rubbing against the stator exists when machine operation occurs at, or while passing through, a rotor flexural mode. Rubbing of seals and other closely controlled clearance areas may take place, with a resultant loss in efficiency and functionality. In extreme cases major damage to the inside parts of the machine may occur, and in some cases to the supporting structure as well.

In order to define typical options available for the positioning of the rotor's flexural frequencies that are applicable to many different machine types, some examples will be discussed.

First, consider a single shaft, two bearing gas turbine where the operating speed range includes the first flexural rotor mode of vibration, of the type shown in Fig. 7.5. The rotor span between the bearings when evaluated in rigid bearings has its lowest critical speed at 4,874 rpm, and the unit is required to operate up to a maximum speed of 6,200 rpm. The rotor is supported on journal bearings attached to a stiff and massive stator. The ratio of stator mass to rotor mass is 8. Machine stator is supported on a structural steel framework that is determined to have a natural frequency of 1,500 cycles/min. Damping in the steel framework is assumed to result in peak amplification factor ($1/2\xi$) of 10, implying a value of 5% of critical damping. The rotor may be considered detached from driven machinery as far as lateral vibrations are concerned.

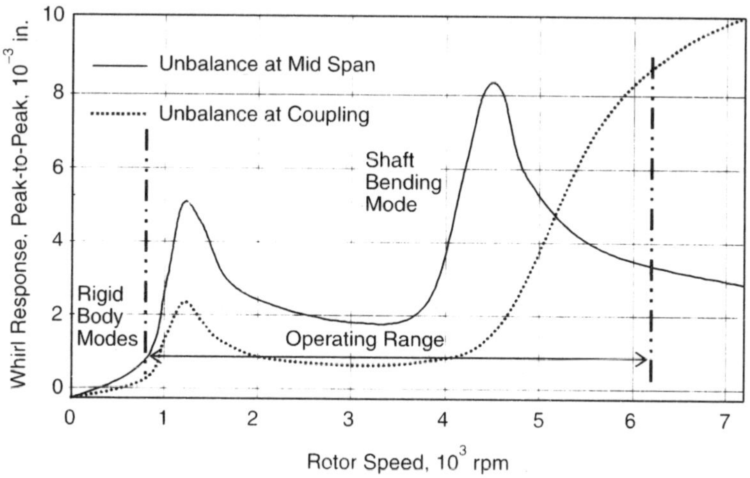

Fig. 7.6. Rotor Response from Unbalance

The first step in the design process is to calculate the undamped critical speeds, and is determined to be 3,890 rpm for the first mode and 7,580 rpm for the second mode. Of special significance is the response to rotor unbalance, especially since machine operation takes place through the first critical speed. The shape of the modes dictates that unbalance at mid span portion of the rotor will be of interest for the first mode. Unbalance located at the 1/4 and 3/4 span length from either bearing will result in high deflections in the rotor when operation of the machine is near the second mode of vibration. Unbalance or misalignment at the flexible coupling will also be sensitive in the second mode. Coupling unbalance, however, will have little effect in the first mode. If operation were required through the second critical speed on this machine, then special attention would be needed to balance the coupling. Unbalance here could cause excessive whirl at the overhung coupling, together with high dynamic loads in the bearing at the higher operating speed. Rotor response is shown in Fig. 7.6.

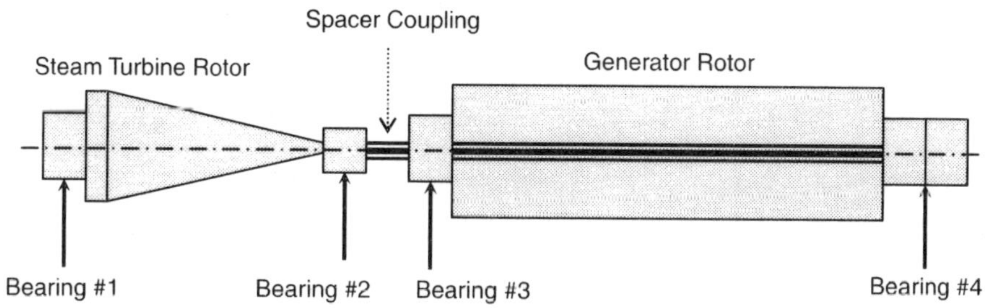

Fig. 7.7. Steam Turbine/Generator

Next, consider the case of a power generation steam turbine rotor with two spans and a total of four bearings, connected by a spacer type of coupling (Fig. 7.7). Individual spans will have their own stiffness and inertia properties, giving rise to a number of critical speeds, each critical speed characterized by a dominant amplitude in the span of the next higher stiffness. The journal bearings are supported in sturdy pedestals anchored to a massive concrete foundation. Due to the substantial characteristics of the mount, the effective spring and damping constants will be due to the journal bearing's lubricant film, explored in detail in section 7.5 on bearings.

The first three undamped critical speeds with bearing constant $K = 2.5 \times 10^6$ lbs/in are 1,550, 2,090 and 4,550 rpm. The spacing between the second and third critical speeds is large enough to accommodate the constant running speed of 3,000 rpm needed for 50 Hz power generation, or of 3,600 rpm for 60 Hz electric power. Note that most electric power equipment manufacturers prefer a margin of 10% above and below the operating speed for vibration free operation. Corresponding mode shapes are plotted in Fig. 7.8.

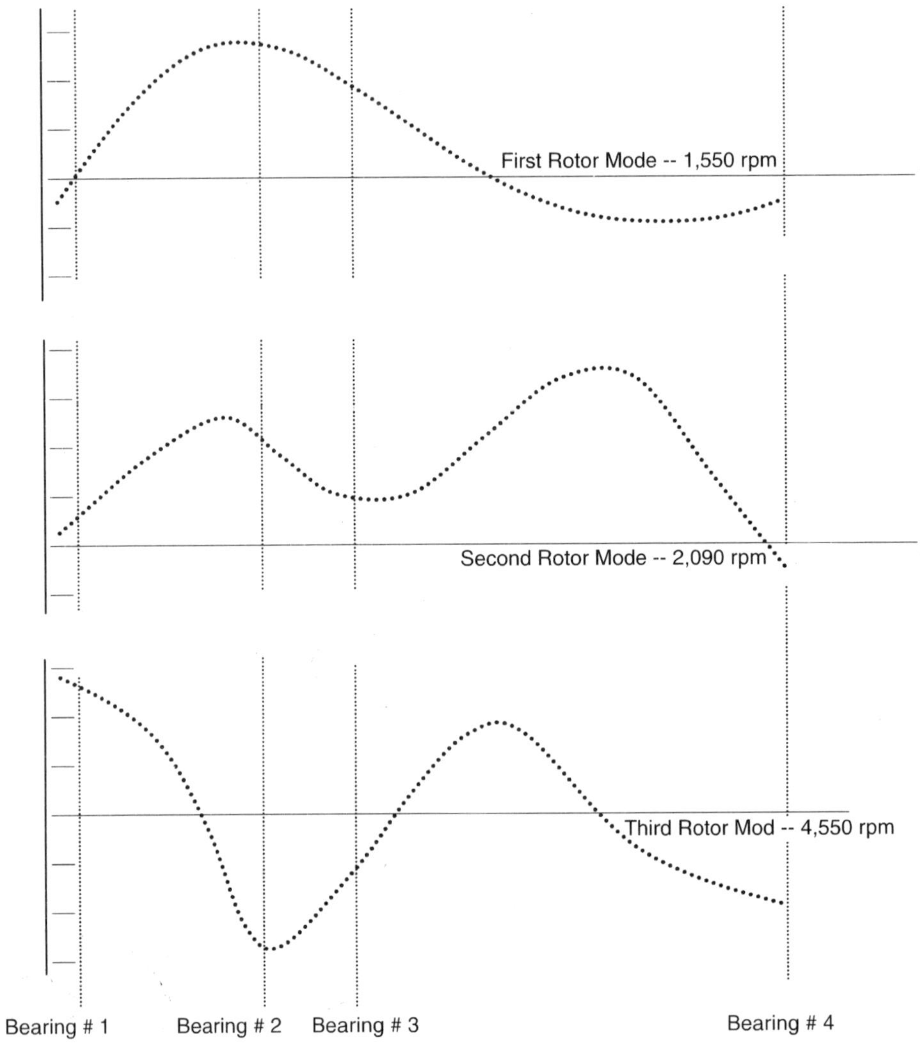

Fig. 7.8. Mode Shapes for Steam Turbine Rotor

Rotor response due to unbalance will depend on the location and the amount of unbalance. Sensitivity to unbalance is related to the fact that the generator rotor has low stiffness relative to the support stiffness provided by the journal bearings and the associated supporting structure. The second critical speed is responsive to unbalance in the turbine span. To significantly excite the third critical speed requires a moment unbalance in the generator.

7.5 BEARINGS

Journal bearings have been used for a long time in all types of rotating machines. High levels of damping in the fluid film of the bearing made possible successful operation of flexible rotors at high speed. Dynamic characteristics of the lubricant film in the form of its stiffness and damping are in a sense a newer development.

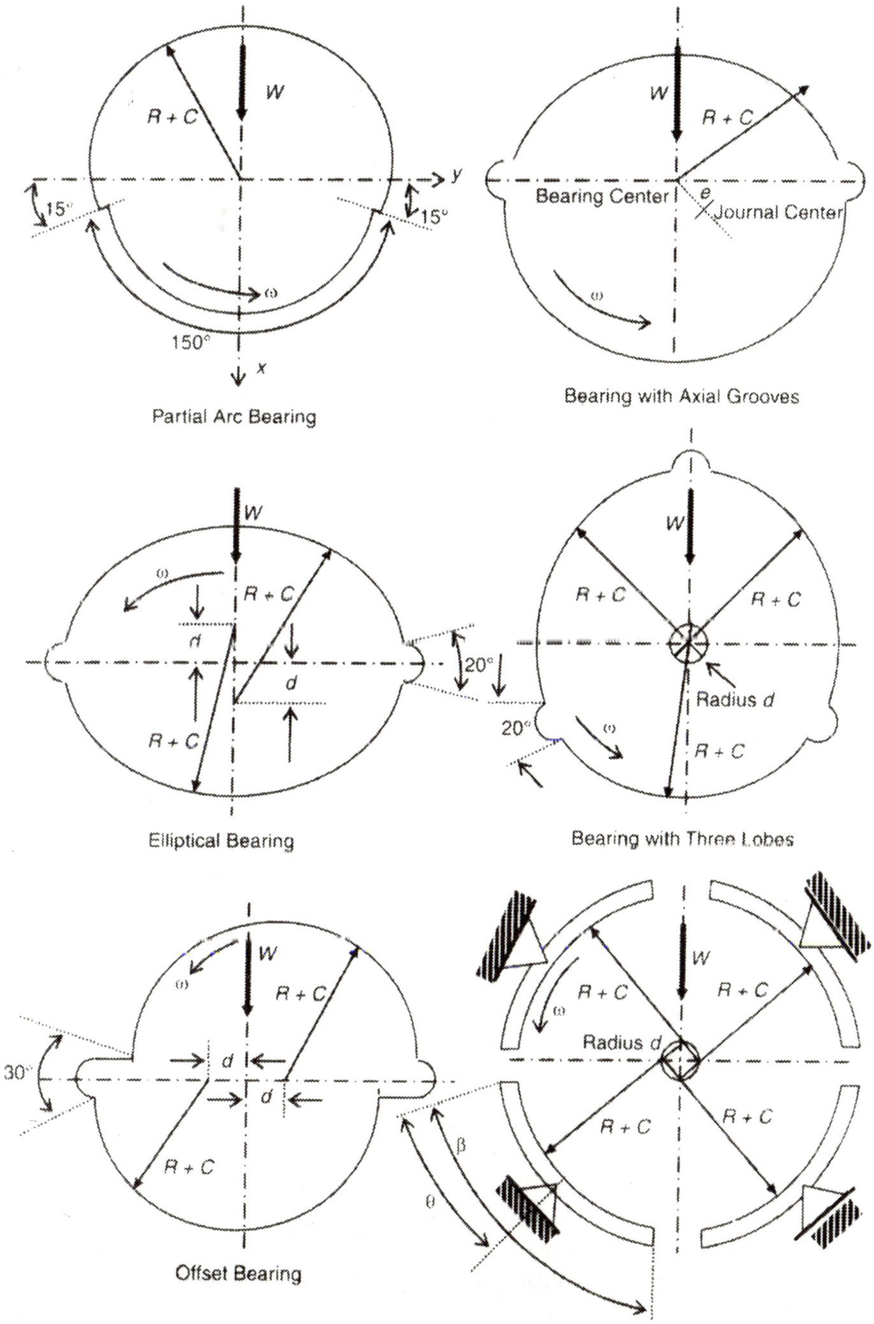

Fig. 7.9. Journal Bearing Types

Many different types of bearings are used to support a rotating shaft. Some representative types are shown in Fig. 7.9. The partial arc and the grooved bearings are adaptations of the plain cylindrical bearings.

With a fixed geometry a cylindrical bearing tends to be an unstable form of support. The elliptical, three-lobed and offset cylindrical bearings are capable of providing some pre-load, which is a function of the ratio of diameter D and the clearance C. A value of 0.5 for the pre-load is widely used. With the pre-load bearings tend to operate with a greater minimum film thickness. When the direction of the steady load on the bearing is varying, the three-lobed configuration is preferred over the offset and the elliptical types.

Bearings with fixed geometry under some conditions exhibit unstable operation. The tilting pad bearing has much better stability characteristics, and may be selected if there is a possibility for instability. Load direction may be towards the center of a pad or between two pads, and can be designed to provide proper level of pre-load.

Experimental determination of the stiffness and damping characteristics of some bearing types was first attempted by Hagg and Sankey (1986). Applying selected values of steady and rotating loads, and measuring the elliptical orbit of the journal, stiffness (K_1 and K_2) and damping constants (B_1 and B_2) were calculated relative to the major and minor axes of the ellipse. Fig. 7.10 shows these constants plotted in a dimensionless form as a function of Sommerfeld's number S for the plain cylindrical bearing. Axes 1 and 2 are independent of the horizontal and vertical directions, $L/D = 1$ and:

$$S = \left(\frac{\mu N}{P}\right)\left(\frac{R}{C}\right)^2 \tag{7.8}$$

$$P = \frac{W}{LD} \tag{7.9}$$

where C = radial clearance

R = journal radius

d = pre-load dimension

m = pre-load = d/C

W = steady load

m = absolute viscosity of lubricant

N = rotor speed

P = bearing pressure

L = axial length

D = journal diameter

This experimental approach has its drawbacks: it does not give information about the cross coupling terms, which are often a significant portion of the stiffness and dynamic characteristics of a bearing. By definition, cross stiffness and damping terms of the lubricant film relate shaft journal motion in a direction perpendicular to the direction of the force. Later evaluations were based on a numerical solution of the lubricant film's generalized equation, known as Reynold's equation. Rotor amplitude is assumed to be sufficiently small, so fluid forces may be replaced by their gradients around the steady state operating eccentricity. The forces are then proportional to the vibratory displacement and velocity, where the stiffness and damping coefficients provide the constant terms. Relative to x and y coordinates as shown in Fig. 7.9, the fluid film forces may be expressed as:

$$F_x = -K_{xx}X - B_{xx}\frac{dX}{dt} - K_{xy}Y - B_{xy}\frac{dY}{dt}$$
$$F_y = -K_{yx}X - B_{yx}\frac{dX}{dt} - K_{yy}Y - B_{yy}\frac{dY}{dt} \quad (7.10)$$

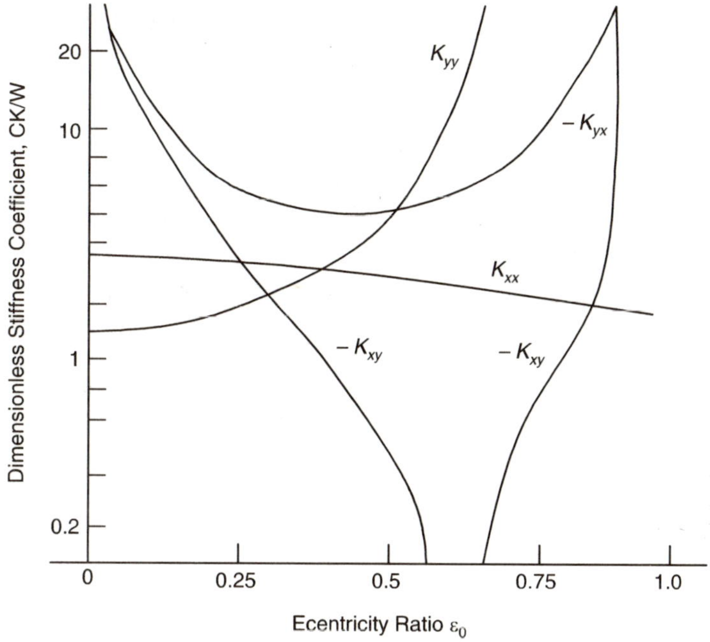

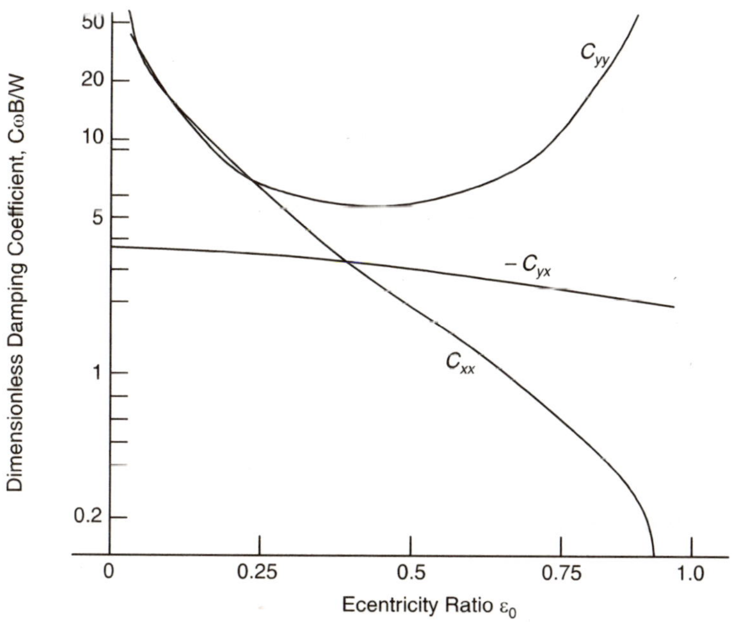

Fig. 7.10. Plain Cylindrical Bearing Stiffness and Damping Coefficients, $L/D = 1$ (*ref.* Lund et al, 1978)

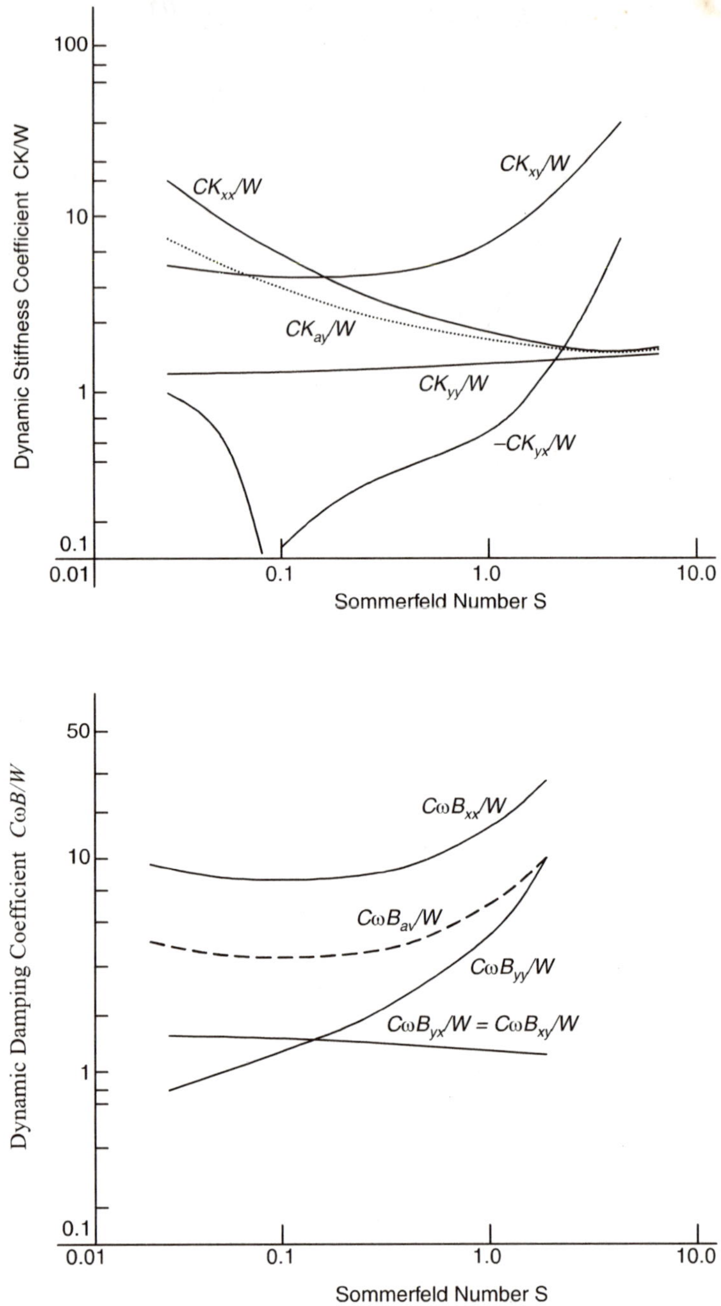

Fig. 7.11. Two Axial Groove Bearing Stiffness and Damping Coefficients, $L/D = 1$ (*ref.* Lund et al, 1978)

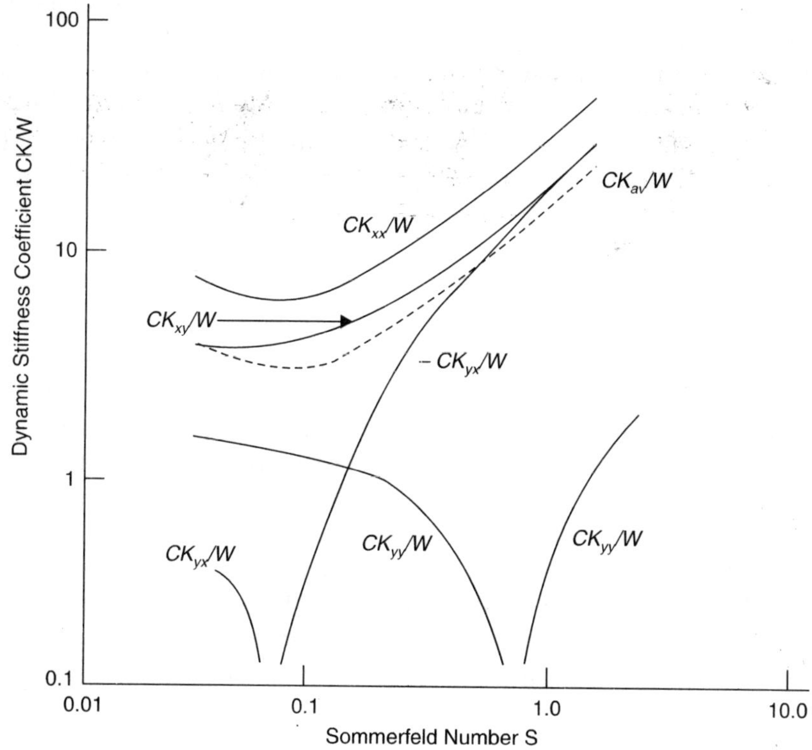

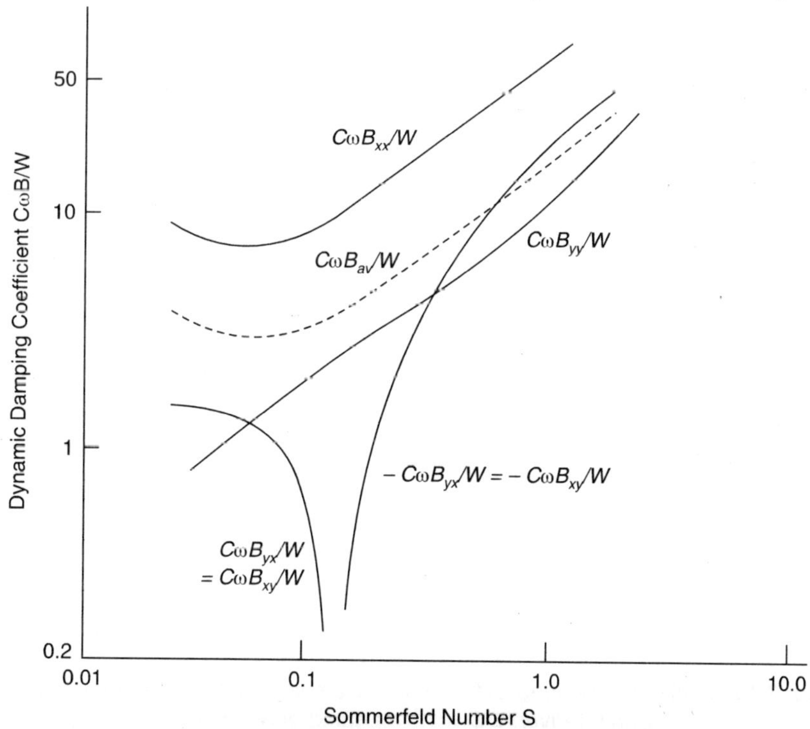

Fig. 7.12. Elliptical Bearing Stiffness and Damping Coefficients, $L/D = 1$ (*ref.* Lund et al, 1978)

Lund and Thomsen (1978) give tabulated values of all the coefficients in (7.10) as a function of Sommerfeld's number. Four of the bearing types illustrated in figure 7.9 are considered, with *L/D* ratios of 0.5 and 1.0. Figures 7.10 to 7.12 depict these coefficients as dimensionless numbers as a function of Sommerfeld's number for three geometric bearing types and *L/D* = 1. Note that some coefficients assume negative values under certain circumstances. Use of this data permits calculations for determining the effects of journal bearing characteristics on the stability of the rotor system, and also the steady state response to unbalance. The cross coupling coefficients are responsible for destabilizing effects in the fixed bearing geometry.

If the bearings are known to operate without instability in a speed range, response to unbalance can be determined more conveniently by using an axi-symmetric model for the bearings. The information thus available is often times useful for many design tasks. An average value of the stiffness and damping coefficients from the set of eight parameter values simplifies the analytical work. The assumption in this case will be that the shaft trajectory will be circular instead of elliptical, with the circle radius equal to the average of the semi-major and minor radii of the ellipse. An alternate method calls for performing two sets of calculations, one using maximum values of the stiffness and damping coefficients from the full set, the other using minimum values. The results will then provide upper and lower bounds for natural frequencies and response to unbalance. Average bearing oil film forces in the radial and tangential directions for angular velocity w and circular whirl radius *r*a are given by the following expressions:

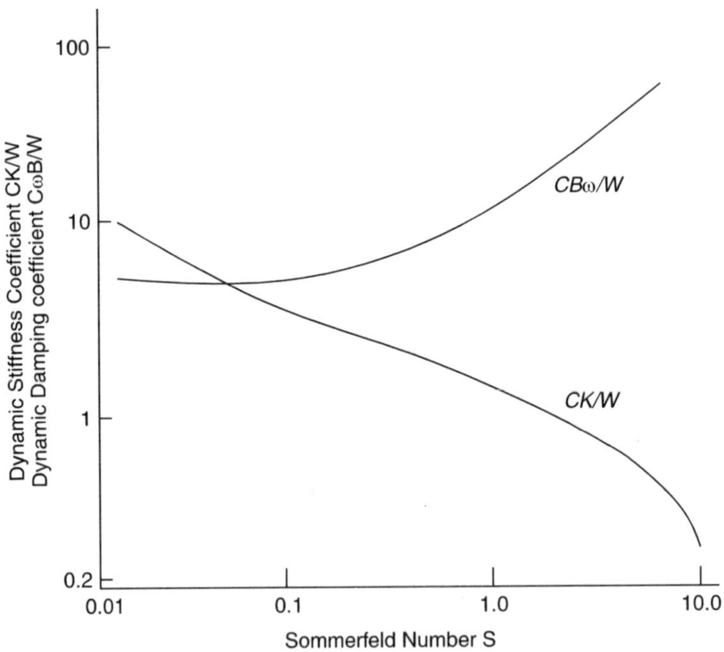

Fig. 7.13. Tilting Pad Bearing Stiffness and Damping Coefficients, Four Pads, Central Pivot, *L/D* = 1. (*ref.* Lund et al, 1978)

$$\frac{F_r}{r_a} = K_{av} = \frac{K_{xx} + K_{yy}}{2} + \frac{\omega(B_{xy} - B_{yx})}{2} \tag{7.11}$$

$$\frac{F_t}{r_a} = \text{w}B_{av} = \frac{-K_{xy} + K_{yx}}{2} + \frac{\omega(B_{xx} + B_{yy})}{2} \tag{7.12}$$

Data for tilting pad journal bearings with 4, 5, 6 and 12 pads is obtained from Lund (1964). The pivots are centrally located on the pads, with several values of L/D. Effects of pre-load and pad inertia is also included. Fig. 7.13 shows data for four pads, with zero pre-load and L/D =1. Stability in a tilting pad is excellent, which is borne out by the absence of cross-coupled terms. A four pad bearing will be symmetric in the x and y directions, hence the generally two direct stiffness and two direct damping coefficients are reduced to one stiffness and one damping terms.

If the pivot on the pad is offset by a small amount the hydrodynamic performance is improved (ref. Orcutt, 1967), but the bearing is limited to operation in one direction of rotation. The author mentions that laminar flow assumption for fluid film bearings is appropriate when using hydrocarbon oils with high kinematic viscosity. Turbulence may set in if lubricant viscosity is low (water, liquid metals) and if rotor speed is high, resulting in Reynold's number to exceed 1,500. For bearing films the expression for Reynold's number is:

$$Re = \frac{\pi DNC}{\mu/\rho} \tag{7.13}$$

where r = lubricant density and m/r = lubricant kinematic viscosity.

Rolling element bearings are applied in a number of applications, primarily due to its advantages of small size, high load carrying capacity, low friction and heat generation and ability to operate at full load and zero rotor speed. Fig. 7.14 shows examples of ball and roller bearings. Grease packing is common on some class of machines. Bearing performance improves substantially if spray lubrication in the form of a mist is used. This serves also to reject any heat developed. Major disadvantages of the rolling element bearing are rubbing or sliding contact between the rolling elements, races and cages, with life limiting consequences. Rolling elements also have inherently low damping features, and if the rotor dynamic aspects of the machine's system require it, some damping elements must be built into the system.

The interaction between rolling element bearings and the rotor is special when looking at the dynamics of the system. Significant factors are:

- Due to lack of damping, systems operating through a critical speed will require attenuation of critical speed response amplitudes. To accommodate this need, squeeze film dampers or dampers made of a resilient material are typically used.
- Rolling element bearings are free of destabilizing forces common in hydrodynamic bearings, such as oil induced whip.
- Differential thermal expansion among the moving and non-moving components of the bearing may result in compression and premature fatigue. To avoid this the bearings must be designed with radial clearance between the elements and the races. But if radial load is reduced during certain operating cycles, bearing life is compromised due to skidding of the elements. Effective softening of the bearing support stiffness may also develop due to larger than normal clearances, with a consequent reduction in natural frequencies and critical speeds. Other disadvantages as a result of larger clearances (and bi-linear stiffness of the bearing) are reported to cause anomalies in vibration response to unbalance, such as peaks at whole number multiples of the

critical speed and hysteresis (Ehrich and O'Connor, 1967). One possibility to overcome the problem is to machine lobes in one of the races, so at least some parts of the bearing circumference provide roller contact with no clearance. At the same time the race in these close contact areas elastically deforms to permit the rolling elements to move along without applying a large compressive load.

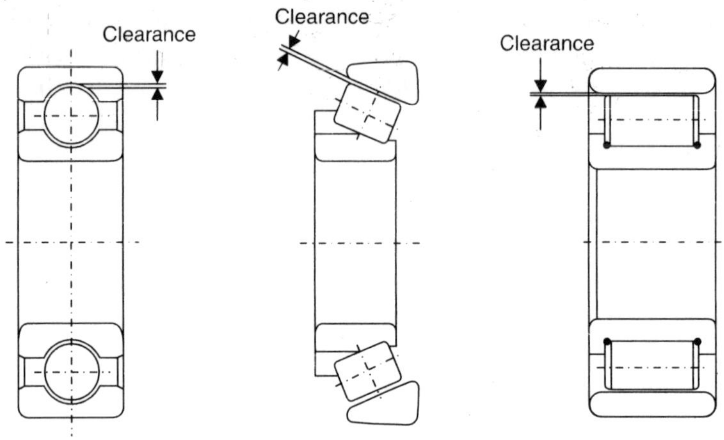

Fig. 7.14. Ball and Roller Bearings
(*Courtesy: SKF Bearings*)

As in hydrodynamic sleeve bearings, stiffness of the rolling element bearing is required for rotor dynamic calculations. Stiffness level plays a major role when dealing with a stiff rotor. Due to its geometric complexity and the number of parts involved, its stiffness is not easily calculated. Deformation is due to three factors: deflection through

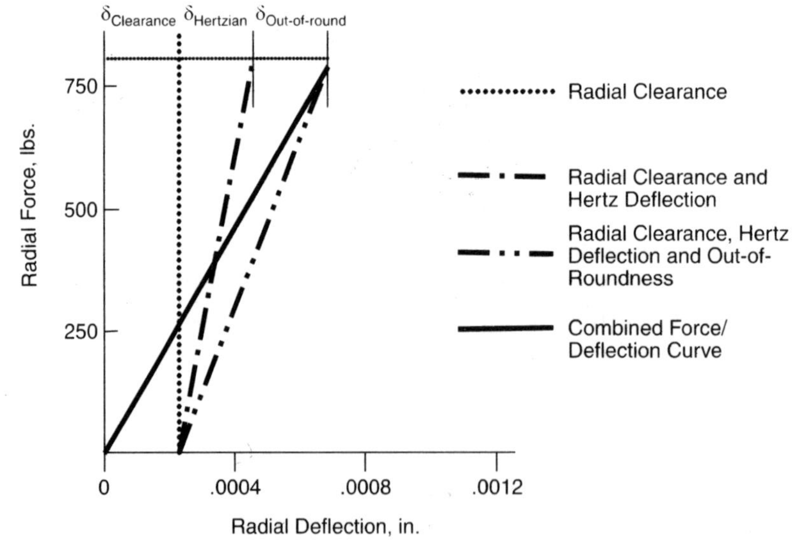

Fig. 7.15. Rolling Element Bearing Stiffness Calculations

the radial clearance, elastic compression in the rollers and deformation of bearing races from a circular to an oval shape. Compression of rollers at the points of contact with the races may be determined from equations developed by Hertz (see example problem # 7.5). The change in the profile produces a change in load distribution used for roller deformation, so an iterative procedure is required for the two combined effects. Because of the presence of the radial clearance, the overall stiffness curve is not linear. Fig. 7.15 shows a linear approximation of the overall characteristic, with slope of the force/deflection being 1.0×10^6 lbs./in. If in the design process it is determined that correlation between calculated and measured rotor critical speeds is not obtained, an adjustment to this stiffness value may be required.

- Unlike journal bearings, rotating elements of the bearing are fixed to the rotor. The rollers and the cage rotate at approximately half the rotor speed as an assembly. As a result problems may occur during balancing and in calculating non-synchronous response due to slight variations in component dimensions, but for the most part cause minor problems.

Magnetic bearings are used in high speed rotors to avoid stability problems of journal bearings and life limitation problems of rolling element bearings. The magnetic bearing levitates the rotor through magnetic forces set up by electromagnets that are opposed. Magnetic bearings use displacement measurements between the rotor and the magnet to actively control forces acting on the rotor. Forces proportional to relative displacement yield effective bearing stiffness, and forces proportional to relative velocity yield damping. Thus, both stiffness and damping are adjustable. There are no cross coupled destabilizing stiffness terms in magnetic bearings, so the bearing is stable. But the bearing can become unstable if looked at as a classical linear sampled data feedback control system. Magnetic bearings have a lower stiffness value than journal bearings with oil film, and its use markedly increases rotor length and diameter at the bearing location.

Optimized stiffness and damping coefficients for magnetic bearings have been developed by Black and Barett, et al. Size limitations, however, have prevented its wider usage. Nevertheless, its ability to overcome destabilizing forces offers some scope for special applications.

7.6 SEALS, DAMPERS AND COUPLINGS

To reduce rotor response due to unbalance at a critical frequency accepted practice calls for putting in a damper device. Squeeze film dampers installed at bearing supports achieve this objective in unstable engines and compressors. The outer race of a rolling element bearing is mounted in the inner element of the damper. A fluid film fills the cavity between the inner and outer damper parts. The inner damper element does not rotate, so there is no fluid rotation to develop destabilizing cross coupling stiffness terms. Dampers develop radial stiffness since the inner element has an orbital motion as it moves with the rotor center. A spring type of device may be used to obtain concentricity of the shaft with the damper, and to improve on stable operation. The spring device is not needed if the damper is not located in a bearing support. Oil is supplied at the center of the damper through holes or grooves. Leakage of oil at the ends is controlled by O rings or piston rings. Stiffness and damping coefficients of the squeeze film damper can be obtained from the generalized Reynold's equation theory by using either the long or the short journal bearing assumption, whereby the tangential or axial variation of oil pressure may be ignored to facilitate closed form solution. If the end seals perform their function flawlessly, the long journal bearing theory is used. For unrestricted flow of damper oil at the end, the stiffness and damper characteristics are determined by using the short journal bearing theory. Marmol and Vance provide useful information on the design and analysis of the damper using Reynold's equation.

Due to the number of variables participating in the squeeze film damper, as also the number of parts, the design process needs careful attention. Experimental verification needs to back up the theoretical concepts in the design at every stage of its development. Practical considerations will no doubt play a major role, but location of the damper in terms of its usefulness within the machine is significant. The effectiveness of the damper increases when placed at a location where relative radial deflection between the rotating and static parts is going to be highest. If the damper is placed at a nodal location of a mode to be suppressed, the effectiveness of the damper will be diminished.

Although the squeeze film has been used very successfully in a number of turbo-machines for vibration control, particularly in aircraft jet engines, it may be expensive for many applications. A convenient alternative in this respect is a damper made of a resilient material, such as an elastomeric compound. The material generally has a non-linear force/deflection property, compressing at a higher pace for smaller loads and at a reduced level as the impressed force increases. Thus, the stiffness is low for small applied loads, and increases exponentially as the load increases. The material, properly dimensioned, fills in the annular space between the bearing's outer raceway and the inner surface of the bearing housing. Elevated temperature, contamination from oil or other machine fluids and repeated load cycles have a tendency to reduce its effectiveness, hence it may require periodic replacement.

Seals are used in compressors, pumps and other machinery to isolate the process fluid from the bearing lubricant. It is also used to prevent leakage to the environment. Fig.7.16 depicts a floating seal construction. Seal segments consist of spring loaded inner and outer halves. Spring pre-load causes the ground and finished external faces of the segments to be in contact with its housing. Rotation of the segments is prevented by a shear pin. Oil is supplied between the segments, then flows axially along the shaft. Most of the oil is recovered, but some of it may be lost to the process fluid stream. The seal is of interest from rotor dynamic considerations because the oil in the annulus provides a form of support for the shaft. But the seals lack load-carrying capability. Fluid forces developed from the relative position and motion in the seal is reacted by the friction at the contact face. Spiral-grooved face on the seal provides a refinement for primary gas sealing in applications calling for high pressures (ref. Sedy, 1979 and 1980). The grooves create a separating force between the two mating faces, causing the seal to operate at tight clearances between the rotating and stationary faces, resulting in a low leakage rate.

Labyrinth seals are used in several ways. In a multi-stage centrifugal compressor, for example, the shaft seal labyrinth may be used to restrict leakage flow along the shaft to the back side of the preceding impeller. To alleviate axial force due to pressure difference between the discharge and inlet pressure on an in-line machine, a balance piston is used to exert force in the other direction, and the balance drum labyrinth controls leakage to the outside. Generally, the fluid forces do not interfere with the rotor's natural frequency or with the damping system. However, high fluid swirl activity due to lack of concentricity between the rotating and stationary parts has been known to create substantial destabilizing forces.

Couplings are used to mechanically attach sub-assemblies or sub-systems of rotors. Connections with splines that include a combination of spline teeth and pilot diameters provide the effect of a solid shaft. The design requires a high degree of precision in the manufacture of the components and during assembly to ensure proper alignment between the components. Inadequate provision for shimming may lead to undesirable vibration of the rotor. In a statically redundant rotor support system using three or more bearings, relatively flexible shafting on one or both sides of the spline joint will relieve the support reaction forces, which potentially could excite vibration in the machine's stator.

To permit axial motion and small amounts of misalignment between mating components, spline joints sometimes use crowned teeth and crowned teeth tip. Lubrication of the spline may be necessary to reduce hysteretic whirl (Marmol and Vance). Hooke's joint falls in this class of couplings in the sense that contact is present only at two pairs of orthogonal points, while accommodating only angular misalignment. Double-hinged joints of this type do not transmit angular motion at a constant angular velocity, and have a tendency to excite torsional vibrations at two times the speed of the rotor.

The spacer form of coupling is used to permit a complete range of misalignments and axial, angular and radial eccentricity between components. For placement of torsional vibrations at particular speeds in a given application, grooves of specific depth and length are sometimes machined to obtain the proper polar cross sectional moment of inertia and corresponding torsional stiffness.

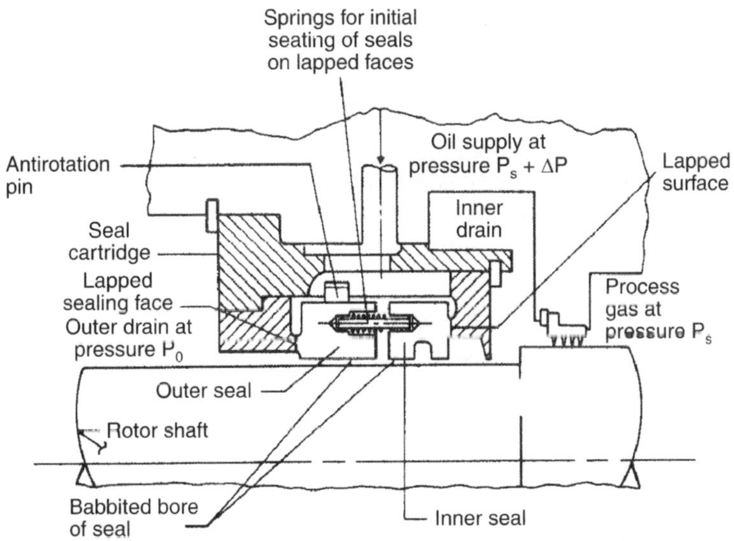

Fig. 7.16. Floating Contact Seal (*ref.* Kirk, 1983)

7.7 PROCEDURE FOR CRITICAL SPEEDS

The direct stiffness method or the transfer matrix method may be used for calculating free and forced response of a rotating system. Both procedures are developed using force-displacement and force-velocity equations, together with momentum principles. The transfer matrix method, an extension of the Holzer method for torsional vibrations, is well suited for analyzing rotating systems. Improvements to the system by Myklestad (1956) and Lund (1964) have made it suitable for computations of lateral vibrations of rotating beam structures. It does not require storage of large system arrays, and is discussed next in detail.

The method utilizes a sequential procedure, starting with boundary conditions at one end and proceeding along fixed locations in the structure to the other end. Elastic and inertia properties of the system are picked up as the calculation sequence proceeds for any shaft speed. Satisfaction of the boundary conditions at the other end for a specified speed value yields the natural frequency of vibration.

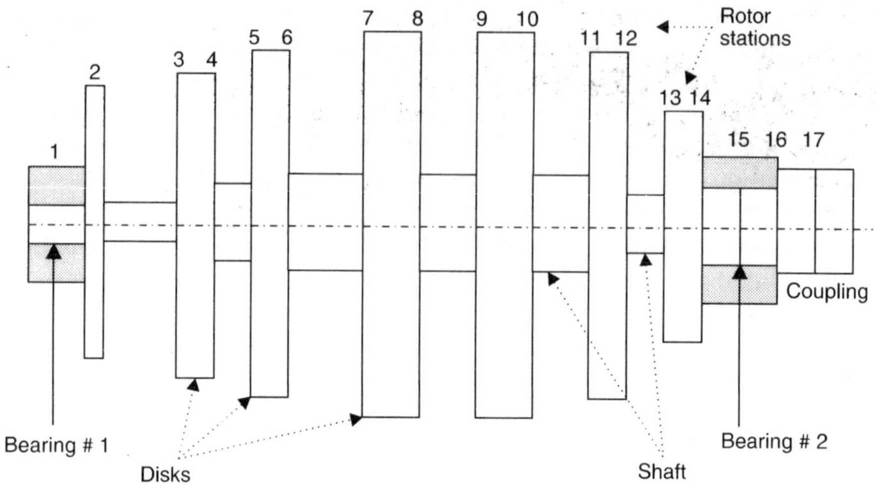

Fig. 7.17. Rotor Model for Lateral Vibrations

Fig. 7.17 illustrates a rotor model for these calculations. Several points along its length are selected to represent the geometric configuration of the system in terms of stiffness and inertia of the rotor and its supports, and are labeled from 1 to n from one end of the system to the other. The state vector at a point defines displacements and forces in the element, and is expressed as:

$$\{z\} = [u\ v\ \theta\ \phi\ M_y\ M_x\ V_y\ V_x]^T \tag{7.14}$$

The forces and displacements are as shown in Fig. 7.18. Positive internal forces act in positive coordinate directions on beams whose cross sections have a positive outward normal.

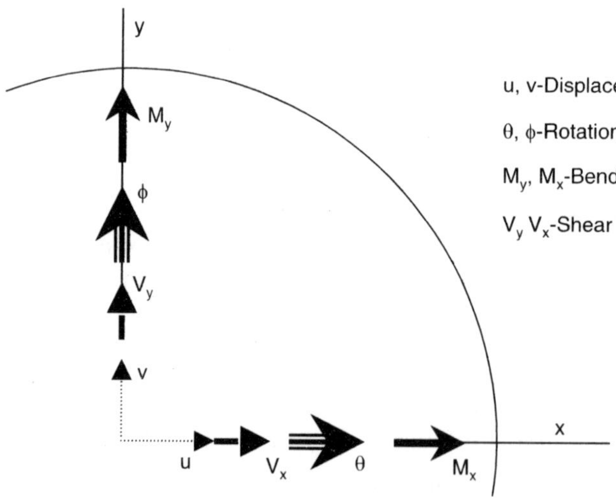

Fig. 7.18. Forces and Displacements on a Beam's Cross Section

Displacement u is along the horizontal x direction and v along the vertical y direction. Corresponding rotations are θ and ϕ. M_y and M_x represent bending moments and V_y and V_x give shear forces in the beam. The 'state' on the left hand side of a rotor station will be different from that at the right side, primarily due to the presence of inertia mass or support stiffness at the station. Superscripts R and L denote the right and left sides of the station i.

The rotor stations are connected by elastic beam sections, and are modeled to exclude their mass. Shear deformation of the beam is included, but it's internal damping is ignored. Two different transfer matrices are used. A point transfer matrix provides transfer of the state vector at a station from its left side to its right side. A typical rotor point may include a rigid disk and a bearing support, so the point transfer matrix will include the disk's inertia properties and the bearing's dynamic stiffness and damping coefficients. The equations of motion for this rigid disk with a bearing support will be:

$$\begin{bmatrix} m & 0 & 0 & 0 \\ 0 & m & 0 & 0 \\ 0 & 0 & I_d & 0 \\ 0 & 0 & 0 & I_d \end{bmatrix} \begin{Bmatrix} d^2u_i/dt^2 \\ d^2v_i/dt^2 \\ d^2\theta_i/dt^2 \\ d^2\varphi_i/dt^2 \end{Bmatrix} + \begin{bmatrix} c_{xx} & c_{xy} & 0 & 0 \\ c_{yx} & c_{yy} & 0 & 0 \\ 0 & 0 & 0 & \Omega I_a \\ 0 & 0 & -\Omega I_a & 0 \end{bmatrix} \begin{Bmatrix} du_i/dt \\ dv_i/dt \\ d\theta_i/dt \\ d\varphi_i/dt \end{Bmatrix} \tag{7.15}$$

$$+ \begin{bmatrix} k_{xx} & k_{xy} & 0 & 0 \\ k_{yx} & k_{yy} & 0 & 0 \\ 0 & 0 & 0 & 0 \\ 0 & 0 & 0 & 0 \end{bmatrix} \begin{Bmatrix} u_i \\ v_i \\ \theta_i \\ \varphi_i \end{Bmatrix} = \begin{Bmatrix} V_x^R - V_x^L \\ V_y^R - V_y^L \\ M_x^R - M_x^L \\ M_y^R - M_x^L \end{Bmatrix} + [m\,\Omega^2]\,\mathrm{Re}(e^{iWt}) \begin{Bmatrix} a_{x'} + a_{y'} \\ -a_{x'} + a_{y'} \\ 0 \\ 0 \end{Bmatrix}$$

I_d represents the mass moment of inertia about its lateral axis and I_a about the polar axis. $a_{x'}$ and $a_{y'}$ give the mass center component in the rotating reference frame. Ω represents the shaft spin velocity. The disk and bearing do not have an axial length, so displacements on the right and left hand sides of the point will be identical. Hence:

$$[u_i^L\ v_i^L\ \theta_i^L\ \phi_i^L] = [u_i^R\ v_i^R\ \theta_i^R\ \phi_i^R] \tag{7.16}$$

For free vibration the solution of (7.15) will be of the form $\{z_i\} = \{Z_i\}e^{\lambda t}$, and may be expressed as:

$$\begin{Bmatrix} u \\ v \\ \theta \\ \varphi \end{Bmatrix} = \begin{Bmatrix} \bar{u} \\ \bar{v} \\ \bar{\theta} \\ \bar{\varphi} \end{Bmatrix} \tag{7.17}$$

where $\lambda = \sigma + i\omega$ represents an eigen value, ω being the damped natural frequency and σ the damping coefficient. Substitution of (7.17) into (7.15) and rearrangement of terms provides the expression:

$$\{Z_i\}R = [P_i]\{Z_i\}L \tag{7.18}$$

where:

$$[P_i] = \begin{bmatrix} [I] & [O] \\ [B] & [I] \end{bmatrix} \tag{7.19}$$

Here $[I]$ is a 4 x 4 unity matrix, $[O]$ is a null matrix and $[B]$ is defined by:

$$[B] = \begin{bmatrix} 0 & 0 & -\Omega I_a\lambda & I_d\lambda^2 \\ 0 & 0 & I_d\lambda^2 & \Omega I_a\lambda \\ b_{yx} & b_{yy} + m\lambda^2 & 0 & 0 \\ b_{xx} + m\lambda^2 & b_{xy} & 0 & 0 \end{bmatrix}$$

The terms $b_{yx} = c_{yx}\lambda + k_{yx}$ and others similar to it provide point transfer matrix for a rigid disk or bearing point.

For a beam with uniform cross section the field transfer matrix connects two rotor stations i and $i + 1$. Elements of the transfer matrix will include shaft material Young's modulus E, shear modulus G, shear deformation factor κ, cross section area A and moment of inertia I and length of shaft segment L between the stations. Since the shaft's mass inertia is included with the disk, the equilibrium requirements in the horizontal plane are:

$$(V_x^R)_i - (V_x^L)_{i+1} = 0$$

$$(M_y^L)_{i+1} - (M_y^R)_i + (V_x^L)_{i+1} L_i = 0 \tag{7.20}$$

From Timoshenko's beam theory expression for deflections in the horizontal plane are given by:

$$u_{i+1} = u_i + \phi_i L_i + (M_y^L)_{i+1} \frac{L_i^2}{2EI_i} + (V_x^L)_{i+1} \frac{L_i^3}{3EI_i}(1 + 3\varepsilon_i) \tag{7.21}$$

$$\phi_{i+1} = \phi_i + (M_y^L)_{i+1} \frac{L_i}{EI_i} + (V_x^L)_{i+1} \frac{L_i^2}{2EI_i} \tag{7.22}$$

where $\varepsilon = EI/(\kappa GAL^2)$ and κ defines modal stiffness shape factor. Similar relations may be written for the vertical plane motion, with some changes to accommodate the sign requirements. Rearranging the terms in (7.21) and (7.22) and including requirements from (7.20) produces the expression:

$$\{Z_i + 1\}^L = [F_i]\{Z_i\}^R \tag{7.23}$$

where

$$[F_i] = \begin{bmatrix} [S] & [T] \\ [O] & [W] \end{bmatrix} \tag{7.24}$$

$$[S] = \begin{bmatrix} 1 & 0 & 0 & L \\ 0 & 1 & -L & 0 \\ 0 & 0 & 1 & 0 \\ 0 & 0 & 0 & 1 \end{bmatrix}$$

$$[T] = \begin{bmatrix} L^2/2EI & 0 & 0 & -L^3(1-6\varepsilon)/6EI \\ 0 & -L^2/2EI & -L^3(1-6\varepsilon)/6EI & 0 \\ 0 & L/EI & L^2/2EI & 0 \\ L/EI & 0 & 0 & -L^2/2EI \end{bmatrix}$$

$$[W] = \begin{bmatrix} 1 & 0 & 0 & -L \\ 0 & 1 & L & 0 \\ 0 & 0 & 1 & 0 \\ 0 & 0 & 0 & 1 \end{bmatrix}$$

and $[O]$ is a null matrix.

Equation (7.24) provides the field matrix for an elastic shaft segment. Pre-multiplication of the rotor's state vector at the left end of the segment just to the right of station i by this matrix determines the rotor's state at the right end of the segment, just to the left of station $i + 1$.

For a single shaft machine the system transfer matrix is obtained with the help of the point and field transfer matrices, which sequentially transfer the rotor's state from the left end at station 1 to the right end at station n. Sequential application of the state and field transfer matrices gives the following expressions:

$$\{Z_1\}^R = [P_1]\{Z_1\}^L$$

$$\{Z_2\}^L = [F_1]\{Z_1\}^R$$

or

$$\{Z_n\}^R = [M]\{Z_1\}^L \tag{7.25}$$

where

$$[M] = [P_n] \prod_{i=n-1}^{1} [F_i][P_i] \tag{7.26}$$

provides the overall system transfer matrix.

The procedure may be readily generalized for multiple shafts. In the two shaft system of Fig. 7.19 the shafts are denoted by A and B. The system matrix proceeds as for a single shaft, as also the sequential calculations, until it reaches the interconnecting member at the gear and pinion set between the two shafts. A transfer across the gear set requires an inter connecting transfer matrix.

For a specified shaft spin speed the system transfer matrix is generated. Boundary conditions at the left and right ends of the system are then imposed. Typical end conditions for the horizontal planes may be as follows: Free condition – $V_x = 0$, $My = 0$; pinned condition – $u = 0$, $M_y = 0$; fixed condition – $u = 0$, $\phi = 0$. Similar conditions are set up for the vertical plane. As a convenience in calculations, the system matrix may be rearranged and partitioned after the introduction of the boundary conditions so the zeros appear as a subset, as shown below:

$$\begin{Bmatrix} \{z^R\} \\ \{0\} \end{Bmatrix} = \begin{bmatrix} [M_{11}] & [M_{12}] \\ [M_{21}] & [M_{22}] \end{bmatrix} \begin{Bmatrix} \{z^L\} \\ \{0\} \end{Bmatrix} \tag{7.27}$$

Then from the lower portion of (7.27) a homogeneous expression is obtained:

$$[M_{21}(\lambda)]\{z^L\} = \{0\} \tag{7.28}$$

and a non-trivial solution for $\{z^L\}$ exists only if the coefficient determinant vanishes. Corresponding values for λ provide the system eigen values.

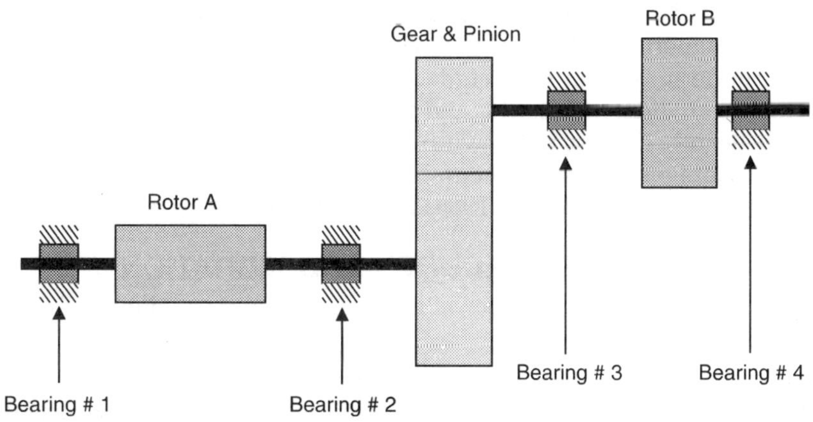

Fig. 7.19. Two-Shaft Rotating System

For an undamped rotor with isotropic supports the orbits of the natural modes of whirl are circular relative to the fixed reference frame. Because of the axial symmetry the order of the system equations may be reduced to half. The point and the field transfer matrices may then be reduced from 8 x 8 to 4 x 4 matrices.

7.8 FORCED VIBRATION RESPONSE

For systems with linear damped bearings the steady state unbalance response is determined by using the extended state vector and its associated transfer matrix. The extended state may be defined as the state of (7.14) with a unit element adjoined to form:

$$\{Z\}_{9\times1} = [U\ V\ \Theta\ \Phi\ M_y\ M_x\ V_y\ V_x\ 1]^T \tag{7.29}$$

The terms in this expression are complex amplitudes of the response, with $u = \mathrm{Re}(Ue^{i\Omega t})$, etc. The field and point transfer matrices are appropriately modified to be 9×9 matrices to include the unit element. Additionally, the point transfer matrix will also include elements for the unbalance at the proper stations. Thus, the expression for a the matrices are:

$$[P_i]_{9\times9} = \begin{bmatrix} [P_i]_{8\times8} & \{G\} \\ [0...0] & 1 \end{bmatrix} \tag{7.30}$$

where :

$$\{G\} = \begin{Bmatrix} 0 \\ . \\ 0 \\ im_i\Omega^2\,(a_{x'} + ia_{y'}) \\ -m_i\Omega^2\,(a_{x'} + ia_{y'}) \end{Bmatrix}$$

and

$$[F_i]_{9\times9} = \begin{bmatrix} [F_i]_{8\times8} & \{0\} \\ \{0\}^T & 1 \end{bmatrix}$$

For multi-shaft systems the system unbalance response is determined by successively utilizing the unbalance excitation from each shaft. Total response is then the superposition of separate shaft responses. For a specified spin speed of the shaft the system transfer matrix is determined as before, but the extended state relations are used.

The inclusion of structure support dynamics adds an extra element, which is not a straightforward use of the transfer matrix method, unless the coupling through the support structure between support points is negligible. A single pedestal mass with damping and stiffness to the ground may then be included at each support point. These masses may be treated as additional single point shafts, and the analysis may proceed as if it were a multi-shaft system. If pedestal stiffness is known from experimental or analytical methods, it may be replaced by a dynamic stiffness element. More discussion on supports for rotating shafts may be obtained from Wang and Lund (1984) and Li, et al. (1981).

7.9 ROTOR ASSEMBLY PROCEDURES AND VIBRATION

Most rotors are assembled from a number of components, and the danger for high vibrations exists from many sources.

One method of controlling relative concentricity between rotating parts is to use rabbets, sometimes with a shrink fit. Splines and curvic couplings are other methods. In all cases the assembly needs to be locked up with a single tie bolt or multiple individual bolts at some radial location. A normal force is made available at the contact surfaces, and because of friction the parts are prevented from relative motion. Internal forces, such as thermal conditions, fluid flow forces and rotor unbalance act to dislodge the mating parts from their seating. If the tension in the bolts is not large enough to overcome the unseating forces, a permanent shift in the relative position of the parts could result in a shift in unbalance.

Any relative motion at the contact surface results in destabilizing energy dissipation, which can lead to unstable vibration of the rotor. The effect is sometimes referred to as hysteretic whirl, and is discussed in more detail in section 7.12.

Axial asymmetry may be caused when keyways are used to transmit torque between components in an assembly, causing dissimilar bending stiffness about two orthogonal planes. This can result in unstable operation in a rotating machine.

7.10 ROTOR OPERATIONAL PROCEDURES

A rotor operating in a temperature varying condition is subject to unbalance. Thermal conditions can cause problems in two ways. In a rotor assembled with components made of different materials, variance in the coefficient of thermal expansion will cause unequal growth. In addition, if thermal gradients are present along the radial or axial directions in the machine, thermally induced growth or shrinkage will be even more severe.

Significant thermal distortions are known to occur in steam and gas turbines when the machine is started or when shutdown. Gravitationally forced cooling air tends to cool the lower half of the machine at a faster pace than the half above the shaft centerline. This causes substantial thermal gradients and unequal shrinkage in a rotor, particularly if the span between the bearings is large, and thermal bow is the consequence. One method to avoid this problem is to keep the rotor spinning at half the speed of full load condition for a period of time through a turning gear mechanism. If a machine is restarted during a cooling period when the machine is not at a uniform temperature, similar thermally induced problems may be expected.

The design of a rotor must include accommodation for introduction of balance weights, both during manufacturing and in the field. During production, even when individual rotating parts or sub-assemblies are dynamically balanced, the need arises to balance the assembled rotor. The final trim balance is done in several ways. At planes selected in the machine from engineering analysis balance weights are placed, which may take the form of washers under the nut. On turbines with known blade weights, the balance may be executed through exchange of blades at different tangential positions. Disks with substantial outer rims can have selected amounts of material removed at a location determined by the balance machine. This procedure has the drawback that once the material is removed, it will be final, hence adequate care must be taken before deciding on this method. Also, the ground edges should not lead to stress risers where a fatigue failure might initiate.

7.11 STABILITY CONSIDERATIONS

Rotating machines experience vibrations arising from a number of causes. Rotor unbalance is just one of them, although a more persistent one. Fluid flow forces, misalignment and maneuver loads in aircraft gas turbine engines are other important sources. There is also another class of vibration that mostly takes the form of flexural motion in the transverse direction at a rotor's natural frequency that is different from the running speed. The vibrations may be sub-synchronous, occurring at a frequency below the shaft speed, or super critical, when the frequency is above the shaft speed. A major feature of these non-synchronous modes of vibration is that the amplitude increases sharply with an increase in speed, and further acceleration may need to be aborted to avoid damage to the machine's components. Thus, such unstable vibrations are damaging and expensive, since they force delays in operation and production.

Modification in engineering design may be required to overcome the problem, since they constitute a restraint on the capabilities of the rotating machine.

The mechanisms of instability may be classified by their characteristics, and by their effects on machine operation. Whirling and whipping of the shaft, stick-slip rubs and chatter are some commonly observed phenomena.

When a shaft experiences whirling and whipping, a tangential force normal to the direction of motion is developed. The magnitude of the force is proportional to the deflection, and develops at some threshold speed when the force is large enough to offset the stabilizing external damping forces (ref: Ehrich, 1992). Fig. 7.20 shows the forces developed in a whirling shaft. In polar coordinates the tangential force may be written as:

$$2m\omega(dr/dt) + c\omega r - F_\theta = 0 \tag{7.31}$$

where a constant rate of whirl is assumed. The whirling is based on the premise that some physical feature in the machine creates an exciting force that is normal to the radial deflection of the shaft and opposes the damping force, which tends to diminish the whirling motion. Often this disturbance force is proportional to the radial deflection. As the shaft speed increases to a threshold limit, the increased shaft radial deflection causes the disturbance force to reach a point where it balances the damping force to set up the condition for the onset of instability. Irrespective of the shaft spin speed, the whirling speed at the onset of instability equals the shaft's natural or critical frequency. The direction of the whirl may be in the same direction as the shaft rotation (called the forward whirl), or opposed to it (backward whirl), depending on the direction of the destabilizing force. In the unstable condition the shaft center trajectory can be proved analytically to take the form of an exponential spiral up to a limit. Non-linear characteristics in the machine system will limit the growth of amplitude in the form of additional dissipating energy. Thus, the amplitude of whirl increases sharply with time before stabilizing at a steady state limit, and may be accompanied by damage to the machine's internal components.

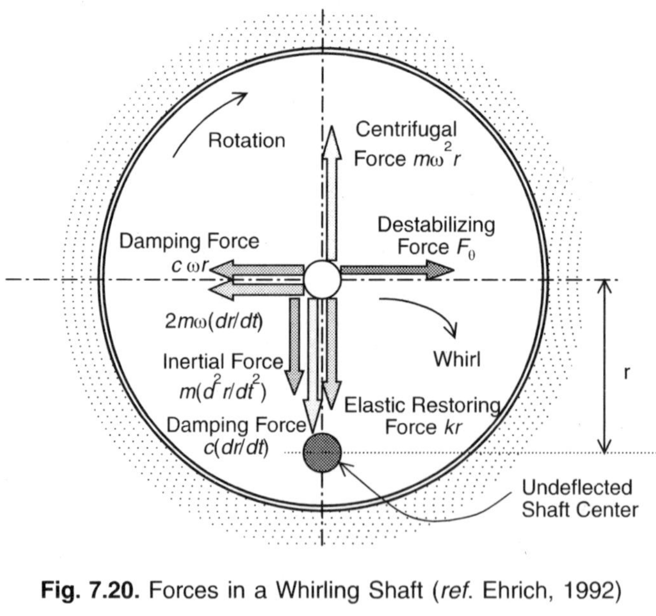

Fig. 7.20. Forces in a Whirling Shaft (*ref.* Ehrich, 1992)

Stability problems vary considerably in intensity. An operating unit may be marginally stable and operate for a long time without experiencing excessive vibrations. At many installations a fleet of identical machines may be operating satisfactorily from vibration considerations for many years, with the exception of a single machine. The machines may be operating basically on similar load cycles, handling the same fluid and work identical schedules. Under the circumstances it would be difficult to explain the dynamic experiences of the single machine with the high vibration signature. In such a unit, comparatively small changes often suffice to reduce the size of the problem. Changing the operating temperature of the fluid, going from one type of bearing to another or changing bearing pre-load may be enough to resolve the problem.

7.12 SELF EXCITED VIBRATIONS

Self excited vibrations in a system begin by themselves without any warning, initiating spontaneously. The amplitude increases until some non-linear effect limits any further increase. The excitation source for these vibrations is a uniform source of power associated with some inherent mechanism within the machine system. The destabilizing force sustaining the motion is derived from and controlled by the motion itself, and the force disappears when the motion stops.

Whirling and whipping instability arises from many sources:

- Internal rotor damping (hysteretic whirl), first reported by Newkirk (1924) and Kimball (1924), can be explained by using the force diagram of Fig. 7.21. Bending in the shaft due to some radial direction sets up elastic restoring forces in a direction parallel and opposed to the direction of deflection. Internal friction in the shaft material also causes a phase shift in the developed stress, which results in a tangential force component. This sets up the condition for the shaft to whirl. The out-of-phase force from the bending stress increases as the shaft accelerates and reaches a point when the forces from the machine's damping system is exceeded to set up the condition for the onset of instability. Besides friction arising from relative motion between the molecules of the material, loose mating components also are responsible for this phenomenon. Bolted joints, shrink fitted assemblies and rabbets are prime examples. If the bolts are not tightened to the proper torque setting, or adequate interference fit does not exist, relative motion between the components will provide cause for hysteretic whirl in the system. Studies by Gunter (1966), Bolotin (1964) and Vance and Lee (1973) have established that hysteretic whirl occurs at shaft speeds above the first critical speed. But once whirl sets in, the induced critical whirl frequency is about half the onset speed.

  Hydrodynamic journal bearings have been a major cause of unstable operation in rotating machines (Newkirk and Taylor, 1925). Referring to Fig. 7.22, a shaft rotating in a journal bearing filled with a lubricant has a nominal radial deflection. Because of the shaft's eccentricity and the average fluid velocity (about half that of the shaft surface), the lubricant develops pressure which is not symmetric about the line of radial deflection. Also, fluid pressure on the upstream side of the clearance region is higher than on the downstream side. This causes the resultant hydrodynamic force on the shaft to have a tangential component in the direction of rotation of the shaft, tending to whirl the shaft in the same direction. When this force exceeds the stabilizing damping forces, unstable operation will be the consequence. The stiffness and damping characteristics of the fluid film play a major role in determining the development of whirl. As the speed increases and the eccentricity ratio gradually becomes smaller, the cross coupling stiffness and damping terms become about equal. The speed ratio at which whirl will develop is given by $k_{r\theta}/c\omega$, where $k_{r\theta}$ is the film stiffness in the cross coupled radial and tangential directions. For plain journal bearings this ratio is about one-half. For a flexible rotor the threshold speed at which instability occurs is approximated by $\omega_n/(k_{r\theta}/c\omega)$, where ω_n is the rotor critical speed. Thus, a flexible rotor supported in plain journal bearings usually experiences sub-synchronous unstable motion when shaft speed is greater than twice the lowest rotor critical speed.

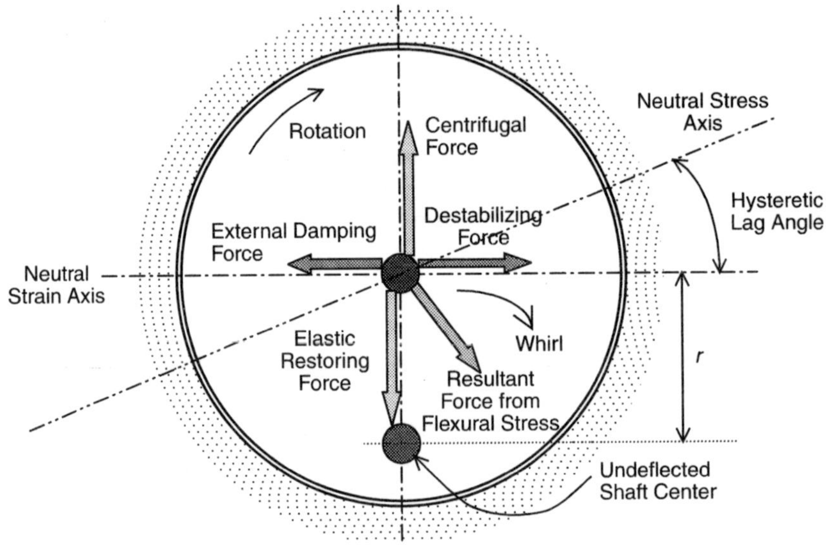

Fig. 7.21. Internal Rotor Damping Causing Hysteretic Whirl (*ref.* Ehrich, 1987)

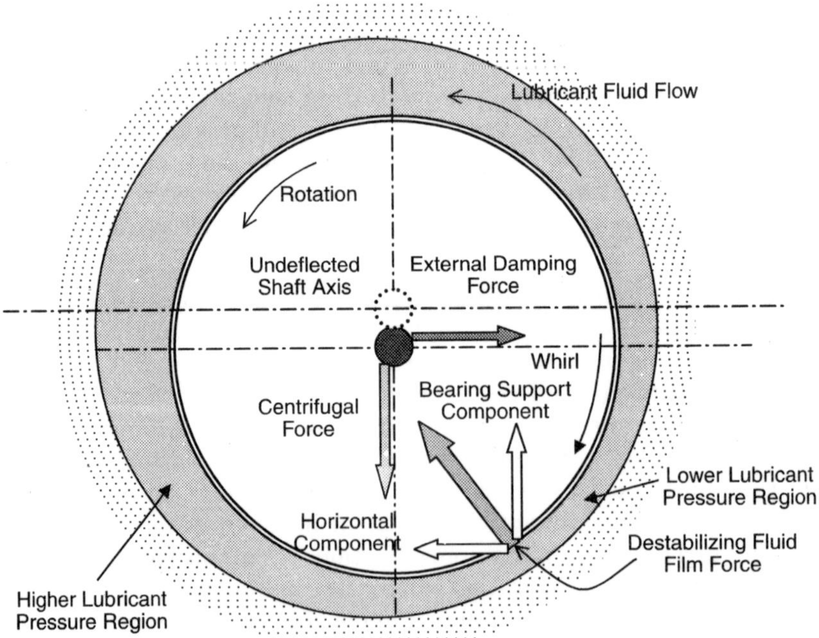

Fig. 7.22. Bearing Lubricant Film Causing Shaft Whirl (*ref.* Ehrich, 1992)

- Steam and gas turbines require close control of radial clearance between the tip of the blades and the casing. The objective behind it is to prevent leakage of steam past the rotating blades without permitting the blades to rub against the case, and thus maximize operating efficiency. In many cases, however, it is difficult to maintain concentricity between the rotor and the stator, with the result that the tip clearance is not uniform along the full circumference. This can happen when the rotor lacks adequate stiffness, or due to larger than normal shaft span between the supporting bearings, causing the rotor to sag downwards. Overhung stages past a bearing support (aircraft jet engines are an example) also tend to exaggerate this situation. From aerodynamic considerations the result of the unequal space around the circumference is to set up a tangential force at the blade tip in the direction of rotation of the rotor. As the clearance distribution becomes more pronounced due to changes in speed or power setting, tangential forces may become large enough to overcome the energy dissipating forces of the damper, with the rotor whirling as a consequence. To overcome the problem, during the design stages bearing and shaft stiffness must be adequate to avoid excessive radial deflection of the shaft. Fig. 7.23 illustrates the forces acting on the rotor as a result of uneven tip clearance. Thomas (1988) offered an explanation for the destabilizing mechanism, with the reaction force of the form:

$$K_{r\theta} = T_{\beta}/(DL) \tag{7.32}$$

where T = torque developed at the stage

D = mean blade diameter

L = blade length

β = change in thermodynamic efficiency per unit change in clearance.

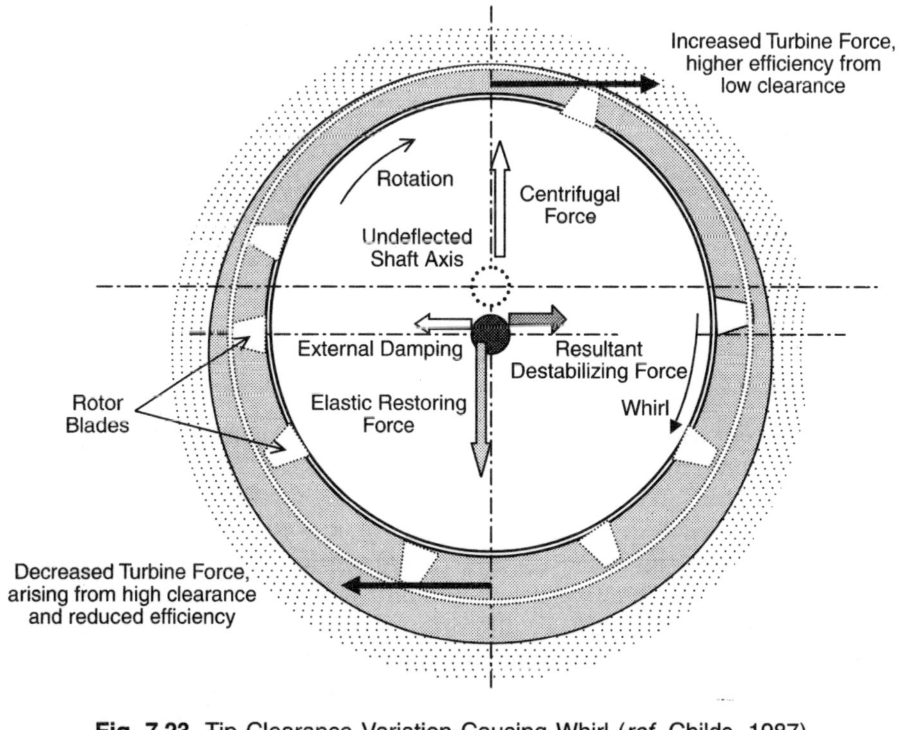

Fig. 7.23. Tip Clearance Variation Causing Whirl (*ref.* Childs, 1987)

Subsequently, Alford (1965) offered similar explanations on stability of compressor rotors. Childs (1986) has done analysis of forces developed on a pump's impeller shroud.

Many examples of unstable rotor vibrations besides those given here can be cited. Aircraft engine propeller whirl has been recognized as a direct consequence of shaft whirl. When the plane of the propeller rotates because of the slope of the shaft's center line, incremental axial velocities are induced (Taylor and Brown, 1984, Houbold and Reed, 1981). At high air speeds destabilizing forces are set up, and can overcome damping forces to cause destructive whirling of the system.

Fluid trapped in a hollow rotor can act as a high-speed centrifuge to set up destabilizing forces. Oil from bearing sumps and steam condensate may inadvertently collect and be trapped in the cavity of the rotor. When the shaft is operating with some nominal radial deflection, the fluid flies out, but does not remain in one fixed location. The surface of the shaft carries along some fluid with it in the direction of rotation. This advance sets up a force component in the tangential direction; the force is the source of instability because it induces forward whirl. The threshold speed for instability will be above the critical rotor speed.

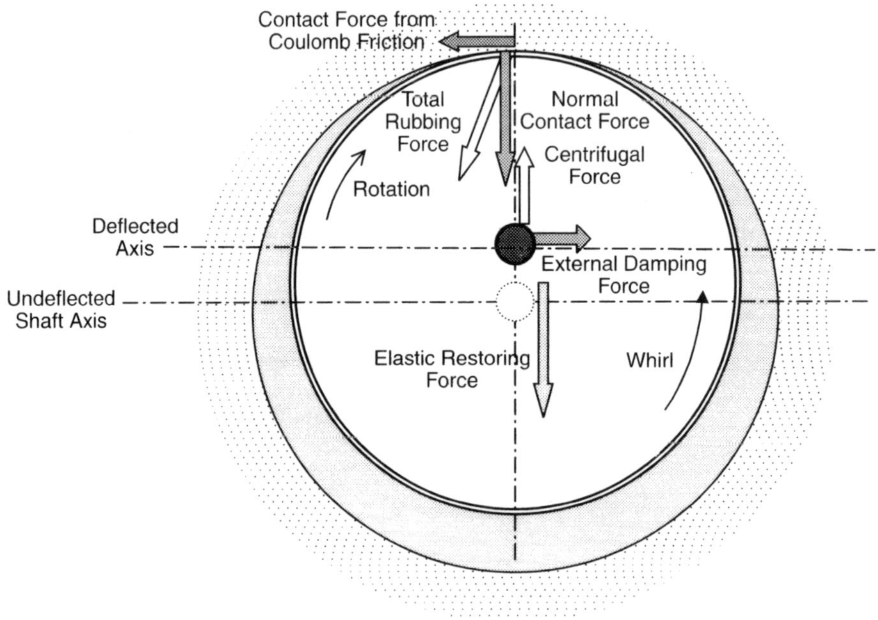

Fig. 7.24. Dry Friction Rub Causing Shaft Whirl (*ref.* Ehrich, 1987)

Rubbing of a shaft against the stator case takes place in an environment where there is no oil or other fluid lubricant, so the rubbing is essentially dry. The dry rub sets up conditions for the shaft to whirl. Dry rub can also take place in a bearing that is starved of the lubricant, or with the loss of clearance in a journal bearing, or due to a loss of radial clearance in a labyrinth seal, or between the blade tip and casing shroud of a turbo-machine (see section 4.15). The mechanics of the rub and consequent shaft whirl can be understood from Fig. 7.24. Coulomb friction due to the rub between the rotating and static parts introduces a tangential force component equal to the product of the radial force and the coefficient of friction. The force induces the shaft into a whirling motion, and the associated centrifugal force produces greater radial deflection, which in turn causes an even greater radial force and whirl producing tangential friction force. Note that the friction force will be in the opposite direction of shaft rotation, thus causing a backward whirl of the shaft. Begg (1983) indicates that the frequency of whirl is generally less than the critical speed. A similar situation is encountered in an axial thrust bearing, where due to the slope of the shaft contact takes place between the rotating and stationary disks at a pivot point. Remedies to alleviate the shaft whirl due to rubbing will range from stiffening of the shaft to reduce radial deflection to improving assembly techniques to maintain concentricity between rotor and stator. If the rub persists during operation, lubrication of the contact surfaces should alleviate the problem. If lubrication is not possible, another remedy would be to use an abradable lining material on the stator to permit disengagement between the interfering components.

Yet another source of excessive rotor vibration has been suggested by Vance. Transmission of torque in high-speed rotors is accompanied by a rotational twist in the shaft. If that twist results in an angular deflection of the torque vector, then the torque will have a small radial deflection, which is nearly proportional to the radial deflection. The radial torque vector induces bending in the shaft and a deflection that is normal to the radial deflection. When this tangential component is large enough to overcome the energy dissipating forces in the system's damping mechanism, the threshold of instability is crossed and the shaft whirls.

7.13 PERIODIC SYSTEM CHARACTERISTICS AND INSTABILITY

An analytical method to investigate the problem of dynamic instability in rotating machines is based on a system of linear differential equations with periodic coefficients, with the period in time t. A specialized form of this equation, called Mathieu's equation, can be used to represent many different physical problems in rotating machines:

$$\frac{d^2y}{dt^2} + [a - 2q\cos(2t)]y = 0 \tag{7.33}$$

where a is a constant and q is periodic in t. Treatment of the theory behind this equation is given by McLachlan (1977). Stability induced by periodic variation in a rotating machine's system parameters such as stiffness and inertia can be evaluated using this approach. In particular, asymmetric shaft cross sections and bearing characteristics, pulsating torque and fluctuating axial compressive loads affect the lateral vibration stability of a rotating shaft.

When the stiffness characteristics of a shaft or stator are significantly different in two orthogonal planes, self-excited vibrations are likely to be a problem. An asymmetric shaft rotating at a speed away from whirl speed acts as a periodic change in stiffness in the plane of its radial deflection, causing instability in specific speed ranges determined by the degree of asymmetry. Unstable operation may be expected at one-third and one-half of critical speed, and at critical speed. But with even larger asymmetries in the shaft (or the support) stiffness, generalizations may not be possible.

Eshelman and Eubanks (1980) carried out experiments to determine the effect of cyclically pulsating shaft torque and its effects on rotor stability. A varying torque was superimposed on a constant level torque in a rotating shaft. This caused the natural frequency of the shaft in the transverse direction to shift. The researchers investigated conditions under which the shaft whirled at the first critical speed, varying speed and pulsating torque. Certain combinations of the steady and pulsating torques caused instability in narrow speed bands at discrete fractions of the critical speed.

In a comparable manner, pulsating axial loads are also known to cause lateral instability in a rotating shaft. When the axial load approaches the critical buckling load for the rotating shaft, the natural frequency of vibration in the transverse direction reaches zero. Since the pulsating compressive load causes a loss in lateral stiffness in the shaft, a corresponding reduction in the frequency is the consequence, as noted by McLachlan (1977).

Dry friction was mentioned before, when the shaft rotating at a high velocity encounters instability due to contact between the rotating and static components of the machine. When the shaft is just beginning to turn breakaway friction produces higher level of friction, causing another group of instability in shafts - chatter, or stick-slip. Periodic motion of the rotor's point of contact is self-excited as a result of negative damping, because motion is taking place in the direction of rotation, and increases the relative contact velocity. Rotor motion counter to contact velocity will reduce relative velocity and increase friction force, thus again reinforcing the periodic motion. In the absence of proper flow of cooling fluid at the cutting bit, dry friction in machine tools exhibit a similar characteristic, and is known as chatter. In both cases the vibration is more torsional in nature than lateral, but vibrations in the transverse direction can also take place.

7.14 METHODS TO ALLEVIATE VIBRATION INSTABILITY

Rotating machines experience instability due to the presence of a tangential force set up by some physical characteristic related to the machine's design or operation. The tangential force arises due to a deflection of the rotor in the radial direction, thus providing a coupling between the two directions. If adequate steps are taken in the design stage of the machine, the tendency towards unstable operation will be diminished.

Stiffness of the shaft is an important consideration. Destabilizing forces generally become more pronounced when machine operation takes place above the first critical speed, since their sensitivity towards unstable operation increases with the shaft's flexibility. Thus, one possibility is to maintain the operating speed of the machine below the first critical. As the speed ratio w/wn increases, the tendency for the shaft to whirl goes up. Keeping shaft span between bearings low helps to increase the critical frequency of the rotor, and consequently the operating speed of the machine can be raised without encountering instability problems.

External damping plays a significant role in eliminating unstable operation in a rotating machine. Care must be taken to ensure that damping is external to the rotating system, so the energy dissipating forces in the damper depend on the relative motion between the rotor and the stator. Adequately designed bearing supports play a major role in this respect. In the preliminary design system parameters require adjustment in order to position the natural frequencies of whirl as required. Bearing and support stiffness plays a crucial role, together with support location and mass and stiffness distribution of the rotating components. For this purpose a plot of the natural frequencies of vibration against a particular design parameter provides useful information about the sensitivity of that parameter. Fig. 7.25 illustrates a plot of critical speeds as a function of identical support stiffness at both ends. When a number of such plots are generated for variables such as distance between bearings and weight of rotor, a comparison of the design from shaft vibration considerations is possible. Comparison can also be made with test measurements during machine operation to improve the mathematical model with the help of critical speed maps. Use of energy maps for design evaluation has also been suggested (Ehrich, 1987). Potential (or strain) energy and kinetic energy distributions in major sub-assemblies in the machine for a particular vibration mode provide useful information. Since strain energy varies with the square of displacement and kinetic energy varies with the square of velocity, the energy distribution gives important data about the level of dynamic activity in different portions of the machine system. When a system design parameter is changed, its impact on the vibration characteristics can be studied from the energy distribution.

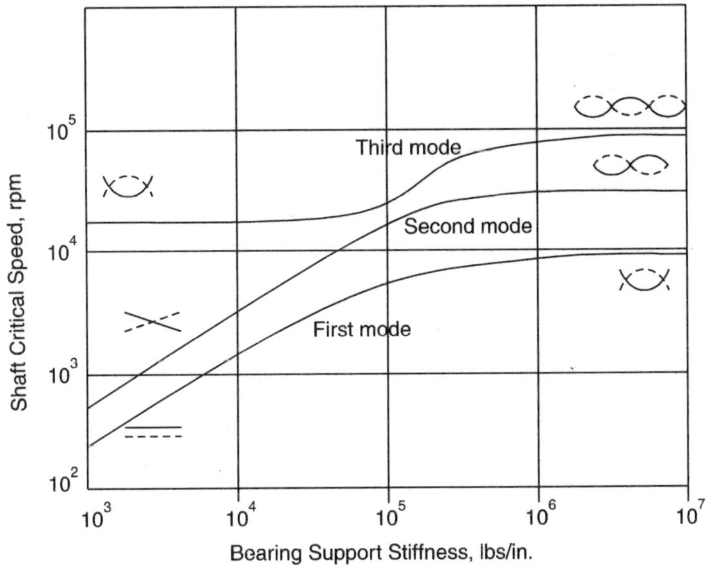

Fig. 7.25. Critical Speed Map

7.15 ROTOR-BEARING SYSTEM DESIGN EXAMPLE

Shafts are used in many different ways in all types of mechanical equipment as power shafts, line shafts and cam shafts, among others (Fig. 7.26). An axle is a stationary member used to support rotating wheels, idler gears and pulleys. A stub, or head, shaft is generally integral with an engine, motor or other prime mover, and has a projection to facilitate connection with other shafts. A jack shaft, or counter shaft, connects a prime mover with a line shaft or a machine. Flexible shafts may be used to transmit motion between two components where the rotating axes are at angle with each other.

Typically a shaft is subjected to fluctuating loads, so considerations of stress concentration and endurance strength play a major role in its design. Of equal importance, and sometimes more so, is the consideration of shaft rigidity, or stiffness. Excessive lateral deflections can cause premature wear and failure of the bearings, while also reducing the critical speed, which can cause unacceptable shaft vibrations when operating at or near this speed.

Generally, shafting of 3 to 3½ in. diameter is made of cold drawn carbon steel round bars. If toughness, shock resistance and strength are of prime concern, alloy steel bars (for example, 1347, 3140, 4150, 4340, 8650) may be more suitable. Heat treatment may also be specified to produce desirable properties. Surface wear resistance requires a carburized grade steel (1020, 1117, 4820, 8620, among others). Cost control can be obtained by using a plain carbon steel, if possible. Nitriding, cyaniding, flame hardening and other case hardening methods provide greater surface wear resistance.

Shafts larger than 3½ in. diameter call for machining from hot rolled carbon steel. Adequate stock must be removed in the machining process to eliminate decarburized scale caused by the hot rolling process. Larger diameter shafts are often times forged and machined to the required size from .45% carbon steel.

Shaft diameter of a rotor system can be selected once the distribution of bending moments and torque along its axial length is available. The diameter of each section of the shaft will depend on the stresses due to the combination of bending moment and torsion. The location of the maximum stress in an application is sometimes not difficult to tell. In general, however, a systematic approach calls for drawing shear and bending moment diagrams to locate the points of maximum moments. An example is considered next to illustrate the approach.

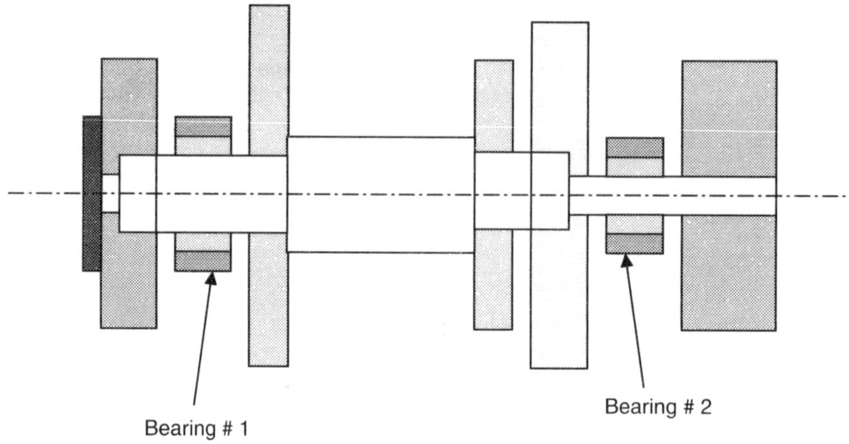

Fig. 7.26. Rotor-Bearing System Design Example

7.16 EXAMPLE PROBLEMS

Example Problem 7.1: A 24.0 in. diameter pulley *B* weighs 145 lbs and receives 33.5 HP at 325 rpm from a shaft located at an angle of 45° below the pulley (Fig. 7.27). A 19.25 in. pitch diameter gear *C* weighing 57.5 lbs. delivers 35 percent of the power horizontally at the top to gear *G*, and a 12.375 in. pitch diameter gear *E* weighing 28 lbs. delivers the remaining power to another gear *H* downward to the left at an angle of 30° below the horizontal. Both gear sets have a 20° pressure angle. The system is required to have corrosion protection for long-term effects. The shaft is supported in bearings at *A* and *D*. Determine the torque transmitted by the gears and the pulley, and obtain the bending moment distribution in the shaft.

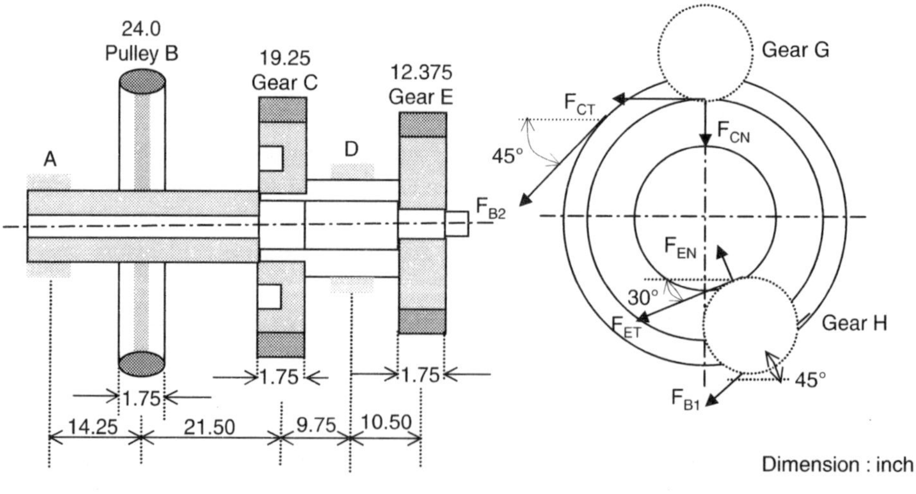

Fig. 7.27

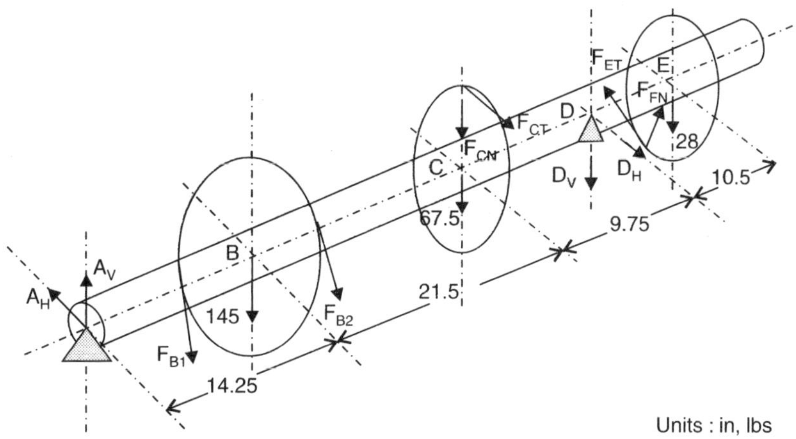

Fig. 7.28

Solution: The torque acting on the shaft between *B* and *C* is:

$$T_B = 63025(\text{hp})/\text{rpm} = 63025 \times 33.5/325 = 6496 \text{ in-lbs.}$$

$$Tc = 63025 \times .35 \times 33.5/325 = 2274 \text{ in-lbs.}$$

The remaining power is delivered by gear *E*. Hence torque between *C* and *E*:

$$T_E = 63025 \times .65 \times 33.5/325 = 4223 \text{ in-lbs.}$$

The torque on the pulley is:

$$(F1 - F2) \times D_B/2 = 6496 \text{ in-lbs.}$$

Hence, $F_1 - F_2 = 6496/12 = 541.4$ lbs.

Assume belt tension ratio $F_1/F_2 = 3$ for a flat belt, with adequate initial tension.

Then: $F_1 = 812.1$ lbs., and $F_2 = 270.7$ lbs.

Total bending force at pulley *B* is: $F_B = F_1 + F_2 = 1082.7$ lbs. Reactive tangential forces on gears *C* and *E* are:

$$F_{CH} = F_{CT} = T_C/r_C = 2274/9.625 = 236.2 \text{ lbs}$$

$$F_{ET} = T_E/r_E = 4222/6.1875 = 682.5 \text{ lbs}$$

The normal force tending to separate contact at the gear teeth in gears *C* and *G* and mating gears *E* and *H* are:

$$F_{CV1} = F_{CN} = F_{CT} \times \text{Tan}(20^\circ) = 236.3 \times 0.364 = 86.0 \text{ lbs}$$

$$F_{EN} = F_{ET} \times \text{Tan}(20^\circ) = 682.3 \times 0.364 = 248.4 \text{ lbs.}$$

Referring to Fig. 13.53, sum the forces in the horizontal and vertical planes:

$$A_H - F_B \sin(45^\circ) - F_{CT} - D_H + F_{ET}\cos(30^\circ) - F_{EN}\cos(60^\circ) = 0$$

$$A_V - F_B\cos(45^\circ) - 145 - F_{CN} - 57.5 - D_V + F_{EN}\cos(30^\circ) + F_{ET}\cos(60^\circ) - 28 = 0$$

Bearing reactions are obtained from static equilibrium conditions (sum of forces and moments is zero) in each plane. Bending moment and shear force diagrams can now be prepared for the horizontal and vertical planes, shown in Fig. 13.54.

Maximum bending moment occurs at pulley *B*. Total bending moment at *B* is:

$$= [(9748.4)^2 + (11086.5)^2]^{1/2} = 14764.6 \text{ in-lbs}$$

Note that at pulley *B* the shaft also transmits a torque load of 6496 in-lbs.

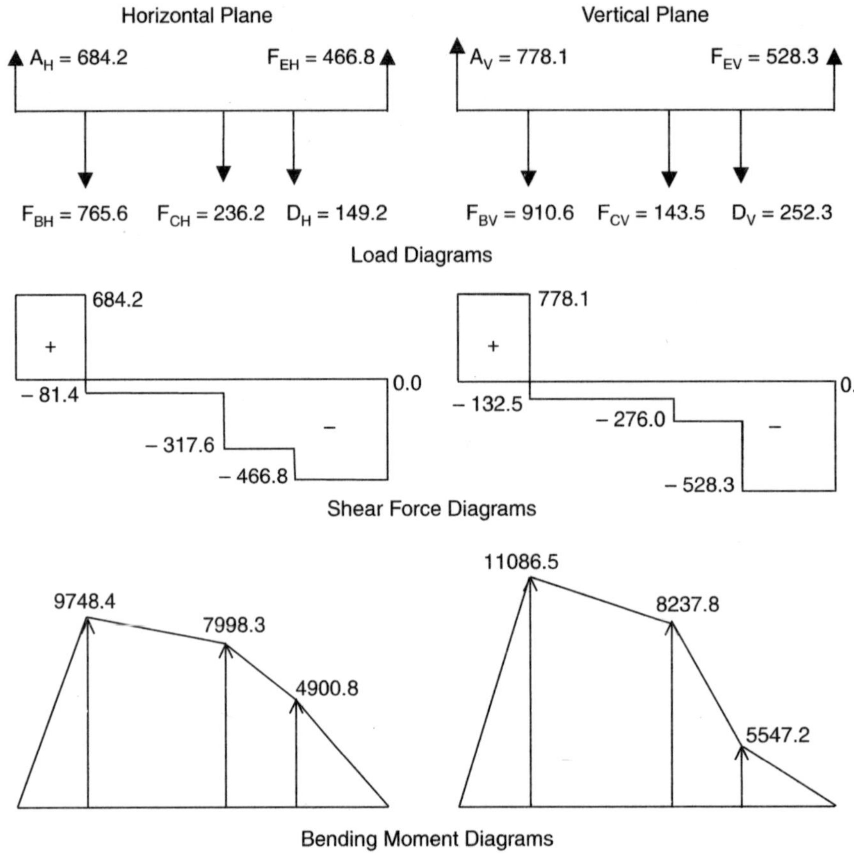

Fig. 7.29

Example Problem 7.2: In problem 7.1 assume the shaft is to be designed for maximum loads. The shaft is to be machined from C1040 steel. Determine the shaft diameters. Assume the shaft is hollow, with $D_i/D_o = 0.5$.

Solution: For a hollow shaft the normal, shear and combined stresses are:

$$\sigma_x = \frac{32\,M}{\pi D_0^{\,3}[1-(D_i/D_0)^4]} \tag{13.38}$$

$$\tau = \frac{16T}{\pi D_0^{\,3}[1-(D_i/D_0)^4]} \tag{13.39}$$

$$\tau_{max} = \frac{0.5S_{yp}}{N} = \frac{16\sqrt{M^2+T^2}}{\pi D_0^{\,3}[1-(D_i\,/\,D_0)^4]} \tag{13.40}$$

where N is a safety factor.

Normalized C1040 steel has the following material properties:

$$S_u = 85{,}000 \text{ lbs/in}^2 \qquad S_{yp} = 50{,}000 \text{ lbs/in}^2$$

Endurance limit for C1040 may be assumed to be: $S_n' = 0.5S_u = 42{,}500$ lbs/in^2. If the shaft has profiled keyways at B and C, K_f (bending) = 1.6 and K_{fs} (torsion) = 1.3. For the keyway at E assume K_f (bending) = 1.3 and K_{fs} (torsion) = 1.3. For operation in a non-corrosive environment and a survival rate of 95 percent, the corrosion factor will be assumed at C_R = 0.869, and for a machined surface C_F = 0.77. Also, since the shaft will be larger than 1/2 in. diameter and smaller than 2 in. diameter, assume C_S = 0.85. The appropriate endurance limit is then obtained from:

$$S_e = (1/K_f)C_R C_S C_F S_n' = (1/1.6) \times 0.869 \times 0.85 \times 0.77 \times 42{,}500$$

or $$Se = 15{,}108 \text{ lbs/in}^2$$

The factor of safety N is assumed to be 3.75. Since the load is to be gradually applied, $K_{sb} = K_{st}$ = 1.0. Stress concentration at the pulley is due to the profiled keyway and light press fit will be K_t = 1.6 for shafts subjected to fluctuating bending loads, and K_f = 2.5 for fretting corrosion. Thus, for the keyway and press fit combination,

$$K_f K_t M_B = 2.5 \times 1.6 \times 14764.6 = 59058 \text{ in-lbs.}$$

$$\tau_{max} = (.5 \times 15108)/3.75 = 2014.4 = [(16/\pi D_o^{\,3}) \times (M^2 + T^2)^{1/2}]/[1 - (D_i/D_o)^4]$$

Substituting the values for moment and torque at location B, then $D_o^{\,3}$ = 160.2, outer diameter D_o = 5.432 in. Select D_o = 5.50 in. and inner diameter D_i = 2.75 in. Shaft segment between A and B experiences $M_{AB} = [(9748.4)^2 + (11086.5)^2]^{1/2}$ = 14764.6 in-lbs and $(M^2 + T^2)^{1/2}$ = 16131 in-lbs. using Kf = 1.6 for corrosion. Then $D_o^{\,3}$ = 69.60, outer diameter D_o = 4.113 in. Select D_o = 4.125 in. and inner diameter D_i = 2.0625 in.

Example Problem 7.3: Shaft deflections must be calculated in order to establish minimum permissible clearance between the pulleys, gears and the assembly enclosure. Explain the numerical integration method. How is the fundamental natural frequency of vibration of the shaft related to its deflections?

Solution: Many methods are available for determining the deflection along the length of a shaft subject to a set of operating loads. The numerical integration approach is based on the following differential equations:

Load equation: $$\frac{d^4y}{dx^4} = \frac{w}{EI}$$

Shear force: $$\frac{d^3y}{dx^3} = \frac{V}{EI}$$

Bending moment: $$\frac{d^2y}{dx^2} = \frac{d\theta}{dx} = \frac{M}{EI}$$

Slope: $$\frac{dy}{dx} = \theta$$

By considering finite distances along the shaft length (x direction), the above equations can be integrated through numerical summation. As an example, $y = \Sigma\theta\Delta x$. The deflections are obtained by using the following step by step procedure, using a tabular format:

- Place a station number at each shear force location in the shaft, as also at each change in its cross-section. When long sections of the shaft with constant section are encountered, place stations at intermediate points. (Column # 1) .
- Enter station numbers on alternate lines of the table, and force on the same line as the station number. Upward forces are positive, downward forces are negative. (Column # 2).
- Calculate vertical shear at each station by summing forces in column # 2, and enter shear values in column # 3.

- Enter the axial distance of the station from the previous station on the same line as station number. (Column # 4).
- Calculate bending moment at each station by multiplying shear force with axial distance. The product when added to the previous entry in column # 5 gives bending moment at that station.
- In column # 6, on a line between stations enter area moment of inertia I at the section between the two stations.
- In column # 7 enter product EI on the same line as I (E is material Young's modulus).
- In column # 8 enter the value of (M/EI). The values are entered both at the station location as well as between stations.
- In column # 9 enter the average value of M/EI on the line in between stations.
- Calculate slope at each station by entering running total of the product of column # 9 and shaft segment length (column # 4). Enter in column # 10 on same line as station numbers.
- In column # 11 enter arithmetic average of slope from column # 10, in between stations.
- Calculate deflection increment by multiplying column # 11 with column # 4, and enter on the next lower line from average slope value. Enter in

  column # 12 on same line as average slope.
- Total all deflection increments between bearing reactions, reverse the sign and divide by distance between the bearings. This gives the integration constant. Multiply this constant by shaft spans of column # 4 and enter in column # 13 on same line as deflection increments.
- Add deflection increments of column # 12 and the integration constant in column # 13. Enter the sum as part of a running total in column # 14 on the same line as station number.

In this sequence of calculation the bearings are assumed to be rigid, hence the deflection at the bearing supports is zero. When shaft loads are separated into horizontal and vertical components, total shaft deflection is obtained by calculating the square root of the sum of the squares of the horizontal and vertical deflections at each station.

Practical difficulties in the manufacturing and assembly of a built-up rotor do not permit the center of mass of a symmetric rotating system to coincide with the actual center of rotation. Additionally, static deflections due to weight of the components and constant lateral loads cause a further shift of the mass center from the axis of rotation. As the rotating shaft's speed increases, the kinetic energy of the attached masses also increases. When the kinetic energy reaches the potential energy of the deflected shaft, the vibrations in the shaft will be at a peak. The rotor speed at which this disturbance occurs is called the fundamental frequency of vibration, or critical speed, of the shaft. Equating the kinetic energy of the rotating masses to the potential energy of the deflected shaft results in an equation that will define the critical speed of the shaft. This equation is called Rayleigh's equation. Thus the lowest, or fundamental, frequency of vibration for a shaft simply supported at two points is given by:

$$f_n = 187.7 \left[\frac{W_1 y_1 + W_2 y_2 + W_3 y_3 + \ldots + W_n y_n}{W_1 y_1^2 + W_2 y_2^2 + W_3 y_3^2 + \ldots W_n y_n^2} \right]^{1/2}$$

where W_n is the weight of the rotating mass in pounds and y_n is the static shaft deflection due to W_n.

Example Problem 7.4: Explain the practical significance of torsional stiffness of a rotating shaft.

Solution: An adequate shaft design calls for it to transmit power uniformly with a steady motion. If a shaft permits excessive angular motion, it contributes to torsional as also lateral vibrations. This can affect performance of gears, couplings and bearings. As a rule, torsional deflection in camshafts for reciprocating machines should be limited to 0.5° over its full length, regardless of its length.

Torsional deflection is given by:

$$\theta = \frac{TL}{JG}$$

where J is the polar moment of inertia of the shaft. For a round hollow shaft it is given by the expression: $J = [\pi/32]\,[D_o^{\,4} - D_i^{\,4}]$. For shafts having other than round cross sections table 7.1 is presented as a designing aid.

When the geometric configuration of a rotor bearing system gets more complex, finite element representation is probably desirable. Oftentimes it is required to find the impact on the dynamic behavior of the rotating system due to relatively small changes, for example, in the bearing sump area, material properties of a component or temperature related variation in a characteristic. In such a case parametric evaluation of the problem provides quantitative results, which may be used as the basis for making design decisions. Experimental results to confirm analytical predictions may also be necessary for the purpose. Accelerometer readings at bearing locations can be obtained in most cases with relative ease. Dynamic strain gages provide useful data, but are more difficult to install. Relative closure data between the stator and the rotor at discrete locations around the periphery using telemetry may also be required.

Example Problem 7.5: A state of tri-axial stress occurs when two bodies with curved surfaces are compressed against each other. This causes point, or line, contact to change to an area contact, causing the state of stress in both bodies to change to 3-dimensional.

Such situations arise in the contact of mating gears, a wheel against a rail and in rolling element bearings. Computing the stress levels resulting from loading one body against another may prevent surface failures.

Consider two solid spheres of diameters d_1 and d_2 with a force F. A circular contact area of radius a is obtained. If the elastic constants (Young's modulus and Poisson's ratio) of the two spheres are E_1, μ_1, E_2, and μ_2, radius a is given by the following expression:

$$a = \sqrt[3]{\frac{3F[(1-\mu_1^{\,2})/E_1] + [(1-\mu_2^{\,2})/E_2]}{8[(1/d_1) + (1/d_2)]}}$$

According to Thomas and Hoersch, stresses in the x, y and z directions are principal stresses, are defined by the coefficient A:

$$A = \frac{4a}{\pi}\,\frac{(1/d_1) + (1/d_2)}{\pi[(1-\mu_1^{\,2})/E_1] + [(1-\mu_2^{\,2})/E_2]}$$

Then for an element on the z-axis, the stresses are:

$$\sigma_x = \sigma_y = A\left[(1+\mu)\left(\frac{z}{a}\cot^{-1}\frac{z}{a} - 1\right) + \frac{1}{2}\,\frac{a^2}{a^2 + z^2}\right]$$

$$\sigma_z = -A\left(\frac{a^2}{a^2 + z^2}\right)$$

Table 7.1. Expressions for Torsional Stiffness and Shear Stress in Rotating Beams

Section	Stiffness Constant K ($\theta = TL/GK$)	Maximum Shear Stress
	$K = \dfrac{\pi d^4}{32}$ d = solid shaft diameter	$\tau = \dfrac{16T}{\pi d^3}$
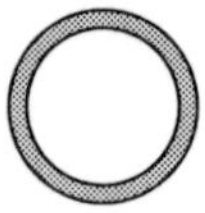	d = outer diameter, d_i = inner diameter $K = \dfrac{\pi[d^4 - d_i^{\,4}]}{32}$	$\tau = \dfrac{16Td}{\pi[d^4 - d_t^{\,4}]}$
	r = mean radius, t = thickness $K = \dfrac{2\pi r l^3}{3}$	$\tau = \dfrac{3T}{2\pi r l^2} + \dfrac{1.8Tt}{(2\pi r t)^2}$
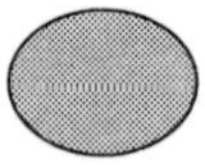	a = length of major axis, b = length of minor axis $K = \dfrac{\pi a^3 b^3}{a^2 + b^2}$	$\tau = \dfrac{2T}{\pi a b^2}$
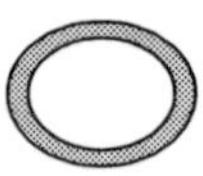	Outer Ellipse: a = length of major axis, b = length of minor axis Inner Ellipse: a_1 = length of major axis, b_1 = length of minor axis $K = \dfrac{\pi a_1^{\,3} b_1^{\,3}}{a_1^{\,2} + b_1^{\,2}}[(1+q)^4 - 1]$	$\tau = \dfrac{2T}{\pi a_1 b_1^{\,2}[(1+q)^4 - 1]}$ $q = (a - a_1)/a_1 = (b - b_1)/b_1$
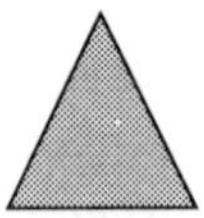	$K = \dfrac{b^4\sqrt{3}}{80}$ b = length of equilateral triangle side	$\tau = \dfrac{20T}{b^3}$
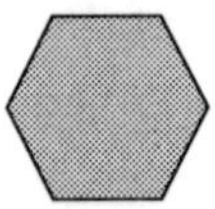	$K = 2.69b^4$ b = length of uniform hexagon side	$\tau = 1.09T/b^3$
	$K = \dfrac{2t_1 t_2 (a - t_2)^2 (b - t_1)^2}{a t_2 + b t_1 - t_2^{\,2} - t_1^{\,2}}$ a, a_1 = length of outer/inner larger side	$\tau = \dfrac{T}{2t_1(a - t_2)(b - t_1)}$, or $\tau = \dfrac{T}{2t_2(a - t_2)(b - t_1)}$ b, b_1 = length of outer/inner shorter side
	$K = 0.1406b^4$ b = length of side of square	$\tau = 4.80T/b^3$

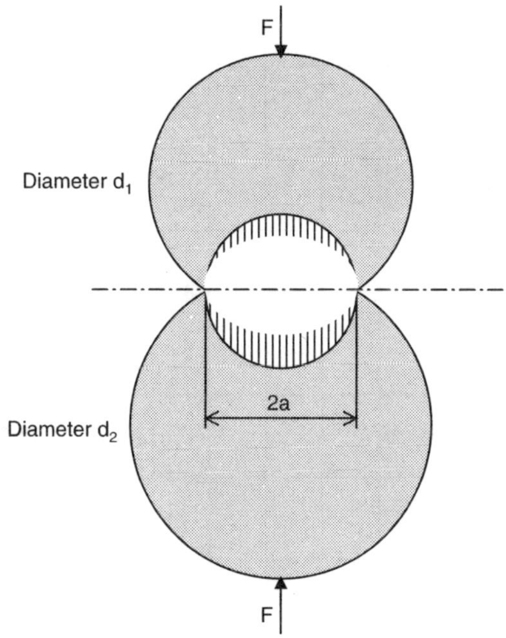

Fig. 7.30. Pressure Distribution Between Two Spheres

The equations given above also apply to the contact of a sphere and a plane surface or to a sphere and an internal spherical surface. For a plane surface use $d = \infty$. For an internal surface the diameter is expressed as a negative quantity.

A ball bearing's inner raceway may be considered to be the segment of a sphere. The rolling elements have a mating surface. Hence, the two contacting surfaces may be idealized as spheres loaded against each other. If the raceway diameter = 2.25 in., ball diameter = 0.775 in. and contact force F = 15 lbs., determine the contact area and stress levels. Use $E_1 = E_2 = 30 \times 10^6$ lb/in^2, Poisson's ratio = 0.3 for both bodies.

Solution: The pressure within each sphere has a semi-elliptical distribution, as shown in Fig. 7.30. Maximum pressure occurs at the center of the contact area, and is given by the expression:

$$P_{max} = \frac{3F}{2\pi a^2}$$

Fig. 7.31 provides variation in the stress components for a distance $3a$ below the surface. Note that shear stress τ reaches a maximum value slightly below the surface. Maximum shear stress is considered to be the leading cause of surface fatigue failure in contacting elements. A crack initiating at the point of maximum shear stress below the surface and lubricant pressure flowing into the crack region may be enough to dislodge metal chips. In this example when the contacting surfaces are cylindrical the contact area is a narrow rectangle of half width b, given by the equation:

$$b = \sqrt{\frac{2F[(1-\mu_1^2)/E_1]+[(1-\mu_2^2)/E_2]}{\pi l[(1/d_1)+(1/d_2)]}}$$

where l is the length of the contact area. The pressure distribution across the width $2b$ is elliptical, and maximum pressure is given by:

$$P_{max} = -\frac{2F}{\pi bl}$$

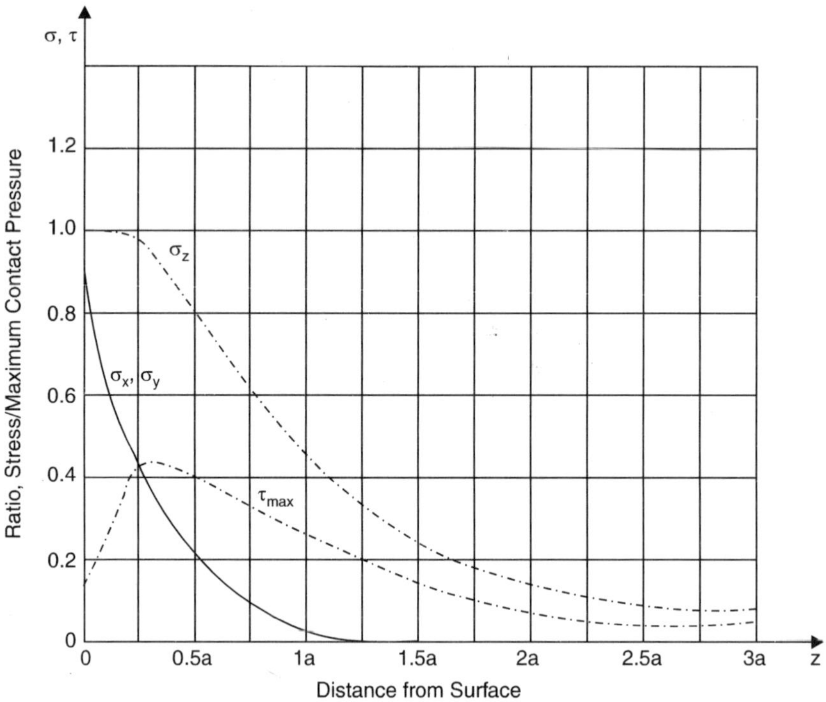

Fig. 7.31. Stress Components Below the Surface of Contacting Spheres

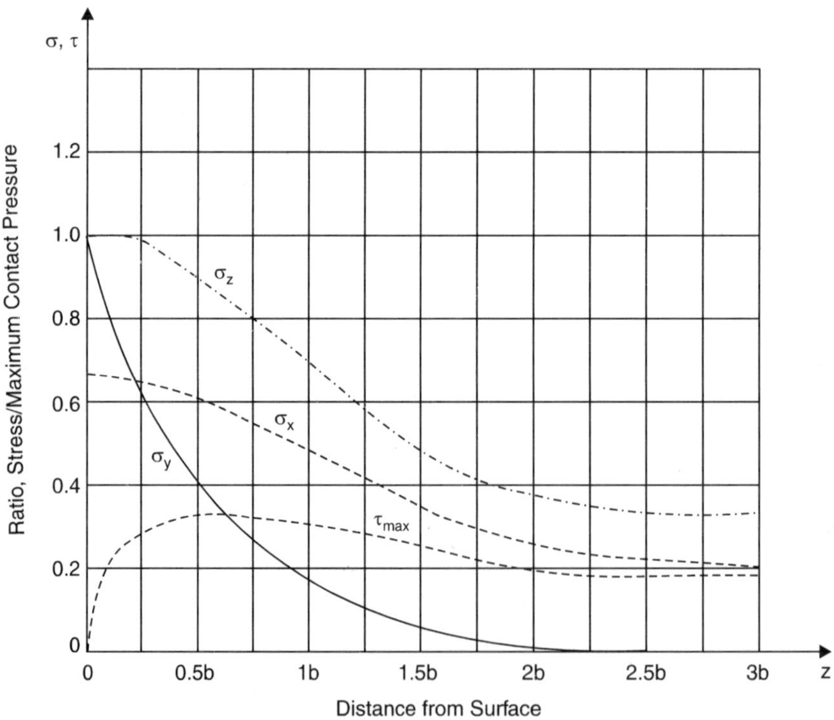

Fig. 7.32. Stress Components Below the Surface of Contacting Cylinders

When applied to a cylinder and a plane surface, for the plane use $d = \infty$. For an internal cylinder d is negative.

To evaluate stresses, select the origin of a reference system at the center of the contact area with x parallel to the axes of the cylinders, y perpendicular to the plane formed by the two cylinder axes and z in the plane of the contact force. For elements on the z-axis, principal stresses σ_x, σ_y and σ_z exist. Fig. 7.32 shows a plot of the stresses for depths up to $3b$ below the contact surface. Also, three different shear stresses are created, the largest of which is given by:

$$\tau_{yz} = \frac{\sigma_y - \sigma_z}{2}$$

This shear stress is also shown in Fig. 7.32, labeled τ_{max}. It reaches a peak value slightly below the surface, similar to what is seen in the case of the contacting spheres.

REFERENCES & BIBLIOGRAPHY

Alford, J., "Protecting Turbomachinery from Self-Excited Rotor Whirl", Journal of Engineering for Power, 87(4):333-334, October, 1965.

Barrett, L. E., Allaire, P. E., Gunter, E. J., "Proceedings of the Conference on the Stability and Dynamic Response of Rotors with Squeeze Film Bearings", University of Virginia, Charlottesville, May 1979.

Barrett, L. E., Allaire, P. E., Gunter, E. J., "Optimum Bearing and Support Damping for Unbalance and Response and Stability of Rotating Machinery", Journal of Engineering for Power, 100(1):89-94, 1978.

Begg, I. C., "Friction Induced Rotor Whirl - A Study in Stability", 83-DET-10, ASME, New York, September, 1983.

Bentley, D. E., "The Re-Excitation of Balance Resonance Regions by Internal Friction", 82-PET-49, ASME, New York, September, 1982.

Bolotin, V. V., "Non-Conservative Problems of the Theory of Elastic Stability", Pergamon Press, New York, 1964.

Booker, J. F., Huebner, K. N., "Application of Finite Element Methods to Lubrication, An Engineering Approach", Journal of Lubrication Technology, 94(4):313-323, October, 1972.

Booser, E. R., "Handbook of Lubrication", Vol. 2, CRC Press, Boca Raton, FL. 1983.

Brown, R. D., Hart, J. A., "A Novel Form of Damper for Turbomachinery", CP2443, NASA, Washington, DC, June, 1986.

Cataford, G. F., Lancee, R. P., "Oil Free Compression on a Natural Gas Pipeline", 86-GT-293, ASME, New York, June, 1986.

Childs, D. W., "Force and Moment Dynamic Coefficients for Pump Impeller Shroud Surfaces", *Proceedings of Fourth Workshop on Rotor Dynamic Instability Problems in High Performance Turbomachinery*, NASA, CP, Texas A & M University, College Station, TX, June, 1986.

Childs, D. W., "Identification and Avoidance of Instabilities in High Performance Turbomachinery", *Von Karman Institute for Fluid Dynamics*, lecture, Rhode-Saint Genese, Belgium, January, 1987.

Choi, Y. S., Noah, S. T., "Non-Linear Steady State Response of a Rotor Support System", Journal of Vibrations, Acoustics, Stress and Reliability in Design, 169(3):255-261, July, 1987.

Crede, C. E., Ruzicka, J. E., "Theory of Vibration Isolation", Shock & Vibration Handbook, 3d ed., Mc-Graw Hill, New York, 1988.

Den Hartog, J. P., "Mechanical Vibrations", Dover Publications, New York, 1985.

Deutschman, A. D., Michels, W. J., Wilson, C. E. "Machine Design - Theory and Practice", MacMillan Publishing Co., New York, 1985.

Dimarogonas, A. D., "Newkirk Effect: Thermally Induced Dynamic Instability of High Speed Rotors", 83-GT-26, ASME, New York, April, 1983..

Ehrich, F. F., "Self-Excited Vibration", Shock & Vibration Handbook, 3d ed., Mc-Graw Hill, sec. 5, 1987.

Ehrich, F. F., "The Role of Bearing Support Stiffness Anisotropy in Suppression of Rotor Dynamic Instability", DE-Vol. 18-1, ASME, New York, 1989.

Ehrich, F. F., "Handbook of Rotordynamics", Mc-Graw Hill, New York, 1992.

Ehrich, F. F., O'Connor, J. J., "Stator Whirl with Rotors in Bearing Clearance", Journal of Engineering for Industry, 89(3):381 - 390, August, 1967.

Eshelman, R. L., Eubanks, R. A., "Effects of Axial Torque on Rotor Response: An Experimental Investigation", 80-WA/DE-14, ASME, New York, December, 1980.

Fowlie, D. W., Miles, D. D., "Vibration Problems with High Pressure Compressors", 85-Pet-28, ASME, New York, September, 1985.

Gunter, E. J., "Dynamic Stability of Rotor Bearing Systems", SP-113, NASA, chapter 4, Washington, D. C., 1966.

Gunter, E. J., Barrett, L. E., Allaire, P. E., "Design of Non-Linear Squeeze Film Dampers for Aircraft Engines", (ASME) Journal of Lubrication Technology, 99(1):57-64, January, 1977.

Hagg, A. C., Sankey, G. O., "Some Dynamic Properties of Oil Film Journal Bearings with Reference to Unbalance Vibration of Rotors", Journal of Applied Mechanics, 23(2):302-306, June, 1986.

Hildebrand, F. B., "Advanced Calculus for Applications", Prentice-Hall, New Jersey, 1986.

Houbolt, J. C., Reed, W., H., "Propeller Nacelle Whirl Flutter", Institute of the Aerospace Sciences, 61-34, New York, January, 1981.

ISO 1940/1, "Mechanical Vibrations - Balance Quality Requirements of Rigid Rotors - Part I, Determination of Permissible Residual Unbalance", International Organization for Standardization, Geneva, Switzerland, 1986.

Kirk, R. G., "Labyrinth Seal Analysis for Centrifugal Compressor Design - Theory and Practice", *Proceedings International Federation for the Theory of Machines and Mechanisms 1986 International Conference on Rotor Dynamics*, pg. 589-596, Japan Society of Mechanical Engineers, Tokyo, September, 1986.

Kirk, R. G., Mondy, R. E., Murphy, R. C., "Theory and Guidelines to Proper Coupling Design for Rotor Dynamics Consideration", 83-DET-93, ASME, New York, September, 1983.

Kimball, A. L., "Internal Friction Theory of Shaft Whirling", General Electric Review, 27(4): 244, April, 1924.

Kramer, E., "Instabilities of Rotating Shafts", Proceedings of the Conference on Vibration of Rotating Systems, Institution of Mechanical Engineers, London, pg. 230-246, February, 1982.

Li, D. F., Gunter, E. J., Barrett, L. E., Allaire, P. E., "The Dynamic Analysis of Multi-Level Flexible Rotor Systems Using Transfer Matrices and Component Mode Synthesis", Final Report UVA/628145/MAE81/102, University of Virginia, Charlottesville, 1991.

Linn F. C., Prohl, M. A., "The Effect of Flexibility of Support upon the Critical Speeds of High Speed Rotors", Transactions Society of Naval Architects and Marine Engineers, 59:536-553, 1971.

Loomis, A. L., "Compressed Air & Gas Data", Ingersoll Rand Co., Woodcliff Lake, NJ, 1989.

Lund, J. W., "Spring and Damping Coefficients for the Tilting Pad Journal Bearing", Transactions American Society of Lubrication Engineers, 7(4):342-352, 1964.

Lund, J. W., Thomsen, K. K., "A Calculation Method and Data for the Dynamic Coefficients of Oil Lubricated Journal Bearings", *Topics in Fluid Film Bearing and Rotor Bearing System Design and Optimization* , ASME, New York, pages 1-28, 1978.

Mabie, H. H., Reinholtz, C. F., "Mechanisms and Dynamics of Machinery", John Wiley & Sons, New York, 1987.

Marmol, R. A., Vance, J. M., "Squeeze Film Damper Characteristics for Gas Turbine Engines", ASME Journal of Mechanical Design, 100(1):139-146, January, 1982.

McLachlan, N. W., "Theory and Applications of Mathieu Functions", Oxford University Press, New York, 1977.

Muszynska A., "Partial Lateral Rotor to Stator Rubs", Paper # C281/284, Institution of Mechanical Engineers, London, 1984.

Muszynska A., "Whirl and Whip-Rotor/Bearing Stability Problems", Journal of Sound & Vibration, 110(3):443-462, 1986.

Myklestad, N. O., "Fundamentals of Vibration Analysis", Mc Graw Hill, New York, 1956.

Newkirk, B. L., "Shaft Whipping", General Electric Review, 27: 169 - 178, 1924.

Newkirk, B. L., Taylor, H. D., "Shaft Whipping due to Oil Action in Journal Bearing", General Electric Review, 28(8): 559-568, 1925.

Orcutt, F. K., "The Steady State and Dynamic Characteristics of the Tilting Pad Journal Bearing in the Laminar and Turbulent Flow Regimes", Journal of Lubrication Technology, 89(3):392 - 404, 1967.

Pincus, O., Sternlicht, B., "Theory of Hydrodynamic Lubrication", Mc-Graw Hill, New York, 1984.

Roark, R. J., Young, W. C., "Formulas for Stress and Strain", 5th ed., Mc-Graw Hill, New York, 1989.

San Andres, L. A., Vance, J. M., "Effect of Fluid Inertia on Finite Length Squeeze Film Dampers", 86-TC-4D-1, American Society of Lubrication Engineers, Park Ridge, IL, October, 1986.

Taylor, C. F., "The Internal Combustion Engine in Theory and Practice", Volume 2, The M. I. T. Press, Cambridge, MA., 1997.

Taylor, E. S., Browne, K. A., "Vibration Isolation of Aircraft Power Plants", Journal of the Aeronautical Sciences, 6(2): 43-49, 1984.

Thomas, J. J., "Instabile Eigenschwingungen von Turbinenlaufern, Angefacht durch die Spaltstromungen Stopfbuschen un Beschaufelungen", AEG-Sonderdruck, 1958.

Timoshenko, S., Young, D. H. and Weaver, W., "Vibration Problems in Engineering", John Wiley & Sons, Inc., New York, 1974.

Vance, J. M., "Torque Whirl-A Theory to Explain Non-Synchronous Whirling Failures of Rotors with High Torque Load", Journal of Engineering for Power, 100(2):235-240, April, 1988.

Vance, J. M., Laudadido, F. J., "Experimental Measurement of Alford's Force in Axial Flow Turbomachinery", 84-GT-140, ASME, New York, June, 1984.

Vance, J. M., Lee J., "Stability of High Speed Rotors with Internal Friction", 73-DET-127, ASME, New York, September, 1973.

Wachel, J. C., Tison, J. D., "Vibrations in Reciprocating Machinery and Piping Systems", *Proceedings of the 23rd Turbomachinery and Piping Systems*, Texas A & M University, College Station, TX 1994.

Wachel, W. D., "Non-Synchronous Instability of Centrifugal Compressors", 82-Pet-22, ASME, New York, September, 1982.

Wang, Z., Lund, J. W., "Calculations of Long Rotors with Many Bearings on a Flexible Foundation", C291/84, *Proceedings of Third International Conference on Vibration in Rotating Machinery, Institute of Mechanical Engineers Conference Publication,* York, England, 1984.

Zhou Ren-Mu, "Instability of Multi-Stage Compressor K 1501", CP2443, NASA, Washington, DC, June, 1986.

Advanced Analytical Methods 8

8.1 INTRODUCTION

Conflicting demands of modern reciprocating machinery design, in particular the quest for high ratings without penalties, require more in-depth analysis than hitherto. Representation of geometric configuration has to be more precise, less sweeping assumptions must be made and calculations more nearly approximating the actual machine operations must be carried out. An increasing disparity of design variants, coupled with a compressed time scale for total design and development, call for more rapid solutions. At the same time market realities and competitive pressures require that the machinery operate at peak efficiency, deliver maximum operating reliability, possess nearly limitless life duration and have minimum development and manufacturing costs.

Various levels of analytical methods are available for carrying out a dynamic and strength evaluation of the crankshaft, cylinder block, bearings, piston, rings, connecting rod and valve system. Computerized procedures may feed information to, or take information from, classical, finite element and modal methods. At the highest level a finite element method includes load sharing effects of the flexible crankshaft within a flexible machine block interconnected by non-linear oil films. This model includes representation of the complete crankshaft, and goes on to determine the dynamic stress characteristics throughout the machine cycle. Fatigue assessment may then be performed to calculate the fatigue safety factor. This can be used for failure investigation or detailed design optimization and verification.

Numerous tools are available on digital computers for the solution of vibration problems. They fall into two basic categories: (*a*) numerical analysis of dynamic systems that are too complex to perform by hand calculation and (*b*) experimental applications requiring rapid acquisition and processing of structural input and response data. Analytical applications encompass general purpose programs such as finite difference methods, finite element techniques, boundary element methods, statistical energy analysis and special purpose programs for vibration isolation and machinery analysis.

8.2 HIERARCHY OF METHODS

Computer aided engineering tools are used throughout the design process. During conceptual design phase of the machine emphasis is on performing rapid design iterations using robust programs having undemanding input data requirements. The computer codes generally have substantial data available for assuming numerical values for key factors related to design, operation and production of the components and assembly. The programs yield easily interpreted results, and can be incorporated into computer-aided design (CAD) based process. Crankshaft concept is often fundamental to the machine layout, and also interacts with the design of other components, thus underlining the need for rapid iterations during the early design stage.

At the other extreme of the design process, when basic machine design parameters are better defined, sophisticated tools are needed to perform detailed calculations. The crankshaft is a finely optimized component from considerations of weight, stress, fatigue and manufacture, encountering substantial bending and torsional stresses. Its design affects engine noise, cylinder block forces, oil film in the bearings, materials and production processes.

A modular approach may be useful to develop a suite of programs at various levels of sophistication. Some levels commonly associated with the design and development of the crankshaft are:

1. CAD parametric drawing and interactive low-mesh density finite element stress analysis.
2. Detailed finite element analysis. Loads calculated for this model use a statically determinate model of the crankshaft and torque derived from a forced and damped torsional vibration model. An initial torsional vibration analysis is carried out.
3. Finite element dynamic stress analysis using load data of level 2.
4. Finite element dynamic stress analysis using loads calculated from a statically indeterminate model of the crankshaft. Torsional vibration calculation is identical to that of levels 2 and 3.

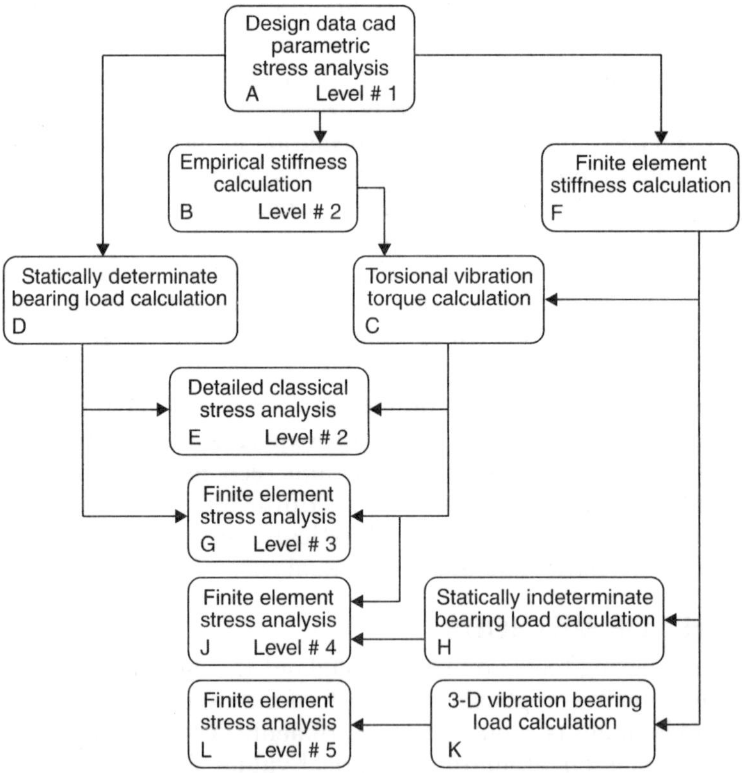

Fig. 8.1. Information Flow Through Different Levels of Analyses

5. Finite element stress analysis, using loads from a full 3-D vibration model of the crankshaft. Fig. 8.1 shows the various calculation modules and the routes through them to produce the hierarchy shown.

Lower level cases often times use fewer elements compared with the higher level finite element models.

Oil film calculations to obtain bearing loads are an essential and integrated part of crankshaft design, and will be used for stress analysis. Various levels of refinement are also available in oil film hydro-dynamic calculations.

8.3 PARAMETRIC CAD EVALUATION

This is the lowest level technique for the design of a crankshaft, and is often integrated into the CAD based design method. The method provides an initial layout of the crankshaft based on fundamental machine data, such as:

- Machine configuration
- Firing order (for internal combustion engines)
- Bore × Stroke
- Reciprocating machine type and duty
- Crankshaft and bearing materials
- Number of counterweights
- Gas or fluid characteristics (for pumps and compressors)
- Rate of flow
- Maximum cylinder pressure
- Reciprocating and rotating weights

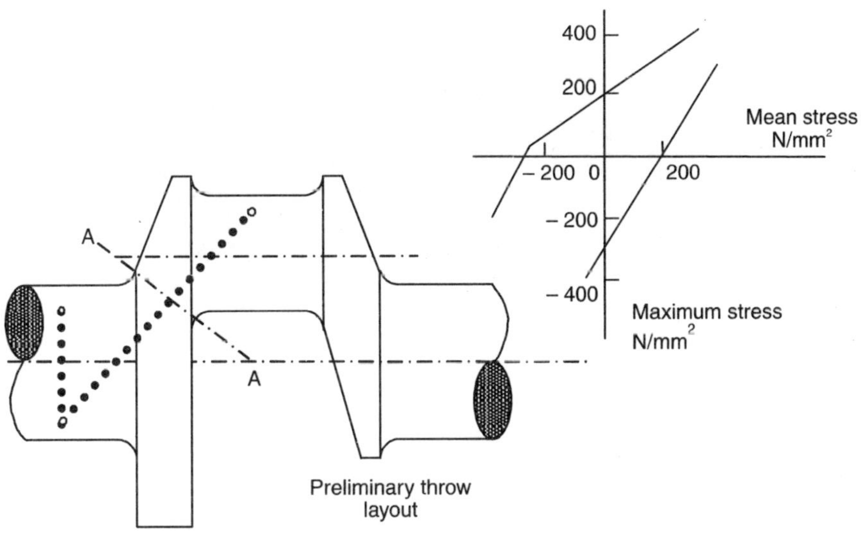

Fig. 8.2. Parametric Design Evaluation of Crankshaft

Before proceeding further, this initial layout must be assessed by an interactive classical analysis program, labeled A in Fig. 8.1. Pseudo-static gas and inertia forces and bending moments are calculated. A statically determinate load sharing analysis is used to determine the bending moment distribution, since crankshaft stiffness characteristics are still unknown. There is, therefore, no bending moment transfer between adjacent crank bays. A full torsional analysis and damper design is too detailed for use at this level, where the aim is to present the user with unambiguous comparative results. Torque values of similar previous machinery design are used. Section properties and stress concentration data are taken from the CAD geometry, and combined bending and torsion stresses compared with fatigue limits of the selected material on a Goodman diagram (Fig. 8.2). This calculation is generally performed at three speed and load conditions: maximum torque, maximum power and no load over speed. A step by step evaluation of the calculated parameters helps in the understanding of the design process.

8.4 CLASSICAL ANALYSIS METHODS

Examination of detailed classical analytical methods for determining the lateral vibration characteristics will be useful before exploring the finite element method. A mathematical model using discrete mass and elastic beam to represent the crankshaft for calculating natural frequencies of vibration and mode shapes has been used extensively, and provides a high degree of accuracy. The method bears some resemblance to the Holzer method for determining torsional frequencies. Holzer's approach requires torsional inertia and stiffness characteristics of individual crankshaft elements. A finite element model with a low mesh density may be used for this purpose. Besides torsional stiffness, the model can determine the full three-dimensional properties of the structure in terms of mass, centroid, inertia and stiffness. Geometric features like bearing fillet cannot be included in the model. Figure 8.3 shows a typical low mesh density model of a crank web. Defining the geometry can be done directly from the CAD system, through a digitizer or input by hand using component dimensions.

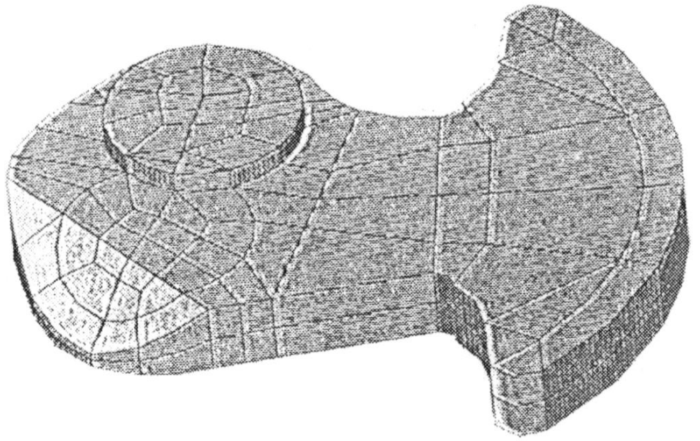

Fig. 8.3. Low Density Finite Element Model of Crank Web for Stiffness Determination

Three load cases are used for crank web stiffness evaluation (Fig. 8.4). Torque is applied to the model in the form of displacements at the nodes of the main bearing cut plane. These nodes are free to move out of plane. Restraint is applied on the crank pin cut plane, and this plane is restrained from displacement in its own plane but is free to move out of plane. Figure 8.4*a* shows the loading in diagrammatic form. The program calculates reaction torque due to the applied displacements, from which web stiffness is calculated. Displacements are applied as a matter of choice, but the program may also be set up to determine displacements as a consequence of applied torque. In-plane bending is also applied to the web as a displacement on the nodes of the main bearing's cut plane, Fig. 8.4*b*. Restraint is applied on the pin cut plane, with the requirement that all nodes on the cut plane remain in the original plane; the plane is free to move axially. A single node in the model is restrained in the axial direction to prevent rigid body motion. Stiffness is determined from the calculated reaction bending moment. For out-of-plane bending moments displacements are applied in a similar manner to the main bearing cut plane. Restraint is applied to the pin cut plane in the axial direction. A single node is fully restrained to prevent rigid body motion. Figure 8.4*c* shows the procedure for load application.

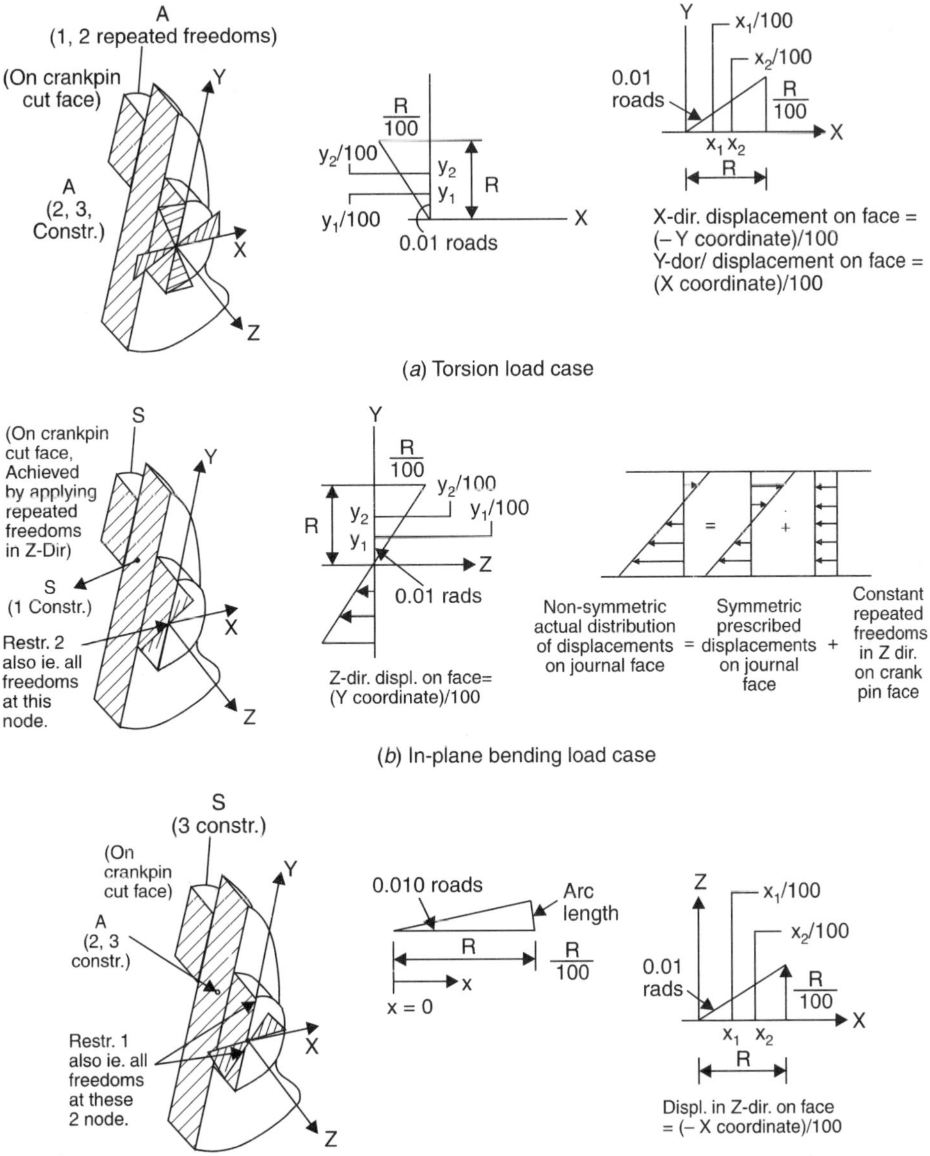

(*a*) Torsion load case

(*b*) In-plane bending load case

(*c*) Out of plane bending load case

Fig. 8.4. Crank Web Stiffness Calculation - Load Definition

With the assumption that the crankshaft can be represented by a linear system with negligible damping, Myklestad's method can be used for calculation of the transverse direction vibrations. The crankshaft is represented by n discrete masses connected by flexible elements, or springs, without mass. A schematic representation of a three bearing, four throw crankshaft is shown in Fig. 8.5.

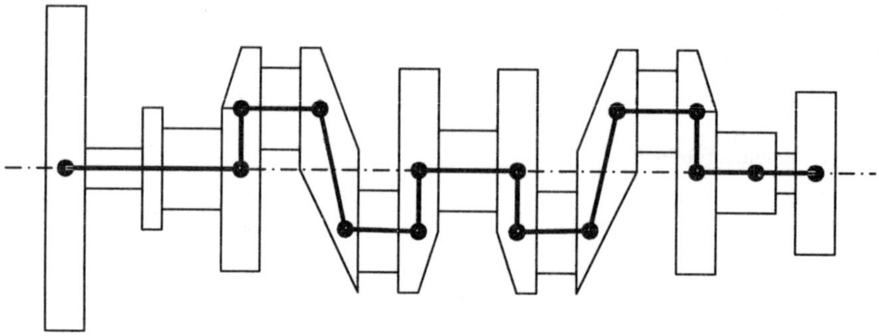

Fig. 8.5. Discrete Mass, Flexible Spring Model of Crankshaft

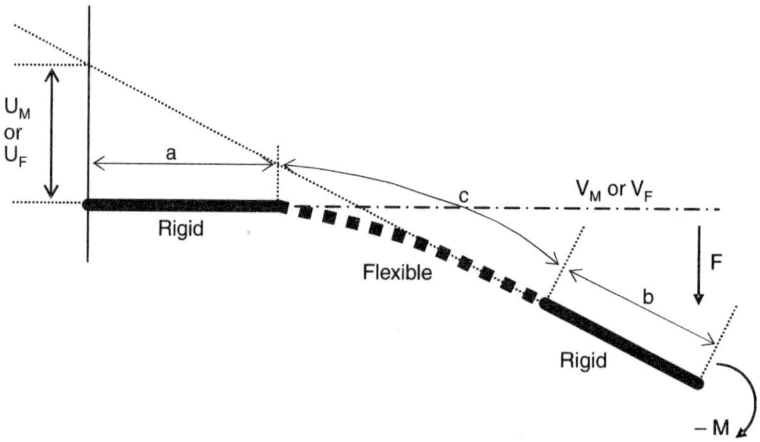

Equations for Deflection Coefficients

$$UF = (1/EI)\{bc(c/2 + a) + c^2/2(c/3 + a)\}$$

$$UM = (1/EI)\{c^2/2 + ac\}$$

$$VF = (1/EI)\{bc + c^2/2\}$$

$$VM = c/EI$$

$$CFLEX = c/(E \times AREA)$$

$TFLEX = \{7c(p^2 + r^2)\}/[2Gp^3r^3]$ for rectangular section of dimensions p, t

or $TFLEX = c/GI$ for circular cross section

Fig. 8.6. Crank Pin and Main Journal Stiffness Evaluation

To understand Myklestad's procedure, consider an isolated element in the plane of axes X and Y at an angle α to the shaft reference axis X. Alternatively, the element may be taken in the plane of axes Y and Z to obtain a more complete representation of a three dimensional crankshaft. Following the method of Myklestad, displacements, forces and moments applied to the right hand end of the element are assumed in the reference frame of the shaft. A transformation matrix in terms of angle α is used to transform displacement, forces and moments to the local coordinate axes of the element. Element inertia and stiffness due to the translation and rotation from the crankshaft's coordinate system are taken into account. Displacements, forces and moments are then calculated at the left hand end in the element's local coordinate axes.

Experimental measurements on one-piece cast or forged crankshafts indicate that, generally, cylindrical sections of crank pins and main journals are flexible for only a part of the length. This arises from the stiffening effects of the adjacent crank web or balance weight. At one end of the crankshaft the flywheel provides additional stiffening of the adjacent journal, and at the other end the pulley may also be expected to do the same to a somewhat lesser extent. Allowance for shear deflection could also be included, but experience suggests it is not necessary since web bending deflections are more significant. Figure 8.6 provides dimensions and expressions for bending, torsional and axial flexibility of the elements. Crank pins and journals connect with points taken to be inside the rigid mass of a web or balance weight; hence if only the pin or journal is flexible, appropriate values for a, b or c can be assumed.

Natural frequency calculations use the principle of superposition. But displacements, forces and moments assume complex values to account for the phase difference associated with gyroscopic terms. With gyroscopic effects vibration is no longer confined to a plane; a whirling motion ensues. To allow for bearing constraints it is necessary to determine intermediate values of the deflections at each journal. Mass inertia arising from piston and connecting rod is located at the center of the crank pin.

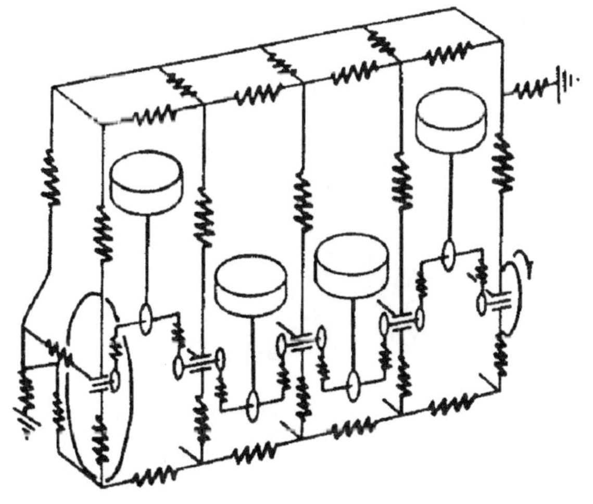

Fig. 8.7. Mass and Spring Simulation of a Complete System

A minimum number of masses are required to calculate natural frequencies with sufficient accuracy, but it is convenient to include a greater number to get a more complete picture of the modal shape of the vibrating crankshaft.

Figure 8.7 shows a mass and spring simulation of a complete machine system, including the crankshaft, connecting rods, pistons and cylinder block. The model is restrained from rigid body motion in all three directions in the same manner as in the actual installation.

8.5 FINITE ELEMENT METHOD

The finite element method is a major analytical tool for studying the behavior of a conceptual model of a dynamic system. A spatial distribution of discrete points throughout a region is used to define three components of displacement and three components of rotation. A small but finite (hence the term *finite* element) volume of the region identifies a sub-set of grid or node points associated with it. Thus, an area where a two-dimensional stress or strain is assumed could be represented by a triangular element with at least three grid points or a quadrilateral element with at least four grid points. A three dimensional field could be represented by trapezoidal, pentagonal or cubic element with at least four, five or eight node points. Figure 8.8 illustrates a solid element model of a throw of a crankshaft using trapezoidal elements.

A major function of a finite element program is to convert a conceptual model of a system that might be an assembly of structural elements (rods, beams, plates and shells) into a mathematical model that can be manipulated and solved by the computer. Engineering description of the reciprocating machinery, including material properties and geometric configuration, is converted into a mathematical model in terms of component and system mass, damping and stiffness matrices. 'Elements' are used to represent behavior in a small but finite volume of components and sub-assemblies. Other features to describe loads, constraints and special modeling requirements enhance the overall model. Commercially available structural mechanics programs include discrete elements such as springs, masses and dampers, as well as whole items like beams, plates, shells and solids.

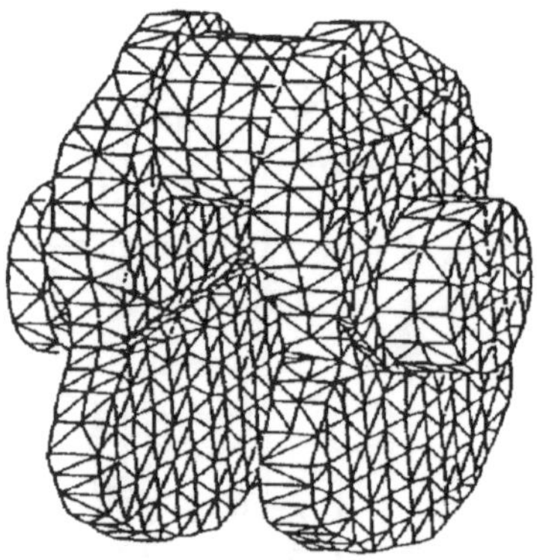

Fig. 8.8. Solid Element Model of One Throw of Crankshaft

The finite element method (Zienkiewicz and Taylor, 1989, Hughes, 1987) uses a class of approximation techniques based on weighted residuals to transform partial differential equations representing the behavior of a volume into a set of algebraic equations that can be solved on the computer. The technique is similar to the Ritz procedure of using an assumed displacement function to determine, for example, the natural frequencies of a beam. The method uses a set of basis functions to approximate the behavior in the finite volume, in a Ritz way. Undetermined basis coefficients associated with a set of Ritz vectors can then be determined by minimizing the error associated with satisfaction of the equations of mathematical physics in the finite volume. The volume's behavior must satisfy the equations of equilibrium, strain-displacement relations and stress-strain relations. Element formulations are based on assumed deflections, defined by a set of shape functions, in the volume. Strains are calculated from the displacements, and stresses obtained from material property values. The approach calls for developing a relationship between element forces and displacements.

Displacement based elements tend to be overly stiff. An alternate approach is based on an assumed stress field (Pian), where the shape functions are chosen to satisfy equilibrium conditions at each grid point. Strains are calculated using the inverse of stress and strain equations. The assumed stress coefficients are determined by minimizing the error in satisfying the strain-displacement relations in an integral sense.

A hybrid approach (Pian and Chen, 1982) assumes displacements on the boundary and stresses, strains or a combination of the two in the interior of the region. This leads to elements that perform well over a wide range of loads, boundary conditions and component geometry. The elements predict verifiable strains in an arbitrarily shaped element, do not become overly stiff and lock and converge to the solution rapidly for a small number of elements.

8.6 SPATIAL MODEL CREATION

The objective in a finite element analysis is to obtain a turnkey solution. Given an unambiguous geometry, loads and constraints, the analysis must determine the behavior of the component or assembly to a specified degree of accuracy. Achievement of this goal generally calls for significant engineering judgement for the analysis of the reciprocating machinery's structure. Prior knowledge of its performance and operating characteristics is necessary in gauging the precision of the calculated results. Automated analysis cannot be realistically expected for the built up assembly, even for static analysis.

Generally creating and verifying a baseline finite-element model, one that gives acceptable accuracy for static analysis, develops the finite element model for dynamic analysis. The spatial model, however, still has to perform for dynamic analysis.

For static analysis the mesh must be refined (or the polynomial order of the shape function increased) until element size is small enough to adequately represent strain gradients. Pre- and post-processor sub-routines in the software are available to identify areas of high strain gradients and repetitively re-mesh and re-analyze until the solution error is reduced to an acceptable limit. High strain gradients occur at grid points where load is applied, at constraints and where component geometry is changing rapidly. Distribution of spatial grid points in dynamic analysis has other requirements. The modal content associated with a specified range of frequencies must be included in the model. If the spatial distribution is not sufficient to represent a mode of interest, then the response will not include that mode.

The library of elements available in most finite element programs will contain, at a minimum, the following element types: scalar masses, springs and dampers, rods, beams, four noded quadrilateral shells, three noded triangular shells, eight noded hexagonal solids, six noded pentagonal solids and four noded tetrahedral solids. For higher order interpolation of local behavior, and for improved geometry definition around contours, element edges may include one or two mid-side nodes. Special situations require added capabilities in the basic element types. For example, a shrink fit between two mating parts may require a gap type of element with interference and stiffness specifications.

Modeling objects and analytical algorithms permit the user to study the dynamic performance of the system. A structural assembly of piston, connecting rod and crankshaft can be studied in detail when subjected to a sudden application of a load. A case in point is the gas pressure load in internal combustion engines immediately following ignition of the air-fuel charge. Section 8.4 discussed modeling of the structural system using discrete parameters. To model continuous systems requires additional skills and knowledge associated with element capabilities and performance. An understanding of an element's effects, capabilities and behavior facilitates representation of the geometry while improving the quality of the results.

A number of options are available for distribution of mass in the model. Concentrated mass elements are convenient for representing flywheels and discs, mainly because their weight and inertia values are more readily available and can be located at a discrete point. Distributed mass associated with continuous components may be calculated within the program, since the element volume, location and material density is known. The program then calculates the mass coefficients associated with the physical displacement degrees of freedom. If a component's mass matrix is independently determined by experiments, it may be directly introduced into the model. Rotary inertia terms are not generated for rotational degrees of freedom, and this may result in a singular mass matrix. Eigenvalue extraction can still be obtained by using a method that does not require transformation to a standard form.

Stiffness properties of an element depend on its type. Beams, rods and spars require area, location of its centroid and moment of inertia properties, together with its material properties, in the stiffness calculation. A continuous type of element has its stiffness calculated within the program. Shear deformation, stiffening due to centrifugal force and spin softening capabilities are available in many computer programs.

Damping may be introduced in the form of damping elements or in the distributed form of structural, modal or proportional damping. Distributed damping removes energy from the system and damps the response of specific modes in the modal formulation. The structural, or complex, damping coefficient multiplies the stiffness matrix to generate an equivalent viscous damping matrix at a specified angular frequency ω. Structural damping that is proportional to the stiffness or mass matrix is useful, since it leads to a diagonal damping matrix as well as diagonal mass and stiffness matrices when real eigen vectors are transformed from physical to modal coordinates.

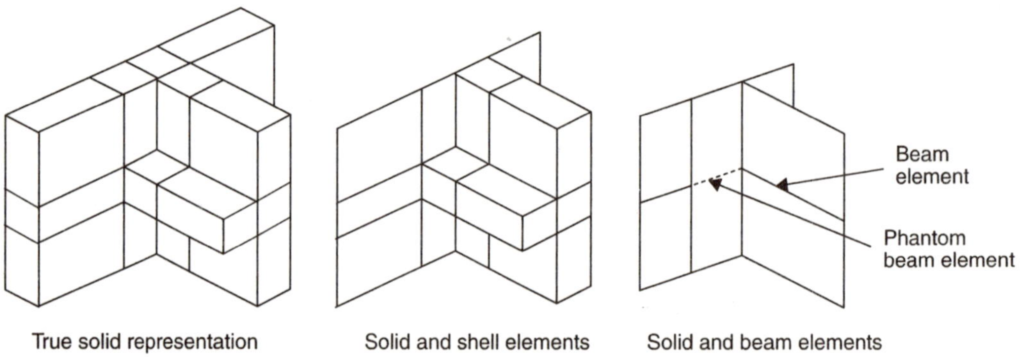

Fig. 8.9. Modeling of Joints Between Differing Element Types

Geometric approximations are inherent in the generation of finite element models, and are more severe for the simpler element types. Hence, more approximations occur in shell and beam element models than in shell and solid element models. As a general rule, fewer higher order elements yield better results than a larger number of lower order elements. Major errors arise from non-inclusion of fillets and bolt holes. Bolted joints always demonstrate some non-linearity, sometimes throwing the validity of the calculated results into question when using the linear behavior assumption. Lack of relevant material properties at operating temperature and dimensional differences between drawing specifications and finished product are beyond the analyst's control.

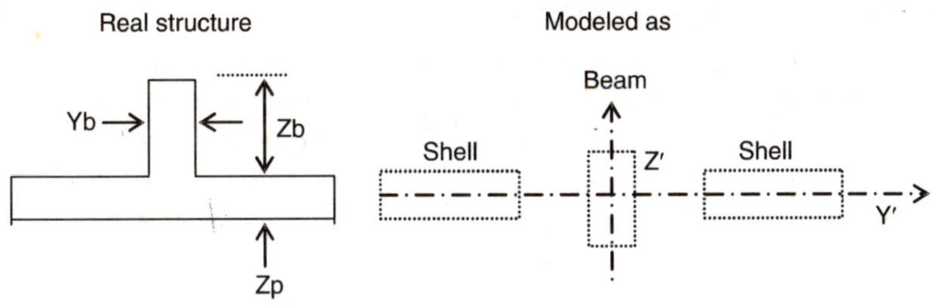

PROPERTIES OF BEAM

Area = $Yb \times Zb$

Constant for Torsion $C = B.A^3\{16A/(3B)[1 - (1/12)(A/B)^4]\}$

where $A = Yb/20$

$B = (Zb + Zp)/20$ If $A > B$, reverse A and B

Moment of Inertia, $Y'Y'$ axis $\{Yb.(Zb + Zp/1)^3\}/12 + \{Yb.(Zb + Zp/2)Zb^2\}/4$

Moment of Inertia, $Z'Z'$ axis $\{(Zb + Zp/2)(Yb)^3\}/12$

Fig. 8.10. Guidelines for beam properties to simulate stiffening ribs

When more than one element type is used in a model, problems due to non-matching of rotational degrees of freedom may be encountered. Consider, for example, the situation shown schematically in Fig. 8.9 to represent a joint in a cylinder block meeting a sidewall. In the shell and beam model a beam is used for the stiffening rib. Now center nodes of shell elements have three degrees of freedom while the beam has six. Torsional stiffness of a joint is an important effect of a beam; tying the rotational capabilities may be achieved by the use of phantom beams with carefully selected property specifications, as shown in Fig. 8.9. Beam element properties must consider rib/plate combinations. Figure 8.10 provides guidelines for determining beam cross section area and torsional constant. Similar problems in joining shell elements to solid elements may be overcome through the use of overlaid shells.

8.7 TIME DEPENDENT BOUNDARY CONDITIONS

Besides the structure's mass, stiffness and damping, time dependent loads or displacements or a combination of the two must be included in the overall model. Many computer programs permit definition of time dependent spatial set of boundary conditions in the form:

$$\mathbf{f}(\mathbf{x},t) = \mathbf{p}(\mathbf{x})\mathbf{v}(t) \tag{8.1}$$

where $\mathbf{p}(\mathbf{x})$ is a spatial distribution and $\mathbf{v}(t)$ is a set of time dependent amplitudes associated with the spatial distribution of the loads. Instead of the time domain, boundary conditions may be specified in the frequency domain, in which case $\mathbf{f}$ and $\mathbf{v}$ will depend on frequency ω instead of the time parameter t.

A special procedure to directly enter time-dependent displacement of the structure's supports is not available in many programs. Acceleration and spin velocity about an axis may be used for this purpose. Mass converts an acceleration into a force, so a large force together with a large mass will provide time dependent support conditions. Rigid body motion of the large mass may be removed by using a modeling procedure that permits displacement of the reference degrees of freedom to be set to zero in such a way that constraint forces are not generated.

8.8 SHOCK AND VIBRATION MODEL

Finite element programs are capable of executing a number of different procedures for simulating and predicting the response of structural dynamic systems to time and frequency dependent loads. Many algorithms and options are available for extracting real modes and frequencies or for transforming the solution to modal coordinates. Algorithms available with most commercial programs are for transient analysis for a set of prescribed initial conditions, steady state frequency response and random response. Response spectrum analysis is discussed in section 8.9.

Transient analysis yields response in the time domain for a given set of time dependent forces and initial conditions. Equations of dynamic motion are integrated in the physical coordinates or in the transformed modal coordinates after extraction of a set of eigen vectors. A numerical procedure such as the Newmark-beta or Wilson-theta methods (Bathe, Wilson) is used to transform the set of differential equations to a set of difference equations that are solved recursively at a sequence of discrete user defined time steps. The choice of time step in direct transient analysis has the effect of filtering out modes whose period cannot be adequately represented by the time step.

Steady state response in the frequency domain is determined by solving the equations of motion for a set of specified frequency dependent loads. When damping is present the response will be complex rather than real. The response of the system will be determined at a set of user specified frequencies.

Random vibration analysis uses statistical methods. When the excitation of the system appears to be random in time it is convenient to describe the temporal response of the system statistically rather then deterministically. Random response is treated as a data recovery procedure applied to the results of either direct or modal frequency response. It can be applied to any variable describing the behavior of the system, including displacements, velocities, accelerations, forces and stresses. Data input requires a set of factors that relate a set of frequency dependent forces.

8.9 RESPONSE SPECTRUM ANALYSIS

Response spectrum analysis, or shock response spectrum, is widely recognized and used for the design of reciprocating machinery subjected to time dependent excitation of the machine's mounting system. The procedure is applicable to machines mounted on vehicles (construction machinery, automobiles), ships and other transport systems. As a military tank traverses over a ploughed field at 25 miles per hour, the diesel engine's suspension system may be subjected to an acceleration level of 10 g's or more in a wide range of frequencies. During a storm pumps and compressors mounted aboard a ship will experience similar levels of floor excitation. Analytical methods are available in many commercial finite element programs. Implementation of response spectrum analysis is associated with modal formulation for transient analysis. In the formulation base motion of a reference point is described. The modal participation factor for each modal coordinate is determined. A modal transient analysis is performed for each mode to determine the maximum values of the modal coordinate. Maximum modal responses are combined to obtain a set of maximum values for the set of physical degrees of freedom. Algorithms for combining shock response spectrum analyses for each modal coordinate use the square root of the sum of squares method or the sum of absolute values.

8.10 NEED FOR COMPLETE CRANKSHAFT AND CYCLE MODEL

Problems are encountered when a single crank segment of the full crankshaft is modeled and analyzed for several 'peak' loading conditions. It is difficult to determine the correct boundary conditions for a model with cut planes at the center of the bearings while maintaining a reasonable level of precision. The cut plane is generally too close to the fillets, and the results obtained from the analysis are at best doubtful. A full model of the crankshaft solves the problem of boundary conditions at the cut planes.

Depending on the level of analysis discussed in section 8.2, three methods may be identified for applying the loads on a crankshaft during engine operation. The lower level method *D* in Fig. 8.1 does not include load-sharing effects between the crankshaft and the cylinder block due to flexibility and deformation. Levels *H* and *K* (Fig. 8.1) do take into consideration stiffness and relative deformation between mating components. In the highest level of analysis (*K*) an additional distinction is that the complete three-dimensional vibration motion of the crankshaft is taken into account. At mid levels *D* and *H* only torsional motion of the crankshaft is included. Note that at all three levels the finite element method is used.

Figure 8.11 illustrates the complete finite element representation of a crankshaft. The fillet between the web and crank pin, and the web and the journal, is a critical location with rapid stress build up. Mesh density, therefore, is critical in the region. Calculated results need to be backed up by experimental tests to establish adequacy of the mesh density and of the total model. Generally, good agreement is obtained by using four or five properly shaped high order elements around a typical fillet cross section. A coarse model may be used for the full geometry, and a local model of the fillet area alone for accuracy in stress prediction. The approach also depends on the level of sophistication of the analysis. A coarser model may use three elements (or elements without mid-side nodes) around a fillet cross section.

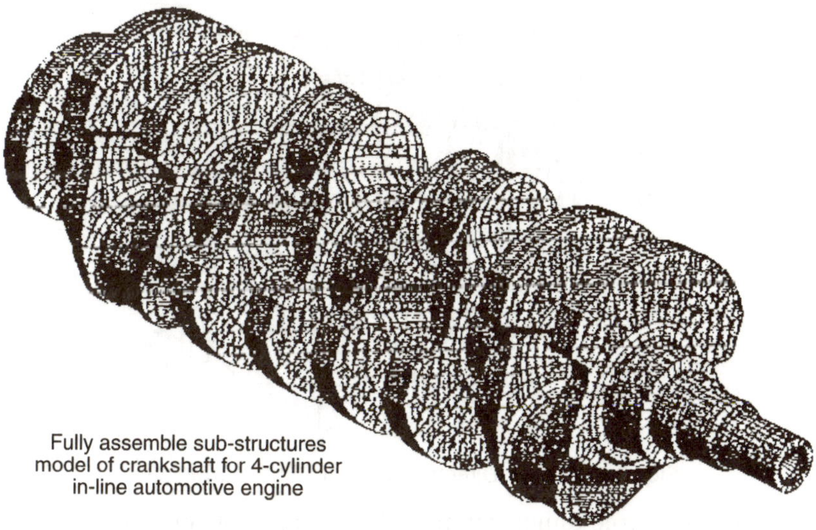

Fig. 8.11. Finite Element Representation of a Crankshaft

The second critical area on a crankshaft is the oil hole break out. This feature may be ignored in the initial finite element mesh. If the oil hole breakout is radial to the cylindrical surface, the finite element method may be used for determination of nominal stress, then combined with the appropriate stress concentration factor from handbooks to predict peak stress. If the breakout occurs at an oblique angle, then a boundary element local model may be required to calculate the stresses in the region. These alternate methods have been found to provide good correlation with strain gage results; also, cost in terms of model preparation and computer time goes down substantially.

For a solid crank web geometry, mesh generation packages are available, starting from the basic configuration defined from the CAD system. Pre-processors are also available for the generation of a three dimensional mesh for more complex crank geometry where lightening drillings or cast hollows are included. In selected regions user defined mesh size is available. The feature comes in handy if prior knowledge of the state of stress in portions of the component is known. If access to a super computer with nearly inexhaustible processor and memory capability is available, then the number of node points and elements in the mathematical model present a lesser obstacle.

With a more refined model of the crankshaft (level H of Fig. 8.1), loading data will be at a more advanced level over the statically determinate analysis. The following effects are included:

- The full three dimensional model of the crankshaft is used.
- Crankcase flexibility is included, both in the linear and rotary senses, the data normally being obtained from a separate finite element analysis of the crankcase.
- Lubricant film stiffness is included. The film's dynamic characteristics are evaluated from the bearing configuration, oil viscosity, clearance and shaft speed. As the speed changes, film characteristics are adjusted, using a piece-wise linear approach.

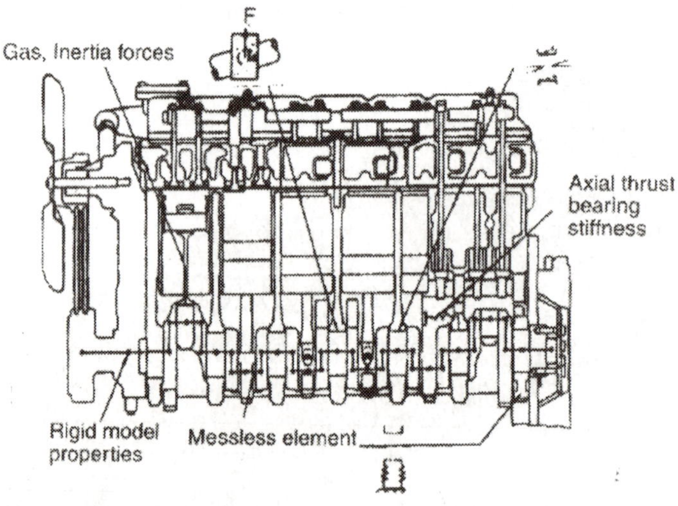

Fig. 8.12. Three Dimensional Load Sharing Model

Figure 8.12 shows a three-dimensional load sharing model of a machine. It is not easy to consider a multitude of loading conditions. Common practice is to consider several 'worst case' scenarios. In order to analyze the crankshaft through a complete machine cycle, four unit loads (two each of forces and moments, or force couples in orthogonal directions) are applied to the full model at the bearings. From these four loads any combination of load direction and magnitude can be developed. The procedure has the advantage that analysis of complete machine cycles for a range of operating conditions may be studied as a post processing exercise. Software to carry out the necessary superposition of load cases is available. When unit forces are applied to a finite element model of a crankshaft they are in the form of a parabolic oil film pressure distribution on the crank pin and the journal, spread over 120°. This approach results in an unrealistic distribution for forces acting midway between the directions of the unit load cases. However, studies on different forms of pressure distribution over bearing surfaces indicate that fillet stress is relatively insensitive to the local form of loading. Investigation of edge loading of the bearing due to misalignment and effects of tolerance on bearing clearance may also be studied with the statically determinate procedure.

8.11 CORRELATION BETWEEN THEORETICAL AND EXPERIMENTAL RESULTS

The assessment of vibration characteristics of a structural system may be split into two stages. The first stage calls for analysis validation by standard checks, such as ensuring that rigid body mode frequencies are at least two orders of magnitude lower than the first non-rigid body mode. The second stage of assessment calls for comparison with experimental data. Frequency correlation can be readily done; mode shapes, however, cannot be so easily compared due to the practical difficulty of visual identification. Forced vibration provides a more reliable key. Accuracy will depend on true representation of the mathematical model, correct application of applied forces and appropriate values of material and structural joint damping.

Instead of simulating all excitations experienced by an operating machine, loads may be selected on the basis of overall requirements. The type of excitation may be of a nominal form (for example, sinusoidal force applied at a point), or use representative force vectors applied to all main bearing housings. To determine the effective forces at the main bearings, one approach is to develop a transfer function of the piston and connecting rod assembly independent of the crankcase. This procedure leads to a reduced model size while permitting a more rapid assessment of design changes in piston, connecting rod and bearings. Velocity is derived at selected locations in the full system due to a combination of forces at other points. The velocity at each location would then be a linear combination of forces at other selected locations and corresponding transfer terms between each pair of points. Selected locations would include points on crankshaft, crankcase, piston and connecting rod. All terms will be complex in nature since both magnitude and phase angle come into play.

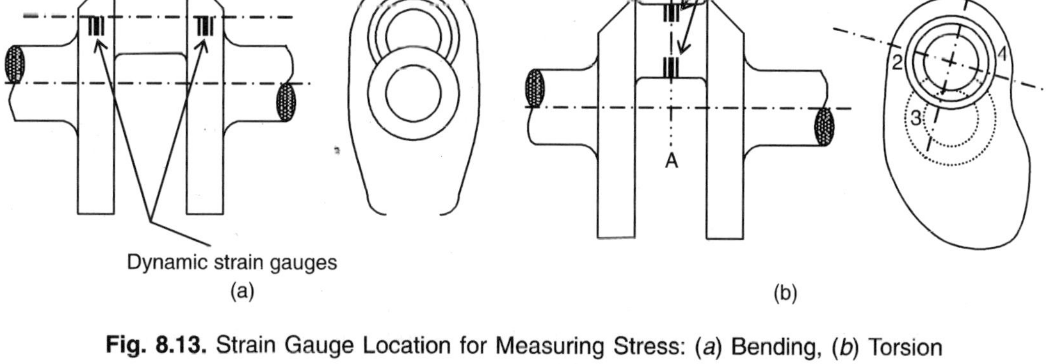

Fig. 8.13. Strain Gauge Location for Measuring Stress: (*a*) Bending, (*b*) Torsion

Coupled torsion and bending vibration stresses in the crankshaft may be experimentally measured. Figure 8.13*a* shows the location of bi-axial strain gauges on the outer surface of each of the crank pins of a V-8 diesel engine (displacing 14.3 liters) for measuring torsional and bending stresses in y and z directions, with the gauges bonded symmetrically about the pin's center line.

Axis 1-3 does not coincide with the z direction, since the lead wire hole coincides with the z axis. Hence these gauges are inclined at an angle of $\pi/9$ radians to the z axis. Figure 8.13*b* shows the location of gauges for measuring bending stress in the crank web. The engine is coupled to an eddy current dynamometer by an intermediate shaft that includes a rubber coupling.

The three-dimensional forced vibration analysis of the shafting system is performed by the transfer matrix method. Figrue 8.14 shows the schematic representation of the full shafting system. Shaft segments are assumed to be elastically supported by the main and thrust bearings of the crankshaft, and by the dynamometer bearings. External forces are applied at the crank pins in the normal and tangential directions. Shaft segments are divided into several smaller segments. Using the finite element approach, transfer matrices are determined for each segment to describe stiffness and inertia characteristics. At each shaft station motion in the axial, torsional and bending in two lateral directions is considered. The state vector is made of the following components: 3 displacements, 3 rotations, 2 bending moments, 3 shear forces, 1 torque and the unity term, making a total of 13 components. The transfer matrices and the transfer equation between the left hand side of the system and the right hand side is expressed in the conventional manner. When free boundary conditions at both ends are assumed, values of forces and moments in the state vectors at the left and right ends are zero. Values for all three terms for displacement only are not known, providing a set of complex linear equations for the prescribed boundary conditions. Numerical precision of the predicted results improves by using non- dimensional terms for the elements of the transfer matrix and state vector before the matrix multiplication process. Wakabayashi provides a complete description of three dimensional analysis of forced vibration by the transfer matrix method for crankshafts. The method includes calculation of coupled bending and torsional vibration stress.

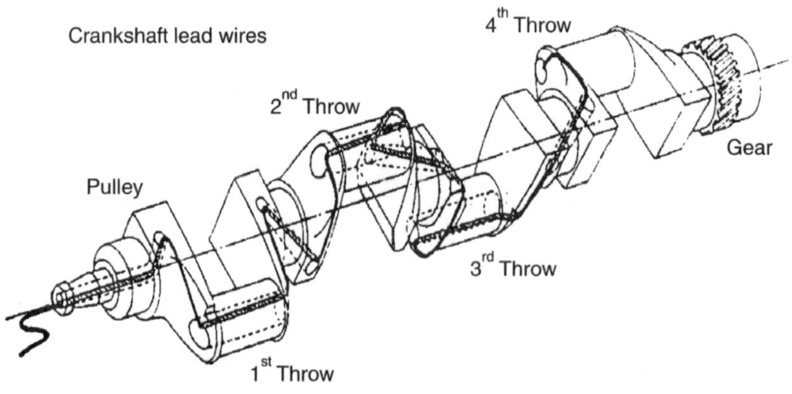

Fig. 8.14. Full Crankshaft With Lead Wire Arrangement for Strain Gauges

Figure 8.15 provides a comparison between calculated and measured torsional vibration stress at the center of crank pin for 8th order vibration. Similar comparisons of the bending vibration stress in the y and z directions are shown in Fig. 8.16 and 8.17. Agreement in the predicted and measured peak torsional stress is good, with the 4th crank pin experiencing 14 MPa at 1,550 rpm. Bending stresses in both planes agree between 60 and 80 percent. Differences arise primarily due to the variations in bearing in bearing stiffness characteristics.

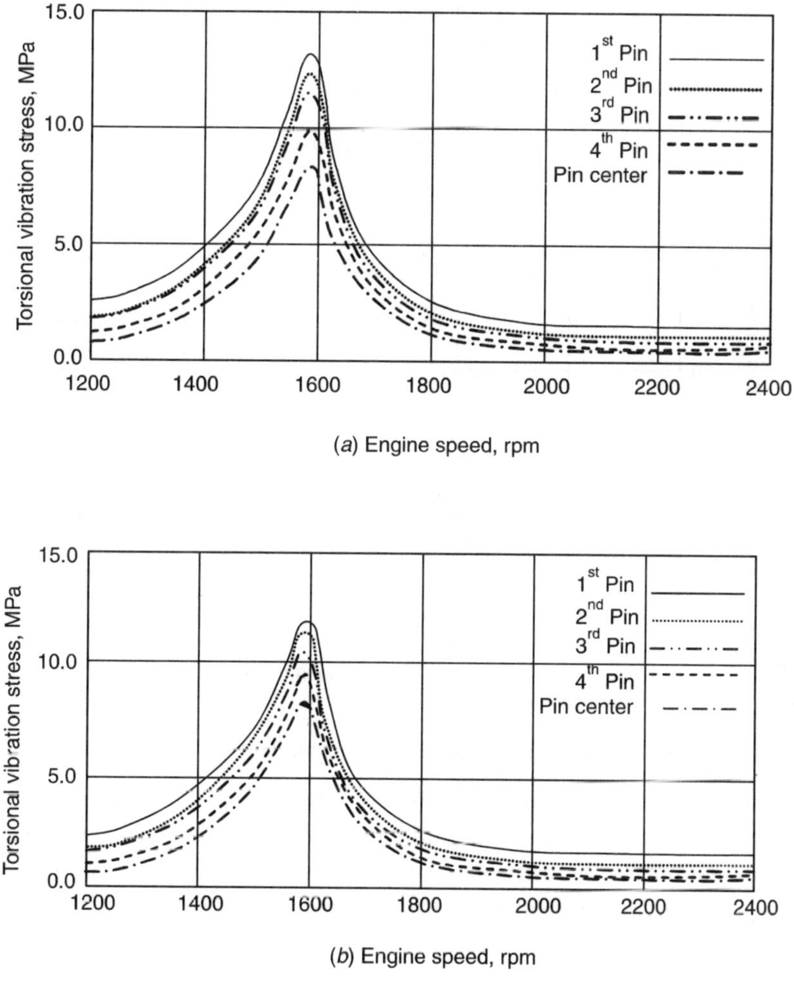

Fig. 8.15. 8th Order Torsional Vibration Stress (*a*) Calculated, (*b*) Measured

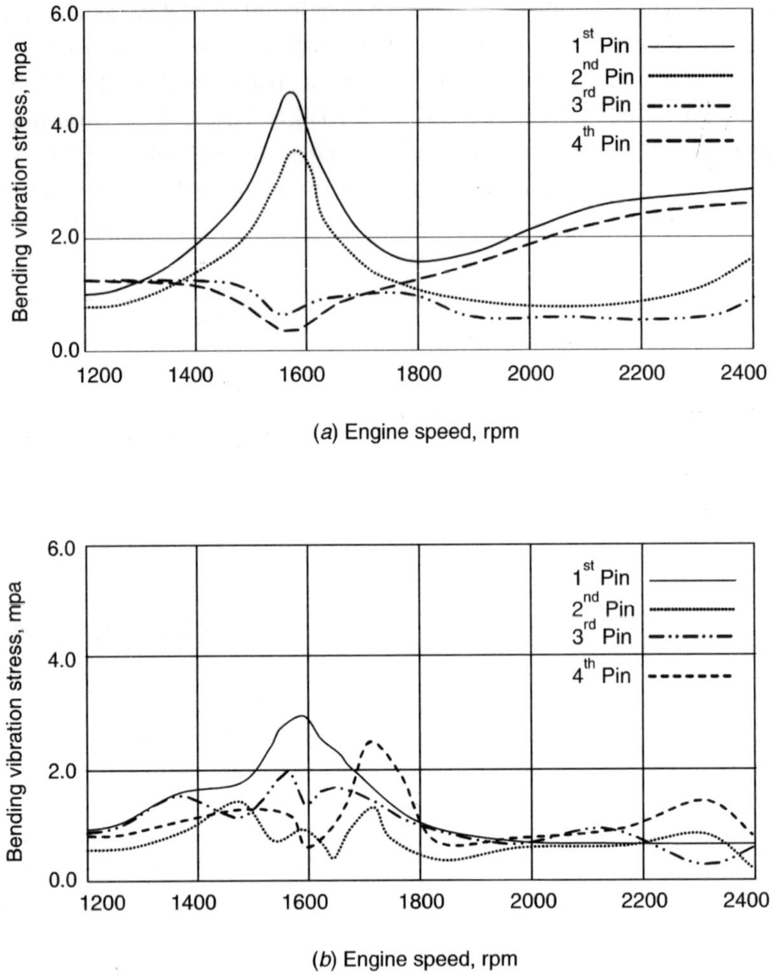

Fig. 8.16. 8th Order Bending Vibration Stress, *y*-Direction (*a*) Calculated, (*b*) Measured

Bending stresses in the 1st and 3rd pins actually reach a minimum stress level at 1,550 rpm. As bending action increases in the main journal and the crank pins, stiffness values of the lubricant film change substantially in a given speed range. Torsional vibration stress calculations are independent of bearing stiffness, and show better correlation with measured stress.

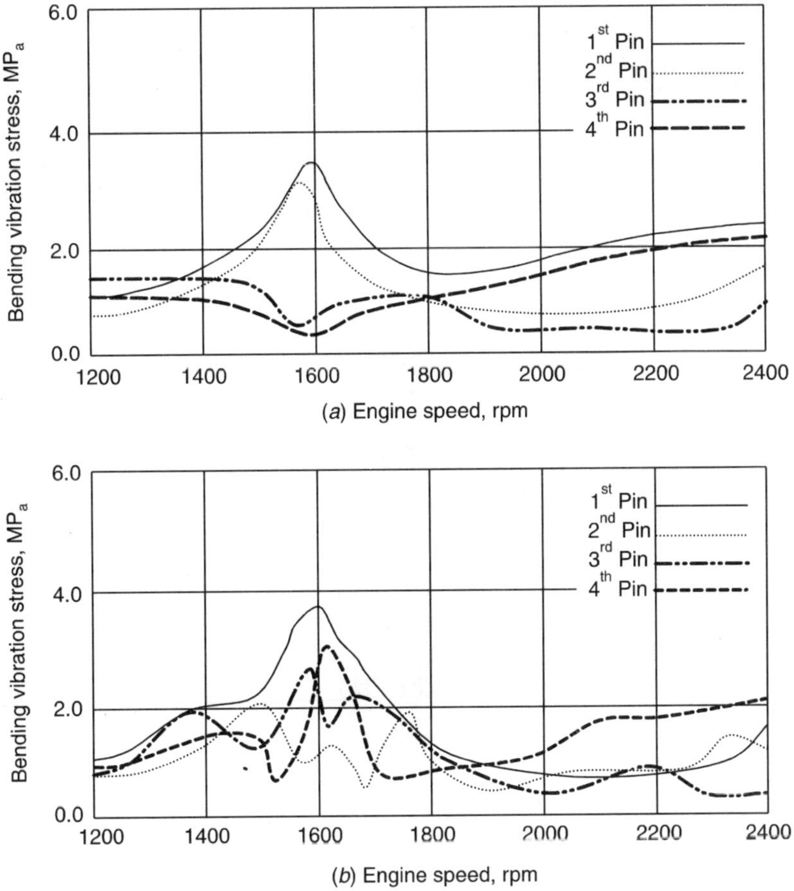

Fig. 8.17. Calculated (*a*) and Measured (*b*) 8th Order Bending Vibration Stress, *z*-Direction

8.12 EXAMPLE OF FLEXIBLE FLYWHEEL MOUNT

Analytical and test investigation of a crankshaft and flywheel system installed in a four-cycle, six-cylinder in-line gas engine revealed bending resonance in the 250 to 350 Hz frequency range for low to medium speeds ranging from 3,000 to 4,500 rpm. Spectrum components for orders 4, 4½ and 5 were noted to be prominent around 4,000 rpm. To compound the problem, order 3 rapidly increased around 6,000 rpm. The last indication may be attributed to degradation of linear behavior of the system in the high-speed region, caused by a sharper rise in the second order component of the inertial force from reciprocating masses.

Initial efforts to reduce the bending vibration of the system concentrated on increasing rigidity of the crankshaft and its support system. Bearing beams and increased bulkhead plate thickness were adopted, but success in eliminating bending resonance out of the frequency range proved elusive.

To tune the frequency, a flexible flywheel illustrated in Fig. 8.18 was used. The flywheel mass is connected to the crankshaft through a flexible plate. The plate is elastic in the bending direction, being rigid in torsion to avoid affecting torsional modes. The plate, therefore, acts as a spring between the flywheel's mass and the crankshaft. To suitably tune the flexible plate so that resonance falls outside the problem frequency domain, the bending mode was split into three categories: resonance consisting mainly of flywheel flap mode below 200 Hz, resonance of flywheel longitudinally below 200 Hz and resonance of crankshaft vibration mode above 900 Hz. Second and third order bending stress in crank pin # 4 for the conventional and flexible flywheels is displayed in Fig. 8.19. Third order peak at 5,400 rpm is eliminated in the flexible flywheel, while also reducing the second order component above 6,000 rpm. Figure 8.20 provides spectral engine vibration comparison, and Fig. 8.21 gives engine noise comparison. Thus, reduction in vibration as well as noise level was accomplished. Chapter 12 provides more discussion on vibration and noise.

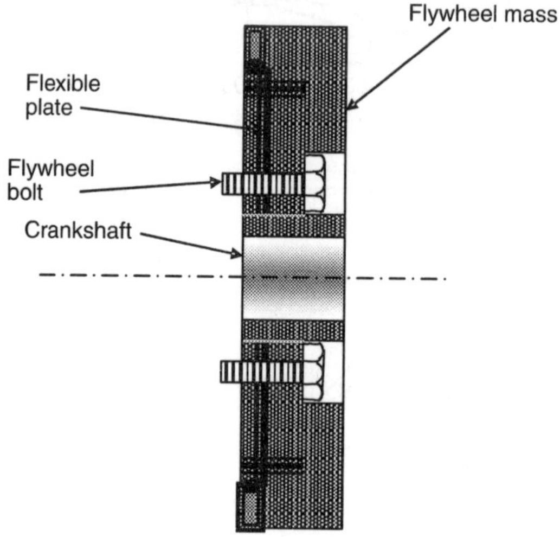

Fig. 8.18. Flywheel With Flexible Plate (in Bending Direction)

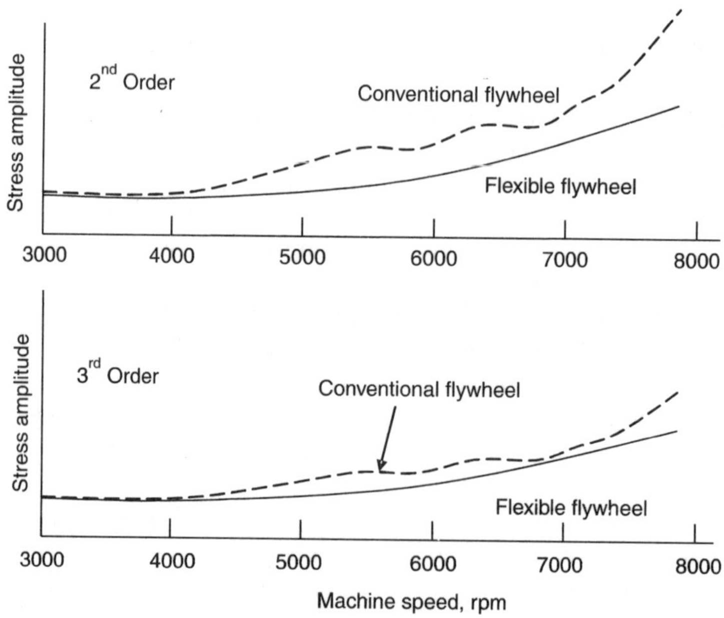

Fig. 8.19. Bending Stress in # 4 Crank Pin

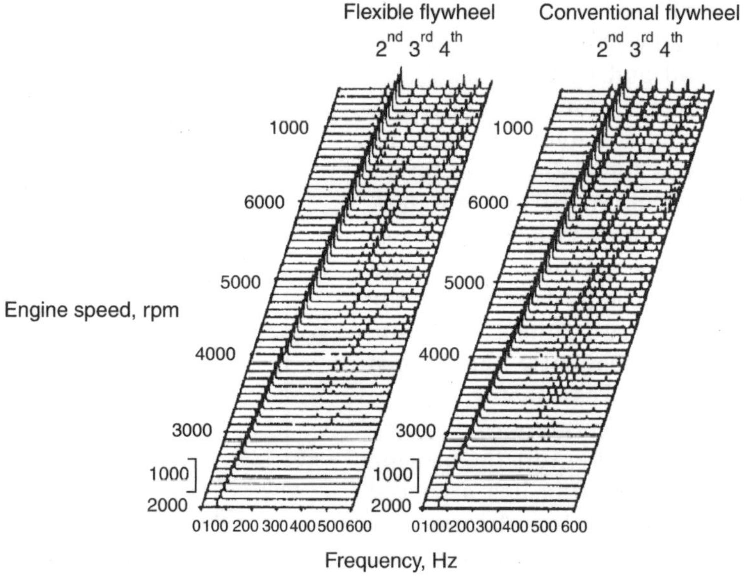

Fig. 8.20. Comparison of Engine Vibration

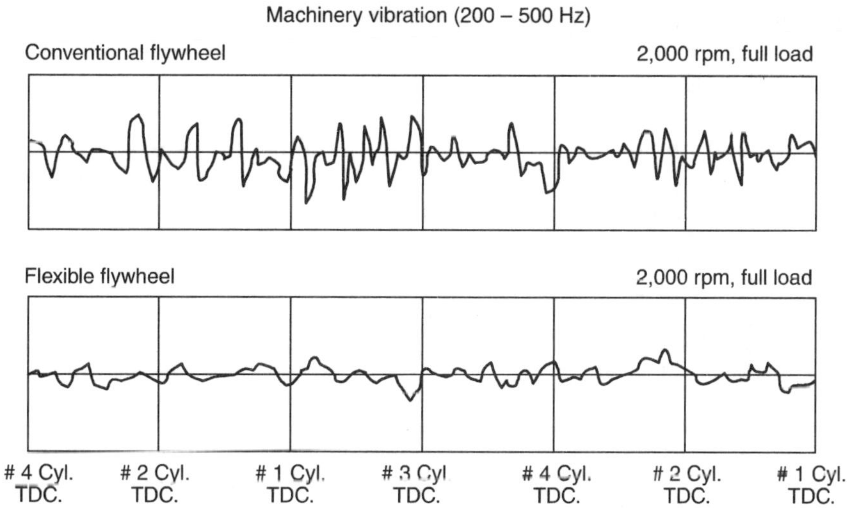

Fig. 8.21. Comparison of Engine Noise

8.13 BEARING ORBITAL MOTION

Bearing calculations are performed to determine peak lubricant film pressure from which it may be sized, and to specify the clearance range. If the bearing operates with a below than normal clearance, excessive heat build up can cause film breakdown. If the clearance is too large, loss of stiffness and bearing action will result.

Bearing oil film calculations can be approached with several degrees of complexity. The mobility method (Booker, 1971) of calculation suffices for many applications. The method is based on the short bearing assumption, and uses the p-film theory. The oil film is assumed to be present only in half the circumference of the bearing around the load vector. Since the journal is static, journal load and hydro-static forces must be in equilibrium. Journal motion is predicted by balancing the dynamic external journal loads and the hydro-dynamic forces induced by the movement of the journal. Leakage through the sides is accounted for. Disadvantages of this method arise from inadequate accounting of oil feed features in the bearing, since the procedure assumes the oil film to be uninterrupted. Consequently, the method may provide acceptable results for a plain cylindrical bearing or a circumferential groove bearing. When the lubricant enters the bearing from a hole, or when a bearing has a partial groove, the accuracy of the results may be compromised. The Integrated Lubrication and Dynamics Model (Vijayraghavan, 1989, Woods, 1989, Paranjpe, 1990) is more refined in its approach, and uses a numerical solution of the full 2-D Reynolds equation. Using the finite-difference technique, the position of the rupture boundary is determined explicitly within Reynold's equation. Journal center orbit under dynamic loading is determined by solving the equation of motion. The method is advantageous in that oil groove shape, size and location can be accurately modeled. Elastic deformation of the bore and journal can also be accounted for. Cavitation, or pockets of vacuum, is taken into account if they exist. This phenomenon can cause film collapse and implosion, as also damage to the bearing surface. Figure 8.22 shows the trajectory of the journal in a bearing and Fig. 8.23 illustrates its instantaneous hydro-dynamic pressure. At each time step the calculation executes the following sequence:

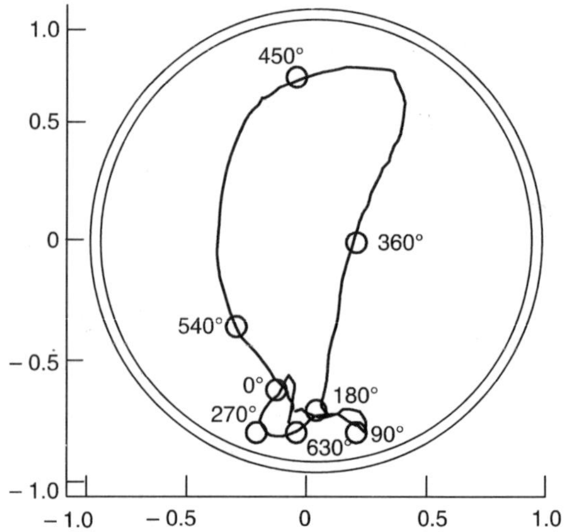

Fig. 8.22. Predicted Trajectory of Journal in its Bearing

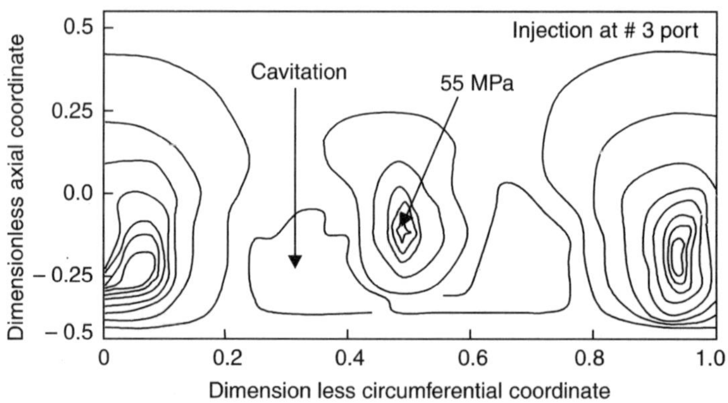

Fig. 8.23. Instantaneous Hydro-Dynamic Pressure Contours in a Journal Bearing

1. Solve Reynolds equation from velocity, density and bearing geometric data.
2. Integrate pressure over the journal surface, and balance it with inertia and pressure load.
3. Determine new journal position.
   Return to step 1.

This procedure is computationally intensive, taking several hours on a computer workstation. The mobility method, on the other hand, takes a few minutes.

Computer software does not employ sophisticated analysis; it applies the equations of motion to complex systems. Friction, flexible bodies and exciting forces can be included in the procedure. Mathematical models can also be associated with CAD and FEM software. An example of an application is shown in Fig. 8.24 and 8.25. Here it is required to calculate the forces on the shorter connecting rod journal for evaluation of the bearing's lubricant film thickness. The dynamics model is set up, and forces on the connecting rod journal are calculated and plotted in Fig. 8.25.

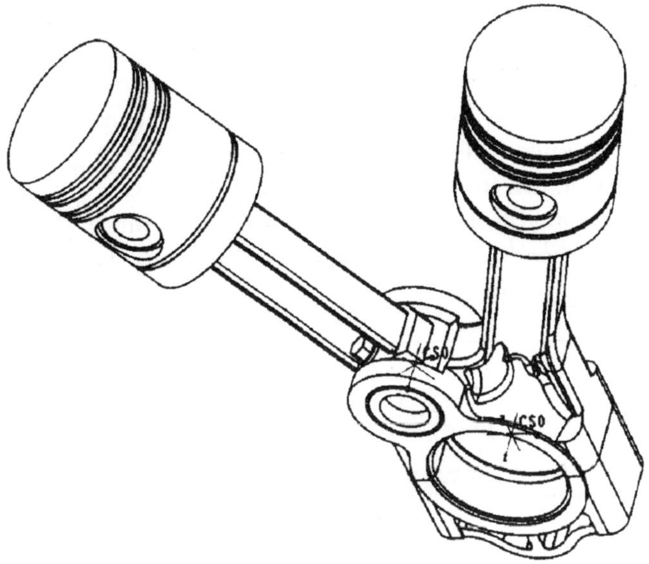

Fig. 8.24. Piston, Connecting Rod Assembly for a Large Stationary Engine

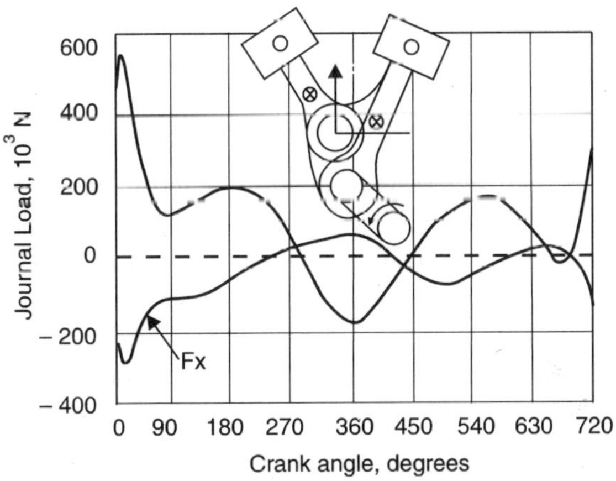

Fig. 8.25. Force Components on Connecting Rod Journal

Computer software to simulate the kinematic motion of slider and crank mechanisms are available. ADAMS™, DAD™ and Working Model™ have found wide acceptance and usage. Motion and loads between components can be simulated to simplify the dynamic analysis of the assembly's components.

8.14 GUDGEON PIN BEARING

The bearing at the small end of a connecting rod of a reciprocating machine operates in an environment where lubrication conditions are often times severe. This results in abnormal wear, seizure and also cracking. A possible remedy, demonstrated here, calls for use of full-floating piston pins.

Friction force can be measured by floating the pin from the connecting rod with a spring and measuring the spring displacement. Phase difference between the rotational oscillation of the pin and the oscillating motion of the connecting rod then provides a measure of the friction force. The measuring device may be provided with roller bearings to minimize extraneous sources of friction. Strain gauges are mounted on a beam linking the connecting rod to a holder fixed to the piston pin, as shown in Fig. 8.26. When friction force F_b is present at the pin boss, an opposite force *Fb′* acts on the beam, trying to move the pin with the boss, as shown in Fig. 8.27. Hence, the beam bends in proportion to this opposing force, and so the friction force is determined from the strain gauges.

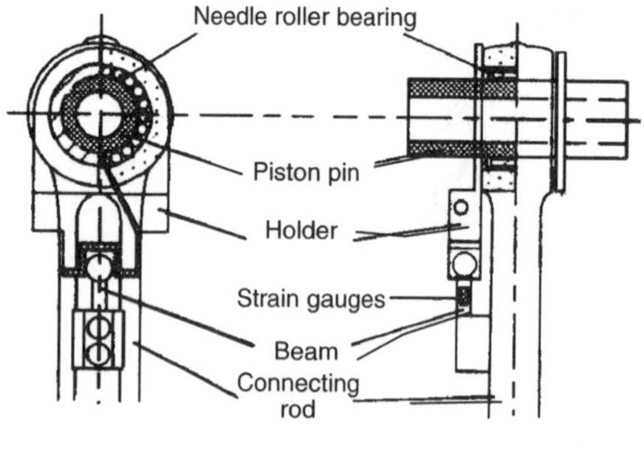

Fig. 8.26. Friction Force Measuring Arrangement

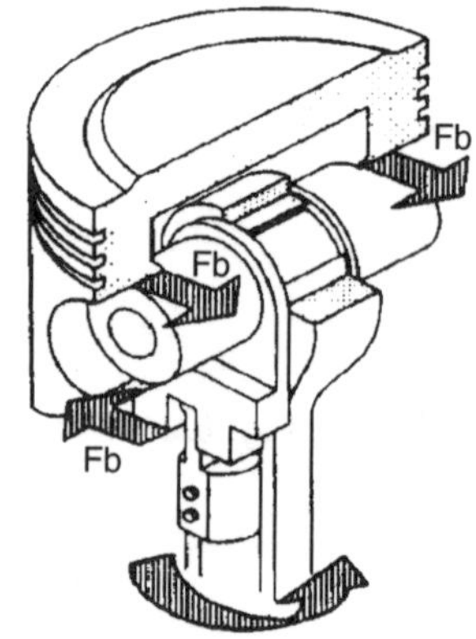

Fig. 8.27. Forces on Pin

From the friction force readings the coefficient of friction can be calculated, using inertial force and cylinder pressure as the load. Friction forces acting on the piston of an automobile engine (86 mm bore, 78 mm stroke) is used here for illustration. The results are plotted in the form of Stribeck diagrams in Fig. 8.28, together with the friction force wave form for comparison. Crank angle varies from 90° before top dead center to 160° after top dead center. Circled numbers on the friction force curve correspond with those on the friction coefficient curve. Overall profile of the Stribeck curve suggests that the coefficient of friction decreases as the load decreases up to the compression top dead center, then rises suddenly and moves into the zone without fluid lubrication. Then, just before 90° after top dead center the curve reaches a level portion, gradually rising again till the crank angle is 160° after top dead center. At full load in particular, the coefficient of friction increases, accompanied by an expansion of the non-fluid lubrication zone. This implies that even though load capacity of the bearing increases due to the increased sliding speed, the oil film remains thin. One reason for this may be due to inadequate flow of lubricant oil to the bearing. Bearing clearance also plays a role. Under partial load there may not be an apparent change in the friction force when the bearing clearance is changed. However, at full load, between 90° before top dead center and 180° after top dead center, the friction force varies substantially, peaking just after compression top dead center, when operating with a reduced clearance. From this observation, it may be inferred that further reduction of bearing clearance for the purpose of reducing noise will come at the expense of higher friction loss and lower efficiency.

8.15 CYLINDER BORE DISTORTION

Distortion of cylinder bores (or liners, when used inside the bore) can have a major impact on the operation of the machinery. Wear characteristics and consumption of fuel and lubricating oil are adversely affected when cylinder walls experience a large amount of deformation. To achieve the ideal condition for function and durability, the profile of the walls or liners should be as circular as possible. Manufacturing tolerances alone can cause the bore to deviate from this condition. Deviation from the ideal geometry by thermal and mechanical loads induced during operation will exacerbate the situation. Distorting loads arise from many sources:

- Mechanical load when assembling cylinder head and gasket with cylinder block.
- Differential thermal expansion of the cylinder head and block. The relative expansion becomes even more severe when materials with differing thermal expansion coefficient are used for the two components.

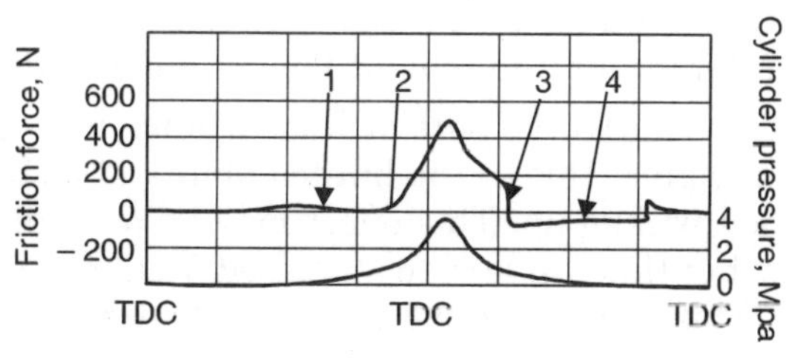

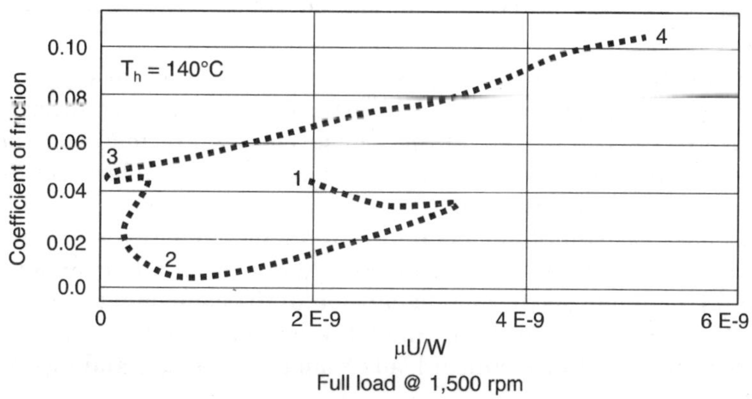

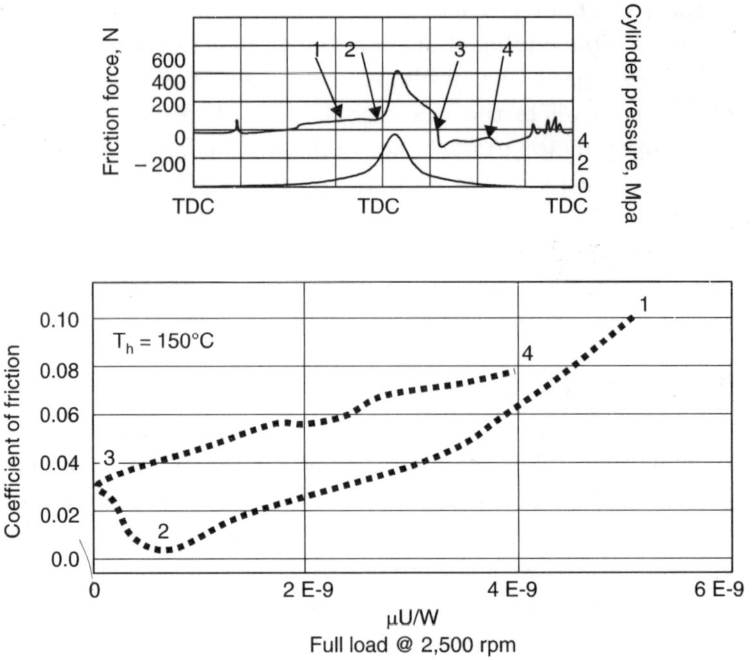

Fig. 8.28. Piston Pin Bearing Friction and Stribeck Diagram

- Pressure build up inside the cylinder barrel. In the case of internal combustion engines, gas pressure rises dramatically following ignition.
- Temperature gradients during operation. Significant temperature gradients also exist between liners and cylinder block and between cylinder head and block, arising primarily from different air and coolant flow in the cylinder block and the head.

Any non-circular distortion in the bore or liner will have a significant influence on the operational behavior of an engine. Due to the limited ability of piston rings to conform to the bore shape, oil consumption will go up. This effect may be partially offset by increasing the tension in the piston rings, but this will lead to increased friction losses and hence to increased specific fuel consumption. Thus, the incentive to reduce distortion in the cylinder bore exists. As an added advantage, low distortion liners or cylinder bores lead to shorter break-in time periods and to reduced wear; and reduced oil and fuel consumption will lower particulate and hydro-carbon emissions.

Bore deformation measurement may be carried out in different ways. The deformations can be determined indirectly by measuring the external contour, or by directly measuring the internal bore surface. As a first step, static measurements are made on the cylinder block without assembling the cylinder block, from which manufacturing tolerances during machining of the block are established. Bore geometry is then measured after assembling the cylinder gasket and head, enabling evaluation of bolt force influence on bore distortion. In the last step, bore measurements are made on the assembled cylinder block and cylinder head filled with heated cooling fluid. Differential thermal expansion of the components can be determined for varying temperature levels of the fluid. Since the engine is not operating, distortion from combustion temperatures and mechanical loads from gas pressure and inertia are not known at this stage.

Cylinder deformation during operation can be determined by using direct and non-contacting measurement. Eddy current inductive sensors continuously measure the gap between two sliding components. By installing eight (or more) sensors on a piston, the cylinder contour is scanned as it moves axially in the cylinder bore. Figure 8.29 illustrates the method of installing the sensors using a central support and tube, with the sensors spaced 45° apart. Since gap between the wall and piston ring is of interest, the sensors are located in the land between two rings. To avoid effects from piston distortion, the inductive sensors are fixed inside carrier tubes. The central support and tubes are made of a material with a low coefficient of thermal expansion, and the tubes are free to float within the local area. Lead wires are taken out through a link rod and rocker mechanism to a fixed external mechanism. The measured signal is amplified and recorded.

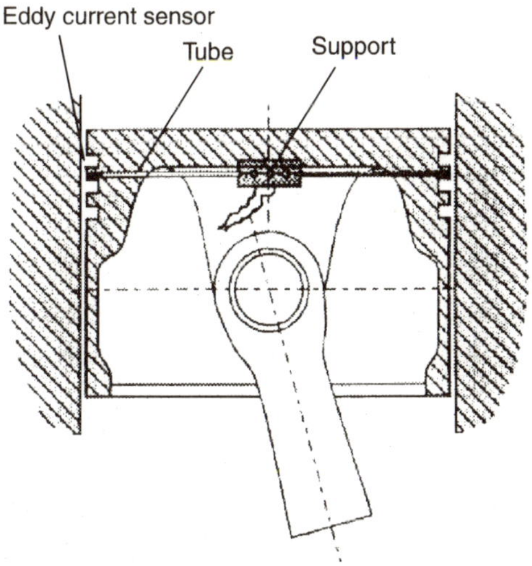

Fig. 8.29. Eddy Current Sensor for Measuring Gap between Bore and Piston Ring

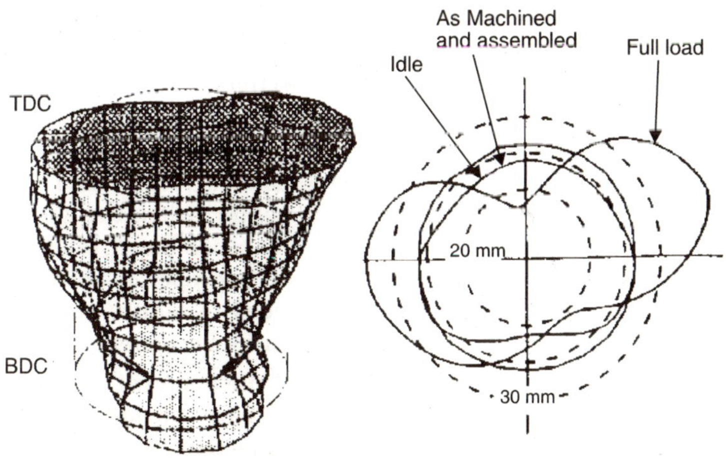

Fig. 8.30. Cylinder Bore Distortion

Fourier analysis of the measured signal determines deviation from the ideal cylinder bore profile. To calculate appropriate Fourier orders, the number of sensors must be twice the order to be determined. Thus, eight sensors will provide data up to the fourth order. Assignment of the overall deformation characteristics to the various orders provides valuable insight into the causes of deformation.

Figure 8.30 shows the influence of various operating conditions on bore deformation at full load and at idle condition. Measurements from machined and assembled bore are also provided for comparison. The measured liner is in an engine with wet liners made of cast iron, and the cylinder head is made of aluminum. When operating at part load thermal stress in the cylinder head is lower. This causes lower level of distortion in the longitudinal direction.

Dunavesky (1990) studied the effects of cylinder bore distortion on oil consumption. The objective is to match bore distortion with the elastic deformation of the piston ring. From these studies the equation defining the limit of bore distortion at which the clearance between piston ring and bore can be kept to zero is proposed:

$$A\omega = \frac{1.52\,(Qr^3)}{ht^3\ E(\omega^2 - 1)} \tag{8.2}$$

where $A\omega$ = Limit of bore distortion

ω = distortion order

h = piston ring height

t = piston ring thickness

r = piston ring radius

E = elastic modulus of piston ring material

Q = piston ring tension

Extensive research has been done on this subject for automobile engines. Geometric configuration of the cylinder block undoubtedly controls the distortion, but thermal distortion is the biggest contributor. Abe (1995) conducted a study of an engine with siamese cylinders in a 4-cylinder in-line engine using the finite element method of analysis and experimental measurements. Second and third order thermal distortion were determined to be more significant. Fourth order distortion from cylinder head clamping reduced as a result of decreased seal face pressure as the head tries to lift from the block. Gas pressure affects bore distortion marginally, but it may increase the second order distortion.

8.16 PISTON MOTION AND SKIRT FRICTION

The skirt of a piston is apt to be placed under a severe load as it is pushed against the cylinder wall by the side force near the compression top dead center, when the sliding velocity is low. Requirements to reduce engine size and weight, on the other hand, tend to reduce the skirt area and clearance, causing the piston to assume an improper posture in a region where lubrication conditions are far from ideal. Hence, the need to reduce piston skirt friction and improve lubrication for increased engine durability and fuel economy can be readily seen.

Figure 8.31 shows the force acting on the skirt of the piston of an automobile engine (86 mm bore, 78 mm stroke) due to friction. Two speed points are considered, 1,500 rpm and 2,500 rpm, with the engine operating under full load. Friction force at the lower speed displays non-fluid friction characteristics, with peaks at the top and bottom dead centers. Also, the force is greater during the expansion stroke than in the others, and reaches its maximum value during the former half of the expansion stroke. At the higher speed level the peaks at the top and bottom dead centers are not so pronounced, and friction characteristics change to the type representative of full lubrication regime. The markedly increased side thrust on the piston presumably causes high friction force during the expansion stroke.

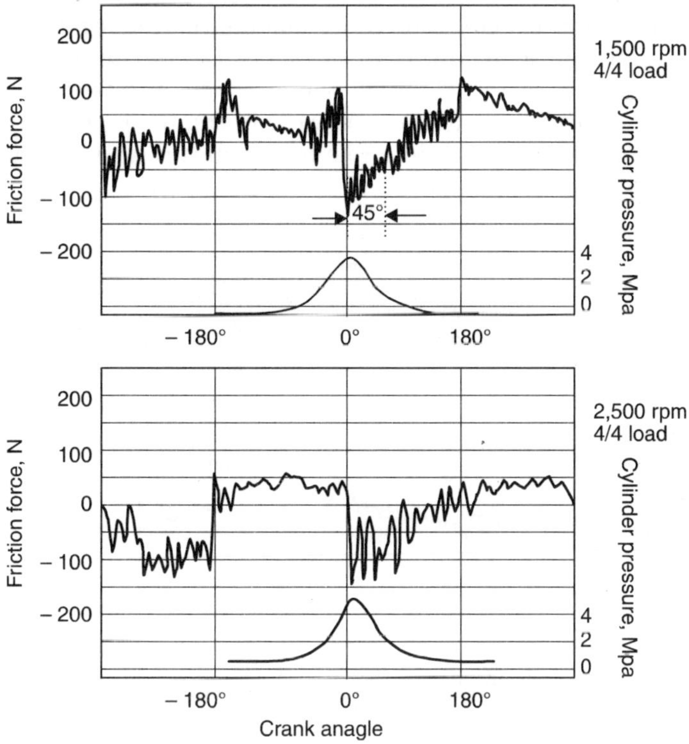

Fig. 8.31. Friction Force on Piston Skirt

Piston slap motion also affects the friction force. Figure 8.32 shows measured results of piston slap motion on the thrust side where gap sensors TFU and TFL are located in the front of the skirt. The rate of wear in this region is the greatest. Piston slap motion at the center of the skirt on the thrust side is included as measured by sensors TU and TL. The mechanics of slap motion during the expansion stroke may be described as follows. The lower portion of the skirt on the front side contacts the cylinder after the piston reaches the top dead center of the compression stroke. The contact area increases to the upper portion of the skirt center. The time point when TFU and TFL contact the cylinder is shown by arrow '*D*', and occurs when a large friction force acts during the first half of the expansion stroke. The increase in the friction force is attributed to the metallic contact between the cylinder wall and the rigid front portion of the skirt on the thrust side of the piston, near the piston side wall. Sensor readings at both speed points are similar, except for sensors TFU and TU at 2,500 rpm, and are indicated by arrow '*B*'. It may be surmised that the large contact pressure at the front skirt portion on the thrust side occurs at the lower speed point, but not at 2,500 rpm. From these observations it can be deduced that piston slap motion at the front part matches with contact of the skirt in the same region.

8.17 PISTON RING DYNAMICS

Piston ring flutter is a major cause of wear in both the rings and in the seating grooves. Unusual wear pattern is important for controlling the flow of oil in the piston, ring and cylinder bore (or liner) system. Oil scraped by the first, or top, ring is an important part of the total oil consumption in reciprocating machines. Diesel engine manufacturers carry out substantial amount of research work to reduce and control lubricant consumption, since oil constitutes a major contributor to particulate emission in the exhaust gas. Oil consumption is also affected by blow-by, since it is a source of corrosive combustion products and abrasive soot particles that are responsible for lubricant oil degradation and wear of engine components. The dynamics of both the piston and the rings play a leading role in the matter, and require careful attention during the development phase.

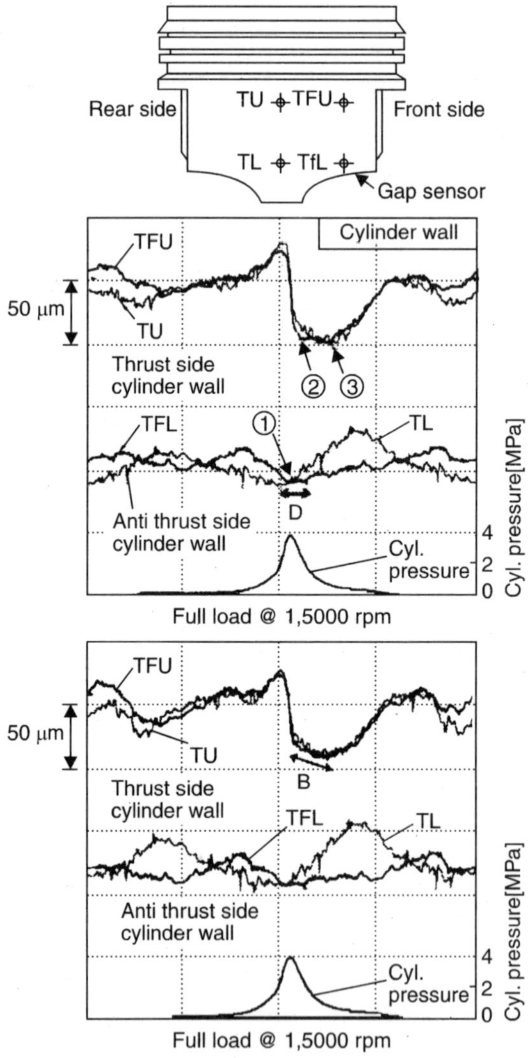

Fig. 8.32. Piston Slap Motion

Axial and angular direction dynamic motion of piston rings depend on a number of factors:

- Static twist in the ring. Wear pattern of the piston ring and groove change after a standard machine operating run. The worn profile of the upper running surface of the ring is primarily determined by the tilt angle of the lower surface of the seating groove during operation.
- Relative angle between the sides of ring and the sides of mating groove.
- Contact pressure between the ring and cylinder bore, and between the rings and piston grooves.
- Condition of lubrication between cylinder surface and ring.

Contact pressure and relative sliding motion between two mating surfaces will determine the wear pattern in both components. Axial and torsional vibratory motion of the rings will arise from forces and moments on the sides of the rings. The loads are generated by gas pressure, pressure in the oil between the ring and the groove and pressure arising from asperity contact between the ring and the groove. Figure 8.33 illustrates various pressures on a typical ring. Friction force on the ring's running surface generally has negligible effects on the dynamics of the ring. Overturning moments generated from the radial pressure on the ring's running surface due to increased gas pressure does not affect ring flutter. The profile of the running ring's surfaces change with the ring twist, which alters the overturning moment generated by the radial pressure on the ring's running surfaces. This effect can be coupled into the angular momentum equation by including an estimated term based on equilibrium conditions.

The profile of the worn upper wedge of the top ring is critical for oil consumption in diesel engines. The ring scrapes up oil on the liner, at the side opposite to the thrust, during late compression stroke. This top ring scraping is due to the combined effect of the tilt of the piston and the ring's dynamic twist. An estimate of this oil up-scraping rate provides an indication of the oil consumption value. With the understanding of the mechanisms of the top ring and groove wear and the worn top ring's up-scraping rate, potential to optimize the profiles of the ring and the groove to minimize oil loss exists.

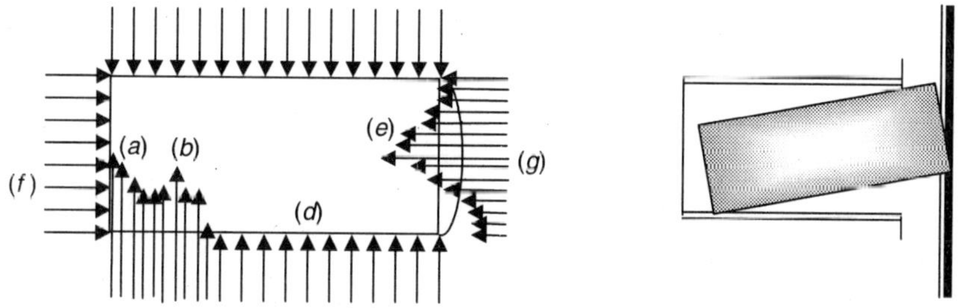

(*a*) Asperity contact pressure
(*b*) Oil squeezing
(*c*) Gas pressure (upper land)
(*d*) Gas pressure (lower land)
(*e*) Ring/liner lubrication
(*f*) Gas pressure from region behind ring
(*g*) Friction

Fig. 8.33. Pressure Sources on a Piston Ring

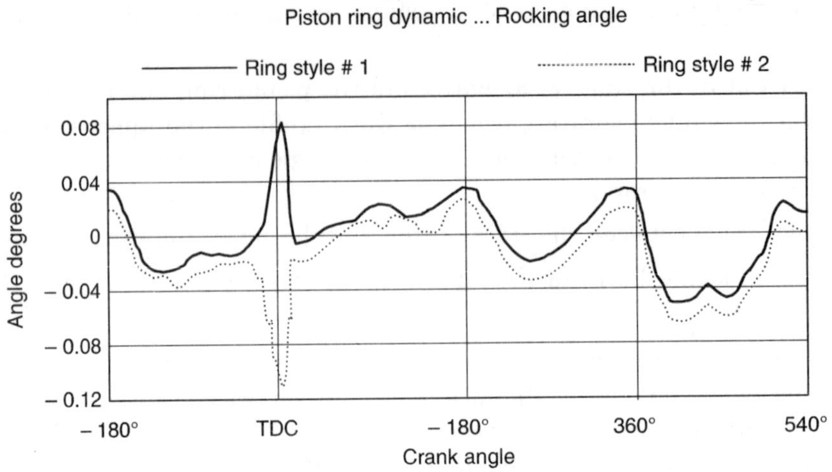

Fig. 8.34. Rocking Motion of a Diesel Engine Piston (*Courtesy:* Ricardo Corp.)

A negative static twist in the second (bottom) ring can result in flutter during the late part of compression and early part of the expansion stroke. As the ring flutters, gas is blown down through the clearance between the ring and the groove, and the pressure in the land area also drops. This may be beneficial for oil consumption control. But if a reverse flutter occurs in the top ring with a positive static twist at a high operating speed, oil may be induced to flow into the combustion chamber, causing loss of oil.

Lateral and tilting motion of the piston also alters the ring's running surface profile. The motion arises from the forces, moments and thermal loads acting on the piston. Elastic and hydrodynamic actions between the piston skirt and the be without breaks, or the twisting effect in the ring may be ignored. A few representative results from ring dynamics calculations for a diesel engine are shown in Fig. 8.35, showing gas pressure between the first and the second rings. Figure 8.36 shows the axial movement of the rings within the groove.

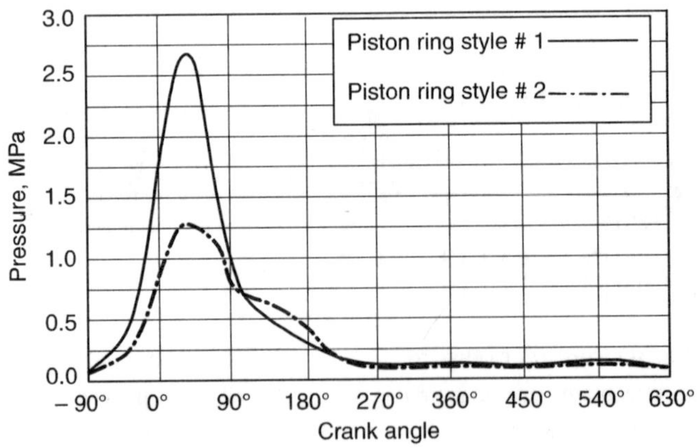

Fig. 8.35. Gas Pressure Between First and Second Piston Rings (*Courtesy:*Ricardo Corp.)

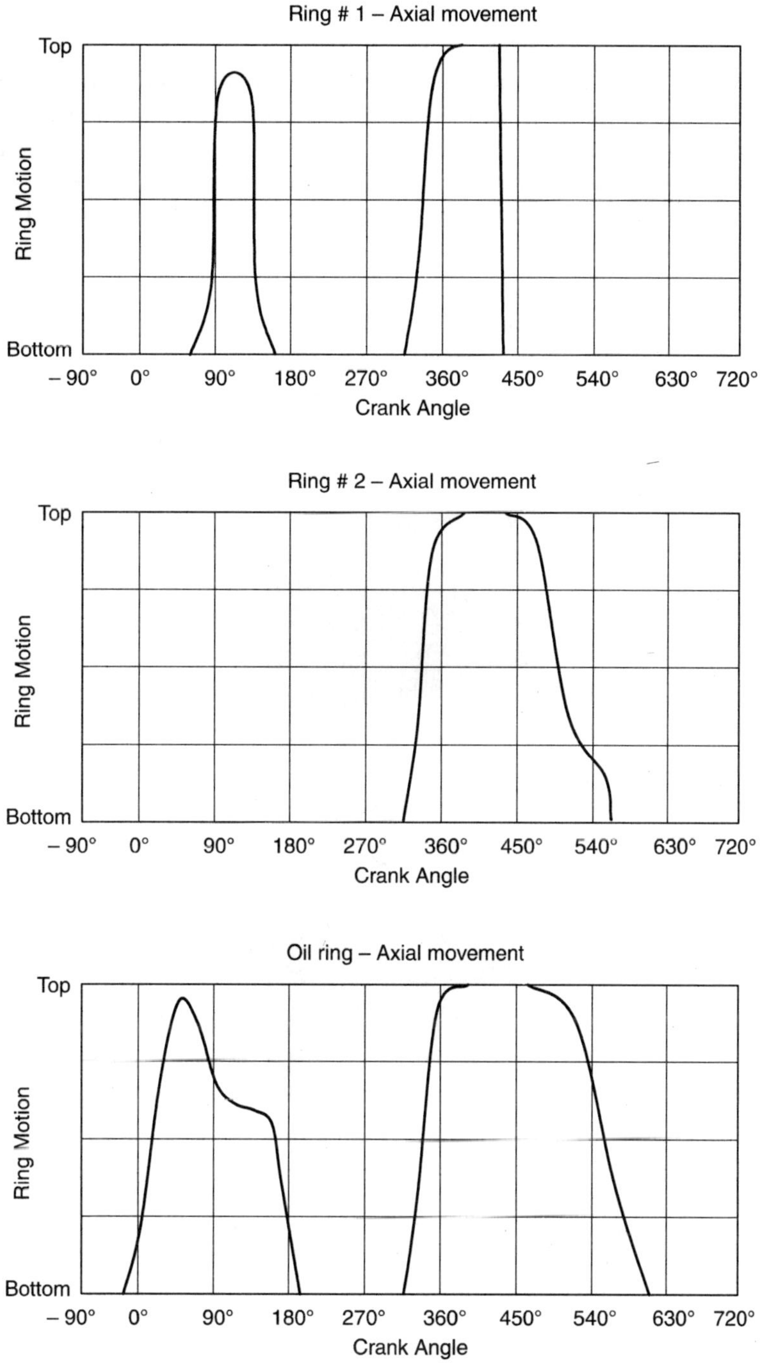

Fig. 8.36. Axial Movement of Piston Ring Inside the Groove (*Courtesy:* Ricardo Corp.)

8.18 CYLINDER HEAD

Cracking of cylinder heads in large diesel engines is a common problem. Cracks at the valve bridge arise in some applications when the engine operates under certain conditions. As an illustration, consider the case of a class of diesel engines installed at nuclear power plants to serve as standby power generators. The engine is a four stroke direct injection diesel with 343 mm bore, 419 mm stroke, and is required to start and quickly go to full load in the event of a loss of off-site power. The engine is started with compressed air, and is required to reach full load within 90 seconds. Several cylinder heads experienced cracks on the flame face surface, initiating between the pressed-in valve seat inserts. Smaller radial cracks around the counter bores for exhaust valve seat inserts were also present, but they do not preclude satisfactory operation of the engine. Cracks through the wall, on the other hand, increase the risk of coolant leaking into the combustion chamber, and are unacceptable.

A finite element model of the cylinder head (Fig. 8.37) was developed to determine if operating temperatures, thermal and dynamic stresses were high enough to cause cracking. Sensitivity studies were also conducted to examine the effect of engine load, changes in deck thickness and assorted design changes to improve durability. Results from the finite element evaluation were backed up by engine tests for a range of engine load and coolant flow rates.

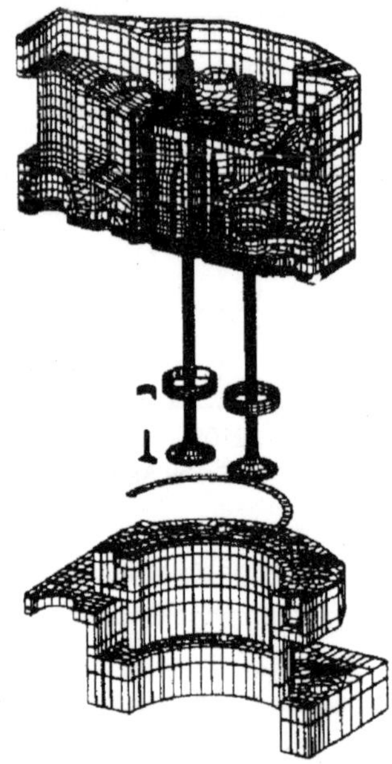

Fig. 8.37. 3-D Finite Element Model of Cylinder Head Assembly

Cracking under the given conditions is normally attributed to low cycle thermal fatigue failure mechanism. The flame face of the cylinder head is exposed to hot combustion gases in the cylinder. Peak temperatures usually occur at the bridge between exhaust valves. Thermal expansion of the bridge section is constrained by the reduced expansion on the opposite surface, where cooler air temperatures prevail outside the cylinder walls. Significant thermal gradients develop at the bridge segment during constant load engine operation, resulting in compressive stresses in the region. When the bridge section reaches a certain temperature level during operation the compressive stresses are large enough to create permanent plastic deformation, and the onset of failure mechanism prevails. As the load is shed, thermal fatigue damage becomes more pronounced. The opposite surface recovers elastically, but tensile stress develops in the bridge segment between the valves, which has undergone plastic deformation. Thermal ratcheting between compressive and tensile states of stress during engine loading and unloading, or starts and stops, results in low cycle thermal fatigue damage, and eventual failure. There are other load sources which aggravate the situation. Cylinder firing events over long periods of continuous operation result in creep. Creep in metal components is a non-linear phenomenon that is driven by four parameters: load (and hence stress), temperature, time duration and creep related material characteristics. In addition to elastic deformation, creep causes plastic deformation, even when the stress is in the linear elastic stress-strain range. As the engine accumulates more cycles damage from creep also increases. The effects of creep become more intense when load cycles are of shorter duration and more severe (going from no-load to maximum load, then back to idle, in a short time span) then when the machine is operating continuously. Consequently, the machine, or the sub-assembly, has a shorter life span.

- Durability of the cylinder head is clearly dependent on operating temperatures.

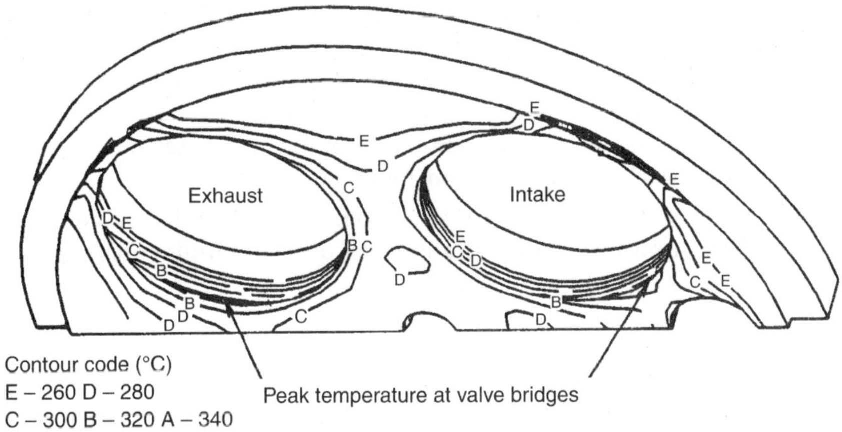

Fig. 8.38. Predicted Thermal Contours on Flame Face of Cylinder Head

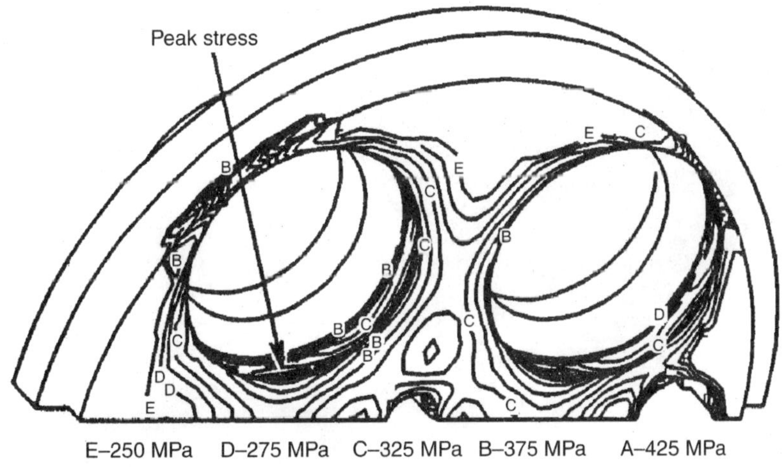

Fig. 8.39. Thermal and Firing Load Stress Contours in Cylinder Head

Fatigue damage will result when operating temperatures produce thermal stresses that exceed the yield strength of the material. For gray iron castings, the temperature limit is set at 380° C, based on service experience of engines of this class. When temperatures exceed 400° C, cracking incident may be considered assured for conventional head designs. Variations in material properties and design parameters (thickness and width of bridge, valve seat insert press fit, among others) will also influence the likelihood of cracking.

Several approaches are available for obtaining reductions in cylinder head operating temperatures and consequent cracking. A sampling list is provided here:

- Coolant flow directed over the bridge section will reduce operating temperature.
- Drilled passages have an advantage over those cast into the component. Coolant flow comes closer to the flame face surface, while controlling variations in metal thickness, inherent with casting defects. Also, heat transfer across the drilled surface is improved.
- Application of thermal barrier coating to insulate the cast iron fire deck reduces flame face metal temperatures.
- Machining scallops, or slots, into the fire deck reduces stress build-up at critical locations.

Adherence to proper operating procedures and maintenance practices are essential in maintaining low rates of failure. A single incident of overload in the form of cylinder over-pressurization, over heating or temporary loss of coolant has the potential to produce sufficient fatigue damage to cause cracks in the cylinder head.

Figures 8.38 and 8.39 provide the results from the finite element analysis. Predicted temperature and stress contours on the flame face of the cylinder head are for 10% over maximum power engine operating condition.

8.19 CYLINDER BLOCK

A heavy-duty engine for commercial vehicles introduced in Japan achieves high power output, low weight and good fuel consumption rate. An innovative feature is the use of dry liners, which is uncommon in naturally aspirated V configurations. This makes it possible to increase the bore size and displacement, while maintaining the bore pitch.

The 90° V type cylinder block (Fig. 8.40 and 8.41) is made of cast iron, and has an integral crank case. The concept of dry liners for cylinder blocks is difficult because of problems

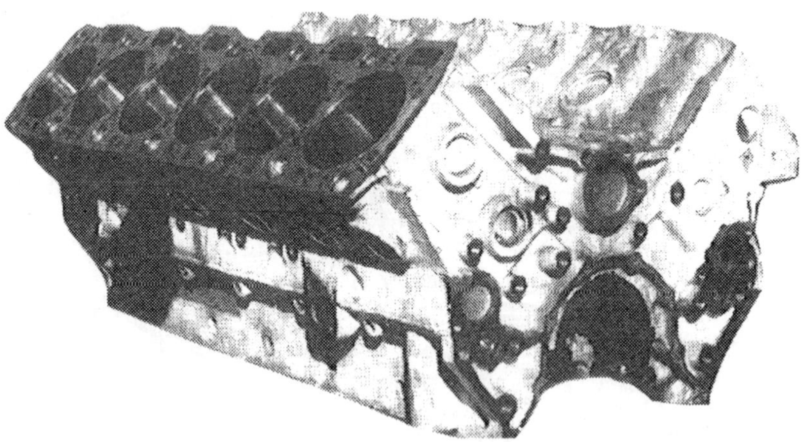

Fig. 8.40. Cylinder Block

encountered during casting. Associated difficulties range from removal of sand from inside the water jacket to ensuring wall thickness around the bore. Core fixing, sand stripping and other casting related processes require improvement for the purpose. The pay-off comes in the form of reduced cylinder block length and total engine weight. Maintaining the stiffness of the cylinder block is important in controlling structural vibrations of the engine system, and to suppress noise. Steel cups and ribs are cast into the block to realize greater stiffness of the basic structure. Optimum size and location of the stiffeners is achieved through a finite element and modal analyses of the block. Figure 8.42 shows the finite element representation. Since the cylinder bores are exposed to high temperatures, holes are drilled between the bores for effective cooling. The liners are made of cast steel with thin walls, and are loosely fitted for easy insertion into the block. To avoid distortion of the bore, the walls between the cylinder head bolts are arranged in a straight line.

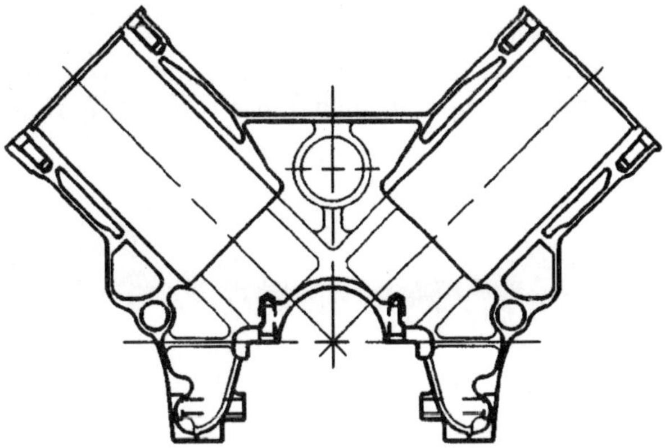

Fig. 8.41. Side View and Cross Section of Cylinder Block

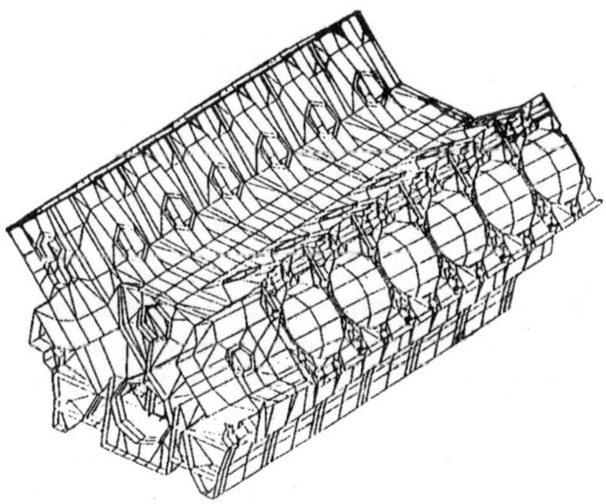

Fig. 8.42. Finite Element Model of Cylinder Block

Arranging the intake and exhaust push rods in the cylinder head side by side permits greater flexibility in the intake port design; as a consequence, the volumetric efficiency improves. Cooling efficiency between the valve seat and nozzle is improved by increasing the coolant flow rate to the area. Conventional steel laminated cylinder head gaskets, with five laminated sheets may be used to obtain optimum surface pressure balance (Fig. 8.43). Rubber grommets are needed for coolant and oil passages. Six fixing bolts per cylinder ensure uniform axial force distribution and adequate sealing.

Fig. 8.43. Cylinder Head Gasket

8.20 CONNECTING ROD

The design of a connecting rod is a fine example where detailed finite element analysis, backed up by strain gauge measurements, helps to withstand increased inertial and gas pressure loads. Figure 8.44 shows the finite element model of a connecting rod used in a diesel engine for commercial vehicles.

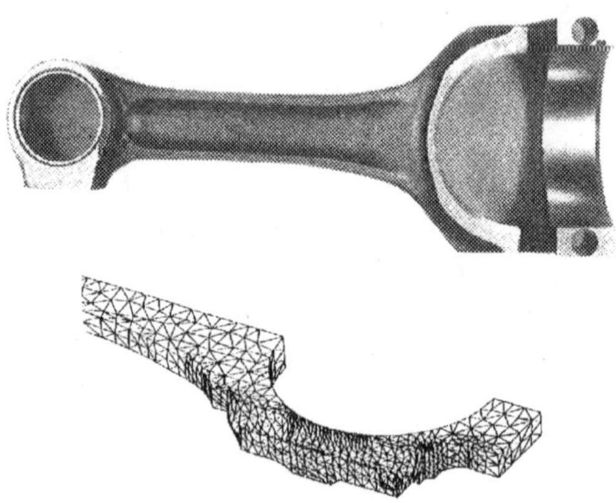

Fig. 8.44. Connecting Rod and Finite Element Model

Crankshaft and connecting rod designs must be closely coordinated for a number of reasons. Considerations for kinematic motion, load transmission, form and fit of the two components of the slider and crank mechanism does not need further elaboration. Bearings at both ends of the connecting rod receive lubricant oil through holes drilled in the crankshaft. Migration in unbalance quality due to the boring of holes in the crankshaft is a common experience in reciprocating machines. It may be dealt with by increasing the weight of bolted balance devices. Manufacturing and assembly process may be improved by positioning the balance device with a knock-pin drive instead of a conventional fit-in. Axial oil seals with slingers help prevent lip damage during assembly, and the use of fluoro-silicon reduces the wear rate of the rubber material. Section 3.7 provides detailed discussion of loads and stresses in connecting rods.

8.21 VALVE TRAIN SYSTEM

In its simplest form the inlet valve is open and the exhaust valve is closed at the top dead center on the piston's outward induction stroke. Expulsion of the fluid or gases from the cylinder is maximized when the exhaust valve opens as the piston slows down on its approach towards the end of the power stroke. To induce as much fresh charge into the cylinder as possible, the inlet valve starts to open towards the end of the exhaust stroke, when the outgoing column of the fluid in the exhaust port still has sufficient velocity to form a depression in its wake. Valve timing is thus designed to have an overlap period for improved breathing efficiency of the reciprocating machine.

The opening and closing of the valves at the right instant is accomplished through a system of cams, camshaft, rocker arms and return springs. A cam is a rotating member, and imparts reciprocating motion to another member (called follower) in a plane at right angles to the axis of the cam by creating a slip-forming pair. A radial plate type of cam has a circular section with a radially protruding lobe, and is provided with a suitable contour to provide a smooth rise and fall for the predicted motion of the follower. This motion is then relayed to individual poppet valves.

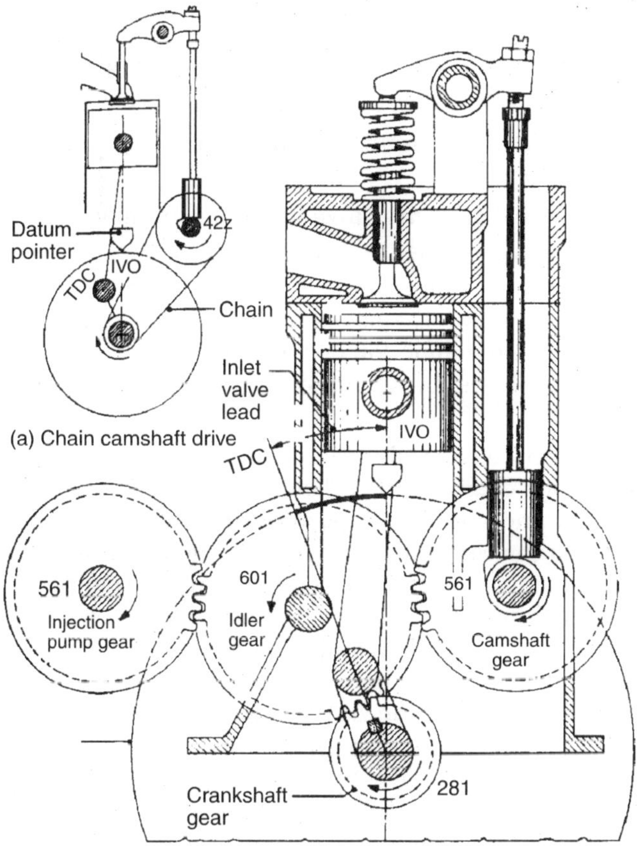

Fig. 8.45. Valve Operating Mechanism (*Courtesy:* Daimler-Benz)

Any back and forth motion of the valve train (valve, rocker, push rod and follower) produces positive and negative inertia forces. Positive forces tend to push the cam follower towards the cam's contour, while negative forces separate them. Negative forces must be reacted by the valve return springs, and so are critical. If these negative forces are large stiff return springs are required to counteract the undesirable jump-off response to the negative forces.

Unfortunately, the same spring load is added to the inertia load during periods of positive acceleration and deceleration. This will increase stress in the components, and adds to the wear and damage of the heavily loaded flanks of the cam and contact faces of the follower. The motion of the follower is controlled positively by its own inertia acting on the cam contour during the first part of its outstroke and the second part of the return stroke. During the opening and closing nose-to-follower phase contact is maintained solely by the return spring stiffness. The following list provides some criteria for good design practices in valve actuating mechanisms:

- Minimum valve train inertia
- Minimum rubbing velocity between cam and follower - helps avoid lubrication problems
- Minimum contact pressure between cam and follower - helps avoid surface scuffing
- Maximum rigidity of camshaft bearing mount
- Maximum stiffness of cam, follower, push rod and rocker arm
- Minimum length of push rod
- Minimum length of rocker arm - minimizes rocker moment of inertia
- Minimum valve lift - keeps cam acceleration and deceleration low
- Minimum cam profile lift acceleration and deceleration - improves accuracy of motion transfer from cam to valve
- Minimum spring load - prevents valve jump at high machine speed.

A high spring rate is a poor substitute to compensate deficiencies in cam and actuating mechanism design. It only serves to increase stress in the components. Opening and closing lift motion of the cam can be accurately relayed to the valve at relatively low camshaft speed. At increased speeds, however, dynamic forces will distort the transmitted motion from the cam to the shaft. Figure 8.45 shows the layout of a valve mechanism operated by an overhead camshaft.

The timed rotational drive from the crankshaft to the camshaft may be transmitted by a roller chain, inverted tooth chain, toothed belt or gear train. Roller and tooth chains are useful to drive complicated and awkwardly positioned camshafts. Variations in the distance between the crankshaft and the camshaft are caused by substantial expansion in the cylinder head and cylinder block during operation. An automatic tension adjustment mechanism is needed to take up the slackness for more positive drive control. A three-cornered chain configuration is illustrated in Fig. 8.46 for a diesel overhead camshaft engine. The arrangement calls for a pair of double-sided vibration dampers, providing closer chain control on the drive side of the chain. In a four-cycle engine the valves open and close once for every two revolutions of the crankshaft. Hence, the camshaft turns once for every two revolutions of the crankshaft. This is accomplished by doubling the number of sprockets on the camshaft drive compared with that on the crankshaft side. A second short chain is used to drive the oil pump, and also to reduce its speed relative to the crankshaft. The driver conveys the torque to the driven sprocket wheel through the taut side of the chain, which alone takes the load. Thus, when the engine is accelerating, the slack side of the chain is free to whip between the sprockets. As the mechanism accumulates a large number of load cycles the sprockets wear and the chain stretches, thus increasing the amount of slackness on the non-drive side. This may eventually cause the rollers to jump over the camshaft sprocket teeth, thereby altering valve timing. Hence, the chain tension mechanism is normally located on the slack side of the chain.

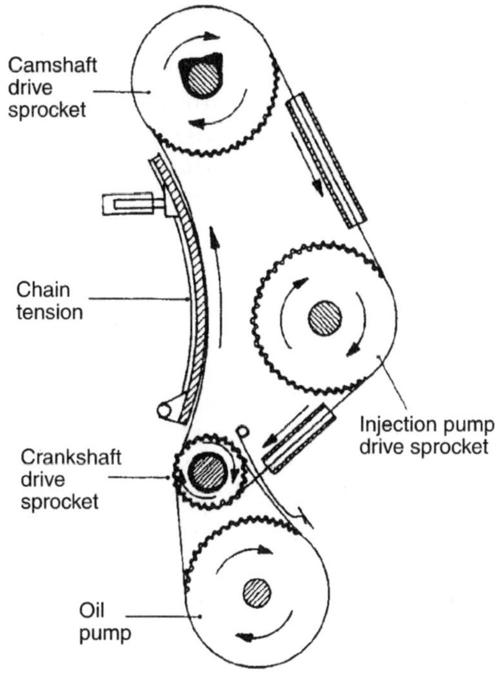

Fig. 8.46. Overhead Camshaft Chain Drive (*Courtesy: Daimler-Benz*)

A synchronous belt drive is an endless strap with internal cogs that engage into identically shaped grooves on the driven pulley's periphery, providing a positive and timed drive. Figure 8.47 shows the construction of a timing belt drive. The coefficient of thermal expansion for the belt material is smaller than for the overall machine system. Thus, the belt tension increases as machine temperature rises from off-load to full load condition. When high loads or high speeds are to be sustained for long operating periods the involute gear train is perhaps the most suitable method of driving the camshaft. Helical involute gears may be used to ensure continuous tooth contact and to overcome tooth pitting and wear.

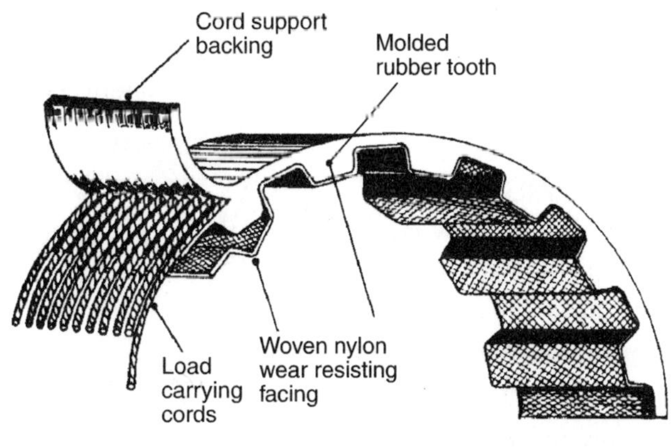

Fig. 8.47. Timing Belt Drive

An evaluation of the dynamic characteristics of the valve train system of a reciprocating machinery is necessary from a number of considerations. Opening and closing of the intake and exhaust valves at the right point in the operating cycle is crucial if the machine is to operate at its expected power and efficiency level. If the inertia, stiffness and damping properties of the components of the valve train system preclude this operation due to time or phase lag, loss in power and operating efficiency will be the resulting penalties. Generally, at lower speed levels inertia and damping forces due to the mass and damping of the components in the valve train system are not substantial to alter the timing sequence of the valves. As the machine operating speed increases, dynamic loads in the rocker arms, tappets, springs, drive mechanism and associated components will cause interference in the opening and closing timing of the valves.

Dynamic analysis using a finite element model of the valve train system helps identify the natural frequencies of vibration and associated mode shapes. All components of the complete valve train system must be included in the model. Thus, if a multi-cylinder diesel engine uses four valves per cylinder, all four valves, together with rocker arms, push rods and return springs for all cylinders must be represented in the dynamic model. A discrete-mass and elastic spring type of representation will be sufficient for most applications. Elastic beams and supports may be used to simulate rocker arms and the camshaft. If a chain or belt type of drive is utilized, stiffness (force per unit deflection) of the mechanism must be included. Chain tension adjustment devices, together with its spring, must have representation in the model, as also chain guides to prevent twisting motion. Inertia of sprocket drive wheels on the crankshaft and camshaft must be included in the dynamic model.

Since the crankshaft drives the camshaft, exciting force frequencies acting on the system can be identified. Hence, in two-stroke internal combustion engines, both shafts operate at the same speed, so the valve train system will be in resonance if it's natural frequency coincides with the speed of the crankshaft. Single acting compressors and pumps will follow the same trend. Valve opening and closing sequence in four-stroke engines, on the other hand, occurs once for every two crank turns, consequently, the resonant frequency of the valve train will be one-half of crank speed. In double acting pumps and compressors, if the same camshaft operates intake and exhaust valves on both sides of the cylinder, resonant conditions will prevail at twice the operating speed of the crankshaft. Excitation forces arising from other sources must be taken into consideration. For example, acoustically induced pressure waves in suction and discharge piping have their own frequency and intensity characteristics. Chapter 11 provides a complete discussion on this subject. If the calculated frequencies of the valve train lie in close proximity of exciting force frequencies, modifications of component, or system, design may be required.

8.22 EXAMPLE PROBLEMS

Example Problem 8.1: A single throw crankshaft of the type depicted in figure 6.6 is to be mathematically simulated using the finite element method. Details of the shaft are as follows:

Main journal outer diameter $D_s = 2.25$ in.

Main journal inner diameter $d_s = 0.75$ in.

Crank pin outer diameter $D_c = 2.00$ in.

Crank pin inner diameter $d_c = 0.625$ in.

Main journal half-length = $b/2 = 1.25$ in.

Crank pin length = $a = 1.875$ in.

Web thickness = h = 0.75 in.

Total bay length = l = 5.875 in.

Crank pin offset from main journal = R = 2.00 in.

Use web chamfers CD, $C'D'$; D is located 0.5 in. from inner cheek.

Web width W_c = 2.310 in.

Web width W = 2.435 in.

Web width W_s = 2.560 in.

Use fillet radius = 0.090 in. between crank pin/web and journal/web.

Shaft material (steel) average properties (70° F to 500° F temperature range): Young's modulus E = 28.5 × 10^6 lbs./in^2, Poisson's ratio ν = 0.30, density r = 0.28 lbs./in^3.

Solution: The solid model of the crankshaft is shown in Fig. 8.48. Solid model generation is available in most commercially available finite element computer programs. Hence, cylinders, webs and fillets can be readily created. One procedure calls for coordinate points to be first set up at key locations in the geometry of each sub component. Key locations may be generally described as those needed to define circles or arcs, end points of a straight line and plane surfaces. For arcs and circles at least three points are needed, one at the two ends of the arc and one at the center of the arc, with all three locations lying in the plane of the circular arc. Alternatively, a circular arc can also be defined by the center, radius of curvature and the angle of the arc in the defined direction. Straight lines can be readily generated by connecting its end points.

Plane, cylindrical and other curved surface areas may be created in a number of ways. A straight or curved line segment may be swept over a domain. Three or more lines along a closed contour can also create an area. For example, to create a quarter cylinder two 90° circular arcs at opposite ends and two straight lines connecting the corresponding pair of end points of the arcs are needed. Another method calls for sweeping the straight line along a trajectory describing the circular arc; the same can also be achieved by sweeping a circular arc along a straight line trajectory. Thus, depending on the capabilities of the software, a number of procedures are available to achieve a given geometric configuration.

Volume generation represents the next step in the chain. Here also there are a number of techniques available. An area may be swept over a prescribed path. A more common approach is to specify all areas that define the boundaries of the volume, then apply the volume generation command. It is not necessary to define the area surfaces in a specific sequence, however, all areas defining the boundaries of the volume must be present.

When a void is present in a volume Boolean operations may be the preferred choice. Insertion of holes in the main journal and crank pin are adaptations of this method. Two solid cylinders need to be generated as separate entities, one for the outer region and the other for the inner hole. Execution of the Boolean command permits subtraction of the inner hole from the outer cylinder. This method provides a hollow cylinder with considerable ease, and saves a number of steps if all of the areas bounding the hollow cylinder were to be defined for volume definition. Oil supply holes can be located in the same manner. This method may also be used for the chamfers. After the web is generated, definition of the cutting plane for the chamfer is required. Subtraction in the Boolean method will then create two volumes on both sides of the cutting plane. Note that the cutting plane does not have to be of the same size as the web's plane, it may extend in any direction. The smaller wedge on the other side of the cutting plane may then be deleted.

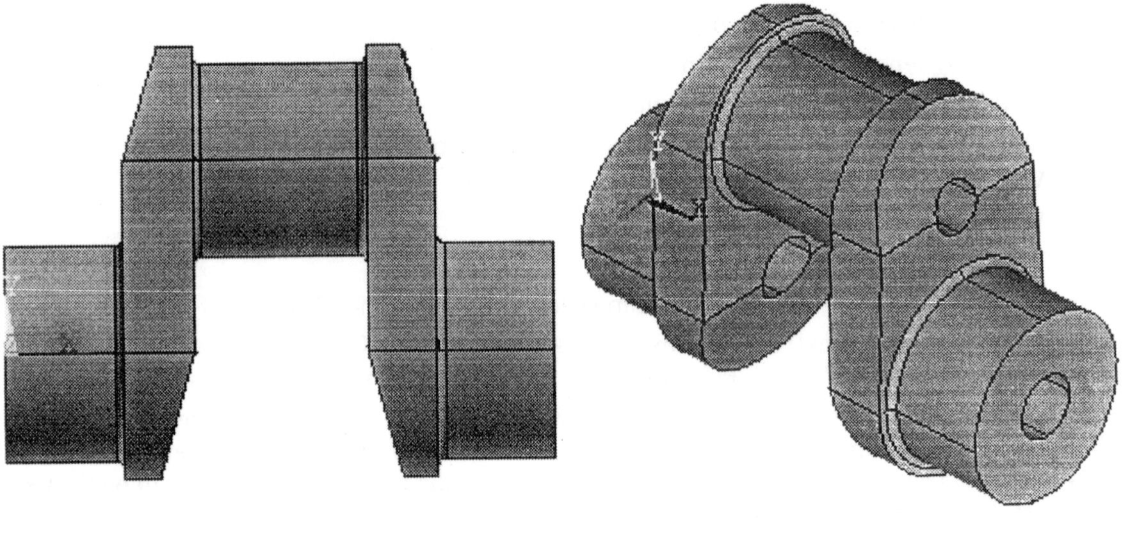

Fig. 8.48. Solid Model of Crankshaft

Assembly of all the hollow cylinders, fillets and webs requires gluing, or bonding, of the pieces to each other. The objective is to have a common surface area at mating locations. The process will then yield one continuous volume for the crankshaft.

When more than one bay of identical design is called for in a crankshaft, a procedure for reflection and copying comes in useful. The second and subsequent bays may be oriented at a different angle relative to the first bay. Definition of a new set of coordinate system for each bay is required. The origin of the new coordinate axes will have the same offsets along the three principal directions as in the design of the crankshaft. Rotation of the axes in the new system generally will not be needed. Copying the first bay geometry into the second bay's own coordinate reference frame will provide solid model of the crankshaft with two bays. If the second bay's geometry happens to be a mirror image of the first bay, provision is also available in many software packages to perform a reflection. The process can be repeated as many times as the number of bays in the crankshaft's design.

Once an adequate solid model of the crankshaft is available, the next step is generation of the finite element mesh. Before proceeding with the meshing, the program will expect information on material characteristics, element type to be used and its attributes. Brick (or 3-dimensional structural solid) elements require a node point at each of its corners, thus requiring a minimum of eight nodes per element. Each node has three degrees of freedom of translation in the x, y and z directions. In some finite element computer programs this element type may be capable of analyzing problems involving plasticity, creep, swelling, stress stiffening, large deflection and large strain. Solution output associated with this element type is in two forms, the nodal solution and the element solution. The primary nodal solution is in the form of displacements, temperatures and pressure for each degree of freedom, and also provides reaction solution at constrained nodes. Element solutions are in the form of values calculated at the centroid. Stress directions may be parallel to the element coordinate system.

Eight-noded brick elements are appropriate where geometric features of the component do not change rapidly. On the other hand, fillet radii are substantially smaller than the radius of the journal or crank pin, and eight-noded brick elements will experience poor shape characteristics. If in addition to the corner nodes the brick element also has one or two mid-sided nodes, meshing efficiency and accuracy of the calculated results generally improves dramatically. Prisms with six nodes (two triangular faces at opposite ends connected by three rectangular faces) are useful in the fillet region. Tetrahedrons consisting of four nodes and four triangular faces sometimes do not offer a high level of numerical precision.

When a component is subjected to varying temperature distribution over a wide range, it may be necessary to provide material properties at a number of temperature points. For meshing a region the number of divisions on each edge line must be specified. As a rule, where the geometry changes rapidly the division size should be smaller. All of the edge line divisions do not have to be equal in size. Thus, if one end of a line is connected to a rapidly changing geometric area compared to the other end, the divisions may be specified to change according to a predefined geometric ratio. The objective should be to obtain a coarser mesh where the geometry does not have sharp features, and a finer mesh where the geometry is changing rapidly. Figure 8.49 provides a finite element representation of the crankshaft.

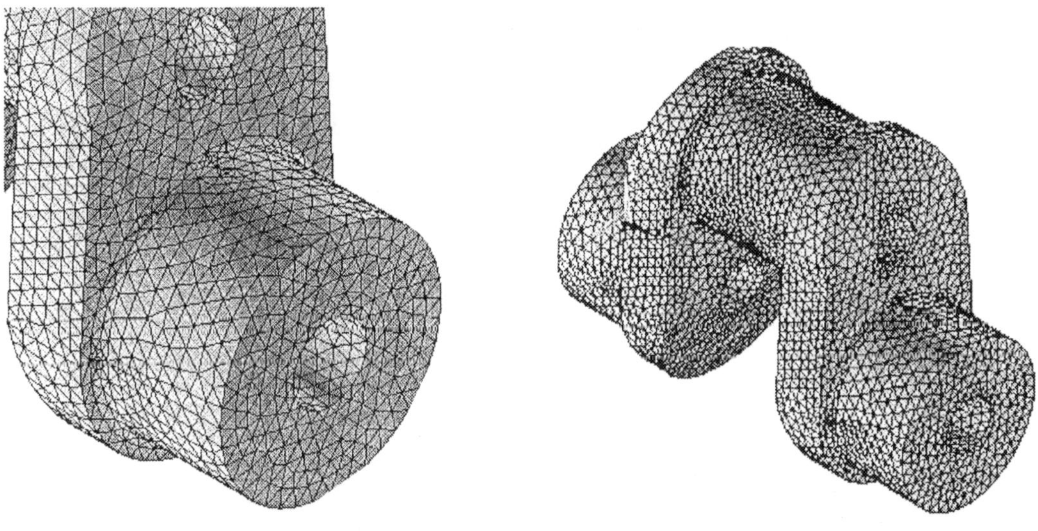

Fig. 8.49. Finite Element Mesh of Crankshaft

Example Problem 8.2: A static load analysis of the crankshaft described in example problem 8.1 is to be performed. Consider the following set of loads:

Crank pin pressure load on upper half of pin = 150 lbs/in^2

Main journal pressure load on lower half of both journals = 200 lbs/in^2

Shaft operating speed = 2,000 rpm

Shaft torque = 350 in-lbs.

Solution: Static analysis is used to determine displacements, stresses, strains and forces in components caused by loads that do not induce significant inertia and damping effects. Steady loading and response conditions are assumed; consequently, the loads and structure response are presumed to vary slowly with time. Load types that can be included in a static analysis are:

- Externally applied forces and pressures
- Steady state inertial forces arising, for example, from gravity or rotor speed

- Imposed non-zero displacements
- Temperatures for thermal strain.

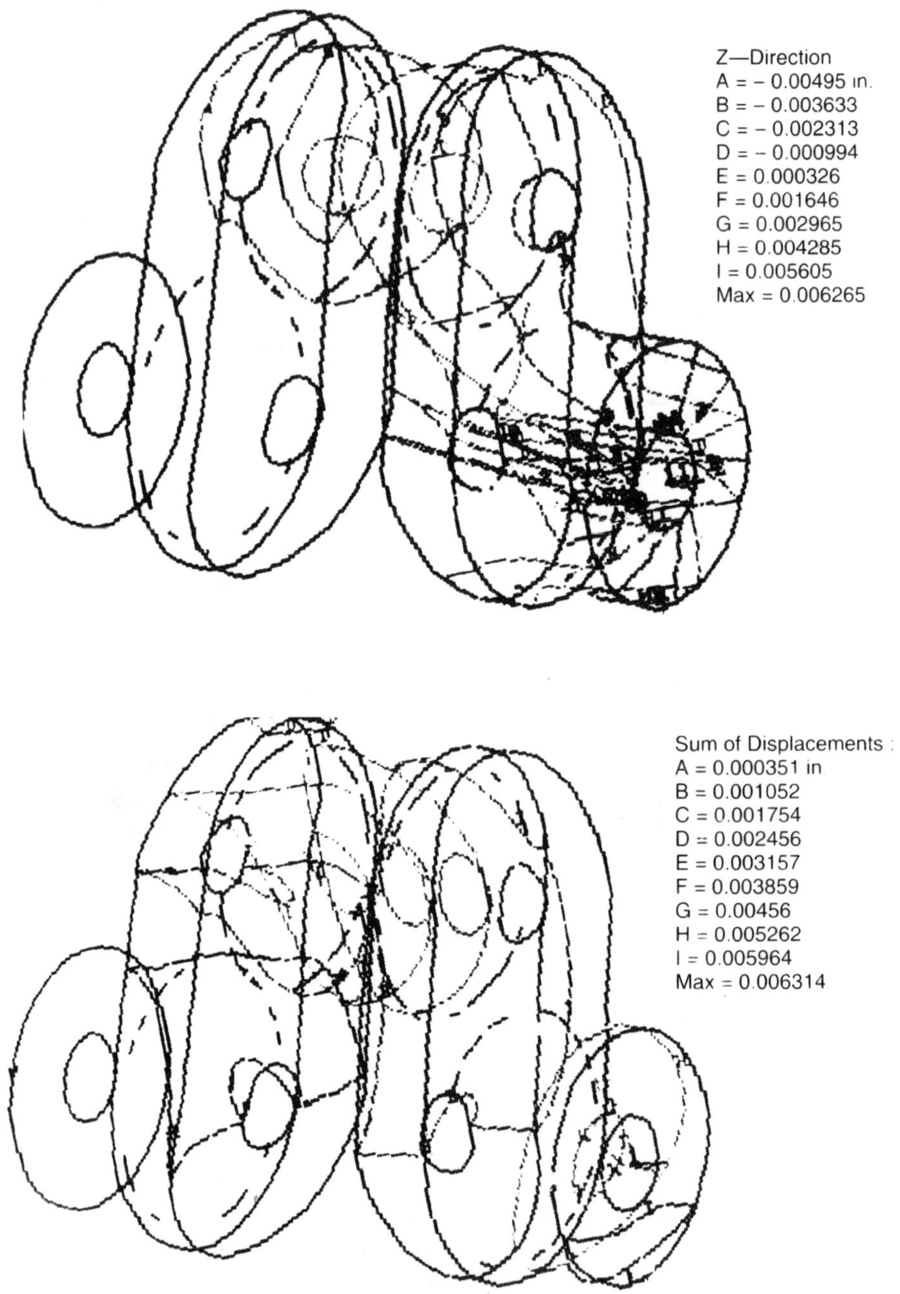

Fig. 8.50. Static Analysis Results-Displacements

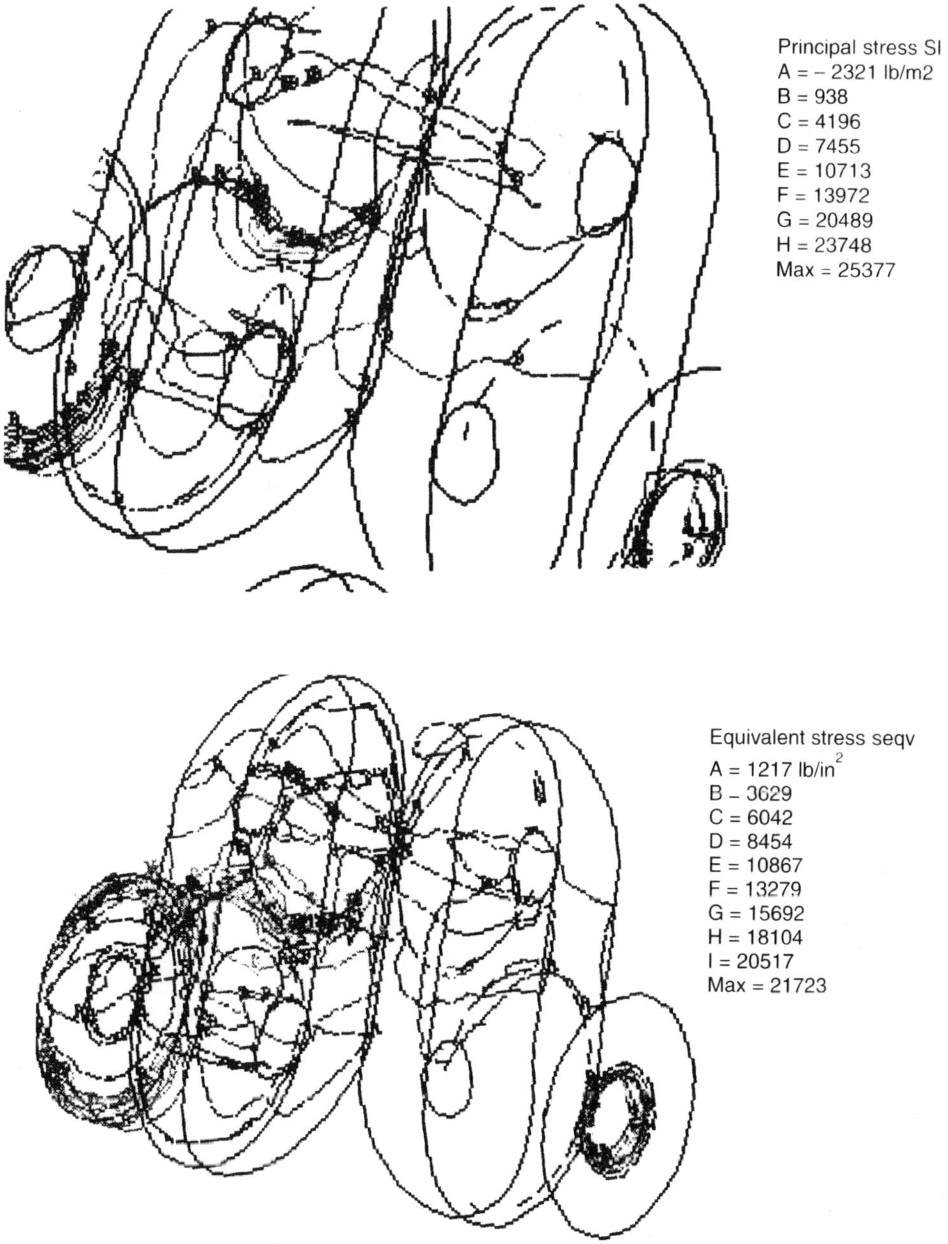

Fig. 8.51. Static Analysis Results-Stresses

Nodal loads (force loads or displacement constraints) may be defined at the nodes, and are not directly related to the element. Element loads (surface loads, body loads and inertia loads) are associated with a particular element, even if the loads are described at the nodes. Surface loads (pressure for structural elements, convection for thermal elements) may be applied at node points of an element face or to the face of an element. Inertia loads are applicable when the elements have mass, and are typically entered in the form of angular velocity or acceleration of the structure.

The primary unknowns, nodal degrees of freedom, calculated in a static structural analysis are displacements. Other quantities, such as strains, stresses and reaction forces are then derived from the nodal displacements. Static analysis may be of linear and non-linear type. Non-linear analyses include plasticity, stress stiffening, large deflection, large strain, hyper-elasticity, contact surfaces and creep.

Static analysis may also be used to perform inertia relief calculations to determine accelerations that will counterbalance the applied loads. The procedure may be thought of as an equivalent free-body analysis. The results from the analysis consist of translational and rotational accelerations required to balance the applied loads, and can be used by other computer codes to perform kinematic studies. Reaction forces at the constraints will be zero because the calculated inertia forces balance the applied forces. The minimum number of displacement constraints must be specified - those required to prevent rigid body motion. Three or fewer constraints are required for 2-dimensional models, and six or fewer constraints are necessary for 3-dimensional models. Additional constraints, such as those required to impose symmetry conditions are permitted, but reaction forces must be checked for null values at all constraints to make sure that the model is not overly constrained for inertia relief.

Results from the static analysis of the single throw crankshaft are shown in Fig. 8.50 and 8.51 in the form of displacement and stress contours.

Example Problem 8.3: Perform a modal analysis of the crankshaft of example problem # 8.1. Determine natural frequencies of vibration and associated mode shapes in the 0 to 4,500 rpm speed range. Assume following values for the attached reciprocating masses:

Weight of piston, fluid and small end of connecting rod = 4.25 lbs.

Weight of large end of connecting rod = 2.25 lbs.

Length of connecting rod = 7.75 ins.

Assume rigid bearings at journals and crank pin.

Solution: Modal analysis is used to determine vibration characteristics-natural frequencies and mode shapes - of a machine component. It can also serve as a starting point for another more detailed dynamic analysis routine, such as a transient dynamic analysis, a harmonic response analysis or a spectrum analysis. Modal analysis can also be performed on components subjected to a steady state of stress, such as a spinning disc or a compressor blade. Modal cyclic symmetry feature permits reviewing mode shapes of a cyclically symmetric structure by modeling only a sector of component, such as a flywheel.

Only linear behavior is valid in modal analysis. Non-linear elements, for example, a gap or an interference between two mating components, is treated as linear, with its stiffness calculated based on its initial status. Material properties of the components may be linear, isotropic or orthotropic, and constant or temperature dependent. Young's modulus (or stiffness characteristics in some form) and density (or mass) must be defined for a modal analysis.

A specific frequency range must be provide for the extraction of modes associated with the natural frequencies. Mode shapes may be normalized to the mass matrix or to unity. For subsequent spectrum analysis the mode shapes must be normalized to mass matrix. Maximum modal response of each mode is obtained by multiplying the mode shape by the mode coefficient. When damping is included in the mathematical model the calculated eigenvalues are complex. Results from the modal analysis of the crankshaft are shown in Fig. 8.52.

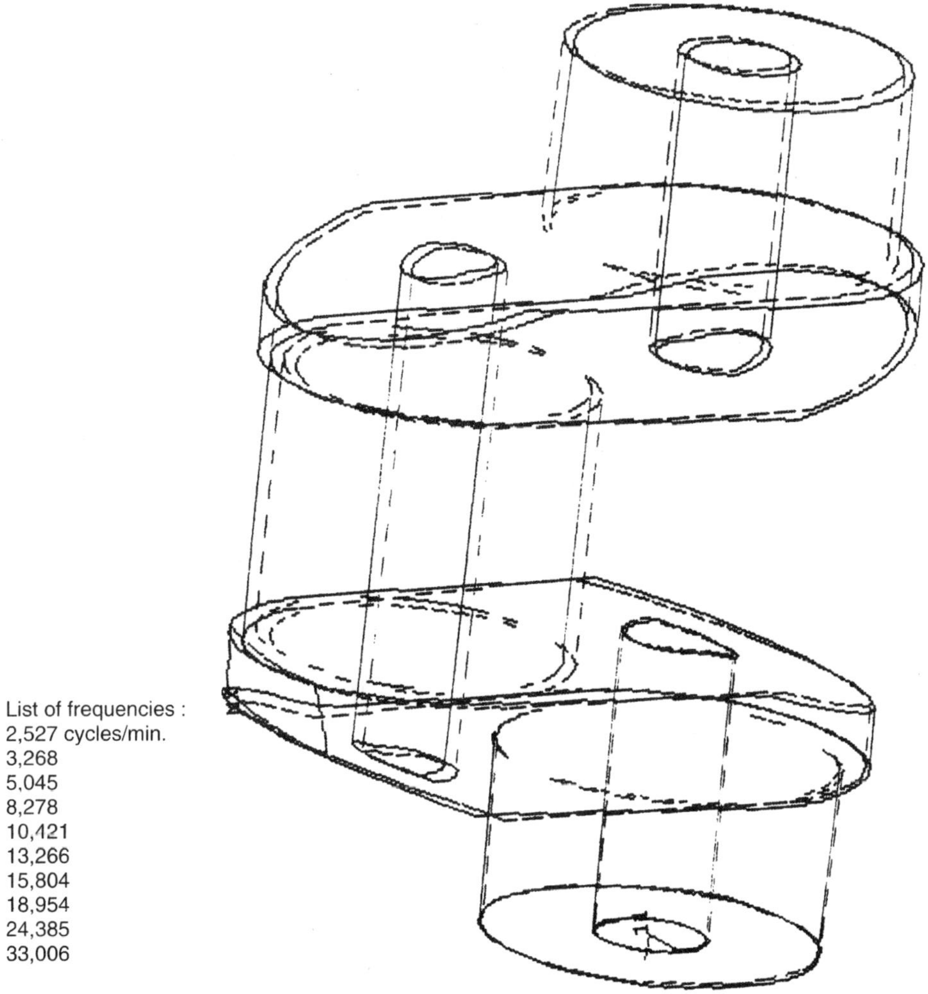

Fig. 8.52. Results from Modal Analysis - Fundamental Mode Shape

Example Problem 8.4: A harmonic response analysis of the single bay crankshaft of example problem 8. 1 is to be performed. The analysis is to be carried out in the 30 Hz to 70 Hz frequency range and torque amplitude of 350 in-lbs. Since the finite element model is large, the problem size may be reduced to calculate the harmonic solution. A limit of 700 degrees of freedom may be imposed, so the program selects nodes where there is high mass concentration and high levels of potential and kinetic energies in the overall model.

Solution: Any sustained cyclic load will produce a sustained cyclic response in a machine component. Harmonic response analysis is capable of predicting the sustained dynamic behavior of a machine system or component. The procedure determines steady state response of a structure to loads that vary sinusoidally, or harmonically, with time. The structure's response is calculated at several frequencies in order to obtain a graphical plot of some response quality (usually displacements) against frequency. 'Peak' responses are then identified on the plot, and stresses reviewed at those frequencies. The procedure thus verifies if the design will successfully overcome resonance, fatigue and other deleterious effects of forced vibrations.

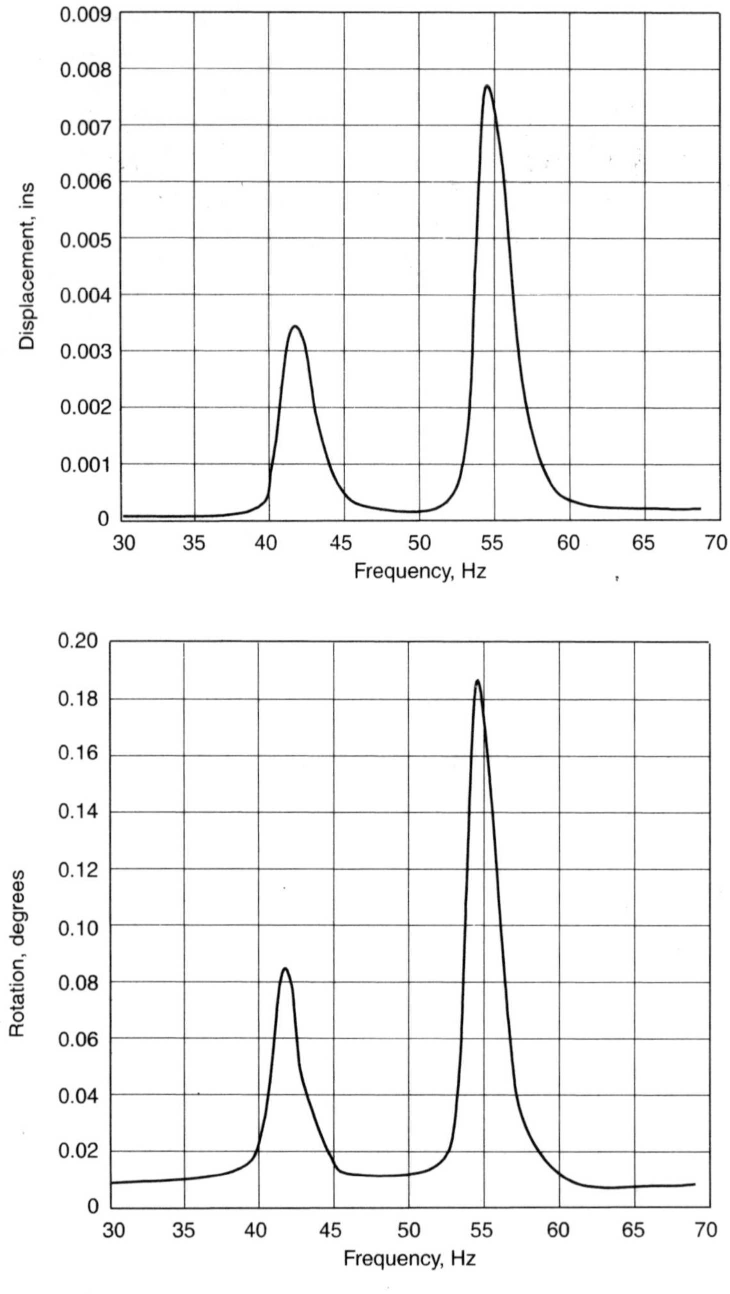

Fig. 8.53. Results from Harmonic Response Analysis - Deflections at Crank Pin Center

Some restrictions apply to harmonic response analysis. Transient vibrations present at the beginning of the excitation are not accounted for in a harmonic response analysis. All applied loads must be sinusoidally time varying, and must have the same frequency. Non-linear geometric features are not permitted.

Any of these restrictions may be overcome by performing a transient dynamic analysis, with harmonic loads expressed as time-history loading functions. Transient dynamic analysis determines time varying displacements, strains, stresses and forces in a machine component or assembly as it responds to any combination of static, harmonic and general time dependent loads. The time scale of the loading is such that inertia or damping effects are of significance in the solution. If inertia and damping effects are not important, static analysis may be used.

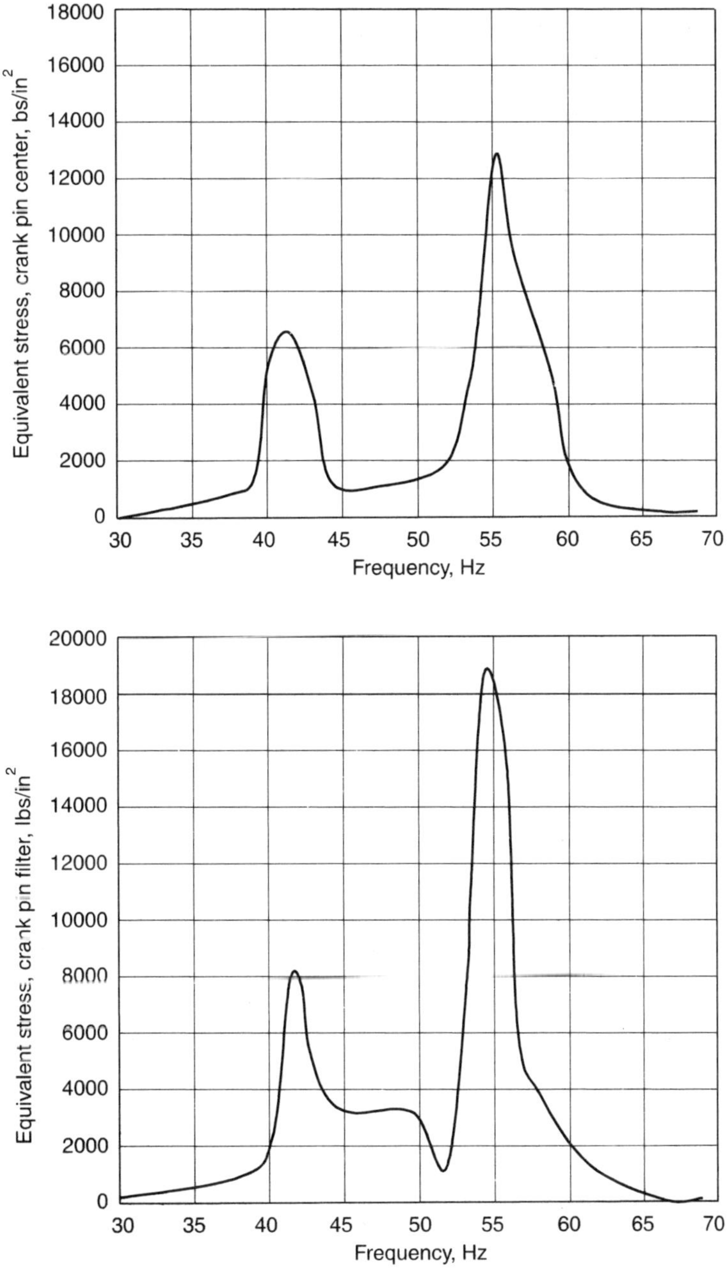

Fig. 8.54. Results from Harmonic Response Analysis - Stresses

The basic equation of motion in a transient dynamic analysis at any given time *t* can be thought of as a set of 'static' equilibrium equations, where inertia and damping forces are taken into account. The time increment between successive time points is referred to as the integration time step.

A transient dynamic analysis is more complex than a static analysis because it generally requires more computer resources, while also calling for better practical understanding of the dynamic behavior of the machine system. Preliminary work done prior to the analysis to understand the physics of the problem helps. Analysis of a simpler model made up of masses, beams and springs provides good insight. Often times the simpler model may be all that is needed to determine the dynamic response of the component or the assembly. When non-linear geometric features are included in the mathematical model, their impact on the structure's response needs to be understood by performing a static analysis prior to the dynamic analysis. In many cases non-linear aspects need not be included in the dynamic model. The dynamics of the problem can also be evaluated and understood by performing the modal analysis. Natural frequencies and mode shapes provide key information on how the structure responds when those modes are excited. Natural frequencies are also useful for determining the correct integration time step. When two or more modes are closely spaced, interaction between the modes may be expected. For example, when two closely spaced modes have most of the dynamic activity confined to two different portions of the overall structure, excitation at the first mode frequency is likely to result in large motion in the regions where both modes are mostly active.

Spectrum analysis uses the results from a modal analysis to combine a known spectrum to calculate displacements and stresses. It is mostly used in place of a time-history analysis to determine the response of structures to random or time dependent loading conditions. Conditions of this type are present during, for example, earthquakes and ocean wave loads. A spectrum is a graph of spectral value versus frequency, which captures the intensity and frequency content of time-history loads. A response spectrum represents the response of single degree of freedom system to a time history loading condition. It is a graph of response versus frequency, where the response might be displacement, velocity, acceleration or force. If damping is to be specified, it may be in the form of frequency dependent damping ratio, so that the program calculates effective damping at each frequency. The damping ratio may also be constant for all frequencies.

Results from the harmonic response analysis in the form of deflections at selected locations in the crankshaft are plotted in Fig. 8.53, and stresses are plotted in Fig. 8.54.

REFERENCES & BIBLIOGRAPHY

Abe, S., Suzuki, M., "Analysis of Cylinder Bore Distortion During Engine Operation", SAE Paper # 950541, 1995.

Aoki, H., "Effects of Power Plant Vibration on the Quality of Sound Inside a Passenger Compartment During Acceleration", Nissan Technical Review, Nissan Motor Co., 1986.

Booker, J., "Dynamically Loaded Journal Bearings: Numerical Application of Mobility Method", Institution of Lubrication Technology Vol. 93, No. 2, 1971.

Booker J., "Design of Dynamically Loaded Journal Bearing", ASME Design Engineering Conference, Fundamentals of the Design of Fluid Film Bearings, ASME, pages 31-44, 1979.

Broch, J. T., "Application of Bruel & Kjaer Measuring Instruments to Mechanical Vibration and Shock Measurements", Bruel & Kjaer Instruments, Inc., Cleveland, 1992.

Collacott, R. A., "Mechanical Fault Diagnosis", Chapman & Hall, London, 1987.

Dunavesky, "Analysis of Distortion of Cylinders and Conformability of Piston Rings", Tribology Transactions, Vol. 33, pages 33-40, 1990.

Gale, N. F., Widener, S. K., Sui, P. C., Zhang. H., "Analytical Engine Design Methods: A Review", SAE Paper # 950806, 1995.

Goetzke, M. B., Chen, J., Scott, D. I., "Computer Aided Analysis of General Motors EMD 710G Series Engine Components", ICE Vol. 9, ASME, 1989.

Heath, A. R., "The Measurement and Prediction of Flywheel Whirl", Institute of Mechanical Engineers, Auto Congress, 1987.

Hodgetts, D., "Vibration of A Crankshaft", Institute of Mechanical Engineers, Paper # C99/81, 1981.

Hughes, T. J. R., "The Finite Element Method, Linear Static and Dynamic Analysis", Prentice-Hall, Englewood Cliffs, N. J., 1987.

Ide, S., "Improvement of Engine Sound Quality Through A New Flywheel System Flexibly Mounted to The Crankshaft", SAE Paper # 900391, 1990.

Jeon, H., Tsuda, K., "Theoretical Analysis of the Undamped, Coupled Torsional - Axial Vibration of Marine Diesel Engine Shaftings", Bulletin of the M. E. S. J., Vol. 4, No. 7, P41, 1989.

Koch, F., Fahl, E., Haas, A., "A New Technique for Measuring the Deformation of Cylinder Bores During Engine Operation", SAE Paper # 950540, 1995.

Lacy, D. J., "Computers in Analysis Techniques for Reciprocating Engine Design", Institute of Mechanical Engineers, # C14/87, 1987.

Law, B., "Prediction of Main Bearing and Crankshaft Loading in Reciprocating Engines", CIMAC Paper # 6358C/424A, 1983.

Law, B., "Prediction of Crankshaft and Flywheel Dynamics", Institute of Mechanical Engineers, Paper # C382/046, pages 487 - 498, Perkins Technology, 1989.

Maas, H., "Calculation of Crankshaft Plain Bearings", 9th CIMAC Conference, Paper # A-22, Stockholm, 1981.

Maas, H., "Calculation of Crankshaft Plain Bearings", 11th CIMAC Conference, Pages 43-68, Vol. 1, Barcelona, 1985.

McNamara, P. M., Trevelyan, J., "Crankshaft Stress Analysis - The Combination of Finite Element and Boundary Element Techniques", Institute of Mechanical Engineers, # C430/030, 1991.

Nakayama, K., Yasutake, Y., Takiguti, M., Furuhama, S., "Effect of Piston Motion on Piston Skirt Friction of a Gasoline Engine", SAE Paper # 970839, 1997.

Paranjpe, R. S., Goenka, P. A., "Analysis of Crankshaft Bearing Using A Mass Condensing Algorithm", STLE Transactions, Vol. 33, pages 333-344, 1990.

Pian, T. H. H., Chen, D. P., "A Hybrid Approach for Displacements in the Interior of a Region", Intl. J. Num. Method in Engg., 18:1679-1684, 1982.

Prakash, V., Aprameyan, K., Shrinivasa, U., "An FEM Based Approach to Crankshaft Dynamics and Life Estimation", SAE Paper # 980565, 1998.

Raft, A., "Finite Element Study of a Composite Material Sump Pan of an I. C. Engine", SAE Paper # 950942, 1995.

Rajakumar, C., Rogers, C. R., " The Lanczos Algorithm Applied to Unsymmtric Generalized Eigenvalue Problem", International Journal for Numerical Methods in Engineering, Vol. 32, pages 1009 - 1026, 1992.

Saito, J., Odajima, M., Harada, T., "Isuzu New V8-V12 PE-Series Diesel Engines", SAE Paper # 950525, 1995.

Shizuo, A., Suzuki, M., "Analysis of Cylinder Bore Distortion During Engine Operation", SAE Paper # 950541, 1995.

Spectral Dynamics Corporation, "Machinery Analysis - Why Signature Ratio?", Technical Publication # RTA-4, San Diego, September, 1994.

Suhara, T., Ato, S., Masaaki, T., Furuhama, S., "Friction and Lubrication Characteristics of Piston Pin Boss Bearings of an Automotive Engine", SAE Paper # 970840, 1997.

Vertin, K. D., Haller, C. L., Lubnow, T. S., "A Root Cause Investigation of Cylinder Head Cracking in Large Diesel Engine Standby Power Generators", SAE Paper # 950518, 1995.

Vijayaragahvan, D., "New Concept in Numerical Prediction of Cavitation in Bearings", Ph. D. Dissertation, 1989.

Wakabayashi, K., Honda Y., Kodama, T., Iwamoto, S., "The Characteristics of Bending Vibration Stress Coupled with Torsional Vibration of Automotive Diesel Engine Shaftings", SAE Paper # 950543, 1995.

Wakabayashi, K., Iwamoto, S., Matsumoto, N., "Analysis of Vibrations of Reciprocating Engine Shaftings by the Transfer Matrix Method - The Second Report, Calculation and Comparison with Measured Data of Torsional Vibration", Bulletin of the M.E.S.J., Vol. 8, No. 1, Pages 40-50, 1980.

Wakabayashi, K., Seki, T., Iwamoto, S., Ito, H., "Analysis of Vibrations of Reciprocating Engine Shaftings by the Transfer Matrix Method-The First Report, Analysis of Forced Vibration of a Crankshaft", Bulletin of the M.E.S.J., Vol. 8, No. 1, Pages 27-39, 1980.

Wilkinson, J. H., "The Algebraic Eigenvalue Problem", Clarendon Press, Oxford pages 515 - 569, 1988.

Woods, C. M., Brewe, D. E., "The Solution of the Elrod Algorithm for a Dynamically Loaded Journal Bearing", ASME Journal of Tribology, April, 1989.

Yeo, S. D., Koo T. K., Chang, S. H., "Torsional Vibration Characteristics of a Crankshaft with a Rubber Damper", SAE Paper # 891232, 1989.

Yunus, J., Shah, M., Pawlak, T. P., Cook, R. D., "Solid Elements with Rotational Degrees of Freedom: Parts 1 and 2", International Journal for Numerical Methods in Engineering, Vol. 31, pages 575-610, 1991.

Zienkiewicz, O., C., Taylor, R. L., "The Finite Element Method, Volume I: Basic Formulation and Linear Problems, 4th Edition, McGraw-Hill, London, 1989.

Vibration Measurement and Balancing 9

9.1 INTRODUCTION

Vibration measurement in reciprocating and rotating machines is a prerequisite for fault diagnosis, performance verification and condition monitoring. Advances in sensors, signal processing procedures and data storage devices have simplified evaluation of dynamics of motion of the rotating shaft. Improvements in electronics and application of microprocessors to vibration analysis has further enhanced capabilities of engineers to obtain vibration free operation of machines. The fast Fourier transform (FFT) is the foundation of most vibration data processing, and diagnostic and modal analysis techniques have taken advantage of FFT capabilities for widespread use in data reduction of vibration signature of rotating machines.

Since reciprocating machines have both reciprocating and rotating components, vibration measurement is generally confined to the main structure of the machine at the bearing caps or pedestals. Vibrations in rotating machines with rigid rotors are also measured at the surface, where vibration levels are indicative of excitation forces generated by the rotor because of unbalance, thermal bow or other sources. Machines with flexible rotors may be set into different modes of vibration as it accelerates through one or more critical speeds to reach the operating regime. In this case, vibration amplitude measured on a structural member may not be indicative of the vibration of the rotor. For example, a flexible rotor may experience very large flexural displacements to cause failure, even though dynamic activity at the bearing cap is low. Hence, in this case it is essential to measure the vibration of the shaft directly. A group of machines like axial flow compressors and fans may have quasi-rigid rotors, where vibration amplitudes measured on the bearing cap are still indicative of lateral activity in the shaft.

Torsional vibration velocity can be measured using a toothed wheel and a fixed sensor. The signal generated by the teeth of the wheel passing the fixed sensor has a frequency that equals the number of teeth multiplied by shaft speed. If the shaft is experiencing torsional vibrations, the carrier frequency will exhibit frequency modulation, or a change in frequency, since the time required for each tooth to pass the fixed pickup varies. Other methods for indirect measurement of torsional vibrations use strain gauges to indicate instantaneous torque. The strain gauges are usually mounted at an angle of 45° to eliminate influence of shaft bending on torque measurements. Such techniques, however, are not common diagnostic tools, since strain gauge signals are difficult to take from a rotating shaft. Measurement of torsional vibrations is, at best, a tricky proposition, because the oscillatory motion is superimposed on the steady rotational motion of the shaft. In comparison, instruments that convert shock or vibratory motion into an optical, mechanical or, most commonly, an electrical signal that is proportional to the experienced motion can more readily measure lateral vibrations of a rotating shaft.

Higher and lower harmonics of the rotor speed, where characteristic frequencies are an integral or fractional multiple of the machine's operating speed, are typical of reciprocating engines, compressors and pumps. For instance, multiple cylinders experience firing of the air-fuel mixture at various crank angle locations in an engine, causing vibration peaks at several frequencies. 'Ghost' components are sometimes observed in vibration spectra obtained from measurements in machines coupled to a gearbox, where vibration frequencies occur in multiples of the number of teeth on a gear. Presence of a number of blades, support struts and obstructions in the flow path of a turbine or compressor has similar effects on the vibration signature of a machine.

A rotor is considered balanced if it rotates about its principal axis without wobbling, assuming that there is no dynamic motion due to the elasticity of the rotor and the lubricant in the bearings. When the journal axis of rotation does not coincide with its principal axis, a net radial force acts on the rotor due to centrifugal acceleration. The process of effecting essential coincidence between the principal inertia axis of the rotor and the shaft axis is called dynamic balancing. Besides unbalance, there are other sources of vibration in a rotating machine, some of which may be aggravated by unbalance under certain circumstances, and so may be of significance. Balancing of rigid rotors (one that does not experience distortion) is important because it comprises the majority of balancing work done in industry. When machine operation takes place in a speed regime where bearing support stiffness at both ends is essentially rigid, shaft elasticity also comes into play. The overall problem in which dynamic action and interaction of rotor elasticity, bearing and support elasticity and damping are taken into account is called flexible rotor, or modal, balancing.

9.2 VIBRATION SENSORS

Vibration measurement instruments fall into three broad categories: transducers to measure the vibration set up in the machine, electronic devices to process the signal and analyzers to calculate spectral blocks of data and display the results. Within each category, associated components are needed to condition, process, record and store the signal.

Piezoelectric transducers, accelerometers and gauges are commonly used to measure vibration signals. Generally exhibiting high elastic stiffness, a piezoelectric material electrically responds to force, and produces an electric charge proportional to the force applied to it in its linear elastic range. Motion transducers (or pickups or sensors) come in two basic types. A fixed reference type has one terminal attached to a fixed point, the other to the point whose motion is to be measured. A mass-spring (or seismic) transducer has a mass and spring system attached to a base, and the base is mounted on the vibrating structure; dynamic movement is inferred from motion of the mass relative to the base. A piezoelectric accelerometer (Fig. 9.1) is a linear seismic transducer that produces an electric charge proportional to the applied acceleration through the base. The piezoelectric crystal producing the charge acts as the spring. In a properly designed accelerometer the mass and the frame have high stiffness coefficient, the crystal has relatively little mass, and viscous damping exists only between the frame and the mass. For a given application practical considerations may not fulfill all of these assumptions. At higher frequencies of excitation, for example, inertial forces may deform the housing and change the performance of the seismic system by distorting the indicated frequency.

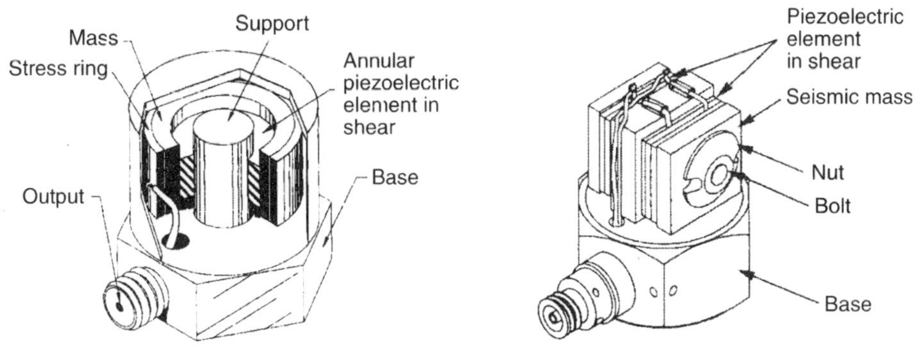

Fig. 9.1. Seismic Piezoelectric Accelerometers (*Courtesy*: Bruel & Kjaer, Endevco)

A force gauge is a mechanical impedance measurement device to relate the force applied to a structure of an operating machine. Force gauges use piezoelectric transducing elements to generate an output charge or voltage proportional to the applied force. Amplitude response, signal conditioning requirement and environmental effects associated with force gauges are similar to those related to piezoelectric accelerometers. Force gauges for use with a vibration excitation are attached at one end to the structure and the other end to a vibration exciter.

For shock excitation force gauges are usually built into the head of a hammer. Interchangeable faces of various materials to control the generated waveform of the shock pulse may be used on the head of the hammer. A hard material produces short duration, high amplitude shocks with fast rise and fall times. Shocks of short duration have a broad frequency spectrum. A softer material, on the other hand, causes longer, lower amplitude shocks with a narrow frequency spectrum. Impedance heads combine the functions of a force gauge and an accelerometer in a single instrument. They are convenient for measuring driving point impedance.

The laser Doppler vibrometer uses the Doppler shift of a laser light, which is scattered from a vibrating test object, and produces a real time analog signal output proportional to the instantaneous velocity. Benefits of the system arise from lack of mounting and mass loading effects, and are not affected by environmental factors. Measurement may be done from a remote standoff distance.

The electro-optical measurement system tracks displacement of a light-dark target. The target provides discontinuity in the intensity of reflected light from an object, and the lens of a photo processing equipment forms the image. Electrons are emitted in proportion to the intensity of the light, causing an electronic image to be generated in real time.

Accelerometer preamplifiers are needed to convert a weak signal at high impedance from a typical piezoelectric transducer into a voltage signal at low impedance. Voltage signal is less prone to the influence of electromagnetic noise and other effects. Pickup of noise in a high impedance circuit increases with cable length, so it is advantageous to mount the preamplifier close to the transducer. A line drive amplifier, with its miniaturized thick film circuits, can be attached or even included in the transducer. The output cable from a preamplifier is less subject to noise problems. Line drive preamplifiers have restrictions on dynamic and frequency range compared with a general-purpose charge preamplifier.

Signal conditioning is required to limit frequency band of the signal, to integrate it (to velocity or displacement) and to adjust its gain. High and low pass filters are required to remove extraneous low and high frequency signals and to restrict measurement to the frequency range of interest. Extraneous low frequency signals arise from thermal transient effects, strain in accelerometer base and vibrations transmitted through the foundation from other sources. At the high end of the frequency spectrum resonance arising from the accelerometer's own geometry must be filtered. The signal in the input amplifier is not affected by this high and low pass filtering.

Many criteria for evaluating operating machinery vibrations are based on velocity, and it is often preferable to express it in terms of displacement. It may also be desirable to determine the total motion of a rotating shaft by adding displacement of the bearing housing, measured with an accelerometer, to the relative displacement of the shaft in its bearing, measured with proximity probes. Acceleration signals can be electronically integrated to yield velocity and displacement; an accelerometer combined with an integrator can produce a velocity signal that is valid over a frequency range of three decades (1000:1).

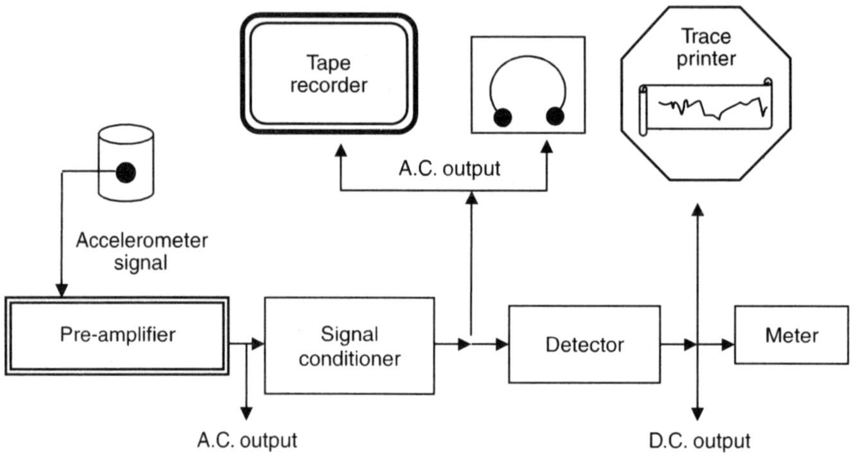

Fig. 9.2. Vibration Measurement System

Detectors are used to extract parameters that characterize a signal, typical ones being arithmetic average, mean square and root mean square values. A full wave rectifier obtains the signal magnitude, and with a smoothing circuit, produces the arithmetic average. Mean square and root mean square values are independent of phase relationships, and are preferred for describing stationary signals. Mean square values are directly additive when two signals of different frequency bands are added together.

Vibration meters receive signal from a transducer and process it to display relevant vibration parameters. For measurement on rotating machines a frequency range of 10 Hz to 10 kHz is desirable. The lower frequency limit includes any sub-harmonic components due to factors such as oil whip in plain bearings. The upper limit includes higher order frequencies in multi-cylinder internal combustion engines, tooth meshing frequencies and their harmonics in gears and frequencies associated with faults in rolling element bearings.

Evaluation of vibration signatures can be done at a later date if the signal is stored on instrumentation tape recorders using direct, frequency modulation or digital techniques. Ability to record low frequency signals is the prime difference between the methods. Direct recordings cannot record down to a direct current signal, 25 Hz being the lower limit. In contrast, frequency modulation method can go down to a direct current signal, which represents a constant deviation of the carrier frequency.

Figure 9.2 provides a block diagram of a typical vibration measuring system consisting of a preamplifier, signal conditioner, detector and indicating meter. Vibration meters often combine all of these elements in a single unit.

9.3 MEASUREMENT TECHNIQUES

Careful pretest planning can save a lot of time and aggravation in the measurement of vibrations, and ensures that the most useful data is obtained. Starting with a clear definition of the test's objectives, factors that need important consideration must be identified. For example, estimates of the characteristics of the motion to be measured must be made, including the range of frequency, amplitude, dynamic motion, duration and principal direction of activity. This information provides a basis in the selection of the measurement equipment. Previous experience on a similar type of equipment may provide a guide in the estimation of signal characteristics. Preliminary measurements may also be helpful. Environmental conditions, locations where measurements are to be made, system components and method of data transmission are other factors to be taken into account.

Selection of the parameter to be measured (displacement, velocity, acceleration, strain) may be predetermined by specifications or standards. At high frequencies acceleration measurements provide the highest signal output. A transducer of small size and mass suffices to measure acceleration, which is a consideration for smaller components. The results are useful in load and stress analysis. Velocity measurement is widely used on rotating machinery because the velocity spectrum is generally more uniform in the intermediate frequency range. If vibration readings are to be correlated with acoustic measurements, sound pressure level is proportional to the velocity of the vibrating surface. Displacement measurements are particularly important if the vibrating part must not come into contact with other components in the machine, and for relative motion between the rotor and the stator. The magnitude of the displacement provides an indication of the stress experienced by the component. Also, at low frequencies acceleration and velocity parameters may be too small to be useful.

A number of factors need to be considered in the selection of transducers, including characteristics of the transducer, motion type and environmental issues. Conflicting situations often arise. As an example, sensitivity requirement would suggest that a transducer of larger size must be selected, but mass considerations would call for a small size in order to reduce inertia loading on the component. In special cases even the smallest transducer may result in an unacceptable load. Then a device that does not come into direct contact with the test surface may be the right choice.

A transducer may be mounted by screw, cement type of epoxy or attached with a bracket to the test surface. The method of mounting affects the resonance frequency, and hence, the useful range of frequency of the transducer. Also, ability to withstand high levels of vibration, operating temperature, stability and repeatability of the method of mounting should be reviewed. A stud-mounted transducer offers the highest resonance frequency of any mounting methods, and accepts high vibration levels without loosening from the test surface. When stud mounting is not possible, a transducer may be bonded to a test surface by a thin layer of cement (epoxy or dental type). The method provides good frequency response when the accelerometer is directly attached to the test surface. Magnetic mounting is useful in measuring low acceleration levels. A hand held transducer provides the poorest performance, but it comes in useful in making a rapid survey of a test surface by changing measurement location. A probe with one end threaded into a transducer is pressed against the test surface. Measurement accuracy is not available with the procedure, and the frequency range is restricted between 20 and 1,000 Hz.

If a transducer is to be used under unusual temperature conditions calibration must be performed in the temperature range in which it will operate. One method calls for a direct comparison of the generated output with the output from another standard transducer when both transducers are attached to a vibration exciter and subjected to the same excitation. The free-fall calibration method calls for the transducer to fall freely for a short period under the influence of gravity. The peak signal is measured for a value of 1 g. Calibration of a complete vibration measurement system may be done by a field calibrator. This is a portable device containing a small shaker. The accelerometer can be mounted on the shaker and subjected to a known displacement, velocity or acceleration at a fixed frequency. The calibrator is provided with an internal oscilloscope and a stable built-in accelerometer in a feedback loop that controls the electro-dynamic exciter.

Data transmission between the transducer and the electronic instrumentation is generally through a cable, but the aerospace industry relies on telemetry. The choice of a suitable cable depends on the application, amplifier type and environmental conditions. Cable jackets may be made of silicon rubber, poly-vinyl-chloride or Teflon, depending on the temperature range to which it is subjected. The cable must be tied down near the accelerometer and at regular intervals to avoid cable whip and induced cable noise. Light and flexible cables prevent loading the transducer and the structure under test. Cables should be generally treated with the same care as transducers in vibration measurement tests.

9.4 SIGNAL PROCESSING

FFT analysis, digital filtering and optimization are the methods used for processing of the signal. Commercial software programs employing these digital techniques are available and widely used. Analog-to-digital (A/D) converters serve to convert a continuous signal into a sequence of digital numbers to represent the instantaneous value of the signal at specified time increments. It may be possible to regain the original analog signal by the reverse process using a digital-to-analog (D/A) converter. Time increments are normally uniform, representing a constant sampling frequency. Time intervals may also be selected on other bases, for instance, uniform increments in shaft speed for order tracking.

A/D conversion is done through the sample and hold procedure. The instantaneous value of the analog signal is sampled at the instant of each pulse of the sampling clock, and holds that analog voltage constant until the A/D conversion process is complete. The process is then repeated. For multiple channel conversion it is common to use a single A/D converter multiplexing between channels, with the sample and hold circuits for all channels synchronized. The output of the A/D converter is a binary integer number with $2N$ possible values, where N is the number of bits. The actual dynamic range of the measurement may be limited by factors other than the least significant bit, such as noise level and linearity in the analog parts of the system.

Discrete sampling in the time domain corresponds in the frequency domain to a periodic repetition of the spectrum, with a periodic spacing equal to the sampling frequency. If it is desired to obtain the correct frequency spectra, it is necessary to ensure the analog signal does not contain frequency components greater than half the sampling frequency, and this is achieved by using appropriate anti-aliasing filters with steep characteristics. Their application makes it possible to use up to 80 percent of the theoretically available spectrum, but it may be accompanied by some phase distortion in the vicinity of the cut-off frequency.

An ideal band-pass filter transmits that part of the input signal within its band pass, and completely attenuates all other frequency components. An analysis may be performed over a frequency range by using a single filter with a tunable single filter swept over the entire frequency range. For general vibration analysis it is common to use tunable filters, with the center frequency tuned manually, or synchronized with the X position of the pen on a graphic recorder. Alternatively, the center frequency may be synchronized with an external signal, for instance, by a trigger pulse once per revolution of a rotating shaft. The filter then becomes a tracking filter, which may be used to identify the component corresponding to a designated harmonic, or a multiple of the synchronizing signal. An adapter to multiply or divide the frequency is needed to generate a tuning frequency from the original signal. Digital filter analyzers have superseded banks of filters, each with their center frequency. A digital filter can process a continuous digitized signal, and provide another modified signal as an output. The relationship between the output and input signals is expressed as a difference equation, with properties similar to those of a differential equation used in describing an analog filter. Figure 9.3 illustrates the difference between a 3 dB bandwidth and the effective noise bandwidth, the first being most relevant when separating discrete frequencies, and the second being used for random signals.

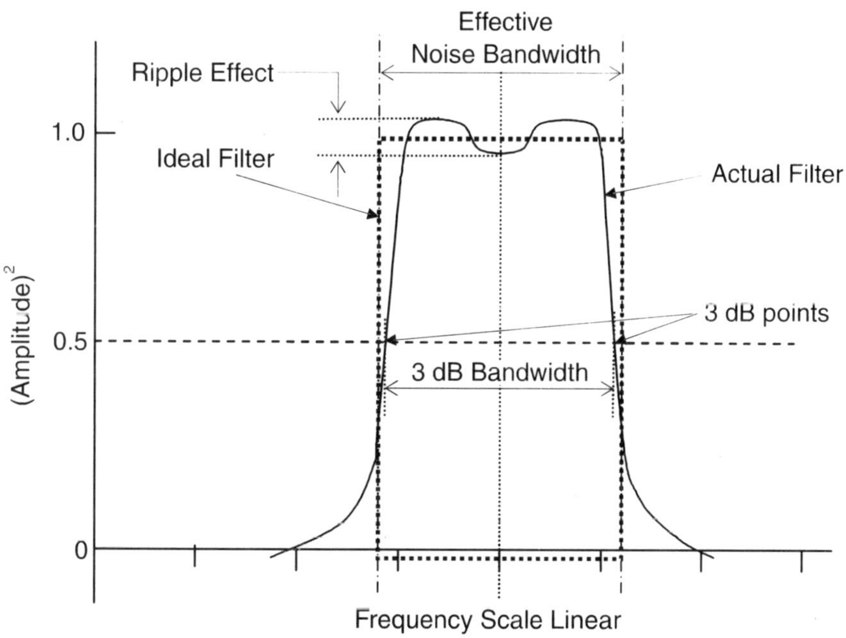

Fig. 9.3. Band Width Definitions for a Practical Filter Characteristic

Externally measured vibrations on a machine casing are usually the result of internal forces acting on a structure, whose frequency response function modifies the result. Because the structural response functions vary over a wide dynamic range, it is advantageous to depict the vibration spectra on a logarithmic amplitude axis. This applies particularly when the vibration measurements are used as an indicator of machine condition, and thus, internal forces and stresses. Another aspect to be considered is the dynamic range. The signal from an accelerometer, including the preamplifier, can have a valid dynamic range of 120 dB. A wide range of this magnitude can only be accommodated on a logarithmic amplitude axis. An exception where a linear amplitude scale is preferable to a logarithmic scale is for representation of relative displacement signals using proximity probes. This parameter often needs comparison with rotor dynamic and bearing calculations, and the dynamic range for relative shaft vibrations does not need depiction on a logarithmic scale.

Filter analyzers can cause significant delays to influence the speed of analysis. The filter and the detector introduce delay on the same order as the response and averaging times. In a swept frequency analysis, the delays mean that the recorded spectrum departs from the true spectrum, as shown in Fig. 9.4. In the selection of paper speed with an X-Y recorder errors in the level of peaks and valleys and the frequency offset due to the delay must be limited.

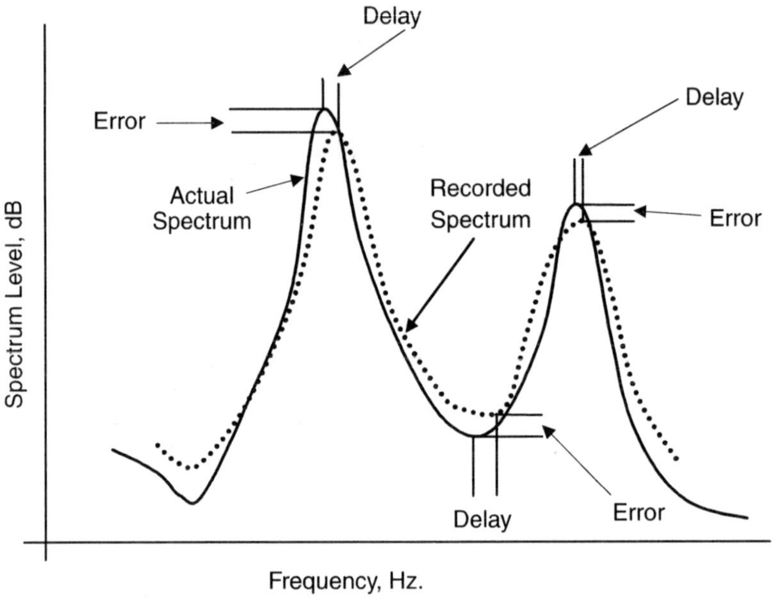

Fig. 9.4. Delay and Error between Actual and Recorded Spectra in Swept Frequency Analysis

The FFT algorithm is an efficient way to calculate the discrete Fourier transform (DFT), which is a finite approximation of the Fourier integral transform. For real valued time signals, the FFT algorithm makes use of the following form of equation:

$$S_x(m) = \Delta t \sum_{n=0}^{N-1} x(n\,\Delta t)\,\exp\,(-j2\pi m\;\Delta f n\;\Delta t) \tag{9.1}$$

$$x(n) = \Delta f \sum_{m=0}^{N-1} S_x(m\,\Delta t)\,\exp\,(j2\pi m\;\Delta f n\;\Delta t) \tag{9.2}$$

The equations shown above give spectrum values $S_x(m)$ at N discrete frequencies $m\Delta f$, and give the time series $x(n)$ at N discrete time points $n\Delta t$. Fourier transforms are integrals of continuous functions, while DFT equations are finite sums, but otherwise have similar properties. Figure 9.5 graphically illustrates the difference between the Fourier integral and the corresponding DFT. The DFT concept is used to generate from a physical signal a unique set of sine and cosine terms at appropriate frequencies. The set is unique because of the orthogonal nature at different frequencies of the sine functions, of the cosine functions and of the sine functions compared to cosine functions.

A typical non-stationary signal resulting from measurements made during a machine run-up or coast-down can be analyzed by dividing it into a series of short quasi-stationary time periods (often overlapping), with the speed roughly constant in each step. The signal may be analyzed with a tracking filter tuned to a specific harmonic of the operating speed, and the results recorded as a function of speed. If a phase meter is included between the filtered and the tracking signals, phase and the amplitude may be recorded against speed to provide a Bode plot.

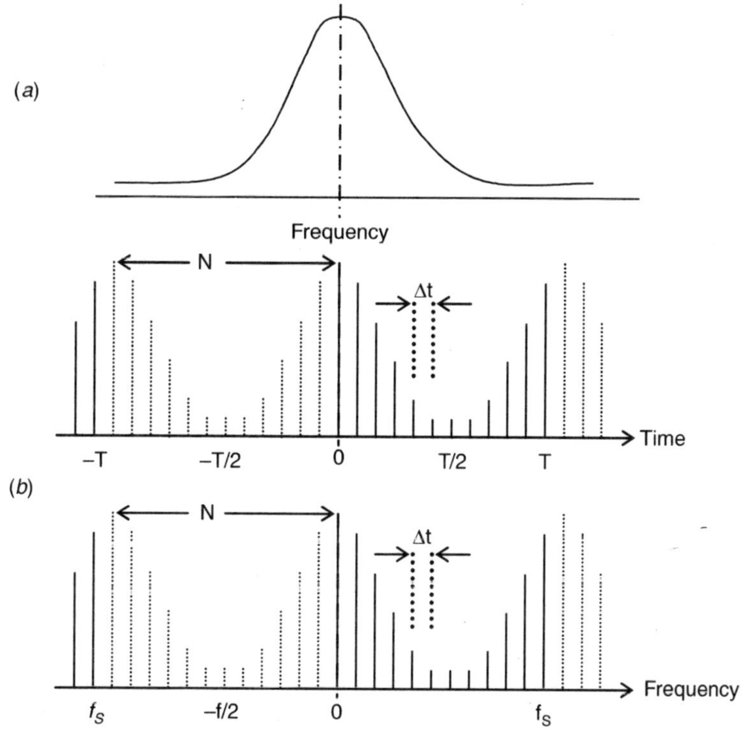

Fig. 9.5. Comparison of (*a*) Fourier and (*b*) Discrete Fourier Transforms

Using an FFT spectrum analyzer, the behavior of several harmonics may be studied simultaneously. A scan analysis on an FFT analyzer with a long memory permits this operation; a short Hanning window is scanned through the record to view successive instantaneous spectra on the display screen. Varying the step length may change scan speed. An effective way of representing a scan analysis is by a 'water fall' or 'cascade' plot, as shown in figure 9.6. The third dimension of the plot may be time or speed; for a simple scan analysis it is usually time. If the spectra are arranged at equal intervals of shaft speed, harmonically related components (their base following radial lines) can be readily separated from constant frequency components (with base line parallel to speed axis). Note that the cascade plot has characteristics of a Campbell diagram. During a slow startup or coast down of a rotating machine it may be possible to obtain vibration spectra at constant shaft speed. Each spectrum is then a windowed section of a continuously varying signal with a small speed change within the window length. As a consequence, the peak corresponding to each harmonic is not always localized in one line; in particular, the higher harmonics are likely to be spread over progressively more lines.

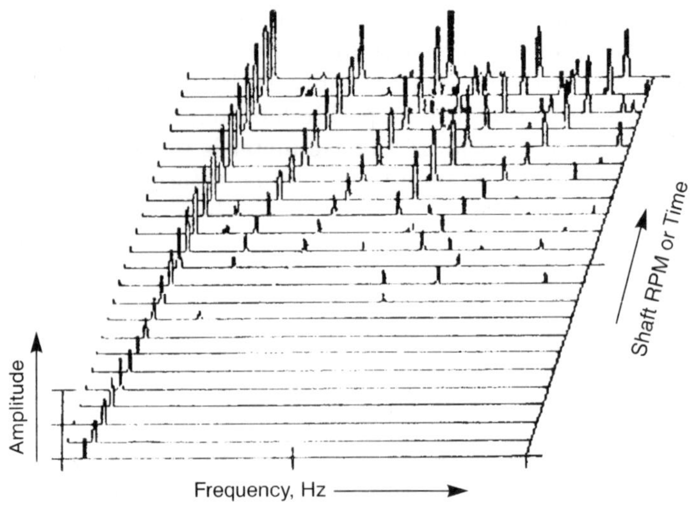

Fig. 9.6. Water Fall Plot of Rotating Machine Vibration Spectra

9.5 MODAL ANALYSIS

Natural frequencies, damping factors, modal vectors and modal scaling of a linear time invariant system can be determined by the process of modal analysis. Modal parameters can also be determined by finite element analysis. Experimental methods provide verification and correction of the calculated results. If a finite element model does not exist, experimental results serve as the model for future evaluation for structural modifications.

The success of an experimental modal analysis process depends on having specific goals for the test. Every phase of the process is affected by the established goals. Modal analysis theory has not changed for a long time; however, its application to experimental measurements has changed substantially. Certain aspects of transform relationships have taken on renewed importance, since digital forms of integral transforms are in constant use. The theory uses a more thorough understanding of how the structural parameters of mass, damping and stiffness relate to the impulse response function (time domain), frequency response function (Fourier or frequency domain) and transfer function (Laplace domain). The primary concern in the acquisition of data for the formulation of a modal model relates to digital signal processing. Conversion of analog signals into a corresponding sequence of digital values must accurately describe the time varying characteristics of the inputs to, and responses from, a system. Digital signal processing helps to condense the amount of measured data, estimate the modal parameters and reduce the influence of noise in the measurement procedure.

The frequency response function is the prime measurement to be made in an experimental modal analysis. A measurement model is needed to permit estimation of the frequency response function from measured input and output data in the presence of noise (representing errors). The Fourier transform is used for this computation, and is performed digitally using a fast Fourier transform algorithm. Several factors contribute to the quality of actual measured frequency response function estimates. Errors arise from overloading the input, extraneous signal input and strong electric or magnetic fields nearby. Nonlinear effects shift energy from one frequency to many new frequencies in a manner that may be difficult to recognize. One way to reduce this effect is by choosing a randomly different input signal for each of the contributing averages.

Signal averaging is widely used in structural signature analysis where the event is a revolution of a rotating shaft. Since the frequency response function is expressed in terms of system properties of mass, damping and stiffness, the functions may also be considered to be constant. Hence, when formulating the frequency response function by using appropriate algorithms, the estimate is intrinsically unique, as long as the system is linear and noise is eliminated. The estimate will be valid whether the input is stationary, non-stationary or deterministic. The uniqueness aspect of the frequency response function permits greater flexibility in the testing procedure. Each function is derived as a result of a separate test, or as the result of different portions of the same continuous test. In either case the estimate of the frequency response function is the same, as long as the time-history data for the auto and cross power spectra are acquired simultaneously. Since the Fourier transform is a linear function, signal averaging may be of synchronous, asynchronous or cyclic type. Input excitation to the system may be of the random signal or deterministic type. The choice of input depends upon the characteristics of the system and of the parameter estimation, as also the expected utilization of the data. For a linear system all input forms should give the same expected value. Many signals are appropriate for use in modal analysis. Figure 9.7 shows a typical test configuration when an impact form of excitation is used on a component.

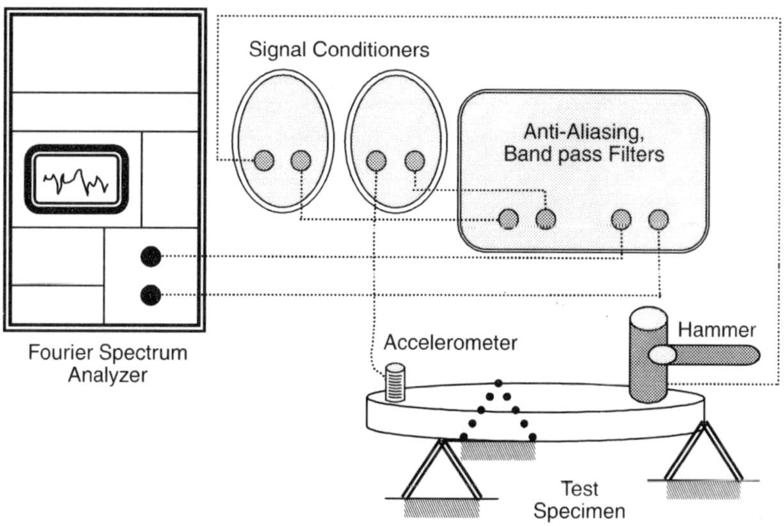

Fig. 9.7. Test Configuration for Impact Excitation

Modal parameter estimation, or identification, uses numerical techniques to separate the contributions of individual modes of vibration in measurements such as frequency response functions. The concept calls for estimation of individual single degree of freedom contributions to the multiple degree of freedom measurement. Modal participation vectors are a result of the estimation algorithms, and relate how well each modal vector is excited from each of the reference locations in the measured data. The process is done in multiple stages. The common characteristics of different modal parameter estimation algorithms can be more readily identified by using a matrix polynomial model rather than a physically based mathematical one. Using a polynomial form for the frequency response function, a linear equation with unknown coefficients can be developed which is valid at each measured frequency. In parallel, a time domain representation is also formulated. Most of the modal parameter estimation starts from a general matrix polynomial formulation that is justifiable based upon the underlying matrix differential equation. The process yields essentially the same characteristic matrix polynomial equation for both time and frequency domain data. The roots, or modal frequencies, of the matrix characteristic equation in the time domain are in the z domain and must be converted to the frequency domain, while the roots in the frequency domain are already in the desired domain. The modal frequencies can be found using a number of numerical techniques. Using a companion matrix approach the roots of the matrix characteristic equation can be found as eigen-values. Eigen-vectors that are found from the solution utilizing the companion matrix might be useful for modal parameters, unless scalar coefficients are used. Modal participation vectors from the eigen-vectors of the companion matrix approach are generally found in a second stage solution of the process. For nearly all modal identification problems a large amount of redundancy exists, because of which the data space is filtered, or sieved, in order to obtain a reasonable solution in the minimum amount of time.

9.6 TORSIONAL VIBRATIONS

Since the shaft is rotating, torsional vibration measurements are more difficult than lateral vibrations. The procedure for signal analysis is similar to that used for lateral vibrations. Torsional response in the form of rotations and strains can be measured at intermediate points in a system. Sensors, however, cannot be placed at a nodal point, so it is advisable to perform Holzer calculations to determine mode shapes prior to selecting sensor location.

Strain gauges, available in a variety of size and sensitivity, can be placed anywhere on the shaft. Calibration may be set up to directly indicate instantaneous torque by applying a torque load on the drive shaft statically. Stress and torque may also be calculated from strength of materials theory. Strain gauges are usually mounted at an angle of 45° in order to eliminate shaft bending effects on torque measurement. The signal must be processed by a bridge amplifier unit which is arranged to compensate for temperature related variations.

A slip ring is needed to obtain the vibration signals from a rotating shaft. Wireless telemetry is also available; a small transmitter mounted on the rotating shaft at a convenient location broadcasts signals to a nearby receiver. Torque transducers are available for measuring oscillating rotations in the shaft, but they must be inserted in the drive line, and run the risk of altering the dynamic characteristics of the system.

The velocity of torsional vibrations may also be measured with a toothed ring mounted on the rotor and a sensor. The ring must be rigid and of light weight construction to prevent distortion of the rotor's inertia characteristics. If space and operating conditions permit it, a number of this arrangement may be placed along the length of the shaft. Simultaneous measurement of torsional vibrations at more than one point on the rotor will provide useful information on its dynamic behavior. If a damper is included in the rotating system, characteristics of the damping mechanism can also be readily evaluated with the aid of these measurements.

9.7 SOURCES OF UNBALANCE

Unbalance in rotating machines arises from a number of sources, including lack of symmetry, lack of homogeneity in material, distortion at service speed and eccentricity. Foundry casting of components oftentimes causes the core to shift inside the cavity, resulting in thinner walls on one side and greater thickness on the opposite side. Rough finish on forging surfaces will also result in asymmetry of the component. Blowholes in cast rotors, inclusions in rolled or forged parts, slag inclusions and variations in the crystalline structure as a result of variations in material density cause the component to lose its homogeneous nature. Turbine and blower fan blades have a geometrically complex configuration. At service speed centrifugal forces and moments cause the blade to assume a new profile due to elastic effects in the material. Rotating parts in built-up rotors have a mismatch at the mating surface due to manufacturing tolerances, causing the assembly to lack symmetry.

Machining processes are a major source of unbalance. Unmachined portions of a casting or forging that cannot be made symmetrical and concentric with respect to the shaft axis represents a potentially severe problem during balancing. Built-up rotors made of several parts have a radial offset due to manufacturing tolerances, thereby introducing unbalance.

At the bearing a journal may not be concentric, creating asymmetry of the rotating system. Misalignment in the bearings is another source of unbalance. Plastic deformation sometimes causes parts to shift within an assembly. Cavitation in pumps and turbulence in compressors are examples of hydraulic and aerodynamic sources of unbalance. Thermal gradients in many instances are responsible for bowed rotors in power generation steam and gas turbines and in paper mill rolls.

Including design features to control rotor unbalance can minimize balancing problems. Large amounts of unbalance require large corrections. If the corrections are to be made by removing metal additional cost is incurred, and also affects part strength. If corrections are made by adding material, space requirement for the balance shots may pose a problem.

Design limitations often impose restrictions that preclude features for unbalance correction. For instance, electrical machinery design requires that one coil be placed at a greater radius than the other coils in certain armatures. If addition of a compensating mass to take care of the associated unbalance is impractical, the machine's frame and rotor will be required to absorb the resulting dynamic load. Rotating parts should be designed for inherent unbalance, whether the component is balanced or not. When service speed is low and effects of a reasonable amount of unbalance can be tolerated, the balancing operation may be eliminated. If functional reasons call for distribution of certain masses in the rotor to be asymmetric, counterbalanced masses must be designed into the assembly.

9.8 MOTION OF UNBALANCED ROTORS

Consider two cases of a rotating structure in soft bearings, as shown in Fig. 9.8. In one case the rotor has an unbalance mass at its mid plane, causing the rotor to be statically unbalanced. As the shaft turns, centrifugal force on the mass will cause the centerline of the shaft to travel in a circular path. In the other case the rotor has two unbalance masses at opposite ends, and are offset 180° to each other. The two unbalance weights will offset each other statically. Also, as the shaft rotates, centrifugal force on the masses will cancel each other, however, an overturning moment will be present due to the axial offset of the masses. The resulting trajectory of the shaft's centerline will be in the form of two cones, and the apex of the cones will be at mid plane due to the symmetric nature of the rotor and the unbalances. When the two forms of unbalance are combined, the apex of the cones will be moved away from the center of gravity. More generally, an apex will not be present, because the shaft will move in a complex combination of motions shown in Fig. 9.8. A random combination of magnitude, axial and phase angle location of the static and couple unbalance will lead to a condition referred to as dynamic unbalance.

Flexibility in the supports of the rotor will add to the motion of the rotor in the lateral directions. When support flexibility at the two ends is not symmetric, the lateral motion will also have differing characteristics, as shown in Fig. 9.9. Here it is tacitly assumed that the rotor is essentially rigid, with the dynamic activity confined to the supports and no flexure in the shaft.

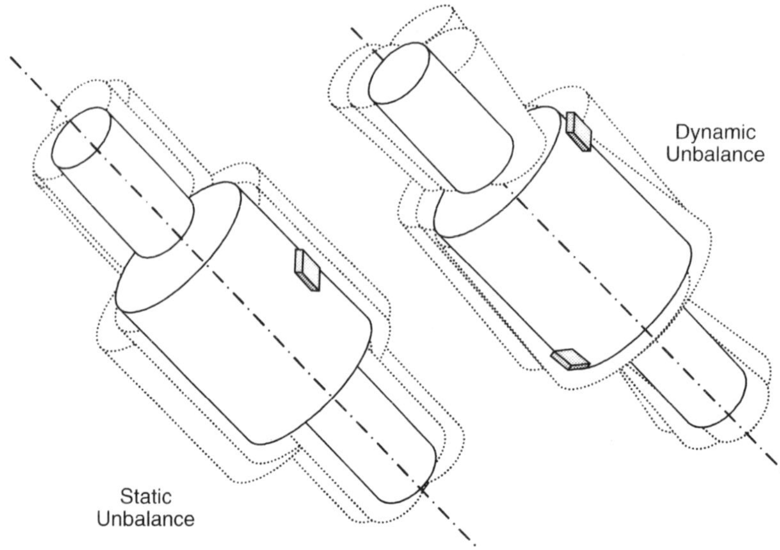

Fig. 9.8. Static and Couple Forms of Unbalance

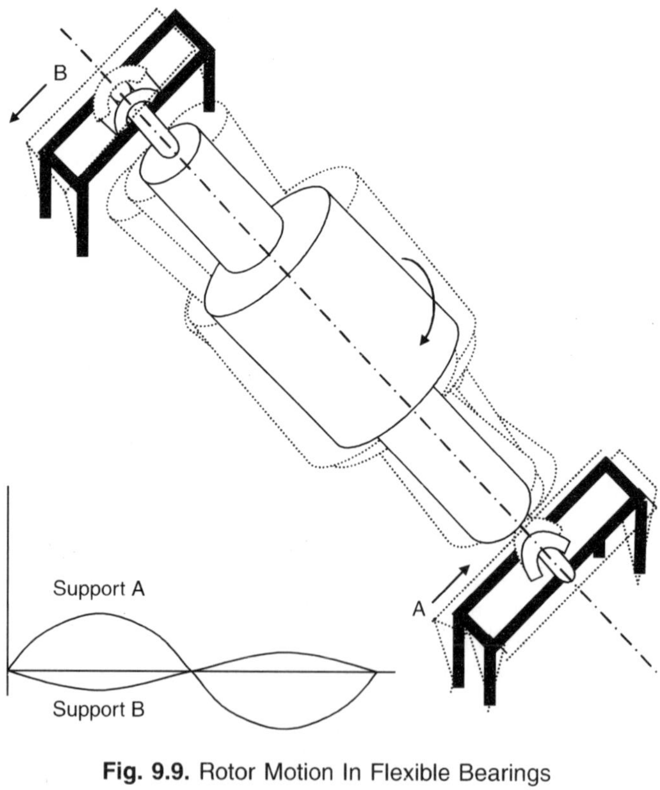

Fig. 9.9. Rotor Motion In Flexible Bearings

To understand the effects of unbalance on a flexible rotor, consider a single mass disk mounted on an elastic shaft. The mass center of the disk is displaced from the center of the elastic shaft, resulting in an unbalanced rotor (Fig. 9.10). The total rotor effective unbalance is equal to the product of the disk's mass and the unbalance eccentricity. The rotating unbalance force on the disk causes the shaft centerline to precess, or orbit, about the geometric centerline. When the bearing supports are rigid, the motion may be referred to as synchronous precession with a circular orbit. If this rotor is mounted on flexible bearing supports, total motion of the disk will include motion of the end supports.

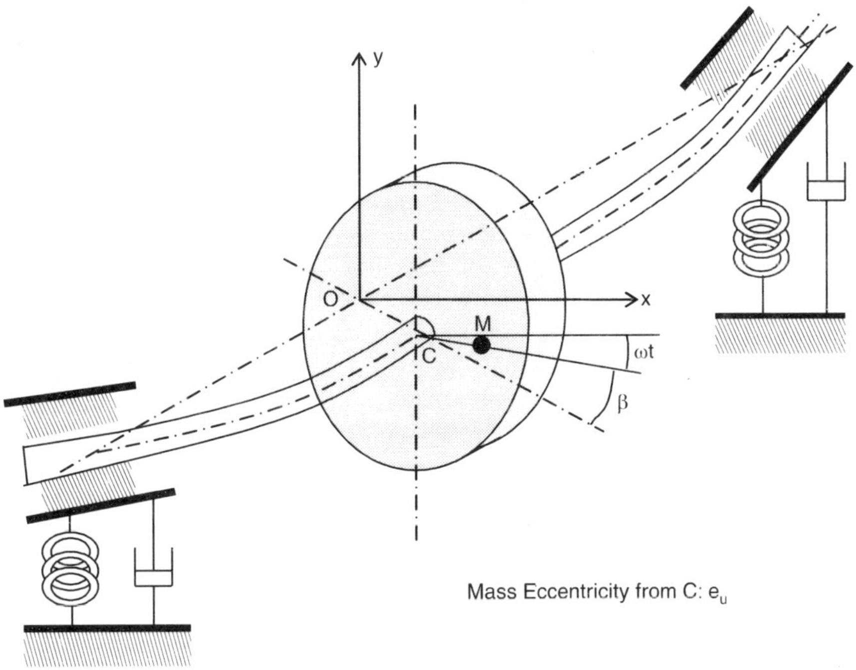

Fig. 9.10. Deflections of Single Mass Elastic Shaft in Flexible Supports

9.9 BALANCING MACHINES

Two types of centrifugal balancing machines are in use to determine the size of the required weight and its angular location in each correction plane, the soft bearing and the hard bearing machines. Soft bearing machines permit the idealized free rotor motion, but usually the activity is restricted to the horizontal plane. In addition, the bearings and its attached components move with the rotor, thus adding mass to the rotor at the location. The increased mass of the rotor will reduce the displacement of the bearings, and output from the device sensing the unbalance will correspondingly reduce.

Hard bearing balancing machines are basically of a similar construction, except that their bearing supports are significantly stiffer in the horizontal direction. This causes horizontal plane related critical speeds of the machine to be several orders of magnitude greater than that for a comparable soft bearing machine. This type of machine is designed to operate at speeds well below the horizontal critical speed. The output from the sensing elements attached to the bearing supports is directly proportional to the centrifugal force reaction resulting from unbalance in the rotor. Since the output is not affected by bearing mass, rotor weight or inertia, a definite connection between unbalance and sensing element output can be established, hence eliminating the need for calibration of a given rotor.

Centrifugal balancing machines may also be categorized by the type of unbalance to be indicated (static or dynamic) and work piece shaft axis attitude (horizontal or vertical). Trial and error balancing machines are of the soft bearing type, and do not directly indicate unbalance in weight; they only provide displacement or velocity of shaft lateral motion at the bearings. Balancing of the job calls for a lengthy trial and error procedure, even if it is one of an identical batch. Also, the machine cannot identify balance correction for a specified plane. Field balancing generally does not have access to a microprocessor, so the method falls into this group. Balancing machines that are calibrated for a balanced rotor from an identical group can be instrumented to permit plane separation. However, for the first piece the trial and error procedure cannot be avoided.

Soft bearing machines may also be provided with integral electronic unbalance compensator type of instrumentation, where a balanced master rotor is not needed. Separation of balance planes and calibration is achieved by using precisely weighed calibrating shots temporarily attached in each of two correction planes. Microprocessor based software determines the influence coefficients. This class of soft bearing balancers may also be provided with electrically driven vibratory shakers fitted to the rotor supports.

Another class of permanently calibrated balancing machines uses hard bearings. Calibration is set up for rotors falling within a range of rotor weight and size in a selected speed range. Unbalance indication is accomplished in the first run by including an analog or digital computer with the instrumentation. Calculations for determining correction weight size and location are based on the following parameters:

- Distance from left correction plane to left support
- Distance between correction planes
- Distance from right correction plane to right support
- Radial location of the correction weights in both planes

The instrumentation then determines the magnitude and angular location of the correction weights in both planes.

Precise definition of the rotor assembly's configuration, capacity and performance is needed in selecting the balancing machine. Details about rotor weight, length, diameter, journal diameters, service speed, application and whether the rotor is rigid or flexible are required to evaluate the suitability of the balancing machine. A manufacturer's selection must be evaluated not only for compliance to the purchase specifications but also on the machine's features, operation and conformance with applicable standards.

9.10 CORRECTION METHODS

Static balance of a work piece may be performed using gravity. The principle is illustrated in Fig. 9.11, where a disk is suspended at its geometric center by a string. Orientation and amount of the disk's tilt gives an indication of the balance mass to be placed to bring it in the horizontal plane. If the weight of the disk precludes this method, it may be mounted loosely at its hub on a vertical mandrel. The balance weight may be split into two equal halves and placed on opposite faces to reduce overturning effects from centrifugal forces during operation. When the disk is attached to a shaft, permitting the shaft to roll on horizontal knife edges gives an indication of unbalance location when it comes to rest.

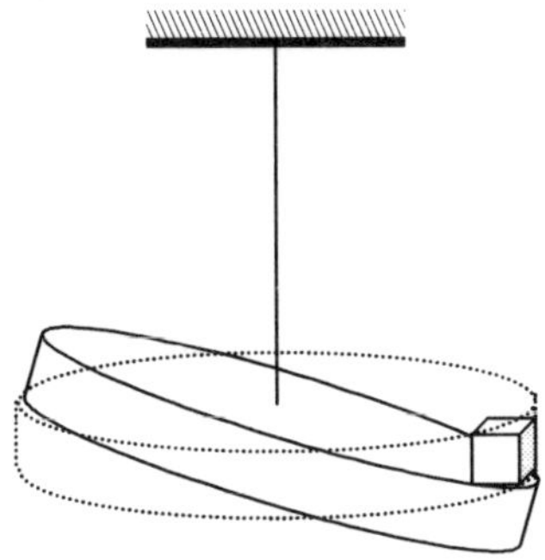

Fig. 9.11. Static Balance Using Pendulum Principle

Balance correction may be done by adding weights or removing material. In special cases shaft axis may need relocation. The selected procedure must ensure that adequate capacity exists to permit additional correction, because initial correction estimates may not permit reduction within tolerance limits. When carried out carefully, common methods usually result in achieving a 10:1 reduction. Under more closely controlled conditions for weight and angular position, reduction in the range of 20:1 is not uncommon. If the maximum initial unbalance does not bring the rotor within permissible limits by a selected method, a second correction procedure may be required to reduce the residual unbalance.

Correction masses may be added in more than one way. It is sometimes difficult to apply wire solder while ensuring that its center of gravity is at the desired location. Errors may also be induced by variations in wire diameter. Bolted or riveted washers are useful if only moderate balance quality is required. Cast iron or lead weights in incremental weights provide large initial unbalance correction, if suitable design features in the rotor permit adequate fastening means. Resistance welding may be used for attaching large correction weights if localized thermal distortion in the rotor from the heat of the welding is kept to a minimum.

Material may be removed from a rotor by drilling a hole to a measured depth for unbalance correction. A depth gauge and switch may be used on the drill spindle to assure intended material removal with a high degree of accuracy. Milling, shaping and fly cutting permit accurate removal if the surface of the rotor from where metal is to be removed is machined, permitting proper measurement of the size of cut. The method is preferable when larger correction is called for. Grinding is essentially a trial and error method of correction, because it is difficult to estimate the amount of metal removed. The method is restricted to design of rotors that do not permit other means of correction. A good example is the shaft connecting the low pressure turbine with the fan module in aircraft gas turbine engines. The shaft is thin walled, axial spacing between bearing centerlines is large and aircraft operating cruise conditions may call for operation close to the first critical speed. Since both ends of the shaft are splined, the only provision for balance correction is land areas where metal is removed through grinding.

Location of the two radial planes is not important from mathematical considerations, since the overturning couple does not take into account axial location of the balance planes. Practical considerations call for the planes to be located axially as far as possible to gain greatest leverage, thus minimizing the magnitude of the masses. Balance planes must also be easily accessible for attaching the correction weights, especially if the operation is to be done in the field. A typical approach is the provision of an otherwise nonfunctional rib to which correction masses may be added, or an annular protrusion in the balance planes from which material can be machined or ground away at the appropriate circumferential location. Since assembly and disassembly is not necessary, the method is advantageous. The process has been adapted to automatic balance correction where material is removed from the rotor by laser melting while rotating in the balance machine. Metal removal procedures are subject to the limitation that balance correction cannot be undone when rebalance of the rotor is required. Also, care must be exercised to avoid introduction of stress risers in highly stressed areas of the rotor in the course of removing material.

If run-out measurements indicate that a rotor's geometric centerline does not coincide with the axis of rotation, a procedure known as mass centering may be useful in reducing unbalance effects. The rotor is mounted in a balanced cage or cradle, which itself is installed in the balancing machine. The rotor is adjusted radially with respect to the cage until the unbalance indication is within limits. The procedure causes the principal inertia axis of the rotor to be brought into essential coincidence with the axis of the balanced cage. Center drills guided along the axis of the cage provide means for establishing an axis in the rotor about which it is in balance. Subsequent machining operations on the rotor, however, may nullify some of the benefits of the mass centering procedure.

An individual turbine stage assembly with blades in place is best balanced by exchanging the location of two blades from their initially assigned dovetail slots. Each blade on the turbine disc is pan weighed before assembly. After a run on the balance machine, vibration indication from the sensor and weight distribution of the blades around the wheel will permit identification of a pair of blades to be switched from their location. More than one such sequence of events may be needed to bring the assembly within tolerance limits.

Balancing of rotors made of two or more individually balanced components or sub-assemblies is often performed to obtain maximum reduction in vibration from unbalance. The complete machine may also be run under service conditions as part of the balancing procedure. Assembly balancing may be necessitated by conditions related to assembly and operation procedures. A balanced flywheel mounted on a balanced crankshaft may not be balanced as an assembly. When connecting rods and pistons are added, more unbalance is likely to be introduced. The unbalance effects can only be reduced by balancing the reciprocating machine as an assembly.

The inner and outer raceways of anti-friction bearings sometimes do not have concentric surfaces. Eccentricity of the races makes it desirable to balance the rotor on its own bearings, with the stator supporting the rolling element bearings. In this respect bearing alignment and preload also become important factors.

9.11 ROTOR CLASSIFICATION

Classification of rotors as rigid or flexible depends on the dynamic behavior of the rotating system in its operating speed range. If analytical methods indicate that potential energy in the bearings is over 80% of the system's total strain energy, the rotor is considered to be rigid. Rigid rotors may be balanced through placement of correction weights in two planes in its given operating speed zone. But if the strain energy due to bending in the shaft exceeds 20% of the system total, and the rotor operates through one or more critical speeds, then it is regarded as a flexible rotor. A two plane balancing may not be sufficient for a flexible rotor, and additional weights may be needed along the length of the rotor to control vibration amplitude at speed.

Documentation published by the International Standards Organization (ISO) (1994) on the subject discusses classification of rotors and balance quality requirements for many different applications. Rieger (1986) also provides a deeper insight on the subject. Five basic groups of rotors from balancing perspective are presented by ISO:

1. *Rigid rotors:* Balanced in any two axial planes, staying in balance in the complete speed range.
2. *Quasi-flexible rotors:* Not fully rigid, but can be adequately balanced in a low speed balancing machine; balance will be maintained in the full speed range.
3. *Flexible rotors:* Cannot be balanced in a slow speed balancing machine, and will require one or more high speed trim plane corrections.
4. *Flexible-attachment rotors:* Rotors of class 1, 2 or 3 have attached to them, or contain within them, flexible components.
5. *Single speed flexible rotors:* Special case of class 3, with balancing of rotor restricted to a single speed.

Class 1 and 2 rotors may be adequately balanced at one speed in one or two planes. Flexible rotors in class 3 and 4 require the modal technique or the influence coefficient method.

In reality, no rotor is completely rigid. The expression may be applied to specific rotors operating at speeds well below their fundamental flexible natural frequency. The rotor is considered flexible if it does not hold balance after being balanced as a rigid body. Figure 9.12 shows a rotor with attached disks. Disks 3 and 7 have unbalance of U_3 and U_7, while the rotor mass center is displaced from the axis of rotation by distance e_u at disk 5. When this rotor is placed in a balancing machine, either the hard or soft bearing type, a two-plane correction will be indicated. Balance correction weights ω_3 and ω_7 are placed at planes 3 and 7 at radii R_3 and R_7. As a rigid body, the rotor now has the appearance of being balanced. Rotor unbalance is represented by $Me_u = U_5$. The rigid rotor balance condition of $\Sigma(U_3 + U_5 + U_7) = 0$ is satisfied, so the system is in static equilibrium. When operated as shown at a sufficiently low speed, the

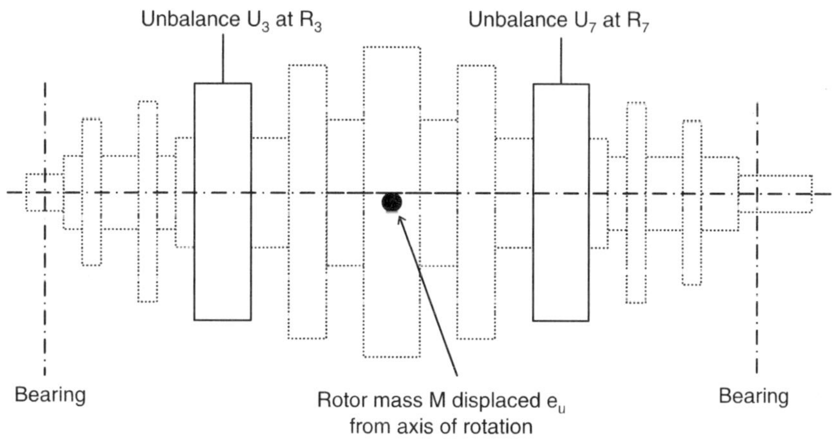

Fig. 9.12. Rotor with Attached Disks

balance will be maintained. As the speed increases centrifugal forces from the unbalance distribution and elastic effects in the shaft cause the shaft to deflect and bow. The effects become more pronounced as the speed exceeds 70% of the critical speed. Additional balancing is required to control vibration amplitudes. Unbalance distribution in the rotor may come in many different variations. Some commonly encountered types include continuous unbalance along the length of the rotor, unbalance due to a bow and moment unbalance from an overhung disk.

9.12 BALANCING OF RIGID ROTORS

Static single plane balancing of a rotor does not cause coincidence of the shaft's running axis with its principal inertia axis; it only ensures that the two axes have a common point. Thus, when the shaft is turning, dynamic balance is not achieved. This can only be obtained by rotating the principal axis about the center of gravity in the longitudinal plane, characterized by the shaft axis and the principal axis. The rotation can be accomplished by modifying the geometry of the journals (an impractical proposition), or by adding or removing metal from the mass of the rotor in the longitudinal plane. Note, however, that adding or removing material disturbs the static balance already achieved. To avoid this possibility, two masses of equal magnitude are added or removed, one on each side of the principal axis to maintain static balance, and one in each of two radial planes to eliminate overturning effects.

For balancing of machines operating in the field, the rotor is already in place in its own bearings. Suitable instrumentation is placed on the shaft, bearing housing or foundation to monitor the vibrations. A number of procedures are employed, with preference of one method over another depending on geometric configuration of the rotor and bearing system.

The single plane, single speed balancing procedure is simple. After vibration measurement from an initial run, a trial weight is placed on the shaft. The vibration response of the rotor at the same speed is again recorded. The change in vibration divided by the trial weight provides an influence coefficient, representing rotor response due to the trial weight at the given speed. Note that the influence coefficient is a vector with amplitude and phase, the latter providing the angle between the forcing function and response. For a linear system application of the balance correction should theoretically reduce amplitude of motion at the given speed to zero. The procedure, called 'one-shot' balancing, is similar to trim balancing and is widely used in many different situations. However, problems in the form of higher vibrations at other speeds and points along the rotor may be encountered.

High vibration at other speeds may be avoided through linear regression, which provides the least-square error method of balancing. Instead of a single speed point, vibration measurements are recorded over a number of points. Linear regression procedure will reduce large responses at other speeds. The magnitude of correction weight is smaller than the single-speed influence coefficient method, and ensures that correction is not limited to a single speed. The method is useful if the shaft has a bow or a run-out.

When the rotor is long, applying correction in one plane may result in high vibrations at another location due to overturning moment effects. Applying correction weights and calculating influence coefficients in two planes at the same time proves more useful. A trial weight is placed on the first plane, and response at first and second planes is recorded. Corresponding influence coefficients are calculated. In the second step the weight is removed from the first plane and placed on the second plane. After measuring the response two more influence coefficients are determined, from which a 2×2 matrix of complex influence coefficients (equivalent to a 4×4 matrix of real numbers) is obtained. Inversion of the matrix will provide magnitude of correction weight and phase angle location for the planes.

As in the case of single plane balancing, vibration readings at multiple speed points may be recorded for the two-plane method. Generalized influence coefficients are then determined using the least-square error procedure. The method provides useful correction weight values even when a rotating system has to pass through a number of critical speeds during operation. Also, the method is preferred when prior knowledge of the system's geometry and rotor dynamic characteristics is not available. Initially developed by Goodman, improvements were reported by Palazzolo and Gunter. When balance planes are not properly selected, the least-square error method may lead to correction weights of magnitude that are impractical to use. Limitations may not be placed on the size of balance weights in this method.

9.13 BALANCING OF FLEXIBLE ROTORS

When a rotor operates through multiple natural frequencies, the modal method of balancing is most suitable. Critical speeds and corresponding mode shapes of the system must be known, either from analytical predictions or from experimental measurements. Weights placed on the rotor at an axial location will be proportional to the relative mode shape deflection at that point. Modal distribution for each natural frequency is used to individually balance the rotor for that mode. The objective is to reduce dynamic activity in modes with high amplification factors without upsetting the rotor's balance at other modes. Balance corrections to be applied on the rotor may be computed by using the influence coefficient method to predict modal influence balancing coefficients. Darlow (1989) provides significant contributions on the combined modal and influence coefficient methods, and the procedure is referred to as the unified method of rotor balancing. Note that rotor mode shapes become quite complex above the second critical speed.

Sometimes it is not possible to obtain an accurate measurement of the phase angle. Blake reports the method of applying trial weights at three different locations to determine the magnitude and location of unbalance. The method has been successfully applied to fans where the speed cannot be held steady due to aerodynamic or other disturbances. Compensation for non-linear effects can be included in this graphical procedure, referred to as the three trial weight method of balancing. The method may be extended to balance flexible rotors through multiple critical speeds (Foiles and Gunter, 1982). A modal distribution of weights is placed on the rotor, and vibration amplitude measured at critical speeds. An FFT analyzer with ability to hold peak values may be used for this purpose. Each mode may then be separately balanced without affecting the balance of the rotor in other modes. The method proves to be advantageous when amplification factors are high, and when it is difficult to balance the rotor by the least-square error method. Drawbacks arise from the need to know the rotor's mode shapes, and three runs must be made for each critical speed to be balanced. After this procedure, the generated modal sensitivity may be used for additional improvements by making only two runs.

Two machines may be sometimes individually balanced, but when they are mechanically coupled, high vibration levels ensue. Trim balance at the coupling is required. One procedure calls for rotation of one half of the system by 180° at the coupling, and then taking another set of vibration readings. From the measurements the new position of the coupling may be computed to obtain acceptable vibrations.

Balancing of rotors without trial weights is reported by Palazzolo and Gunter (1977). The Nyquist's plot of vibration amplitudes is used to determine the amplification factor for various modes, from which modal unbalance eccentricity can be obtained. Phase angle for the modal weight distribution is also determined from the polar plot to balance the rotor at particular critical speeds. Rotor modal mass and mode shapes for the natural frequencies of interest must be known beforehand. Balance predictions from this procedure may be used for initial trial runs, then refined by the influence coefficient method of balancing.

9.14 BALANCING CRITERIA

ISO/DIS standard 10816 (ANSI Counterpart S2.12, Part 6-199X) establishes procedures and guidelines for measurement and classification of mechanical vibrations of reciprocating machines. Vibration measurement and evaluation criteria provided in this standard are recommended for securing safe and reliable operation of machines, and to avoid problems with auxiliary equipment attached to the main structure. In general, the standard calls for vibrations on the main structure of the machine. A number of components are mounted on the machine. Heat exchangers, governors, pumps, filters, gauges and piping components are some examples. Vibration guidelines based on experience with similar machines are useful to ensure damage is not sustained in the mounted attachments. Vibration readings on the main structure, however, are not a true indicator of the state of dynamic stress of the components within the machine.

Vibration of shipboard hull arises from excitation by the propeller shaft and propeller blades. Vibration is normally highest in the aft parts of the ship, with displacement amplitudes sometimes exceeding 0.10 in. At higher shaft speeds hull vibrations tend to be lower. With shipboard machinery mounted on the deck, hull and bulkhead stiffeners, vibration amplitudes at the equipment supports is substantially the same as in the hull. Acceptable values for diesel engine driven merchant ships of 100 meters in length or more are provided in ISO 6945. The shaded area in Fig. 9.13 shows acceptable vibration levels for shipboard structures. Additional related information is also available in ISO 2631, ISO 4867 and ISO 4868.

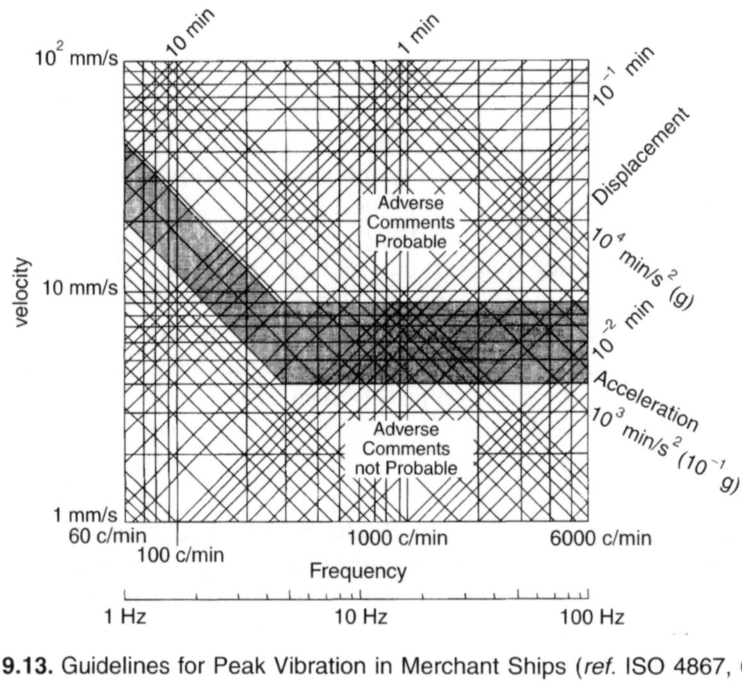

Fig. 9.13. Guidelines for Peak Vibration in Merchant Ships (*ref.* ISO 4867, 6945)

The Hydraulic Institute specifies acceptable vibration limits for horizontal clear liquid pumps. The limits apply only if the pump is restricted to the following operating conditions: (1) Operation is non-cavitating and non-separating, (2) the suction pipe provides straight uniform flow, (3) piping arrangement avoids undue stress on the pump and (4) shaft coupling is aligned according to manufacturer's instructions. If the vibrations in the pump exceed specified values, the machine system should be inspected for possible defects and correction. Often, changes in the level of dynamic activity over a period of time are more significant than the recorded vibration amplitude. Vibration levels in excess of specified limits may be acceptable if there is no increase over a sustained period of operation. Evidence of damage in the bearings or increased noise level, however, would be reasons for closer investigation.

Vibration criteria are available from the Compressed Air and Gas Institute for compressors in air and gas service, with operating speeds up to 15,000 rpm. A change in the vibration pattern of an operating compressor, as a rule, is more important than the measured level at any time. Thus, a substantial change in the vibration pattern for the same operating conditions should be promptly examined to determine its cause.

Standard ISO/WD 8579-2 describes the procedure for determining vibrations in enclosed gear reductions units operating within its specified design speed, load and temperature bounds for acceptance testing. Lubrication requirements from the manufacturer must be followed. A set of two vibration measurements are required on units provided with journal bearings, one on the shaft, the other on the bearing housing. Proximity probes may be used to measure shaft vibration relative to the housing for frequencies ranging from 0 to 500 Hz. The transducers must be as close as practical to the bearing, and measurements must be made in three orthogonal directions, one of them parallel to the shaft axis. Housing vibrations may be measured with a seismic type of accelerometer in frequencies ranging from 10 to 10,000 Hz. The standard provides a method for classifying gear boxes based on shaft displacement and housing velocity measurements, together with a subjective assessment of acceptable vibration levels for typical applications.

9.15 EXAMPLE PROBLEMS

Example Problem 9.1: Consider a shaft of uniform diameter D and length L supported between two bearings, assumed to be ideal rigid supports. The rotating beam has natural frequencies of vibration (called eigen-values), and corresponding to each natural frequency is a mode shape (or eigen-vector), as depicted in Fig. 9.14. The displacement mode shape does not represent actual shaft deflection, it represents a normalized shape that the shaft assumes at that frequency. When the shaft accelerates in speed, unbalance or shaft bow will cause the shaft to bow outwards as the speed approaches the shaft critical speed. Actual whirl radius will depend on magnitude and location along shaft axis of unbalance, ratio of rotor speed to critical speed, and amount of damping in the system. Considering a shaft of following dimensions, explain the characteristics and significance of the mode shapes.

Shaft length,	$L = 75.0$ in.
Shaft diameter,	$D = 3.125$ in.
Material Young's Modulus,	$E = 30.0 \times 106$ lbs/in^2
Material density,	$r = 0.283$ lbs/in^3

Solution: The first three critical speeds of this shaft are $\omega_n = (K_n/2\pi) \times [EIg/wL^4]^{1/2}$, where $K_1 = 9.87$, $K_2 = 39.5$ and $K_3 = 88.8$ and w = weight per unit length. Then:

$$\omega_1 = 2.648 \times 10^3 \text{ rpm}, \omega_2 = 1.0597 \times 10^4 \text{ rpm and } \omega_3 = 2.3823 \times 10^4 \text{ rpm}$$

Corresponding mode shapes are shown in Fig. 9.14.

Critical speeds of a shaft represented by discrete lumped masses will be considered next. Figure 9.15 shows a three-station discrete lumped-mass representation of the uniform shaft. Weights of $W/4$ are located at the ends of the shaft, $W/2$ at the center.

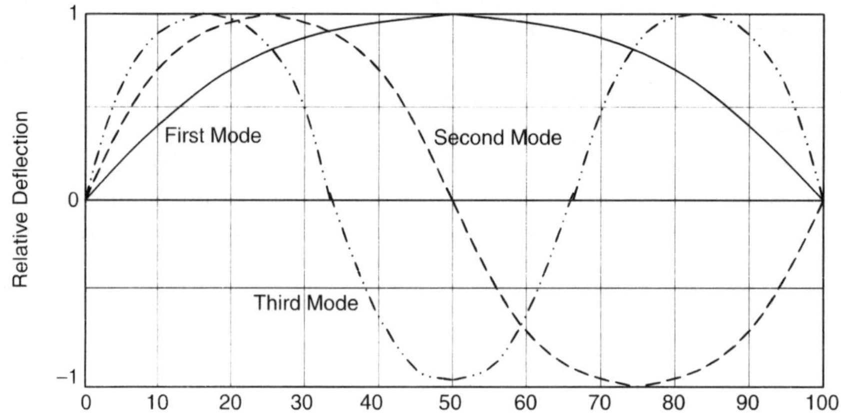

$\omega_1 = 2.648 \times 10^3$ rpm; $\omega_2 = 1.059 \times 10^4$ rpm; $\omega_3 = 2.383 \times 10^4$ rpm

Fig. 9.14. Mode Shapes of a Uniform Beam

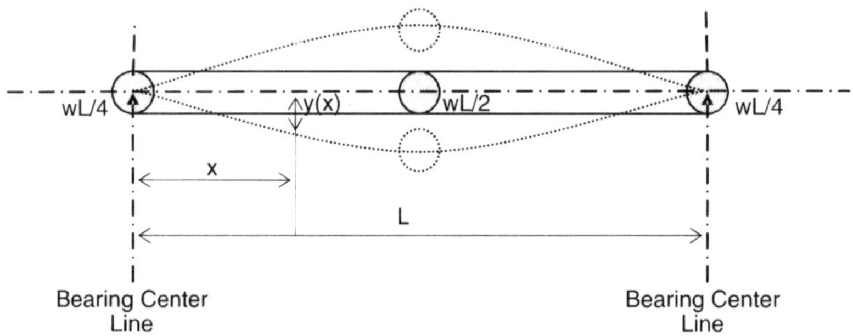

Fig. 9.15. Three Station First Critical Speed Representation of Uniform Shaft

Note that this model is valid only for the first rotor critical speed.

The first critical speed of lumped-mass representation gives $w1 = (6.93/2p)$ ´ $[EIg / WL3]1/2$ = 2,656 rpm, where $W = wL/2$. Thus, this three-station representation of the uniform shaft provides first critical speed with a deviation of only 0.3 %.

The second and higher critical speeds of the lumped-mass representation are considered next. To approximate the characteristics of the uniform beam for the second mode, five mass stations are required (Fig. 9.16). The ends have lumped weights of $wL/8$, center stations having lumped weights of $wL/4$. This results in a model that approximates the second rotor critical speed to less than 0.7 percent error. Here it should be noted that closed form solutions for the second and higher order critical speeds are not available, computerized finite element method is needed for this purpose. Note that the deflected mode shape of the shaft crosses the spin axis twice. The point along the shaft at which zero displacement occurs for a particular critical speed is called a nodal point. Figures 9.16 and 9.17 show the lumped-mass model and mode shape for the second and third modes.

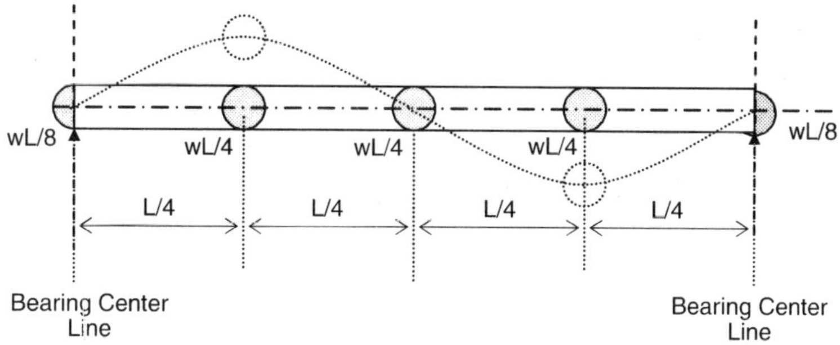

Fig. 9.16. Five Station Second Critical Speed Representation of Uniform Shaft

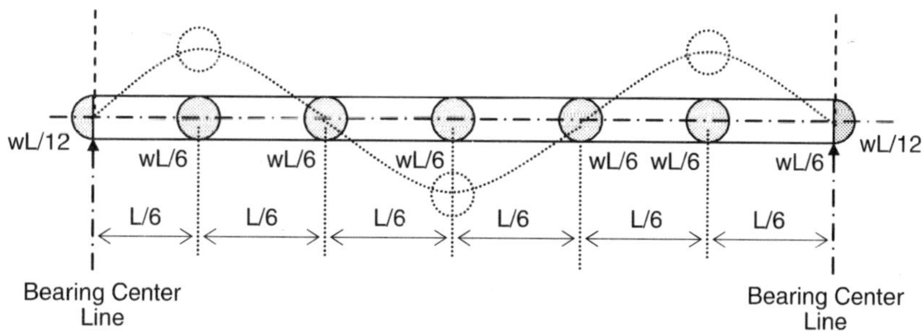

Fig. 9.17. Seven Station Third Critical Speed Representation of Uniform Shaft

Example Problem 9.2: Rotors are classified as either rigid or flexible according to their dynamic behavior at operating speeds. The classification may be determined by performing a critical speed analysis of the rotor-bearing system. If the potential energy in the bearings is over 80 percent of the system's total strain energy, then the rotor may be classified as rigid in that operating speed range. A rigid-body rotor may be balanced in two arbitrary planes. The rotor will then maintain balance throughout its operating speed range. Fig. 9.18 shows illustrations of a disk and a cylinder with two-plane couple form of unbalance. Illustrate the two-plane rigid body balancing procedure using the influence coefficient method.

Solution: Practical considerations must be included for proper selection of the planes to balance a rigid-body rotor with unbalance. On long cylindrical rotors the balance planes may be at the ends of the rotor. On a turbocharger shaft assembly balance planes may be opposite disk faces of the impeller. Measurements may be made of bearing forces or supports (as in a hard bearing balancing machine); readings of the rotor's displacement, velocity or acceleration, or of the casing or the foundation may also be used for this procedure. A third transducer, called key phasor, is required to establish a timing reference mark on the shaft to produce a phase angle reference of peak vibration with respect to the timing mark.

Vibration Readings:	*Amplitude*	*Phase Angle*
Initial Reading – Plane 1:	$Y_1 = 7.4 \times 10^{-3}$ in.	$\alpha_1 = 83°$
Initial Reading – Plane 2:	$Y_2 = 6.9 \times 10^{-3}$ in.	$\alpha_2 = 234°$
Trial Balance Weight:		
Plane 1:	$U_{t1} = 15$ grams	$\alpha_{t1} = 249°$

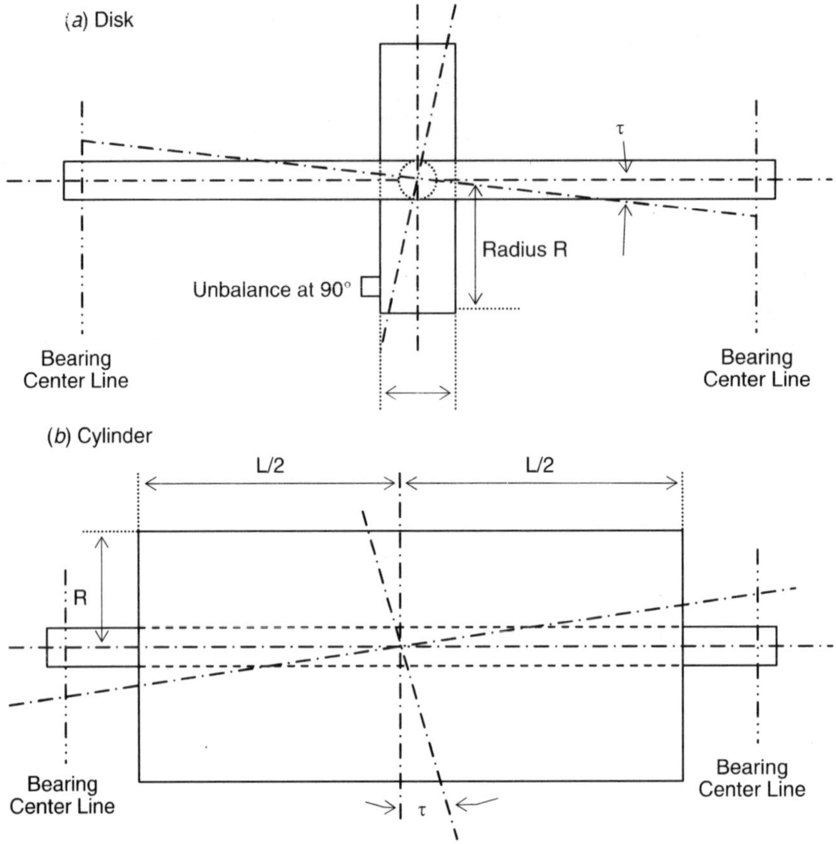

Fig. 9.18. Two-Plane Couple Unbalance in: (*a*) Disk, (*b*) Cylinder

Resulting Vibrations:

Plane 1:	$Y_{11} = 7.8 \times 10^{-3}$ in.	$\alpha_{11} = 138°$
Plane 2:	$Y_{21} = 3.4 \times 10^{-3}$ in.	$\alpha_{21} = 267°$

Trial Balance Weight: (Note – Plane 1 trial weight must be removed)

Plane 2:	$U_{t2} = 20$ grams	$\alpha_{t2} = 90°$

Resulting Vibrations:

Plane 1:	$Y_{12} = 8.7 \times 10^{-3}$ in.	$\alpha_{12} = 48°$
Plane 2:	$Y_{22} = 9.8 \times 10^{-3}$ in.	$\alpha_{22} = 138°$

Balance Correction Weight:

Plane 1:	$U_{b1} = 21.38$ grams	$\alpha_{b1} = 158.5°$
Plane 2:	$U_{b2} = 17.80$ grams	$\alpha_{b2} = 250.1°$

Example Problem 9.3: The influence coefficient method is used here to balance a flexible, single disk rotor. The rotor (sometimes referred to as a Jeffcott rotor when weight of shaft is comparably small) may be balanced by operating the rotor through various speeds below, at or above the critical speeds and recording the appropriate amplitude and phase angle. It is assumed the shaft is straight with no appreciable run-out or thermal bowing, since the influence coefficients would not be repeatable otherwise. Figure 9.19 shows the relationship between displacement and phase of the disk.

Solution: The procedure involves measurement of the rotor's synchronous amplitude and phase at one or more speeds. The rotor is then stopped and a known trial weight is placed on the rotor at a specific radius and angle measured from a reference mark on the disk, preferably placed so as to reduce rotor response. Also, the magnitude of the trial weight should create a rotating unbalance load of about 10 percent or less of the static rotor weight. A large rotating unbalance force can result in permanent shaft bowing, while also creating non-linear effects in the balancing procedure. Since influence coefficients do change with the level of unbalance, a flexible rotor badly out of balance may require several balancing runs to obtain a low vibration reading. Rotor response is then measured at the same speed as the initial peak-to-peak vibration reading.

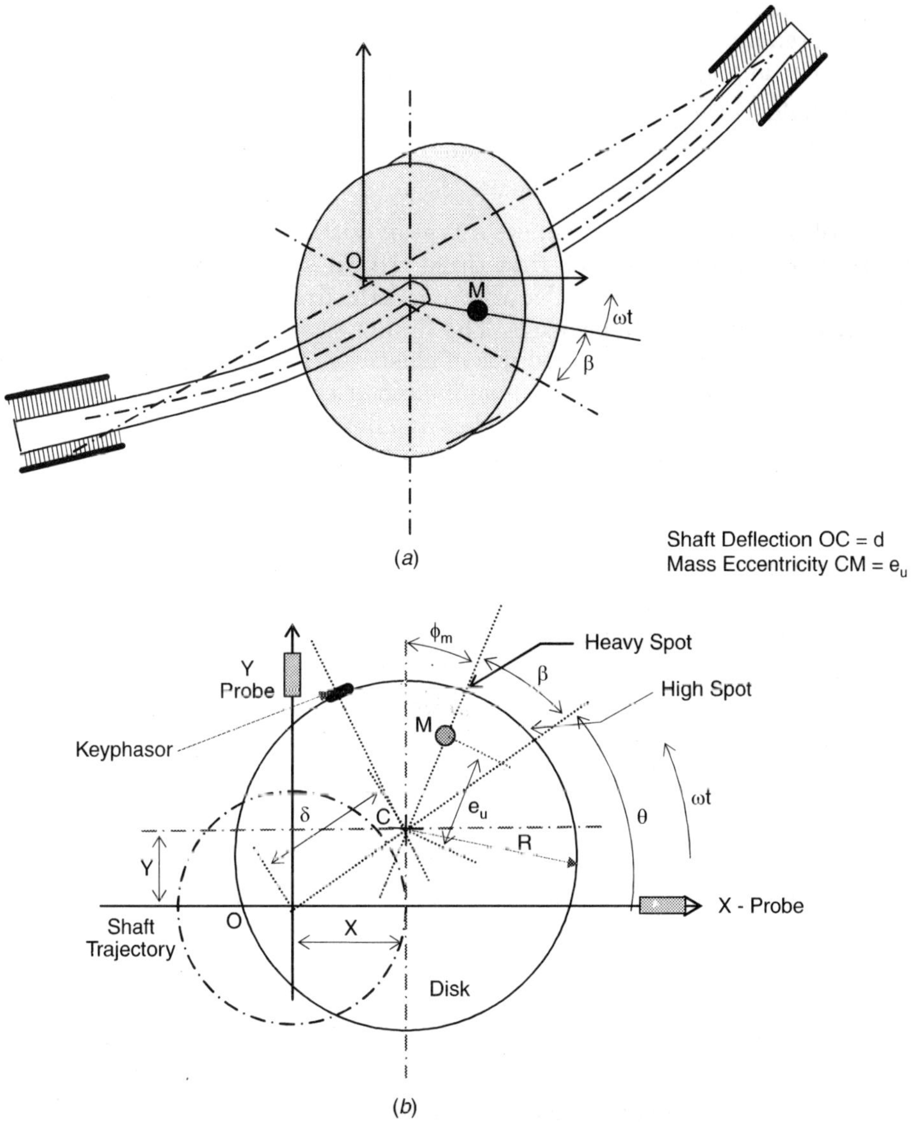

Fig. 9.19. (*a*) Single Disk Jeffcott Rotor, (*b*) Cross Section of Jeffcott Rotor Showing Disk Displacement/Phase Relationship

Rotor Weight:	W = 1475.0 lbs., rotor speed: N_1 = 7250 rpm
Initial Vibration Reading:	$Y_1 = 2.75 \times 10^{-3}$ in. at angle: $\phi_1 = 285°$
Trial Weight:	W_{t1} = 0.185 oz. at radius: r_{t1} = 11.0 in. at angle $f_{t1} = 45°$ from timing mark against rotation.
New rotor response:	Y_3 = 2.275 mils at angle: $\phi_3 = 224°$
Balance Correction Weight:	W_{b1} = 2.17 oz.-in. *Angle:* $\phi_{b1} = -5.38°$

The following procedure is for trim balancing. When the balance weight is left in the rotor, since the overall rotor amplitude is reduced, the resulting balance correction weight is referred to as a trim balance correction. For the case where a trial weight is welded to the rotor, it may not be convenient to remove this weight. However, if response is much larger than the original reading, the trial weight must be removed and another trial weight, preferably smaller, at another angular location must be attached.

The trim balance correction weight in the example problem above is calculated to be:

Correction: U_{tr1} = 1.79 oz.-in.

Angle: $\phi_{tr1} = -66.4°$

Example Problem 9.4: Figure 9.20 shows a typical critical speed map of a uniform shaft for various bearing stiffness values. The first three critical speeds are plotted as a function of bearing flexibility parameter $[\delta_s/(g\omega^2)]^{1/2}$, where ω_1 is the first critical speed with rigid supports and δ_s indicates flexibility of the bearings. The first critical speed $\omega_1 = \pi_2(gEI/WL^3)^{1/2}$, where g = acceleration due to gravity, E = modulus of elasticity, I = moment of inertia and W and L are total span weight and length. For a given value of ω_1 the first and second critical speed ratios decrease continuously with increasing bearing flexibility and the associated mode shapes transforming gradually. For larger values of bearing flexibility the two modes assume the first

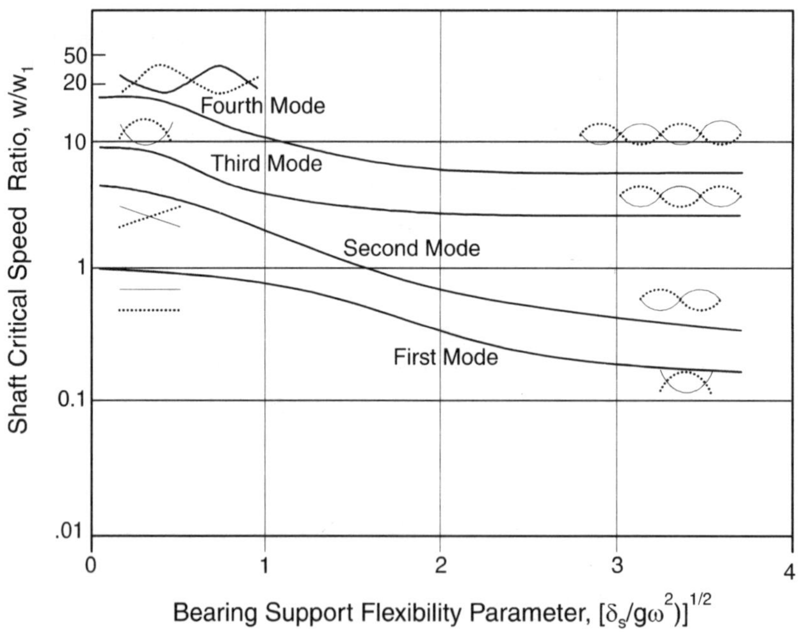

Fig. 9.20. Critical Speed Map of Uniform Flexible Shaft

and second rigid body modes. The ratio between the two critical speeds is 3.5 for large values of bearing flexibility, which is the theoretical value for the rigid body modes. For very large values of bearing flexibility the two rigid body critical speeds both become zero.

To understand the proper method of balancing such a rotor up to the third mode certain concepts such as modal eccentricity, modal weights and rotor response need to be understood. The procedure for balancing a rigid body rotor to take care of the first two rigid body modes has been discussed earlier using the two-plane influence coefficient method and the least-square-error method. The general procedure and method to balance the third mode arising from rotor flexibility is to be explored. Modal eccentricities of a uniform flexible rotor will be considered next.

Consider the uniform shaft of example problem # 9.1.

Shaft Length:	$L = 75.0$ in.
Shaft Diameter:	$D = 3.125$ in.
Material Young's Modulus:	$E = 30 \times 106$ lbs/in^2
Material Density:	$r = 0.283$ lbs/in^3 (128.4 gm/in^3)
Shaft weight	= 162.8 lbs (73866 gm)

It is required to find the modal unbalance eccentricity vectors for the system using the first and the third modes assuming the following unbalance distribution:

$\mathbf{U}_1 = 125.0$ gm-in at $x_1 = L/4$

$\mathbf{U}_2 = -250.0$ gm-in at $x_2 = L/2$

$\mathbf{U}_3 = 125.0$ gm-in at $x_3 = 3L/4$

Solution: The normal mode vectors for the for the indicated unbalance locations are:

$$\phi_1 = \begin{bmatrix} 0.707 \\ 1 \\ 0.707 \end{bmatrix} \qquad \phi_2 = \begin{bmatrix} 1.0 \\ 0 \\ -1.0 \end{bmatrix} \qquad \phi_3 = \begin{bmatrix} 0.707 \\ -1 \\ 0.707 \end{bmatrix}$$

Modal weight for the uniform beam for the three modes is $W_1 = W_2 = W_3 = W_{total}/2 =$ 36933 gm. Modal unbalance eccentricities ei are calculated to be:

$$e_1 = (1/36931) \times [.707\ 1\ .707] \times \begin{Bmatrix} 125 \\ -250 \\ 125 \end{Bmatrix} = -0.00198 \text{ in.}$$

$$e_2 = (1/36931) \times [1\ 0\ -1] \times \begin{Bmatrix} 125 \\ -250 \\ 125 \end{Bmatrix} = 0.0 \text{ in.}$$

$$e_3 = (1/36931) \times [.707\ -1\ .707] \times \begin{Bmatrix} 125 \\ -250 \\ 125 \end{Bmatrix} = 0.01156 \text{ in.}$$

Modal unbalance eccentricity for the third mode is nearly 6 times that for the first mode, indicating that rotor response at the third critical speed will be considerable.

Example Problem 9.5: Calculate the rotor response in example problem 9.4 due to the specified unbalance for the first and third critical speeds and prescribed amplification factors of $A_1 = 8.0$ and $A_3 = 3.0$.

Solution: Rotor critical speeds were determined earlier. The first three critical speeds are: $\omega_1 = 2.648 \times 10^3$ rpm, $\omega_2 = 1.0597 \times 10^4$ rpm, $\omega_3 = 2.3824 \times 10^4$ rpm. At the first critical speed maximum rotor response at the center of the shaft is:

$$\{\mathbf{Y}_1\}_{x=L/2} = 2A_1e_1\{\phi_1\} \angle{-90°} = 2 \times 8 \times (-0.001983) \times 1 \angle{-90°} = 0.03173 \angle{-270°} \text{ in.}$$

At the third critical speed $\omega_3 = 2.3824 \times 10^4$ rpm the rotor response at the center of the shaft is calculated to be:

$$\{\mathbf{Y}_3\}_{x=L/2} = 2A_3e_3\{\phi_3\} \angle{-90°} = 2 \times 3 \times 0.01156 \times 1 \angle{-90°} = 0.0693\text{Ð-}90° \text{ in.}$$

The balance correction weights thus result in considerable dynamic motion in the shaft in the first mode and even greater in the third mode. The balance distribution may be expressed in modal components by:

$$\{\mathbf{U}\} = U_{m1}\{\phi_1\} + U_{m2}\{\phi_2\} + U_{m3}\{\phi_3\}.$$

Example Problem 9.6: Determine the modal unbalance components U_{mn} in example problems # 9.4 and 9.5 due to 150 gm-in unbalance at $x = L/2$ (Fig. 9.21) and at $3L/4$ locations on the shaft. This single unbalance may be represented as a modal distribution located at three planes (three balance planes will be used).

Solution: For 150 gm-in unbalance at $x = L/2$ at $\angle 0°$:

$$\{U\} = \begin{Bmatrix} 0 \\ 150 \\ 0 \end{Bmatrix} = U_{m1}\begin{Bmatrix} 0.707 \\ 1 \\ 0.707 \end{Bmatrix} + U_{m2}\begin{Bmatrix} -1 \\ 0 \\ 1 \end{Bmatrix} + U_{m3}\begin{Bmatrix} -0.707 \\ 1 \\ -0.707 \end{Bmatrix}$$

Solving the simultaneous equations gives $U_{m1} = 75.0$ gm-in, $U_{m2} = 0$, $U_{m3} = 75.0$ gm-in, and the corresponding modal unbalance eccentricities are $e_1 = 0.00406$ in., $e_2 = 0.0$ in. and $e_3 =$ -0.00406 in.

When the 150 gm-in unbalance is located at $x = 3L/4$ at $\angle 0°$, then:

$$\{U\} = \begin{Bmatrix} 0 \\ 0 \\ 150 \end{Bmatrix} = U_{m1}\begin{Bmatrix} 0.707 \\ 1 \\ 0.707 \end{Bmatrix} + U_{m2}\begin{Bmatrix} -1 \\ 0 \\ 1 \end{Bmatrix} + U_{m3}\begin{Bmatrix} -0.707 \\ 1 \\ -0.707 \end{Bmatrix}$$

resulting in $U_{m1} = 53.04$ gm-in, $U_{m2} = 75.0$, $U_{m3} = -53.04$ gm-in.

Example Problem 9.7: The condition for modal balancing a rotor at the first critical speed is given by the expression $[\phi_1]^T [\{\mathbf{U}\} + \{\mathbf{U_{bj}}\}\{\phi_j\}] = \mathbf{M_1e_1} = 0$. The mode shapes are orthogonal for a uniform beam if a symmetric axial mass distribution in the rotor is assumed. The condition for the balance of the first mode will then reduce to the form:

$\phi_1^T \phi_1 [\{\mathbf{U}_{m1}\} + \{\mathbf{U}_{b1}\}] = 0$, or $\{\mathbf{U}_b\}_1 = \mathbf{U}_{b1}\phi_1 = -\mathbf{U}_{m1}\{\phi_1\}$. The flexible rotor in example problems 9.4 to 9.6 has been balanced rigidly. Calculate the balancing components for the first and third modes.

Solution: Components for the balancing of the first mode are: $\phi_1^T[\{\mathbf{U}\} + \mathbf{U}_{b1}\phi_1] = 0$, or:

$$\{0.707 \quad 1.0 \quad 0.707\} \times \left[\begin{Bmatrix} 125 \\ -125 \\ 125 \end{Bmatrix} + U_{b1}\begin{Bmatrix} 0.707 \\ 1.0 \\ 0.707 \end{Bmatrix}\right] = 0$$

Thus, $-73.25 + 2.414U_{b1} = 0$, or $U_{b1} = 30.343$ gm-in. Hence, balance distribution for first mode is $\{U_{b1}\} = 30.343\{\phi_1\} = \{21.45 \; 30.34 \; 21.45\}^T$ gm-in. The new unbalance distribution now becomes:

$$\{U\}_{\text{current}} = \begin{Bmatrix} 125 \\ -250 \\ 125 \end{Bmatrix} + \begin{Bmatrix} 21.45 \\ 30.34 \\ 21.45 \end{Bmatrix} = \begin{Bmatrix} 146.45 \\ -219.65 \\ 146.45 \end{Bmatrix}$$

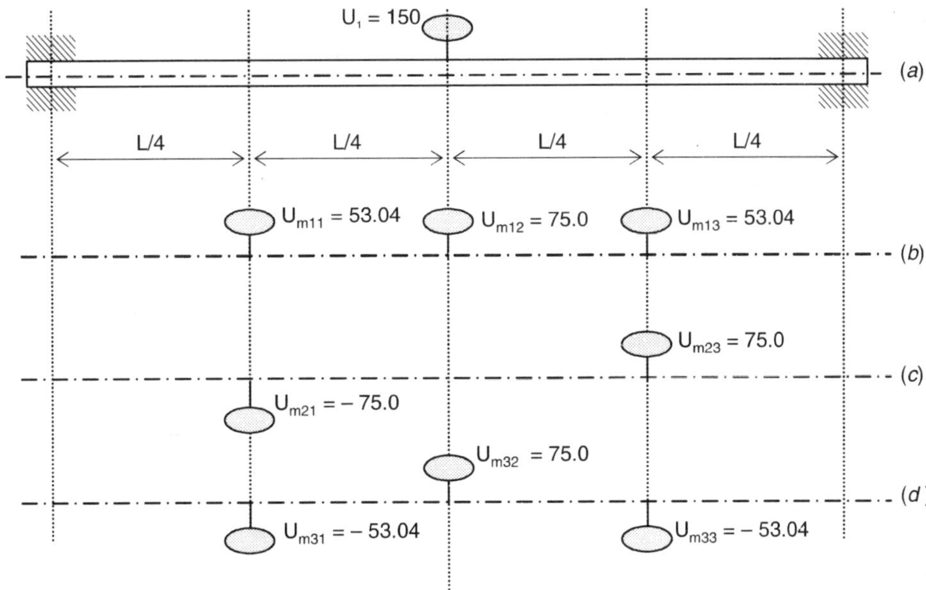

Fig. 9.21. Single-Plane Unbalance of Flexible Shaft Represented as Modal Distribution
(*a*) Uniform Rotor with 150 gm-in at Shaft Length $x = L/2$ at $\angle 0°$
(*b*) Equivalent First Mode Unbalance Distribution
(*c*) Equivalent Second Mode Unbalance Distribution
(*d*) Equivalent Third Mode Unbalance Distribution

But this configuration is not balanced as a rigid rotor, even then it holds the balance for operation through the first two critical frequencies. For operation through the third mode the balancing component, the condition is $\phi_3^T[\{U\} + U_{b3}\phi_3] = 0$, or:

$$[0.707 \quad -1 \quad 0.707] \times \left[\begin{Bmatrix} 146.45 \\ -219.65 \\ 146.45 \end{Bmatrix} + U_{b3} \begin{Bmatrix} 0.707 \\ 1.0 \\ 0.707 \end{Bmatrix} \right]$$

Hence, $426.80 + 2U_{b3} = 0$, or $U_{b3} = 213.40 \angle 180°$ gm-in. For the third mode the balance components are $\{U_{b3}\} = 213.40\{\phi_3\} = \{-146.45\ 219.66\ -146.45\}^T$ gm-in. The final unbalance distribution then becomes:

$$\{U\}_{\text{final}} = \begin{Bmatrix} 146.45 \\ -219.65 \\ 146.45 \end{Bmatrix} + \begin{Bmatrix} -146.45 \\ 219.65 \\ -146.45 \end{Bmatrix} = \begin{Bmatrix} 0 \\ 0 \\ 0 \end{Bmatrix}$$

Thus, after the balancing of the third mode the rotor is fully balanced. This procedure is recommended for balancing of flexible rotors in a specified speed range.

Example Problem 9.8: Figure 9.22 shows a thin disk mounted on a uniform shaft. The shaft has a set bow. When the rotor is placed on a pair of knife-edges and slowly rotated, the disk centerline describes an orbit. The bow is a vector quantity requiring radial bow displacement and phase lag angle from the shaft timing mark, producing an equivalent unbalance whose magnitude is proportional to weight of the rotor and radius of the orbit. Rotor data is as follows:

Rotor weight:	$P = 13.5$ lbs
Distance between bearing center-lines:	$L = 64.375$ in.
Shaft outer diameter:	$Od = 1.375$ in.
Shaft inner diameter:	$Id = 0.75$ in.
Shaft material Young's modulus:	$E = 28.5 \times 106$ lbs/in^2
Damping coefficient:	$C = 0.525$ lbs-second/in.
Eccentricity of unbalance:	$e_{u1} = 0.0015$ in.
Unbalance phase angle:	$\phi_{m1} = 15.0°$ (lag)
Shaft bow vector:	$d_{r1} = 0.001$ in.
Shaft bow phase angle:	$\phi_{m2} = 100.0°$ (lag)
Balancing speed:	$N_{b1} = 1600$ rpm.

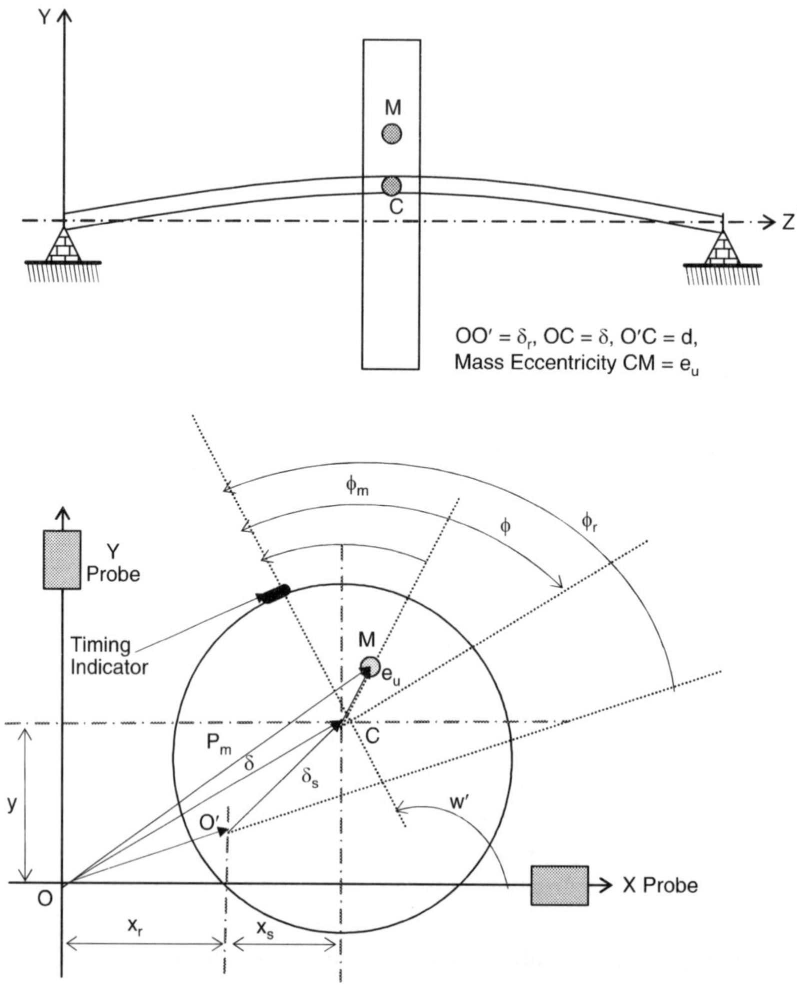

Fig. 9.22. Single Disk Flexible Rotor with Bowed Shaft

It is required to balance the rotor using the standard influence coefficients method, the run-out subtractions method, the bow vector compensation method and the least square error method (*ref*: Nicholas et al, 1976).

Solution: The equivalent unbalance due to the bow in the shaft is: $U_u = (13.5 \times .001 \times 453) = 6.12$ gm-in. For a simply supported uniform shaft the stiffness is given by:

$$K = (384EI/5L^3) = [384 \times 28.5 \times 10^6 \times \pi \times (1.375^4 - .75^4)]/[5 \times 64.375^3 \times 64]$$

$$= 1312.15 \text{ lbs/in.}$$

Then $\omega_{cr1} = (K/M)^{1/2} = (1312.13 \times 386/13.5)^{1/2} = 193.69$ rad/sec, or $N_{cr}{}^1 = 1849.7$ rpm. Critical damping $C_c = 2M\omega_{cr} = 2 \times 13.5 \times 193.69/386 = 13.55$ lbs-sec/in, damping ratio $\xi = C/C_c = 0.525/13.55 = 0.0387$ and amplification $A_{cr} = 1/2\xi = 12.903$. In complex notation shaft unbalance $e_{u1} = 0.0015e^{-i15^\circ}$ in. $= 0.038e^{-i15^\circ}$mm and shaft bow is $\delta_{r1} = 0.001e^{\,i100^\circ}$ in. $= 0.025e^{-i100^\circ}$ mm. The balancing speed ratio $f = 1600/1849.7 = 86.5$ percent of the critical speed. Phase angle $\phi = \tan^{-1}(2\xi f)/(1 - f^2) = 14.9°$.

In the method of standard influence coefficients, the coefficients for unbalance are:

$$c_u = (f^2 e^{-i\phi})/[(1 - f^2)^2 + (2\xi f)^2]^{1/2} = 2.872 \angle{-14.9^\circ}$$

$$c_r = e^{-i\phi}/[(1 - f^2)^2 + (2\xi f)^2]^{1/2} = 3.839 \angle{-14.9^\circ}$$

Total initial response at the given balance speed is:

$$Y_1 = c_u e_u + c_r \delta_r = (2.872 \angle{-14.9^\circ}) \times (0.0015 \angle{-15^\circ}) + (3.838 \angle{-14.9^\circ}) \times (0.001 \angle{-100^\circ})$$

$$= 0.00431 \angle{-29.9^\circ} + 0.00383 \angle{-114.9^\circ}$$

$$= 0.006015 \angle{-69.4^\circ} \text{ in.} \quad \text{or} \quad 0.153 \angle{-69.4^\circ} \text{ mm}$$

Attaching a trial weight on the rotor results in response of $Y_{tr} = c_u(e_u + e_{tr}) + c_r\delta_r$. The corresponding influence coefficient is obtained by subtracting the initial response Y_1 from the response with the measured trial weight ($2.5 \angle 180°$ oz.) response of $0.0024 \angle{-85.5^\circ}$ in. due to etr. Then: $c_u = (Y_{tr} - Y_1)/\mathbf{e_{tr}} = 1.508 \angle{-59.2^\circ}$.

The standard influence coefficient method assumes that the rotor responds only to mechanical unbalance. Balance correction is obtained from:

$$e_b = -(Y_1/c_u) = -(6.014 \angle{-69.4^\circ})/(1.507 \angle{-59.2^\circ}) = 0.00399 \angle{-190.2^\circ} \text{ in.}$$

If the balance radius is 15 mm (0.59 in.), the balance correction weight is:

$$U_b = we_b = w_bR = 13.5 \times .00399 = 0.0539 \text{ lb-in} = 24.40 \text{ gm-in.}$$

Hence, $w_b = 24.4/0.59 = 41.3$ gm at an angle of 190.2° from the timing mark.

But the rotor is balanced at 1600 rpm, and will have high response at the critical speed. At the critical speed $c_u = c_r = A_{cr} \angle{-90^\circ} = 12.903 \angle{-90^\circ}$, and the amplitude is:

$$Y_{cr} = (e_u + \delta_r)A_{cr} = (1.5 \angle{-15^\circ} + 1.0 \angle{-100^\circ}) \times 12.905 \angle{-90^\circ} = 24.18 \angle 42.9^\circ$$

Response at the critical speed will be:

$$Y_{cr,\,b} = c_u(e_u + e_b) + c_r\delta_r = 12.905 \angle{-90^\circ} \times (1.5 \angle{-15^\circ}$$

$$+ 3.99 \angle{-190.2^\circ}) + 1.0 \angle{-100^\circ} \times 3.838 \angle{-14.9^\circ}$$

$$= 51.86 \angle 102.9^\circ \text{ mils} = 1.317 \angle 102.9^\circ \text{ mm.}$$

Total response due to the factors is high and is out of phase with the bow vector, leading to the high response. Therefore, the influence coefficient method is not acceptable.

The run out subtraction method assumes rotor response of the form $Y \approx c_u e_u + \delta_r$. So $e_b = -e = -(Y - \delta_r)/c_u = -(6.014 \angle{-69.4^\circ} - 1.0 \angle{-100^\circ})/(2.872 \angle{-14.9^\circ}) = 2.40 \angle 121.3^\circ$ gm-in. The balance correction is then $U_b = e_b w_b$ = (4.80 ∠121.3°) and the correction weight w_b = 4.80/0.59 = 8.129 gm. The amplitude at critical speed is:

$$Y_{cr,\,b} = 12.905 \angle -90^\circ \times (1.5 \angle -15^\circ + 2.40 \angle 121.3^\circ) = 21.55 \angle 173.1^\circ \text{ mils} = 0.547 \text{ mm}$$

In the bow vector compensation method the influence coefficient for the bow vector is obtained from the expression $c_r = c_u / f_b^2$, where $f_b = N_b/N_{cr}$ = 1.0 at the critical speed. With this procedure, since $Y_{cr,\,b} = c_u(e_u + e_b) + c_r\delta_r = c_{cr}(e_u + e_b + d_r) = 0$, the coefficients for shaft bow and unbalance become equal. The balance correction then becomes $e_b = -(e_u + \delta_r)$. For speeds other than the critical speed (ϕ_b = 0.865), c_r = 2.872/(.865²) = 3.839 ∠– 14.9°. The amplitude is obtained from $Y_b = c_u(e_u + e_b + d_r/f_b^2) \neq 0$, and the balance eccentricity vector then takes the form of $e_b = [Y_1/c_u + \delta_r(1 - 1/f_b^2)]$. For the given values of shaft bow measured at the lower speed and the ratio f_b, the value of the balancing eccentricity is e_b = 2.78 ∠– 112.7° mils, or 0.0706 ∠– 112.7° mm.

It is difficult to distinguish between true shaft bow and a constant shaft residual run-out vector. To overcome this, multi-speed balancing based on minimization of rotor response over a speed range has been developed using the least-square-error method. In this procedure the rotor low-speed run-out vector is measured. Rotor response is then measured at several speed points. A trial weight is next attached to the rotor, and new response is then measured at the same speeds. The single mass Jeffcott rotor with shaft bow and unbalance considered earlier will be balanced by the Least-Squared-Error method by measuring vibration amplitude below and above rotor critical speed, at 1600 and 2000 rpm.

Vibration amplitude and the corresponding phase angle at three speeds points are:

$$Y_{1600} = 4.84 \times 10^{-3} \angle -83^\circ \text{ in.}$$
$$Y_{1850} = 14.32 \times 10^{-3} \angle -147^\circ \text{ in.}$$
$$Y_{2000} = 7.84 \times 10^{-3} \angle -179^\circ \text{ in.}$$

Compensation by the slow roll vector from constant shaft run-out is δ_r = 0.001 ∠– 100° in. Subtracting it from the vibration amplitudes gives:

$$Y_{1c} = 4.84 \times 10^{-3} \angle -83^\circ - 0.001 \angle -100^\circ = 3.895 \times 10^{-3} \angle -78.7^\circ$$
$$Y_{2c} = 7.84 \times 10^{-3} \angle -179^\circ - 0.001 \angle -100^\circ = 7.712 \times 10^{-3} \angle -6.3^\circ$$

The trial weight is 0.826 gm at a radius of 1.0 in., or $U_{tr} = 1.824 \times 10^{-3} \angle 167^\circ$ in-lbs. Vibration readings after the placement of trial weight are:

$$Y_{tr1600} = 3.89 \times 10^{-3} \angle -75^\circ \text{ in.}$$
$$Y_{tr1850} = 9.47 \times 10^{-3} \angle -148^\circ \text{ in.}$$
$$Y_{tr2000} = 3.72 \times 10^{-3} \angle -182^\circ \text{ in.}$$

Influence coefficients at 1600 and 2000 rpm are:

$$c_1 = 4.71 \angle -245.7^\circ \text{ and } \mathbf{c_2} = 9.33 \angle -173.3^\circ \text{ mils/gm.}$$

Complex conjugates of these influence coefficients are:

$$c_1^* = 4.71 \angle 245.7^\circ \quad \text{and} \quad c_2^* = 9.33 \angle 173.3^\circ \text{ mils/gm.}$$

Balance correction by the least square error method in a single plane for two speeds is:

$$U_b = -(Y_{1c}c_1^* + Y_{2c}c_2^*)/(c_1c_1^* + c_2c_2^*) = -(49.767 \angle 167^\circ)/33.169 \angle 180^\circ$$
$$= 0.826 \angle -193^\circ \text{ gm.}$$

Example Problem 9.9: Figure 9.5 shows an overhung rotor with radial unbalance and skewed disk. The plane of the disk is inclined from the vertical by a small angle. Balancing this rotor is more complex than a conventional Jeffcott rotor symmetrically situated between bearings due to gyroscopic moments generated by the skew. If the disk skew is out of phase with the unbalance the effects combine to increase the amplitude of motion at critical speed. On the other hand if disk skew is in phase with disk unbalance, the gyroscopic moment will help to offset the radial unbalance force, thus reducing overall vibration amplitude. Provide a procedure for balancing the rotor.

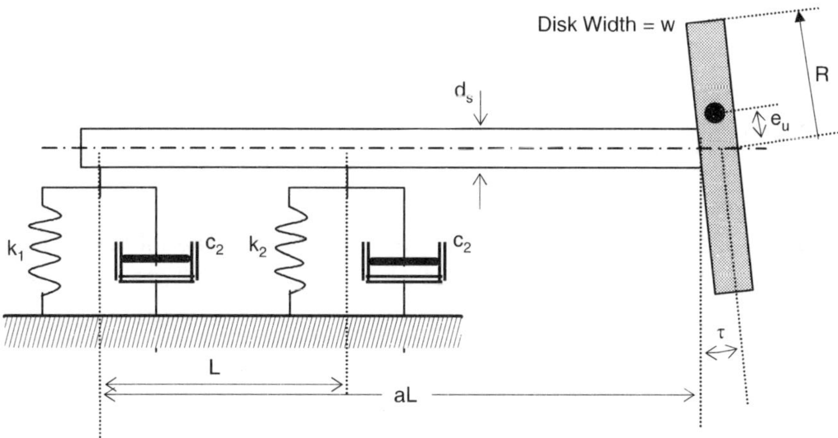

Fig. 9.23. Overhung disk with skew and unbalance

Solution: The recommended procedure for balancing an overhung rotor in the first attempt is to single-plane balance the disk by using influence coefficients and vibration amplitudes obtained from the near and far bearings. This will produce a compromise balance that will prevent excessive vibrations in one bearing. If the resulting vibration is still not low, a trial balance couple using two weights 180 degrees out of phase to each other and separated by an axial distance should be attached to the disk.

Similar balancing can also be achieved by the two-plane balancing procedure in which trial weights are placed on the two flat faces of the disk and vibrations for the near and far bearings recorded for two trial runs.

Vibration monitored at the near and far planes is assumed to be a function of two unknown values of unbalance. A trial balance is placed in each plane and resulting amplitudes of motion are measured. In this calculation positive sign represents lag angles for both vibration and balance weight measures against the direction of rotation.

Measurements for the slow roll vector (from constant shaft run-out) are:

$$Y_{srn} = 0.8 \times 10^{-3} \angle 256° \text{ in.} \qquad Y_{srf} = 0.95 \times 10^{-3} \angle 238° \text{ in.}$$

Note that vibration and trial weight angles are measured against rotation from the vibration probe. Initial vibration readings at near and far locations are measured to be:

$$Y_{in} = 1.95 \times 10^{-3} \angle 175° \text{ in.} \qquad Y_{if} = 3.25 \times 10^{-3} \angle 142° \text{ in.}$$

Compensated near and far vibration amplitudes are obtained by subtracting the slow roll vectors:

$$Y_n = 1.99 \times 10^{-3} \angle 151.6° \text{ in.} \qquad Y_f = 3.48 \times 10^{-3} \angle 126.2° \text{ in.}$$

The vibrations at the two planes are assumed to be functions of the two unknown unbalances U_1 and U_2:

$$Y_n = c_{n1}U_1 + c_{n2}U_2 \qquad Y_f = c_{f1}U_1 + c_{f2}U_2$$

A trial weight is first attached in the near plane and the amplitude of motion is measured. Leaving it in place, attach a second weight in the far plane and measure the resulting amplitude of motion.

First trial weights in the two planes are:

$$U_{1t} = 7.5 \angle 83° \text{ gm} \qquad U_{2t} = 5.2 \angle 180° \text{ gm}$$

Vibration readings at near and far locations after placing the trial weights in the two planes are measured to be:

$$Y_{n1} = 2.42 \times 10^{-3} \angle 179° \text{ in.} \qquad Y_{f1} = 2.29 \times 10^{-3} \angle 356° \text{ in.}$$

$$Y_{n2} = 3.19 \times 10^{-3} \angle 248° \text{ in.} \qquad Y_{f2} = 2.87 \times 10^{-3} \angle 169° \text{ in.}$$

The set of four influence coefficients may be expressed in the following form:

$$c_{n1} = (Y_{n1} - Y_n)/U_{1t} = 0.150 \angle 150.4° \qquad c_{f1} = (Y_{f1} - Y_f)/U_{1t} = 0.701 \angle 62.7°$$

$$c_{n2} = (Y_{n2} - Y_n)/U_{2t} = 0.758 \angle -81.9° \qquad c_{f2} = (Y_{f2} - Y_f)/U_{2t} = 0.458 \angle 71.4°$$

The balance correction weights are given by:

$$\begin{Bmatrix} U_{b1} \\ U_{b2} \end{Bmatrix} = -\begin{vmatrix} U_1 & U_{1t} \\ U_2 & U_{2t} \end{vmatrix} = -\begin{vmatrix} c_{n1} & c_{n2} \\ c_{f1} & c_{f2} \end{vmatrix}^{-1} \begin{Bmatrix} Y_{n2} \\ Y_{f2} \end{Bmatrix}$$

The balance correction weights are:

$$U_{b1} = (Y_{f2}c_{n2} - Y_{n2}c_{f2})/(c_{n1}c_{f2} - c_{f1}c_{n2})$$

$$U_{b2} = (Y_{n2}c_{f1} - Y_{f2}c_{n1})/(c_{n1}c_{f2} - c_{f1}c_{n2})$$

Correction weight and location at near plane is calculated to be:

$$U_{b1} = 5.77 \angle 120.9° \text{ gm against rotation}$$

and the correction weight and location at far plane is:

$$U_{b2} = 3.19 \angle 141.76° \text{ gm against rotation.}$$

Vibration readings after placing correction weight in near plane are measured to be:

$$Y_{n1} = 0.6 \times 10^{-3} \angle 271° \text{ in.} \qquad Y_{f1} = 0.55 \times 10^{-3} \angle 4° \text{ in.}$$

Vibration readings after placing correction weight in far plane are measured to be:

$$Y_{n2} = 2.7 \times 10^{-3} \angle 298° \text{ in.} \qquad Y_{f2} = 1.5 \times 10^{-3} \angle 139° \text{ in.}$$

REFERENCES AND BIBLIOGRAPHY

American Petroleum Institute, "Special Purpose Gear Units for Refinery Service", API Std. 613, 4th edition, December, 1994.

American Petroleum Institute, "Centrifugal Compressors for General Refinery Service", API Std. 617, 5th edition, April, 1994.

American Petroleum Institute, "Centrifugal Pumps for General Refinery Service", API Std. 610, 8th edition, December, 1994.

Bentley, D. E., "Polar Plotting Applications for Rotating Machinery", Vibrations Institute, Clarendon Hills, ILL, 1982.

Bishop, R. E. D., Parkinson, A. G., "On the Isolation of Modes in the Balancing of Flexible

Shafts", *Proceedings of the Institute of Mechanical Engineers*, 177:407-423, 1963.

Bishop, R. E. D., Parkinson, A. G., "Second Order Vibration of Flexible Shafts", *Proceedings Royal Society of London*, 259:1-31, 1966.

Compressed Air and Gas Institute, "In-Service Standards for Centrifugal Compressors", Cleveland, OH, 1993.

Cooley, J. W., Tukey, J. W., "An Algorithm for the Machine Calculation of Complex Fourier Series", Mathematics of Computation, 19(90):297-301, 1985.

Darlow, M., "Flexible Rotor Balancing by the Unified Balancing Approach", Rensselaer Polytechnic Institute, Troy, NY., 1989.

Den Hartog, J. P., "Mechanical Vibrations", 4th ed., Mc-Graw Hill, New York, 1985.

Foiles, W., Gunter, E. J., "Balancing a Three Mass Rotor with Shaft Bow", University of Virginia, Charlottesville, 1982

Goodman, T. P., "A Least-Squares Method for Computing Balance Correction", Journal of Engineering for Industry, 86(3), pages 273 - 279, 1984.

Hydraulic Institute, "Acceptable Field Vibration Limits for Horizontal Pumps", 14th edition, Centrifugal Pump Applications, 1994.

ISO 11340, "Mechanical Vibration - Methods and Criteria for the Mechanical Balancing of Flexible Rotors", 1994.

ISO 1925 and ANSI S2.7, "Balancing - Vocabulary", 1990.

ISO 1940 and ANSI S2.19, "Balancing Quality of Rotating Rigid Bodies", 1986.

ISO 2371 and ANSI S2.38, "Field Balancing Equipment - Description and Evaluation", 1982.

ISO 2372, "Mechanical Vibration of Machines with Operating Speeds from 10 to 200 rps-Basis for Specifying Evaluation Standards", 1989.

ISO 2954, "Mechanical Vibration of Rotating and Reciprocating Machinery-Requirements for Instruments for Measuring Vibration Severity", 1984.

ISO/DIS 2953, "Balancing Machines - Description and Evaluation", 1984.

ISO 4867, "Guidelines for Evaluation of Vibration in Merchant Ships", after ISO 6954, 1993.

ISO 6945, "Guidelines for Evaluation of Vibration in Merchant Ships", 1987.

Military Standard, "Mechanical Vibration of Shipboard Equipment (Type 1-Environmental and Type 2 - Internally Excited)," MIL-STD-167-1 (SHIPS), May, 1984.

Palazzolo, A. B., Gunter, E. J., "Multimass Flexible Rotor Balancing by the Least-Squares Error Method", Vibration Institute, Clarendon Hills, Ill., 1982.

Rieger, N., "Balancing of Rigid and Flexible Rotors", SVN 12, The Shock and Vibration Center, 1986.

Schneider, H., "Balancing Technology", 4th edition, Carl Schenck, AG, Darmstadt, Germany, 1991.

Stadelbauer, D. G., "Fundamentals of Balancing", 3rd edition, Schenck Trebel Corp., Deer Park, NY., 1990.

Winkler, A. F., "High Speed Rotating Unbalance, Coupling or Rotor", Proceedings Vibration Institute, Nassau Bay Conference, Clarendon Hills, Ill, 1983.

Applications from Computational Fluid Dynamics

10

10.1 INTRODUCTION

Unsteady flow of air and fluid through the intake manifold, past the valve ports into the cylinder and out through the exhaust has a major impact on performance and emission of noise and pollutants in reciprocating machinery. Design approaches applied to flow paths in naturally aspirated or supercharged systems substantially affect the induction of air or fluid into the cylinder. Internal combustion engines often use helical ports to induce swirl into the gas inside the cylinder to enhance combustion rate, making valves and ports complex fluid dynamic arrangements. Limitations on volumetric efficiency of a cylinder, brought about by effective valve area, can be mostly removed by an adequate design of the manifolds. At the end of the cycle the machine relies almost entirely on the operation of the exhaust gas dynamics to purge the cylinder of the combustion products in internal combustion engines and pressurized gas or fluid in the case of pumps and compressors. Design of the exhaust system in turbo-charged engines is often vulnerable to lack of sufficient energy available at the turbine inlet, over the full spectrum of operating conditions, to drive the compressor which then produces the required power boost.

A reciprocating machine may be used for a number of applications. For instance, a 2.5 liter naturally aspirated internal combustion engine might be used in a family type sedan, a sports car or even a four wheel drive multi-purpose vehicle. Each of these applications requires different characteristics from the engine and a different flow path configuration of the intake and exhaust systems. Consequently, even if new power plants are not designed often by an automobile manufacturer, it is necessary to go through a major redesign of the manifold systems for individual applications.

When a fluid flows along a pipe at low speed the velocity at any point in the flow does not vary with time. Transfer of heat and momentum across the flow occurs only by molecular transport, because the flow is laminar. When the velocity is increased a speed is reached at which the laminar flow becomes unstable, vorticity is generated wherever there is shear and the flow becomes turbulent. At any point the flow velocity now varies randomly about the local mean. The variation of velocity is produced by the passage of vortices carried by the flow. Various scales can be defined to characterize the size of the largest, smallest and average vortices. A large scale transport may be of magnitude of 1 mm, while the smaller eddies are characterized by the order of magnitude 0.1 mm. In a reciprocating machine the situation is more complex. During a cycle the mean velocity at any point is continuously changing. If the mean flow pattern remained identical from cycle to cycle, it would still be possible to apply the definition of intensity in order to determine values at particular points in the cycle by collecting and averaging data over a large number of cycles.

Measurement of instantaneous velocities in a turbulent flow has been by the hot wire anemometer in the past. It is based on the principle that flow normal to a short wire heated by an electric current changes the rate of heat loss from the wire, consequently changing the current required to maintain it at a constant temperature. If the wire is small enough, the response is sufficiently fast to permit measurement of velocity variations in the turbulent flow. Typical hot wire dimensions are 2 mm long and 10mm in diameter. Such wires, however, cannot survive the heat of combustion in an engine, and can only be used in a motorized system. Further, they cannot distinguish flow direction and so give false results if local velocities vary significantly from the mean velocity, as is frequently the case in engines. A more recent Laser Doppler anemometry procedure uses a laser beam split into two parallel beams, which are then made to converge and cross at the point where measurement is required. Since the beams are coherent, crossing produces an interference transverse to the intersection patch. If a particle carried in the local flow crosses the interference pattern, the light scattered by it can be observed, and the time of transit over a number of interference bands gives a measure of the component of velocity normal to them. This procedure can be used during combustion also.

Application of computational fluid dynamics theory in the design of manifolds for the inlet and exhaust of the fluids can be used to maximize machine performance while minimizing noise and pollutant emissions. Manifold components can be engineered to attenuate the amplitude of pressure waves, and can be designed to act on specific frequency components; the challenge then is to obtain the desired vibration and noise spectrum without producing deleterious effects on the management of flow pressure waves which enhance overall machine performance. Unsteady flow in the manifolds can also have impact on emission levels. Optimization of valve events in modern internal combustion engines assists in meeting legislated levels of unburned combustion while maintaining acceptable performance levels. Transport of chemical species and energy through an exhaust system to a catalytic converter must also be considered in determining its location: close coupled catalysts give rapid light-off but have a detrimental effect on exhaust system tuning by compromising the dynamics of pressure waves.

Pressure wave phenomena dominate the processes described above. Inviscid one-dimensional models of the gas flow have been successfully applied as design and analysis tools. Numerical techniques are required to simulate unsteady flows in the machine. The equations of one-dimensional unsteady flow are introduced, and in conjunction with normally encountered boundary conditions, solved by the method of characteristics and finite volume technique to provide a mathematical procedure for simulating flow conditions. Computer programs based on this approach are available for use both by the power plant and component manufacturers.

The process of tuning the intake and exhaust manifolds is based on linearized wave equations to estimate natural frequencies of vibration of the system. Analytical predictions simulating manifold arrangements may then be compared with measurements. The procedure may then be further applied to investigate various tuning features and to consider strategies for variable geometry manifolds and effects of variable valve timing.

10.2 FLOW IN INTAKE AND EXHAUST MANIFOLDS

A schematic diagram of a turbo-charged and inter-cooled spark ignition engine is shown in Fig. 10.1. Air is drawn into the intake manifold, passes through a filter, then regulated by a throttle valve. Fuel is injected into the air stream. This mixture enters the cylinder when the air intake valves are open, where it is burned. At the end of the expansion stroke, after the exhaust valves open, combustion products enter the exhaust manifold. In modern engine designs a small percentage of the exhaust gas is re-circulated around the intake manifold for preheating the cold incoming air. The remaining exhaust gases progress through an emission control and noise suppression devices before releasing into the atmosphere. Thus, all three gaseous, liquid and solid phases of the working media are exposed to the manifolds. Note that in diesel and direct injection spark ignition engines the fuel may be directly admitted into the cylinder.

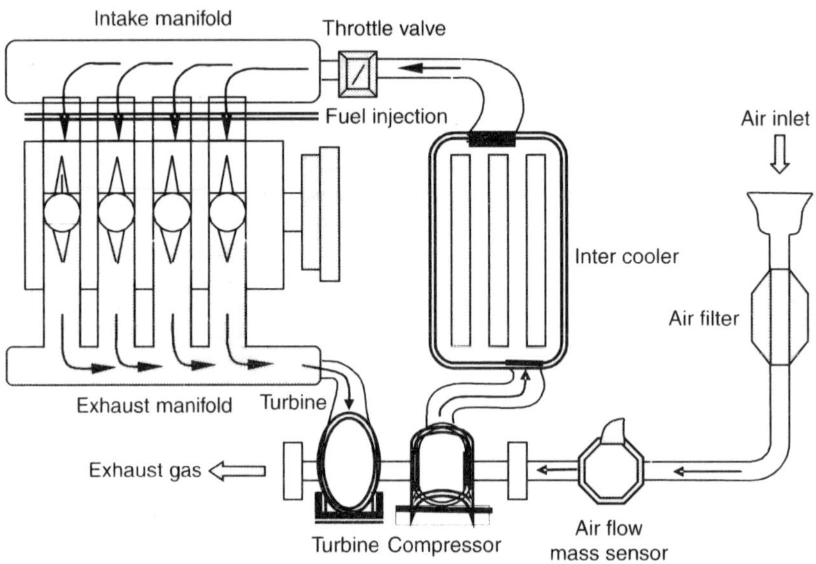

Fig. 10.1. Air Flow System of Supercharged, Inter-cooled Engine

Most of the material passing through the piping systems is in the gaseous phase. In the intake manifold air is the major component, some moisture may accompany the air, and liquid or vapor fuels are mixed in the region of fuel injectors or carburetors. Some products of combustion may also be present as a result of back flow from the combustion chamber when the intake valve is open. In small amounts blow-by gas, forced past the piston rings into the crank case during combustion, may also be vented into the intake manifold. Exhaust manifolds will be exposed mainly to the products of combustion, but may be mixed with some air and fuel passing through the cylinder during valve overlap periods. Since nitrogen is the main component of air, both manifolds will mostly have it, passing through the combustion process unaltered except for a small amount of oxides of nitrogen (NO_x) pollutants due to the high combustion temperature. Consequently, flow in the manifolds can be simulated by solving the equations for gas flow.

Fuel injectors in the intake manifold produce a spray of fuel droplets suspended in the air stream. Some of the fuel, however, may impact and coalesce on the manifold walls to form rivulets. To avoid this the fuel may be introduced at a point chosen to give a better distribution to all cylinders. Fuel may also be injected at more than one point for a more even distribution, thus permitting manifold design for better volumetric efficiency. Note that characterization of liquid fuel in a manifold requires three-dimensional simulation of atomization, surface impingement, heat transfer and evaporation, a far more complex task then calculation of motion of the gas in the manifold. Solid material is present in exhaust manifold in the form of sub-micron soot particles generated during combustion, having little effect on gas flow in the manifold.

10.3 GAS DYNAMICS AND MACHINE PERFORMANCE

Treating the intake and exhaust manifolds as a network of pipes permits one-dimensional gas dynamics simulation and subsequent evaluation of the performance of the reciprocating machine. The network can be evaluated for unsteady flow in the pipes by assuming the flow can be represented by a single average velocity at any cross section. The procedure neglects the effects of turbulence and viscosity to establish the velocity profile of the flow in the pipe. Effects of friction can be included through the use of a friction factor. When the pipe configurations are reasonably straight and uniform, as in many internal combustion engines, the approach yields satisfactory results and correlation with measured data. If the curvature of the pipes gets more pronounced, the results are less appropriate. The simplified procedure is also limited at pipe junctions and at locations where the manifold flow pattern tends to display multi-dimensional characteristics. The difficulties may be overcome by using empirical coefficients based on test measurements.

Pressure waves occur in manifolds because of the periodic nature of the induction and exhaust processes in the machine cycle. When acting on the compressible fluid in the intake manifold, the process generates pressure waves travelling at the local velocity of sound in all directions relative to the fluid. Large amplitude waves are non-linear in character, but small amplitude waves behave in a mathematically linear fashion. Linear behavior means the combination of two interacting waves is obtainable by summing the effects of each of them when acting by itself. In the exhaust manifold of internal combustion engines gas temperature is high, and the wave action is generally non-linear, requiring more complex methods of calculation. In four stroke engines the shape of the volumetric efficiency/engine speed characteristic is governed by the geometry of the intake system, while exhaust piping will have only secondary effects due to the small valve overlap period.

The basic requirement of an intake manifold is to provide a high pressure level at the intake valve so that mass flow is maximized at any machine speed. The pressure level must be maintained in a period between bottom dead center of the induction stroke and a point just before the closing of the valve. This enables trapping a large amount of the fluid at high pressure, and prevents any significant back flow into the manifold. Supercharging and turbo-charging by a compressor is one way towards this objective, if care is taken to assure pulsation effects from the forced induction does not adversely affect stable operation of the compressor. Use of an inter cooler isolates the compressor from the rest of the manifold, when it is possible to make use of gas dynamic tuning effects (Sato, 1987).

After the intake valve opens, descent of the piston during the induction stroke reduces pressure in the cylinder, causing a rarefaction wave to be propagated into the intake pipe. The rarefied wave travels to the end of the intake pipe when it is reflected as a compression wave. Arrival of the peak of the compression wave at the right instant as the intake valves are closing augments filling of the cylinder with the fluid.

Residual waves are produced in the intake pipe after the valve closes, creating additional peaks in the volumetric efficiency curve. When the residual waves and the open period pulsation combine, a peak in a residual wave from the previous wave can coincide at the start of the following intake stroke, the resulting depression pulse is enhanced, thereby creating a peak in the volumetric efficiency curve (Fig. 10.2). This effect can occur at more than one machine speed as different residual waves fill the closed period. Broome (1989) proposed the following equation:

$$(2n-1)\theta_t + \theta_d = 720 \tag{10.1}$$

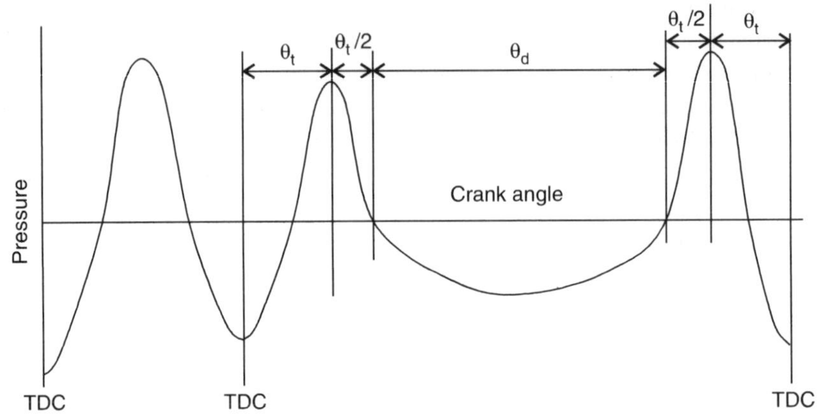

Fig. 10.2. Effects of Wave Tuning on Cylinder Pressure

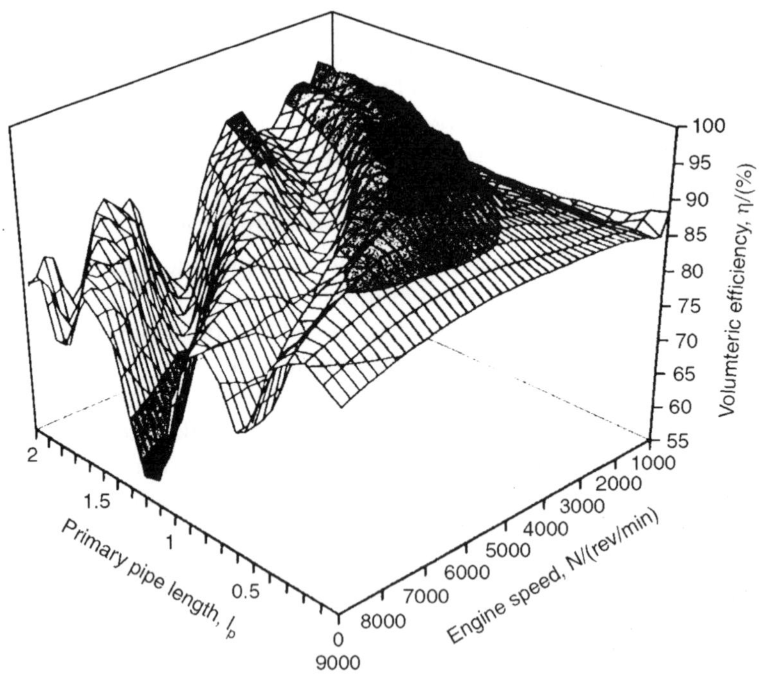

Fig. 10.3. Variation in Volumetric Efficiency due to Machine Speed and Primary Pipe Length

where θ_t represents the angle of crank rotation required for an acoustic wave to travel from the cylinder to the end of the pipe and back, defined by:

$$\theta_t = 360N \times (2l_p)/60 \tag{10.2}$$

where N = machine speed, rpm, and l_p = length of pipe. The value of θ_d is given by the duration of the suction pulse in crank angle degrees.

Figure 10.3 gives a three dimensional plot to show the impact on volumetric efficiency due to variations in pipe length and machine operating speed for a single cylinder engine. Volumetric efficiency may be maximized for a number of combinations of pipe length and machine speed, so performance of a fixed geometry inlet manifold is compromised when operation takes place at other than design condition. To achieve peak efficiency over a larger speed range it would be necessary to vary the length of the intake pipes. Note, however, that valve timing, non-uniform pipe cross sectional area, dependence of wave propagation speed on local fluid velocity and pressure and pipe end effects render simpler acoustic analogies subject to a margin of error in the prediction of manifold tuning speeds.

10.4 UNSTEADY FLOW OF GAS IN A PIPE

Simplification of an unsteady flow of a compressible fluid to a single space dimension is acceptable if the intake manifold of a reciprocating machine is mostly in the form of straight pipes. At pipe bends, junctions and chambers located just before entry into the cylinder through the intake valves effects of multi-dimensional flow manifest themselves as increased pressure drops.

In the special case where there is no heat transfer or flow friction the entropy level of the fluid friction remains constant. The passage through a gas stream of a pressure wave's characteristics does not change its entropy level. In the case of flow through a uniform cross section pipe, the continuity and momentum equations are given by:

$$\frac{\partial \rho}{\partial t} + u\frac{\partial \rho}{\partial x} + \rho\frac{\partial u}{\partial x} = 0 \qquad (10.3)$$

$$\frac{\partial u}{\partial t} + u\frac{\partial u}{\partial x} + \frac{1}{\rho}\frac{\partial p}{\partial x} = 0 \qquad (10.4)$$

For isentropic conditions $ds = 0$, and changes in pressure may be related to density variations from: $dp = (a^2)d\rho$. Here a is the speed of sound obtained from the expression $a^2 = \kappa RT$, κ being the ratio of specific heats at constant pressure and constant volume and R is the specific gas constant. Substitution into (10.3) and (10.4) yields the classical acoustic wave equation:

$$\frac{\partial^2(\Delta p)}{\partial t^2} = a^2\frac{\partial^2(\Delta p)}{\partial x^2} \qquad (10.5)$$

In terms of velocity and density, the wave equation may also be expressed as:

$$\frac{\partial^2(\Delta u)}{\partial t^2} = a^2\frac{\partial^2(\Delta u)}{\partial x^2} \qquad (10.6)$$

$$\frac{\partial^2(\Delta \rho)}{\partial t^2} = a^2\frac{\partial^2(\Delta \rho)}{\partial x^2} \qquad (10.7)$$

D'Alembert obtained solutions for the wave equation based on the assumption that individual waves moving in the positive or negative x-direction exist, with their form unaffected by interaction with each other. If p_e is the excess pressure above the undisturbed level, D'Alembert's general solution may be expressed in the form:

$$p_e = f(x - at) + g(x + at)$$

where f and g denote longitudinal component pressure waves. The velocity field induced by the wave is parallel to the direction of propagation. The resultant pressure at any section is the sum of excess pressure in individual waves, giving rise to their description as linear waves. It can also be shown that all portions of the wave move at the same propagation speed a without distortion of its form. A proportionality constant between excess pressure and fluid component velocity in the direction of wave propagation u exists, expressed by:

$$p_e = \pm \rho a u \tag{10.8}$$

In non-linear waves this simplicity is lost because the speed of sound is no longer constant. When the fluid is a perfect gas, and only isentropic change in state is considered, the requirement $p = (\rho^{\kappa}) \times$ constant is satisfied and wave speed is given by:

$$\frac{2}{\kappa - 1}\, da \pm du = 0 \tag{10.9}$$

This equation relates the change of speed of sound to the velocity along the propagated waves. Variation of properties of the flow at a fixed point in the pipe, with the time needed to be determined; this calls for investigation of the state of the fluid across the characteristics.

The method of characteristics can be used as a graphical technique to calculate flows in the manifold. The method is illustrated for the isentropic case with zero friction losses using equation 10.9. Note that the positive sign in (10.9) relates to waves travelling from left to right, called the λ characteristics, while the negative sign applies to waves moving in the opposite direction and are called β characteristics. The relationship between a and u along a rightward propagating wave from point 1 to point 2 along the wave is obtained by integrating (10.9) and evaluating between the two points:

$$a_2 - a_1 + \frac{\kappa - 1}{2}(u_2 - u_1) = \text{Constant} \tag{10.10}$$

Rearranging the terms gives the expression:

$$a_2 + \frac{\kappa - 1}{2} u_2 = a_1 + \frac{\kappa - 1}{2} u_1 = \lambda \tag{10.11}$$

defining the characteristic:

$$\lambda = a + \frac{\kappa - 1}{2} u \tag{10.12}$$

for a rightward propagating wave. Likewise, for a leftward propagating wave:

$$a_2 = \frac{\kappa - 1}{2} u_2 = a_1 - \frac{\kappa - 1}{2} u = \beta \tag{10.13}$$

defining the characteristic:

$$\beta = a - \frac{\kappa - 1}{2} u \tag{10.14}$$

Equations (10.12) and (10.14) require a and u to be constant along the path of a wave. Parameters λ and β are numerically equal to the speed of sound a when gas velocity is zero. Note that a and u may be expressed in terms of λ and β by:

$$a = \frac{\lambda + \beta}{2} \quad \text{and} \quad u = \frac{\lambda - \beta}{\kappa - 1} \tag{10.15}$$

The physical significance of these results may be seen by considering waves in the $x - t$ plane of Fig. 10.4. The rightward travelling wave from point L at speed $u + a$ arrives at point i at time $n + 1$. Similarly, the leftward travelling wave from point R arrives at point i at time $n + 1$. If u is non-zero, then point L is not equidistant from point i as point R.

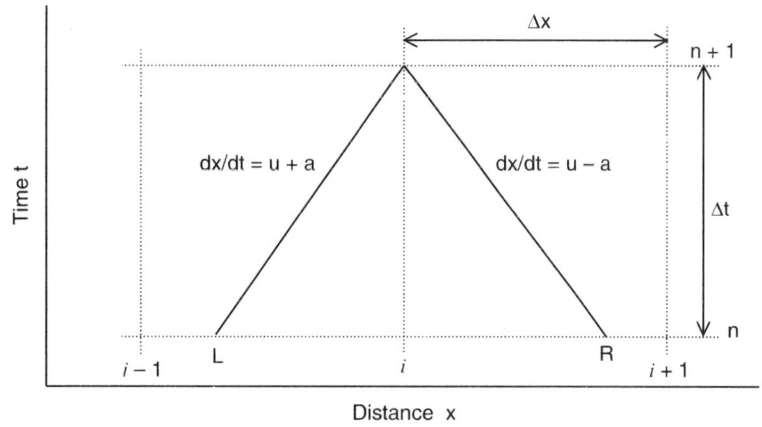

Fig. 10.4. Position of Wave Flow Characteristics for Isentropic Case, No Friction

When the general solution of the flow equation takes into account variations in entropy, the energy equation must also be solved in addition to the wave characteristics defined by (10.9). The additional equation introduces a new characteristic called a path line, representing transport of fluid properties at the local particle velocity. Inclusion of pipe wall friction and heat transfer in the analysis modifies the λ and β characteristics, as do changes in pipe cross sectional area. Hence $d\lambda$ and $d\beta$ are not zero, and the corresponding position of wave flow characteristics is depicted in Fig. 10.5. Changes in entropy along wave characteristics are obtained from the path line characteristics. Change in entropy along the path line is due only to pipe wall friction and heat transfer in the governing equations. Entropy level of the gas may change across a path line due to variations introduced at pipe boundaries.

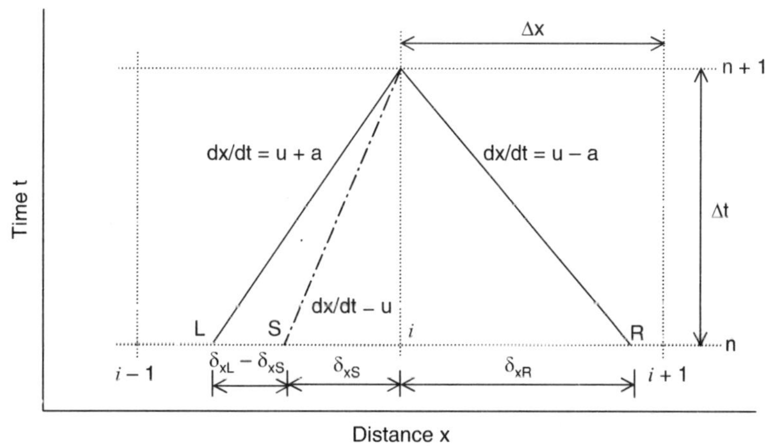

Fig. 10.5. Position of Wave Flow Characteristics for Non-Isentropic Case with Friction

Unsteady flow calculations by the method of characteristics are supplanted by the finite difference, or finite volume, methods. A number of distinct advantages accrue, chief among them:

- better conservation of flow properties (mass, energy, etc.)
- higher order accuracy

- faster solution using a computer
- capability to deal with entropy discontinuities and real gases.

Several procedures are available. The simplest finite volume scheme resembles the method of characteristics as it attempts to trace back all three characteristic paths from $[i\Delta x, (n + 1)\Delta t]$ on to the time level $n\Delta t$ as shown in Fig. 10.5. Due to linear interpolation of the characteristics and time the method provides first order accuracy in space. The scheme is not based on governing equations in conservation law form, hence discontinuities such as shock waves cannot be conveyed with the right speed, or evaluate correct jump conditions across them. To overcome these shortcomings the time step is subject to constraints imposed through stability considerations. This requires the wave information to travel not more than one mesh length in one calculation time increment.

A commonly used procedure is based on the Taylor series of expansion for the state vector, requiring evaluation of the Jacobian matrix. If the equations are solved in a two step approach this matrix does not need to be calculated, while still retaining second order accuracy. The first step is based on space centered differences about the point $(i\Delta x, n\Delta t)$, while the second step is a midpoint leap frog calculation with time differences centered on the point $[i\Delta x, (n + 1)\Delta t]$. Thus, important phenomena in the behavior of real waves are protected, and accuracy of predicted results enhanced, by the use of the procedure.

An important aspect of the calculation is boundary conditions at the two ends of the pipe that must be adequately simulated. Typical boundaries are: open ends, closed ends, in-flow boundaries, valve boundaries, junctions, pulse converters, orifices, valves, throttle bodies, turbines, compressors and inter coolers. A common feature is that they are treated as if the flow is quasi-steady, meaning that the boundary characteristic is not a function of the frequency of the incident wave form. The assumption is more appropriate for a simple boundary such as a closed end than for a complex volume such as a compressor. Thus, the physical size of the boundaries can be neglected compared with the length of the pipe connected to it, and the equations of steady flow can be applied locally to the small region. Based on physical characteristics of each boundary at the end of a pipe, the conditions can be expressed mathematically for both subsonic and sonic outflow through the type of boundary. The defining feature of a simple open end boundary is that the pressure at pipe exit is equal to that outside it. The effect is to invert an incident wave and hence an incident compression wave is reflected as a rarefaction wave. At a closed end boundary flow velocity is zero, and an incident compression wave is reflected as a wave of equal magnitude. Valves and ports influence performance of the reciprocating machine by the period they are open and the area available for flow. Note that besides transferring fluid into and out of the cylinder, they also facilitate air motion inside the cylinder. Thus, use of a helical air inlet passage provides swirling motion to the air inside the cylinder of diesel engines for improved combustion.

10.5 PARAMETER MEASUREMENT

Flow characteristics of reciprocating machine components may be measured using steady flow rigs. Unsteady flow rigs for evaluating performance of manifolds under dynamic conditions are also widely applied, obtaining data for simulation programs and quasi-steady boundary conditions. Inlet and exhaust manifolds are modeled using wave action computer codes, where geometrically complex components like valves, junctions, throttles and sensors are connected with pipe elements in which the flow is assumed to be one-dimensional. The actual flow is locally multi-dimensional and separated in the boundary regions, thus delineating the extent of continuous computational domain of each piping segment. A complete treatment of the boundaries would require a multi-dimensional model of the unsteady flow, with means to characterize turbulence in the fluid in the system. A simpler alternative lies in assuming flow in the boundary regions is quasi-steady and applying experimentally measured empirical coefficients at specific boundaries to maintain the effects of turbulence in the system.

When gas flows through a valve, separation and re-circulation regions develop a vena-contracta where the actual cross sectional area of the gas stream is less than the area of the orifice. This can be readily simulated in a one dimensional model as an empirical data. Measured data can provide effective valve area or a discharge coefficient, from which variation of pressure drop with mass flow rate across the device can be calculated. As an example, consider subsonic flow through an exhaust valve. Figure 10.6 shows flow through an exhaust valve into an exhaust port, with a diverging section immediately downstream of the valve. Further downstream an orifice plate measures mass flow rate of the air. If downstream stagnation pressure is taken as p_{OB}, the effective area represents the resistance to flow presented by the physical system between the valve and section *B*. On the other hand if downstream pressure is taken to be p_A, the resulting value of effective area represents flow resistance of the combined valve and section of the pipe up to section *A*. Since pipe section at *A* is larger than at *B*, $p_A > p_B$, if friction related pressure drop is ignored. For the same mass flow rate measured at the orifice plate and the same upstream stagnation pressure p_{oc} gives a greater effective area than if p_B was used as the downstream pressure. It is important to ensure the data obtained from a steady flow rig represents the pressure ratio/mass flow characteristics of the system being analyzed. The flow rig should be designed such that pressure values measured are representative of conditions and characteristics in the immediate vicinity of the valve. Flow rigs are of two types. A 'blowing' rig has a high pressure gas supply connected upstream of the device to be tested; in 'vacuum' or suction rigs gas flow is sucked through the device. When flow is from a reservoir (or cylinder), pressure measurement is similar in both type of rigs.

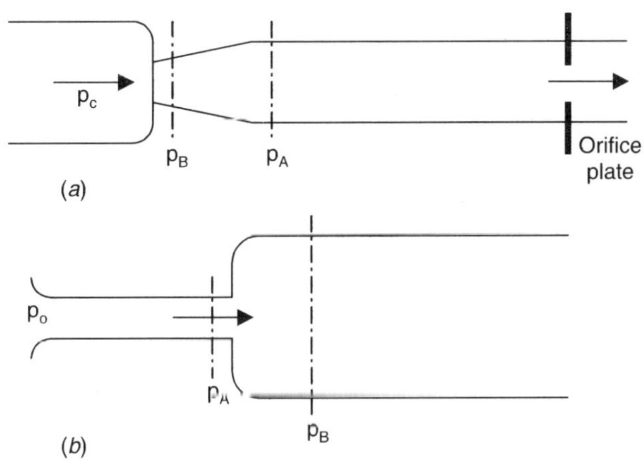

Fig. 10.6. Flow Measurement Through Valves: (*a*) Exhaust Valve, (*b*) Inlet Valve

Yoshitomi (1989) suggests a method of evaluating gas dynamic performance of a manifold. An actual manifold and cylinder head are used to make up the pulse generator. The crankshaft is not driven, with the pistons locked in the cylinder bore to act as plugs (Fig. 10.7). The head and manifold are attached to a large receiver to damp out pressure fluctuations, which in turn is connected to a high pressure air supply. Valves are fitted to the supply line to maintain pressure at a specific level, together with an orifice plate to measure mean flow rate. The inlet camshaft in the cylinder head is driven through a variable speed electric drive. The rig, using a standard automobile engine, is provided with pressure transducers, thermocouples and an encoder, and the data collected through an acquisition system. The manifold, however, is of a modular form to permit measurements for a range of designs. Note that tuning effects will not be fully representative of the engine because the pistons are not moving. Also, effects of the low temperature of the exhaust will not be exact as in an actual operating engine. But the rig does permit tuning natural frequencies of manifold designs to be assessed in order to establish accuracy of the flow. Modular manifolds may be built from combinations of sections of pipe lengths and plenum volumes, enabling a broad range of natural frequencies to be covered. Measured and predicted pressure traces are shown in Fig. 10.8 at two operating speeds, 1,000 and 3,000 rpm.

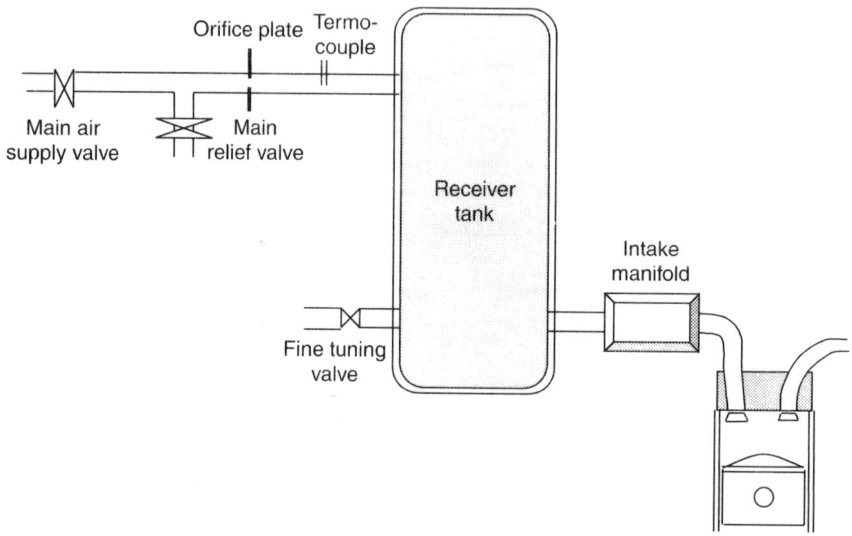

Fig. 10.7. Measurement of Dynamic Characteristics of Flow Through Valves

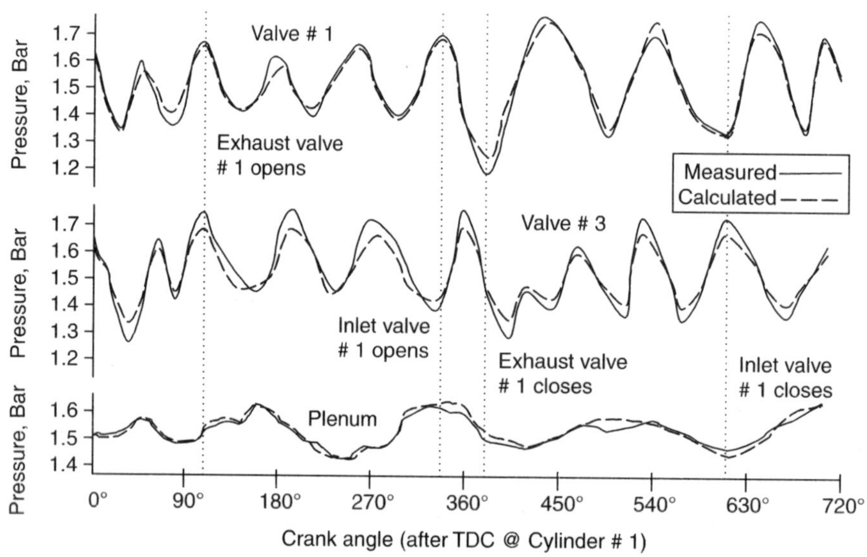

Fig. 10.8. Measured and Calculated Pressure Variations in Inlet Manifold

10.6 MANIFOLD DESIGN ASPECTS

The fundamental requirement is to synthesize a system to meet all requirements within a cost constraint. The difficulty in applying this philosophy is the wide range of parameters encountered in the design of the manifold. Figure 10.9 defines the geometric parameters affecting the tuning of an inlet manifold. Certain ground rules in relating to basic flow velocities help in deciding the basic pipe cross sections. Design procedures promulgated from the use of linear approaches provide additional information. In conjunction with experimental data from a similar machine fitted with selected manifold configurations, comparison can be obtained with simulated results. Thus, primary pipe length, plenum volume and secondary pipe segments can be varied to collect data at wide open throttle. When wave attenuation effects are not included for the predicted amplitude of pressure waves, larger than normal variations may be expected with traces measured at the intake valve. Lumped attenuation coefficients dependent on mass flow rate will generally provide better amplitude match between predicted and measured data for a specific manifold configuration at a given operating speed.

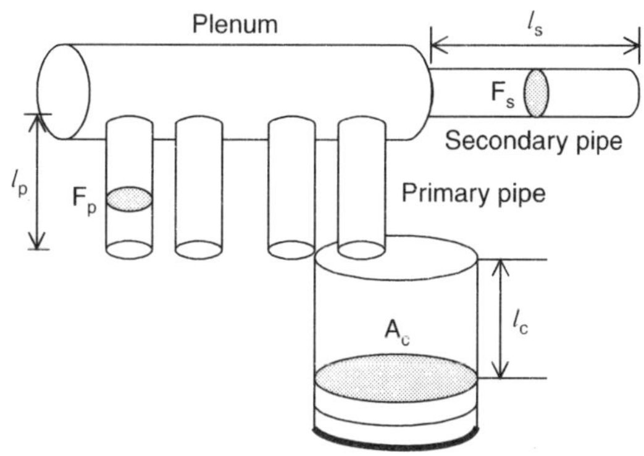

Fig. 10.9. Geometric Parameters of an Inlet Manifold

To obtain correct phasing of the simulated pressure traces it is necessary to use effective pipe lengths in place of geometric lengths. Effective pipe length consists of geometric length and end effects. For example, an open pipe at both ends has effective length about 1.7 times the geometric length due to end effects. Primary pipes are open at both ends, but only for a quarter of the engine cycle. In addition, phasing errors in linear theory result from neglecting effects of mean flow on wave propagation velocity. Figure 10.10 shows a comparison of measured pressure at the intake valve with a trace predicted by a linear simulation over five engine cycles. Engine speed and manifold configuration are for optimized amplitude and phase of the simulated case.

The effects of component dimensions on volumetric efficiency need to be observed closely, chief among them, primary and secondary pipes and plenum parameters. Figure 10.11 provides a predicted volumetric efficiency versus machine speed curve for a four-cylinder engine from which measured traces were obtained. A broad peak at high engine speed and a narrower peak at reduced speed are the main features of the characteristic, and is typical of the simplified manifold configuration.

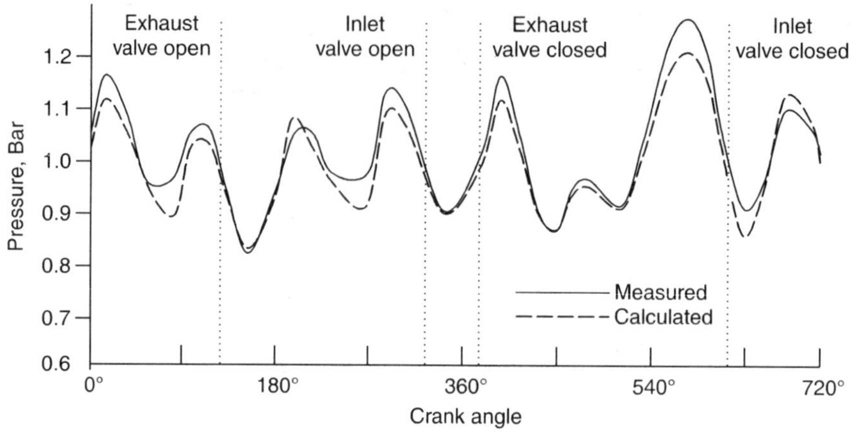

Fig. 10.10. Comparison of Measured Pressure in Manifold with Linear Simulation Values

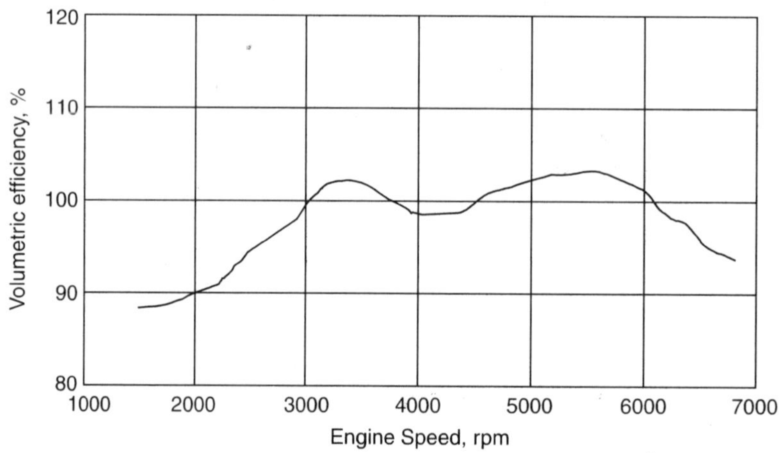

Fig. 10.11. Volumetric Efficiency versus Engine Speed for a Four Cylinder Engine

When primary and secondary pipe dimensions are held steady, an increased plenum volume decreases engine speed at which the lower peak in the volumetric efficiency curve occurs, together with a reduction in the magnitude of the peak. But the higher peak is not affected. The plenum volume can have a profound effect on the idle speed control, which may be a major consideration in selecting its dimensions.

The effect of increasing secondary pipe length is to decrease engine speed at which the lower peak in the volumetric efficiency is placed. Again, the higher speed peak is unaffected by this change. This effect is illustrated by Ohata et al. (1985) for Toyota's Acoustic Induction Control System (ACIS). The system is a variable geometry type, and works by operating a control valve which effectively shortens the secondary pipe length above a particular engine speed to maintain high volumetric efficiency.

Decreasing cross sectional area of the secondary pipe reduces the speed at which the lower speed volumetric efficiency peaks. The Hino EP100-II engine employs this aspect by dividing the secondary pipe into two segments axially, using a flap valve to cover and uncover half of the pipe's cross sectional area at the proper engine speed (Sato, 1987). Together with variations in secondary pipe length it enables the engine to maintain a high volumetric efficiency across the full operating speed range. The results indicate that the lower speed peak in the volumetric efficiency curve is affected by the dimensions of the whole manifold, while the higher speed peak is affected only by the primary pipe and cylinder dimensions in an engine. Examination of the resonance characteristics of the manifold offers a basic insight into the tuning mechanism. Figure 10.12 presents predicted relative amplitude of pressure responses for two different manifold configurations at the open end of a primary pipe that is subjected to a harmonic mass flow rate excitation. Considering only the solid line, the lowest and highest peaks represent the first and second harmonics of the manifold. Peak 2 provides resonance of the pressure in the primary pipe only, cylinder volume being omitted. For the lowest frequency peak 1, pressure in both plenum and primary pipe are at resonance and in phase with each other. At the highest frequency peak both pressures are in resonance but π/2 radians out of phase with each other.

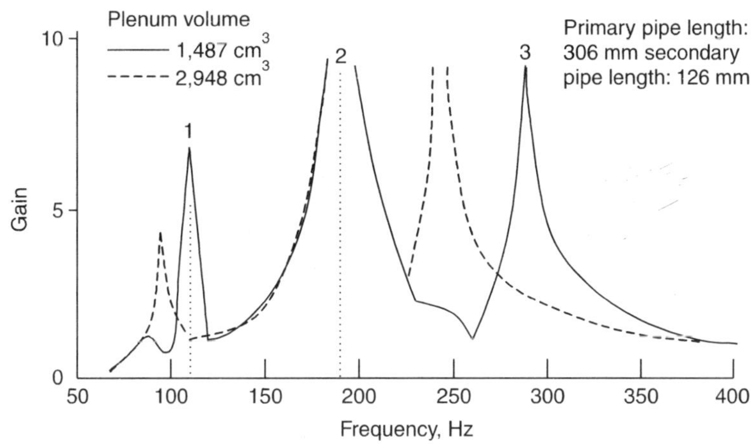

Fig. 10.12. Gain in Pressure at Pipe End 2 due to Sinusoidal Mass Flow Rate

10.7 FLOW SIMULATION IN INLET PORT WITH MOVING VALVE

Multi-dimensional flow simulation of airflow in an inlet port and cylinder using experimentally determined boundary conditions at the valve helps identify a number of problems that are not readily recognized with one-dimensional analysis. 'KIVA' code from Los Alamos National Laboratory (Amsden et al. 1985) is specifically designed for application in internal combustion engines. It is a Lagrangian Eulerian code using cell and vertex flags for computational domains made of arbitrary hexahedrons. Once the outer boundaries of the object are established, the mesh generator proceeds with the task of filling the volume with elements. In the finite difference method, however, each cell layer has to be carefully built, and additional cell layers are progressively added. A wire frame representation of the final mesh is depicted in Fig. 10.13, with the position of the valve at its minimum and maximum lifts.

A moving valve is simulated by translating all cell vertices pertaining to the valve as prescribed by the valve function. The topology of the computational domain is not altered during valve movement since the connectivity of each cell is maintained during the operation. Still, the possibility of distorted or negative volume cells remains.

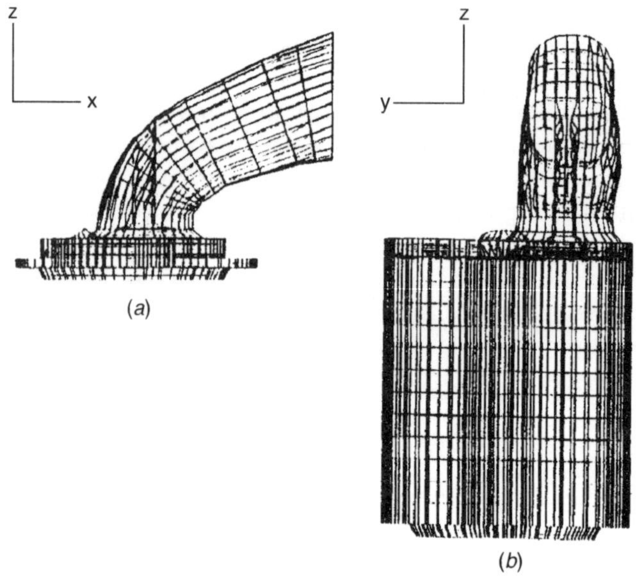

Fig. 10.13. Port and Cylinder Showing (*a*) Minimum, (*b*) Maximum Valve Lift

With moving vertices in the flow simulation, as the boundary of the computational domain moves, the position of internal vertices can be recalculated at each time step using analytical mesh generation. As the piston and the valve move the boundary of the computational domain is continuously distorted. The adopted rezoning process must guarantee that no coordinate line crossings will occur, particularly in the two limiting cases of small valve lift and small space between piston and valve. The procedure works well in two-dimension work when the geometry is not too distorted. However, in three dimensions, as soon as the boundary contours become complicated, negative volume cells may appear with increasing frequency. The problem can be resolved by using an iteration procedure which when applied to a topologically accurate grid yields a mesh which is more adapted to the present flow computation procedure (Taghavi and Dupont, 1989). Limitation in the minimum allowable cell size may restrict valve movement all the way to closure.

Figure 10.14 provides results in the form of velocity field on a horizontal plane, coinciding with the lower surface of the valve. At lower valve lifts (for example, at – 352.5° crank angle (with 0° taken at compression top dead center) the velocity profile is less affected by the presence of the wall than at higher lifts. As the valve lowers wall effects become predominant, and the flow is diverted to portions of the valve circumference that are further away from the wall. Wall proximity creates an axis of symmetry along the longer axis of the cylinder head cavity, superceding effects due to the incoming flow from the intake port.

During the early stages of the intake process a fresh air front advances through the inlet port, through the valve passage outlet and along the walls of the combustion chamber. The region immediately below the valve is reached last by the fresh air front.

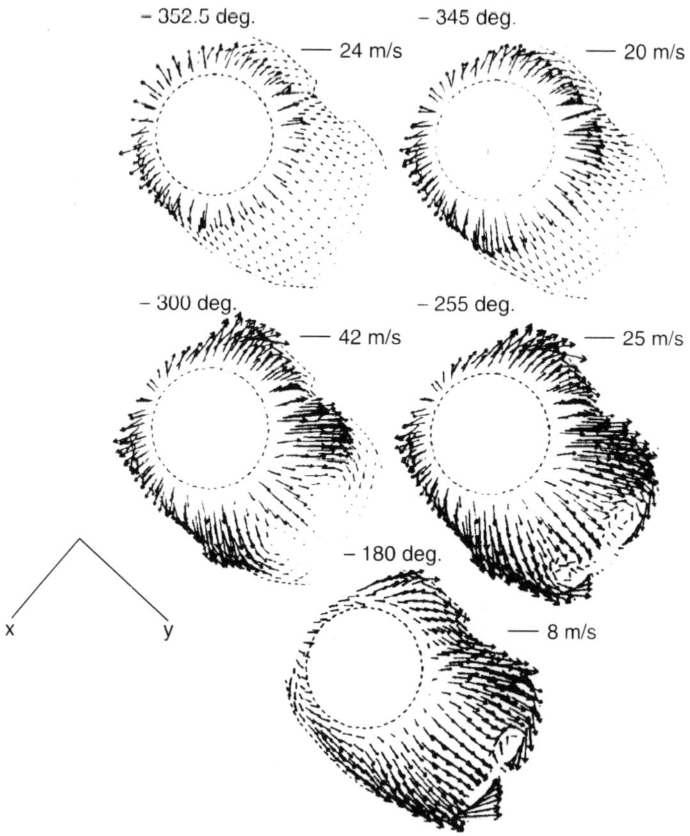

Fig. 10.14. Gas Velocity Field in Lower Plane of Valve

10.8 FLOW PATTERN AT MANIFOLD JUNCTIONS

A detailed and quantitative picture of the flow pattern at manifold junctions (or multi pulse converters) for different flow configurations and mass flow ratios from Laser Doppler Anemometry (LDA) measurements is useful for comparison with numerical algorithm and overall system performance improvement (Flamang and Sierens, 1988). Figure 10.15 shows the pipe arrangement of a typical pipe junction. Pressure loss coefficient L_i is defined as:

$$L_i = (p_1 - p_5)/p_5 \tag{10.16}$$

where $i = 1$ to 4

where pressures p_1 to p_5 are mean static pressure readings in sections at equal distance of 5 times duct diameter from the junction. The loss coefficients are a function of :

$$L_i = f\{(dm_1/dt)/(dm_5/dt), (dm_2/dt)/(dm_5/dt), (dm_3/dt)/(dm_5/dt), M_5\} \tag{10.17}$$

where $M_5 = u/a$ is the Mach number in pipe 5.

The LDA set-up used in the experiment is a dual beam system operating in forward scatter mode. The counter is connected to a data acquisition system. Seeding of the flow is obtained by atomization of oil at the entrance of the pipes. Test rig schematic is shown in Fig. 10.16.

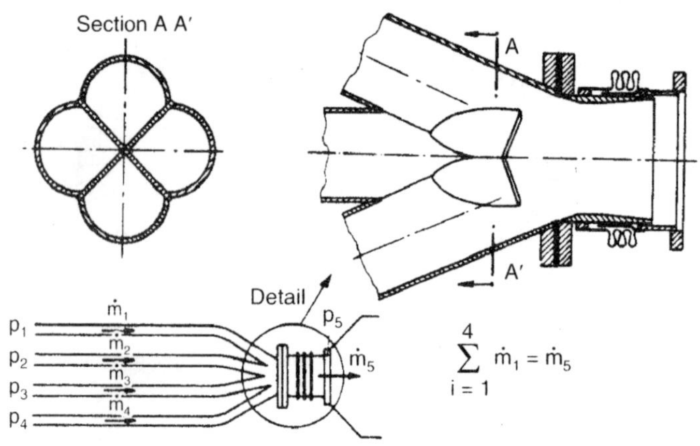

Fig. 10.15. Pipe Arrangement at Junction

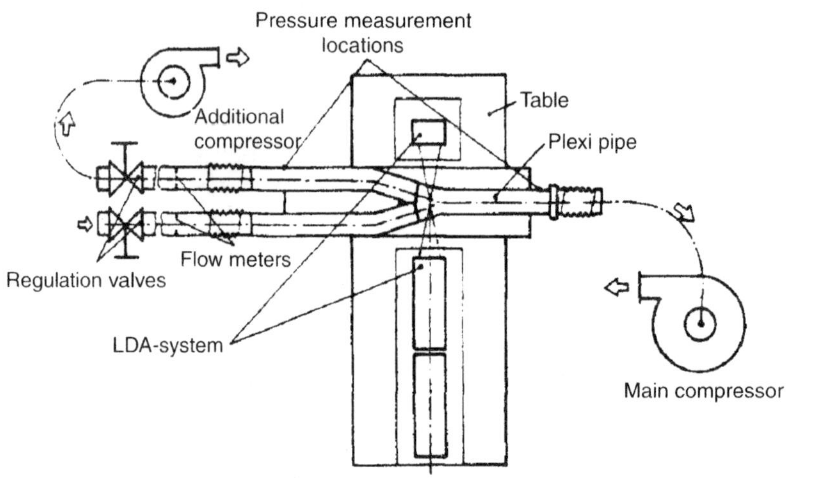

Fig. 10.16. Test Rig Arrangement

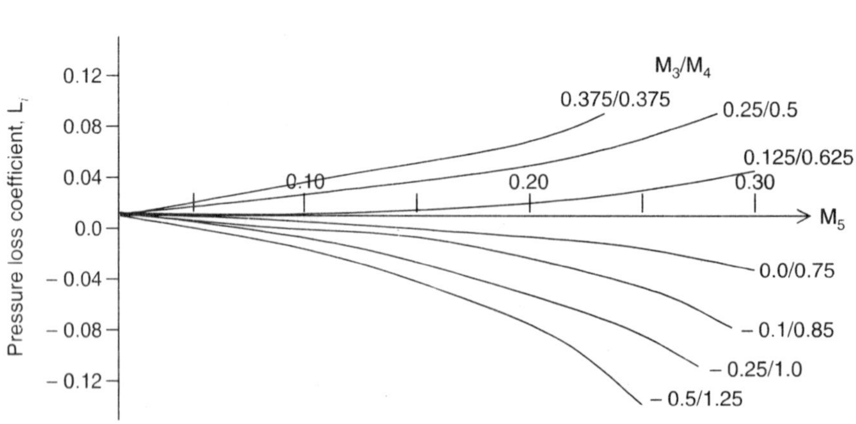

Fig. 10.17. Pressure Loss Coefficient

Pressure loss coefficient L_1 for pipe 1 is plotted as a function of the Mach number M_5 in Fig. 10.17 for a specific mass flow ratio. Loss coefficients for the other pipes follow a similar trend.

Figure 10.18 shows the evolution of the axial flow for the flow configuration of 0.5/0.5/0.5/-0.5 at planes -25, -12 and 17 mm. Back flow velocity occurs only in 25 mm plane, shown in dotted lines.

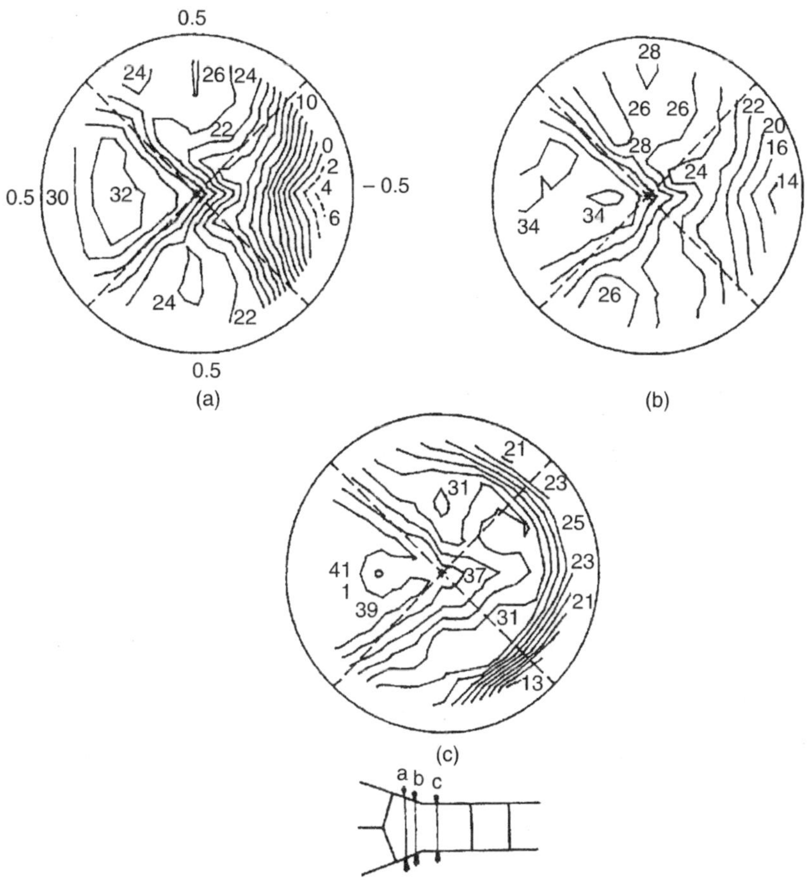

Fig. 10.18. Axial Velocity Profiles: M_5 = 0.0996, Flow Configuration: 0.5/0.5/0.5/-0.5

10.9 TUNING OF MANIFOLDS

Results from computer simulation and test measurements can be used to investigate manifold tuning arrangements. Tuning is substantially affected by the length of the primary pipes in the system. If the pipe is shortened by 10-12% the first peak may not move noticeably, but the second resonance will move to a higher speed, and the magnitude of tuning reduces. The latter is due to the inertia tuning brought about by resonator effects in the manifold. If the pipe is lengthened the inertia peak becomes bigger due to increased mass of air in the pipe, and is accompanied by reduced tuning of frequency. The two tuning modes may even start to merge, resulting in a steady increase in volumetric efficiency and torque over a substantial speed range. But it will also have the drawback of rapid volumetric efficiency drop at higher speeds and consequent reduction in performance.

Computer simulations using linearized wave equations often show higher volumetric efficiency when compared with measured data. Discrepancies may be attributed to exclusion of heat transfer from cylinder walls and ports to the charge air (more prominent at low engine speed), assuming incompressible flow for valve model (more prominent at high engine speed), and presence of residual gases in the cylinder.

Low speed resonance effect can be regained by grouping cylinders whose induction periods do not overlap together at separate plenums. A scheme for such grouping in a V-6 engine is shown in Fig. 10.19. At the confluence of the two secondary pipes a small but noticeable fluctuation

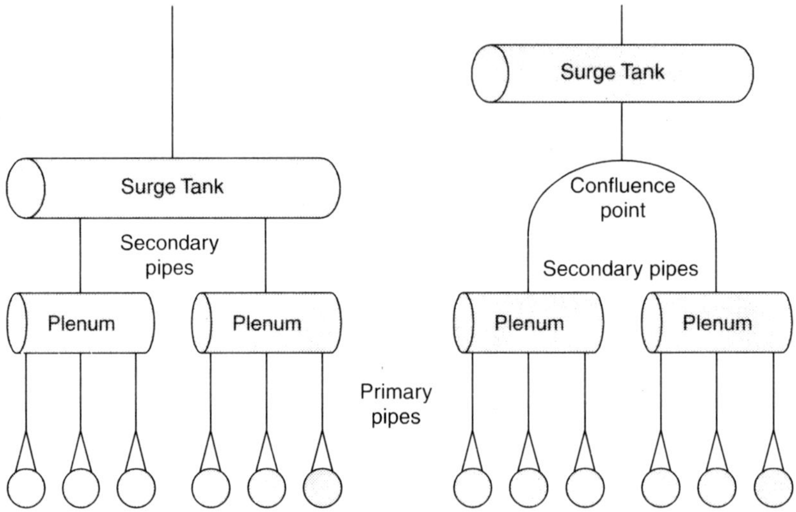

Fig. 10.19. Six-Cylinder Machine Manifold Systems to Obtain Three Cylinder Grouping

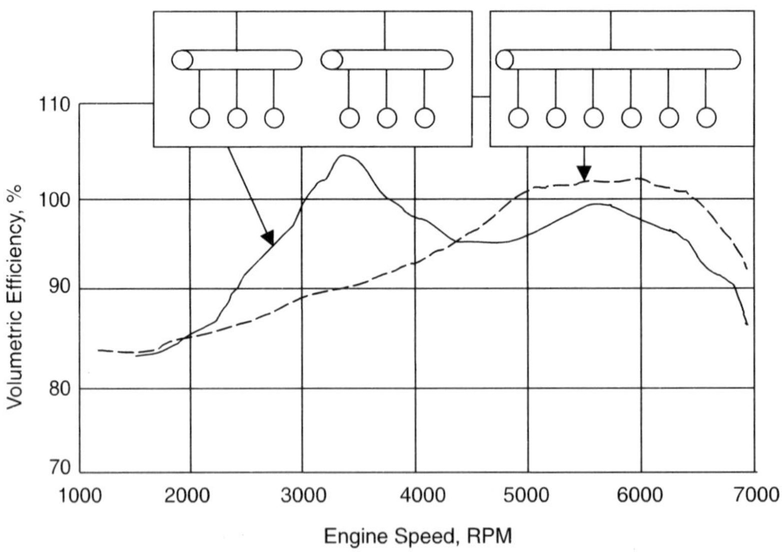

Fig. 10.20. Predicted Volume Efficiency Comparison - Single & Double Plenums

in pressure will occur, since waves of small amplitude and nearly opposite phase arrive simultaneously from each half of the system (*ref*: Hatamura et al, 1987). Consequently, the secondary pipe length is the effective distance from the plenum to the point of confluence. Figure 10.20 provides a comparison of predicted volumetric efficiencies for a six cylinder machine.

10.10 VARIABLE GEOMETRY MANIFOLDS

Volumetric efficiency results provided in Fig. 10.20 indicate that if the double plenum configuration at low speed could be modified to the single plenum version at high speed, the performance of the combined system would be advantageous. This is the concept used in many reciprocating machines behind a variable geometry manifold. At lower operating speed the two halves of the plenum chamber are disconnected, then connected at higher speed. Figure 10.21 shows calculated volumetric efficiency against speed characteristic of a six-cylinder engine with variable geometry manifold, where the plenum chambers are connected by a short pipe containing a valve. The valve is set to open at higher operating speeds, thus forming one continuous pipe, and to close at lower speed, below 3,200 rpm, to form two separate plenums. The geometry gives rise to a low speed peak in the volumetric efficiency curve at 2,700 rpm.

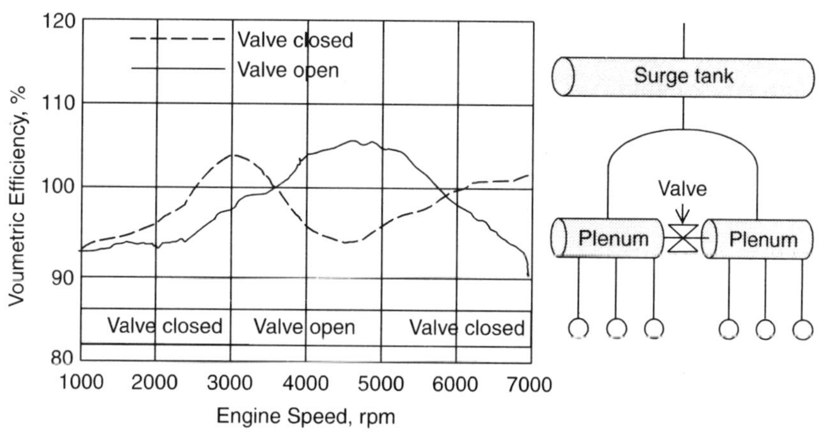

Fig. 10.21. Variable Manifold Geometry - Volumetric Efficiency

It is also possible to exploit the low frequency pulsation of the induction system when divided into two secondary pipe lengths. The change in effective length may be achieved by connecting the two secondary pipes with a shorter pipe. Since pressure waves in the secondary pipe are significantly out of phase in such an arrangement, the point of confluence of the two pipes acts as a pressure node point, just as an open ended pipe to the atmosphere, and thus dictates the effective pipe length.

Increasing the length of a primary pipe lowers its natural frequency. A variable geometry induction system using this property is shown in Fig. 10.22 (Isomura et al. 1987). By opening and closing the valve in the auxiliary primary pipe, the effective length is changed. When the valve is closed, two closed ends are produced in the shorter pipe, and the effect is included in the computations.

On four valve engines with one primary pipe per valve it is possible to close one of the primary pipes with a butterfly valve. Reduction in the effective flow area as a consequence lowers the frequency of the primary pipe resonance. Closing the pipe in a primary pipe substantially lowers the speed at which the peak in the volumetric efficiency curve occurs.

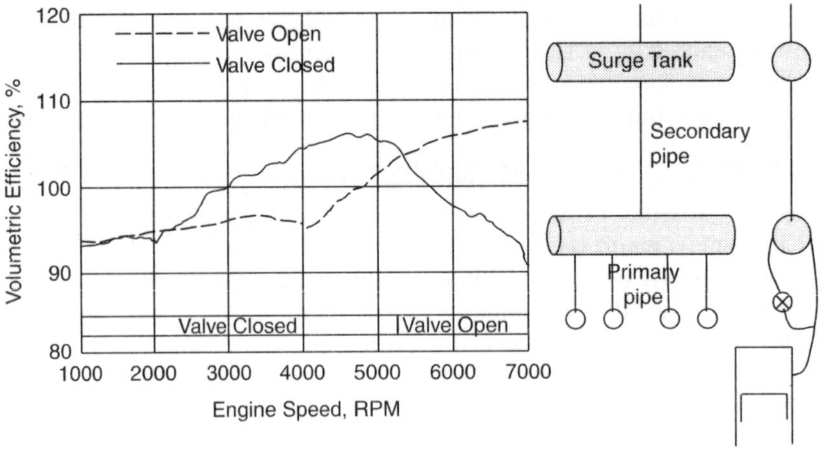

Fig. 10.22. Variable Primary Pipe Length

10.11 VALVE ARRANGEMENT

In addition to higher valve area, the use of multiple valves per cylinder with a pent-roof combustion chamber produces an ordered motion of the incoming charge about an axis perpendicular to the cylinder axis. The tumbling and swirling motion of the charge can be closely controlled with multiple valve arrangements in order to produce the style of structured turbulence in the cylinder; this gives burning rates sought in the design of modern engines. Besides improving the characteristics of the motion of the mixture flowing into the cylinder, a slight raising of the cylinder head has the potential to increase available valve area above that obtained with a flat gas face of the cylinder head.

Woods and Brown (1996) present an analysis of the gains in geometric area experienced through the use of multiple valves, although it is not the only factor which makes the system more desirable. There are several automotive engines that incorporate four valves per cylinder, two for air intake, two for exhaust. Yamaha motor cycles (model FZR 750) are provided with five valve cylinder heads, three inlet valves, two exhaust valves. Aoi et al. (1986) performed an analysis using the following design criteria: minimum clearance between valve seats 2 mm for intake valves, 4 mm for exhaust valves; intake valve opening area 1.3 times that for exhaust valve. The calculations showed that calculated intake valve opening area for a given bore diameter of a five valve configuration is greater than for four, six and seven valve designs. Four and five valve engines of the same bore (57 mm) were built. Steady state flow tests showed a superior mass flow rate through the cylinder of the five valve engine across the lift range for a fixed pressure drop. Volumetric efficiency of the five valve engine proved to be nearly 10% better than the four valve configuration in the 7,000 to 13,000 rpm range. For similar constraints on valve spring design the reduced equivalent weight and lower lift of the five valve design increased the speed of valve bounce occurrence by almost 5%.

Other published data show higher values for geometric area for five valves than for four valves (Ohrenberger and Mann, 1994). Larger web area between the spark plug and valves are obtainable in five valve designs, permitting better water cooling around the plugs for improved knock resistance. Consequent reductions in valve train mass and camshaft drive torque are additional advantages of a five valve arrangement.

Kanazawa et al. (1993) show that measured mean effective intake valve area with four inlet valves is 29% larger than for a cylinder with circular section equal in area, and two inlet valves. The elliptical section increases effective intake valve area by 12% when two valves are used. The increase in the flow area is attributed to reduced interference between the incoming air and cylinder walls.

10.12 TURBULENCE IN CYLINDERS

Steady flow bench measurements provide data for engine simulation models for characterizing turbulence and swirl generation capabilities of a given intake port (Davis and Tabaczynski, 1988). Velocity distribution produced near the valve/head interface as the flow enters the cylinder of any intake port design is unique. The effects of velocity distribution on turbulence and burn rate inside the cylinder needs to be assessed.

During experiments the induction system can be modified from the base configuration by masking the intake valve and inserting flow restricting or directing tubes upstream of the intake valve inside the intake ports. For steady flow measurement of inlet velocity at 10° increments around the intake valve is made with a hot wire anemometer, thus obtaining mass flow rates at the same locations. Typical results for a four cylinder engine are shown in Figs. 10.23 and 10.24.

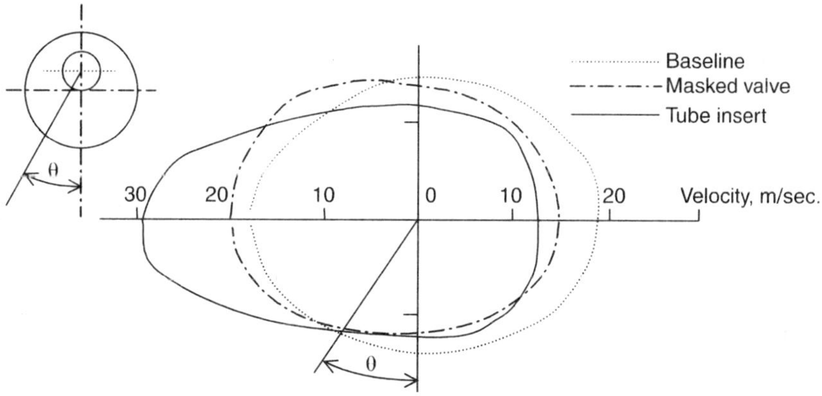

Fig. 10.23. Flow Velocity Distribution Around an Intake Valve

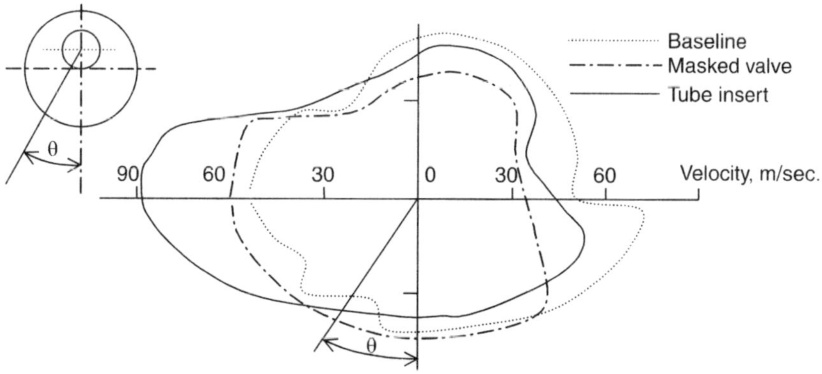

Fig. 10.24. Mass Flow Rate Distribution Around an Intake Valve

Intake generated turbulence is assumed to be proportional to instantaneous mass flow rate and mean kinetic energy, represented by the equations:

$$P_{gen} = [dm_{in}/dt]\{K_{in} - K\}/V \quad (10.18)$$

where: $K_{in} = (C_k V_{in})^2$

and V_{in} is the instantaneous velocity past the intake valve. The proportionality constant is set at 0.15, similar to that of a turbulent jet. Turbulent kinetic energy is also produced due to squish and during compression of the charge, with some dissipation effects also taking place.

In-cylinder peak turbulence may be expected to increase by 20% with the inserted tube over the uniform flow case, as shown in Fig. 10.25. Dissipation effects cause the difference to diminish to about 5% by the time combustion initiates. Significant differences in the predicted intake generated turbulence are noted, and persist up to the time of spark and during combustion. The differences are caused by variations in valve discharge flow coefficient, as also in maximum valve lift.

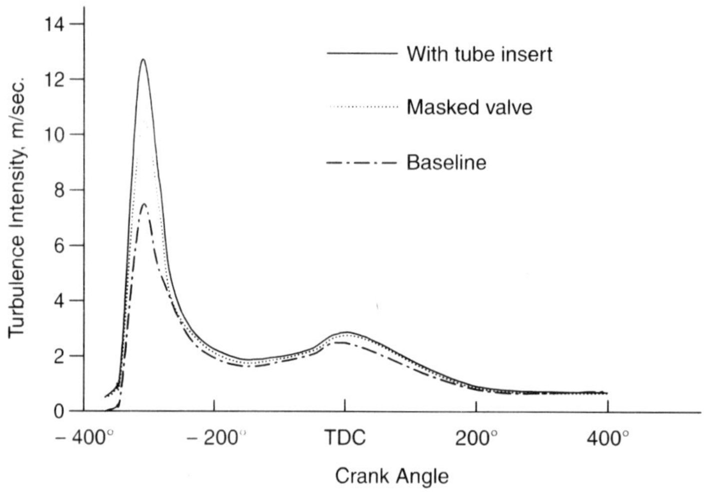

Fig. 10.25. Effects of Tube Insert on Turbulence Intensity

10.13 SCAVENGING PROCESS

Exhaust manifold systems in machines with multiple cylinders range from connecting all cylinders into one manifold, through the type where two cylinders are connected initially and common pipes joined downstream, to the tuned 'banana bunch' manifold where all pipes are brought together at one junction.

Exhaust manifolds must aim to separate exhaust flows of cylinders which are liable to interfere with each other, and to produce beneficial wave reflections to assist in scavenging. To understand the operation consider two primary sub-manifolds connecting cylinders 1 and 4 and cylinders 2 and 3 of a four cylinder engine. The blow down pulse from cylinder 1 travels down the pipe till it meets junction 1, where a sudden enlargement doubles the pipe area at the confluence. A rarefaction wave is then transmitted back to cylinder 1 to assist the scavenging. Smaller pressure waves are transmitted into the other two pipes, one going towards the junction, the other to cylinder 4. If the wave reaches cylinder 4 when the exhaust valve is closed, it will not interfere with the flow through the cylinder. This is likely to be the case if wave travel time is nearly 80° crank angle when machine speed is 3,500 rpm and phase angle between cylinders 1 and 4 is 360°. The wave travelling towards the junction will be partially reflected to cylinder 1 as a rarefaction and partially transmitted through the junction with a reduced amplitude towards cylinders 2 and 3, where it will interfere with the scavenging process of the cylinder firing 180° after cylinder 1.

In two stroke engines the scavenging process takes place when the exhaust valves and inlet ports are open, fresh air then replacing the combustion products. Because the available oxygen for combustion is strongly dependent on the scavenging efficiency, total machine performance heavily relies on the process. Scavenging efficiency is a function of the ratio of air mass supplied to the cylinder and the mass required to fill it at system pressure and temperature. Air retained in the cylinder during scavenging will depend on system geometry (inlet ports, piston bowl, cylinder head, exhaust valves) port and valve timing, pressure and temperature.

A transient three dimensional code (Carapanayotis and Salcudean, 1988) may be based on turbulent kinetic energy and dissipation model for computing effective viscosity, the technique then numerically solving continuity and momentum equations. Boundary treatments assume no transport at cylinder axis, specified flow velocity components, temperature and concentration at inlet and zero tangential velocity at cylinder walls. At the outlet changes in valve area because of valve motion during scavenging is performed in finite steps.

Scavenging characteristics display three recirculation zones in the cylinder. The first vortex is in the piston bowl, the second close to the cylinder wall and the third close to the cylinder axis due to swirl, or the back flow vortex. In the upper part of the cylinder swirl velocities are different. When an engine is equipped with four exhaust valves the swirl velocity is higher then in a two valve configuration. In the two valve geometry air advance to the cylinder head from the inlet port is slower than with four exhaust valves. This may partly be due to the two valves being further away from the cylinder axis, but the effect on overall scavenging performance is not substantial.

10.14 MACHINE SPEED AND SCAVENGING CHARACTERISTICS

Detailed calculations to predict pressure and temperature fluctuations in exhaust manifolds show good agreement with measured results by introducing loss coefficients at junctions. Models based on constant pressure junctions tend to under-predict pressure produced by the main pulses close to the exhaust valve of the cylinder, but over-predict subsidiary pulses. The under-prediction of the main pulse may be due to low resistance to transmission of pressure from sub-manifolds. Pressure pulses in exhaust manifolds of automotive engines may reach a maximum value of 3 bar when operating at 6,500 rpm, underscoring the need to avoid interference between cylinders. Figure 10.26 provides a comparison between measured and calculated exhaust manifold pressures over a speed range from 2,500 to 5,000 rpm.

A high pressure level will blow back through the engine cylinder into the inlet manifold. Some back flow into the inlet manifold does occur when temperature at the inlet valve is raised to between 310° C when the valve opens and around 360° C when substantial flow passes into the cylinder. Comparison between calculated and measured pressure traces shows major characteristics are similar. Primary pressure pulses tend to dominate, with subsidiary ones occurring just before them.

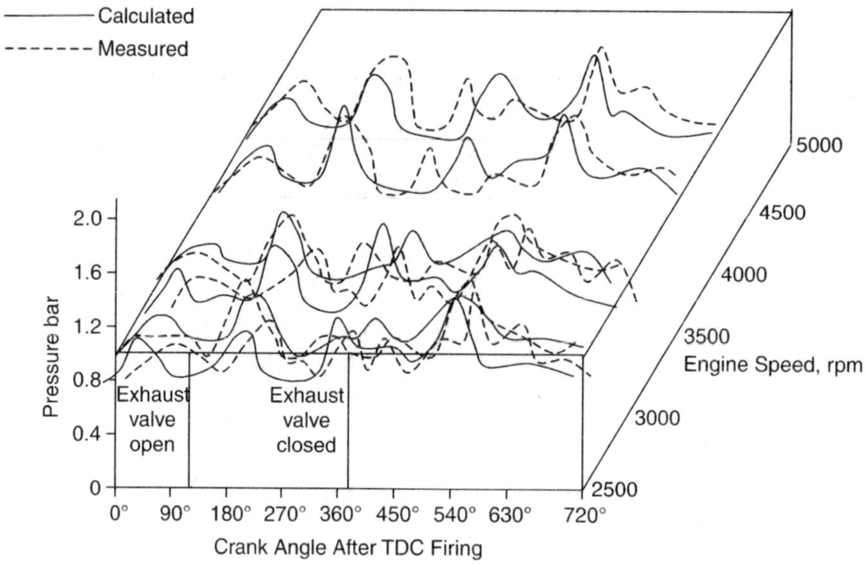

Fig. 10.26. Measured and Calculated Pressure Traces in Exhaust Manifold

10.15 REACTIVE SILENCERS

Reactive silencers abate sound pressure levels by reflecting acoustic power, carried by pressure perturbations, back to the noise source. The mechanism of reflection and transmission of sound waves at geometrical discontinuities are exploited to control acoustic power generated by the source, then carried downstream along the pipe system through interaction between engine noise source and the silencing device.

Reactive silencers are different from absorptive ones that use sound absorbing material to dissipate acoustic energy as heat when pressure waves travel along ducts of an acoustic filter. The simplest configuration that may be introduced in intake and exhaust pipe systems are expansion chambers and column side resonators. Since they represent basic elements used to build up more complex silencer systems through acoustically equivalent schemes, care must be exercised to determine their characteristics. An expansion chamber (Fig. 10.27*a*) provides broad band attenuation, approaching zero for the resonant (or transparency) condition. The frequencies are given by:

$$f_n = na/(2l_e) \tag{10.19}$$

where $n = 1, 2, 3 \ldots$ and l_e represents length of the expansion chamber. Conversely, resonators possess narrow band action, with intense attenuation only at resonant frequencies.

The frequency of a Helmholtz resonator (Fig. 10.27*b*) is:

$$f_o = \frac{a}{2\pi}\sqrt{\frac{F}{l_n V}} \tag{10.20}$$

where F and l_n are cross sectional area and length of neck, respectively, and V is volume of chamber. Frequency of a column resonator (Fig. 10.27*c*) is given by $f_n = na/(4l_r)$, where l_r is length of the resonator column and $n = 1, 3, 5 \ldots$.

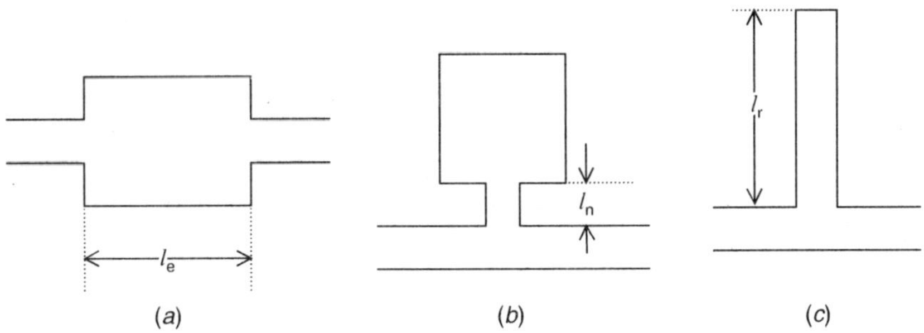

Fig. 10.27. Reactive Silencer Elements: (*a*) Expansion Chamber, (*b*) Helmholtz Side Resonator, (*c*) Column Side Resonator

It is possible to simulate wave motion in silencers with more complex internal structures. Modeling of a muffler is feasible using derived boundary conditions for sudden enlargements, contractions, pipe junctions and open ends when coupled with equations for end effects at discontinuities. Typical internal elements are orifices, axial side branches and flow reversers. Figure 10.28 shows schematic diagrams of silencers with typical internal structures and equivalent

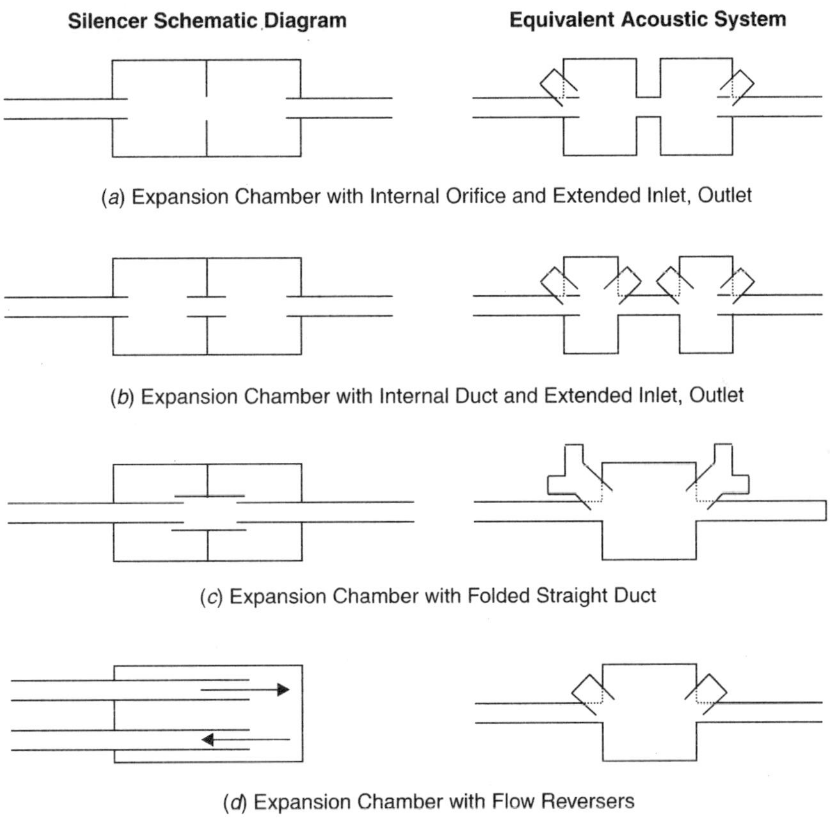

Fig. 10.28. Typical Internal Elements for Silencers

acoustical systems. Gas dynamic modeling of reactive silencer elements is based on existing boundary conditions. Abrupt area change of an expansion chamber may be treated as a sudden enlargement-contraction model. The cavity of a Helmholtz resonator may be represented by an equivalent cylindrical duct with appropriate geometric length and volume to avoid limitations of a lumped parameter approach, and to compute resonant frequencies of the system. Closed end boundary condition is used for the end wall of both the cavity and the resonator. *T* junctions arising in resonators at the interface with tubes can be satisfactorily described by the constant pressure or equal total enthalpy models. Since mean flow in side ducts of figures 10.27*b* and *c* is zero, pressure loss in the junction is not significant; however, a pressure-loss model for the junction may provide greater accuracy for resonator attenuation.

Perforated tube silencers are found in many silencers. Studies have been carried out to investigate effects of perforations using one dimensional and non-linear gas dynamic methods. The dissipative effect of perforated ducts may be introduced by means of an appropriate friction coefficient for short ducts representing orifices. The effect is thus directly included as a source term in conservation equations. Junctions of several pipes may be treated as boundary conditions in the constant pressure model.

10.16 TAILPIPE NOISE

High flow velocity of gas in exhaust mufflers results in generation of gas flow, or aerodynamic, noise. When gas passes through succeeding silencing elements (expansion chambers, flow reversers, perforated pipes, among others) and exhausts from the tailpipe duct as a high velocity jet, significant turbulence is generated. Each geometric discontinuity in the flow path is a site for vortex shedding, and is responsible for self generated broad band noise. The silencer itself behaves as a noise source, and its intensity may sometimes be comparable to that of the whole engine. Special care is required in the design of the silencer to avoid generation of excessive gas flow noise and meet stricter legislative requirements. Gas flow noise can become an even more difficult problem at high flow rates encountered at high operating speeds.

At the exit of the exhaust system calculated pressure is nearly constant because the outflow is practically continuous, except for sporadic flow inversions. At the intake calculated pressure exhibits only small departures from constant atmospheric pressure. Thus, pulse noise radiated from open terminations cannot be calculated from predicted pressure trends. Open terminations of duct systems are generally modeled with simple quasi-steady boundary conditions. In the case of sub-sonic outflow static pressure at the vena contracta is assumed to be constant and equal to ambient pressure.

The sound field generated around the outlet periphery may be evaluated satisfactorily from predicted flow velocity. For small perturbations, if the outlet is considered to be a monopole source, radiating spherical waves have a pressure field defined at distance *r* by the expression:

$$p(r, t) = \frac{\rho_o F}{C_v r} \frac{d}{dt}\left[u\left(t - \frac{r}{a_o}\right)\right] \tag{10.21}$$

where F is cross sectional area of the terminal, ρ_o, a_o are density and sound velocity in the surrounding environment and C is a constant equaling 4 for spherical radiation and 2 for hemispherical radiation. Predicted velocity at the open end may be calculated for discrete crank angles. After decay of a relatively short transient wave motion (lasting around 3 to 10 engine cycles) flow velocity, pressure and temperature are periodic in time. Discrete Fourier Transform procedure may be ideal for evaluating spectral content of a periodic set of values using the Fast Fourier Transform algorithm. Spectral content of the radiated noise may be determined as follows. The set of velocity values need to be interpolated to get a continuous function for differentiation in the time domain. Velocity derivatives corresponding to specific time or crank angle may be evaluated by the FFT algorithm. Once the amplitudes of the spectral components are known, root-mean-square values can be computed to obtain pressure field spectral components from the equation:

$$p_{n(\mathrm{rms})} = \frac{\rho_o F}{4\pi r}\left(\frac{d}{dt}\right)_{n(\mathrm{rms})} = \frac{\rho_o F}{4\pi r}\,\frac{(du/dt)_n}{\sqrt{2}} \tag{10.22}$$

As a last step components of the noise spectrum at each frequency f_n can be expressed in terms of sound level L_p as:

$$L_p = (f_n) = 20 \log \frac{p_{n(\mathrm{rms})}}{2 \times 10^{-5}} \tag{10.23}$$

where the harmonics for a four cycle engine are:

$$f_n = nf_o = n\,\frac{1}{T} = n\,\frac{360(N/60)}{720} = n\frac{N}{120} \tag{10.24}$$

and f_0 is fundamental frequency, T the period of signal and N engine speed, rpm.

Perturbation imposed upstream of the system is 'white noise', where spectral content of the pressure signal is uniformly distributed over a frequency band. It is possible to evaluate gas dynamic response of the muffler to all concurrent excitation frequencies at once. When a steady pattern of standing waves has been reached after decay of the initial transient motion, pressure and velocity oscillations in the system are characterized by a significant amplitude of the spectral components for each frequency of white noise introduced in the silencer. Approach of white noise is rapid when compared with acoustic excitation from a large set of single harmonic pressure oscillations.

Figure 10.29 shows a comparison between an appropriate signal obtained with Δp = 0.0005 bar and random angles of phase ϕ_n and another inappropriate signal, with the

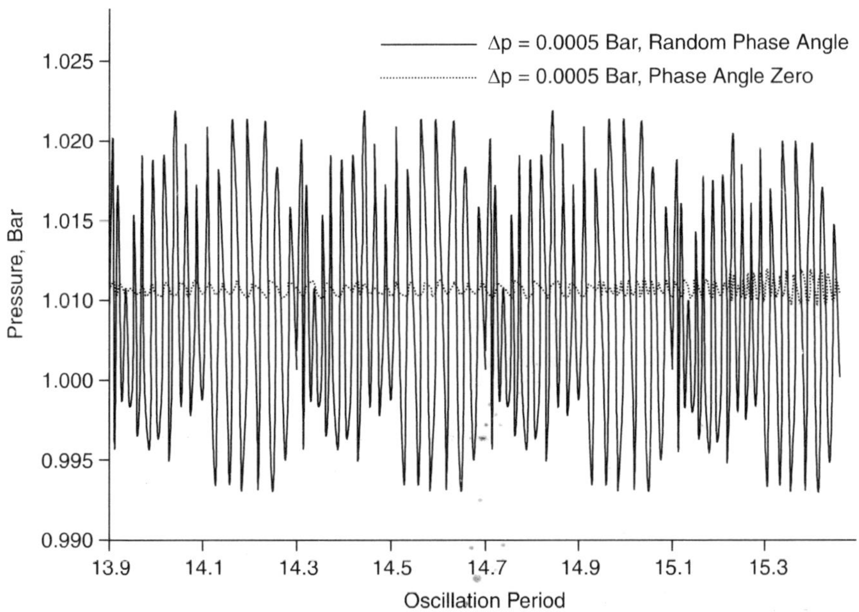

Fig. 10.29. White Noise Pressure Trace from Silencer

same Δp but where the phase angles ϕ_n are set to zero. The second choice gives excessive non-acoustic amplitudes both at the beginning and at the end of the pressure fluctuation. The spectrum of white noise pressure signal over a frequency range of 20 – 2,000 Hz confirms the discrete and uniform quality of the spectral content. The pressure trace of the 'white noise', $p(t)$, may be imposed as a boundary condition at the entry cross section of the silencer. Attenuation curves of complex systems of silencers with perforated pipes and dissipative silencers with sound absorbing material may be calculated using the 'white noise' approach.

The major advantage of silencers using sound absorbing material is good attenuation over a large frequency band, however, the attenuation quality reduces substantially at low frequencies. Also, the absorptive lining material faces rapid erosion due to high mean flow velocities. Hence, use of absorptive silencers in internal combustion engines is limited. Reactive silencers that guarantee lasting service performance are preferred. Dissipative ducts are still found in many complex muffling devices, consequently proper modeling methods for these elements is important if wave motion in the whole intake and exhaust duct systems and the resulting radiated tail pipe noise are to be correctly simulated.

REFERENCES & BIBLIOGRAPHY

Amsden, A. A., Ramshaw, J. D., O'Rourke, P. J., Dukowicz, J. K., "KIVA, A Computer Program for Two and Three Dimensional Fluid Flows with Chemical Reactions and Fuel Sprays", Los Alamos National Laboratories Report # LA-10245MS, 1985.

Aoi, K., Nomura, K., Matsuzaka, H., "Optimization of Multi-Valve Four Cycle Engine Design - the Benefit of Five Valve Technology", SAE Paper # 860032, 1986.

Arcoumanis, C., "Internal Combustion Engines-Combustion Treatise", Academic Press, London, 1996.

Broome, D., "Induction Ram", Parts 1-3, Automotive Engineer, April issue, Pages 130-133, May issue, Pages 180-184, June issue, Pages 262-267, 1989.

Carapanayotis, A., Salcudean, M., "Mathematical Modeling of the Scavenging Process in a Two Stroke Diesel Engine", Basic Processes in Internal Combustion Engines, ICE Vol. 3, Pages 97-105, 1988.

Davis, G. C., Tabaczynski, R. J., "The Effect of Inlet Velocity Distribution and Magnitude on In-Cylinder Turbulence Intensity and Burn Rate - Model Versus Experiment", Basic Processes in Internal Combustion Engines, ICE Vol. 3, Pages 59-67, 1988.

Flamang, P., Sierens, R., "Study of Steady State Flow Pattern in a Multi Pulse Converter by LDA", Basic Processes in Internal Combustion Engines, ICE Vol. 3, Pages 67-78, 1988.

Gosman, A. D., Ahmed, A. M. Y., "Measurement and Multi-dimensional Prediction of Flow in an Axisymmetric Port/Valve Assembly", SAE Paper # 870592, 1987.

Hatamura, K., Morishita, K., Asanomi, K., Shiraishi, H., "Mazda's New V-6 Gasoline Engine and its Innovative Induction System", SAE Paper # 871977, 1987.

Isomura, S., Kurihara, Y., Yamashita, T., Koike, M., "Development of Variable Inertia-Charging System for 1.61 DOHC Engine", JSME Review, **8**, Pages 77-79, 1987.

Kanazawa, S., Akagi, M., Fukumachi, M., Suzuki, M., "Development of Elliptical Intake valve Piston Engine for Motor Cycles", Paper # SAE 930224, 1993.

Lovell, L., "Four Valve Heads", Car Design Technology, Pages 70-73, March, 1992.

Matsumoto, I., Ohata, A., "Variable Induction Systems to Improve Volumetric Efficiency at Low and/or Medium Engine Speeds", SAE Paper # 860100, 1986.

Munjal, M. L., Krishnan, S., Reddy, M. M., "Flow Acoustic Performance of Perforated Element Mufflers with Application to Design", Noise Control Engineering Journal, 40, 159, 1993.

Ohata, A., Saruhashi, H., Matsumoto, I., Imamura, Y., "Acoustic Control Induction System for Six-Cylinder Engines", JSAE Review, Pages 8-15, August, 1985.

Ohrenberger, G., Mann, M., "The Audi Five Valve Cylinder Head Concept", SAE Paper # 945004, 1994.

Payri, F., Torregrosa, A. J., Broatch, A., Brunel, J. P., "Pressure Loss Characterization of Perforated Ducts", SAE Paper # 980282, 1998.

Pearson, R. J., Winterbone, D. E., "A Rapid Wave Action Simulation Technique for Intake Manifold Design"" SAE Paper # 900676, 1990.

Robeson, G., "Cosworth-The Search for Power", Patrick Stephens Limited, London, 1991.

Sato, A., Suenaga, K., Noda, M., Maeda, Y., "Advanced Boost-up In Hino EP 100-II turbo-charged and Charge Cooled Diesel Engine", SAE Paper # 870298, 1987.

Sykes, R., "Development of a Five Valve per Cylinder Variant of the Ford Duratech 2.5 Liter V-6 Engine", *Transactions of I Mech E,* Autotech Conference, Paper # C498/22/174, 1995.

Taghavi, R., Dupont, A., "Multi-Dimensional Flow Simulation in an Inlet Port/Combustion Chamber Assembly Featuring a Moving Valve", Basic Processes in Internal Combustion Engines, ICE Vol. 6, 1989.

Wakisaka, T., Shimamoto, Y., Ishiki, Y., "Induction Swirl in a Multiple Intake Valve Engine-Three Dimensional Numerical Analysis", I Mech E Paper # C40/88, 1988.

Winterbone, D. E., Pearson, D. J., "Design Techniques for Engine Manifolds", Professional Engineering Publishing, London, 1999.

Woods, W. A., Brown, P. G., "Flow Area of Multiple Poppet Valves", *Proceedings Institution of Mechanical Engineers,* Part D, J. Auto Engrg., 10, Pages 347-351, 1996.

Yoshitomi, M., "Investigation of Wave Action Inside Intake Manifold of a Reciprocating Internal Combustion Engine", Thesis, UMIST, Manchester, UK, 1989.

Pressure Pulsations in Compressors and Pumps 11

11.1 INTRODUCTION

The mostly harmonic motion of the piston in a reciprocating pump or compressor results in the undesirable consequence of imparting fluctuating flow velocity and pressure of the fluid entering at the suction end and exiting at the discharge end. Pulsations, or pressure variations, are a major drawback of reciprocating pumps and compressors. Generation of the pressure pulse is a direct consequence of the oscillatory flow from the discharge valve of the machine. Operation of the machine at low pressure levels may not experience substantial variations, but in the case of high fluid pressures, the pulsating flow possess enough dynamic energy to destroy the piping system connecting the reciprocating machine to its destination point for transmitting the gas or liquid. The resulting loads from the pressure pulsations have the potential to cause high vibrations and dynamic stresses in pipes, valves, supports, accessories and machines. Suction and discharge valves can distort the pressure/volume card, reduce compressor capacity and efficiency and shorten valve life as a direct consequence of fluctuations in the pressure. Cavitation problems are also encountered in reciprocating pumps when substantial pressure fluctuations distort the working components.

High levels of vibration in the piping system are the most common manifestation of pressure pulsations. The connection between the acoustical excitation mechanism of reciprocating machines and the piping acoustical response is substantial. Coupling between acoustical and mechanical characteristics and piping mechanical response is also present. The geometric configuration of the complete piping system must be taken into consideration in order to determine deleterious effects of the pulsating pressure in the flow. Acoustic characteristics of the fluid handled by the pump or compressor are included in the evaluation for specific operating conditions. Dynamic characteristics in the form of natural frequencies, mode shapes and response to acoustic excitation is then calculated. Dynamic stresses may then be determined for evaluation of the system's components through appropriate fatigue criteria.

The most practical way of studying gas pulsations is to use an electro-acoustic analogical system where every part is individually adjustable or replaceable. The primary objective in the study is to avoid any resonance between the active systems (the compression cylinders) and the passive systems (the whole piping network); the secondary purpose is to reduce the amplitudes of the remaining pressure pulsations to negligible levels. Theoretically, the means available are: (1) change of diameter and of length of the gas piping, (2) removal of pipe connections or adjunction of additional piping, and (3) use of pulsation snubbers and of orifices at judiciously selected locations. In reality, the possibil-ities are restricted because of the high speed of sound in the gas (1000 to 2000 m/s), because of the very low compressibility of the gas, and because of the very high price of vessels and piping. However, in a large number of applications it has proved relatively easy to reduce a dangerous pulsation (for instance, from 25 percent down to 5 percent) by inexpensive and sim-ple means.

Designers dealing with compressors for very high pressures need to keep in mind at least three basic ideas: (1) safety, (2) large forces (how to apply them), and (3) accessibility. The last two are, of course, chiefly economic, but they are often combined with the aim of safety. For instance, in the design of large compression cylinders, the long through-bolts connecting the base with the cylinder head are an important safety factor. If, by chance, the gas decomposed in the cylinder, these long bolts, acting as springs, would be elastically length-ened by an appreciable amount without significantly increasing the stresses and would allow the gas to escape between the liner and the head. The bolts require equal pre-tensioning with a very high force.

The most severe form of pulsations in the discharge of a reciprocating pump is a consequence of the flow-variation induced pulsations in the pump itself. Gas dampeners work on the principle that they prevent the formation of pulses in the entire system, including the pump, by the simple yet elegant process of smoothening pressure spikes in the flow. Acoustic form of dampers, on the other hand, permit formation of the pulses, then partially filter them out on the downstream side. Use of chokes and tubes as acoustic elements (which are relatively large compared to the size of the pump), the components sometimes add resistance to the flow, and consequently add to the power load.

11.2 PROPAGATION OF WAVE IN FLUID

Consider the motion of a piston in a long pipe filled with a compressible fluid, as shown in Fig. 11.1. The piston, at rest at the end of the pipe in the upper diagram, travels toward the other end at velocity v, and in the process compresses the medium. Density ρ and pressure p are increased in the immediate vicinity of the pressure side of the piston above the normal undisturbed condition, as seen in the lower diagram at time t and piston displacement $x = vt$.

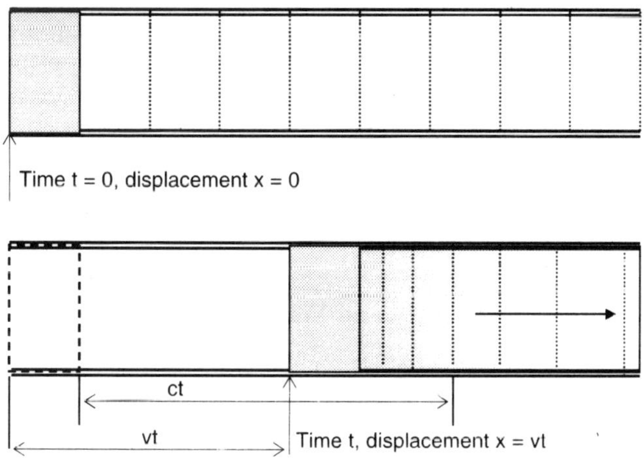

Fig. 11.1. Generation of Waves in a Long Pipe Filled with Fluid

The compressed fluid next to the piston moves forward while compressing the layers next to it. A pressure pulse travels down the conduit at the acoustic velocity, or speed of sound, c of the fluid medium, which defines the rate at which a pressure wave (or noise) propagates. If the piston were to be retracted, as occurs during the return stroke of the machine, the fluid will expand and cause the pressure and density to below the condition at rest. A rarefied pulse now travels down the tube at acoustic velocity c.

Since the fluid is initially at rest, there is no steady fluid velocity. But if some non-zero fluid velocity is present, the pressure pulse will travel at the acoustic velocity relative to the fluid velocity. Generally, the average fluid velocity in piping for reciprocating pumps and compressors is much lower than the pulse velocity c.

The frequency at which a pressure wave propagates down a pipe is primarily dependent on the mass and elastic properties of the gas or liquid itself if it is tacitly assumed that the pipe walls are rigid. Turbulence in the pipe flow may have a small effect on the velocity of wave propagation. Also, the effects of fluid flow friction in the pipe walls are insignificant. In the absence of flow in the radial direction the pulse as considered here travels essentially with a flat front. Such a plane wave in a single dimension provides a valid acoustic model for the frequency range of interest in the reciprocating equipment and its associated piping. This permits pressure and flow properties of the gas or liquid to vary along the length of the conduit, remaining constant along the cross sectional area.

11.3 FLUID FLOW DYNAMICS OF RECIPROCATING MACHINERY

The positive displacement pump generates very specific dynamic liquid flow and pressure conditions, but the response of the system to the pump dynamics is a result of the system operating conditions and specific piping arrangements.

There are three steady-state pressure conditions that affect system liquid dynamic responses: friction-generated pressure from the average liquid flow, vertical head from a column of liquid, and pressure applied by a gas blanket. There are also three sources of dynamic pressure disturbances generated by the positive displacement mechanism: frictional pressure drop, acceleration from the liquid flow variation of the pump, and low-amplitude water hammer-type pressure disturbances that occur each time a pump valve opens or closes.

Reciprocating equipment exhibit certain typical flow variations in the discharge (outlet) and suction (inlet) caused by the rotary motion of the power end driving the displacement elements. Equations for instantaneous velocity are:

$$V_P(\text{ft/s})_q = 0.42s\omega[\sin\theta + (s/4L_c)\sin 2\theta] \quad (11.1)$$

where: $\omega = 2\pi N/60$

$V_s(\text{ft/s})_\theta = \Sigma(V_P D_P^2/D_s^2)$

$V_d(\text{ft/s})_\theta = \Sigma(V_P D_P^2/D_d^2)$

The general Equations for Velocity and Velocity Pressure in pipes are;

$$V = 0.4085 \times Q/D^2 \quad (11.2)$$

$$P_v = 0.0001137 \times Q^2/D^4 \quad (11.3)$$

where A_p = are a of piston, in^2

A_d = area of discharge pipe, in

A_s = area of suction pipe,in^2

V_p = piston velocity, ft/s

V_s = suction (liquid) velocity ft/s

V_d = discharge (liquid) velocity, ft/s

Q = flow rate, gallons/min

Σ = summation (algebraic) of individual cylinder functions

The machine type is responsible for large differences in the shape of the flow pattern (Fig. 11.2). This shape depends on both the ratio of connecting rod length L_c to crank radius r and the loss of displacement on the piston rod end (crank end) of a double-acting pump.

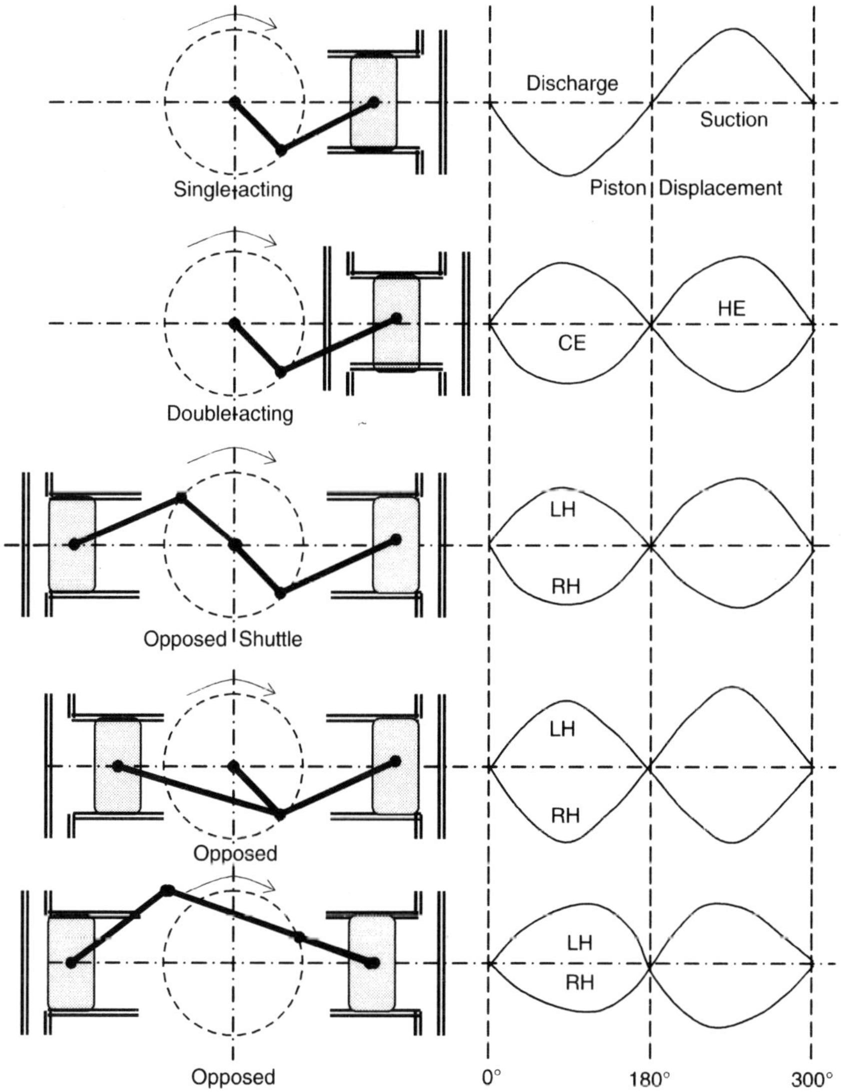

Fig. 11.2. Flow Pattern in Reciprocating Machinery (*CE* = Crank end, *HE* = Head end, *RH*, *LH* = right hand, left hand of opposed cylinder pump)

Conventional machinery design is such that the arrangement and number of cylinders generate different maximum and minimum flow rates. The flow rates are expressed as a percentage of the average as shown in Fig. 11.2, which illustrates one typical pump geometry with specific L_c/r ratio. For instance, a triplex single-acting pump has six points of maximum flow rate per revolution. These points are duplicated in both the discharge and the suction of a pump.

Because frictional pressure drop in a pipeline is a function of the square of the velocity, these flow variations will be converted into pressure variations (pulsation) following the same shape as the flow pattern.

In reciprocating equipment, again because of the typical harmonic motion, the fluid entering and leaving each cylinder experiences a start-and-stop flow with velocities ranging from zero at the beginning and end of a stroke to a maximum at mid-stroke. The velocity variation and its contribution to the instantaneous value of acceleration is determined by the following:

$$a(\text{ft/s}^2)_\theta = 0.042 s\omega\,[\cos\theta + (s/2L_c)\cos 2\theta] \tag{11.4}$$

The accelerations of the liquid at the suction and discharge connections are the result of the cylinder-generated acceleration and are calculated from the expressions:

$$a_s(\text{ft/s})_\theta = \Sigma(a_P D_P^2/D_s^2) \tag{11.5}$$

$$a_d(\text{ft/s})_\theta = \Sigma(a_P D_P^2/D_d^2) \tag{11.6}$$

The liquid has mass, and so is subjected to Newton's laws of motion, force = (mass × acceleration). Converting to pressure:

$$P_{as}(\text{lb/in}^2)_\theta = 0.0069 L\rho a_s \tag{11.7}$$

$$P_{ad}(\text{lb/in}^2)_\theta = 0.0069 L\rho a_d \tag{11.8}$$

11.4 PULSE CHARACTERISTICS

As the pressurized and rarefied pulse travels through a liquid or a gas, portions of the medium undergo a series of compression and expansion from the equilibrium state. When these compressions and expansions occur at a rapid rate, exchange of heat between portions of the gas and it's surroundings does not occur, so the process is mostly adiabatic. The acoustic velocity from the theory of waves is based upon changes in pressure and density under adiabatic conditions and is expressed by:

$$c^2 = \left.\frac{\partial p}{\partial \rho}\right|_s \tag{11.9}$$

For an ideal gas the acoustic velocity is approximated by:

$$c = \sqrt{kRT} = \sqrt{k\frac{p}{\rho}} \tag{11.10}$$

where: c = acoustic velocity

p = pressure

ρ = mass density

$k = c_p/c_v$ = ratio of specific heat

R = gas constant

T = absolute temperature

s = entropy

For incompressible liquids the acoustic velocity can be determined using the adiabatic bulk modulus K.

$$c = (K/\rho)^{1/2} \tag{11.11}$$

where K = adiabatic bulk modulus. Physical properties of some commonly used liquids are shown in table 11.1.

Flexibility of pipe walls made from steel produce an effect that is significant in the case of liquid systems, but is insignificant in gas systems. This is because the bulk modulus of liquids is considerably higher than for liquids. Pipe walls behave in a spring like manner in the frequency range in which one-dimensional pulsation theory applies, and at the same time its inertia effects are negligible. This wall compliance reduces the effective acoustic velocity by a value that depends on the ratio of the fluid and pipe wall bulk modulii. The classical Korteweg correction to the acoustic velocity can be used for thin wall pipes (thickness < 10% of diameter) to account for the pipe wall effect:

$$c' = \frac{c}{\sqrt{1+\dfrac{Kd}{Et}}} \tag{11.12}$$

where d = pipe diameter, inch

t = pipe wall thickness, inch

K = adiabatic bulk modulus, lb/in^2

E = pipe material elastic modulus, lb/in^2

c' = effective speed of sound

If the time varying flow velocity and pressure functions repeat after a fixed period of time, then the flow as a combination of steady and oscillatory flow. The time interval at which the two functions repeat themselves is the period of oscillation. Referring to Fig. 11.3, the reciprocating motion of the piston has constant amplitude at angular frequency w in the combination of steady and oscillatory manner. The motion of the piston is described by the expression:

$$x = X \, . \sin(\omega t) \tag{11.13}$$

A continuous train of pulsating pressure waves will travel along the infinitely long tube. The wavelength λ of the pressure wave is defined by the acoustic velocity c and frequency f:

$$\lambda = c/f \tag{11.14}$$

where c = speed of sound, ft/sec, and $f = \omega/2\pi$ cycles/sec. In the absence of reflections of waves from the other end of the pipe, as occurs if the pipe is very long, Newton's law may be used to obtain the amplitude of the pressure wave:

$$P = \rho c x \omega \sin(\omega t) = \rho c u \sin(\omega t) \tag{11.15}$$

Table 11.1. Properties of Common Liquids

	Liquid Density lb/ft^3	*Bulk Modulus* $10^3\ lb/in^2$	*Acoustic Velocity* ft/sec
	Atmospheric Conditions – Isentropic Bulk Modulus		
Fresh Water	62.3	318	4870
Sea Water	63.9	344	4990
Benzene	54.8	222	4320
Methanol	49.3	144	3680
Ethanol	49.3	155	3810
Turpentine	54.2	223	4360
	At 68° F, 200 lb/in^2 – Isothermal Bulk Modulus		
Propane	30.8	25	2300
Isobutane	35.0	41	2700
N-butane	35.8	53	3000

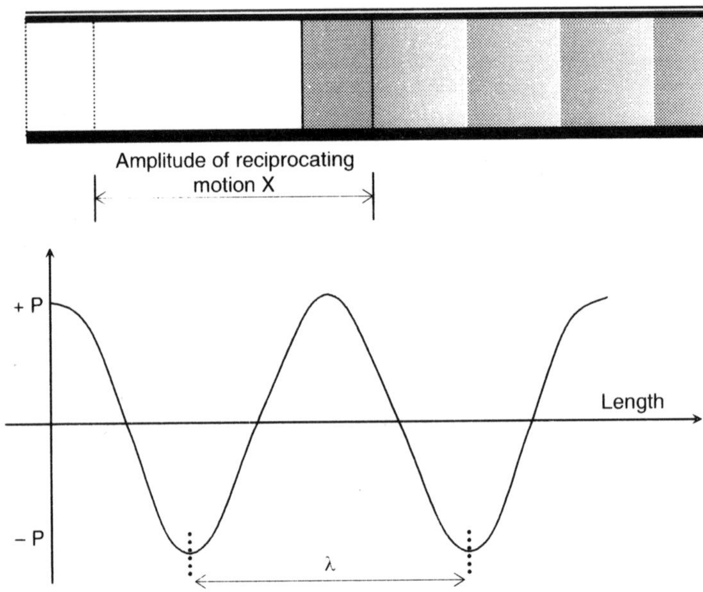

Fig. 11.3. Traveling Pressure Wave in a Long Pipe due to Reciprocating Piston

where ρ = mass density, lb-sec^2/in^4

c = acoustic velocity, in/sec

x = piston displacement amplitude, in

ω = piston circular frequency, rad/sec

$u = dx/dt = x\omega$, piston velocity amplitude, in/sec.

Besides being a function of density and speed of sound, the pressure wave is proportional to the velocity of the piston, u, which explains the primary source of pulsations in reciprocating compressors and pumps. As the piston executes an oscillatory motion, the velocity varies from a positive maximum value to a minimum negative value. The fluctuating pressure values, imposed on the steady state pressure at rest, also correspondingly change from a minimum value to a maximum value.

11.5 ACOUSTIC RESONANCE AND MODE SHAPES

Reflections of sound in any fluid medium are the same as an echo. If a source placed between two large parallel walls emits noise, a series of echoes are generated, each one of diminishing volume. The original sound is caused by a sudden pressure pulse of compressed air created by the emitting source. The pressure wave moves at the speed of sound and reflects from the wall's surface, and occurs due to a change in acoustic impedance. The characteristic acoustic impedance of a one-dimensional pipe at a particular cross section in the pipe is defined by the expression:

$$Z = \rho c/A \tag{11.16}$$

where r = mass density, lb-sec^2/in^4

c = speed of sound, in/sec

A = pipe area, in^2.

In a uniform pipe the impedance value is constant, and the pressure wave travels indefinitely in a controlled manner along its length. When the wave comes up against a discontinuity in the pipe's cross section, the impedance value changes and a reflection of the wave takes place. Common examples of such discontinuities where the impedance value shifts are: closed end in a pipe loop, open end of a pipe segment, change in cross section area of the pipe, branch in pipe circuit, pipe tees, piping instrumentation, such as an orifice, valve or a flow metering device and shift in fluid density or acoustic velocity. A change in the impedance value reinforces the acoustic wave, which may result in resonance. Discontinuity in the pipe causes the wave to be reflected. At open and closed ends of the pipe a complete reflection of the wave takes place.

While a full reflection takes place at zero or infinite impedances in a pipe, intermediate changes in the pipe cross-section will produce an intermediate effect in the form of a partial reflection of the pulse wave. The ratio of reflected and incident pressures P_r/P_i when the cross sectional area increases from A_1 to A_2 is given by the expression:

$$\frac{P_r}{P_i} = \frac{A_1 - A_2}{A_1 + A_2} = \frac{A_1/A_2 - 1}{A_1/A_2 + 1} \tag{11.17}$$

The ratio of transmitted and incident pressures P_t/P_i when the when pipe area is reduced from A_1 to A_2 is given by:

$$\frac{P_t}{P_i} = \frac{2}{A_2/A_2 + 1} \tag{11.18}$$

The equations show the same phase reversal characteristics as discussed for complete reflections of the acoustic wave pulse. When $A_1 > A_2$, pressure ratio P_t/P_i is positive, indicating a compression wave pulse. This is accompanied by an increase in the transmitted pressure level compared with the incident pressure. But when pipe area $A_1 < A_2$, a phase reversal occurs at the change in pipe cross section, and the pressure ratio P_r/P_i is negative, resulting in a rarefied wave pulse. Also, the sudden enlargement of the pipe will lead to an attenuation of the transmitted pressure.

When two waves of the same frequency and in certain phase relationship travel in opposite directions, the result will be a classical standing wave. One of the two waves may be considered as the incident wave and the other as a reflected wave. Superposition of the two waves produces fixed locations for the maximum and minimum deflections, producing a standing wave.

Resonance associated with the distributed system of a constant pipe diameter has a sinusoidal mode shape. Length resonance of some pipe elements is described in terms of a full wave length. Generally, multiple wavelength resonances decrease in severity with the higher multiples, since more energy is required to drive the higher frequency modes.

The full wavelength resonance of a closed-closed or open-open pipe of length L has resonant frequency $f = c/L$. Thus, pipe lengths equal to multiples of wave lengths will be in resonance. Pipe lengths of certain fractions of a wavelength, such as one-half and one-quarter can also be in resonance. A quarter wave length resonance will occur in a pipe with one open end and one closed end. The open end will have a condition of minimum pulsation amplitude (zero) and maximum flow velocity, while the close end will have a condition of maximum pressure amplitude and zero flow velocity. Since $f = c/\lambda$, if $L = \lambda/4$, then $\lambda = 4L$ and $f = c/(4L)$, giving the frequency of a quarter wave length resonance. Resonant frequencies occur at odd multiples (for closed-closed or open-open pipes):

$$f = nc/4L, n = 1, 3, 5, \ldots \tag{11.19}$$

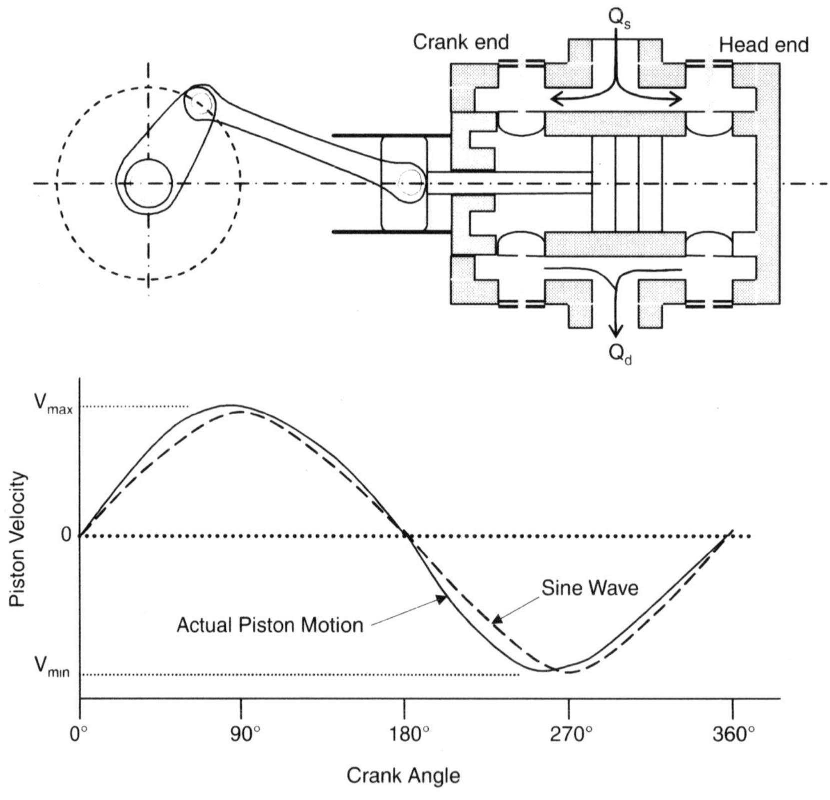

Fig. 11.4. Piston Motion and Velocity for Single-Cylinder Reciprocating Pump or Compressor

In an open-closed pipe the first modes occur in the form of quarter-wave stubs. The stub has its resonant frequencies at odd multiples of the fundamental quarter-wave frequency. A half wave resonance can occur in a pipe having the same conditions at both ends, either opened or closed. In the equation for frequency $f = c/\lambda$, the fundamental resonance mode is determined by substituting $L = \lambda/2$, hence $f = c/2L$. Resonance will occur at integer multiples of the fundamental mode (for open-closed pipes):

$$f_n = nc/2L, \; n = 1, 2, 3,... \tag{11.20}$$

11.6 RECIPROCATING COMPRESSOR AND PUMP SYSTEMS

Many forms of disturbances are observed in the pressure characteristics generated by reciprocating pumps and compressors. The flow modulations come about as a result of intermittent flow through the suction and discharge valves. Geometry of the finite length of the connecting rod is also responsible because of the associated inertia effects.

Figure 11.4 shows a schematic diagram of a single-cylinder compressor. The suction flow, Q_s, enters the cylinder, and the discharge flow, Q_d, exits the cylinder. The motion of the piston may be approximately simulated by a sinusoidal wave in shape, the deviation arising from the limited length of the connecting rod. The acoustic excitation of the compressor is proportional to the flow (Q_s or Q_d) modulation. Since the flow is based on the product of the piston velocity and the piston swept area (Q = Area × Velocity), the profile of the discharge flow at the piston face is also similar.

Also, the suction and discharge valves of one end (head end) of a compressor are never open simultaneously; hence the suction and discharge piping systems are isolated acoustically.

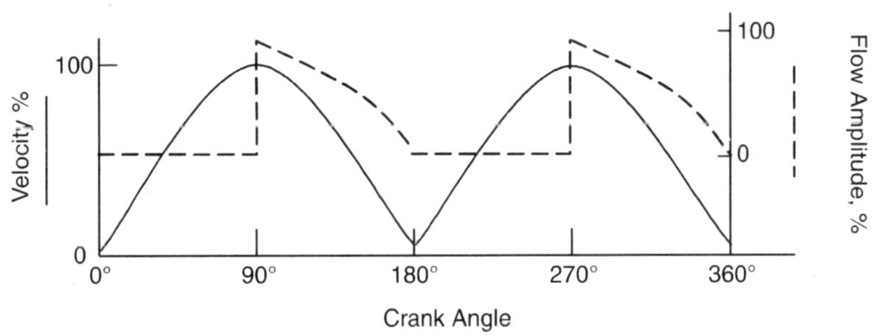

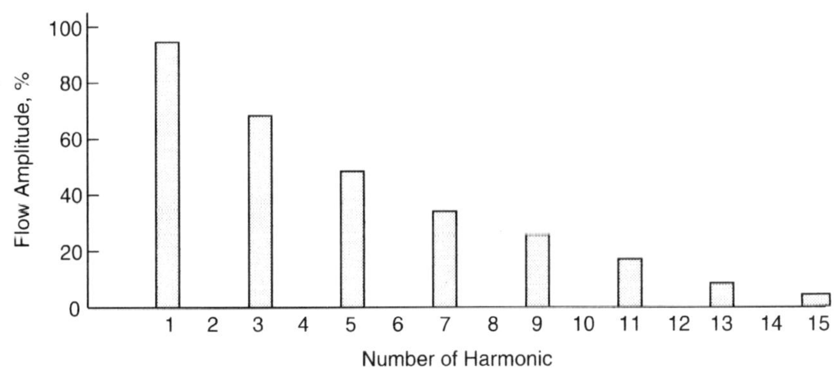

Fig. 11.5. Ideal and Actual Flow versus Time Wave (*upper*) and Flow Frequency Spectrum (*lower*) For Symmetrical Double Acting Compressor with Ideal Valves

Figures 11.5 show the effect of valve action on flow through the discharge valves of a compressor. During the compression stroke the suction and discharge valves are closed. As the pressure in the cylinder increases to the discharge back-pressure, the discharge valve opens, and the flow amplitude time wave through the valve has the shape of a portion of the velocity wave shown by the dashed curve in the figure. As the cylinder reaches top dead center (TDC), the discharge valve closes, and the flow stops. For a 'perfect' double acting cylinder (equal head end and crank end flows, $L/R = \infty$, inertia and stiffness loads absent) the flow versus time contains two flow portions 180º apart in time. The odd harmonics (in the idealized case) cancel, so that the cylinder flow excitations occur at even harmonics of running speed (2 ×, 4 ×, ...) as shown in the figure.

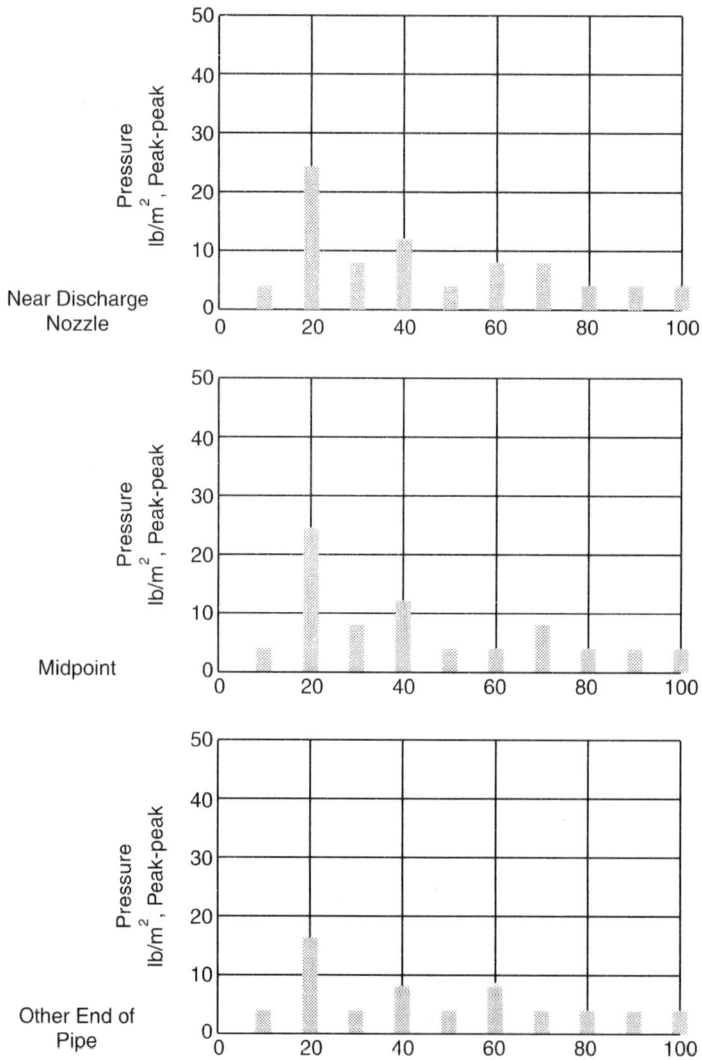

Fig. 11.6. Pulsation Spectra due to Cylinder Discharge Excitation into Long Pipe

All reciprocating machinery have cylinders, piston rods, differences in clearance volumes and finite length connecting rods, so the flow volumes delivered during the forward and backward strokes are not identical. Thus, even in the double acting machine, the cylinder will produce flow excitation at all harmonics of running speed. The cylinder excitation constitutes a flow modulation mechanism consisting of many frequencies, which when injected into the piping, creates response at these various frequencies. Figure 11.6 shows typical pulsation response spectra in a long pipe very close to the discharge nozzle, midpoint and at the other end of the pipe that is excited by a compressor's discharge flow. The pulsation levels are closely the same at all three test points. Note that the relative amplitudes between the various harmonics are the same for both the excitation and the pulsation response.

11.7 COUPLING BETWEEN ACOUSTICAL AND MECHANICAL CHARACTERISTICS

Fluctuating pressures affect many aspects of reciprocating compressor operation. These effects include acoustical and mechanical coupling, changes in performance, cavitation (liquid systems), and valve damage. Rather than the pulsation itself, a point of acoustical and mechanical coupling is present in the system that develops dynamic forces, which in turn produces the vibra-tion. Several geometric discontinuities such as elbows, bends, reducers, change in area, bottles and capped ends are some common force-coupling points in industrial piping systems.

Elbows or bends have differential areas due to outside and inside radii differences, which produce forces in the plane of the elbow. As the angle of the elbow becomes larger, the force at elbows becomes smaller in magnitude. Therefore, the number of elbows in a piping system should be minimized to reduce the number of force coupling points. 45° elbows are preferable to 90° elbows. Where elbows are required, dynamic restraints (clamps) are desirable near the elbows in order to absorb dynamic forces at these locations.

The equation for force acting on the differential area occurring at a reducer is:

$$F = P_{Dynamic} \times A = P_{Dynamic} \times \pi \times (D_2^2 - D_1^2)/4$$

where D_2 and D_1 are the diameters of the reducer at the bigger and smaller ends and $P_{Dynamic}$ is the peak-to-peak fluid pressure acting on the reducing wall. Force, F, is along the axis of the reducer. Since the instantaneous pressure at the reducer is a function of time, the force F varies with time.

An extreme case of area change is that of a capped end. This condition occurs in all vessels, and is particularly important in surge volumes. The net axial force acting on a bottle or surge volume is determined by summing the forces acting on each end of a bottle. For the fundamental closed-closed resonance mode, the force acting on the right end of the bottle of area A is given by the expression:

$$F_{x1} = (P_{static} + P_{dynamic})A \tag{11.21}$$

and on the left end:

$$F_{x2} = -(P_{static} - P_{dynamic})A \tag{11.22}$$

The total force is:

$$F_R = F_{x1} + F_{x2} = 2AP_{dynamic} \tag{11.23}$$

When excitation frequencies occur near the fundamental acoustic resonance mode of a bottle, high shaking force levels may be expected since the dynamic pressures are amplified and the instantaneous pressures at the ends are 180° out of phase.

A similar condition occurs when a volume-choke-volume system is excited at resonance. The dynamic pressures in the two chambers are 180° apart in phase. Hence, the force acting along the axis and proportional to the area of the choke is produced. If the diameter of two equal volumes is D and area $A = \pi D^2/4$ and diameter of the choke connecting the volumes is d and area $a = \pi d^2/4$, and the pressures in the volumes is P_1 and P_2 then the net force $F = -P_1 a + P_2 a = a(P_2 - P_1)$. Pressures P_1 and P_2 are equal in magnitude and out of phase by 180°. Hence, $P_2 = -P_1$ and $F = 2P_1 a$. The frequency of this resonance may be controlled by adjusting the bottle and choke sizes (diameters and lengths). This concept is a fundamental tool for control of bottle shaking force.

11.8 COMPRESSOR PERFORMANCE AND OVERLOAD

Pressure variations in the connecting pipe work arising from the time varying and repeatable flow in reciprocating compressors affect the volume discharge capacity of the unit. Pulsations, like heat transfer, can either increase or decrease capacity. The characteristics are:

1. Higher pressure in the suction passage at maximum cylinder volume increases capacity.
2. Higher pressure in the discharge passage at minimum cylinder volume decreases capacity.
3. Lower pressure in the suction passage at maximum cylinder volume decreases capacity.
4. Lower pressure in the discharge passage at minimum cylinder volume increases capacity.

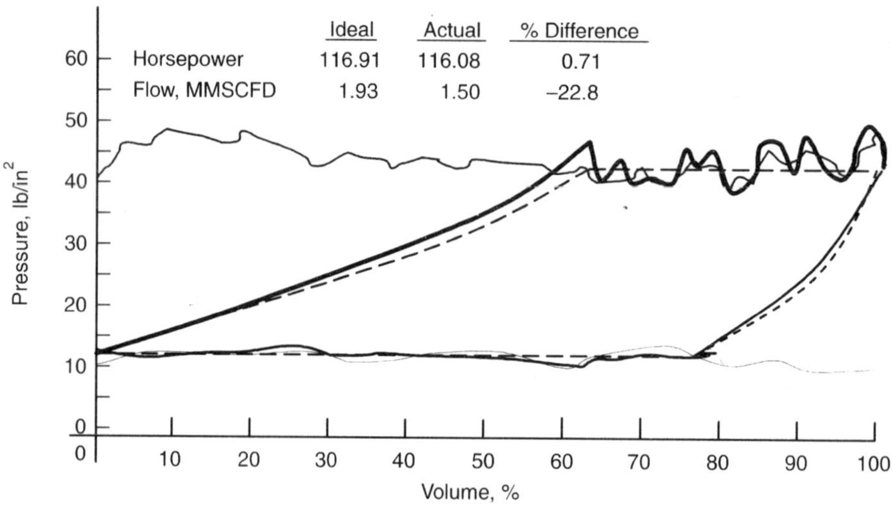

Fig. 11.7. Effect of Pulsation on Pressure-Volume Diagram and Performance

Figure 11.7 shows the effect of high pulsation due to an acoustic resonance in the discharge piping of a cylinder. The PV diagram is significantly distorted from the ideal (dotted) card by the high pulsation levels. The actual horsepower of the card is decreased over that of the ideal, but the flow (MMSCFD) is considerably decreased.

This distortion of the PV card also increases pressure forces acting on the piston rod. If the rod design does not have a sufficient safety factor to handle such an overload, rod failure can occur. In addition, increased pressure level can increase torsional excitation level at individual crank throws, possibly resulting in crankshaft failure if a torsional resonance is excited.

The performance of reciprocating compressors can be significantly affected by piping acoustics and compressor interaction. Compressor horsepower can be greater or less than the ideal horsepower, and capacities can be increased or decreased. The horsepower required to drive a compressor can even be increased sufficiently by pulsation effects such that the driver is incapable of supplying sufficient power to drive the compressor.

11.9 CAVITATION IN LIQUID PUMP SYSTEMS

Cavitation occurs in liquids when the local static pressure falls below, or attempts to fall below, the liquid vapor pressure as seen in figure 11.8. The extent of cavitation depends upon many factors, including nuclei concentration. The nuclei serve as seed for formation of the cavitation bubbles. Abundant nuclei are usually available in the form of dissolved gases, liquid impurities, and surface imperfections.

Cavitation in the liquid end of a reciprocating pump is described as the rapid conversion of the liquid to a vapor, followed by a sudden collapse of the vapor bubble to the liquid phase. It is this sudden collapse that generates microscopic but intense blasts of high-velocity liquid. If cavitation occurs near a metal surface, damage to that surface can be expected, usually in the form of pits or honeycombed surfaces in local areas. These areas are usually on the upper surface of horizontal bores (above the plunger travel) in the liquid end. If pitting occurs at the intersection of bores, the stress risers created can induce early corrosion-fatigue failure. Plungers will sometimes show cavitation erosion on the packing area or on the end of the plunger exposed to the liquid. A ring of pits around a plunger is a sure sign of cavitation. Such conditions result in rapid packing wear.

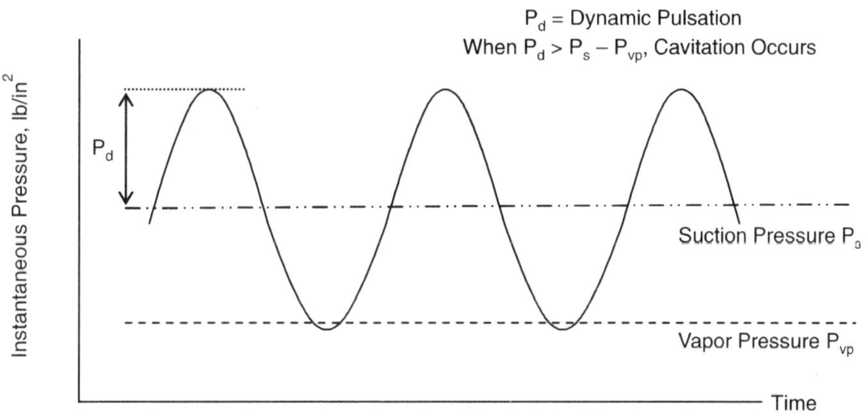

Fig. 11.8. Cavitation Caused by Pulsation

Pressure deficiency at the pump inlet can be the result of a poor suction system liquid that can be described by 'long' or 'small diameter' or 'sharp bends' or 'low head'. Pressure reduction can also be generated by the acceleration of the liquid at the beginning of a plunger or piston suction stroke. In that case, increasing the head on the supply tank or source alone may not cure the problem. The installation of a properly designed suction stabilizer will usually be the only remedy short of a newly designed suction system. It is also believed that air or gas dissolved in the liquid can cause cavitation effects in the sense that some gas can break out with the same results as vapor formation.

Cavitation is usually manifested by liquid knocking or pipe vibration. Examination of the instantaneous pressure at the pump inlet measured with sensitive electronic equipment and displayed on an oscilloscope screen, will show the typical rounded-bottom, sharp-upward-spikes type of waveform instead of the usual symmetrical shape. Figure 11.9 (*upper*) shows a typical waveform for cavitation in a pump's suction. In this case, a great improvement was achieved, as shown in the lower diagram, by the use of a suction stabilizer. Suction pressure is 63 lb/in^2 peak-to-peak, and the pump is *not* equipped with suction stabilizer. Note the typical 'cavitation' waveform with sharp upward spikes and rounded bottoms, accompanied by extreme pressure excursions in an upward or positive direction, while the bottom of the trace is prevented from extending into the negative region by the formation of vapor. In the lower trace the suction pressure is 13 lb/in^2 peak-to-peak. The pump now is provided with the suction stabilizer, and is using the same operating conditions. Note the 'clean' sine wave and high frequency typical of 'good' suction. The low-frequency cycles, over which the high fre-quency is imposed, are the remnant of pump rotation-generated cycles. While this flow-induced pulsation is not serious, it could have been reduced by the use of a properly gas-charged bladder in the stabilizer.

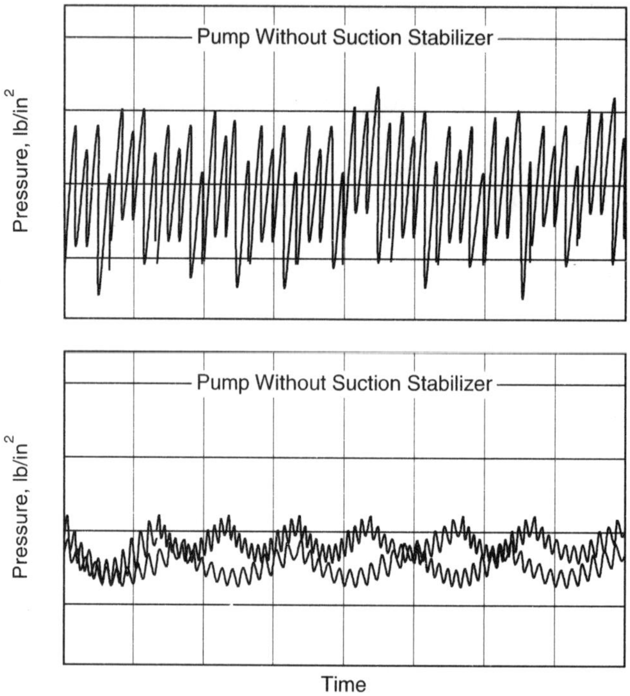

Fig. 11.9. Suction Pressure Trace of Typical Cavitation Waveform

Cavitation can also be caused by the dynamic effect of valves. Valve parameters such as lift, valve mass, spring rate, preload, valve lip area, flow areas, etc., can greatly influence local cavitation at valves and plungers. Valve effects can be simulated with a computer program, which includes these variables and also considers the system acoustics.

Cavitation in reciprocating pump installations can cause many mechanical problems, ranging from high amplitude vibration of the piping and pump to failures of all major components of the machine. Cavitation can occur even in those systems that have ample net positive suction head (NPSH). According to guidelines specified by the Hydraulic Institute Standards, in addition to the net positive suction head required (NPSHR) by the manufacturer, an allowance should be made for inlet piping pressure drop and acceleration head. Acceleration head calculations are an attempt to account for the dynamic behavior of the system using quasi-static assumptions. In practice, these calculations may prove to be inadequate, since they ignore the dynamic acoustical response characteristics of the fluid. Accurate calculations of the pulsation levels in the pump systems must consider the dynamic flow by taking into account all parameters influencing the system substantially, including the pump's fluid end, pump valves, and associated suction piping.

11.10 PULSATION CONTROL IN COMPRESSORS

For reciprocating compressors, API 618 defines the minimum suction and discharge surge volumes required for pulsation control. The formula for calculation of these minimum volumes is contained in paragraph 3.9.2.2.2 of API 618 (1986). Pulsation control in compressor piping systems can be accomplished by application of the basic acoustic elements of compliance, inductance, and resistance. The components may be combined in various configurations in order to achieve pulsation control ranging from attenuation of pulsations to true acoustic filtering. The theory for attenuation and filtering in acoustic fluid systems is similar to that for electrical and mechanical systems, and direct analogies can be made between mechanical, electrical, and acoustical systems. The pulsation suppression devices discussed here range from empty bottles that act as surge volumes to acoustic filters in the form of bottles with internals.

Surge volumes in the form of relatively large empty bottles attached to the suction or the discharge of the compressor are often used to attenuate the pulsations produced by compressors. The volume acts as an acoustic compliance (equivalent to spring flexibility in a mechanical system), isolating the piping fluid from flow modulations induced by the compressor. The attenuation characteristics of the bottle depend on its volume enclosed and the expansion ratio of the attached pipe and bottle diameters.

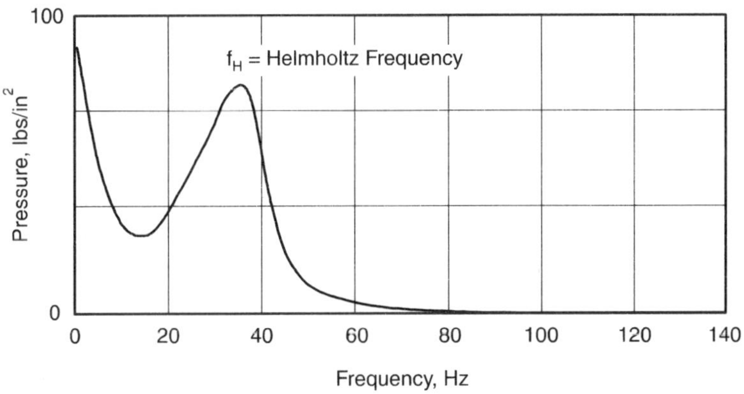

Fig. 11.10. Typical Frequency Response to Flow Excitation of Volume-Choke-Volume Filter

Pulsation amplitudes may be diminished by the surge volume, but may not be adequate for attenuating all pulsation frequencies. Further reduction of amplitudes at higher frequencies may be necessary, especially when chances of mechanical natural frequencies of the piping coinciding with frequencies associated with acoustically induced shaking forces are high. In this respect the volume-choke-volume filter is a more effective pulsation control device. Consisting of two volumes connected by a relatively small diameter pipe or choke tube, the volumes of the two chambers form acoustic compliances, while the mass of fluid in the choke tube becomes an acoustic inductance. The combination produces a low-pass filter for attenuating pulsation frequencies above a specific cutoff limit.

Figure 11.10 illustrates the dynamic pressure transmission characteristics of the volume-choke-volume filter as a function of frequency. The resonant peak at frequency, f_H, is called as the Helmholtz frequency of the two-chambered filter, and will amplify pulsations at that frequency. Filters may also have pass-band for amplifying certain frequencies. The pass-bands are related to the lengths of certain elements such as choke tube length, chamber length and inlet nozzle length.

Consider a volume-choke-volume design where the choke length is L_C, area is A_C and diameter is d_C. The choke connects two volumes of capacity V_1 for the primary bottle and V_2 for the secondary bottle. The equation for the ideal Helmholtz frequency of the filter is:

$$f = \frac{c}{2\pi}\sqrt{\frac{A_c}{L_c'}\left(\frac{1}{V_1}+\frac{1}{V_2}\right)} \tag{11.24}$$

where f = frequency (Hz)

c = speed of sound (ft/sec)

A_c = area of choke tube (ft^2)

L_c = length of choke tube (ft)

$L_c' = L_c + 0.6\, d_c$ (ft)

d_c = choke diameter (ft)

V_1 = volume of primary bottle (ft^3)

V_2 = volume of secondary bottle (ft^3)

Although this symmetrical design is rarely achieved due to layout considerations, the equation describes the relative effect of the bottle and choke diameters. Usual design procedures for low pass filters call for the Helmholtz frequency to be below the lowest pulsation frequency to be attenuated. Effective pulsation reduction can be achieved for frequencies above approximately twice the Helmholtz frequency.

Response to acoustic resonance at the flange of the nozzles connecting a bottle to a reciprocating compressor cylinder generally has a strong response, because the pulsating flow from the cylinder flows directly into the nozzle with no attenuation. The nozzle and the cylinder's internal flow passages, create a resonance similar to a quarter wave, with the region acting as an open end and the surface of the piston, when the valves are open, acting as the closed end. An orifice plate creates a pressure drop and serves as an acoustic resistance, and may be provided in the nozzle to help dampen this response. For the most effective damping of the resonance associated with the nozzle, the orifice plate should be located close to the bottle-end of the nozzle, and may also be placed at the cylinder flange for better control. API 618 sets limits on the allowable pressure drop through an acoustic filter, including orifice plates, in paragraph 3.9.2.2.4.

The type of pulsation suppression device required depends on the degree of pulsation control desired. In certain cases, a surge volume may provide adequate attenuation of pulsations. However, for maximum attenuation over a wider frequency range, an acoustic filter should be designed. For example, consider a volume-choke volume filter consisting of two symmetrical volumes of length L and diameter D and connected by a tube of diameter d and length L. The Helmholtz frequency for the system is:

$$f = (c \times d)/(\sqrt{2} \times \pi \times L \times D) \tag{11.25}$$

where f = frequency (Hz)

c = speed of sound (ft/sec)

11.11 PULSATION CONTROL WITH GAS-TYPE DAMPENERS IN PUMPS

The theory of pulsation control in liquid systems is identical to that in gas compressor systems. Hence, surge volumes, acoustic filters, and orifice plates are also effective pulsation control devices for reciprocating pumps. The pumps can experience many different types of pressure disturbances, and the method for alleviating the problem must take into consideration the nature and form of the problems.

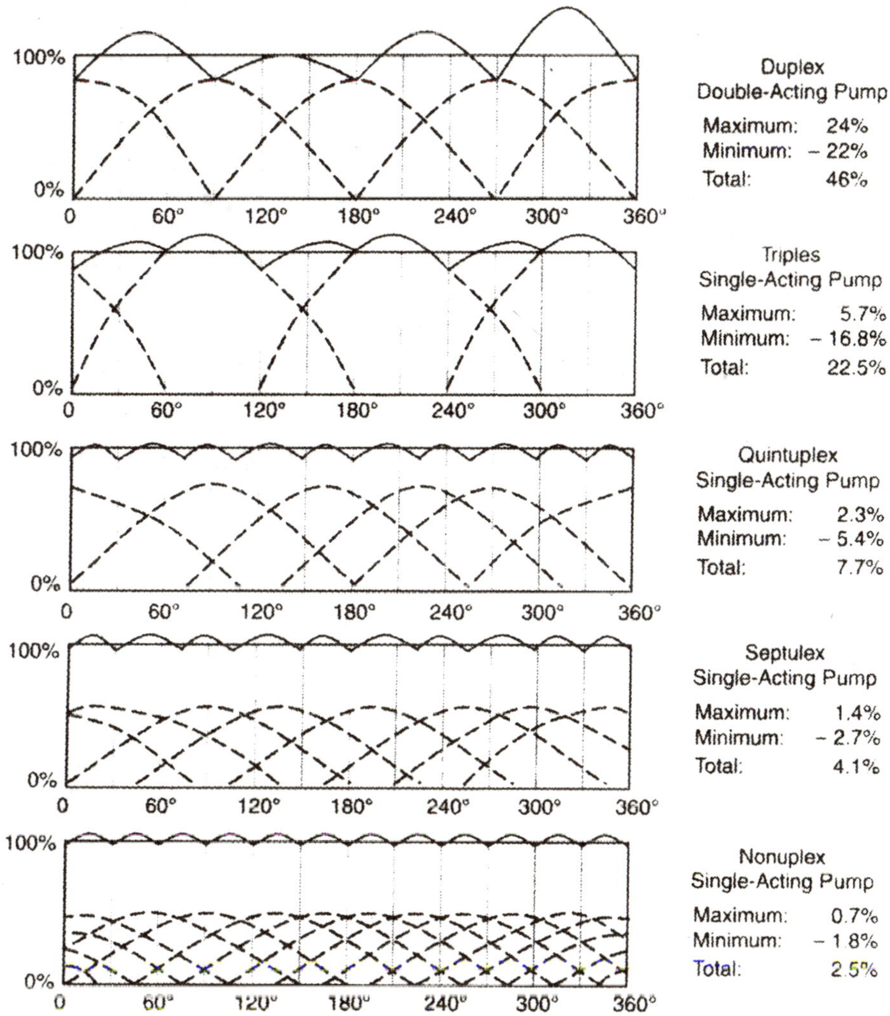

Fig. 11.11. Discharge Velocity Flow Variation in Multi-Cylinder Pumps

Some pulsation control devices operate on the principle of absorbing the oscillatory energy present in the pump flow, with the dampening effect obtained from a compressible gas, liquid or a combination of the two. The acoustic principle is used in some designs in order to filter out certain frequencies in the pulsating flow. Similar to the acoustic method, but requiring a larger quantity of liquid, is the reactive principle of design, where the intermediate choke element is deleted. A confined or an unconfined bladder made from some flexible membrane form of material may be used in some designs. Bladders made of cellular foam have also experienced good operating success. Some dampeners do not use a bladder. The various devices offer the flexibility of being attached as a Tee on a pipe, in series with the line to obtain total passing of the flow through the dampening mechanism, or the flow may be diverted to the bladder while maintaining the flow-through effect.

The discharge-velocity (DV) type of pressure disturbance is caused by restrictions encountered in a long discharge pipeline during varying-flow operating condition. The frequency of the disturbances is related to the pump's operating speed and the number of active cylinders, as shown in Fig. 11.11. These inherent and predictable pulsations are usually the most commonly observed when the disturbing frequencies coincide with the natural vibration frequencies of the attached pipe spans. The pulsations can be almost eliminated by installing gas filled and energy absorbing dampeners shown in Fig. 11.12. The gas used is usually Nitrogen. The design shown on the left (type Gas Confined Appendage Energy GCAE) uses gas compression, has a confined bladder, is attached as an appendage in the form of a Tee to the pipe, and works on the energy absorbing effects arising from compressibility of the gas. A bladder anti-extrusion button allows nearly unlimited high pre-pressure charge to extend effectiveness range, but excludes protection below the pre-charge pressure. The design shown on the right (type Gas Unconfined Flow through Attached GUFA) uses gas compression, has an unconfined bladder, is attached in series between the pump and the attached pipe to obtain total flow through the device and works on the acoustic principle to filter out certain known frequencies. In conjunction with the use of the device, it may also be necessary to modify placement for the pipe supports.

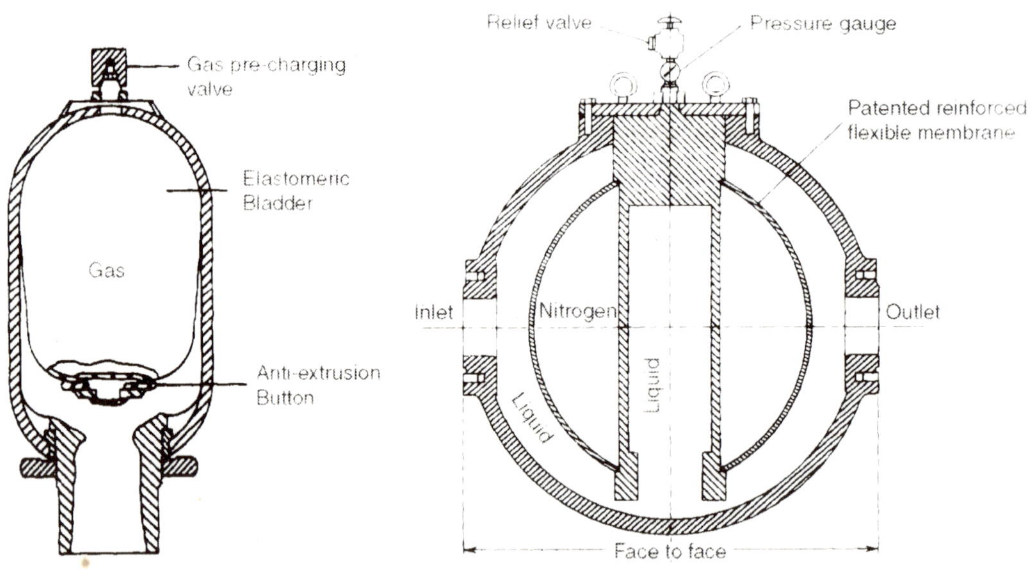

Fig. 11.12. Gas Dampening Pulsation Control Device for Pump Discharge-Energy Absorbing GCAE Type (*left*), GUFA Type (*right*) (*Courtesy*: Greer Inc., White Rock Engineering)

Pulsation problems at the suction nozzle usually occur when the attached pipeline is exceptionally long or has a small diameter. Serious disrupting effects are encountered when flow acceleration is large enough to overwhelm the effects from the velocity, leading to a surging effect at the suction valve opening. The resulting pulsations appear near the start of each plunger suction stroke as rapidly diminishing sinusoidal pressure waves of a frequency unrelated to the pump's operating speed. Poor pump performance is experienced due to the negative excursions in fluid pressure dipping into the vapor pressure zone, thus resulting in cavitation. The smaller gas type of dampener is generally used to alleviate the problem. A suction stabilizer using dual gas and liquid is shown in Fig. 11.13, and is useful for relieving the pump suction of fluid acceleration effects.

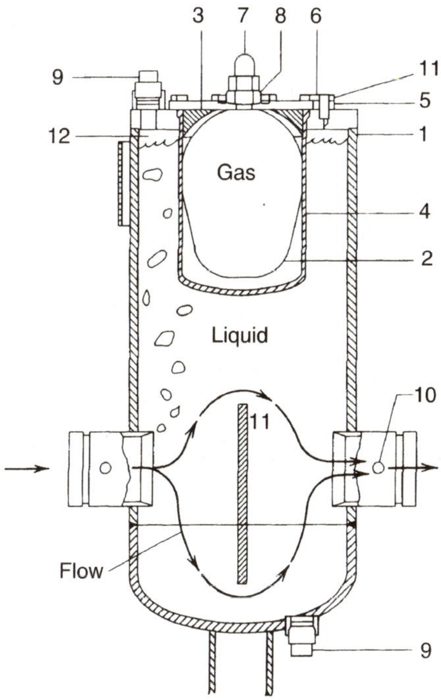

1. Gasing
2. Bladder
3. Retainer
4. Bladder housing
5. Cover plate
6. Housing restraint
7. Pressure gauge
8. Pressure holder
9. Charge and drain plugs
10. Pressure tap
11. Housing pin
12. Fluid level

Fig. 11.13. Suction Stabilizer (*Courtesy*: Greer Inc.)

Gas filled bladder type of dampeners are extensively used for pumping fluids across long pipelines for services such as oil field water injection, drilling-mud circulation and in process systems with high-pressure letdown. Any practical degree of fluid flow induced discharge pulsation attenuation can be obtained by selecting proper dampener sizing and gas pre-charge pressure. High-frequency acceleration generated pulses cannot be attenuated by employing gas bladder dampeners since reaction time is not quick enough to absorb the surge.

Reciprocating pumps employed for transmission of liquids over long distances may require the use of two or more discharge dam-peners (with a total gas volume as required for one dampener) for the following reasons:

1. Because pipeline start-up pressure rise is brought about over a relatively long period (several minutes to an hour or more), there is a period during which the pump pressure is less than the optimum pre-charge pressure of about 60% of the discharge pressure for one dampener. During that time, the pumps would operate without pulsation control. With two dampeners (the second of which is pre-charged to about half of the first), for example, pulsation control is extended well into the critical start-up period.

2. In some cases, it is necessary to pump alternate batches of liquids with widely differing viscosities. It is obvious that the pump pressure required to displace the less viscous liquid is less than that for the more viscous one-sometimes less than the optimum pre-charge pressure for one dampener at maximum pipeline pressure. A second dampener pre-charged to a lower pressure would give protection for a wide range of pipeline pressures.

3. The use of multiple discharge dampeners provides some redundancy and protection in case of the loss of pre-charge or bladder failure in one of the other dampeners.

Combining the features of both types are often effective. The suction stabilizer shown in Figure 11.13, the PASAFE (Pump Acceleration Stabilizer And Flow Equalizer) dampener using closed cell elastomer in Figure 11.14 and the PULSETONE gas-type dampener shown in Figure 11.12 (*right*) are examples.

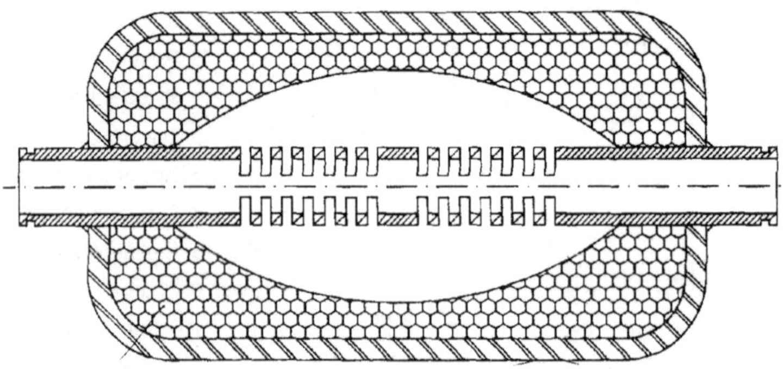

Fig. 11.14. Energy Absorbing Impedance Dampener Using Closed Cell Foam Elastomer

11.12 ACOUSTIC FILTERS FOR PUMPS

Acoustic filters (dampeners) operate on distinctly different principles than gas-type energy-absorbing dampeners. Acoustic filters, as the name implies, do not alleviate flow variations, they filter out the high-amplitude pressure waves generated by the pump before they travel into the downstream system of piping, whereas gas-type dampeners prevent the generation of such waves by the pump by actual reduction of the flow-variations.

Because reciprocating pumps are of considerable interest, it is only fair to emphasize that the acoustic filter, reacting to the deliberate insertion of flow restriction in the device, can impose greater discharge pressure and con-sequently higher pressure pulses in the pump liquid end itself, even beyond those of an undampened pump. This is demonstrated in Fig. 11.15, where the electrical analogy shows that the insertion of resistance alone reduces the down-stream pulsations but increases the amplitude of the upstream pump-generated pulses.

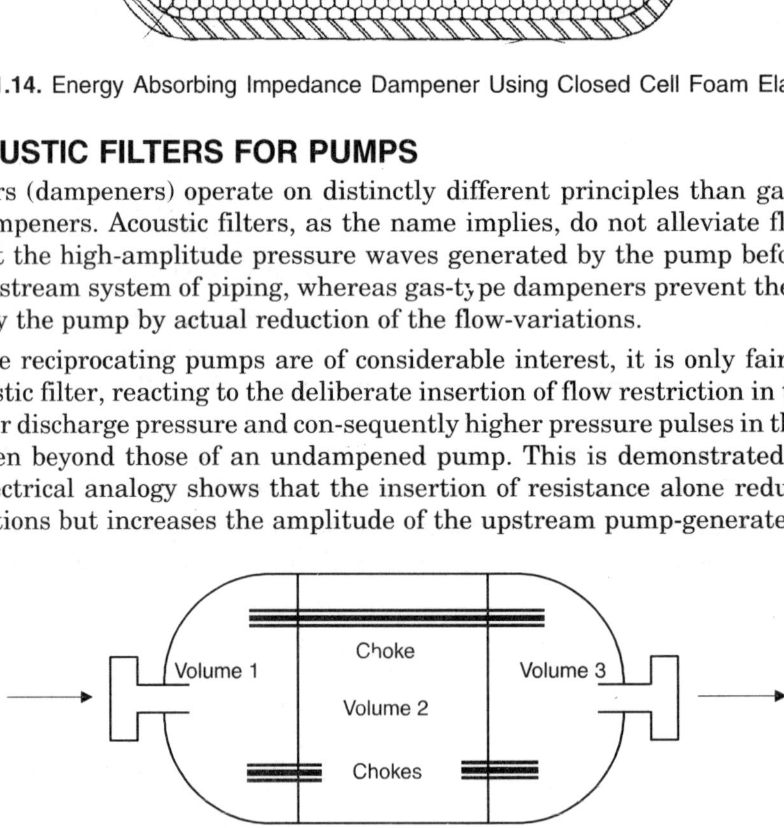

Fig. 11.15. Acoustic Dampener

Acoustic filters are most efficient at only one particular design pressure wave frequency, which militates against their acceptance for systems that require variation in speed or displacement, liquid type and operating temperature. They also become exponentially more efficient at higher frequencies, while gas-type dampeners exhibit an inverse reaction to frequency. This feature precludes the use of acoustic filters on large, low-speed pumps such as slurry pipeline triplex pumps at 75 RPM, which have a basic frequency of 7.5 Hz. This low frequency combined with the great flow rate would dictate an abnormally large and costly acoustic device.

It has been shown that reciprocating pumps usually generate pressure pulsations of several different frequencies in both suction and discharge lines, the most violent and damaging being that related to the basic pump pulse frequency determined by the pump speed times the number of cylinders and usually times a number related to the crank arrangement. Figure 11.11 illustrates various pump-type flow peaks per stroke. These basic frequencies seldom exceed about 50 Hz. Secondary high-frequency pulsations result from the effect of acceleration discontinuities in both the suction and discharge. These are on the order of 75-150 Hz, having no relation to pump speed. Unpredictable pressure pulses are also generated when the acoustic properties of the system result in "noise" from flow-through orifices, valves, regulators, etc. These latter high frequencies can certainly be reduced by the use of a properly designed acoustic filter.

11.13 SIZING OF PULSATION CONTROL DEVICE

Gas dampener size may be established if the degree of allowable residual pulsation (generally between 3 and 6 percent) and the pre-charge pressure. In bladder type of dampeners the pre-charge pressure the pre-charge pressure is the gas charge (usually dry nitrogen) pressure injected into the bladder. Desirable pre-charge pressure is between 60 and 70 percent of the average working pressure, or the permissible bladder design limit. Pre-charge pressures below 25 percent of the discharge pressure should be avoided since the effective bladder size is disrupted.

Equations for determining design parameters for gas dampeners are provided next. The unit for volume used here is US gallons. The PASAFE model is based on the pump flow rate. The suction and discharge units are sized to be equivalent to 10 percent of the flow rate in gallons/min, or 170 percent of flow rate in m^3/hour delivered by the pump. Volume displacement

Table 11.2. Gas Dampener Parameter Selection

Liquid Volume		*Gas Volume*		*Max Gas Pressure*	*Charge Pressure*
Gallons	*Liters*	*Gallons*	*Liters*	*lbs/in²*	*Bar*
20	75	2.6	10	300	20.7
40	150	5	19	250	17.2
80	300	13	49	200	13.8
120	450	28	106	150	10.3
240	900	53	201	125	8.6
400	1500	91	344	100	6.9
600	2250	112	424	100	6.9

of a single-cylinder pump for a given bore and stroke (using inch and mm units is:

$$V_{cyl}(\text{gallons}) = [(\pi/4) \times \text{bore} \times \text{stroke}]/231 \text{ (inch units)}$$

$$V_{cyl}(\text{liters}) = (\pi/4) \times \text{bore} \times \text{stroke} \times 10^{-6} \text{ (mm units)}$$

Variation in pump flow volume is defined by the quantity of liquid the pulsation control unit must be able to accommodate and dispense to stabilize pipeline flow rate. A factor for each pump type is available from integration of measured generated flow data. Assumed are 10° valve sealing delay and 10° liquid compression delay. The resulting factors are nearly all pump applications. Exceptions are high-pressure pumps with large clearance volumes and compressible liquids. Variation dV in pump capacity is obtained by multiplying cylinder displacement by the factors listed below:

Pump Type	*Variation Factor*
Simplex SA	0.600
Simplex DA	0.280
Duplex SA	0.240
Duplex DA	0.160
Triplex SA	0.026
Quintuplex SA	0.016

Gas volume available in the dampener at system operating pressure is obtained from:

$$V_G = (V_{GP} P_{CP})/P_O \tag{11.26}$$

where: V_G = PASAFE specified gas volume at system operating pressure

P_O = system operating pressure

V_{GP} = PASAFE specified initial gas volume

P_{CP} = PASAFE specified nitrogen charge pressure (50% of system pressure or maximum PASAFE charge pressure, whichever is less, lbs/in^2 or bar).

Residual pressure variation resulting from changing frictional pressure drop as a result of downstream flow variation when using the selected PASAFE Unit is calculated using the equation below. Gas volume provided by the PASAFE Unit allows down stream flow variation to be reduced because of its ability to accommodate peak flow and dispense fluid during periods of low pump flow. Pressure pulsations when operating below normal system pressure are less than the pulsations created at normal system pressure, but the observation is not applicable for appendage dampeners when system pressure is below the dampener charge pressure.

$$d_P = P_O - (P_O V_G)/(V_G + dV_{\text{pump}}) \tag{11.27}$$

where: P_O = system operating pressure (lb/in^2 or bar)

V_G = PASAFE gas volume at operation pressure (gal or liter).

Target residual pressure variation (peak-to-peak) is figured as a percent of system operating pressure. The following table lists industrial recommended standards. However, system requirements or economics may require a different target percentage dP to be used, where:

$$\text{Percentage } dP = 100\, dP/P_O$$

and the target values are:

Type Pump	*Target % dP*
Duplex DA	10%
Triplex SA	3%
Quintuplex	2%

Adjustment of residual pressure pulsations can be obtained by modifying the PASAFE unit or by adding an appendage pulsation dampener to augment pulsation control at high system operating pressures. The addition of gas volume from another dampener requires an adjustment to the total gas volume and dampener charge pressure to recalculate the residual hydraulic pressure variation at design operating pressure. Since the PASAFE Unit has a relatively low nitrogen charge pressure, the PASAFE Unit provides pulsation control over a broad operation pressure range. An appendage dampener used to augment the PASAFE performance can be charged to an optimum higher charge pressure. Use the following equation to determine a new gas volume available at system oper-ating pressure.

$$V_G = [(V_G\, P_{CP}) + (V_{GA}\, P_{CA})]/P_O \tag{11.28}$$

where: V_{GP} = PASAFE initial gas volume

P_{CP} = PASAFE nitrogen charge pressure

V_{GA} = Appendage dampener initial gas volume

P_{CA} = Appendage dampener nitrogen charge pressure

The following acoustic filter formulas are intended only to show how all the parameters of an acoustic filter are related. Refer to Fig. 11.15 for a typical acoustic filter configuration.

Effective bulk modulus, lb/ft^2 = $144\, K/[1 + (KD/Eh)(1 - \mu^2)]$

a, acoustical velocity = $(K'g/p)^{0.5}$, ft/sec

C, acoustic capacitance of volume = $Vg/pa^2 = v/K'$, ft^5/lb

L, acoustic inductance of tube = pLc/gB, lb-s^2/f

RA, acoustic resistance of pipeline = P/Q, lb-s/ft^5

P, pressure of rapid valve closure = paQ/gA, lb/ft^2

Then: $R_A = pa/gA$

P_L/P_O, transmission ratio = $[1/(X + Y)]^{0.5}$

where: $X = [(\omega L/R_A)^2 (CR_A^2/L)^{-1}]^2$

$Y = (\omega L/R_A)^2 (CR_A^2/L)^2$

A = flow area of pipeline, ft^2

B = cross-sectional area of tube, ft^2, (trial)

D = pipe inside diameter, in

E – modulus of elasticity of pipe, lb/in^2

f = frequency of pulsation, Hz

g = gravitational constant, ft/s^2

h = pipe wall thickness, in

K = bulk modulus, uncorrected for pipe, lb/in^2

K' = effective bulk modulus

L_T = tube length, ft, (trial).

P = pressure, lb/ft^2

P_L = peak pulsating pressure with filter, lb/ft^2

P_O = peak pulsating pressure Without filter, lb/ft^2

Q = volume of flow, ft^3/s

V = capacitance volume, ft^3, (trial).

ρ = liquid density = lb/ft^3 × 0.53

μ = Poisson ratio

$\omega = 2\pi f$

11.14 VALVE DYNAMICS EFFECTS

In addition to pulsation effects causing overload and distortion of the plunger/piston pressure versus time wave similar to that discussed for compressors, valve dynamics effects can cause large deviations in pressures acting on the pump's working parts.

An overpressure can occur once per cycle which produces dynamic stresses on the power end as well as the fluid end, including the connecting rod, crosshead, crankshaft and working barrels or cylinders and manifold. Large overpressure spikes can cause various problems such as failure of working barrel, crosshead guide and case, crankshaft and connecting rod, bearing damage, and reduced valve life.

A typical valve consists of a spring and a valve plug or plate, which seals against a seat. Since pressure required to open the valve is controlled by the spring force and the differential pressure across it, the valve design can influence the magnitude of over pressure.

The cause of pressure overload (spike) produced by valve effects is described as follows. A differential area is required at the valve seat to seal the liquid, which causes an overpressure. Let A_c define the area of the valve in contact with the cylinder pressure P_c, and A_m the area of the valve in contact with manifold pressure P_m. In order for the valve to open, cylinder pressure must be greater than manifold pressure by the ratio of the areas A_m/A_c, plus an additional amount to overcome the spring force. The area ratio A_m/A_c is typically 1.1 to 1.5 to allow sufficient seating area to control valve impact stresses. This field data indicated an overpressure of up to 150% of the discharge pressure for a quintuplex pump. This valve simulation assumes a perfect seal of the valve-seat contact area. Actual over pressure levels in pumps may be significantly lower due to non-perfect sealing effects and consequent leakage.

The time required for a valve (suction or discharge) to return from a fully open position to its seat is dependent upon the valve mass and return spring properties. Therefore, as the speed of a pump increases, the fixed finite time required for valve closure to occur results in a greater relative valve lag in terms of crank rotation. As the speed increases from 260 to 300 rpm, the intake or suction portion of the cycle decreases from 164 to 153 degrees. The force acting on the pressure side is F_1 and on the spring side is F_2. Forces on closed valve are:

Spring Force:	$F_s = F_{\text{pre-load}} = F_{pl}$
Pressure Forces:	$F_1 = P_c A_c$ (in cylinder)
Then:	$F_2 = P_m A_m + F_s$ (in manifold)
Valve will open when:	$F_1 > F_2$

Hence: $$P_cA_c > P_mA_m + F_{pl}$$

and: $$P_c > P_mA_m/A_c + F_{pl}/A_c$$

The lag in the closing of the discharge valve actually causes backflow through the discharge valve, and the lag in the closing of the suction valve allows backflow through the suction valve. This not only reduces pump capacity, but also changes the flow excitation characteristics of the pump.

11.15 PIPING VIBRATIONS

Since vibratory characteristics of piping spans can be described by the differential equations of motion for uniform beams combined with appropriate boundary conditions, natural frequencies of the piping spans can be calculated. The distinguishing characteristic between classical beam vibration problems and actual piping vibrations is the deviation between the theoretical end conditions and the configuration, those that exist in typical plant piping. While ideal or classical end conditions are never exactly realized in the field, the classical treatment of ideal beams gives a valuable starting point.

The differential equation of motion for lateral vibration in a beam is:

$$d^4y/dx^4 - \beta^4 y = 0$$

where: $\beta = [\mu\omega^2/gEI]^{1/4}$

Solution of the vibration equation (11.23) gives the natural frequency equation:

$$f = (\beta^2/2\pi) \times [gEI/\mu]^{1/2} \tag{11.25}$$

where: m = weight of beam per unit length = γA

γ = density of material, lb/in^3

A = beam cross sectional area, in^2

ω = frequency, radian per second

E = modulus of elasticity, lb/in^2

l = moment of inertia, in^4

g = gravitational acceleration constant, in/sec^2

The effect of piping elbows (L-bends, U-bends, Z-Bends and three-dimensional bends) on natural frequencies may be accounted for by introducing a frequency factor in terms of the length of the span, $\lambda = \beta^2 l^2$. For steel ($E = 30 \times 10^6$ lb/in^2, $\gamma = 0.283$ lb/in^3), then:

$$f = 223 \times \lambda \times k/L^2 \tag{11.26}$$

where k = radius of gyration, inches

L = length, feet.

Energy methods can be used to solve for the natural frequencies of mechanical systems. Basically, this is a method of solving for the fundamental frequency of a system by writing the equations for the maximum kinetic and potential energy and equating the two, since for a force conservative system (one without damping), these are equal. For equilibrium maximum kinetic energy is equal to the maximum potential energy, or $1/2(M\omega^2) = 1/2(kx^2)$. It can be shown that for a non-rotating beam the expression for natural frequency is given by:

$$\omega_0^2 = \frac{\dfrac{EI}{2}\displaystyle\int_0^L \left(\frac{d^2y}{dx^2}\right)^2 dx}{\dfrac{\mu}{2g}\displaystyle\int_0^L y^2\,dx} = \frac{\text{Potential Energy}}{(1/\omega_0^2)(\text{Kinetic Energy})} \tag{11.27}$$

The use of Rayleigh's method to solve for the fundamental natural frequency requires a preliminary knowledge of the deflection curve during vibration. For a given deflection curve $y(x)$ of a beam, the first derivative dy/dx gives the slope, the second derivative d^2y/dx^2 is proportional to the bending moment and third derivative d^3y/dx^3 is proportional to the shear force. But the deflection curve of a pipe segment is not known, since its determination requires solution of the problem by the classical method, consequently Rayleigh's method may not be employed gainfully.

When a concentrated weight is present in a span, it does not add to the stiffness of the beam. The potential energy of the beam will not change consequently, but the added mass will add to the kinetic energy. The natural frequency of a steel beam with a concentrated weight in the span is given by the following expression:

$$f_p = [223 \times \lambda \times k/L^2]/(1 + \alpha^P W)^{1/2} \tag{11.28}$$

where P = concentrated load

W = weight of beam span of length L

f_p = natural frequency

α^P = weight correction factor

k = radius of gyration.

Values for the weight correction factor α depend on the configuration of the piping segment, location and magnitude of attached weights and mounting condition at the ends of the pipe. A cantilever pipe segment of uniform cross-section with a mass located at a quarter-length from the fixed end has $\alpha = 1.1$, increasing to 2.0 if the mass is attached at half-length. For a simply supported section the corresponding values are 1.7 and 3.9. Pipe sections with a two or three-dimensional bends have varying correction factors for first and higher modes of vibration. Finite element form of analysis must be relied upon for precise solutions as the configuration, size, attached masses and end conditions become more complex. When two such concentrated weights are located in a span, Dunkerley's equation may be used to determine the effect of the second weight:

$$f_{P1+2}^2 = \frac{1}{\dfrac{1}{f_{P1}^2} + \dfrac{1}{f_{P1}^2} - \dfrac{1}{f_0^2}} \tag{11.29}$$

High frequency vibrations of piping shell walls can be caused by excitation of tangential and radial mode natural frequencies. A number of theories are available to calculate the natural frequencies and dynamic stresses due to shell wall vibration. Shell theories are among the most accurate and easily applicable. The curvature of the shell couples the flexural and extensional vibrations to complicate the analysis. Calculation of natural frequencies for cylindrical shells of infinite length may be determined from:

$$f_i = [\lambda/2\pi R] \times [E/\{\gamma(1 - \nu^2)\}]^{1/2} \tag{11.30}$$

$$\text{where: } \lambda_i = \frac{1}{\sqrt{12}} \frac{h}{R} \frac{i(i^2-1)}{(1+i^2)^{1/2}} ; i = 2, 3, 4... \tag{11.31}$$

f_i = shell wall natural frequency, Hz

λ_i = frequency factor

R = Mean radius of pipe wall, inch

E = Young's modulus

ν = Poisson's ratio

γ = mass density of pipe material, lbs-second2/in^4

h = pipe wall thickness, inch

i = mode number

For a steel pipe with ν = 0.3, the frequency equation reduces to:

$$f_i = 33749 \times [\lambda/R] \tag{11.32}$$

The dynamic stress in a pipe can be calculated from the classical beam theory, $s = MR/4I$. Bending moment may be expressed as a function of the mode shape by $M = EI(d^2y/dx^2)$. Hence, if $y = Y_o \sin(\pi x/l)$ is the mode shape definition, then the maximum dynamic stress is:

$$s = -\frac{ED}{2}\left[-Y_o \frac{\pi^2}{l^2}\right] \tag{11.33}$$

Vibration induced stress in a piping span that includes bends and vibrates at resonance may be expressed in terms of maximum vibration amplitude and deflection and stress factors discussed above:

$$s = K_d\, y \frac{D}{L^2} (\text{SCF}) \tag{11.34}$$

where: s = dynamic stress, psi

K_d = deflection stress factor for the first two modes of pipe configurations with equal elbow legs

y = maximum vibration amplitude, 10^{-3} inch

D = pipe outside diameter, inch

L = pipe span, feet

SCF = stress concentration factor

11.16 COMPRESSOR MANIFOLD VIBRATION

The manifold system may be considered to include crosshead guide, distance piece, cylinder, suction and discharge bottle and attached piping to the first clamp (Fig. 11.16). Mounting options for secondary pulsation bottles may include horizontal attachment near the grade elevation, vertical attachment to minimize unsupported riser dimensions or fabricated together with the primary filter bottle. Hence, mechanical design of the machine's suction and discharge manifold systems must be carried out in conjunction with piping pulsation analysis.

A general purpose finite element computer program with dynamic analysis capabilities can be used to evaluate the system dynamics. Mass, inertia and stiffness characteristics of the crosshead guide, distance piece and cylinder must be available for the preparation of the mathematical model. Crosshead guide and distance piece may be simulated by flexible beam elements, although the cylinder is generally is rigid in the frequency range of interest. The distance piece and crosshead guide require modeling in sufficient detail to represent their dynamic characteristics adequately by selecting a number of cross-sections along the length. The number of cross sections required to adequately describe a component depends on its complexity. Properties of the weakest cross section will usually be the controlling item. Gusseted or oblong suction and discharge nozzles are treated as piping or beam elements.

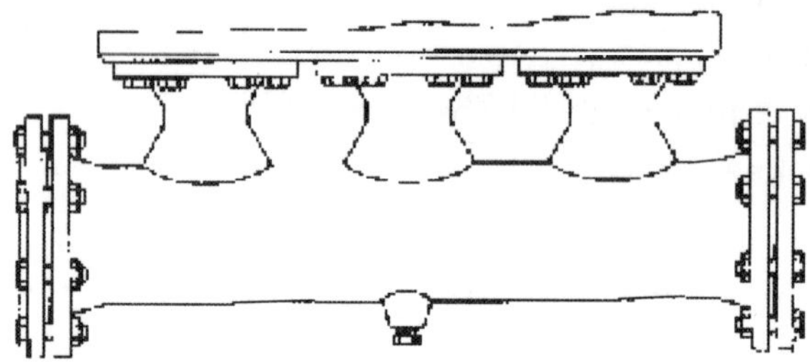

Fig. 11.16. Typical Manifold System

Useful tips for connections between nozzle and bottle branch are available in Welding Research Council bulletin # 297. Equivalent rotational and translation springs may be used for branch connection flexibility, the spring values derived from a separate finite element model of the joint area. Detailed definition of a branch connection such as pad thickness and type enhance accuracy of the calculated results, but may be dispensed with if component drawings are not available. Hemispherical or semielliptical heads are treated as equivalent lengths of pipe to develop the right mass and center of gravity location. Valves, flanges and reducers may also be simulated by special structural elements available in a number of computational codes. Calculated natural frequencies must also be verified by modal testing methods where the manifold system is available.

Once the manifold system's natural frequencies are understood, practical means of controlling the associated dynamic activity must be considered. The two sources of excitation are the synchronous rotor related forces and pulsation induced forces. Dynamic forces arising from rotor speed are generally relevant at the lower harmonics of compressor speed (1x and 2x). Shaking forces induced by pulsating fluid pressure frequencies are controlled by the acoustic design of the suction and discharge nozzles and the attached lines. The natural frequencies of the manifold system must be maintained above the third multiple of compressor speed if possible, and they should not coincide with shaking force frequencies. Hence, practical methods for controlling and placing the dynamic natural frequencies of the compressor manifold system are of interest.

Knowing the dynamic natural frequencies of the manifold system from the reciprocating machine's design, judicious selection of bottle size, nozzle length and wall thickness is necessary. As a rule, long suction nozzles are preferable for convenience in maintenance and installation of the equipment. But if the consequent reduction in the natural frequency and the associated resonance is detrimental to the overall system's structural integrity, the length of the nozzles may require reduction. Heavier nozzle walls may also be considered to improve stiffness and raise frequency, even if thinner walls may be adequate for withstanding fluid pressure.

Supporting system and method for discharge bottles is clearly beneficial for improving the dynamic natural frequencies to the appropriate level. Many compressor installations designed according to API 618 specifications use straps and wedges for supporting for the bottles. The support mechanism can substantially change the low frequency modes, namely, the bottle cantilever, rotary and angular modes of vibration.

A visco-elastic damping material inside the clamp improves surface contact and also reduces the dynamic response. The straps and wedges form of clamping must permit thermal growth to ensure the related stresses are not exacerbated at elevated operating temperatures.

Flexible supports, oriented in the direction of bottle stretch and perpendicular to the shaft axis, offer the advantage of raising the low frequency mode above the troublesome excitation frequencies. The support system is made of steel bar and plate material, and attaches to the cylinder head bolts. Leveling and tightening at operating temperature, and is not intended for alignment needs.

Supports for suction bottles are not required often, except where the bottle is substantially large or the nozzle is extended. Support structures made from steel beam sections usually offer adequate stiffness to obtain the proper natural frequency.

Suction riser designs using elbows pose the difficulty of inserting added flexibility, and at the same time create a force coupling point. Avoiding some of the elbows in suction risers is a good design practice. Where suction bottles are oriented vertically, as is the case for many secondary bottles with the intent to limit the length of the riser, attention must be focused on the mounting details for the vessel. Exceptional cases arise if the total length of the riser is excessive, and consequently flexible, calling for concrete foundations and steel frame structure to provide the necessary support.

The design procedure inherently becomes more reliable and efficient when the analytical predictions can be verified with test measurements. Spending time and effort at as many design stages as possible to obtain pressure, temperature, flow rate, accelerometer and strain gage readings is an investment that pays excellent returns in the form of an improved design. In the preliminary stages of a new machine or system design the difficulties encountered in obtaining verification of calculated results may be alleviated by using similar other designs of a different size. The observed results may then be scaled by using the similarity approach. Experiments may also be performed on scale models.

REFERENCES AND BIBLIOGRAPHY

ANSI/ASME OM-3, "Requirements for Pre-Operational and Initial Start-up Vibration Testing of Nuclear Power Plant Piping Systems", 1981.

API Standard 618, "Reciprocating Compressors for General Refinery Services", American Petroleum Institute, Washington, D.C., February, 1986.

API Technical Data Handbook, Project 44, American Petroleum Institute, Washington, D.C., 1985.

American Welding Society, "Structural Welding Code", Section 10.7, page 104-107, 1972.

Arnold, R. N., Warburton, G. B., "The Flexural Vibrations of Thin Cylinders", Journal Proceedings of Institute of Mechanical Engineers, Vol. 167, 1983.

Blevins, R. D., "Formulas for Natural Frequency and Mode Shape", Van Nostrand Reinhold Co., New York, 1979.

Bloch, H. P. "A Practical Guide to Compressor Technology", McGraw-Hill, New York, 1995.

Carucci, V. A., Mueller, R. T., "Acoustically Induced Piping Vibrations in High Capacity Pressure Reducing Systems", ASME Paper # 82-WA/PVP-8, Pressure Vessel and Piping Conference, 1982.

Fung, Y. C., Sechler, E. E., Kaplan, A., "On the Vibration of Thin Cylindrical Shells Under Internal Pressure", J. Aeronautical Sciences, Vol. 94, 1981.

Harvey, J. F., "Theory and Design of Pressure Vessels", Van Nostrand Reinhold Co., New York, 1985.

Helmholtz, H. L. F., "On the Sensations of Tone", Translation, 2nd Edition, Dover, 1954.

Hydraulic Institute Standards, Hydraulic Institute, Cleveland, OH., 14th Edition, 1983.

Kinsler, L. E., Frey A. R., "Fundamentals of Acoustics", 3rd edition, John Wiley, New York, 1992.

Meirovitch, L., "Analytical Methods in Vibrations", Macmillan, New York, 1987.

Mikacinovic, M., "Vibration Acceptance Criteria of Circular Cylindrical Shells", Pipeline Dynamics and Valves, ASME PVP Vol. 180, page 47-52, Pressure Vessel and Piping Conference, July 1989.

Miller, J. E., "The Reciprocating Pump: Theory, Design and Use", Krieger Publishing Co., Malabar, FL., 1995.

Olson, D. E., "Piping Vibration Experience in Power Plants", Pressure Vessel and Piping Technology Book # H00330, American Society of Mechanical Engineers, New York, 1985.

Pichot, P., "Compressor Application Engineering", Volume I, Compression Equipment, Gulf Publishing Co., Houston, 1986.

Rachel, J. C., "Piping Vibration and Stress", Proceedings of Machinery Vibration Monitoring and Analysis", Vibration Institute, April, 1991.

Roark, R. J., Young, W. C., "Formulas for Stress and Strain", Mc-Graw Hill, New York, 1994.

Sparks, C. R., "Controlling the Effects of Pulsation and Fluid Transients", Report # 160, SGA-PCRC Open Seminar, Southwest Research Institute, November, 1991.

Sparks, C. R., "On the Transient Interaction of Centrifugal Compressors and Their Piping Systems", Journal of Engineering for Power, Vol. 135, page 891-901, October, 1995.

Shigley, J. E., Mitchell, L. D., "Mechanical Engineering Design", Mc-Graw Hill, New York, 1993.

Towne, D. H., "Wave Phenomena", Addison Wesley Publishing Co., 1987.

Tulk, J. D., "Correlations Between Dynamic Stress and Vibration Velocity in Piping Systems", Pipeline Dynamics and Valves, ASME PVP Vol. 180, page 19-23, Pressure Vessel and Piping Conference, July 1989.

USAS B31.7, "Nuclear Power Piping", ASME Code, 1969.

Wachel, J. C., Bates, C. L., "Techniques for Controlling Piping Vibrations and Failures", ASME Paper # 86-PET18, New York, 1986.

Wachel, J. C., "Field Investigation of Piping System for Vibration-Induced Stress and Failures", ASME Bound Vol. # H00219, Pressure Vessel and Piping Conference, June 1982.

Wachel, J. C., Morton, S. J., Atkins, K. A., "Piping Vibration Analysis", 19th Turbomachinery Symposium", Texas A & M University, 1990.

Reciprocating Machinery Dynamics and Noise

12

12.1 INTRODUCTION

Mechanical power produced by reciprocating engines, compressors and pumps has been responsible for a great deal of prosperity and abundance over the last century. Agricultural machinery has increased grain production several folds, electric power generators illuminate the remotest corners of the world and motor boats and snow mobiles provide leisure time activity for everyone. Reciprocating machines have been at the forefront to lead the transformation. But this progress has come at a price. Vast amounts of pollutants are spewed out into the air by reciprocating engines, depleted energy resources and created noise pollution. Noise is now no longer considered to be only an annoying irritant. Medical science has proved that prolonged exposure to high levels of noise results in hearing impairment, sometimes even in a total loss of hearing. Excessive noise can have severe psychological and physiological effects on humans. Cardiovascular, muscular and fatigue effects can be directly correlated to the intensity of exposure to sound levels. Hence, the impetus to make reciprocating machines operate quietly certainly exists.

The connection between mechanical vibration of a component and noise caused by it is easily recognized. Man has used sound for a long time to discriminate between properly functioning machines and those where failure of a component is imminent. Before the advent of accelerometers and spectrum analyzers production line inspectors merely listened to an operating machine as it was run for final inspection. This type of a quality evaluation, however, had several problems. The ambient noise level plays a significant role. A particularly noisy machine can be readily identified when the background noise is low. But if the noise in the background is high it becomes more difficult to hear the tell-tale signature of an unacceptable component. In addition, after a long exposure to the same sound the human faculty of hearing loses its ability to identify discrete noise sources. Individual mood changes may also play a role in applying the acceptance criteria. More recently, sound level and vibration meters are extensively used together with signal processing equipment for production line quality control testing of reciprocating machines and components. The analyses can be made more objectively and definitively, and the data can be stored for further evaluation at a later date.

Analysis of dynamic excitation, structural response and sound radiation characteristics of reciprocating machines helps to identify dominant sound radiating surfaces, and to determine important dynamic features of the internal and external components related to sound transmission and radiation. Evaluation of the data then permits selection of various treatments and modifications for controlling the noise at its source. The analyses could be performed experimentally or analytically. However, complexities and non-linearities make analytical studies time consuming and generally more expensive than experimental methods. Digital signal processing techniques using Fourier analysis type of instrumentation make many of the measurements practical. Data acquired during operating and static tests provide noise source identification, dynamic force generation and structural response. Frequency response of the structure can be measured by applying a controlled force when the machinery is shutdown. Resonant frequencies, their corresponding mode shapes and damping values can then be obtained. The task of defining sound emission characteristics of the reciprocating machine may appear at first sight to be straightforward, but it requires ingenuity and difficult work that is rarely complete. After understanding the noise generation and radiation process, a cure has to be found for the diagnosis. Finite element studies come in useful at this point. Once the cure is found, it has to be acceptable from production and cost considerations, otherwise one may be faced with a terminal case!

12.2 OPERATING PARAMETERS

The amplitude of the root-mean-square of the acoustic pressure wave gives sound pressure. At the threshold of hearing the magnitude of sound pressure $p_o = 2 \times 10^{-4}$ μ bar, or 2×10^{-5} N/m². Sound power per unit area defines sound intensity, and can be computed from sound pressure:

$$I = p^2/\rho c, \text{ W/m}^2 \tag{12.1}$$

where ρ = density of air, kg/m³

c = acoustic speed of air, m/second.

At the threshold of hearing the sound intensity $I_o = 10^{-12}$ W/m². Sound intensity ratio is the ratio of sound intensity I over the reference intensity I_o. The ratio varies from 10^0, the threshold of hearing, to 10^{14}, the threshold of pain. For convenience the exponent of 10 is used as the unit bel. Therefore the thresholds of hearing and pain are 0 and 14 bels, or 0 and 140 decibels (dB). Sound intensity and pressure levels are defined as:

$$SIL = 10 \log \left(\frac{I}{I_o}\right) \text{ dB} \tag{12.2}$$

$$SPL = 20 \log \left(\frac{p}{p_o}\right) \text{ dB} \tag{12.3}$$

Reciprocating machinery noise is often measured at a distance of 1 m from the machine. Only the bare machine is used for noise evaluation. The machine is equipped only with built-in accessories essential for operation, such as flywheel, fuel, oil and water pumps and intake and exhaust manifolds. Sound from the manifolds, flywheel housing, test stand and dynamometer must be minimized to measure the sound level of the basic, or bare, machine. Sound radiation occurs at machine surfaces. These include valve and timing covers, oil pan and other non-structural load carrying panels. Radiation of surface noise is affected by factors like force transmissibility from machine structure to support, area and radiation efficiency, damping and modal density of the machine's radiating surfaces. Hence, the basic machine noise depends on exciting force characteristics, structural response, vibration transmissibility and radiation efficiency of machine's external surfaces.

Since sound pressure level is a physical quantity, it does not consider human response to noise. A weighted average, dBA, approximately simulates the frequency response of the human ear to noise. The dBA level may be referred to as the noise of the source. The advantage of considering changes in noise levels is that changes can be calculated from sound pressure level, provided measurement conditions are maintained consistently. Hence:

$$\Delta L = \Delta(\text{dBA}) = \Delta(SPL) \tag{12.4}$$

where L signifies noise level.

12.3 CORRELATION BETWEEN DYNAMICS AND NOISE

The intensity of a machine's radiated noise level depends on the characteristics of exciting forces and response of its structure. An understanding of the exciting force mechanism, the machine structure and their interaction is the first step towards the goal of reducing the noise. Exciting forces may be expressed in the time domain as a force-time diagram or in the frequency domain as a force-frequency spectrum. Figure 12.1 displays these characteristics for representative reciprocating machines and associated components. Force spectral data can be

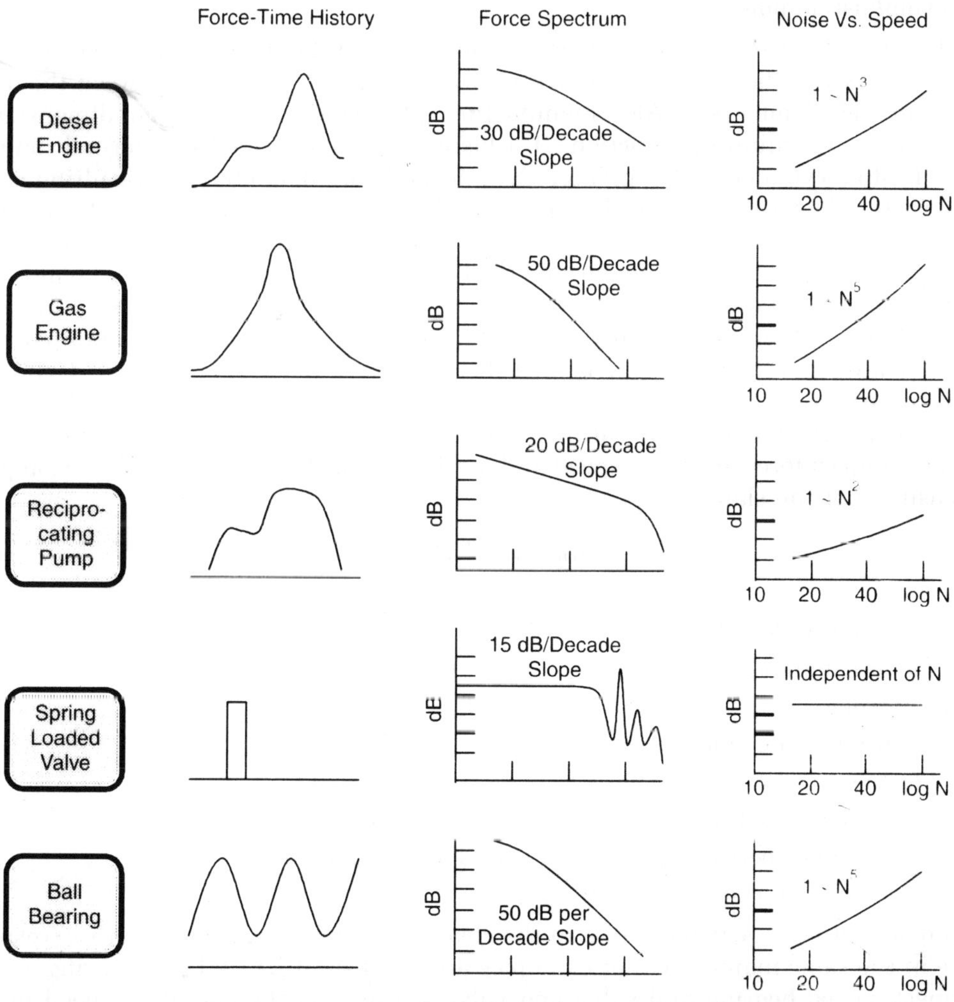

Fig. 12.1. Examples of Force-Time History, Force and Noise Frequency Spectra (*ref.* Clarkson, 1982)

obtained directly from time data by Fourier or harmonic analysis methods. In the time domain as the force function's initial ascent becomes more rapid, high frequency components in the frequency domain will increase. In frequency ranges above 240 Hz the forcing function can be approximated by a straight line, with the slope depending on the rate or impulsiveness of the forcing function. Frequency domain data has the advantage that the effect on output can be estimated, as seen in the noise frequency spectrum of Fig. 12.1. Evaluation of data from a transducer such as a microphone, a load cell, an accelerometer or a strain gage begins with analog time domain signals, and can be reviewed in the time domain on an oscilloscope or a strip chart. Complexity of time domain signals and sensitivity of the dynamic response of physical systems to frequency call for studying the signals in the frequency domain.

Surface noise emits from the vibrating surfaces of the reciprocating machine.

Excitation forces come from combustion, gas or fluid pressure and from impact caused by reciprocating and rotating components. Rapid increase in the cylinder pressure of internal combustion engines following ignition generates large excitation forces. Cylinder gas or fluid pressure applied to the piston and cylinder head due to compression, combustion and expansion is called combustion noise.

Mechanical noise is also a direct consequence of metal to metal impact between moving parts. Since a piston slides in the cylinder, clearance must be available between the diameter of the piston and cylinder bore. Also, manufacturing tolerances in component dimensions have to be provided to facilitate production. If a piston with a smaller than normal diameter is assembled into a cylinder with a greater than normal bore diameter, the resulting clearance will be greater than the nominal design value. Besides motion in the radial direction, the piston will be able to rock about the piston pin within the clearance envelope. Clearance also exists between the piston pin diameter and the bearing at the small end of the connecting rod. This provides additional freedom for the piston to turn about the cylinder axis. Gas pressure load immediately following ignition of the fuel charge will cause the piston to impact against the cylinder wall, and the piston pin against the connecting rod bearing surface. Cylinder bore and piston surface roughness also plays a role. Ross and Ungar determined that the noise generated by piston slap in reciprocating machines is affected by the impact distance, the rate of change of impact force, machine speed and machine load. The following expression provides the intensity of piston slap:

$$I_{P\text{-Slap}} \propto \frac{\rho_{\text{piston}}}{\rho_{\text{cyl. block}}}\left[\frac{Z}{S}\,\frac{B}{t_b}\,\frac{B}{t_1}\right]\left(\frac{\delta}{B}\right)^{4/3} C_m^{\;3} \tag{12.5}$$

where: ρ = material density

B = bore diameter

S = stroke

Z = number of cylinders

t_b = thickness of cylinder block

t_l = thickness of liner

d = clearance between piston and cylinder bore diameters

C_m = mean piston speed

Impact between gear teeth due to the clearance between meshing teeth is responsible for noise in the same manner. Other examples are crankshaft journal and bearing, crank pin and connecting rod bearing and valve and valve seating. As the operating speed increases, impact forces between neighboring components goes up, as also the noise level.

12.4 IDENTIFICATION AND RANKING OF SOURCES

Noise generated by reciprocating machines may be classified into three groups: aerodynamic noise, combustion noise and mechanical noise. Aerodynamic noise includes exhaust gas and intake air noise, as also the noise arising from cooling and auxiliary fans and other airflow. Combustion noise in internal combustion engines refers to noise emitted from the vibrating surfaces of the machine's structure, components and accessories as a direct consequence of the combustion process. Mechanical noise is generated by the vibrating surfaces of the overall machine structure, components and accessories after excitation by the reciprocating components.

Noise may also be classified by the source where it originates. Exhaust system noise includes noise from exhaust gas pulses leaving the muffler or the tailpipe, together with noise emitted by the mechanically vibrating surfaces of exhaust system components. Exhaust system components are also excited by the engine structure's vibration energy. Intake system components generate noise in a similar manner by the flow of fluid or gas in the machine's inlet manifold.

Water-cooled reciprocating machines typically use a radiator as a heat exchanger, with an axial flow fan forcing cooler air through the radiator. Air-cooling is accomplished generally by a centrifugal fan with its shrouding to direct the air across the machine's surfaces. Fan noise consists of discrete frequency tones and broadband noise. Broadband components of fan noise are caused by the shedding of vortices from the rotating fan blades and by turbulence in the fan's air stream. Discrete frequency components arise from pressure fluctuations when a fan blade passes an obstruction in the pressure field. For dynamic balance fan blades are spaced at equal angular intervals, hence the fundamental discrete tone will occur at the fan's blade passing frequency, given by:

$$f_o = (\text{Fan speed, rpm}) \times (\text{Number of fan blades})/60$$

where the frequency is in cycles/sec. Noise intensity of the fundamental tones, as also the strength of its harmonics, will depend on the nature of the pressure impulses generated by the obstruction.

In axial flow fans blade tip speed is the dominating factor affecting the sound emission level. In many applications the cooling fan is driven directly by the main drive shaft at a fixed ratio of engine speed. The speed ratio is selected usually for the most adverse condition (load or temperature). During part load or idle operation, excessive power may be needed to drive the fan, resulting in higher sound generation. Mechanical vibration of blades and shrouding may also add to the noise level.

12.5 NOISE GENERATION IN DIESEL ENGINES

Cylinder gas pressure acting on piston and cylinder head during compression, combustion and expansion of the air fuel mix is known as combustion noise. For engines of comparable size and power output, maximum gas pressure is higher in diesel powered engines when compared with gasoline engines. Consequently, the noise level in diesel engines is higher than in gasoline engines. Figure 12.2 illustrates the comparison of pressure and noise spectra for the two engine types.

Combustion forces are primarily affected by cylinder bore size, fuel rate and ignition delay, which control the engine's power output, rate of pressure rise, maximum cylinder pressure and duration of combustion. Power output determines the low frequency (< 500 Hz) forcing input. Fuel rate and ignition delay determine the high frequency forcing input.

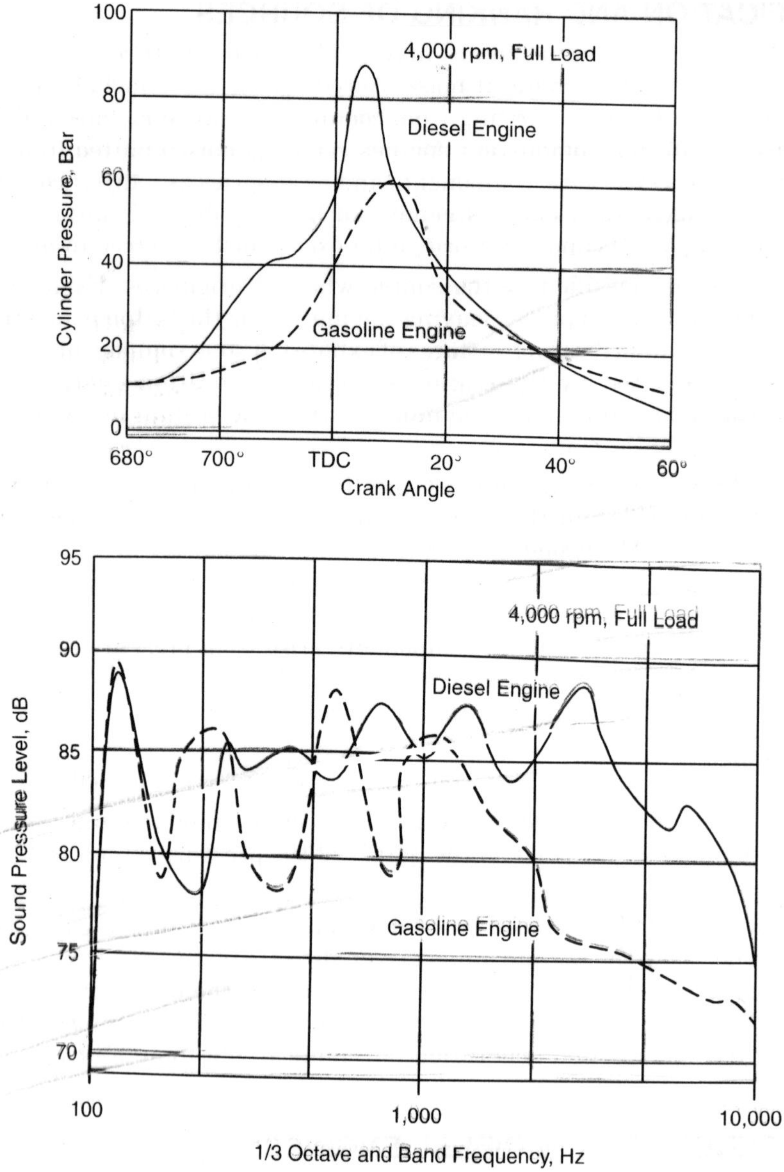

Fig. 12.2. Cylinder Pressure and Noise Spectra for Diesel and Gasoline Powered Engines (*ref.* Society of Automotive Engineers, 1984)

Ignition delay is the most important factor in causing rough combustion noise pulses. Ignition timing is affected by engine intake pressure and temperature, which in turn is affected by engine temperature, oil and water temperature and ambient air temperature. The timing can also be affected by fuel's heat value, rate and timing of fuel injection, extent of fuel and air mixing, compression ratio, combustion chamber configuration, speed and load. Engine combustion noise intensity may be expressed (Anderton, 1980) as follows:

$$I_{comb} \propto B^5 N^n \tag{12.6}$$

where: B = cylinder bore

N = engine speed

n = combustion index, varying between 2 and 5, depending on roughness of the combustion process. In acoustic terminology this equation for combustion noise level may be written as:

$$L_{comb} = 50(\log B) + 10n(\log N) + \text{constant} \qquad (12.7)$$

A delay in ignition timing results in the combustion process taking place more spontaneously. There is less time available for the flame to spread and for combustion of the fuel charge, resulting in a faster pace of increase in the gas pressure. Consequently, the force acting on the piston and cylinder head is also more sudden. The impulsiveness of the gas forces in the time domain causes force and its associated noise to have high frequency components in the frequency domain. To illustrate the effect of combustion roughness on noise, Fig. 12.3 shows a set of pressure curves. All three cases have the same average and maximum cylinder pressures, as well as combustion and total cycle time duration. Curve # 1 shows a gradual pressure increase with time, and produced lower noise level in the high frequency range. Curve # 2 has a sudden pressure change, accompanied by a rapid rate of pressure build-up. Curve # 3 represents the intermediate condition between # 1 and # 2, both in terms of pressure increase and level of high frequency noise components.

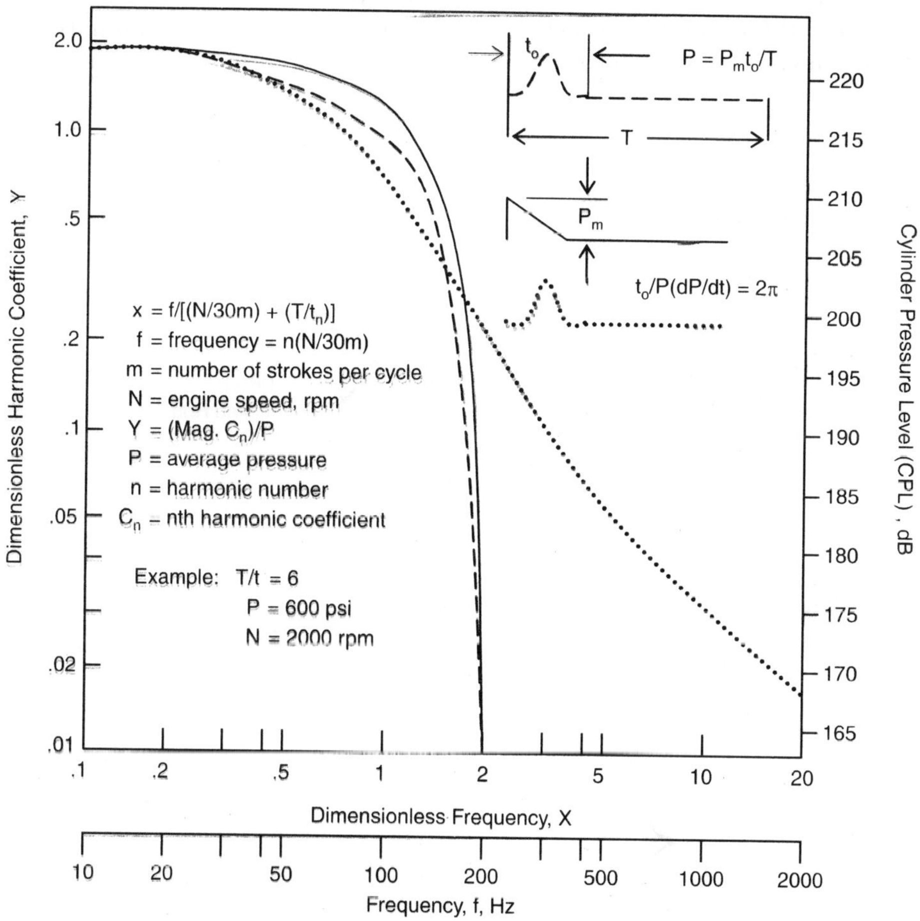

Fig. 12.3. Cylinder Pressure Analysis (*ref.* Wu, 1988)

The forces encountered in a diesel engine are:

- Gas forces from combustion
- Inertia forces due to mass and acceleration of components

  Impact forces from parts moving across gaps with neighboring components of the assembly
- Shock forces as a consequence of misalignment, rough cam profile, shaft unbalance and poor lubrication
- Aerodynamic forces arising from the flow of gas around valves, combustion chamber passages, fan blades and gear teeth.

Figure 12.4 shows the connection between various engine features and their effect on force and noise generation. Combustion noise component can be separated from mechanical noise by comparing it with total noise across a wide range of injection timings at the same engine speed. This is because combustion noise varies considerably with timing, while mechanical noise is independent of it. Where combustion noise dominates the operation, combustion noise may be calculated by measuring cylinder pressure level with a transducer mounted flush with the combustion chamber wall, and by measuring engine structure attenuation. When cylinder pressure and attenuation are at the same frequency and of the same octave band, the difference in cylinder pressure and structure attenuation provides combustion noise level (*ref*: Anderton, 1980).

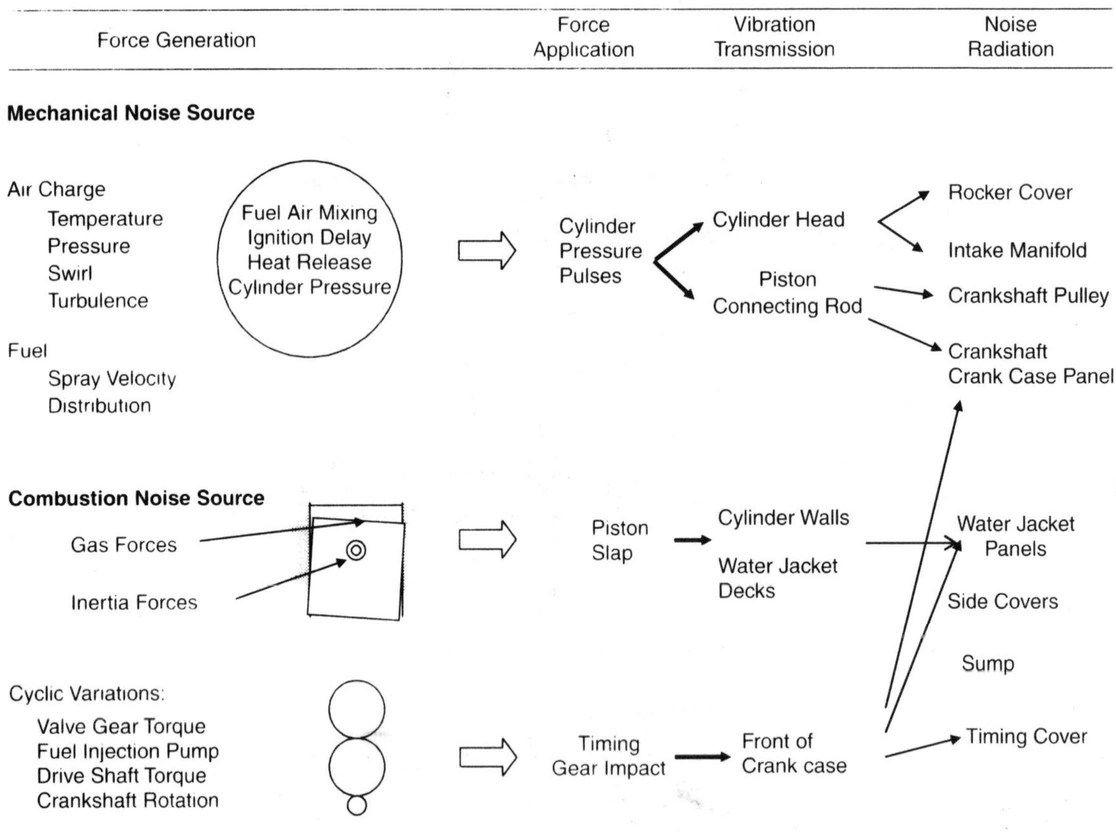

Fig. 12.4. Diesel Engine Noise Factors (*ref.* Russell and Cavanagh, 1980)

12.6 EXCITING FORCES, STRUCTURE RESPONSE AND NOISE

The structure of a reciprocating machine may be divided into internal (for example, connecting rod and piston) and external (block, crankcase) structures. Attenuation of exciting forces by the structure depends on the stiffness and natural frequencies of the components and the machine assembly. Idealized noise radiating models with various vibrating modes for an in-line machine and a V configuration are shown in Fig. 12.5. Expressions for natural frequencies

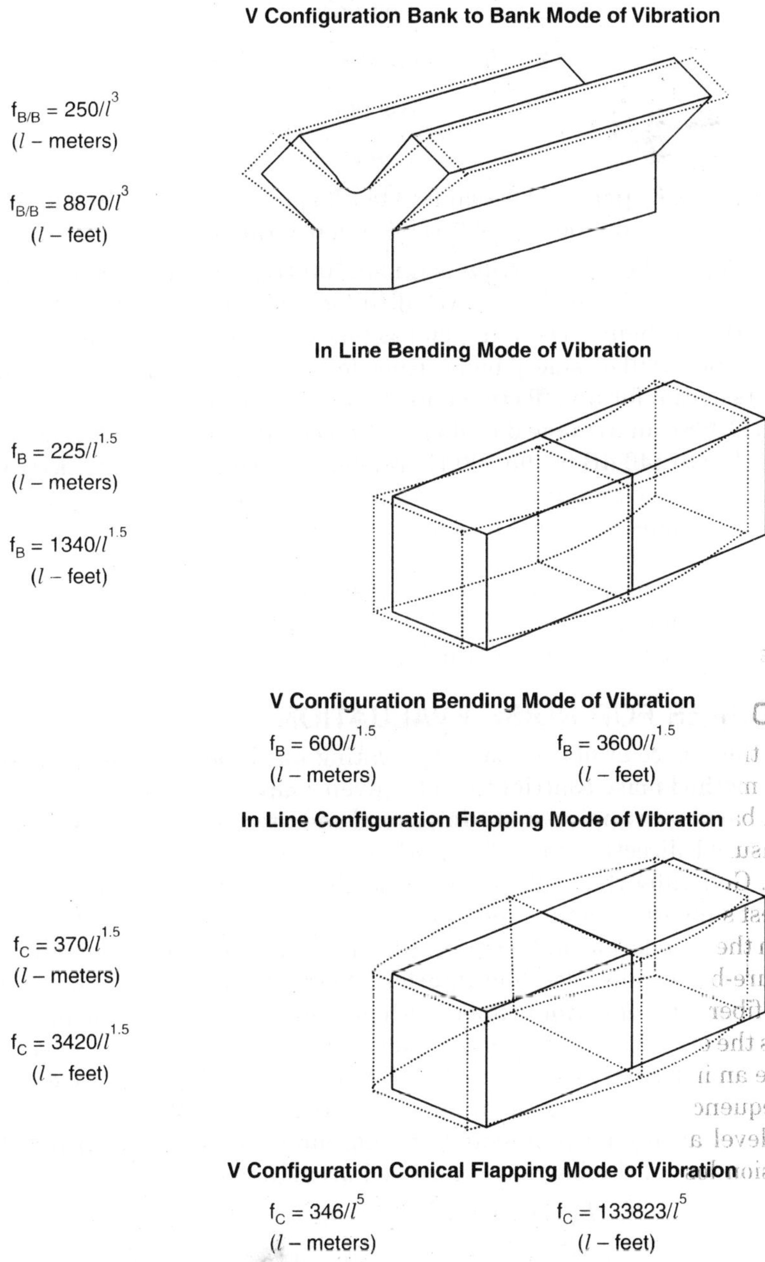

Fig. 12.5. Dynamic Mode Shapes of Machine Block as Noise Source (*ref.* Lalor and Petyt, 1985)

of vibration are shown in terms of length of machine block (Lalor and Petyt, 1985). One mode describes motion between the banks of a *V* configuration, the second involves bending of an in-line type and the third shows conical, or flapping, activity in the side walls. Below 800 Hz (representing the conical mode natural frequency) the machine vibrates essentially as a complete body. Above 800 Hz the cylinder block's side panels show a lot of activity, and the motion of the machine as a whole is not readily recognizable. Natural frequencies of the panels will depend on their size, shape and weight, and lie predominantly in the range f_{p1} and f_{p2}, defined by the expressions:

$$f_{p1} = \frac{45}{l_p^{\,2},\ \text{meter}^2} = \frac{70{,}000}{l_p^{\,2},\ \text{inch}^2}\ \text{Hz} \tag{12.8}$$

$$f_{p2} = \frac{70}{l_p^{\,2},\ \text{meter}^2} = \frac{109{,}000}{l_p^{\,2},\ \text{inch}^2}\ \text{Hz} \tag{12.9}$$

where l_p represents panel span, and is equal to block span divided by the number of cylinders. Thus, a 10 in. span will lie in 700 to 1,090 Hz frequency range.

When the primary forcing function is under 500 Hz, as is the case in many high output diesel engines, the load path within the overall structure is mostly controlled by the stiffness characteristics of the elements. Damping characteristics and resonant frequency will control activity in the thinner walled side panels. Knowledge of the stiffness characteristics of the cylinder block is essential for an effective noise control program of the machine. From studies conducted by Wu (1988), an average diesel engine block attenuates combustion noise (cylinder pressure level) at 130 to 140 dB in the 100 Hz range, or at 80 to 95 dB at 1 kHz when based on 1/10 octave. On a frequency-amplitude type of a plot the response of an engine structure can be approximated by two ramps connected by a nearly flat plateau. The middle plateau lies between 800 Hz and 2,500 Hz. On the low frequency side of the curve the ramp tends to have a positive slope of 50 to 60 dB per decade, while the falling ramp at the other end will have a negative slope of 10 to 20 dB per decade. Increasing the overall cylinder block stiffness will tend to shift the plateau towards a higher frequency range.

12.7 PROCEDURES FOR NOISE EVALUATION

Identification of the source of noise in a reciprocating machine is a principal objective. In the partial exposure method noise contribution of a given source is measured by blocking, through the use of sound barrier wrapping material, sound radiated by all other sources. Since sound pressure is measured directly, noise from other sources must be quantified or reduced to negligible levels. Generally this calls for acoustically treating intake and exhaust manifolds, dynamometer, test support structure and any other noise sources in the test chamber. Vibration isolators between the foundation and reciprocating machine and dynamometer may be necessary to ensure structure-borne noise does not interfere. Sound barrier materials consist of layers of acoustical foam, fiberglass or ceramic fiber, with an outer layer of a dense material. The inner softer layer helps the outer layer from the component vibrations. Such sound barrier treatments generally provide an insertion loss of 10 to 20 dBA. The insertion loss is mostly due to acoustic leaks at high frequencies and poor isolation at low frequencies. With the treatment in place, sound pressure level at any position equals background noise plus unwrapped engine noise minus transmission loss of the barrier. Expressed acoustically:

$$L_1 = L_b + L_e^{\,w} \tag{12.10}$$

where L_b = background noise and Lew = sound level of wrapped engine. If each source is assumed to radiate sound independently of that radiated by other sources, then with one source exposed the sound level is given by:

$$L_2 = L_b + L_s + L_e^{\ w} - L_s^{\ w} \tag{12.11}$$

where L_s = sound level of exposed source and $L_s^{\ w}$ = sound level of exposed source when wrapped. The two equations can be solved only by assuming a value for $L_s^{\ w}$; it may be neglected if it is much less than L_s. The advantage of this method is that it does not require sophisticated instruments or highly trained personnel. However, the procedure is time consuming and requires costly sound barrier material.

Acoustic intensity measurement method relies on finite sum and difference approximations of acoustic pressure and particle velocity, respectively, of two closely spaced sound pressure measurements. The measuring apparatus consists of two microphones separated by a distance Δr and producing two sound measurements $p_1(t)$ and $p_2(t)$. Particle velocity can be approximated from the pressure gradient. Cross spectral density of $p_1(t)$ and $p_2(t)$ can be computed on a two channel Fourier analyzer. Sound intensity is then obtained from the expected value function of the cross spectral densities. Additional treatment of the mathematical theory is available from work done by R. W. Fotter, 1994. Supplementary information about the frequency content of the acoustic intensity and variation of acoustic intensity over the surface improves the accuracy of this method to identify the noise source. Acoustic intensity data may be plotted using line contour plots, raster plots and three-dimensional contour plots. Figures 12.6(*a*), (*b*), (*c*) show measurements made on one side of a diesel engine using the same interpolation algorithm.

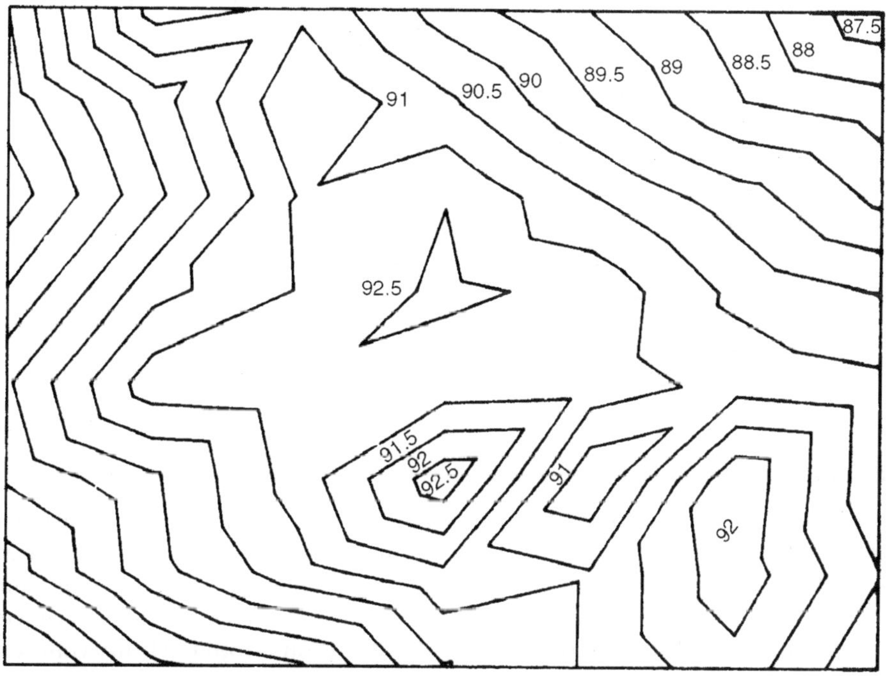

Fig. 12.6. (*a*) Acoustic Intensity Contour Map of Diesel Engine

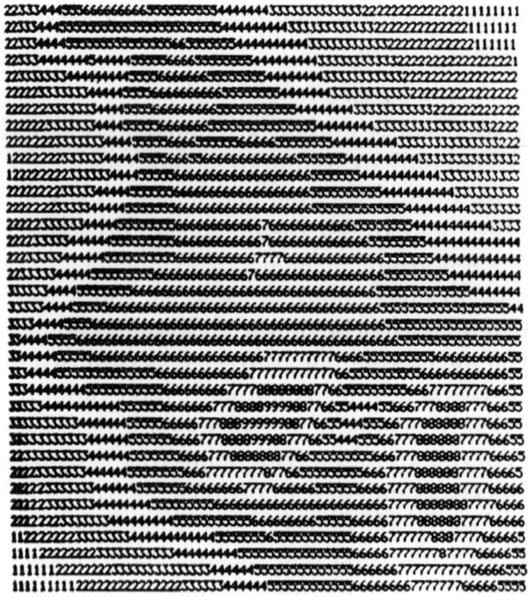

Fig. 12.6. (*b*) Acoustic Intensity Raster Plot of Diesel Engine

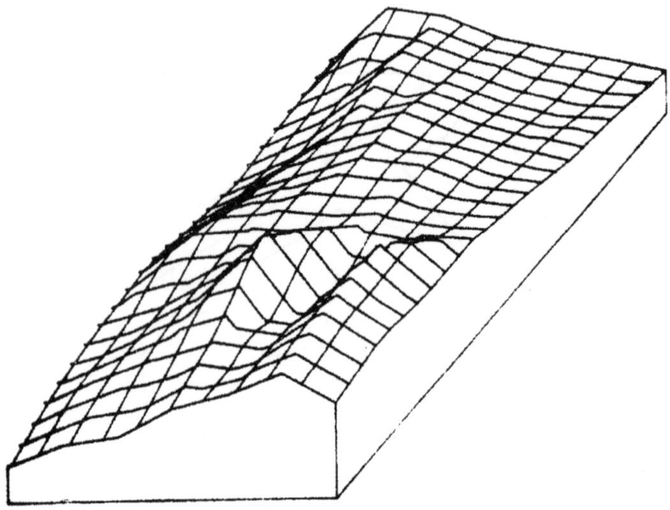

Fig. 12.6. (*c*) Acoustic Intensity Three-Dimensional Contour Map of Diesel Engine

Surface vibration method of sound measurement is available for a few simple radiators, such as a rigid circular piston in an infinite baffle. Combining the closed form analytical solution of the source field with the concept of radiation efficiency permit calculation of sound power radiated by the source. Other techniques that hold promise use the partial coherence function to relate a given input to the output by removing contributions from partially coherent inputs. The phased-array microphone method calls for use of a number of microphones in the far field of the noise generating system.

Varying the relative phase of individual microphones in the array focuses on a specific area of the noise generating system. Incremental variation in the phasing of the array permits scanning all areas of the system, and yields relative contributions of the various areas.

12.8 STATIC TESTS

The purpose of static testing is to obtain frequency response information on the structural response elements of the reciprocating machine. Objectives of the test relate to identifying resonant frequencies, measuring structural damping values, determining resonant mode shapes and measuring frequency response functions. Operational exciting forces are applied at locations where vibration response is of interest. In a noise control task it is often found that relatively low frequency and sufficiently well spaced resonance modes dominate the transmission of structure borne energy and the radiation of airborne sound.

Frequency response is basically a determination of the output of a system in terms of the input to the system. For structural dynamics work this usually takes the form of vibration response in terms of displacement, velocity or acceleration at some location on the structure relative to a force applied at the same or another point in the structure. Complete description of a frequency response function requires both magnitude and phase information, although often only the magnitude of the response is of interest. Force may be applied to the machine structure in several ways. A piezoelectric or an electromagnetic exciter is attached to the structure through a load cell for measuring the applied force. An impulsive force may be applied by a hand held hammer with a load cell attached to it. When measurements are to be made on a weak or lightly damped structural component, attachment of an exciter can significantly alter the structure's characteristics. If the impact technique is not suitable in such a case, a non-contacting acoustic or magnetic exciter may be useful.

Before the advent of digital equipment the swept sine wave method was widely used for frequency response testing. With the traditional analog system exciter signal is a sine wave swept through the frequency range of interest; the rate must be low relative to the response time of the structure. Thus, in structures where damping level is low measurement time is high. Another disadvantage of the method is that on non-linear systems frequency response is also non-linear. A major advantage derives from the fact that energy is concentrated at a single frequency. Hence, signal-to-noise ratio is relatively high, so high force levels can be maintained at a constant level through a feedback loop.

The random excitation method became popular with the availability of digital Fourier analyzers. The signal applied to the exciter is a band-limited random signal with a period much greater than the analysis sample time. The technique is much faster than the swept sine method, because information is being obtained simultaneously throughout the frequency band. Also, it produces the best linear approximation, in the least-square sense, of a non-linear frequency response function. However, the spectrum shape of the signal to the exciter cannot be controlled easily.

Problems frequently arise in the testing of reciprocating machines due to the presence of non-linearities. Clearance at bearings and joints is a prime example. The usual approach is to pre-load the gap with flexible springs. This will ensure that mass-controlled compliance associated with the modes introduced by the flexible springs will not contribute substantially to the overall response in the frequency range of interest. Problems also occur because of the basic non-linear aspects of a reciprocating mechanism-inertia characteristics vary with the crank angle, so frequency response also varies with the crank angle (section 4.6). In internal combustion engines primary gas forces from combustion occur over a small crank angle, so a dedicated piston and crank arrangement will not have a serious problem. However, multi-cylinder engines pose a different problem. In a 90° V configuration, for example, the connecting rods of the opposite banks are attached to the same crank, hence ignition will take place after the crank turns a corresponding angle. A host of other factors also come into play. Non-linear damping characteristics may be encountered internally among machine components. Since damping force varies with gap size and velocity across a joint, impact testing may be ill-suited. A high ratio of peak to root-mean-square level of the excitation signal leads to substantial errors in the measurement.

12.9 MONITORING FOR QUALITY CONTROL

Among other measurable parameters (cylinder pressure, temperature, flow, torque, horsepower, speed) information obtained from a noise and vibration control program of a reciprocating machinery system provides useful insight on the machine's operation. In measuring the machine's sound spectrum, however, it has to be made in a location where background noise is minimal. Such a situation does not commonly occur on a production line. If sound is selected as a measuring parameter for quality control testing, a low ambient anechoic test chamber must be available, and the machines must pass through the chamber one by one for sound test. Most automobile manufacturers now routinely carry out this procedure as part of the overall quality evaluation of the engines.

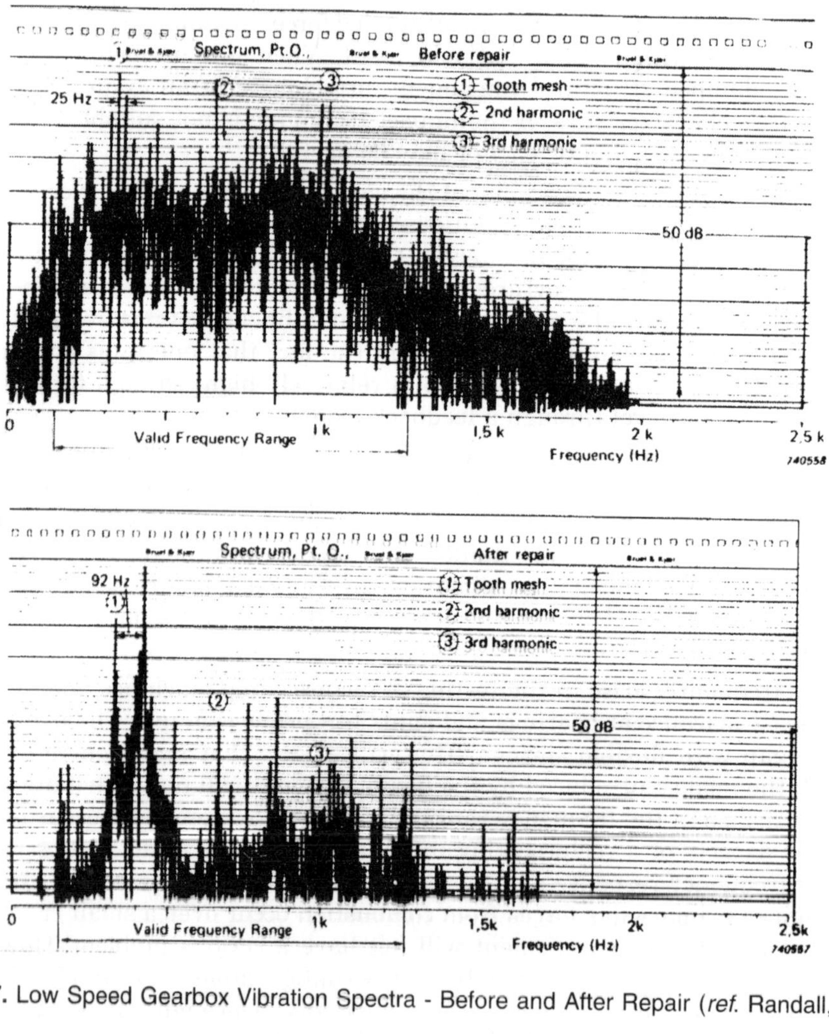

Fig. 12.7. Low Speed Gearbox Vibration Spectra - Before and After Repair (*ref.* Randall, 1995)

As an alternative to sound testing, a machine's vibration spectrum can be evaluated. The correspondence between sound emitted by a defective machine and its vibration levels at certain selected points in the structure must be determined. Evaluation of the machine's vibration signature can then be used for quality control. Although vibration may be measured in terms of displacement, velocity or acceleration, the amplitude and frequency range of interest of most reciprocating machine's vibration spectra make measurement of acceleration most convenient. If the vibration measurement is performed on a production line, the machine must be vibration isolated from the assembly line. This will ensure that interference between assembly line fixtures and the machine structure is eliminated, and only the machine structure's vibration spectrum will then be considered in determining the machine's acceptance. Local structural resonances and its effect on the level of noise and vibration must be taken into account. Since resonant frequencies tend to change from unit to unit, means of detecting and eliminating this parameter is vital. Selection of accelerometer type and its location should take into account the frequency range of interest.

Selected signals under various operating conditions must be carefully checked in order to observe the characteristics of acceptable and unacceptable machines or components. Both time and frequency transforms of the signals should be examined, keeping in mind the analyzer's frequency range, resolution and need for tracking (for order analysis). Visual comparison of the signals and their transforms provide useful information. Figure 12.7 shows the vibration signature of a badly worn gearbox before and after repair (Randall, 1995).

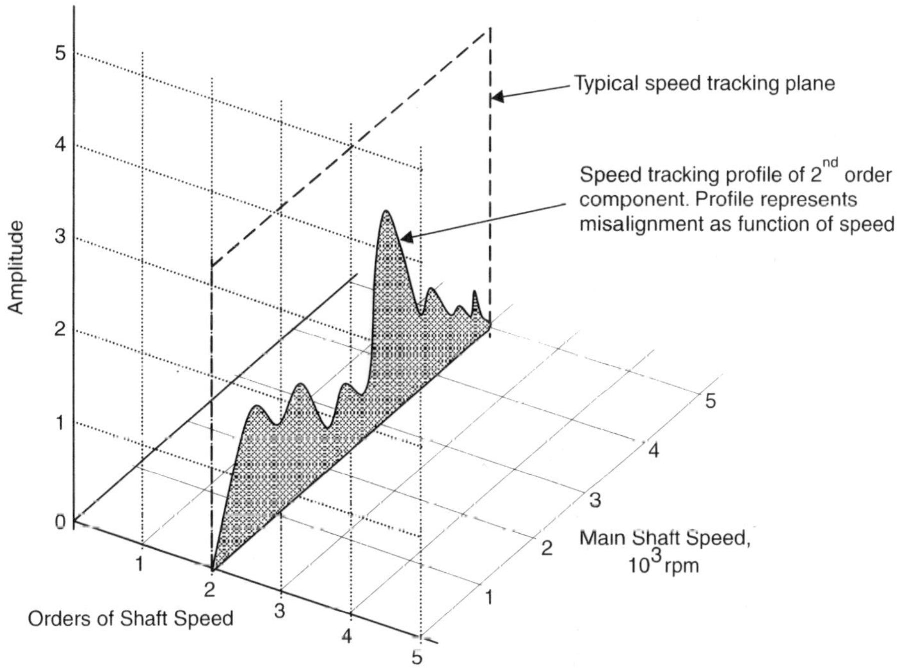

Fig. 12.8. Harmonic Order Analysis (*Courtesy:* Spectral Dynamics Corp.)

For normal operation a machine's characteristic sound and vibration signature will vary around some average spectrum. With limits placed around the average spectrum the procedure can be used as a quality control tool. Many reciprocating machines in a given operating mode will generate harmonically related components, which can yield meaningful information. In a simple frequency analysis, however, separation of the machine's harmonic contents from random noise or vibration may be difficult. Also, small variations in the machine's speed will result in larger variations in the frequency of higher order harmonics. If narrow bandwidths around the harmonic content of a machine's frequency spectrum are established, some form of frequency tracking may be used to ensure that higher order harmonics do not move into the bandwidth when operating speed changes. Figure 12.8 illustrates this concept.

Cepstrum analysis method is used to enhance information made available from an engine's harmonic contents. The cepstrum is defined as the power spectrum of the logarithm of a signal's power spectrum. The unit of amplitude is dB, and time dimension is called quefrequency, meaning spectrum periodicity. Measures of harmonic content and strength of side bands are provided by peak amplitudes. Figure 12.9 illustrates the concept. In addition to processing of signals by a single channel analyzer, a cross spectrum analysis technique between two accelerometer signals improves the forcing function by minimizing the effects of local resonances.

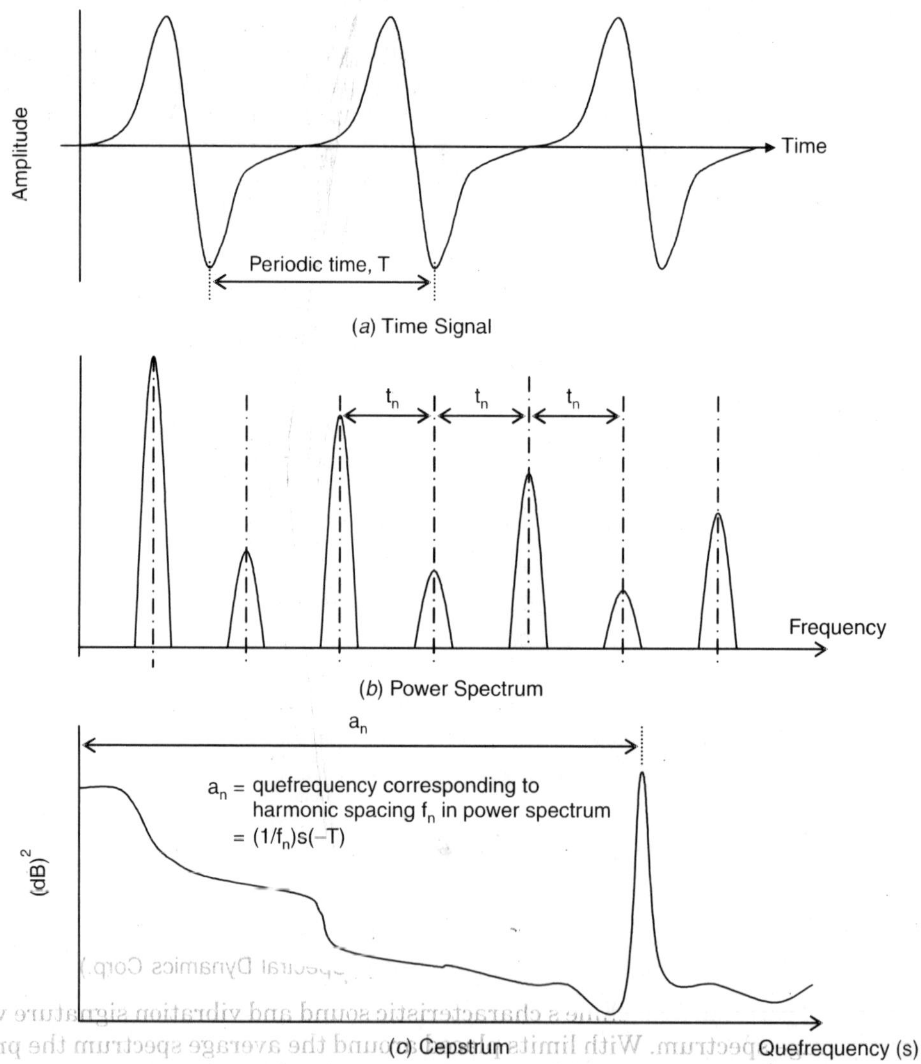

Fig. 12.9. Time Signal, Log Spectrum and Cepstrum of a Periodic Signal (*ref.* Randall, 1995)

Instantaneous acoustic intensity at two different locations on an object may be used for cross spectrum between the two locations. Another technique suggested towards the same objective is meant for high frequency vibration monitoring of gears and bearings. Since a defect causes a broad band frequency spectrum, the resonant frequency will be amplitude modulated at the defect's impact rate (Fig. 12.10). To determine a problem the carrier resonant frequency is filtered, demodulated and spectrum analyzed as shown in Fig. 12.11.

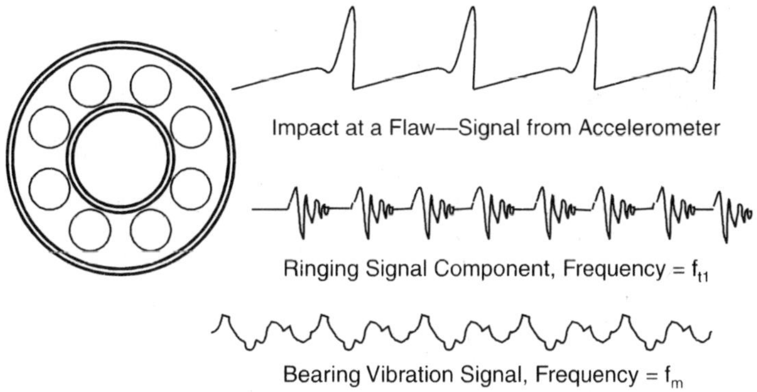

Fig. 12.10. Amplitude Modulation of Transducer Signal due to Flaw in Bearing (f_{t1} = ringing signal, f_{tn} = component vibration signal) (*ref.* Harting and Taylor, 1984)

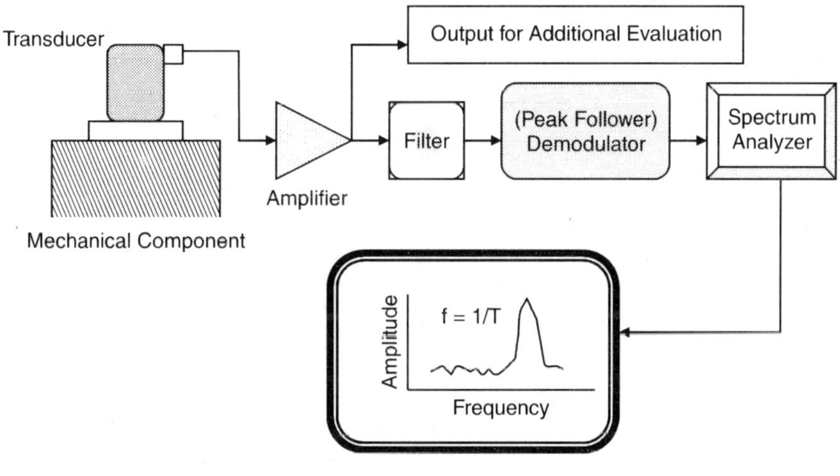

Fig. 12.11. Demodulation and Analysis of Resonance Signal of Fig. 12.10 (*ref.* Harting and Taylor, 1984)

12.10 CRANKSHAFT TORSIONAL VIBRATION AND NOISE

Multiple ignition centers at hot deposits coating the combustion chamber of an engine accounts for rapid and excessive pressure build-up in the cylinder, giving rise to a noise sometimes described as 'thud'. 'Rumble' is associated with high compression reciprocating machinery running for long periods at part load.

Noise emanating from main bearings is of considerable interest. Crankshaft acceleration due to gas pressure torque and elastic oscillations of its elements cause the crank pin and main journal to strike the bearings, generating vibration and noise. Many names have been given to this phenomenon, including 'roughness', 'bearing thrash' and, unfortunately, 'rumble'. Noise generation follows the following sequence:

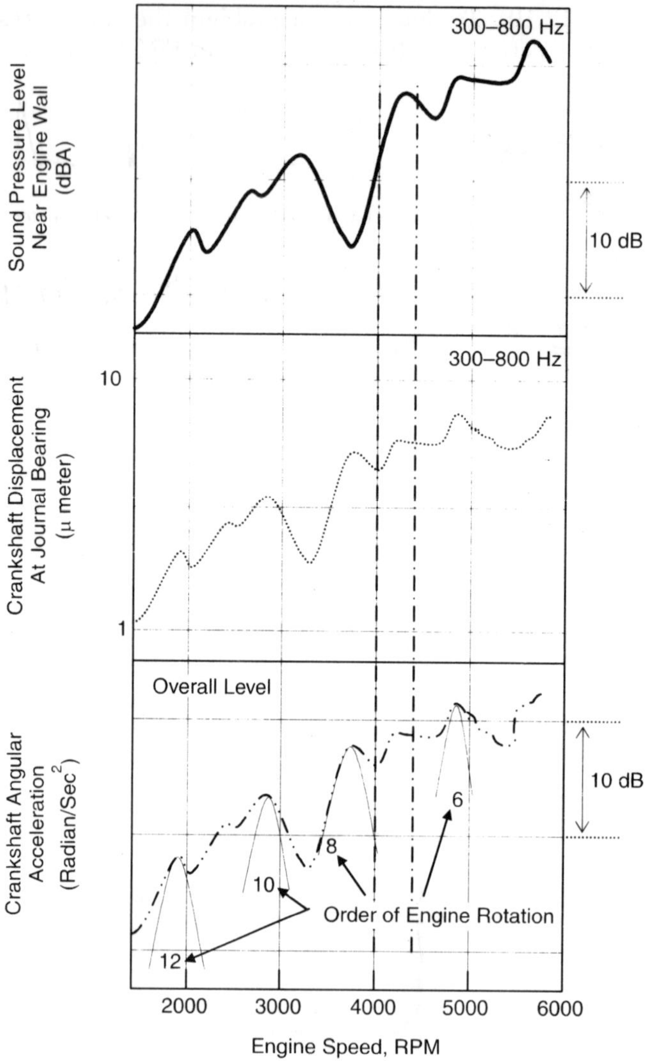

Fig. 12.12. Rumbling Noise and Crankshaft Vibrations (*ref.* Aoki, 1986)

- Variable inertial forces act on the reciprocating mass, pulling the crank pin.
- Variable gas pressure loads are imposed on the piston. In the case of a combustion engine a sudden increase in the gas pressure takes place after the ignition and combustion phase of the cycle. The connecting rod transmits this load and strikes the crank pin.
- The crankshaft accelerates and strikes the main journal bearings. Shock waves transmit to the cylinder block through the main bearings, and noise is radiated from the cylinder block surface. In a 4-cycle machine the process is repeated $n/2$ times, where n is the number of cylinders.

Aoki (1986) reported analytical investigation of the phenomenon, combined with measurements of torsional oscillations of the crankshaft and sound pressure levels. Measurements were made on a water-cooled 4-cycle, 4-cylinder, in-line, gasoline driven automotive engine. Engine displacement is 1.958 liters (81 mm bore x 95 mm stroke), output 144 horsepower at 6,200 rpm, torque 172 N-m at 4,000 rpm. Using displacement sensors, accelerometers and proximity probes, continuous and simultaneous measurements were made of: (*a*) vertical and lateral crankshaft displacements at each journal, (*b*) rotational and axial acceleration at each throw of crankshaft, (*c*) flywheel displacement in all three directions, (*d*) acceleration of crankshaft pulley in all three directions. Figure 12.12 shows sound pressure levels measured near the engine wall and angular acceleration of the crankshaft at the pulley. Rumbling noise peaks in the shaded area around 4,100 rpm, indicating a link between the crankshaft's torsional vibration amplitude and sound pressure level. Interestingly, the author predicted resonance at the sixth order of engine speed of 4,100 rpm using the finite element approach.

Torsional vibrations of a 6-cylinder in-line 4-cycle diesel engine are investigated by Ide, 1990. Noise spectrum of the engine operating at 2,800 rpm at full load revealed order 7 to be of interest. Figure 12.13 shows engine noise and crankshaft torsional vibration signatures, where

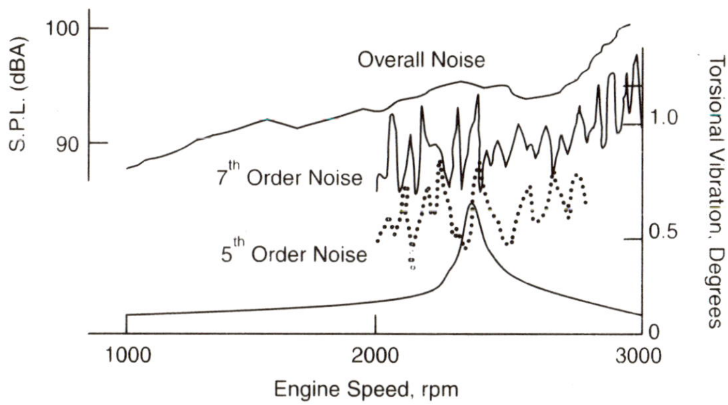

Fig. 12.13. Noise and Crankshaft Torsional Vibrations of Diesel Engine (*ref.* Ide, 1990)

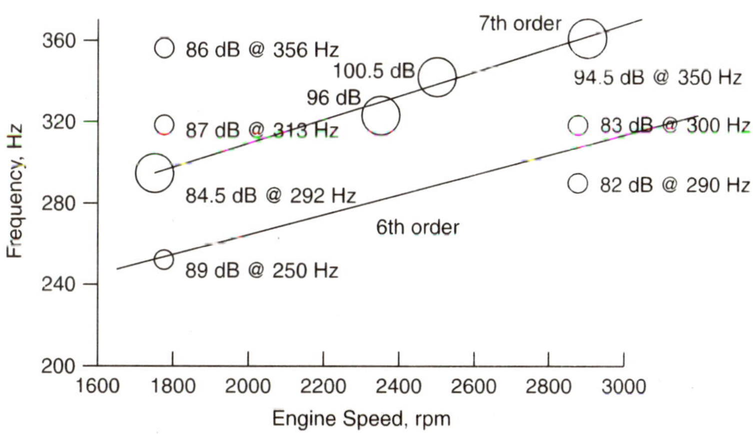

Fig. 12.14. Campbell Diagram Representation of Noise Data of Diesel Engine (*ref.* Ide, 1990)

a peak occurs at order 6 of engine speed. Exciting forces caused by torsional vibration of order n has components of $(n + 1)$ and $(n - 1)$ orders; this explains the presence of orders 7 and 5 components of engine noise peaks when order 6 in torsional vibration appears. Figure 12.14 shows noise speed characteristics of the diesel engine in Campbell diagram format.

12.11 CRANKSHAFT BENDING VIBRATION

Discomforting noise often accompanies bending vibrations of a crankshaft and flywheel system operating at or near maximum load condition. Bending in the crank pin results in angular misalignment with the bearing. Noise level and bending stress increase at high operating speeds, as shown in Fig. 12.15 for an internal combustion engine (Sasaki, 1984). Bending vibration at crank pin # 4 is shown in Fig. 12.16, and engine noise spectrum in Fig. 12.17. Since noise frequency band is composed of harmonic components of engine revolution in the 200 to 500 Hz frequency range, it may be concluded that noise problem is caused by solid propagation of vibration from the engine. As noted in the example of diesel engine, an order n of crankshaft vibration appears as $(n + 1)$ and $(n - 1)$ order of vibration under static condition. Bending stress in crank pin number 4 shows a resonant peak at 270 Hz, the third order of 5,400 rpm rotor speed.

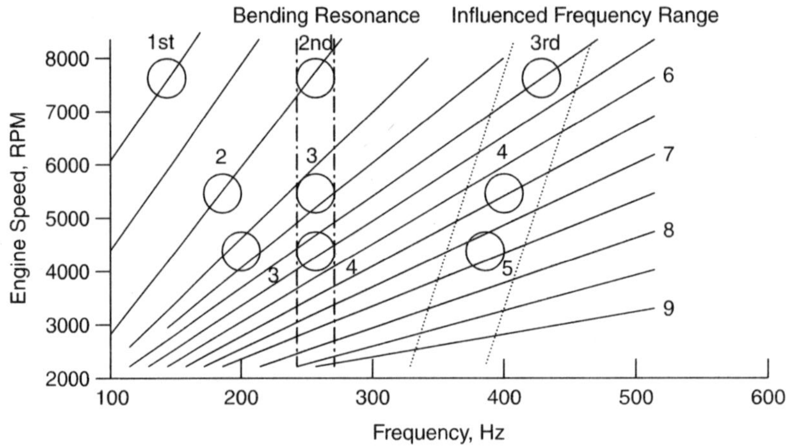

Fig. 12.15. Crankshaft-Flywheel Bending Resonance and Noise (*ref.* Sasaki, 1984)

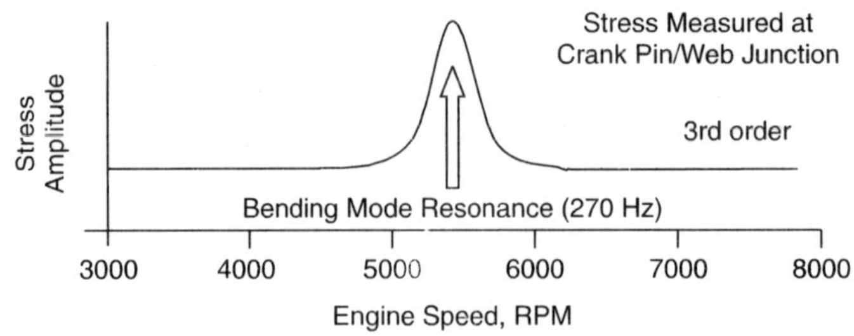

Fig. 12.16. Bending Vibration of Crankshaft at # 4 Crank Pin (*ref.* Sasaki, 1984)

12.12 NOISE CONTROL THROUGH DESIGN MODIFICATION

In a reciprocating machine's structure gas or fluid pressure and mechanical forces occur over a relatively wide frequency range. They are transmitted to the external machine surfaces through two dominant paths: through piston, connecting rod and crankshaft to the crankcase, and through the liner to the head and machine block. To control the noise level of the machine's structure vibration amplitudes of its surfaces and attachments must be reduced. Among major noise contributing factors, the role of each component in causing resonant and forced vibration response must be established. Depending on the outcome of the noise source and severity evaluation, different approaches can be taken to change the dynamic characteristics of a component in order to reduce its response to exciting forces. Figure 12.18 shows the overall noise levels in a V-8 engine and its major noise contributing components in the 900 to 2,500 rpm operating speed range. What is of interest here is that a number of components in a diesel engine exhibit similar overall noise levels. Hence, if a decrease in overall noise is required, investigation and reduction of noise level of every component is necessary. Also, noise levels generally increase with speed of all the components. The effect of load on noise is not as pronounced as speed.

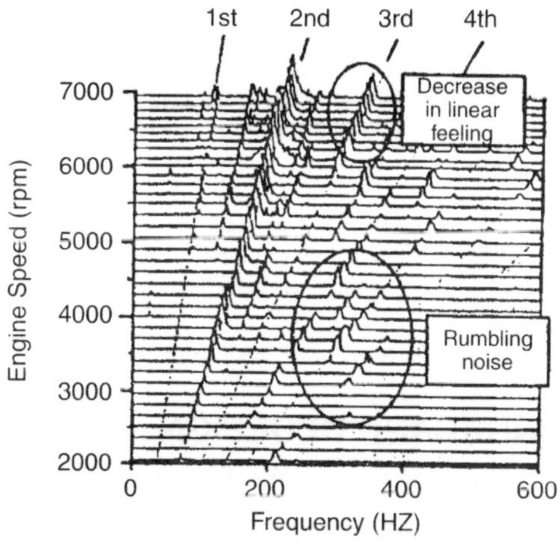

Fig. 12.17. Noise Spectrum of Engine in Figures 12.15 and 12.16 (*ref.* Sasaki, 1984)

To control the noise of each component experimental work must be done for different speed and load conditions. Narrow band analysis and contour maps of overall noise levels must be included. A number of microphones are usually needed for measurements in the full frequency range. Dynamic response of the component is determined over the frequency range where high noise levels exist. Analysis of the component's vibration responses must be made under the same load and speed conditions as those for the noise measurements, including attachment points. Correlation of measured and calculated results helps in identifying the nature of the problem. Specifically, it is of interest to know if a high noise level is associated with a forced or resonant vibration condition. Modification of the component may then be tried to reduce the vibration response. For example, if a sheet metal component has a resonant response, damping may help to control the response. On the other hand if the component responds mainly in a forced mode, as in a cast aluminum component, an isolation system might be more appropriate. A crankcase responding to a forced response might require added stiffness to reduce the noise level. Components need to be examined on an individual basis before deciding on modifications for controlling its noise level.

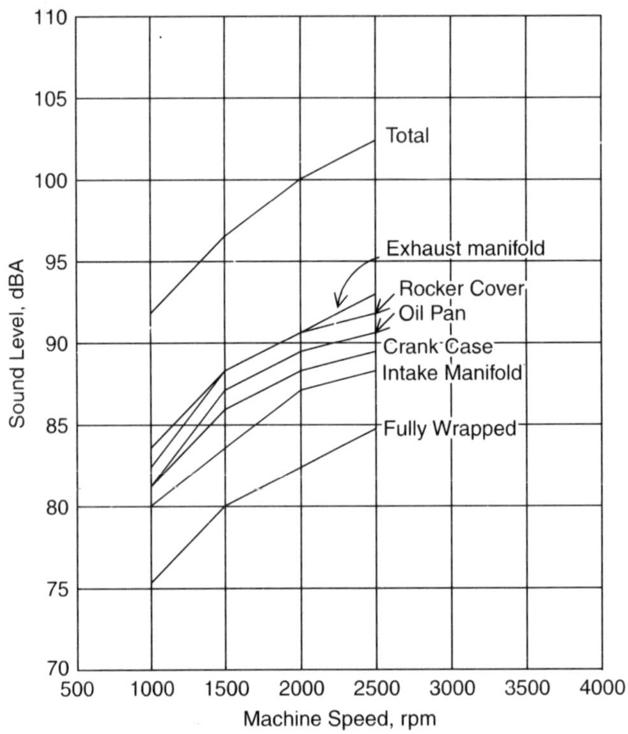

Fig. 12.18. Noise Contribution of Major Components and Surfaces of Diesel Engine

12.13 EXAMPLE OF CRANKCASE REDESIGN

The structure of a crankcase is a major noise contributor on many reciprocating machines. In diesel powered automobiles it is important because of its location in the vehicle. It also transmits vibration to the oil pan. Figures 12.19 and 12.20 represent narrow band analysis of measured sound and vibration data obtained simultaneously on the crankcase of a V-8 engine operating at rated speed and load (Baxa, 1982). The noise and vibration spectra are similar with variations in frequency, primarily because the crankcase is a major contributor to the overall engine noise when measured from the right or the left side. Although a number of peaks are present in the vibration spectra, it is not readily determined if they are due to forced or due to resonant response, or a combination of both. However, an interesting observation to be made is the increase in both the noise and vibration level above 800 Hz. The increase in vibration may be attributed to a higher forcing function or due to a dynamic weakness in the crankcase structure, especially around the oil pan rails. Dynamic response may be analytically or experimentally monitored. The crankcase is quite a stiff structure with few resonances at high frequencies. Vibration peak levels are also quite wide, indicating high modal damping in the system. High modal stiffness and damping are characteristic of automotive crankcases.

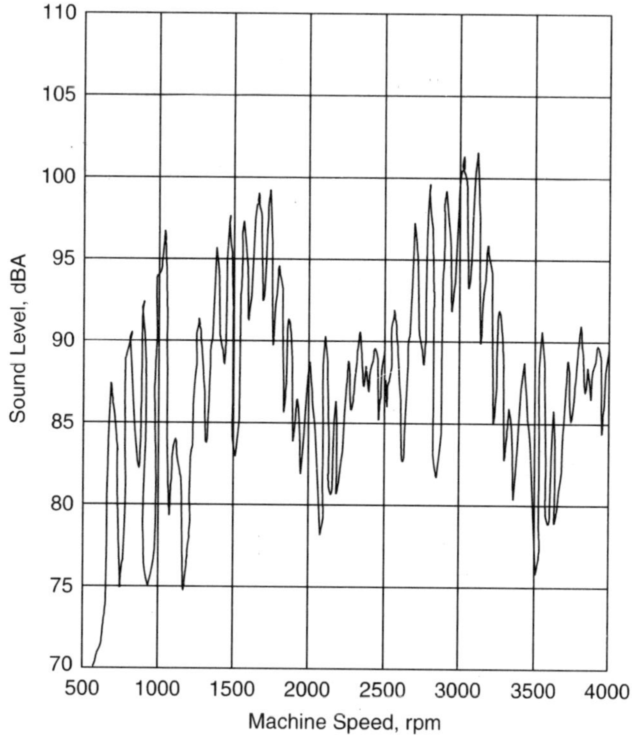

Fig. 12.19. Operating Sound at Side of Engine (*ref.* Baxa, 1982)

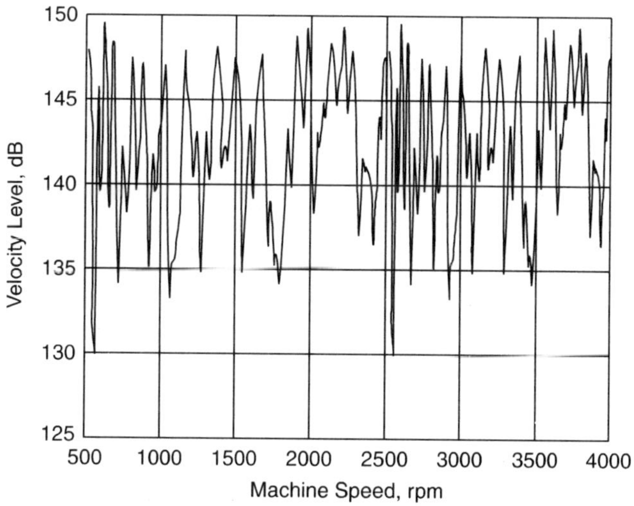

Fig. 12.20. Measured Vibration at Oil Pan Rail of Crankcase (*ref.* Baxa, 1982)

Further investigation of the problem requires an understanding of the relative mode shapes at natural frequencies. Figure 12.21 illustrates the mode shape at 1600 Hz, the fundamental mode, with both sides vibrating in phase. For clarity only one side of the crankcase is shown, although the complete structure was analyzed. The peaks seen in Fig. 12.20 are caused by components inside the crankcase. From figure 12.19 high associated noise levels occur at frequencies below and in the neighborhood of the fundamental mode of vibration. The situation represents the case where operating noise is controlled by the stiffness of the crankcase, so both noise and vibration levels may be reduced by increasing the static stiffness of its structure. Increasing the damping in this example is not useful, since the number of resonances at high frequencies is limited. Introducing more damping into the system would require comparable adjustments to the stiffness, which is not always practical.

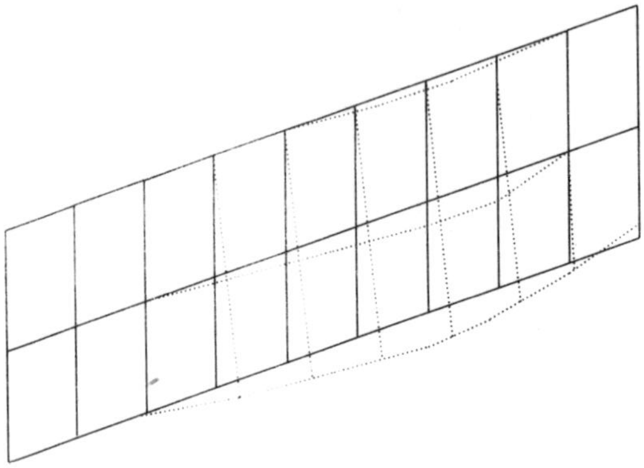

Fig. 12.21. Fundamental Mode Shape of Side of Crankcase-1,600 Hz (*ref.* Baxa, 1982)

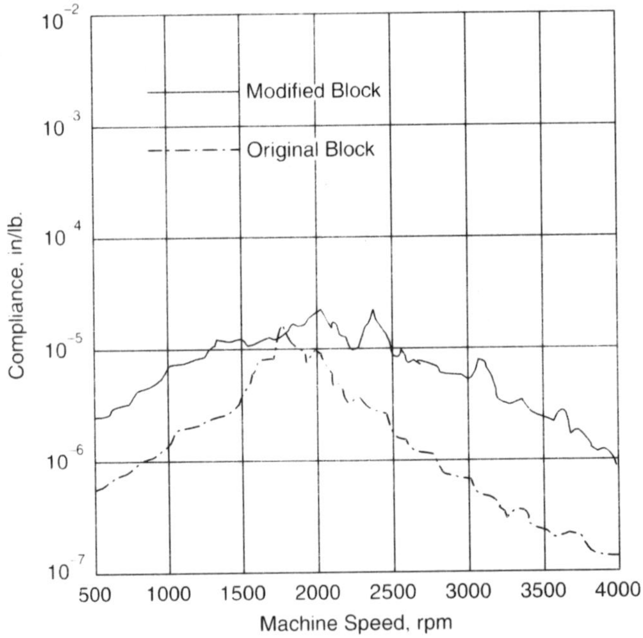

Fig. 12.22. Comparison of Driving-Point Frequency Response (*ref.* Baxa, 1982)

The level of increase in stiffness required for the crankcase may be estimated from the following considerations. The root-mean-square displacement is inversely proportional to $K^{3/4}M^{1/4}$ $\eta^{1/2}$, where K, M and η are modal stiffness, mass and damping respectively. Although derived for a single degree of freedom model, the expression is valid for a plate vibrating in its fundamental mode of vibration, and offers an estimate of the modification required in the structure.

Based on this procedure, if the mass and damping remain the same, an improvement of 10 dB in the stiffness will reduce the overall displacement by 7 dB. From practical considerations this level of reduction may be satisfactory. A modified cylinder block used the required changes at the oil pan mount rails. The resultant frequency response plot is shown in Fig. 12.22, with the response plot of the original block from Fig. 12.20 also included. The static stiffness increased by 8 dB. Figure 12.23 shows a comparison of the sound level. Vibration response has been in both low and high frequency ranges, and similar reductions have been achieved in the sound levels for the given set of operating conditions.

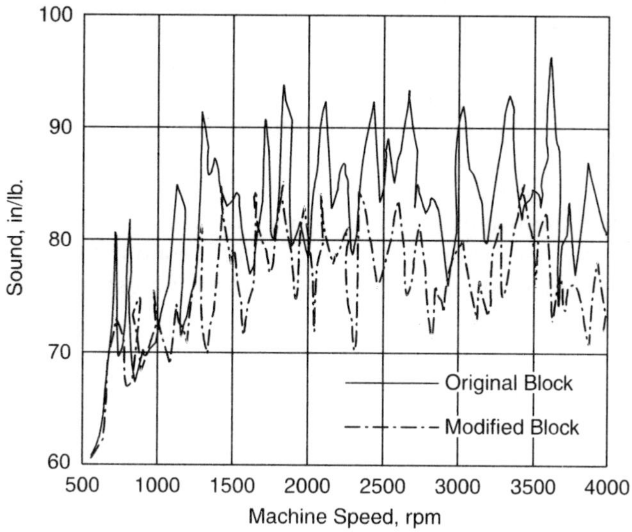

Fig. 12.23. Comparison of Operating Sound at Side of Engine (*ref.* Baxa, 1982)

12.14 DAMPING TECHNIQUES

Damping materials have been used effectively to reduce forced vibration amplitudes in structures, increase fatigue life and to reduce structure borne noise. Damping may not be the ideal solution in many circumstances, but it provides an alternative solution for controlling vibration and associated noise problems in a cost effective manner. Damping becomes advantageous when:

- A component has a number of natural frequencies of vibration in a given frequency band, and the source is exciting many of these vibration modes.
- Modifications to the source or to the load-carrying path cannot be readily made.
- Little improvement is expected by stiffening or reducing the mass.
- Structure's resonances are important to the problem.

An understanding of the basic characteristics of the visco-elastic damping material, as also its configuration, is necessary for a successful application of the technique. The energy dissipated by the damper is proportional to the Young's modulus and the loss factor of the material. The product of the two parameters is called loss modulus, a basic material property.

Two types of damping treatments are commonly used, surface and tuned damping devices. Surface damping is used to solve vibration and noise problems in sheet metal structures, such as an engine's oil pan, and provides high damping capability over wide temperature and frequency ranges. Surface dampers may be subjected to extensional or shear deformation.

Extensional damping, also referred to as free-layer or unconstrained, is more commonly used. It calls for application on one or both sides of a structure, so the material is subjected to tension and compression deformation as the structure undergoes cyclic loading. The performance of any visco-elastic damping material is affected by temperature, thickness of material, frequency, structural damping and method of bonding. The loss modulus of most commercially available damping treatment materials reaches the maximum value in the 70° to 110° F region, hence the effectiveness of the material to reduce vibration and noise also is at a peak in this range. Increasing the thickness of the damping material achieves higher damping up to an extent. If the thickness of the material is much greater than that of the base structural material, the composite system damping asymptotically approaches that of the material itself. For small thickness the increase in the damping is nearly linear. The structural damping coefficient of the base material in some cases is negligible, but in many other cases it is significant and should be included in the evaluation. This is particularly the case in assemblies with joints; some diesel engine sub-assemblies have modal damping values approaching 0.05. Addition of unconstrained damping treatment in small amounts will not help in such structures. The effect of frequency on the performance of damping is secondary to that of temperature. For most applications the effect is small and can be ignored, unless a wide frequency range is of interest.

It is to be expected that the performance of the damping material to control both vibration and sound is as good as the method of attachment to the structure. Some commercially available materials come with their own adhesive, and generally provide easier and better attachments if the component surface permits it. However, a majority of unconstrained damping materials require a structural epoxy or an adhesive layer in order to reach complex surface contours. The adhesive layer should be as stiff and thin as possible. A softer epoxy layer between the damping treatment and base structure will yield to shear deformation, and will detract from the performance of the damper.

Energy dissipation in the damping material is a function of deformation, hence damper efficiency increases when located in areas of the structure that are subjected to large deformations. Detailed knowledge of the vibration mode shapes of the structure is necessary in determining locations where maximum amplitudes and bending stresses occur. Location at a nodal point in the structure will have minimal effects. However, when vibration and associated noise protection is required in a large frequency range, many modes are likely to be present, and partial coverage of the component based on modal activity may not be adequate.

Attempts have been made to broaden the damping performance of damping materials with temperature. Limited success has been achieved by varying the composition of the visco-elastic material and filler type. A more effective approach is to use multiple materials with different properties. For example, if the operating temperature range is – 50° to 150°F, it may be necessary to select one material with an optimum damping value around 0°F and another material at about 100°F. The materials may then be placed on opposite sides of the component wall, or one on top of another, to obtain a wider temperature range of protection.

Shear type of damping treatment (also called sandwich damper) is more efficient than unconstrained layer damping treatment (Collacott, 1987). Since greater analytical effort is required, more precautions are necessary in its application. The rubber like visco-elastic material is constrained by a metal layer. During cyclic bending the metal layer constrains the damping material to deform in shear, and energy is dissipated through shear deformation. Performance of the constrained layer damper depends considerably on the geometry of the component. Maximum shear strain is present in the visco-elastic material only if the constraining layer is sufficiently stiff. At the same time, however, the layer's stiffness should not exceed that of the structure. Hence, the constraining layer is of the material and geometry as that of the structure to be damped.

12.15 TUNED DAMPERS

Tuned dampers have the advantage of low weight, and can be used in structures responding to a number of resonant frequencies. Tuned dampers differ from classical vibration absorbers in that it uses a rubber like material to provide stiffness, which also dissipates dynamic energy in the form of heat.

For simple cross sections the stiffness of the rubber like tuned damper will depend on the mode of deformation. When it is subjected to shear deformation its stiffness is expressed by:

$$k_G = GA/H \tag{12.12}$$

and when it is subjected to tension or compression loads the expression for stiffness is:

$$k_E = \frac{EA}{H}\left[1 + \beta\left(\frac{A}{A'}\right)^2\right] \tag{12.13}$$

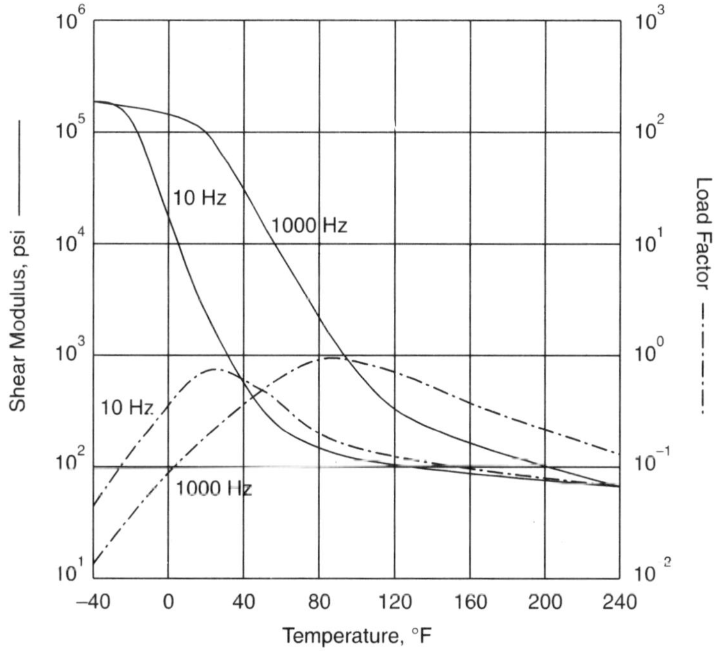

Fig. 12.24. Material Properties of a Typical Tuned Damper

where G = shear modulus

E = Young's modulus

H = height

A = loaded area

A' = unloaded area

β = 1.5 for filled rubber like materials

= 2.0 for unfilled rubber like materials

A correction factor based on the shape of the damper is necessary in (12.13); it arises because Poisson's effects can introduce a bi-axial state of stress in the material. Properties of a typical damper material are shown in Fig. 12.24. Performance of the damper depends on the operating temperature, since its properties vary with it. If the operating temperature is going to be over a wide range, it may be necessary to use the properties at the higher temperature end, since both the loss factor and the modulus vary at a slower pace in the region.

Besides damping, reduction in noise level may also be obtained through proper evaluation of the source. For example, reduction in the exhaust system of an internal combustion engine is generally achieved through proper muffler selection, as also piping lengths and diameters, mounting of exhaust system components and positioning of the exhaust outlet (see sections 10.15, 10.16). On the air intake side an engine's air cleaner will provide significant attenuation of intake air noise. Additional attenuation requires an intake air silence and adequate design, selection and mounting of the components to take care of surface radiated noise. Cooling system noise depends on fan tip speed (see Fig. 12.25) and aerodynamic design of fan blade. A water pump and radiator may be used to enhance cooling efficiency; the radiator core must be of sufficient surface and airflow area. A thermostatically controlled variable speed fan drive will add to the cost, but fan cooling output can be matched to machine cooling requirements. At lighter loads, or when vehicle speed results in adequate air flow through the radiator, fan speed will be reduced, decreasing driving horsepower and output sound level.

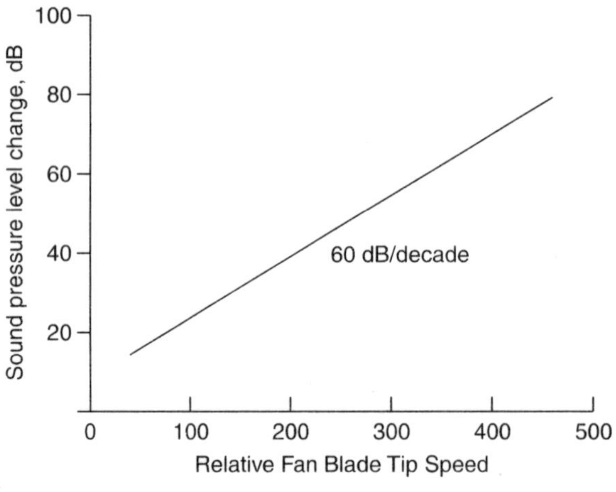

Fig. 12.25. Effect of Fan Speed on Sound Generation

Reduction in reciprocating machine surface radiated noise requires smaller running clearances and machining tolerances of components. Noise generation from piston slap against cylinder walls and shaft journal strike against bearings can be controlled if the component has to travel less before contacting the other surface. Partial or complete machine enclosures can be used to reduce noise transmitted from machine surfaces to the environment, but will add to the cost of the assembly. An effective enclosure incorporating sound absorption materials, sound barrier materials and vibration damping material can attenuate surface radiated noise by 20 dBA. The enclosure will require vibration isolation from the machine and its support structure to prevent the enclosure panels also from becoming noise sources.

REFERENCES AND BIBLIOGRAPHY

Alpini, A., Busso, M., Rsupa, G., Turino, G., "Analysis Techniques of Combustion Noise and Vibrations in Diesel Engines", SAE Paper # 800406, 1980.

Anderson, D., Grover, E. C., Lalor, N., Priede, T., "Origins of Reciprocating Engine Noise-Its Characteristics, Prediction and Control", ASME Paper # 82-WA/DGP-3, 1982.

Anderton, D., "Relation Between Combustion System and Engine Noise", Paper # 790270, SAE *Proceedings* P-80, 1980.

Andrews, S. A., "The Analysis and Mechanism of Engine Crank Rumble", Institute of Mechanical Engineers, ISVR, pages 99-109, 1989.

Aoki, H., "Effects of Power Plant Vibration on the Quality of Sound Inside a Passenger Compartment During Acceleration", Nissan Technical Review, Nissan Motor Co., 1986.

Atkins, K. A., Challen, B. J., "A Practical Approach to Truck Noise Reduction", Paper C131/79 in Noise and Vibration of Engines and Transmissions, *Institute of Mechanical Engineers Conference Publication,* July, 1988.

Baxa, D. E., "Noise Control In Internal Combustion Engines", John Wiley & Sons, New York, 1982.

Beranek, L. L., ed., "Noise and Vibration Control", Mc-Graw Hill, New York, 1991.

Booker, J., "Dynamically Loaded Journal Bearings: Numerical Application of Mobility Method", Institution of Lubrication Technology Vol. 93, No. 2, 1983.

Booker, J., "Design of Dynamically Loaded Journal Bearing", ASME Design Engineering Conference, Fundamentals of the Design of Fluid Film Bearings, ASME, pages 31-44, 1984.

Broch, J. T., "Application of Bruel & Kjaer Measuring Instruments to Mechanical Vibration and Shock Measurements", Bruel & Kjaer Instruments, Inc., Cleveland, 1994.

Clarkson, B. L., "The Social Consequences of Noise", Nominated Lecture, *Proceedings of Institute of Mechanical Engineers*, Vol. 189, August, 1982.

Collacott, R. A., "Mechanical Fault Diagnosis", Chapman & Hall, London, 1987.

DeJong, R. G., Manning, "Modeling of Vibration Transmission in Engines to Achieve Noise Reduction", SAE Paper # 890360, 1989.

Haddad, S. D., "Simulating Piston Slap by an Analog Computer", Journal of Sound & Vibration, Vol. 72, No. 1, pages 79-93, 1987.

Harting, D. R., Taylor, J. W., "Demodulated Resonance Analysis System", US Patent # 3,842,663, October, 1984.

Hodgetts, D., "Vibration of A Crankshaft", Institute of Mechanical Engineers, Paper # C99/87, 1987.

Ide, S., "Improvement of Engine Sound Quality Through A New Flywheel System Flexibly Mounted to The Crankshaft", SAE Paper # 900391, 1990.

Lalor, N., Petyt, M., "Modes of Engine Structure Vibration as a Source of Noise", SAE Paper # 850833, 1985.

Law, B., "Prediction of Main Bearing and Crankshaft Loading in Reciprocating Engines", CIMAC Paper # 6358C/424A, 1983.

Law, B., "Prediction of Crankshaft and Flywheel Dynamics", Institute of Mechanical Engineers, Paper # C382/046, pages 487-498, Perkins Technology, 1989.

Maas, H., "Calculation of Crankshaft Plain Bearings", 9th CIMAC Conference, Paper # A-22, Stockholm, 1984.

Maas, H., "Calculation of Crankshaft Plain Bearings", 11th CIMAC Conference, Pages 43-68, Vol. 1, Barcelona, 1986.

Morse, P. M., Ingard, K. U., " Theoretical Acoustics", Mc-Graw Hill, New York, 1992.

Nashif, A. D., "Application of Damping for Noise Control in Diesel Engine Components", *Damping Application for Vibration Control,* ASME, AMD-Vol. 38, New York, 1980.

Pischinger, F. F., Schmillen, K. P., Leipold, F. W., SAE Paper # P-88, *Proceedings Diesel Engine Noise Conference,* page 39, March, 1987.

Potter, R. W., Mathematical Work, Unpublished, September, 1994.

Randall, R. B., "Cepstrum Analysis and Gearbox Fault Diagnosis", Application Notes, Bruel & Kjaer Instruments, Inc., Cleveland, OH, 1995.

Russell, M. F., Cavanagh, J. E., "Establishing a Target for Control of Diesel Engine Combustion Noise", Paper # 790271, SAE *Proceedings* P-80, 1980.

SAE Engine Sound Level Measurement Procedure: J-1074, SAE Handbook, Vol. 2, Page 35-39, 1984.

Sasaki, Y., "A Study of Engine Caused Abnormal Noise of Automobiles at Acceleration", MHI, Technical Review, Vol. 18, No. 1, 1984.

Snowdon, J. E., "Vibration and Shock in Damped Mechanical Systems", John Wiley & Sons, New York, 1988.

Spectral Dynamics Corporation, "Machinery Analysis - Why Signature Ratio?", Technical Publication # RTA-4, San Diego, September, 1987.

Suzuke, T., "Analysis and Mechanism of Engine Intake Rumble Noise", SAE of Japan Paper # 902246, Toyota Motor Co., 1990.

Tsuge, K., "Study of Noise in Vehicle Passenger Compartment During Acceleration", Journal of Society of Automotive Engineers of Japan, Vol. 39, No. 12, 1985.

Tyrell, R. J., "Engine Noise: Practicalities and Prediction, Part 2-Finite Element Analysis", SAE Paper # 870978, Ricardo, 1987.

Ungar, E. E., Ross, D., "Vibrations and Noise due to Piston Slap in Reciprocating Machines", Journal of Sound and Vibration, Vol. 2, pages 132-146, 1985.

Wu, T., "Discussion of SAE Papers 880405 and 880407", SAE *Transactions* Vol. 77, Page 1408-1409, 1988.

Design Optimization 13

13.1 INTRODUCTION

Resources from nature of all kinds are very large, but not boundless, hence calling for man to use them wisely. Structural optimization of a machine component directs all the synergies at an engineer's disposal towards achieving the objective of conserving those precious resources. The expression 'optimum design' refers to one that meets all specified requirements but with a minimum of expense of certain factors such as weight, surface area, natural frequency and cost, among others. Virtually any aspect of a machine's design can be optimized: dimensions, placement of bearing supports, cost of manufacturing, material property, to mention a few. The primary objective of engineering design is to come up with the best configuration for a given situation; hence, structural optimization is at the very heart of the problem. The traditional approach calls for a series of trial and error to obtain a set of targets. Starting from a concept based on past experience, intuition and creativity, the design is analyzed to evaluate form, function and fit to come up with a solution that satisfies more of the specific requirements. The results of the analysis provide the basis for modifying and developing a new design. Further analysis, evaluation and redesign is repeated till additional gains are not possible. Optimization finds the best possible, or optimal, component or assembly design that meets the varied requirements arising from manufacturing and functional operation. Thus, structural optimization defines the rational establishment of the design of a component that is the best of all possible designs within a prescribed set of objectives and inherent limitations.

Analysis of a structure's design calls for the determination of behavior and response of a component (for example, natural frequencies of vibration, response, and dynamic stress) from a given set of dimensions, material properties and load characteristics. In contrast, in the design optimization process, certain dimensions, load and property data are not known. The aim is to find the minimum, or maximum, value of a predefined function subject to certain limitations and constraints. The predefined function may be weight, deflection, manufacturing cost, among others. The constraints may be geometric in nature, such as thickness or height of a member, or a calculated parameter, such as limits on frequencies or deflections.

Optimum design techniques may be classified on the basis of intuition, optimality criteria or mathematical programming. Design of aircraft engines with reciprocating pistons in the first half of the twentieth century was based on failure occurring at full stress, and represents an example of the use of intuitive methods. Optimality criteria may be considered to work in two steps: (*i*) identifying a set of requirements to be satisfied by the optimum design and (*ii*) developing an iterative design procedure to achieve that optimum. In the mathematical programming based method, structural design is treated as a set of problems to reach extreme values of an objective function (manufacturing cost, for example) in a space involving many multidisciplinary parameters; at the same time, the solution must meet a number of geometrical and behavioral constraints.

Preliminary efforts in structural optimization focus on sizing of a part, requiring, for example, determination of optimum cross sectional dimensions of the part. Depending on the complexity of the machine component, more than one such cross section may be necessary. At this level the domain is fixed and does not change during the optimization process. As a further development of the problem, finding optimal boundaries to fit within a given space while interfacing with other members in the machine assembly must be considered. Boundaries for the structure are established by finding the location of joints of a skeletal structure, or finding optimal values of parameters that define the middle surface of a shell structure. This portion of the procedure may be referred to as shape optimization, where the shape of the domain is not fixed but the topology is. It is mathematically equivalent to the geometry of the part being transformed by bending, stretching and twisting without tearing or gluing points together.

Sizing and shape optimizing may lead to sub-optimal results because they suffer from the problem of not necessarily starting with the optimal topology. To overcome this deficiency topology optimization must be considered. In three-dimensional continua the aim is to determine features such as number of holes and their location. The problem of finding the optimal configuration and spatial sequence of mating parts in the assembly of components also lies in this category. The ideal is to simultaneously optimize size, shape and topology of the structure to define its layout. The procedure may be referred to as optimization at the conceptual stage.

At the analysis stage the design must satisfy criteria based on structural integrity and operability. Thus, if stress build up in the area where a crank web meets the journal of a crankshaft does not meet fatigue life requirements, the fillet's radius of curvature must be increased. The number of such calculated parameters will depend on the machine's complexity, performance expectations and cost. As a result the necessary analytical effort has to be at the corresponding level. It also calls for greater level of tradeoffs between the requirements. If manufacturing cost due to a better finish on a component's surface due to fatigue life expectations goes up, adjustments must be made either to the cost and/or life of the part.

13.2 DESCRIPTION OF OPTIMIZATION PROBLEM

Figure 13.1 shows a cantilever beam, and is used to illustrate various aspects pertaining to design optimization. The beam has a length L and uniform cross sectional height h and width b. A uniform load P is imposed along the full length. The beam is to be designed to have minimum weight, and is subject to three constraints:

1. Maximum stress should not exceed σ_{max}
2. Deflection should not exceed δ_{max}
3. Height should not exceed h_{max}

The weight of the beam represents the objective function, or dependent variable, that is to be minimized. It will be a function of the beam's volume and material density.

Since the defined problem does not mention anything about density and beam length, they are considered to be non-variable, and do not directly enter into the optimization picture. Hence, the beam's height and width are independent design variables that may change in order to achieve the optimum design. Height h has a specified upper limit, serving as a constraint on the design. Also, height and width cannot be zero or negative, so their lower limit would be $b, h > 0.0$. Thus, the width has a single-sided limit, while the height has double-sided limits imposed upon it.

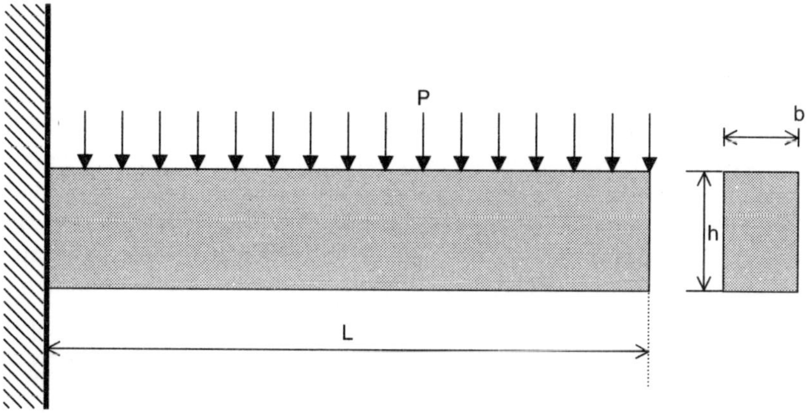

Fig. 13.1. Optimization of Cantilever Beam

Another set of variables is a function of the design variables, and may be grouped as dependent variables. Typically they are response quantities, and may have a maximum or minimum limit. Beam deflection and stress will fall in this group, and have predefined limits dmax and $h$max, respectively. The objective function is the dependent variable to be minimized. It should be a function of the dependent variables. Hence, changing their values should affect the objective function.

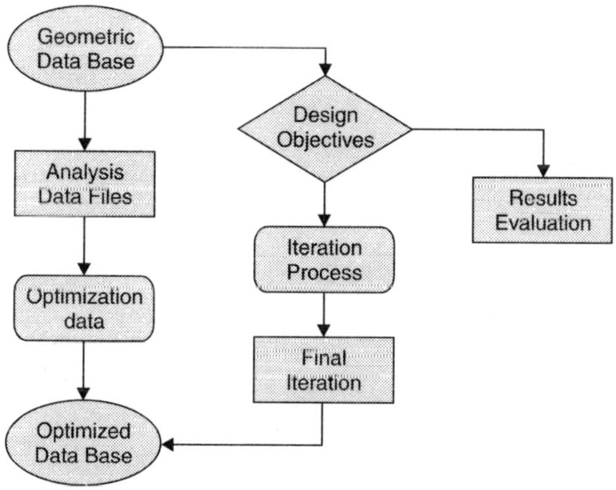

Fig. 13.2. Information Flow during Optimization

The procedure can be understood through the block diagram of Fig. 13.2, showing the flow of information during its execution. The analysis database is separate from the optimization data files. Geometric data on the component is fed into the analysis files to obtain calculated results. Comparison of the results with design requirements provides indications about adjustments to be made to the geometric data, while also building up optimization data. The analysis, evaluation and modification process is continued till a final iteration process produces results that come within acceptable limits of design requirements. Geometric data used in the final iteration is then established in the optimized database for the part, as also for the overall assembly.

13.3 CONSIDERATIONS IN TOPOLOGY OPTIMIZATION

Methods for obtaining topological optimization of a structure range from rigorous mathematically based procedures to the more intuitive approach using engineering judgement. The latter may also be referred as the hard kill/soft kill method. As an example, the length of the crankshaft, and consequently of the multi-cylinder machine, increases by a certain ratio if bigger fillet radii are employed at the crank pins and the journals. The bigger fillet is needed to obtain adequate operating life of the component. But if space requirements for the overall machine dimensions preclude any addition to the length, then the larger size of the fillet must be abandoned. In this example it is tacitly assumed the load carrying capacity of the journal and crank pin bearings, and hence the length of the bearings, cannot be compromised. The upshot in this example is that in order for the machine to be available for the specific application reduced fatigue life will have to be accepted. Hence, reduced fatigue life of the crankshaft is the price to be paid, otherwise the engine cannot be employed.

In many situations minimization of a single function may not be the only goal. Attention must also be directed towards satisfaction of predefined constraints placed on the design. Design configurations that satisfy all constraints may be considered feasible designs. If one or more violations of the constraints are encountered the design would be infeasible. As design sets are generated and if an objective function is defined, the *best* design set is computed and stored. The best set is determined under one of the following conditions:

1. If one or more feasible sets exist the best design set is the one with the lowest objective function value. Put another way, it is the set that most closely agrees with the mathematical goals.
2. If all design sets are infeasible, the best design set is the one closest to being feasible, irrespective of its objective function value.

13.4 OPTIMALITY CRITERIA METHOD

Optimality criteria methods were initially developed as an alternative approach to mathematical programming methods for solving structural optimization problems. The procedures are indirect methods of optimization, and unlike the mathematical programming methods, attempt to satisfy a set of criteria related to the behavior of a structure. The criteria are derived intuitively or rigorously. 'Fully stressed design' and 'simultaneous failure mode design' are examples of intuitive optimality criteria methods. Rigorously derived optimality criteria methods are often based on Kuhn-Tucker optimality conditions. To understand the conditions, consider an optimization problem of the form:

$$\begin{aligned} &\text{Minimize:} && f(x) \\ &\text{such that:} && hj(x) = 0, && j = 1, 2, \ldots n_h \\ &&& g_k(x) + (s_k)^2 = 0 && k = 1, 2, \ldots n_g \end{aligned} \tag{13.1}$$

where $x = [x_1, x_2, ...x_n]T$ are design variables, s_k is an approximate penalized function and n_h and n_g are the number of equality and inequality constraints.

The necessary conditions for minimization of the objective function may be obtained by using the technique of Lagrangian multipliers. The technique replaces the objective function together with the equality constraints with a new objective function in the Lagrangian form $F(x, \lambda)$, where λ_j are the Lagrangian multipliers. Without any loss of generality, then the optimization problem may be stated as:

Minimize: $F(x)$

subject to: $gk(x) \leq 0 \quad k = 1, 2, ...n_g$ (13.2)

According to the Kuhn-Tucker theorem, if x is a local minimum of $F(x)$ satisfying the constraints, then there must exist multipliers nk, such that:

$$\nabla F(x) + \sum_{k=1}^{n_g} \upsilon_k \nabla g_k(x) = 0,$$

$$\upsilon_k g_k(x) = 0, \qquad \text{where } k = 1, 2....n_g \quad (13.3)$$

$$\upsilon_k \geq 0$$

The inequality and the two equalities of (13.3) are called Kuhn-Tucker conditions.

To obtain physical explanation of the theory, consider the Euler-Bernoulli beam problem with a single criterion. In Fig. 13.3 the height of the cross section $z(x)$ and the deflection of the beam $y(x)$ are assumed to be the design variable and response function, respectively. The objective function is given in the form of $\int_0^L \rho(x)\, y(x)\, dx$, where $r(x)$ is a specified weighting on the beam deflection $y(x)$. The lower bound for z is specified as z_o, and the upper bound for the volume as V. The equilibrium equation of the beam is:

$$\frac{d^2}{dx^2}\left(EI(x)\frac{d^2 y(x)}{dx^2}\right) - p(x) \qquad (13.4)$$

together with the boundary conditions.

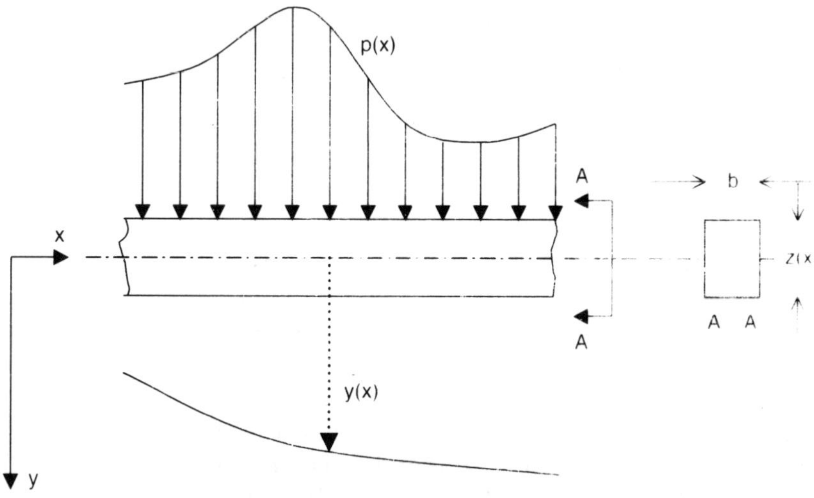

Fig. 13.3. Euler-Bernoulli Beam

The optimal design problem then becomes:

$$\text{Minimize:} \quad \int_0^L \rho(x)\, y(x)\, dx$$

$$\text{Subject to:} \quad \{EI[x, z(x)]y''(x)\}'' - p(x) = 0$$

$$z_0 - z(x) < 0$$

$$\int_0^L bz(x)\, dx \le V \tag{13.5}$$

where L and V are length and volume of the beam, EI is bending rigidity of the beam and superscript (′) represents differentiation with respect to x. By using the Kuhn-Tucker conditions, the optimality criterion may be expressed as:

$$\frac{dEI(x, z)}{dz} \bar{y}'' y'' + g(x) - \Gamma b = 0 \tag{13.6}$$

where $\bar{y}(x)$, $\gamma(x)$ and G are Lagrangian multipliers. Together with (13.6) the other optimality conditions are:

$$\gamma(x)[z_0 - z(x)] = 0; \ \gamma(x) \ge 0$$

$$\Gamma\left[\int bz(x)\, dx - V\right] = 0; \ \Gamma \ge 0 \tag{13.7}$$

Note that the term $\dfrac{dEI(x, z)}{dz} \bar{y}'' y''$ in (13.6) is in fact the gradient of the objective function with respect to the design variable $z(x)$.

13.5 OPTIMIZATION TECHNIQUES

Mathematically rigorous procedures concentrate on a number of calculated parameters. In one procedure derivative information, that is, gradients of the dependent variables with respect to the design variables, is used extensively in order to determine the increment in the design variables required for the subsequent iteration. There may be a number of dependent and design variables associated with a problem, varying over a wide range of design space. Many different tools are available to attain the stated objective:

- The analysis is performed in a closed loop to produce one solution at a time. 'What if' type of scenarios may be studied by using a series of single loop analyses, setting different design variable values before each loop.
- Multiple loops are performed with random design variable values at each loop. A maximum number of loops and a desired number of loops may be used. This tool is useful for studying the overall design space, and for establishing feasible design sets for subsequent optimization analysis.
- Starting from a reference design set, a sweep generation process can obtain several sequences of design sets. Specifically, it modifies one design variable at a time over its full range, using increments in the design variables. Global variational evaluations of the objective function and of all the variables are possible by using this approach.
- Statistical tools may be used to generate design sets at all extreme combinations of design variable values. The technique has similarities with the 'design of experiment' technology that uses a two-level full and fractional factorial analysis. The primary objective is to compute main and interaction effects for the objective function and the state variables.

- At a reference design set gradients of the objective function and the state variables with respect to the design variables are calculated. Local design sensitivities may then be investigated.

There are many reasons for performing multiple optimization calculations. The initial optimization exercise may not find the desired optimum. The design process may call for a start by using one design approach, and optimization may be executed subsequently. The knowledge gained from the first few runs may prompt changes in the design space, and optimize yet again.

Understanding the procedure used by an analytical method or a computer program is always helpful, and is particularly true in the case of design optimization. In the procedure using gradients, partial derivatives of the objective function and the state variables with respect to each design variable are established. A reference design set is defined as the point of evaluation for the gradient. Focusing on the objective function, for example, let the reference set be denoted as $f_r(x)$. The gradient of the objective function is expressed as:

$$\nabla f_r = \left[\frac{\partial f_r}{\partial x_1}, \frac{\partial f_r}{\partial x_2}, \ldots\ldots \frac{\partial f_r}{\partial x_n} \right] \qquad (13.8)$$

With respect to each of the n design variables xi, the gradient is approximated using a forward difference of the type:

$$\frac{\partial f_r}{\partial x_i} = \frac{f_r(x + \Delta x_i e) - f_r(x)}{\Delta x_i} \qquad (13.9)$$

where: e = vector with 1 as its ith component and 0 for all others

$\Delta x = \Delta D(x_i - x_{i-1})/100$

ΔD = percent forward difference step size.

Similar calculations are performed for each state variable.

The first order optimization method calculates and makes use of the derivative information. The constrained problem is transformed into an unconstrained one through the application of penalty functions. Derivatives are formed for the objective function and the state variable penalty functions, leading to a direction for search in the design space. The search direction vector $\mathbf{d}(j)$ for each optimization iteration (j) is devised for the next iteration $(j + 1)$ by applying the following equation:

$$x(j + 1) = x(j) + s(j)\, \mathbf{d}(j) \qquad (13.10)$$

where $s(j)$ is the line search parameter corresponding to the minimum value of the objective function in the direction of $\mathbf{d}(j)$. The solution for $s(j)$ may be based on quadratic fitting technique, with its values ranging from 0 to a percentage of a specified step size of maximum line search. Various steepest descent and conjugate direction searches are performed during each iteration till convergence is reached. Each iteration is composed of sub-iterations that include search direction and derivative computations. Thus, a first order design optimization iteration will perform several analysis loops. The key to the solution of global minimization relies on the sequential generation of search directions and on internal adjustments of the response parameter. For the initial iteration $(j = 0)$ the search direction is assumed to be the negative of the gradient of the unconstrained objective function. The procedure is computationally intensive than other methods, and provides greater accuracy in the calculated results.

Another advanced procedure requires only the values of the dependent variables (objective function and state variables), and not their derivatives. The dependent variables are first replaced with approximations by means of least squares fitting, and the constrained minimization problem is converted into an unconstrained problem using penalty functions. Minimization is then performed for every iteration on the approximated penalized function, which may be considered as a sub-problem, until convergence is achieved or terminated by the computer program. Convergence is assumed when either the present design set or the previous design set or the best design set is feasible, provided that the difference in the calculated values of the objective function and each design variable in two successive sets satisfies the prescribed tolerance level. Each iteration is equivalent to one complete analysis loop. Since the method relies on approximation of the objective function and each state variable, certain amount of data in the form of design sets is needed to initiate the process. Preliminary data can be directly generated from another analytical procedure, or a random design set may be generated.

13.6 APPLICATION TO FULLY STRESSED DESIGN

The fully stressed condition is an intuitive optimality criterion that may be readily applied to truss structures. In general, the optimum lightweight structure is not necessarily a fully stressed one. The reason is that for an indeterminate structure it is possible that some members are not fully stressed to help relieve stresses in other members. A structure is considered to be fully stressed if all members (or elements) of that structure are either at their limit, or have reached their permissible stress level, under at least one other design load condition.

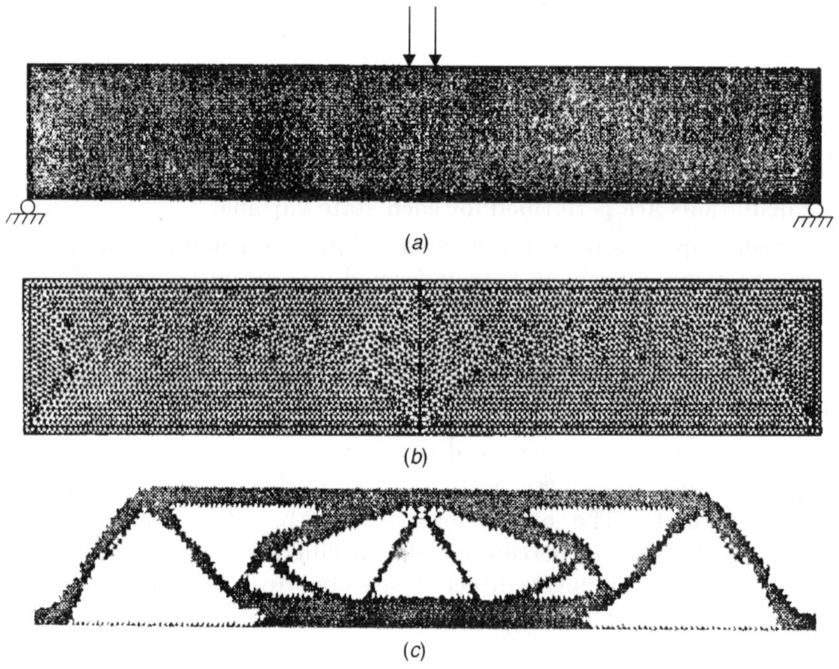

Fig. 13.4. Fully Stressed Optimization of Beam; (*a*) Problem Definition; (*b*) Finite Element Mesh; (*c*) Resulting Topology (*ref.* Sienz, 1994)

The extension of this idea applies to two and three dimension continua, and in combination with the finite element method, provides an alternative approach to structural topology optimization. In this procedure regions having a low stress level are iteratively removed until a quasi-optimum state is reached. Thus, elements which have an effective stress less than a threshold value are removed, or killed, giving rise to the expression 'hard kill method'.

This approach differs from the 'soft kill method' where, instead of removing the elements with low stress level, they are made softer and weaker. This may be carried out, for example, by reducing the thickness of a shell element. In practice, to avoid changing the finite element mesh during the finite element optimization, a small positive value for the Young's modulus may be assigned to the elements that are to be 'removed'. In selecting the small value for the modulus of elasticity care must be exercised to ensure singularities are not caused in the stiffness matrix. This approach is easy to implement in a commercially available finite element software if multiple material properties can be accommodated.

Figure 13.4 shows an example of a simply supported beam used in the floor of an aircraft. For functional purposes an enveloping non-design region is defined around the beam. The fully stressed evolutionary approach has been employed to obtain the resulting optimized topology of the beam.

13.7 NATURAL FREQUENCY OF A ROTOR

Optimization of the crankshaft and attached machinery's system can be accomplished in a number of ways. Section 6.8 used the Holzer method for calculating the torsional natural frequency of vibration, mode shapes and response. Microsoft's 'EXCEL' computer program has been used to perform the sequential calculations. Within this program an option referred to as the 'solver' may be used to obtain the optimal value for an equation in a target cell. The solver

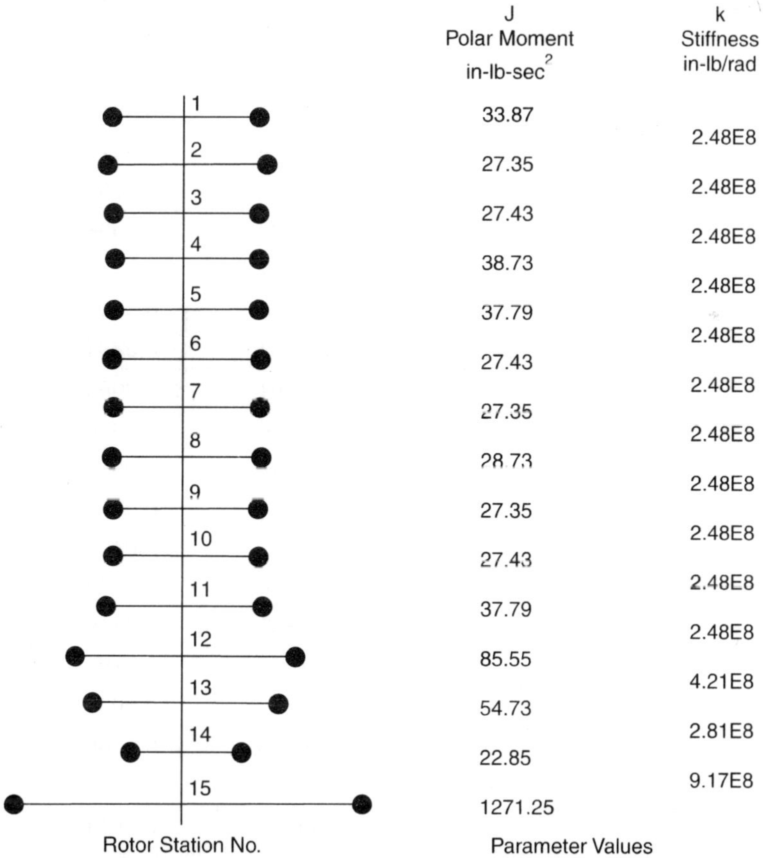

Fig. 13.5. Moments of Inertia and Stiffness Distribution of Diesel Engine Coupled with Gas Transmission Pump

works with a group of cells that are related, either directly or indirectly, to the formula in the target cell. The solver adjusts the values in specified adjustable cells to produce the required result from the target cell's formula. Constraints may be applied to restrict the values the procedure can use in the model. The constraints may also have functional relationship with other cells that affect the target cell's equation.

Table 13.1. Initial Holzer Sequential Calculations for Torsional Natural Frequency of Figure 13.5 Trial Frequency ω_S^2 = 0.026085

Station #	Trial ω_S^2 = 0.026085		β	T	k
1	0.456	0.0119	1.0000	0.0119	
			0.0119	0.0119	1.000
2	0.369	0.0096	0.9881	0.0095	
			0.0214	0.0214	1.000
3	0.370	0.0096	0.9667	0.0093	
			0.0307	0.0307	1.000
4	0.522	0.0136	0.9360	0.0127	
			0.0435	0.0435	1.000
5	0.509	0.0133	0.8925	0.0119	
			0.0553	0.0553	1.000
6	0.370	0.0096	0.8372	0.0081	
			0.0634	0.0634	1.000
7	0.369	0.0096	0.7738	0.0074	
			0.0708	0.0708	1.000
8	0.522	0.0136	0.7029	0.0096	
			0.0804	0.0804	1.000
9	0.369	0.0096	0.6225	0.0060	
			0.0864	0.0864	1.000
10	0.370	0.0096	0.5361	0.0052	
			0.0916	0.0916	1.000
11	0.509	0.0133	0.4445	0.0059	
			0.0975	0.0975	1.000
12	1.153	0.0301	0.3471	0.0104	
			0.0636	0.1079	1.698
13	1.616	0.0422	0.2835	0.0119	
			0.1058	0.1199	1.133
14	0.675	0.0176	0.1777	0.0031	
			0.0333	0.1230	3.698
15	37.533	0.9791	0.1444	0.1414	
		Residual Torque =		0.2644	

Note: Stiffness values are divided by 2.48E8. Polar moments of inertia values are divided by 74.2 to be adjusted, or ω_S^2.

Table 13.2. Optimal Holzer Sequential Calculations Using 'Goal Seek' Method for Target Function Trial Frequency ω_S^2 = 0.03641

Station #	*Trial ω_S^2 = 0.03641*		β	*T*	*k*
1	0.456	0.0166	1.0000	0.0166	
			0.0166	0.0166	1.000
2	0.369	0.0134	0.9834	0.0132	
			0.0298	0.0298	1.000
3	0.370	0.0135	0.9536	0.0128	
			0.0427	0.0427	1.000
4	0.522	0.0190	0.9109	0.0173	
			0.0600	0.0600	1.000
5	0.509	0.0185	0.8509	0.0158	
			0.0757	0.0757	1.000
6	0.370	0.0135	0.7752	0.0104	
			0.0862	0.0862	1.000
7	0.369	0.0134	0.6890	0.0092	
			0.0954	0.0954	1.000
8	0.522	0.0190	0.5936	0.0113	
			0.1067	0.1067	1.000
9	0.369	0.0134	0.4869	0.0065	
			0.1132	0.1132	1.000
10	0.370	0.0135	0.3737	0.0050	
			0.1183	0.1183	1.000
11	0.509	0.0185	0.2554	0.0047	
			0.1230	0.1230	1.000
12	1.153	0.0420	0.1324	0.0056	
			0.0757	0.1286	1.698
13	1.616	0.0588	0.0566	0.0033	
			0.1164	0.1319	1.133
14	0.675	0.0246	– 0.0598	– 0.0015	
			0.0353	0.1304	3.698
15	37.533	1.3666	– 0.0950	– 0.1299	
		Residual Torque =		0.0005	

Note: Stiffness values are divided by 2.48E8. Polar moments of inertia values are divided by 74.2.

In the example problem illustrated in Fig. 13.5 a diesel engine drives a gas pipeline transmission pump. Stiffness and inertia properties of the full drive system are shown in the figure. Using Holzer's technique for the primary torsional natural frequency, the residual torque value (table 13.1) for ω_S^2 = 0.026085 is determined to be 0.2644. In the sequential calculations the requirement for the selected value of ω_S^2 to be a natural frequency is the residual torque value must be zero, or very close to it. In this case there is only one cell value (ω_S^2) that needs to be altered for which the target cell (residual torque) must reach zero. Table 13.2 shows calculations for trial frequency $\omega_S^2 = 0.0364$. The complete procedure can be executed through the following sequence:

- Select the 'goal seek' on 'EXCEL's Tools menu.
- In 'Set Cell' box enter the cell address of the target function, or the residual torque value.
- In the 'To Value' box enter zero.
- In the 'By Changing Cell' box enter the reference for the cell containing the value

The procedure yields the frequency ω_S^2 for which the residual value is nearly zero, from which the torsional natural frequency may be directly obtained.

The problem may also be looked at from a different angle. Consider a situation where the calculated natural frequency is unacceptable from design and operational requirements. To place the machine's torsional vibration frequency in a different speed regime modifications are required in the rotor's geometric configuration. One possibility is to alter the shaft diameter in selected regions. The procedure may be executed in the 'solver' sub-routine of 'EXCEL'. With the solver an optimal value for a formula in the target cell can be achieved by modifying values in a group of cells that are directly or indirectly related with it. The sub-routine alters the values in the specified adjustable cells to produce the required value in the target cell. Constraints may be applied to restrict the values the sub-routine may assume for the adjustable cells to a certain range. The solver uses the generalized reduced gradient non-linear optimization code. Linear and integer problems use the simplex method, with prescribed bounds on the variables. For non-linear problems it is helpful to try different starting values for the adjustable cells, especially if the program finds a solution that is significantly different from expected values. Setting adjustable cell values close to those that are likely to be used in the optimal solution can reduce solution time.

An interesting feature of this program is that it provides information on how sensitive the solution is to small changes in the constraint parameters. For non-linear systems the sensitivity report provides values for reduced gradients and Lagrangian multipliers. A list of upper and lower bounds for adjustable cell values that satisfy the constraints is also generated.

Table 13.3 provides illustration for the example problem of Fig. 13.5. ω_S^2 is required to be greater than 0.03925, and stiffness of the shaft segments between rotor stations 1 through 8 may be permitted to increase by up to 10.0%.

The procedure for the solver calls for the following steps:

- On the tools menu of 'EXCEL' program activate the Solver.
- In the 'Set Target Cell' box enter cell address for the residual torque parameter.
- Since the residual torque must be zero for a critical frequency, activate 'Value Of' and enter zero in the box.
- In 'By Changing Cells' box enter the address of each adjustable cell. In this example the stiffness of shaft segments between stations from 1 through 8 represent adjustable cell parameters.
- In the 'Subject To Constraints' box, enter constraints on values the program is permitted to assign to the adjustable cells. Each shaft segment stiffness will have a constraint, requiring it to be ≤ 1.10.

- Activate the Solver sub-routine.

After successful execution the solution may be saved for evaluation and additional modification if necessary.

Table 13.3. Determining Shaft Diameters in Holzer Calculations To Obtain Required Torsional Natural Frequency Trial Frequency ω_S^2 = 0.039258

Station #	*Trial* ω_S^2 = *0.039258*		β	*T*	*k*
1	0.456	0.0179	1.0000	0.0179	
			0.0164	0.0179	1.0954500
2	0.369	0.0145	0.9836	0.0142	
			0.0293	0.0322	1.0978900
3	0.370	0.0145	0.9544	0.0139	
			0.0419	0.0460	1.0987850
4	0.522	0.0205	0.9125	0.0187	
			0.0589	0.0647	1.0989000
5	0.509	0.0200	0.8536	0.0171	
			0.0746	0.0818	1.0954000
6	0.370	0.0145	0.7790	0.0113	
			0.0847	0.0931	1.0987890
7	0.369	0.0145	0.6943	0.0100	
			0.0944	0.1031	1.0928700
8	0.522	0.0205	0.5999	0.0123	
			0.1055	0.1154	1.0936950
9	0.369	0.0145	0.4944	0.0072	
			0.1120	0.1226	1.0938735
10	0.370	0.0145	0.3823	0.0055	
			0.1173	0.1281	1.0921590
11	0.509	0.0200	0.2650	0.0053	
			0.1214	0.1334	1.0989500
12	1.153	0.0453	0.1436	0.0065	
			0.0824	0.1399	1.6975806
13	1.616	0.0588	0.0612	0.0036	
			0.1267	0.1435	1.1330645
14	0.675	0.0246	– 0.0655	– 0.0016	
			0.0384	0.1419	3.6975806
15	37.533	1.3660	– 0.1038	– 0.1419	
		Residual Torque =		0.0000	

Note: Stiffness values are divided by 2.48E8. Polar moments of inertia values are divided by 74.2.

13.8 REINFORCEMENT OF PLATE STRUCTURES

Plates are one of the more important structural members used in engineering practice. In the problem of topological reinforcement of plates, cavities and ribs need to be introduced. By employing the concept of layers, the model of Fig. 13.6 proves useful in the understanding of the subject.

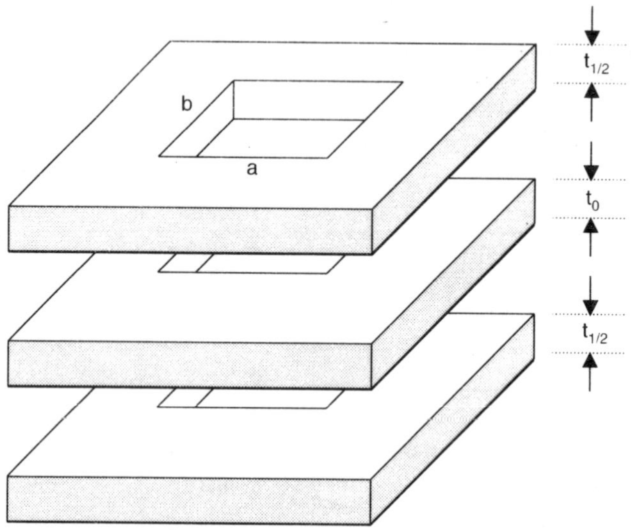

Fig. 13.6. Base Plate with Ribbed Reinforcement

When the objective is to find the stiffest plate, the optimization problem may be expressed as the maximization of total potential energy. By considering total potential energy as the objective function the problem, together with its prescribed boundary functions, may be solved without changing its formulation. In the finite element model a reduced integration rule may be used to evaluate the stiffness matrix associated with the shear strain energy. The shear strain energy terms may be interpreted as penalty functions that force the shear strains to be zero as the plate thickness is reduced. By using the reduced integration procedure, over constraining of the shear portion of the stiffness matrix is alleviated. However, the procedure may also cause the problem of rank deficiency of the stiffness matrix. Consequently, a selective integration approach may be used where the shear portion uses reduced integration, but the remaining terms are fully integrated in order to attain the required rank of the overall stiffness matrix.

The selected plate element should have certain features if it is to perform adequately in the finite element calculations. It should:

- not lock where the plate is thin
- not have spurious zero energy modes
- satisfy the usual completeness and convergence requirements
- not be sensitive to distortion
- be easy to implement

As an illustration of this discussion consider a simply supported square plate with a central circular hole. Length of the plate is 4.0, hole diameter is 1.0, and its thickness is 0.01. Thickness of the external reinforcement layers is 0.03. The plate is subjected to a uniform pressure load. Figure 13.7 shows the finite element mesh for a quarter of the domain, where advantage is taken of the bi-symmetry condition.

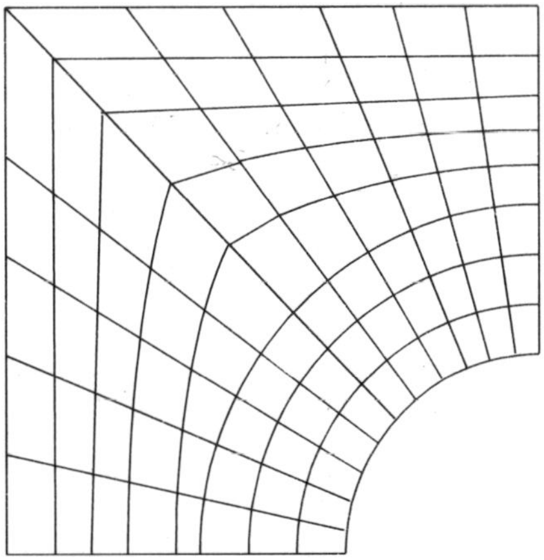

Fig. 13.7. Finite Element Mesh of One-Quarter of Square Plate

Reinforcement is applied over 50% of the surface. Figure 13.8 provides the resulting layout. When the support on the straight edges is of the fully clamped type and reinforcement applied over 30% of the surface, the layout is of the type shown in Fig. 13.9. The reinforcement around the central hole is substantially reduced when the edges are fully clamped.

Fig. 13.8. Reinforcement Layout of Fig. 13.6 for Simply Supported Edges

13.9 INTEGRATED STRUCTURAL OPTIMIZATION

In a competitive environment manufacturers are prone to cut design and development time, as also cost. Rather than follow the traditional process of design engineering, concurrent, or simultaneous, engineering provides an edge over competitors while emphasizing on the quality aspects of the machine. The key to success is a fully integrated process, from the beginning of the design to rapid production and test in a computer-aided design and manufacturing environment.

Considerable speed and efficiency is obtainable by employing optimization techniques in place of the traditional design process. In fact, the nature of the design problem is reversed in a fully integrated optimization approach. Instead of suggesting a design and analyzing it to see if it meets a set of objectives and requirements, an optimization problem is constructed and based on objectives and constraints to find a design.

Topological structural optimization can only provide a rough idea about the boundary surfaces of a machine member. The process is carried out on a fixed domain without the need to change the initial finite element discretization, and they result in an optimal material distribution layout that in some areas may be neither a solid nor a void. Mechanical components have a shape defined by segmental boundary curves. Conventional shape optimization methods yield the optimal form of these boundary curves. Unlike topology optimization where only global objectives (average stress, frequency, etc.) can be dealt with, shape optimization methods permit definition of local objective functions. As a rule, design optimization requires more considerations than global objectives. Information obtained from topological optimization is not sufficient; more precise fulfillment of the requirements calls for the use of more conventional shape optimization methods, such as the boundary variation method.

From these considerations an integrated design and optimization system may be constructed with two main modules, one for topology of the structure, the other to look at its shape aspects. Due to fundamental differences between the two, an interfacing link is required to identify and interpret patterns and to generate boundary curves from the topology optimization module. This may be achieved through curve smoothing and computer pattern recognition techniques. Computer pattern recognition automatically constructs explicit meaningful descriptions of physical objects from images, usually given as arrays of density variation. The process is illustrated in Fig. 13.10.

Fig. 13.9. Reinforcement Layout of Fig. 13.6 for Fully Clamped Edges

A procedure called 'the bubble method' includes elements of conventional boundary variation method for shape optimization while taking into account the effect of topology (Eschenauer, 1994). The idea is to iteratively position new holes, or bubbles, in a structure to change the topology class, followed by an ordered shape optimization. The objective is to determine the optimal shape of the current bubble and external boundaries.

For optimization problems with simple global functions (for example, total potential energy), the optimum bubble position can be determined analytically by using variational methods. A bubble in a given initial position may be analytically guided to the optimal position. The process follows the following steps:

1. Starting from an initial design reference domain, including the objective and constraint functions, conventional shape optimization using boundary variation is carried out.
2. A new hole is inserted, changing the class of topology. The optimum position of the bubble is determined analytically or numerically.
3. Shape optimization is then carried out to find the optimal shape of the bubble and to determine its influence on other variables.
4. Steps 2 and 3 are repeated until some convergence criterion is satisfied.

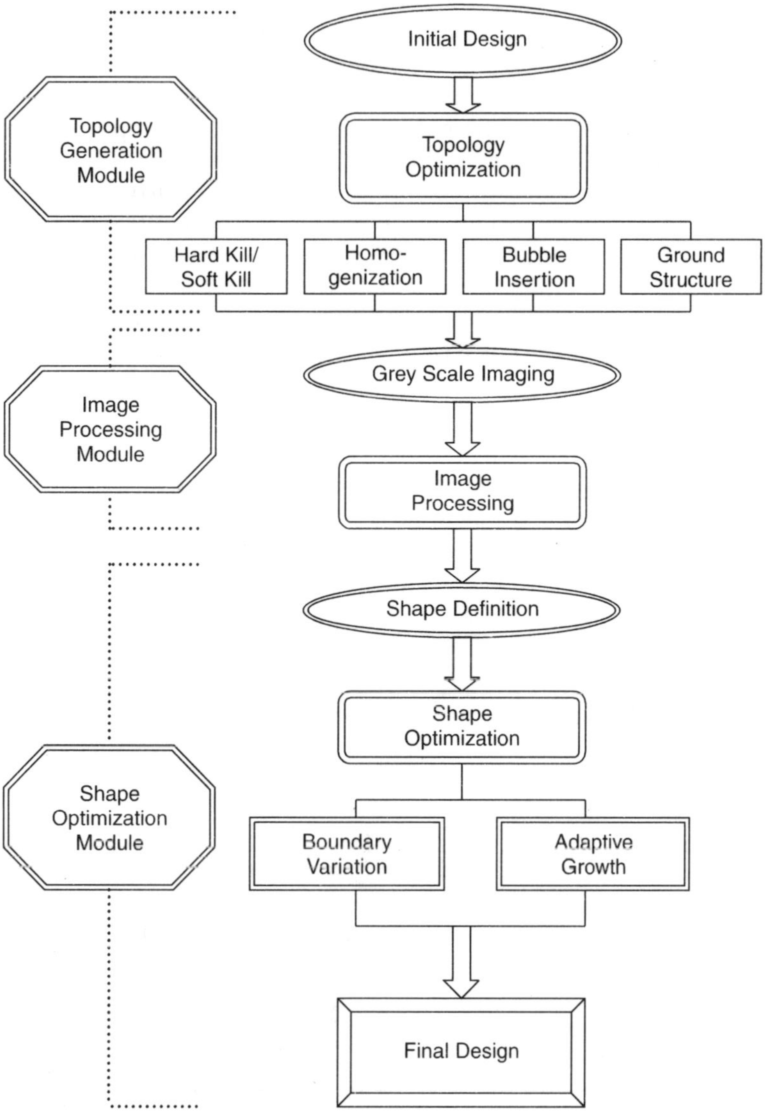

Fig. 13.10. Integrated Structural Optimization (*ref.* Sienz, 1994)

The outcome of the topology optimization module is in the form of an array, which from a design point of view, is not entirely satisfactory. The components will have jagged boundaries due to the profile of the elements. Boundaries of the structure have to be extracted. Depending on the value of the volume constraint the resulting optimum topology can vary from a truss-like configuration to a closed solid shape. Selection of the control points on boundary segments can be done manually or automatically (*ref*: Maute, 1995).

The shape optimization module may be considered to be a method for post-processing of results from the topology optimization module. The output of the topology optimization module is an optimal layout with rough boundaries and areas where material properties are differing. Shape optimization leads to an optimized detail design where additional design constraints are met. Geometrical restrictions associated with manufacturing and assembly requirements are considered, and other performance criteria (buckling of slender members, for example) are taken into account. If necessary, sizing optimization may also be carried out at the same time. The procedure is performed with mathematical programs working in conjunction with finite element analysis. Intuitive as also mathematically derived rigorous optimality criteria may be used to enforce restrictions on the design. Experimental photo-elasticity techniques may also be applied. For instance, bolted joints in machine components sometimes require a number of fasteners. In this type of design the number of bolts, their size and location, or shape, of holes can be optimized from experimental results. The same observation can also be made about plates and shells.

13.10 CONNECTING ROD EXAMPLE

Optimization of a connecting rod is used to provide illustration of the process. For simplification the connecting rod is assumed to be symmetric at the crank pin and the gudgeon pin ends. Consequently, the cross section of the rod between the two ends will be uniform. Taking advantage of this symmetry, only one quarter of the rod is modeled on a finite element program. Figure 13.11 provides details about the geometric configuration, boundary conditions, material properties, loading and other related data.

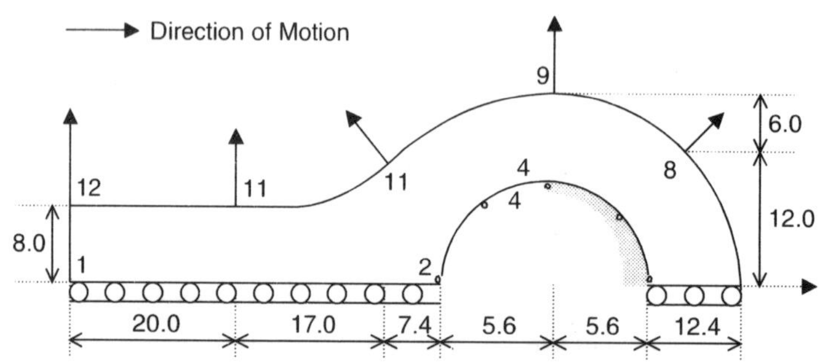

Fig. 13.11. Connecting Rod Data (*ref.* Sienz, 1994)

The objective in this optimization exercise is to minimize the material volume. The connecting rod is subjected to a maximum stress constraint imposed at points 1, 3, 4, 5, 6 and 12. The initial and optimum shapes are shown in Fig. 13.12, together with the corresponding finite element mesh and stress contours. The volume of the initial design is 2,904 mm^3, which compares with 1,072 mm^3 for the optimized design. Corresponding maximum stress values are calculated to be 668.2 N/mm^2 and 1198.9 N/mm^2. Figure 13.13 provides the shape and stress contours for the complete connecting rod.

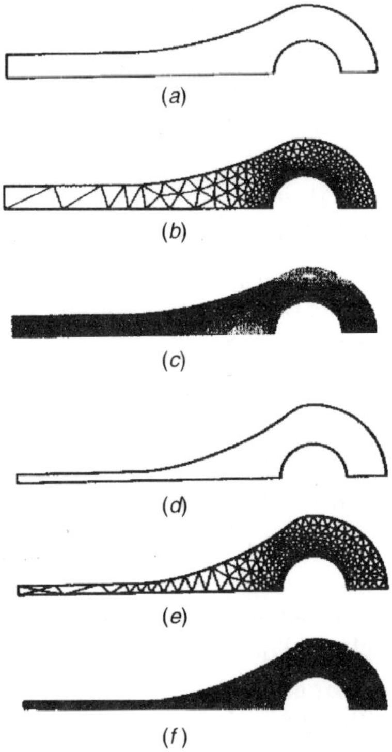

Fig. 13.12. Shape, FE Mesh and Stress Contours of Initial Design (*a, b, c*), and Optimum Design (*d, e, f*) (*ref.* Sienz, 1994)

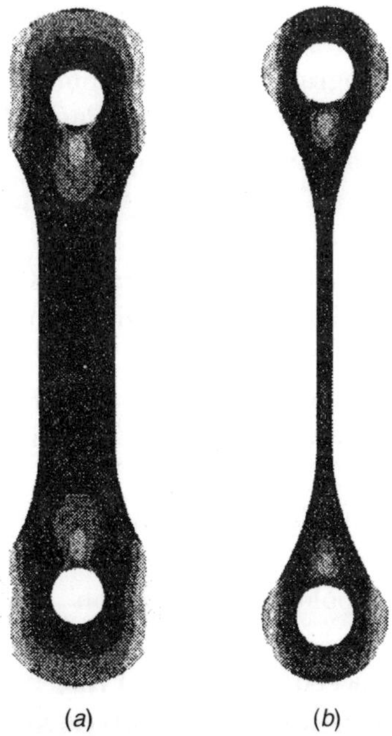

Fig. 13.13. Stress Contours of Complete Connecting Rod: (*a*) Initial Design, (*b*) Optimum Design (*ref.* Sienz, 1994)

REFERENCES AND BIBLIOGRAPHY

Azegami, H., "A Proposal of a Shape Optimization Method Using a Constitutive Equation of Growth", JSME International Journal, 33(1), Pages 64-71, 1990.

Bathe, K. J., "Finite Element Procedures in Engineering Analysis", Prentice-Hall, Englewood Cliffs, 1982.

Bayo, M., "A Modified Lagrangian Formulation for the Dynamic Analysis of Constrained Mechanical Systems", Computer Methods in Applied Mechanics and Engineering, Vol. 71, Page 183-195, 1988.

Biot, M. A, "Mechanics of Incremental Deformation", John Wiley & Sons, New York, 1965.

Broch, J. T., "Application of Bruel & Kjaer Measuring Instruments to Mechanical Vibration and Shock Measurements", Bruel & Kjaer Instruments, Inc., Cleveland, OH. 1988.

Chen, J. L., Tsai, W. C., "Shape Optimization by Using Simulated Biological Growth Approaches", AIAA Journal, 31(11), Pages 2143 - 2147, 1993.

Clough, R. W., Penzien, J., "Dynamics of Structures", McGraw Hill, New York, 1985.

Cook, R. D., "Concepts and Applications of Finite Element Analysis", 2nd Edition, John Wiley & Sons, New York, 1981.

Eschenauer, H. A., Kobelev, V. V., Scumacher, A., "Bubble Method for Topology and Shape Optimization of Structures", Structural Optimization, 8, Pages 42 - 51, 1994.

Gutkowski, W., Mroz, Z., "Proceedings of the Second World Congress of Structural and Multidisciplinary Optimization", Zakopane, Poland, 1997.

Hemada, M., Seguchi, Y., Tada, Y., "Shape Determination Problems of Structures by the Inverse Variational Principle", Bulletin of Japan Society of Mechanical Engineers, 10, 23 - 184, 1980.

Hildebrand, F. B., "Advanced Calculus for Applications", Prentice-Hall, New Jersey, 1986.

Imgrund, M. C., "ANSYS Verification Manual", Swanson Analysis Systems, Inc.,1992.

Irons, B. M., "A Frontal Solution Program for Finite Element Analysis", International Journal for Numerical Methods in Engineering, Vol. 2, No. 1, Pages 5-23, 1980.

Jiang, F., Rodgers, G., "Combined Lagrangian Multiplier and Penalty Function Finite Element Technique for Elastic Impact Analysis", Computers & Structures, Vol. 30, Page 1219-1229, 1988.

Kohnke, P. C., "Large Deflection Analysis of Frame Structures by Fictitious Forces", International Journal for Numerical Methods in Engineering, Vol. 12, No. 8, Pages 1278-1294, 1978.

Kuhn, H. W., Tucker, A. W., "Non-Linear Programming", *Proceedings of the Second Berkeley Symposium on Mathematical Statistics and Probability*, edited by J. Neyman, Page 481-492, 1951.

Law, B., "Prediction of Crankshaft and Flywheel Dynamics", Institute of Mechanical Engineers, Paper # C382/046, Pages 487 - 498, Perkins Technology, 1989.

Maute, K., Ramm, E., "Adaptive Topology Optimization", Structural Optimization, 10, Pages 100-112, 1995.

Olhoff, N., Rozvany, G. I. N., "Proceedings of the First World Congress of Structural and Multidisciplinary Optimization", Vol. 1, Pergamon Press, Goslar, Germany, 1995.

Powell, M. J. D., "An Efficient Method for Finding the Minimum of a Function of Several Variables Without Calculating Derivatives", Computer Journal, Vol. 7, Pages 155-162, 1964.

Przemieniecki, J. S., "Theory of Matrix Structural Analysis", Mc-Graw Hill, New York, 1968.

Rajakumar, C., Rogers, C. R., "The Lanczos Algorithm Applied to Unsymmetric Generalized Eigenvalue Problem", International Journal for Numerical Methods in Engineering, Vol. 32, pages 1009-1029, 1992.

Rajan, S. D., Belegundu, A. D., "Shape Optimal Design Using Fictitious Loads", AIAA Journal, 27(1), Pages 102 - 107, 1989.

Robeson, G., "Cosworth - The Search for Power", Patrick Stephens Limited, London, 1991.

Sienz, J., "Integrated Structural Modeling, Adaptive Analysis and Shape Optimization", C/ph/181/94, University College of Swansea, England, 1994.

Sienz, J., Hassani, B., Hinton, E., "Structural Optimization", *Advanced Finite Element Solution Techniques in Innovative Computer Architecture*, edited by Bugeda G., Papadrakakis, M., CIMNE, Barcelona, 1996.

Roark, R. J., Young, W. C., "Formulas for Stress and Strain", 5th ed., Mc-Graw Hill, New York, 1989.

Sigmund, O., "Design of Material Structures Using Topology Optimization", Ph. D. Thesis, Technical University of Denmark, 1994.

Steven, G., Querin, O. M., Guan, H., Xie, Y. M., "Proceedings of the Australasian World Congress of Structural and Multidisciplinary Optimization", Oxbridge Press, Sydney, Australia, 1997.

Suzuki, K., Kikuchi, N., "Generalized Layout Optimization of Three Dimensional Shell Structures", *Geometric Aspects of Industrialized Design,* edited by V. Komkov, SIAM, 1991.

Timoshenko, S., Young, D. H. and Weaver, W., "Vibration Problems in Engineering", John Wiley & Sons, Inc., New York, 1974.

Zienkiewicz, O. C., "The Finite Element Method", McGraw-Hill Company, London, 1985.